# Cu-Interconnects, Glass, and AI-Assisted Simulation for Chiplets and Heterogeneous Integration

John Lau • Kuo-Ning Chiang

# Cu-Interconnects, Glass, and AI-Assisted Simulation for Chiplets and Heterogeneous Integration

Springer

John Lau
Unimicron Technology Corporation
Palo Alto, CA, USA

Kuo-Ning Chiang
Department of Power Mechanical Engineering
National Tsing Hua University
Hsinchu, Taiwan

ISBN 978-981-95-6890-1 ISBN 978-981-95-6891-8 (eBook)
https://doi.org/10.1007/978-981-95-6891-8



This Springer imprint is published by the registered company Springer Nature Singapore Pte Ltd.
The registered company address is: 152 Beach Road, #21-01/04 Gateway East, Singapore 189721, Singapore

# Preface

There have been strong demands for lower power consumption, faster speed, higher density, finer pitch, and higher bandwidth in optical/electrical interconnects used for artificial intelligence (AI) and networks in a data center. The adoption of direct Cu-interconnects (by hybrid bonding) and co-packaged optics (CPO) has been expected for both high-performance computing (HPC) driven by AI and high-bandwidth and high-speed communication networks in a data center. Glass substrate provides better electrical performance than the organic substrate or silicon substrate and is ideal for HPC/AI. The connections between chiplets (bridges) are the most important tasks in chiplet design and heterogeneous integration. The electronic packaging (EP) community has long relied on design-on-simulation (DoS) technology for developing new packaging structures. However, this traditional approach faces significant challenges, particularly in ensuring the trustworthiness and consistency of simulation results, which are often highly dependent on the individual researcher's domain knowledge and background in the finite element method (FEM). This reliance results in a lengthy design cycle that hinders the ability to meet critical time-to-market demands for increasingly complex packaging structures, such as system-in-package (SiP), chiplet, and heterogeneous packaging. To overcome these substantial limitations and achieve more consistent, trustworthy predictions, artificial intelligence (AI) and machine learning (ML) algorithms have been introduced, leading to the development of AI-assisted design-on-simulation (AI-DoS) technology.

For the next few years, we will see more implementations of hybrid bonding, CPO, glass substrates, bridges, and AI-assisted design-on-simulation technology. Unfortunately, for most of the practicing engineers and managers, as well as scientists and researchers, these hybrid bonding, CPO, glass substrates, and bridges, along with the methods for ensuring the trustworthiness and consistency of simulation results, are not well understood. Thus, there is an urgent need, both in industry and research institutes, to create a comprehensive book on the current state of knowledge of these hybrid bonding, CPO, glass substrates, and bridges and to establish dependable modeling and simulation frameworks. This book is written so that readers can quickly learn about the basics of problem-solving methods and understand the trade-offs inherent in making system-level decisions.

There are nine chapters in this book, namely: (1) Advanced Packaging, (2) Cu-Cu Hybrid Bonding, (3) Glass Packaging, (4) Heterogeneous Integration of PIC, EIC, and Switch (CPO), (5) Chiplet Connections: Bridges, (6) Fatigue and Fracture Models for Electronic Packaging, (7) AI and Machine Learning Algorithms, (8) Modeling and Design-on-Simulation Technology, and (9) Integrating AI with Design-on-Simulation for Advanced Packaging.

Chapter 1 presents the 2D, 2.1D, 2.3D, 2.5D, 3D, 3.3D, and 3.5D IC integrations. Chiplet communication such as bridges embedded in build-up package substrate and fan-out epoxy molding compound (EMC) with RDLs (redistribution layers) will be examined. The high-volume products by Cu-Cu hybrid bonding will be presented. High-bandwidth memory (HBM), customized HBM (cHBM), CoWoS (chip on wafer on substrate), and CoPoS (chip on panel on substrate) are the key elements of HPC products driven by AI will be discussed. CoWoS-S, CoWoS-R, and CoWoS-L will be systematically presented. Glass-core substrates and glass-core interposers will be provided. 3D heterogeneous integration of photonic IC (integrated circuits) (PIC) and electronic IC (EIC) driven by AI and high-speed communications will be briefly presented. Some recommendations will be provided.

Chapter 2 presents the recent advances and trends in the direct Cu-interconnects by Cu-Cu hybrid bonding. Emphasis is placed on the design, materials, processes, fabrications, reliability, challenges, advantages and disadvantages, and examples of Cu-Cu hybrid bonding. Specially, the works by Sony, Samsung, Applied Materials, EV Group, Imec, Daicel Corporation, High Energy Accelerator Research Organization, Toray Industries, Entegris, National Yang Ming Chiao Tung University, Industrial Technology Research Institute, MediaTek, National Tsing Hua University, BASF, STMicroelectronics, CEA-Leti, Binghamton University, IBM, TAZMO, LINTEC, Institute of Science Tokyo, TSMC, Seoul National University of Science and Technology, YOKOHAMA National University, Tokyo Electron Kyushu Limited, SK Hynix, Tokyo Electron Limited, Toray, DISCO, University of Tsukuba, Kurita Water Industries, and Unimicron will be presented and discussed. Also, some recommendations will be provided.

Chapter 3 presents two problems of flip chip on glass-core package substrate. The first problem deals with the flip chip on glass-core package substrate with microbumps and the other deals with the flip chip on glass-core package substrate with Cu-Cu hybrid bonding. Emphasis is placed on the solder joint reliability due to the glass-core substrate. Some recommendations will be provided. The advantages and disadvantages of glass-core substrate vs. glass-core interposer, through-glass via (TGV), and redistribution layers (RDLs) will be briefly mentioned first.

Chapter 4 presents the on-board optics (OBO), near package optics (NPO), and CPO. Emphasis is placed on the trends and outlooks of heterogeneous integration of chiplets such as the PIC, EIC, silicon photonic and application specific IC (ASIC) switch with or without bridges on various CPO substrates such as organic, silicon, and glass. Various forms of 3D heterogeneous integration of PIC and EIC will be presented. Some recommendations will be provided.

Chapter 5 presents the connections (bridges) between chiplets. Emphasis is placed on the bridges embedded in (or above) the organic build-up package substrate, bridges embedded in fan-out epoxy molding compound (EMC) and redistribution layers (RDLs), and hybrid bonding bridges. Finally, thermal-mechanical simulations of a silicon bridge connecting two chiplets with microbumps w/o underfill are presented. Some recommendations are provided.

Chapter 6 presents the fundamental solid mechanics principles necessary for ensuring the mechanical integrity and long-term reliability of electronic packaging. It covers the complex behavior of diverse materials, such as brittle silicon and viscoplastic solder alloys, which exhibit elasticity, plasticity, creep, and stress relaxation. This chapter also introduces key constitutive models, including the Chaboche model for cyclic plasticity and the Anand or hyperbolic sine models for time-dependent viscoplasticity in solders. Reliability is addressed through fatigue life prediction using models like the strain-based Coffin-Manson and the energy density Models (e.g., Darveaux), which account for creep. Furthermore, it examines brittle fracture (relevant for silicon, assessed via the maximum normal stress theory) and interfacial delamination. Numerical analysis of delamination is facilitated by methods such as the modified virtual crack closure technique (MVCCT), which calculates the energy release rate for various fracture modes.

Chapter 7 presents the critical role of artificial intelligence (AI) and machine learning (ML) algorithms in the field of electronic packaging (EP), with a focus on enhancing reliability and design efficiency. ML leverages large datasets to automatically learn patterns and develop predictive models, forming the basis for AI-assisted design-on-simulation (DoS). The main application in EP is supervised regression, used for predicting continuous values such as reliability life cycles (cycles to failure), warpage, and extracting material constitutive parameters (e.g., for Chaboche and Anand models). ML provides predictions in seconds to minutes, significantly accelerating the design process compared to time-consuming experiments or simulations. This chapter introduces various algorithms, including ANN, RNN, SVR, KRR, GPR, random forest (RF), and CNNs, and outlines essential training methodologies. These include data preprocessing (e.g., RobustScaler and Min-Max Scaler), k-fold cross-validation to prevent overfitting, and grid search for hyperparameter tuning. Efficiency is boosted by utilizing GPU-based parallel computing for faster training and optimization of complex neural networks.

Chapter 8 presents design-on-simulation (DoS) technology as a critical methodology in electronic packaging, replacing time-consuming and costly design-on-experiment (DoE) for reliability assessment. DoS utilizes the finite element method (FEM) to analyze stress, strain, and fatigue failures resulting from CTE mismatch, particularly in solder joints. A major focus is on overcoming the inconsistencies and mesh-size dependency inherent in traditional FEM. Solutions include the finite volume-weighted Method, which accurately characterizes strain near singular regions. The document also covers predicting complex solder reflow geometry using methods such as the energy-based method, which is accurate for arbitrary pad shapes. To boost efficiency in complex 3D modeling, techniques such as the equivalent methods (rule of mixtures) and the multipoint constraint (MPC) technique are detailed, allowing for accurate local-global analysis while significantly reducing computational time. Crucially, AI-assisted DoS employs equation-informed neural networks (EINN) alongside Bayesian inference to rapidly and robustly extract coefficients for constitutive models, such as the Anand and Chaboche models.

Chapter 9 presents the strategic value of AI-driven design-on-simulation (AI-DoS) in electronic packaging, offering a cost-effective alternative to laborious experimentation while mitigating the limitations of traditional FEM approaches. AI-DoS methodology involves validating FEM simulations to generate large datasets, which are then used to train supervised machine learning (ML) algorithms (like ANN, KRR, random forest (RF), and GPR). Once established, these AI models can rapidly predict component reliability life cycles (e.g., WLP fatigue life), often in milliseconds to seconds, significantly shortening the design cycle. This chapter also explores strategies for handling limited training data, such as ensemble learning, which combines multiple algorithms to improve prediction accuracy and robustness. Finally, it analyzes empirical models, endorsing the use of coefficients derived from simulation data for the modified acceleration factor (AF) equation, as this excludes uncertainties inherent in physical testing.

For whom is this book intended? Undoubtedly, it will be of great interest to three groups of specialists: (1) those who are active or intend to become active in the research and development of Cu-Cu hybrid bonding, CPO, glass packaging, the mechanical behavior of packaging, AI technology and machine learning algorithms, modeling and simulation, and the coupling of AI techniques with finite element analysis; (2) those who have encountered practical Cu-Cu hybrid bonding, CPO, glass packaging, and investigations into mechanical performance, with emphasis on fatigue and cracking issues, and how to ensure the simulation results and wish to understand and learn more methods for solving such problems; and (3) those who have to choose a reliable, creative, high-performance, high-density, low-power consumption, and cost-effective technique for their products. This book can also be used as a text for college and graduate students who have the potential to become our future leaders, scientists, and engineers in the electronics and optoelectronics industries.

We hope that this book will serve as a valuable reference source for all those faced with the challenging problems created by the ever-increasing interest in Cu-Cu hybrid bonding, CPO, glass packaging, and crack and reliability life prediction. We also hope that it will aid in stimulating further research and development on key enabling technologies and more sound applications of Cu-Cu hybrid bonding, CPO, and glass packaging, particularly in enhancing the trustworthiness and consistency of simulation results to support the development of more competitive products. The organizations that learn how to design and manufacture Cu-Cu hybrid bonding, CPO, glass packaging, and how to apply AI-assisted design-on-simulation technology in their semiconductor packaging systems have the potential to make major advances in the electronics and optoelectronics industry and to gain great benefits in cost, performance, functionality, density, power, bandwidth, quality, size, and weight. It is our hope that the information presented in this book may assist in removing roadblocks, avoiding unnecessary false starts, and accelerating the design, materials, process, manufacturing development, and reliability of Cu-Cu hybrid bonding, CPO, glass packaging, and AI-assisted design-on-simulation technologies.

Palo Alto, CA, USA John Lau
Hsinchu, Taiwan Kuo-Ning Chiang

# Acknowledgments

The development and preparation of *Cu-Interconnects, Glass, and AI-Assisted Simulation for Chiplets and Heterogeneous Integration* was facilitated by the efforts of a number of dedicated people. We would like to thank them all, with a special mention of Mrs. Suganya Kothandaraman and Mrs. Mahalakshmi Saravanan, Springer Nature Scientific Publishing Services (P) Ltd., for her unswerving support and advocacy. Our special thanks also go to Ms. Jasmine Dou, Springer, who helped make our dream of this book a reality by effectively sponsoring the project and solving many problems that arose during its preparation. It has been a great pleasure and a fruitful experience to work with all of them in transforming our messy manuscripts into a very attractive printed book.

The material in this book has been derived from many sources, including individuals, companies, and organizations. We have attempted to acknowledge the assistance we received through citations in the appropriate sections of the text. It would be quite impossible for us to express our thanks to everyone concerned for their cooperation in producing this book, but we would like to extend our due gratitude. Also, we would like to thank several professional societies and publishers for permitting us to reproduce some of their illustrations and information in this book, including the American Society of Mechanical Engineers (ASME) conference proceedings (e.g., International Intersociety Electronic Packaging Conference) and transactions (e.g., *Journal of Electronic Packaging*), the Institute of Electrical and Electronic Engineers (IEEE) conference proceedings (e.g., Electronic Components and Technology Conference and Electronics Packaging and Technology Conference) and transactions (e.g., Components, Packaging, and Manufacturing Technologies), and the International Microelectronics and Packaging Society (IMAPS) conference proceedings (e.g., International Symposium on Microelectronics) and transactions (e.g., *International Journal of Microcircuits and Electronic Packaging*).

John would like to thank his former employers, ASM (HK), Industrial Technology Research Institute (ITRI), the Hong Kong University of Science and Technology (HKUST), the Institute of Microelectronics (IME), Agilent, and HP, for providing him with excellent working environments that have nurtured him as a human being, fulfilled his need for job satisfaction, and enhanced his professional reputation.

Also, John would like to thank Dr. Don Rice (HP), Dr. Steve Erasmus (Agilent), Prof. Dim-Lee Kwong (IME), Prof. Ricky Lee (HKUST), Dr. Ian Yi-Jen Chan (ITRI), and Mr. Lee Wai Kwong (ASM) for their kindness and friendship while he was at their respective organizations. Furthermore, John would like to thank Mr. Tzyy-Jang Tseng (Chairman of Unimicron Technology Corporation) for his trust, respect, and support of his work at Unimicron. Definitely, John would like to thank his eminent colleagues (the enumeration of whom would not be practical here) at Unimicron, ASM, ITRI, HKUST, IME, Agilent, EPS, HP, and throughout the electronics industry for their useful help, strong support, and stimulating discussions. Working and socializing with them has been a privilege and an adventure. John learned a lot about life and advanced packaging from them.

Lastly, John would like to thank his daughter Judy and wife Teresa for their love, consideration, and patience by allowing him to work peacefully on this book. Their simple belief that I am making a contribution to the electronics industry was a strong motivation for me. Thinking that Judy and her supportive husband (Bill) have been doing very well in the electronics industry and their two adorable kids (Allison and James) have been raising in a happy and loving environment, and Teresa and I are in good health, I want to thank God for His generous blessings.

Kuo-Ning would like to express his heartfelt gratitude to his parents, his wife, Tracy, his son, Jeffrey, and his daughter, Jacqueline, for their enduring support and understanding throughout his research and professional journey. Their constant encouragement has been a continual source of motivation. I also wish to extend my sincere thanks to National Tsing Hua University for providing an inspiring and supportive environment that has greatly nurtured my research. The university's steadfast commitment to academic excellence has played a vital role in shaping the foundation of this work. I want to thank my former employers. At the MacNeal-Schwendler Corporation, my experience in developing MSC/NASTRAN provided me with deep insight into how finite element and supercomputing theories can be effectively integrated with practical engineering applications. At the National Center for High-Performance Computing in Taiwan, I had the privilege of collaborating with my colleagues on the design and construction of several supercomputers. This experience provided me with valuable insight into developing system architectures optimized for software performance.

My deepest appreciation also goes to all the students who have been part of my laboratory over the years. Their dedication and contributions have led to numerous significant research achievements, many of which are referenced in this book. I am also sincerely grateful to the National Science and Technology Council of Taiwan, SRC-USA, and the supporting companies for their generous research funding throughout my academic career. Finally, I would like to thank my current Ph.D. students, Qinghua Su and Y. T. Su, for their careful and thoughtful proofreading of my manuscript.

**Competing Interests** The authors have no competing interests to declare that are relevant to the content of this manuscript.

# Contents

# About the Authors

**John Lau** was the CTO from July 2019 to July 2021 and has been a Senior Special Project Assistant at Unimicron in Taiwan and Palo Alto since August 2021. Prior to that, he was a Senior Technical Advisor at ASM Pacific Technology in Hong Kong for 5 years; a specialist at the Industrial Technology Research Institute in Taiwan for 5 years; a Visiting Professor at Hong Kong University of Science and Technology for 1 year; the Director of the Microsystems, Modules, and Components Laboratory at the Institute of Microelectronics in Singapore for 2 years; and a Senior Scientist/ MTS at Hewlett-Packard Laboratory/Agilent in California for more than 25 years. He earned a Ph.D. in Theoretical and Applied Mechanics from the University of Illinois at Urbana-Champaign, an M.A.Sc. in Structural Engineering from the University of British Columbia, a second M.S. in Engineering Mechanics from the University of Wisconsin at Madison, and a third M.S. in Management Science from Fairleigh Dickinson University in New Jersey. He also holds a B.E. in Civil Engineering from National Taiwan University. With more than 40 years of R&D and manufacturing experience in semiconductor packaging, John has published more than 530 peer-reviewed papers (serving as principal investigator for 385), holds 52 issued and pending US patents (serving as principal inventor for 31), and has authored 24 textbooks. John is an elected IEEE Fellow, IMAPS Fellow, and ASME Fellow and has been actively participating in industry, academic, society meetings, and conferences to contribute, learn, and share his expertise.

**Kuo-Ning Chiang** received his Ph.D. from the Georgia Institute of Technology in the United States. He is a Chair Professor at National Tsing Hua University in Hsinchu, Taiwan. After graduating from Georgia Tech, he worked for four years as a senior researcher at MSC/NASTRAN, a world-famous finite element system. From 2010 to 2013, he served as the General Director of the National Center for High-Performance Computing, which is the National Strategic Research Center of Taiwan. He has received outstanding research awards from the Ministry of Science and Technology of Taiwan three times and has published more than 450 technical papers in international journals and conference proceedings. He has been granted more than 50 invention patents. Among the major awards he has received are the

Excellence in Mechanics Award from ASME-EPPD (2022) and the Outstanding Sustained Technical Contribution Award (2020) from IEEE-EPS. He served as the Editor-in-Chief of *IEEE Transactions on Advanced Packaging*, Editor-in-Chief of *Journal of Mechanics*, and Senior Area Editor of *IEEE Transactions on Components, Packaging and Manufacturing Technology*. Currently, he serves as the Academic Editor of *Materials* (SCI) and Associate Editor of the *Journal of Electronic Packaging* (SCI). He is a Fellow of the IEEE, ASME, STAM, AIIA, iMAPS, and AAIA, and an Academician of the International Academy of Engineering (IAE). He has made significant achievements in simulation-based science and technology.

# Chapter 1
# Advanced Packaging

**Abstract** In this chapter, 2D, 2.1D, 2.3D, 2.5D, 3D. 3.3D, and 3.5D IC integrations will be presented and updated. Chiplet communication such as bridges embedded in build-up package substrate and fan-out epoxy molding compound (EMC) with RDLs (redistribution layers) will be examined and updated. The high volume products by Cu-Cu hybrid bonding will be presented. High bandwidth memory (HBM) and customized HBM (cHBM) and CoWoS (chip on wafer on substrate) and CoPoS (chip on panel on substrate) are the key elements of high-performance computing (HPC) products driven by artificial intelligence (AI) will be discussed. CoWoS-S, CoWoS-R, and CoWoS-L will be systematically presented. Glass-core substrates and glass-core interposers will be provided. 3D heterogeneous integration of photonic IC (integrated circuits) (PIC) and electronic IC (EIC) driven by AI and high-speed communications will be presented. Some recommendations will be provided.

## 1.1 Introduction

Fifteen years ago, the boundary and task between the semiconductor, packaging, and carrier/PCB (printed circuit board) were very clear and each one was doing their jobs properly as shown in Fig. 1.1. During the TSMC's investor conference for the third quarter of 2011 when Dr. Morris Chang (founder of TSMC), without any advance warning shocked everybody by announcing his company would move into the packaging and testing field. The first product would be CoWoS, which integrates logic computing and memory chips by mounting them on a silicon interposer and then placing them directly on a package substrate. Today, the industry calls CoWoS 2.5D IC integration.

Since then, the boundary and task between the semiconductor, packaging, and carrier/PCB become blurry and there are overlapping between the semiconductor and the packaging, and between the packaging and carrier/PCB as shown in Fig. 1.2. It can be seen that both the semiconductor and packaging are working on, e.g.,



J. Lau, K.-N. Chiang, *Cu-Interconnects, Glass, and AI-Assisted Simulation for Chiplets and Heterogeneous Integration*,
https://doi.org/10.1007/978-981-95-6891-8_1

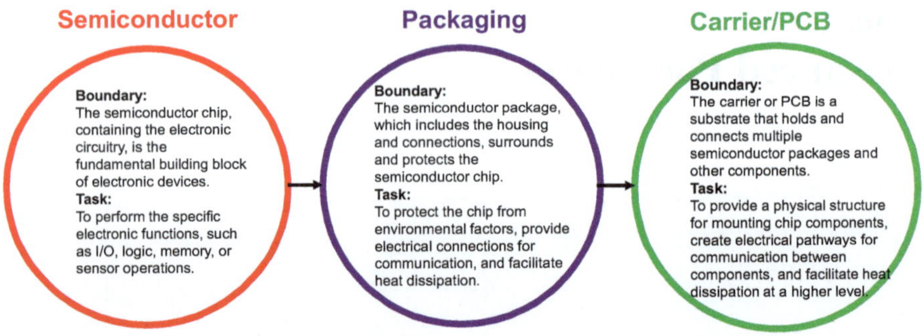


**Fig. 1.1** Boundaries and tasks among semiconductors, packaging, and carrier/PCB

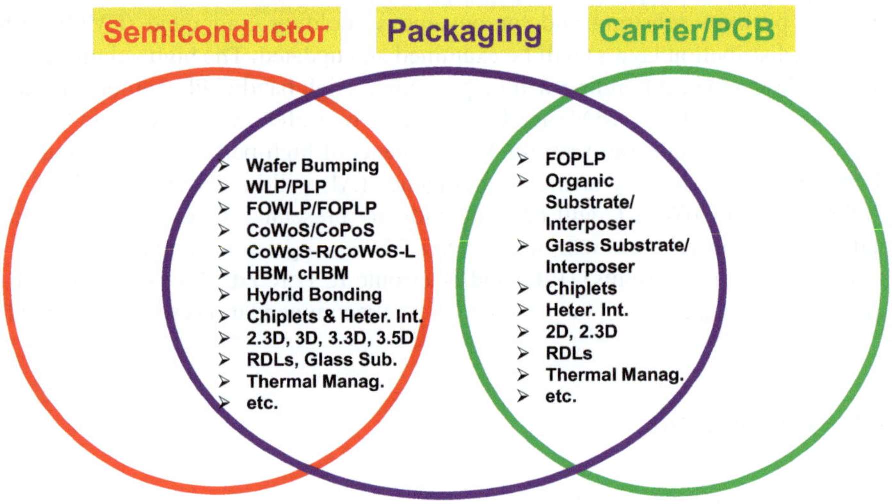


**Fig. 1.2** Overlapping boundaries and tasks among semiconductor, packaging, and carrier/PCB functions

wafer bumping, WLP (wafer-level packaging), PLP (panel-level packaging), FOWLP (fan-out wafer-level packaging),

FOPLP (fan-out panel-level packaging), CoWoS, CoPoS, chiplets and heterogeneous integration, 2.D, 2.3D, 3D, 3.3D, 3.5D, RDLs (redistribution-layers), HBM, cHBM, glass substrate/interposer, thermal management, etc. Also, both the packaging and carrier/PCB are working on FOPLP (fan-out panel level packaging), organic substrate/interposer, glass substrate/interposer, chiplets and heterogeneous integration, 2D, 2.3D, RDLs, thermal management, etc. In this study, the focus is on most of the overlapping between the semiconductor and the packaging.

Four years ago, the recent advanced and trends in advanced packaging have been given [1]. In the past four years, because of the strong drive of AI, there have been

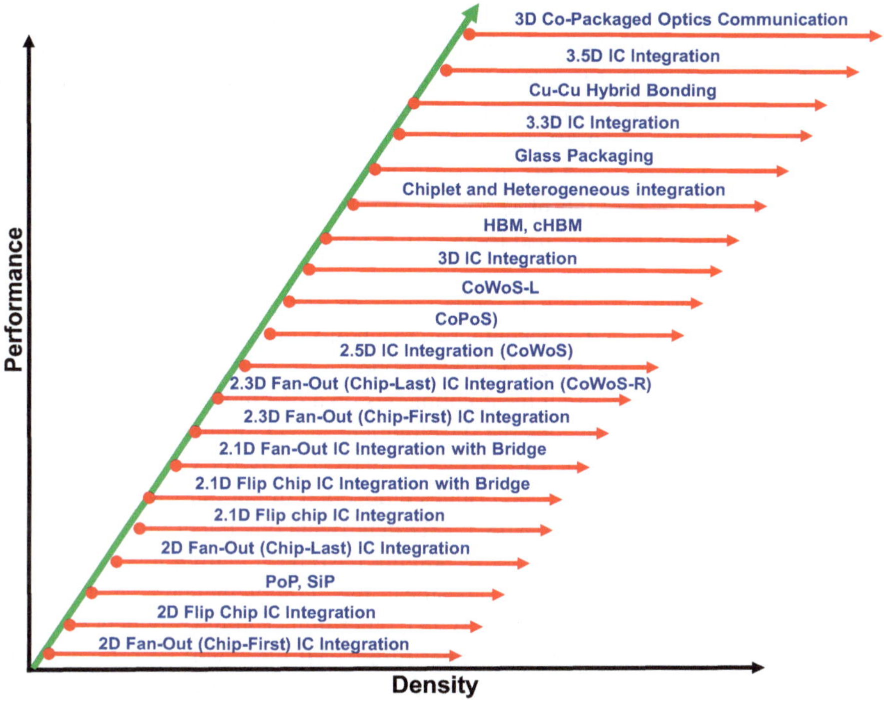


**Fig. 1.3** Advanced packaging ranking according to their density and performance

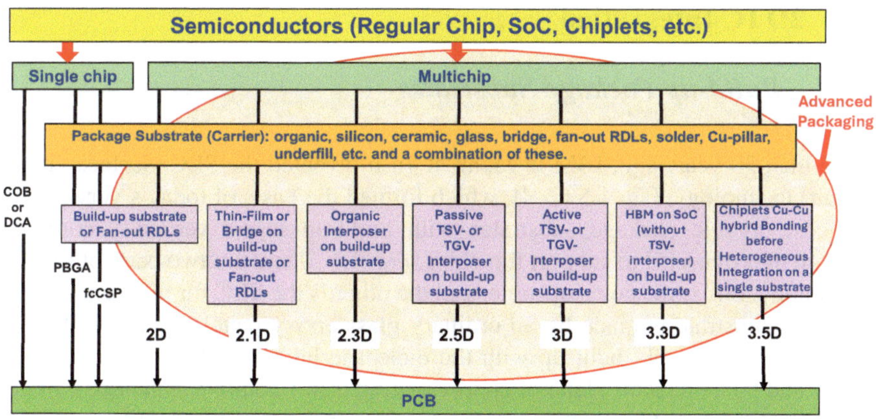


**Fig. 1.4** Groups of advanced packaging: 2D, 2.1D, 2.3D, 2.5D, 3D, 3.3D, and 3.5D IC integration

many new developments and new products being shipped for HPC. In this study, these new developments and products will be presented.

Just like in [1], advanced packaging is defined, Figs. 1.3 and 1.4, as from the 2D IC integration all the way to the 3.5D IC integration. Depending on applications, their electrical performance and interconnect density ranking, and the groups of packaging are schematically shown in Figs. 1.3 and 1.4. It can be seen that for 2D, the multichip are attached to a build-up substrate or fan-out RDLs. For 2.1D, the multichip are attached to a thin-film layer or bridges on a build-up substrate or fan-out EMC with RDLs. For 2.3D, the multichip are attached to an organic interposer on a build-up substrate. For 2.5D, the multichip are attached to a passive through-silicon via (TSV)- or through-glass via (TGV)-interposer on a build-up substrate. For 3D, the multichip are attached to an active TSV- or TGV-interposer on a build-up substrate.

For 3.3D, the multichip such as the HBM or cHBM are attached directly on top of the SoC (system on a chip) then attached to a build-up substrate without any TSV or TGV interposer. For 3.5D, Cu-Cu hybrid bonding some of the chiplets before the heterogeneous integration of all the components on a common substrate.

In the past four years, besides the development of 2D through 3.5D IC integrations, other developments such as HBM vs. cHBM, CoWoS vs. CoPoS, bridges embedded in build-up substrate vs. bridges embedded in fan-out EMC with RDLs, CoWoS-S vs. CoWoS-R vs. CoWoS-L, 3D heterogeneous integration of PIC and EIC (co-packaged optics), and glass packaging have been getting lots of tractions. As for Cu-Cu hybrid bonding, besides the complementary metal-oxide-semiconductor (CMOS) image sensor (CIS), other high-volume products have also been shipped or to be shipped in the near future. All these will be presented and discussed in this study. Some recommendations will be provided.

## 1.2 2D IC Integration

### *1.2.1 Build-up Package Substrates*

More than 30 years ago, IBM in Japan at Yasu invented the SLC (surface laminar circuit) technology, Fig. 1.5 [2–4], which formed the basis of today's very popular low-cost organic package substrates with build-up layers vertically connected through microvias [5] to support flip chips, Fig. 1.6. There are two parts of the SLC technology: one is the core substrate, and the other is the SLC for the signal wiring. The core substrate is made by an ordinary glass epoxy panel. However, the SLC layers are sequentially built up with the dielectric layers made of photo sensitive epoxy and the conductor plane of copper plating (semi-additive technique). In general, a package substrate with twelve layers (e.g., two core-layers and ten build-up layers (5-2-5)) and 10 μm-line width and spacing is more than adequate to support most of the chips. Today, a (12-2-12) build-up package is not non-common.

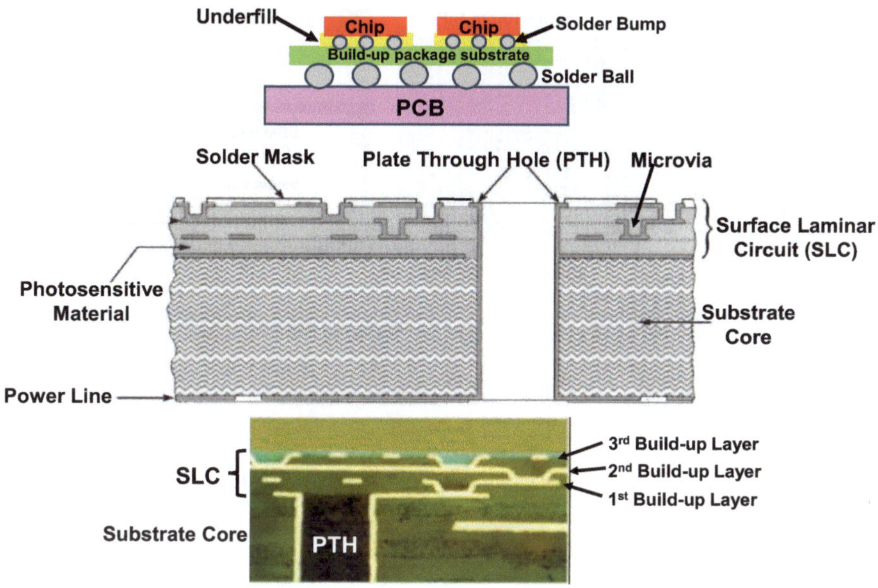


**Fig. 1.5** IBM SLC (surface laminar circuit) technology

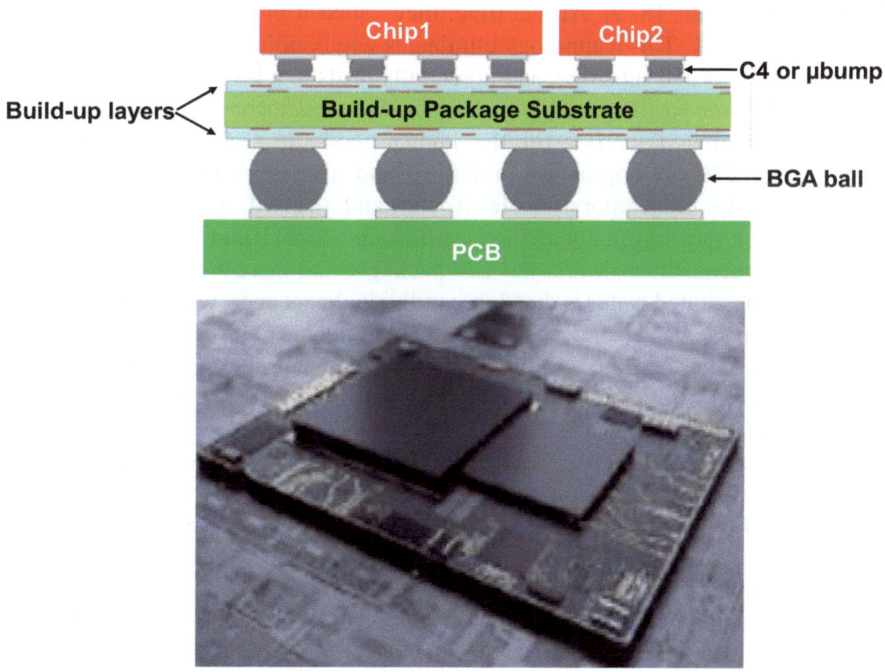


**Fig. 1.6** Flip chips on build-up package substrate

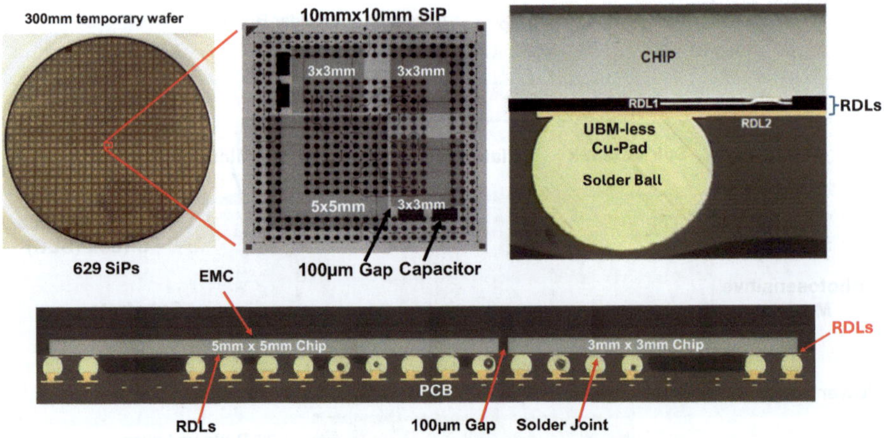


**Fig. 1.7** FOWLP – heterogeneous integration of 4 chips and 4 capacitors

## *1.2.2 Fan-out RDL-Substrates*

The first fan-out wafer-level packaging (FOWLP) U.S. patent was filed by Infineon on October 31, 2001 [6, 7], and the first technical papers were also published (at ECTC2006 and EPTC2006) by Infineon and their industry partners: Nagase, Nitto Denko, and Yamada [8, 9]. At that time, they called it embedded wafer-level ball (eWLB) grid array. This technology eliminates wirebonding or wafer bumping and leadframe or package substrate, and potentially leads to a lower cost, better performance, and lower profile packaging. Alternatively, this technology requires a temporary (reconstituted) carrier for the known-good die (KGD), epoxy molding compound (EMC), compression molding, and the fabrication of the RDLs. Figure 1.7 shows the temporary wafer (round) and a closed-up on one of the packages by the fan-out chip-first with die face-down process [10, 11]. It can be seen that the 300 mm reconstituted wafer has 629 packages and there are 4 chips in a package. The reconstituted wafer is diced into individual packages, where the RDL1 and RDL2 of the package are shown. The thickness of the metal layer of RDL2 (7.5 μm) is thicker than that of RDL1 (3 μm). The thicker metal layer of RDL2 is for the under bump metallization (UBM)-less thick Cu pads to "resist" the Cu consumption from the solder ball reflow and during operation.

The design, materials, process, and fabrication of a heterogeneous integration of four chips by a FOPLP method are investigated in [12, 13], Fig. 1.8. Emphasis is placed on (1) the application of a dry-film epoxy molding compound for molding the chips and (2) the application of a special assembly process called uni-substrate-integrated package for fabricating the RDLs of the FOPLP. The Ajinomoto build-up film is used as the dielectric of the RDLs and is built up by the semi additive process. Electroless Cu is used to make the seed layer, laser direct imaging is used for opening the photoresist, and printed circuit board (PCB) Cu plating is used for making the conductor wiring of the RDLs. The panel dimensions are 508 mm × 508 mm.

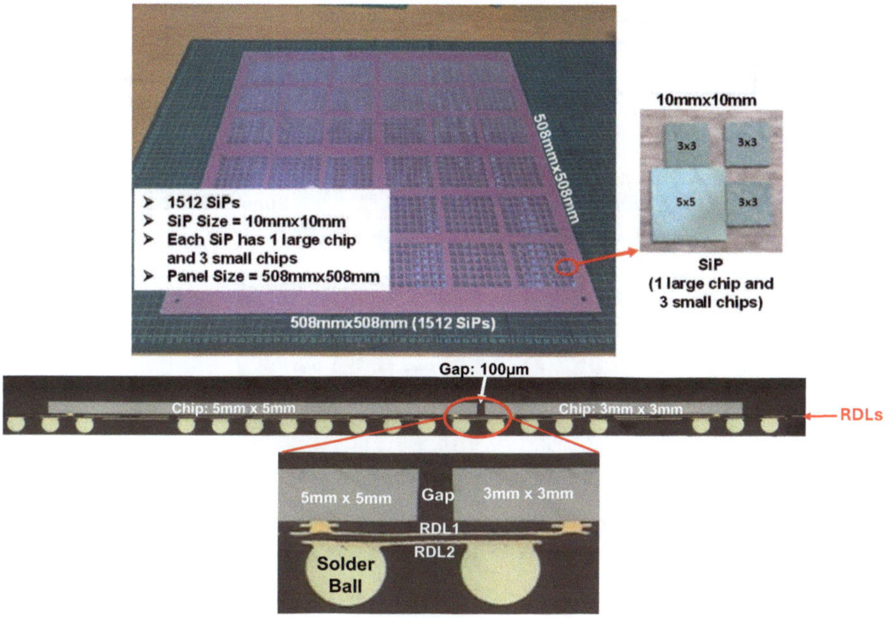


**Fig. 1.8** FOPLP – heterogeneous integration of 4 chips

The package dimensions of the FOPLP are 10 mm × 10 mm. The large chip size and the small chip sizes are, respectively, 5 mm × 5 mm and 3 mm × 3 mm. The uniqueness of this study is that all the processes are carried out by using the PCB equipment.

## 1.3 2.1D IC Integration

### *1.3.1 Thin-Film Layer above Build-up Package Substrate*

Shinko's i-THOP (integrated thin film high density organic package) is shown in Fig. 1.9 [14, 15]. In 2013, Shinko proposed to make thin-film layers on top of the build-up layer of a package substrate. The thickness, line width, and spacing of the thin-film Cu RDLs can be as small as 2 μm. The thin-film Cu RDLs are vertically connected through a 10 μm via. The surface Cu pad pitch is 40 μm, and the Cu pad diameter is 25 μm with a height of 10–12 μm. However, because of the flatness of the build-up substrate, the manufacturing yield of the thin-film layer is not good.

The flatness problem of the build-up substrate could be solved [16] by fabricating the build-up substrate and the thin-film layers separately and combing them together with a photoimageable dielectric or Ajinomoto build-up film. Then, drill vias and fill with Cu through the thin-film layer and the adhesive layer, Figs. 1.10 and 1.11 [16].

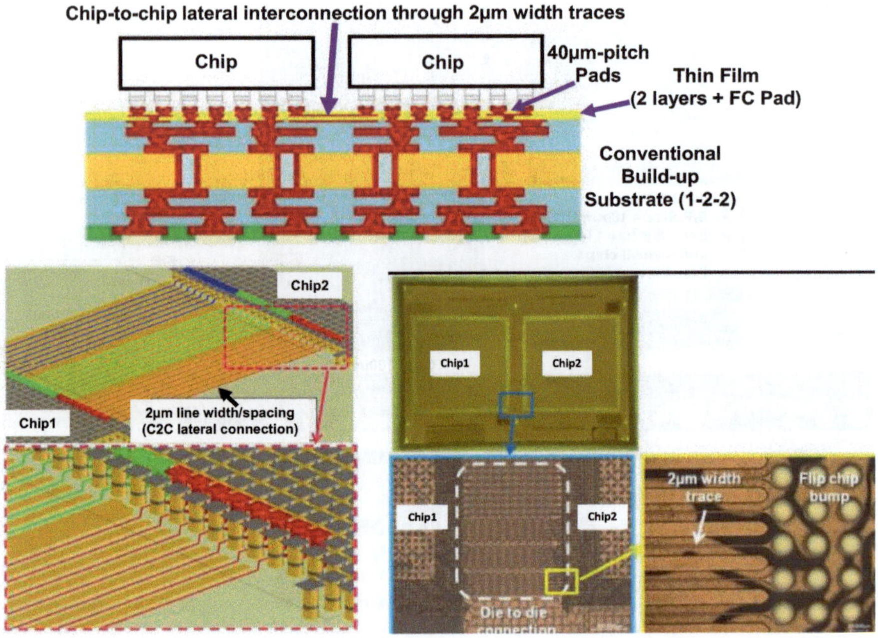


**Fig. 1.9** Shinko's i-THOP

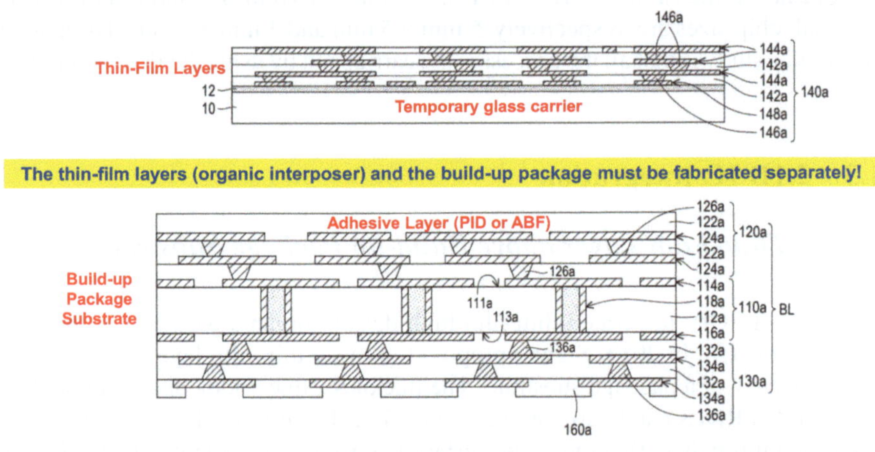


**Fig. 1.10** Fabricate the thin-film layers and build-up package substrate separately

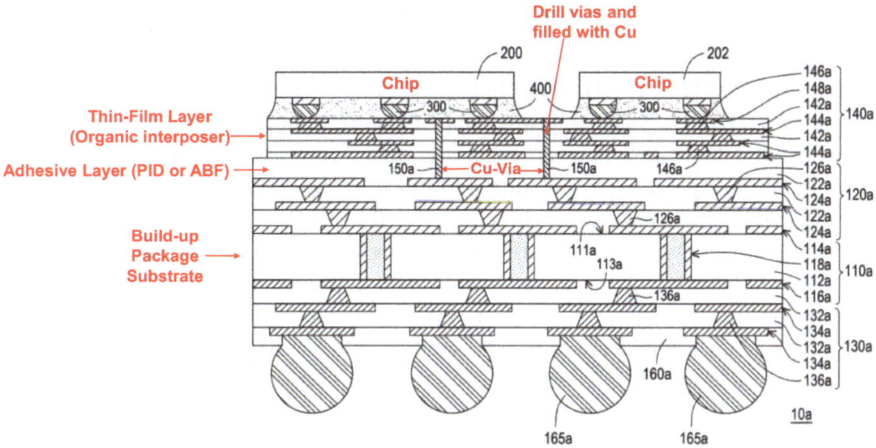


Fig. 1.11 Thin-film layers directly on top of build-up package substrate (2.1D)

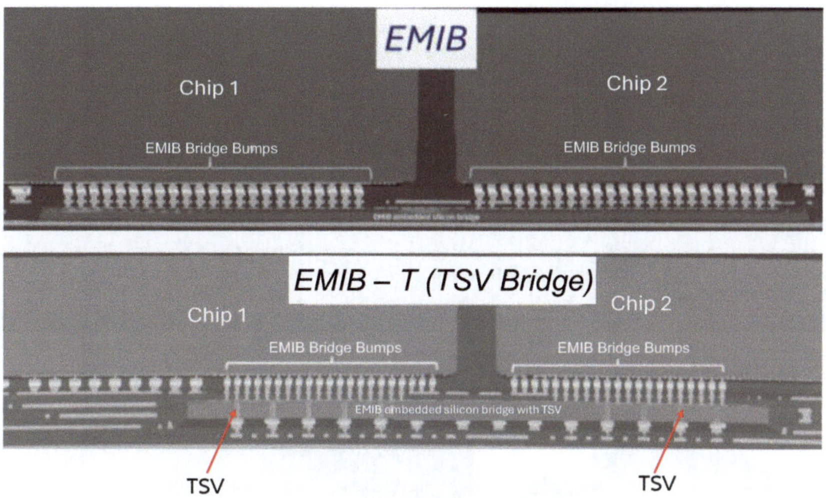


Fig. 1.12 EMIB with TSVs

## 1.3.2 Bridge Embedded in Build-up Package Substrate

During ECTC 2016, Intel published the paper on "Embedded multi-die interconnect bridge (EMIB)—A high-density, high bandwidth packaging interconnect" [17, 18], The EMIB is semi-embedded on a build-up package substrate with cavity. Intel's EMIB has been used in a number of Intel products such as FPGA (Agilex), Ponte Vecchio GPU, and Sapphire Rapids. During IEEE/ECTC 2025, Intel published another paper on EMIB with TSV as shown in Fig. 1.12 [19]. It can be seen that besides RDLs for the horizontal chiplets communications on the EMIB there are also TSVs for the vertical chiplets communications.

### *1.3.3 Bridge above Build-up Package Substrate*

During IEEE/ECTC 2023, IBM published a paper on "Direct Bonded Heterogeneous Integration (DBHi): Surface bridge approach for die tiling" [20], Fig. 1.13. It showed that a Si bridge is microbumping with two chiplets and the Si bridge is above the build-up package substrate. In this case it releases the pressure in making the build-up package substrate with cavity for the EMIB.

UltraFusion is Apple's innovative packaging architecture that interconnects the die of two M1 Max chips to create a SoC with unprecedented levels of stunningly compact design, extensive connectivity, performance and capabilities (shipped in 2022) [21]. This architecture does not combine two M1 Max dies into a single chip package—it also makes the two dies present themselves as a single chip. The interconnection between the two M1 Max dies is by a "silicon bridge" as shown in Fig. 1.14. The chiplets and the Si-bridge are supported by an ordinary build-up package substrate without any cavity.

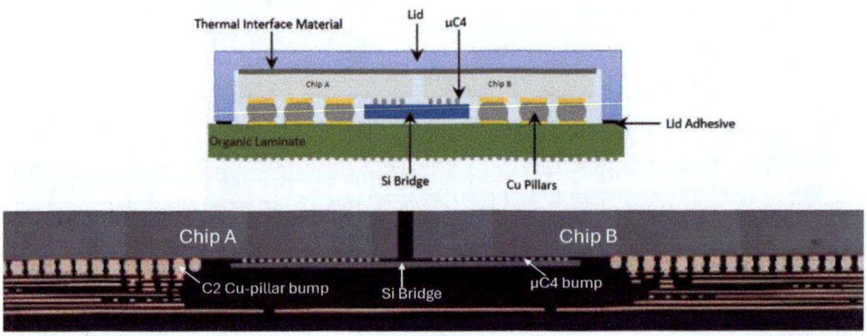


**Fig. 1.13** Direct bonded heterogeneous integration with Si bridge

**Fig. 1.14** UltraFusion with Si bridge

### 1.3.4 *Bridge embedded in Fan-Out EMC with RDLs*

Figure 1.15 shows a structure of chiplets communication through a Si bridge embedded in fan-out EMC with RDLs [22]. It can be seen that the Si bridge is fabricated by a 64 nm process and the minimum pitch of the RDLs is 0.4 μm. After dicing, the Si bridge is embedded in the fan-out EMC with RDLs. The circuits above the EMC are transmitted to its bottom through the through mold via (TMV) and then through the C4 bumps to the package substrate.

## 1.4 2.3D IC Integration

The motivation of 2.3D IC integration is to replace the TSV interposer (2.5D IC integration) with a fan out fine metal L/S RDL- substrate (or organic interposer), Figs. 1.16, 1.17, 1.18, 1.19, 1.20, 1.21, 1.22, 1.23, 1.24, 1.25 and 1.26. The hybrid substrate structure consists of a build-up package substrate [or high density interconnect (HDI)], solder joints with underfill, and an organic interposer, Figs. 1.16, 1.17, 1.20, and 1.24. There are many publications in this area [23–55]: [23–27] are with fan-out chip-first packaging process, while [28–55] are with fan-out chip-last packaging process.

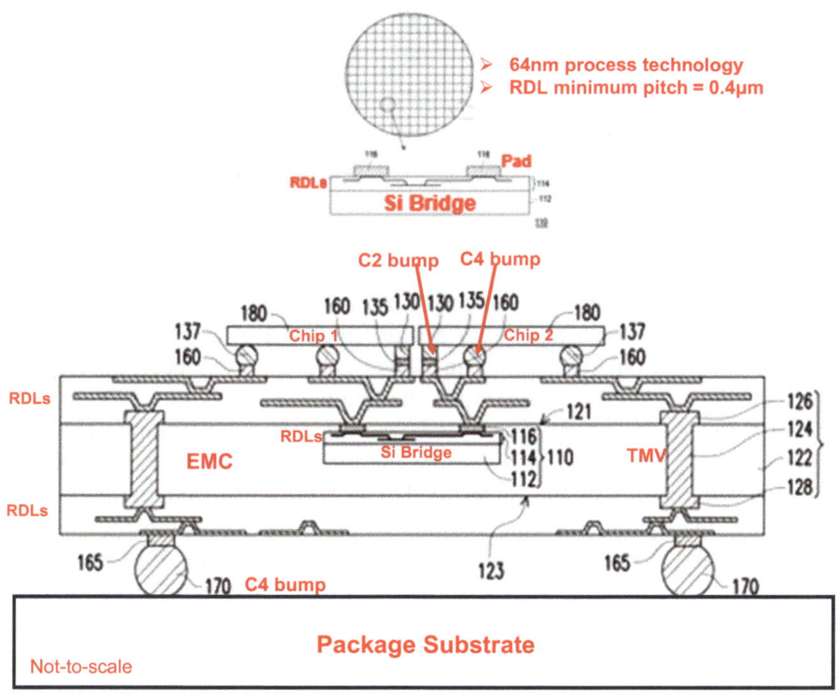


**Fig. 1.15** Bridge embedded in fan-out EMC with RDLs

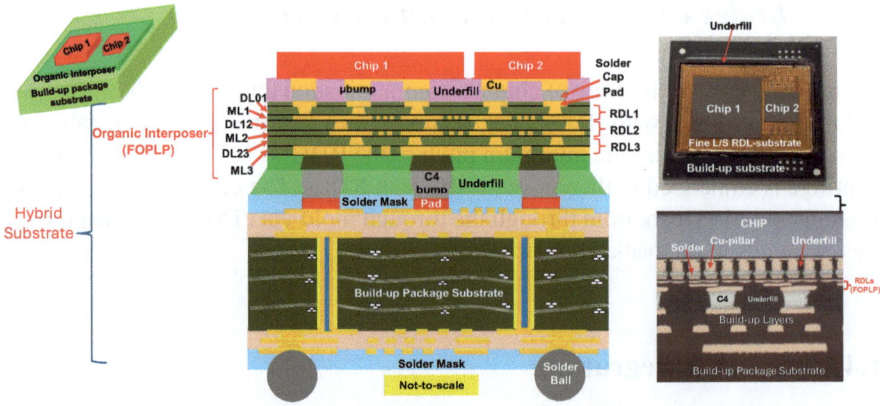


**Fig. 1.16** Hybrid substrate by fan-out RDL-First panel-level packaging (PID)

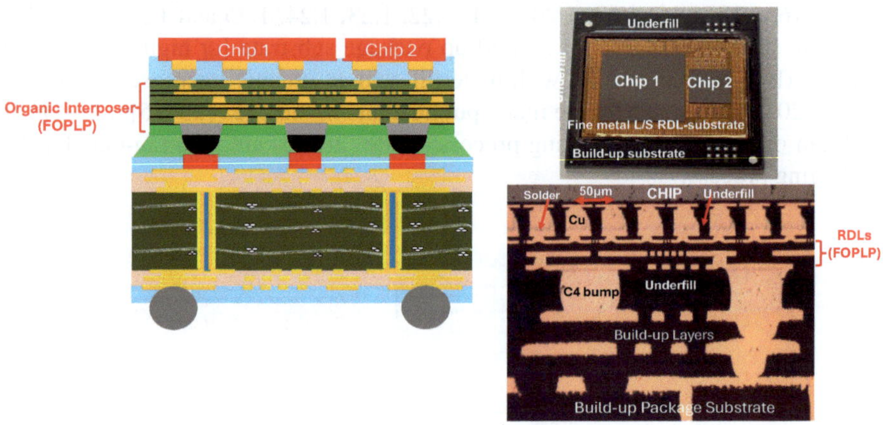


**Fig. 1.17** Hybrid substrate by fan-out RDL-First panel-level packaging (ABF)

The feasibility of the design, materials, process, and fabrication of a heterogeneous integration of two chips with 50-μm pitch on a 2.3D hybrid substrate (Fig. 1.16) by a fan-out RDL first panel-level packaging has been demonstrated [49–52]. In order to increase throughput, the fine metal L/S (2 μm) RDL-substrate has been fabricated on a temporary glass panel (515 mm × 510 mm). The hybrid substrate has been fabricated by soldering the fine metal L/S RDL-substrate on a build-up package substrate and underfilled. The 50-μm pitch microbumped chips are bonded to the hybrid substrate and underfilled. The hybrid substrate has been shown to pass the thermal simulation and the drop test. In [49–52], the dielectric material for the fine metal L/S RDL-substrate is photoimageable dielectric (PID). Recently, the dielectric material has been changed to ABF (Ajinomot build-up film) [54, 55], Fig. 1.17, which leads to much flatter metal traces of the RDLs, as shown in Figs. 1.18 and 1.19.

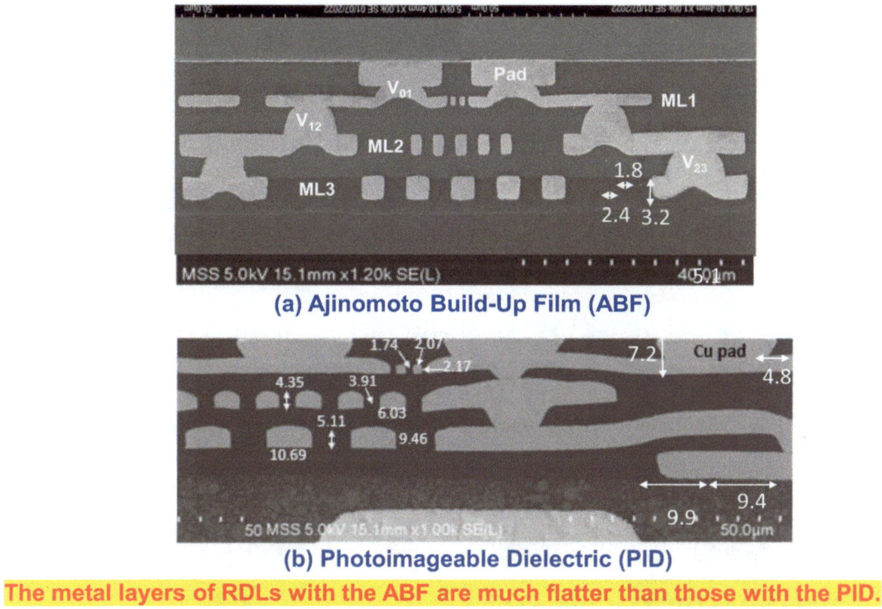


**Fig. 1.18** SEM images of the cross section of the organic interposers with (**a**) ABF and (**b**) PID

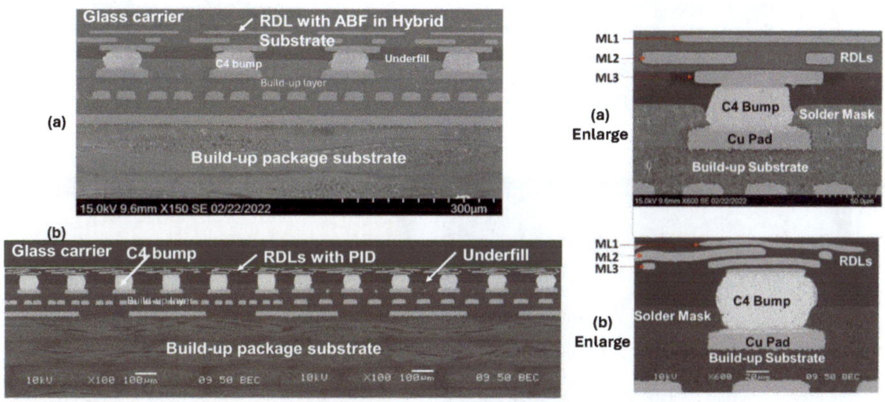


**Fig. 1.19** SEM images of the FOPLP RDLs in the hybrid substrate. (**a**) ABF. (**b**) PID

The fine-metal L/S RDL-substrate and the build-up package substrate (or HDI substrate) can also be combined through an interconnect layer [56–58] into a hybrid substrate. This is very similar to [49–52] except the C4 solder joint and underfill are replaced by an interconnect layer, as shown in Figs. 1.21, 1.22 and 1.23. For more information on the design, materials, process, fabrication, and reliability of the heterogeneous integration of three chips on a hybrid substrate with an interconnect layer by a fan-out RDL-first panel-level package, please read [56–58].

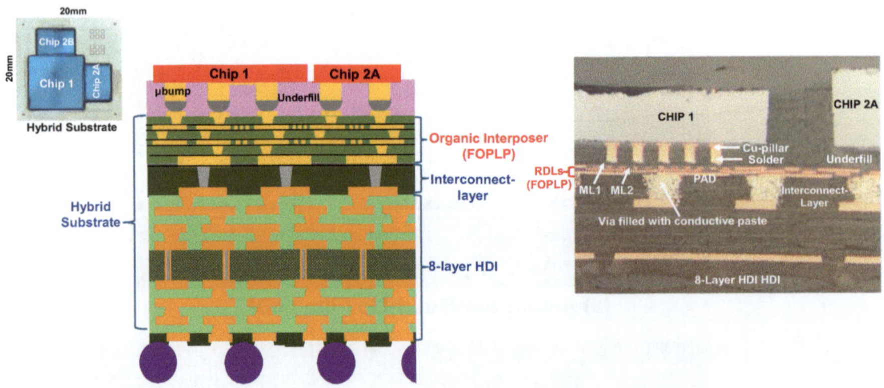


**Fig. 1.20** Hybrid substrate with interconnection-layer for heterogeneous integration

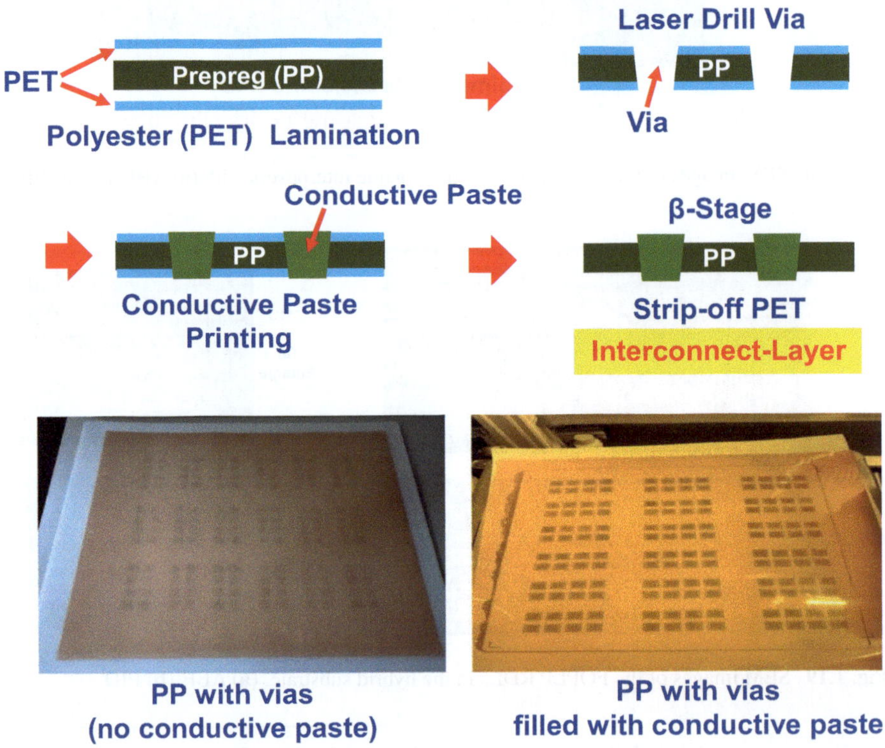


**Fig. 1.21** Process flow of the interconnection-layer

Figure 1.24 shows TSMC multilayer RDL-interposer for heterogeneous device and module integration [59–62]. The structure consists of (1) the chips are attached on an organic or inorganic RDL interposer with microbumps and underfill; (2) the RDL interposer is attached to a build-up package substrate with C4 bump with underfill; and (3) the package substrate is attached to a PCB with BGA (ball grid

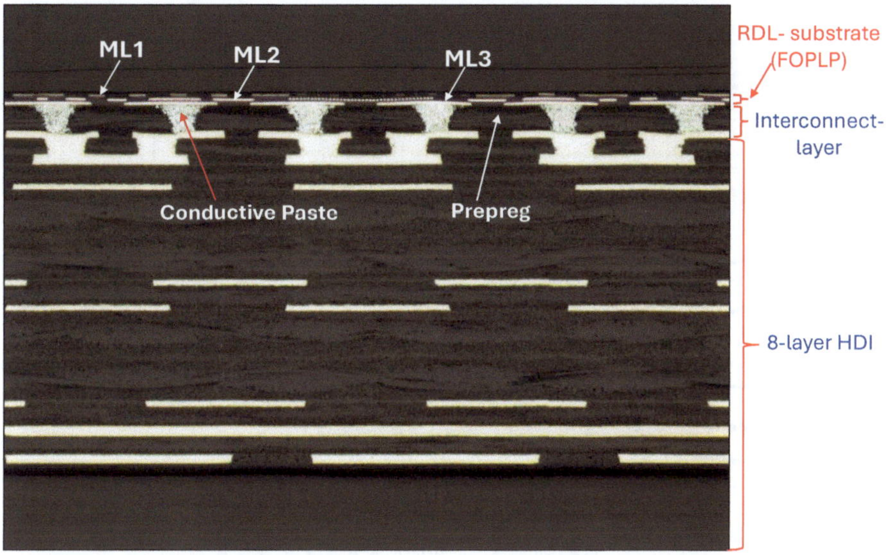


**Fig. 1.22** Typical cross section of the hybrid substrate with interconnection-layer

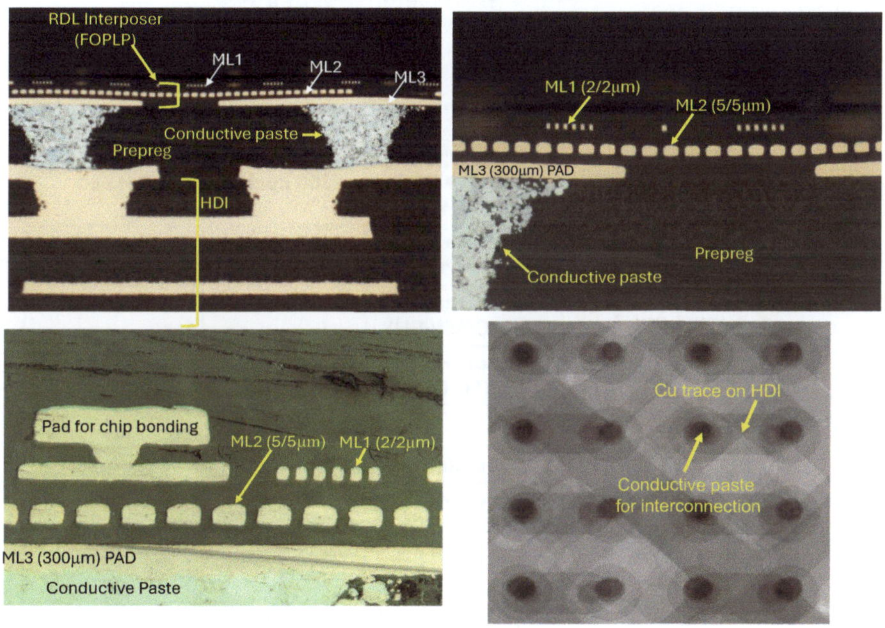


**Fig. 1.23** Close-up views of the hybrid substrate with interconnection-layer

array) solder ball. TSMC called it CoWoS-R. Figure 1.24 shows some images of the assembly. It can be seen that the chips and dynamic random-access memory (DRAM) are attached to a multilayer RDL- interposer with μbumps and then attached to a package substrate with C4 bumps. There are six RDLs with various via

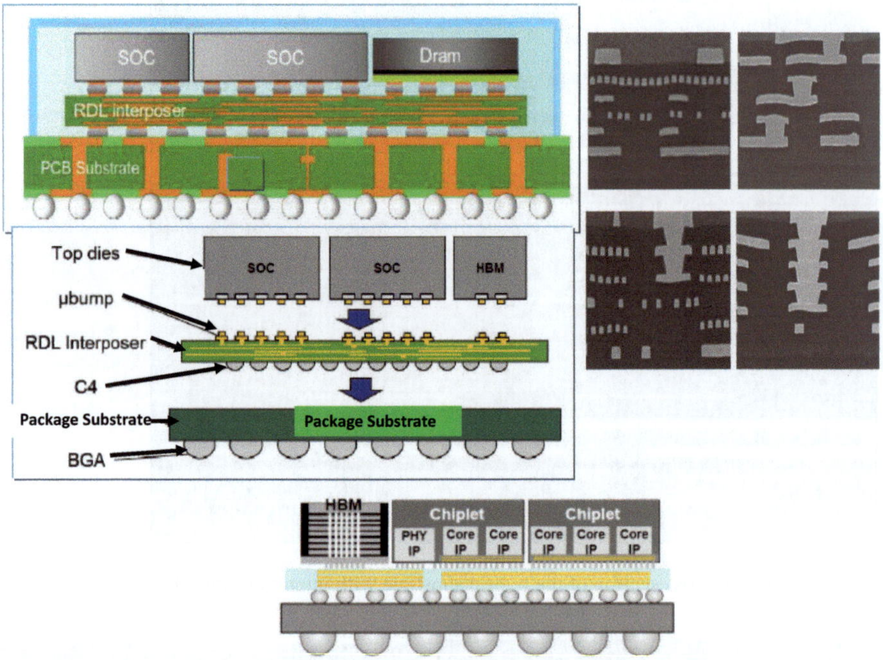


Fig. 1.24 CoWoS-R

(a)

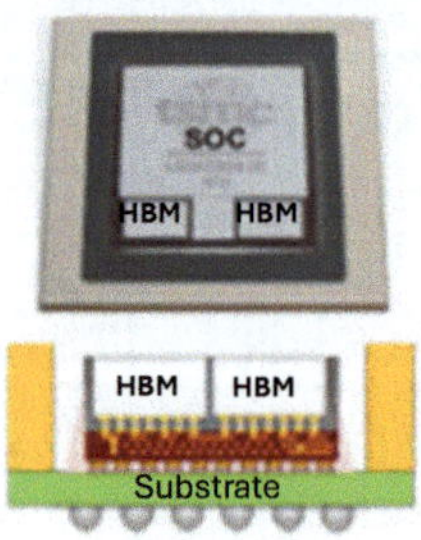


| Test Item | Test Conditions and Results | |
|---|---|---|
| Precon + TCC (-65°C ↔ 150°C) | 1300 cycles | Passed |
| Precon +TCG (-40°C ↔ 125°C) | 2400 cycles | Passed |
| Precon + uHAST (110°C/85%RH) | 264 hrs | Passed |
| Procon + HTS (150°C) | 1000 hrs | Passed |

(b)

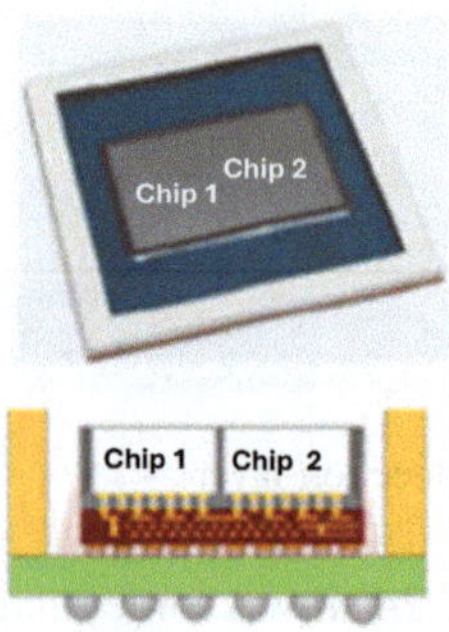


| Test Item | Test Conditions and Results | |
|---|---|---|
| Precon + TCC (-65°C ↔ 150°C) | 1000 cycles | Passed |
| Precon +TCG (-40°C ↔ 125°C) | 2500 cycles | Passed |
| Precon + uHAST (110°C/85%RH) | 264 hrs | Passed |
| Procon + HTS (150°C) | 1000 hrs | Passed |

Fig. 1.25 (**a**) SoC with two HBMs. (**b**) Two chiplets. Reliability data of CoWoS-R

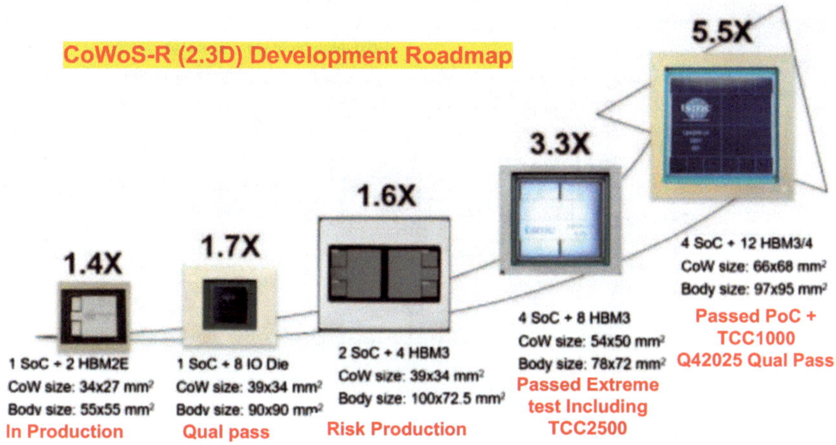


**Fig. 1.26** Roadmap of CoWoS-R

structures such as stager vias, two stacking vias, and four stacking vias. Reliability data are shown in Figs. 1.25 (a and b) [60].

Recently, TSMC integrated both a large amount of high-density integrated passive devices (IPDs) and fine pitch Si-based connection block of convenient IP migration [61]. Their 2.3D IC integration (CoWoS-R) development roadmap is shown in Fig. 1.26 [62]. It can be seen that the case of one SoC with two HBM2es (organic substrate size = 34 × 27 mm$^2$ and build-up package substrate size = 55 × 55 mm$^2$) is in production.

It should be pointed out that, unlike all the other 2.3D IC integrations, TSMC's RDLs (as much as 9 layers) on the organic interposer are built on a temporary 300 mm Si-wafer with the 64 nm semiconductor process technology.

## 1.5 2.5D IC Integration

### *1.5.1 CoWoS or CoWoS-S*

Because of the very high-density, high I/Os, and ultrafine pitch requirements such as the field programable gate array (FPGA), graphics processing units (GPU), central processing units (CPU), HBM, etc. even a fourteen build-up layers (6-2-6) package substrate is not enough to support the chips and a TSV (through-silicon via) interposer (2.5D IC integration) is needed [63–110]. Today, CoWoS or CoWoS-S (CoWoS-silicon interposer) is the package of choice for HPC driven by AI as shown in Fig. 1.27.

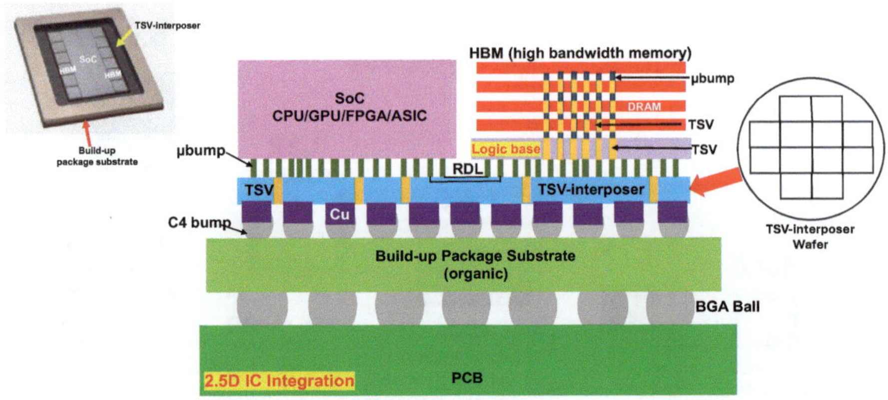


**Fig. 1.27** 2.5D with TSV-interposer fabricated on wafer

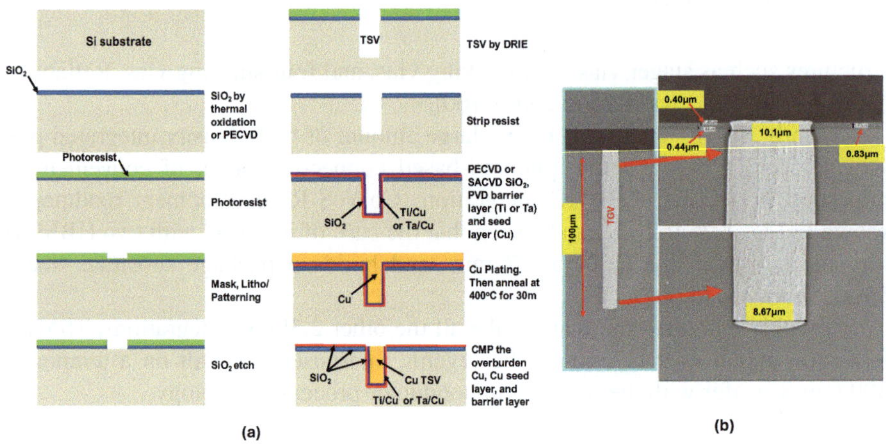


**Fig. 1.28** (**a**) TSV process flow. (**b**) SEM image of a TSV

There are at least two methods in fabricating the TSV: laser and deep reactive iron etching (DRIE). The process flow of making the TSV interposer with DRIE is shown in Fig. 1.28a, and the image of the TSV cross section is shown in Fig. 1.28b [79]. Depending on the linewidth (L) and spacing (S) of the RDLs, there are at least two methods in fabricating the RDLs for CoWoS, as shown in Figs. 1.29 (a and b). Figure 1.29a shows the process flow for the conventional RDLs (L/S ≥ 2 μm) with polymers such as photo-imageable dielectric (PID) or Ajinomoto build-up film (ABF) for the dielectric layer and electrochemical deposition (ECD) Cu for the conductor layer. The polymer is spin-coated on the silicon wafer.

Figure 1.29b shows the other process flow for the RDLs (L/S < 2 μm). In this case, the RDLs are fabricated with the 64 nm process technology. The dielectric layers ($SiO_2$) are fabricated by plasma

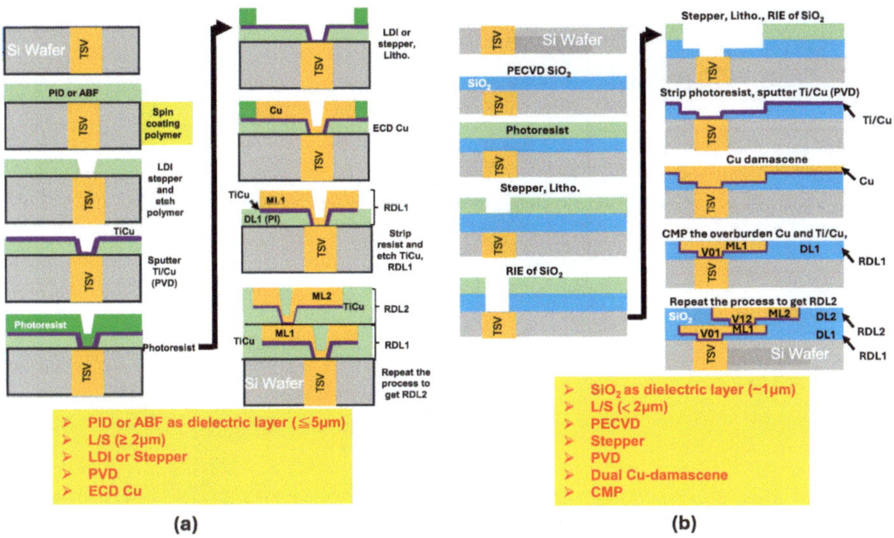


**Fig. 1.29** RDL process flow. (**a**) L/S ≥ 2 μm. (**b**) L/S < 2 μm

enhanced chemical vapor deposition (PECVD). The metal layers are fabricated by dual Cu-damascene and chemical-mechanical polishing (CMP). For example, the scanning electron microscope (SEM) image of the TSV with RDLs is shown in Fig. 1.30 [79]. Today, in all the CoWoS, the RDLs are fabricated by this method and the minimum pitch of the RDLs is 0.4 μm.

On October 20, 2013, Xilinx and TSMC [40] have jointly announced production release of the Virtex-7 HT family with 28 nm process technology, what the pair claims is the industry's first 2.5D IC integration in production. Since then, there are many products use the passive TSV-interposers such as AMD/UMC's Radeon R9 Fury X GPU [111], Nvidia/TSMC's Pascal 100 GPU [112], Graphcore/TSMC's Bow—An IPU (intelligence processing unit) [113], Fujitsu/TSMC's Exascale supercomputer [114], Samsung's I-Cube4/H-Cube/MIoS [115, 116], etc.

The recent passive TSV-interposers specifically used for HPC driven by AI are shown, e.g., in Figs. 1.31, 1.32 and 1.33. Figure 1.31a shows the NVIDIA-A100 double precision Tensor Cores. Combined with 80 GB of the fastest GPU memory, it can reduce a 10-h, double-precision simulation to under four hours on A100. The size of the GPU is 826 mm$^2$ and it is supported by six HBM2s, each consisting of 8 DRAMs (dynamic random access memories) on a logic base chip. These GPU and HBM2s are supported by a passive TSV-interposer and then the TSV-interposer is supported by a build-up package substrate.

Figure 1.31b shows the NVIDIA-H100 Tensor Core GPU. H100 accelerates AI training and inference, HPC, and data analytics applications in cloud data centers, servers, systems at the edge, and workstations. The GPU is supported by 6 HBM3 and each consists of 8 DRAMs and the logic base. All these chips are supported by a passive TSV-interposer, and then on a (4-2-4) build-up package substrate.

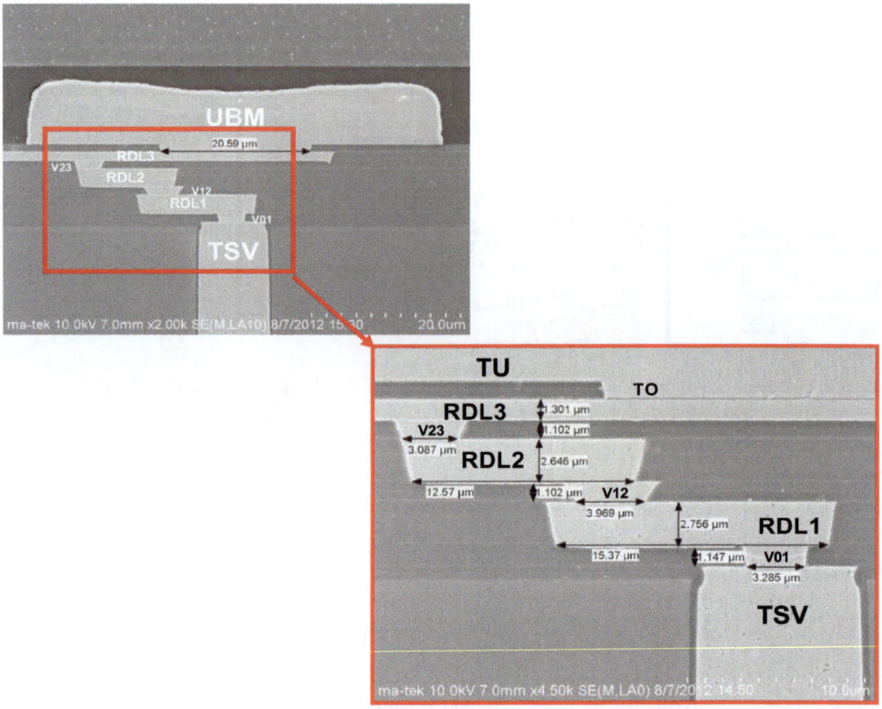


**Fig. 1.30** SEM image of RDLs on top of a TSV

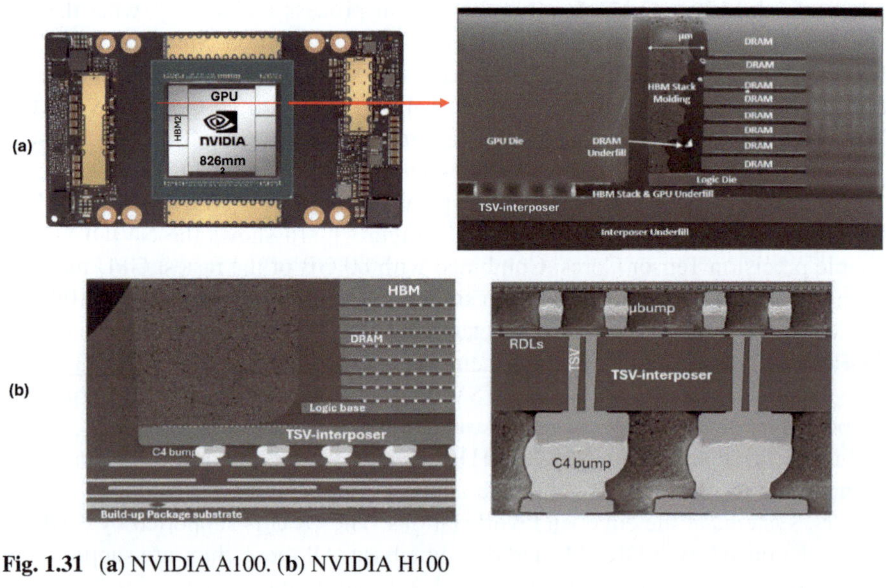


**Fig. 1.31** (**a**) NVIDIA A100. (**b**) NVIDIA H100

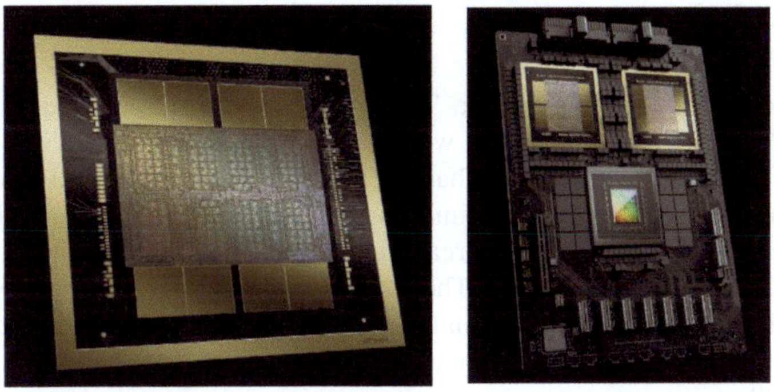


**Fig. 1.32** NVIDIA GB200 Grace Blackwell superchip

**Fig. 1.33** Intel® Gaudi® 3 accelerators

Figure 1.32 shows the NVIDIA-GB200 Grace Blackwell Superchip which is a key component of the NVIDIA GB200 NVL72, connecting two high-performance NVIDIA Blackwell Tensor Core GPUs and an NVIDIA Grace CPU using the NVIDIA® NVLink®-C2C interconnect to the two Blackwell GPUs. Packed with 208 billion transistors, Blackwell-architecture GPUs are manufactured using a custom-built 4NP TSMC process with two-reticle limit GPU dies connected by 10 TB/second chip-to-chip link into a single, unified GPU. The GPU is supported by 8 HBM3es and they are supported by a passive TSV-interposer.

Figure 1.33 shows the Intel® Gaudi® 3 accelerators platform. Intel® Gaudi® 3 AI are built to handle demanding training and inference. It supports AI workloads in data centers or in the cloud—from node to mega cluster, all running on the Ethernet infrastructure. The Intel Gaudi® 3 is supported by 8 HBMs with 1024 GB memory capacity. The Intel Gaudi® 3 and the HBMs are on a passive TSV-interposer and then on a build-up package substrate.

### 1.5.2 CoPoS

During the TSMC North America Technology Symposium on April 23, 2025 [117], TSMC announced that they will change some of the CoWoS to CoPoS as shown in Fig. 1.34. It can be seen that the TSV-interposer fabricated on a Si-wafer (Fig. 1.27) is replaced by the glass-interposer fabricate on a glass-panel (Fig. 1.34). One of the key reasons is that the area efficiency of PLP is better (lower cost) than that of WLP as shown in Fig. 1.35. The size of the glass panel is 310 mm × 310 mm, which strikes a critical balance, small enough to maintain high yield and process control but larger enough to capture area efficiency. The first product is scheduled for Q1 2029

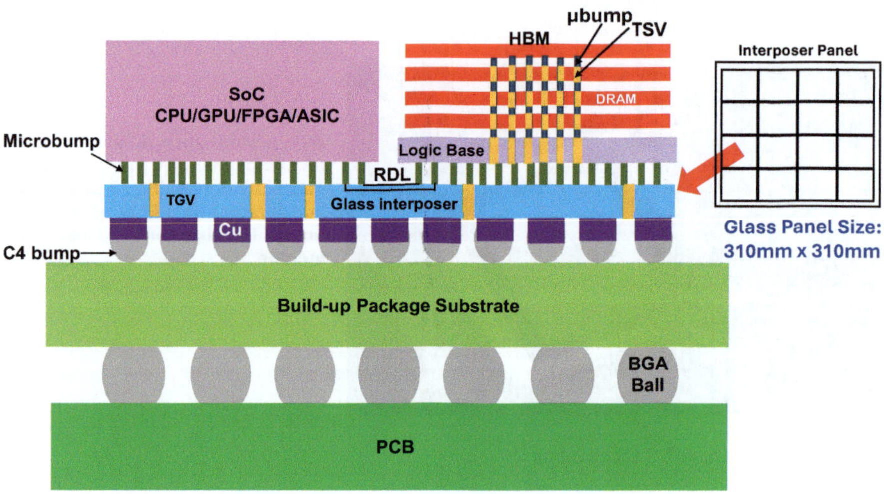


**Fig. 1.34** 2.5D with TGV-interposer fabricated on panel

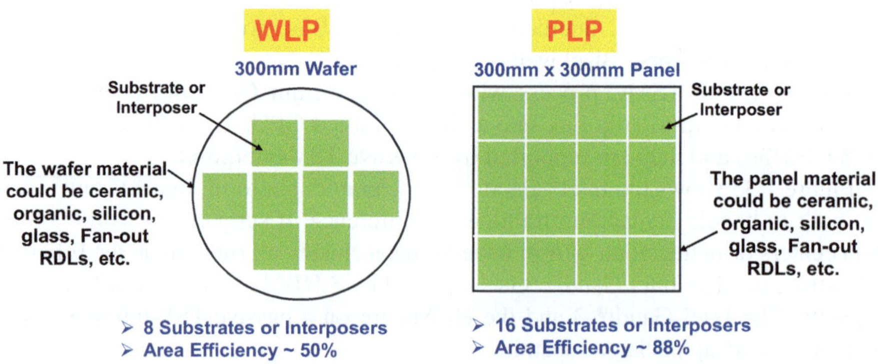


**Fig. 1.35** WLP (Wafer-Level Packaging) vs. PLP (Panel-Level Packaging)

## 1.6 CoWoS-L and FO-EB-T

Because of the great demands of high-bandwidth memory capacity for HPC driven by AI, the number of HBM surrounding the SoC in the 2.5D IC integration (CoWoS-S) increases as shown in Fig. 1.36. Thus, the TSV-interposer is getting bigger and bigger, and their semiconductor manufacturing yield loss of the large-size TSV-interposer is becoming unsustainable.

During IEEE/ECTC 2023, TSMC published two papers [99, 118] showing that when the size of the interposer is ≥ 2500 mm$^2$, then some of the CoWoS-S will be replaced by the local silicon interconnects (LSIs) (Si bridges) embedded in fan-out EMC with RDLs [99, 118] (CoWoS-L) as shown in Fig. 1.37. The new interposer consists of multiple LSIs or Si bridges w/o TSV and global integrated fan-out (InFO) EMC with RDLs to form a reconstituted interposer (RI). The small-size LSI (bridge) inherits all the attractive features of TSV-interposers by retaining sub-micron Cu interconnects, TSV, and embedded deep trench capacitors (eDTCs) to

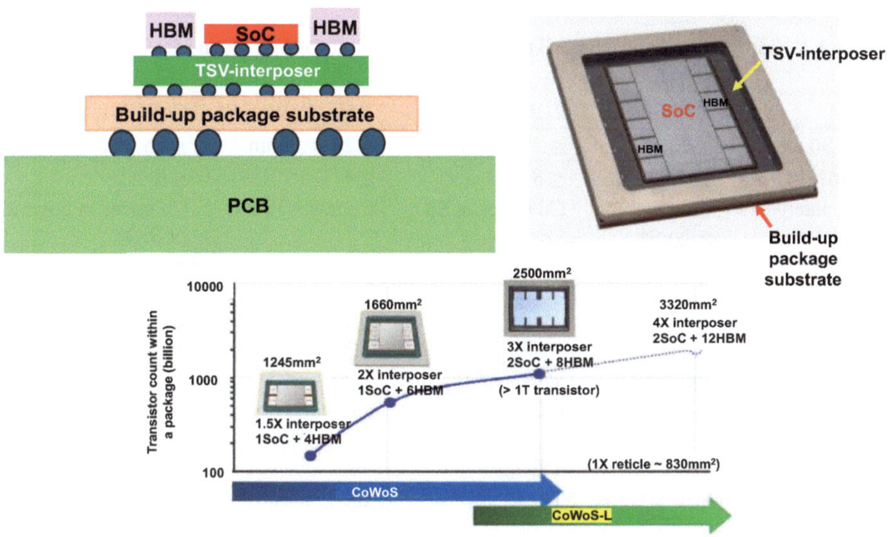


Fig. 1.36 TSMC's TSV-Interposer (2.5D IC integration or CoWoS) roadmap

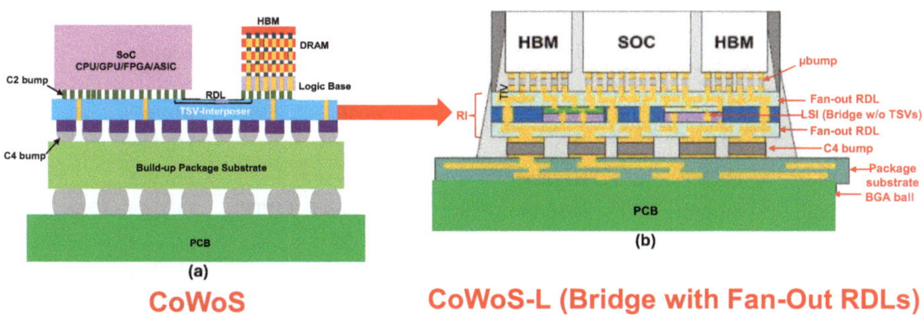


Fig. 1.37 (a) CoWoS-S. (b) CoWoS-L

ensure good system performance, while avoiding the issues associated with one large TSV-interposer, such as yield loss.

During the same conference, SPIL also published two papers called FO-EB-T [119, 120], Fig. 1.38. They are basically the same as CoWoS-L. A comparison between the 2.3D (chip-last), CoWoS-S, CoWoS-R, and CoWoS-L is shown in Table 1.1.

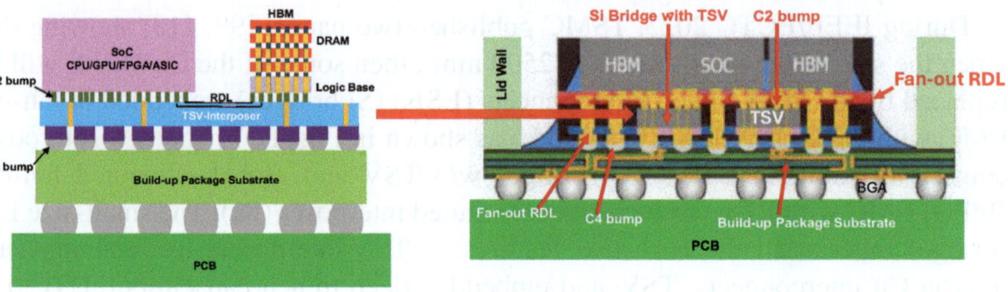


**Fig. 1.38** FO-EB-T

**Table 1.1** Comparison between 2.3D (chip-last), CoWoS-S, CoWoS-R, and CoWoS-L

| Items | 2.3D (Chip-Last) | 2.3D (CoWoS-R) | 2.5D (CoWoS-S) | CoWoS-L |
|---|---|---|---|---|
| SoC Size | $\leq 20 \times 20$mm | $\leq 25 \times 25$mm | $\leq 25 \times 25$mm | $\leq 25 \times 25$mm |
| No. of HBMs | $\leq 6$ | $\leq 8$ | $\geq 4$ | $\leq 8$ |
| RDL interposer | Cu traces in polymer | Cu traces in $SiO_2$ | Cu traces in $SiO_2$ | Cu traces in polymer or $SiO_2$ |
| RDL (L/S) interposer | $\geq 2\mu$m | $\geq 1\mu$m | $< 1\mu$m | $\geq 1\mu$m |
| RDL (Layer) interposer | $\leq 8$ | $\leq 9$ | $\leq 12$ | $\leq 9$ |
| Vertical interconnect | Via in RDL interposer | Via in RDL interposer | TSV in TSV interposer | TSV and via in LSI and via in RDL interposer |
| C4 bump | Yes | Yes | Yes | Yes |
| Underfill (chip/interposer) | Yes | Yes | Yes | Yes |
| Underfill (interposer/sub.) | Yes | Yes | Yes | Yes |
| Total package profile | About the same | About the same | About the same | About the same |
| Package substrate size | Big | Big | Biggest | Bigger |
| Process steps | Less | Less | Most | More |
| Density | High | Higher | Highest | Higher |
| Performance | High | Higher | Highest | Higher |
| Decoupling capacitor | Discrete IPD | Discrete IPD | Embedded DTC | Discrete IPD |
| Cost | High | Higher | Highest | Higher |
| Applications | HPC, data center, etc. | HPC, high bandwidth data center, etc. | HPC, high bandwidth data center, etc. | HPC, high bandwidth data center, etc. |

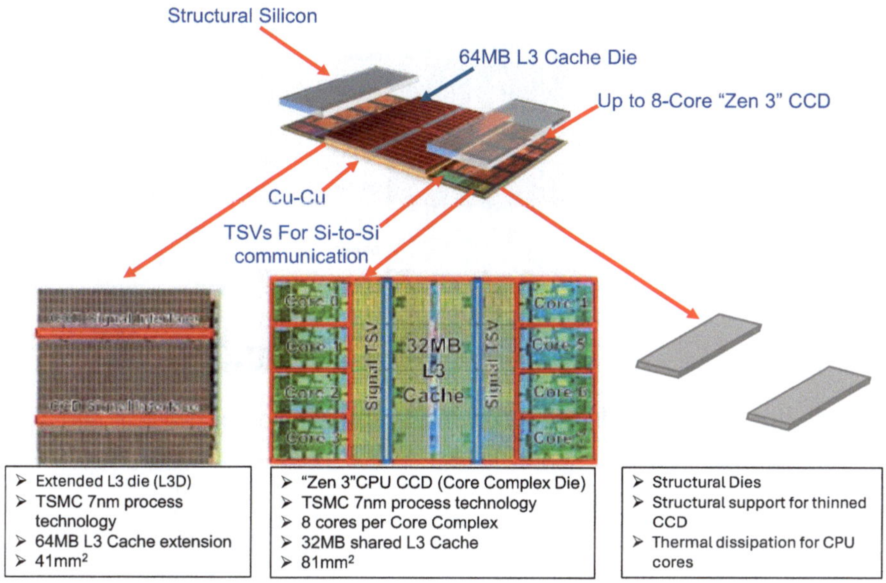


**Fig. 1.39** AMD's 3D V-Cache

## 1.7 3D IC Integration

Figure 1.39 shows AMD's 3D V-Cache chiplet design and heterogeneous integration packaging [121]. The key components of this structure are a bottom compute die, top static random access memory (SRAM) die, and structural dies to balance the structure and provide thermal path for heat dissipation from bottom compute die to the heat sink. The bottom die (81 mm$^2$) is the "Zen 3' CPU which is fabricated by TSMC's 7nm process technology. The top die (41 mm$^2$) is the extended L3 die which is also fabricated by TSMC's 7 nm process technology. The bottom die with TSV is face-down with C4 bumps. The top die is also face-down, which is face-to-back Cu-Cu hybrid bonding to the bottom die.

## 1.8 HBM and cHBM

By vertically connecting multiple DRAMs, HBM has superior performance, bandwidth, capacity, efficiency, and data processing speeds, over conventional DRAM. It is, however, more expensive at around two to three times the price of a standard DRAM. Figure 1.40 shows the HBM, HBM2, HBM3, HBM4 (6th-general HBM), and HBM4E (7th-general HBM). They work with SoC and are a must for HPC applications driven by AI as shown in Figs. 1.27 and 1.34.

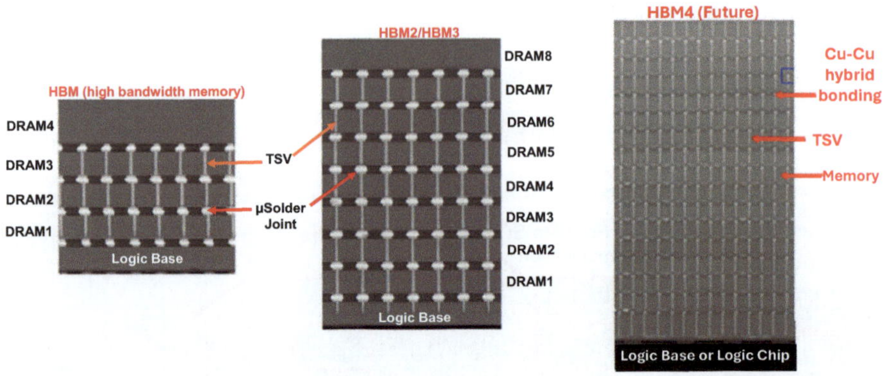


**Fig. 1.40** HBM (high bandwidth memory)

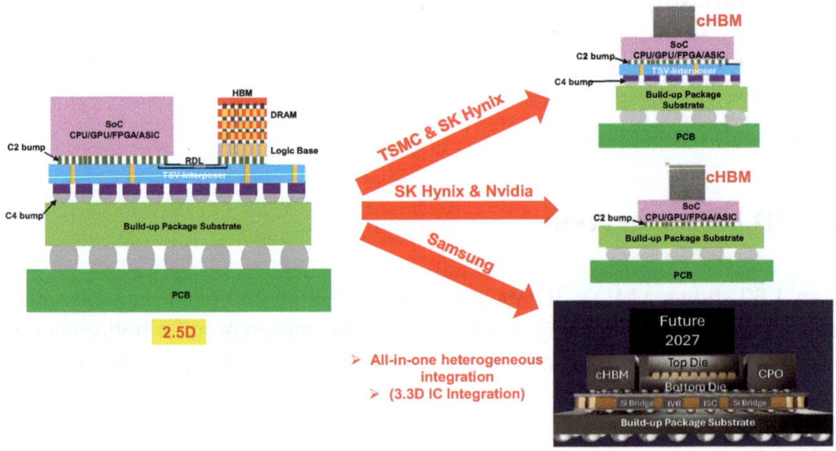


**Fig. 1.41** 3.3D IC integration

Today, the HBM (DRAMs + Logic base) is controlled by the memory houses, e.g., SK Hynix, Samsung, and Micron. Thus, the system houses, e.g., NVIDIA, AMD, Broadcom, etc. just take whatever (HBMs) they are given by the memory houses. System houses' hands are tight.

One and half years ago, things have been changing, the system houses wanted to redesign the HBM (Fig. 1.41) especially the logic base—they have been working with the memory houses to redesign the logic base into a logic chip, so called costume HBM (cHBM). The logic base vs. logic chip is shown below:

- Logic-base refers to the logic die that sits at the base of the HBM stack and manages communication between the DRAM layers and the external world, such as the SoC.

- This logic base is not designed for general-purpose computation or complex calculations; it primarily focuses on data transfer and control.
  - Logic chip is a broader term encompassing the processor or GPU that utilizes HBM for its computational tasks. Essentially, the logic in the HBM stack acts as a translator and controller, while the logic chip is the main processing unit that benefits from HBM's high bandwidth and low latency. (cHBM)
  - This logic chip is the main processing unit that utilizes HBM for its operations. It is the "brain" of the system, performing calculations and executing instructions. (cHBM). TSMC is making the logic chip for SK Hynix and planning to ship the cHBM in 2027.

## 1.9 3.3D IC Integration

As mentioned earlier, because the demanding for more HBMs, the size of the TSV interposer is becoming larger and larger. For large size of TSV interposers, the manufacturing yield is very low which translates to very high cost! One solution is to stack the cHBM on top of the SoC in an array format and then bond the SoC directly on an organic-core or glass-core build-up package substrate without the TSV-interposer, so called 3.3D IC integration, Fig. 1.42. It is scheduled to be in manufacturing in 2027.

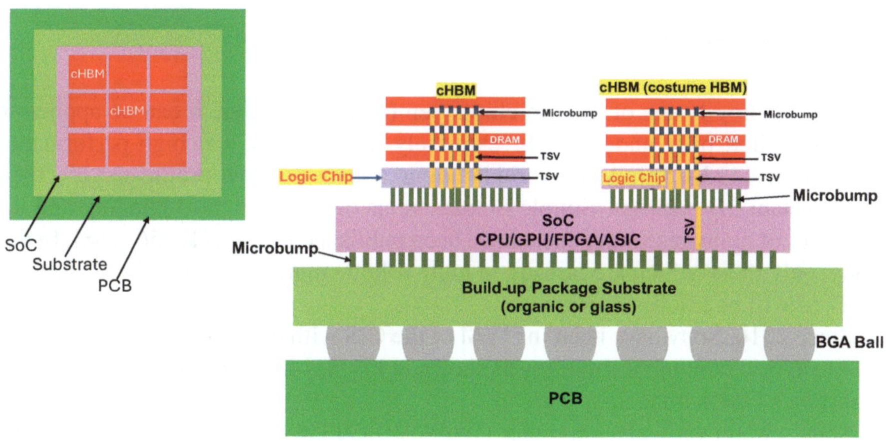


**Fig. 1.42** 3.3D IC integration (no interposer)

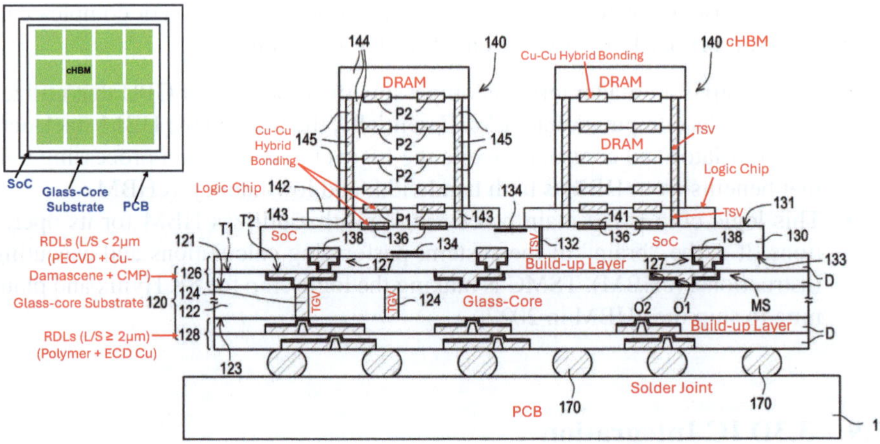


**Fig. 1.43** cHBMs on top of SoC on glass substrate (3.3D IC integration)

Figure 1.43 schematically shows the structure of a 3.3D IC integration [122]. Fourteen cHBMs are stacked on top of a SoC. The DRAMs and the logic chip of the cHBM are Cu-Cu hybrid bonded. Then the cHBM is Cu-Cu hybrid bonded to a glass-core build-up substrate. Finally, the package is surface mount technology (SMT) to the printed circuit board (PCB).

## 1.10 Cu-Cu Hybrid Bonding

Hybrid bonding is one of the flip-chip assembly technologies. It can be applied to very fine-pitch pads and used for extremely high-density and high-performance applications [123–155]. With hybrid bonding, it leads to a low-profile and low thermal resistant 3D package. Simple fundamentals of Cu-Cu hybrid bonding and the summary of many research papers have been published in, e.g., [1, 156–158]. In this study, only those which are already in high-volume manufacturing (HVM) or to be in HVM will be briefly mentioned.

Since 2016, Sony have been in HVM of the CIS with the wafer-on-wafer (WoW) Cu-Cu hybrid bonding technology as shown in Fig. 1.44a [159]. Figure 1.44b shows Samsung's GM% imager.

Sony's wafer-on-wafer-on-wafer (WoWoW) Cu-Cu hybrid bonding technology is scheduled to be in HVM for their high-performance or multi-functionality product in 2026 [160]. The STEM cross section image of WoWoW test vehicle having 1-μm pitch facet-to-face (F2F) and face-to-back (F2B) Cu-Cu connections and 6-μm pitch TSVs is shown in Fig. 1.45. Nicely aligned and bonded Cu pads are observed at both F2F and F2B bonding interfaces.

Since 2022, AMD have been shipping their 3D V-Cache product [161]. The memory chip is Cu-Cu hybrid bonding to the CPU chip. Figure 1.46 shows the

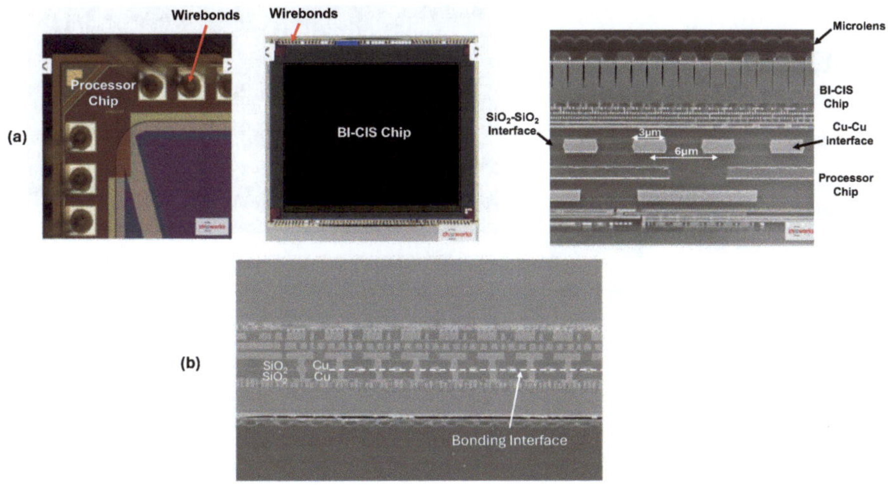


**Fig. 1.44** (**a**) Sony's CMOS image sensor (CIS) with hybrid bonding (WoW). (**b**) Samsung's GM5 imager

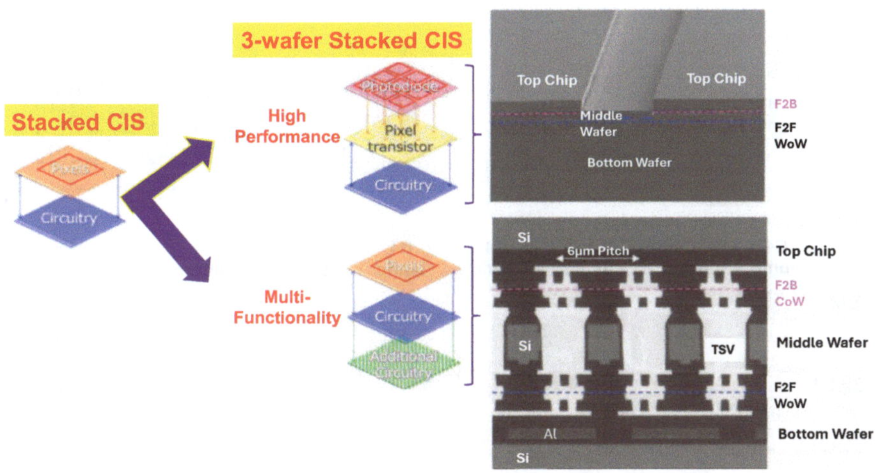


**Fig. 1.45** Sony's WoWoW Cu-Cu hybrid bonding

bonded interface, which is fabricated by TSMC' SoIC (system on integrated chips) technology [162, 163]. The chip-on-wafer (CoW) Cu-Cu bonding pitch is 9 μm. This is the first processing unit product fabricated by Cu-Cu hybrid bonding.

Since 2022, Graphcore have been shipping their intelligence processing unit (IPU) processor [164]. Again, TSMC's SoIC WoW stacking results in a processor chip (bottom) connected to a power-delivery chip (top) by Cu-Cu hybrid bonding (yellow). Signals and power pass through the top chip from solder bumps (grey), Fig. 1.47.

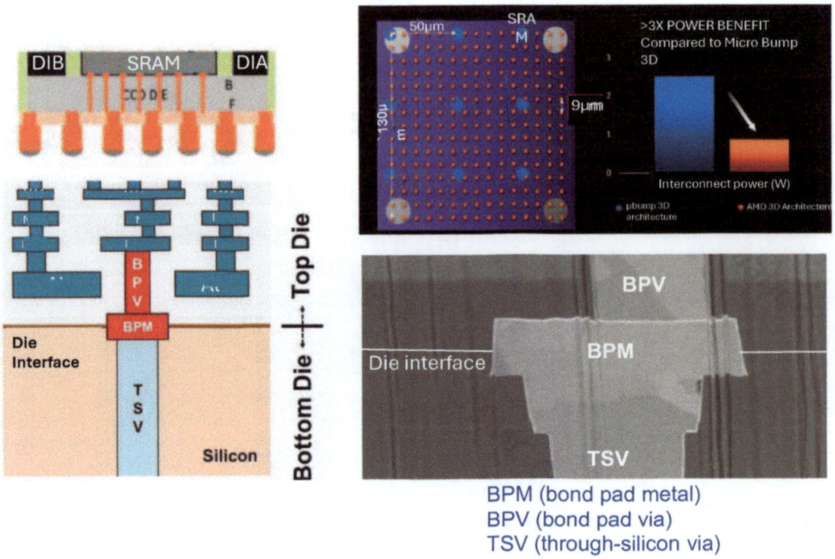


**Fig. 1.46** AMD's 3D V-Cache with TSMC's SoIC technology

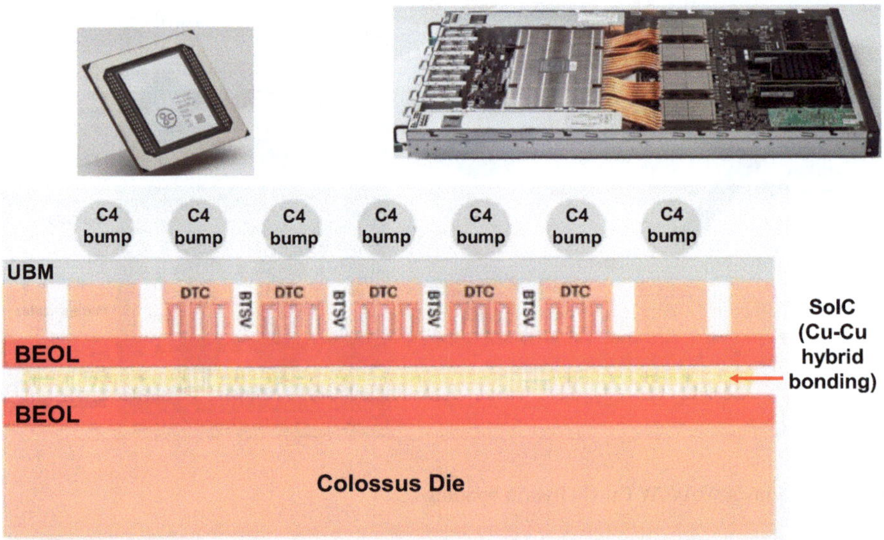


**Fig. 1.47** Graphcore's IPU processor with SoIC

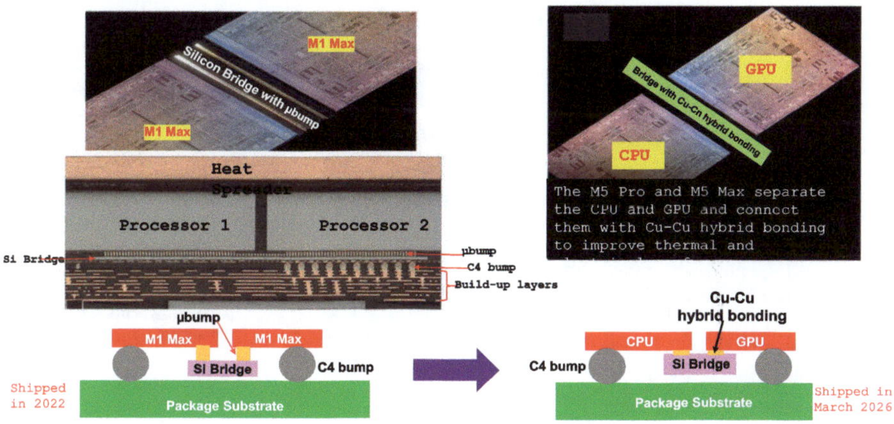


**Fig. 1.48** Apple's M5 (CPU and GPU) are connected by Bridge with hybrid bonding (SoIC)

As mentioned earlier, in 2022, Apple shipped their notebook computer with microbumps connecting the silicon bridge and the processors. In March 2026, Apple shipped their new notebook computers (M5 Pro and M5 Max) with Cu-Cu hybrid bonding connecting the silicon bridge and the GPU and CUP, Fig. 1.48. It is for density, thermal and electrical performance, and low profile.

For Samsung's HBM with Cu-Cu hybrid bonding, the DRAMs are bumpless and the Cu-pad to Cu-pad pairs are bonded with die-to-wafer (D2W) and die-to-die (D2D) methods [165, 166], Fig. 1.49. Figure 1.49a shows the scanning electron microscope (SEM) image of a16H multi stack using the hybrid bonding method. It can be seen that the total package thickness is smaller than the one using the thermocompression bonding (TCB) solder bumped flip-chip method and there is no gap (i.e., gap-less) between the DRAMs. Figure 1.49a also shows the following: (1) there are no voids at the interface, and (2) the Cu diffusion and grain growth at triple points was observed at the Cu interface. One of the challenges of HBM is thermal management. Because there are no gaps between the chips, the vertical thermal resistance of the 16H HBM with the Cu-Cu hybrid bonding is 20% lower than that for the flip-chip case with gaps for the µbumps and underfill (NCF) as shown in Fig. 1.49b. HBM4 with hybrid bonding is scheduled to be shipped in 2027.

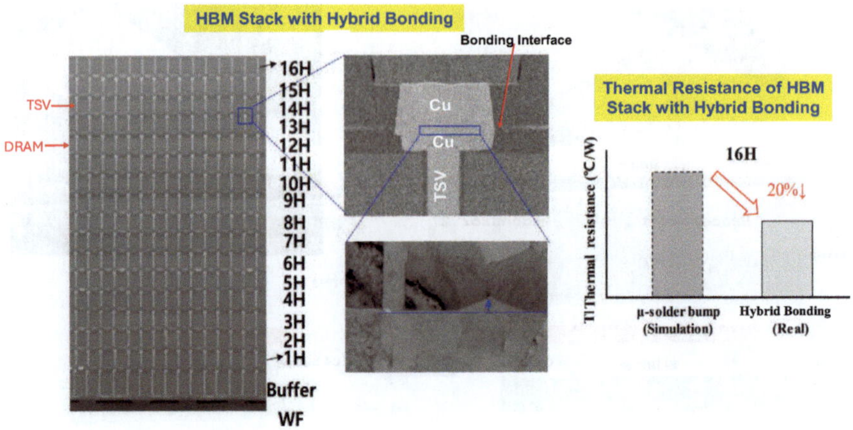


**Fig. 1.49** Samsung's HBM4 with hybrid bonding

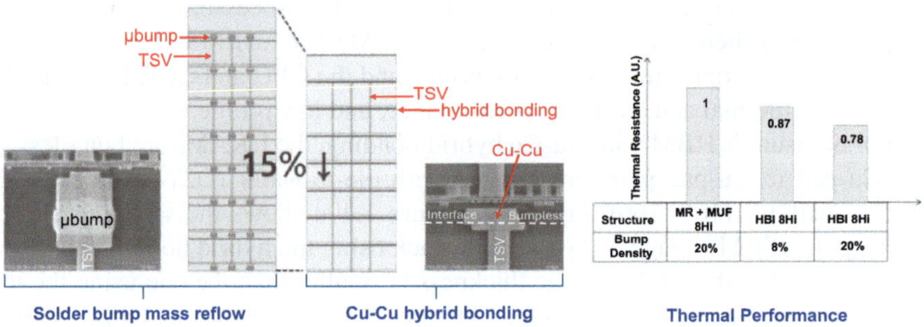


**Fig. 1.50** SK Hynix's HBM4E with hybrid bonding

SK Hynix presented a paper on a chip-to-wafer (C2W) hybrid bonding interconnect technology for higher density and better thermal dissipation of HBM [167]. Their focus is on the demonstration of an 8Hi-HBI (hybrid bonding interconnect) with C2W multi-stack HBM as shown in Fig. 1.50. It can be seen that compared with the solder-bumped mass reflow method, the thickness of the HBM package can be reduced by 15% with the Cu-Cu bumpless hybrid bonding method. SK Hynix also demonstrated that hybrid bonding interconnect (HBI) has a better thermal performance, Fig. 1.50. HBM4E with hybrid bonding is scheduled to be shipped in 2027.

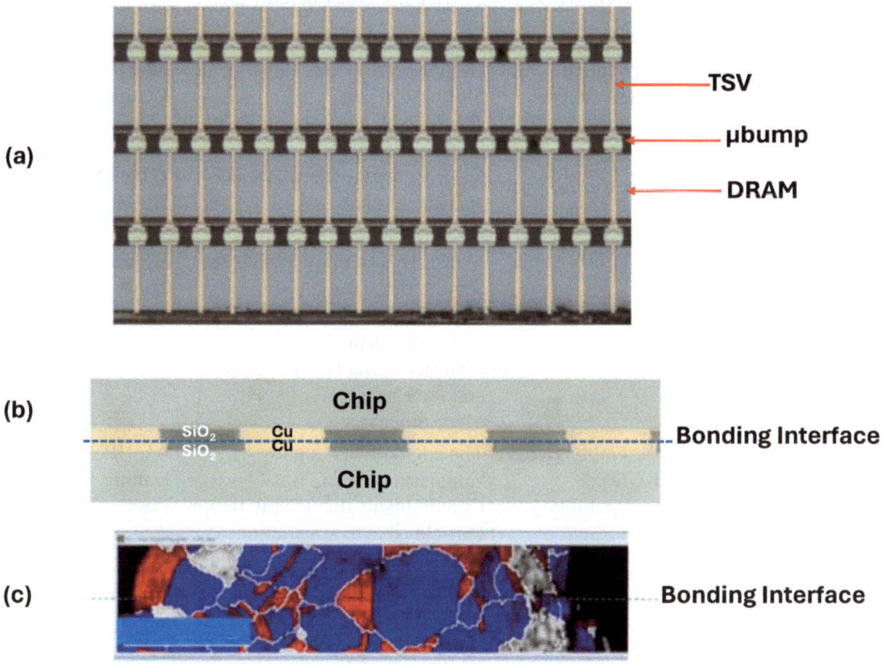


**Fig. 1.51** Micron's HBM4E with hybrid bonding. (**a**) TCB, (**b**) hybrid bonding, (**c**) close-up of hybrid bonding inerface

Micron's hybrid bonding for C2W HBM stacking is shown in [123]. Figure 1.51a shows a conventional stacking of TSV-dies with μbumps solder reflow method for HBM. The μbump pitch can go down to 20 μm. However, with hybrid bonding, the Cu pad pitch can easily go down to 10 μm and there is no bump, i.e., bumpless as shown in Fig. 1.51b. Figure 1.51c shows the electron backscatter diffraction (EBSD) across the bonding interface where two Cu pads are joined together. Due to Cu-to-Cu diffusion, these two Cu pads become one with the Cu crystal growth occurring across the bonding interface. Micron's HBM4 with hybrid bonding is planned to be shipped in 2027.

Today, the bonding methods of Cu-Cu hybrid bonding are CoW, WoW, WoWoW, and reconstructed wafer. A comparison of them is shown in Table 1.2.

**Table 1.2** Comparison of CoW, WoW, WoWoW, and reconstructed wafer

| Hybrid Bonding | Advantages | Disadvantages (Challenges) | Applications |
|---|---|---|---|
| WoW | High throughput<br>Mutual technology<br>High Volume manufacturing | High assembly yield loss<br>Similar die sizes<br>Topography control of the Cu/dielectric surface<br>CMP for metal recess, clean, and flat surface | AI, HPC, Machine leaning<br>Logic, SoC, HBM<br>CIS<br>3D NAND flash<br>3D Monolithic DRAM |
| WoWoW | Very high throughput | Very high assembly yield loss<br>Similar die sizes<br>Topography control of the Cu/dielectric surface<br>CMP for metal recess, clean, and flat surface<br>More process steps | High performance<br>Multi-functionality |
| DoW (CoW) | High assembly yield with KGD<br>Design flexibility | The edge effects due to singulation<br>Contaminants due to singulation<br>Particles due to singulation<br>Higher accuracy pick & place (P&P) machines<br>Slightly larger pads to compensate the P&P tolerance<br>CMP for metal recess, clean, and flat surface<br>Topography control of the Cu/dielectric surface | AI, HPC, Machine leaning<br>Logic, SoC, HBM |
| Reconstructed Wafer | High throughput | More process steps | AI, HPC, Machine leaning<br>Logic, SoC, HBM |

## 1.11 3.5D IC Integration

As mentioned in [168], frontend integration of some of the chiplets (before package heterogeneous integration) with Cu-Cu hybrid bonding can yield a smaller package size and a better performance. Figure 1.52 shows an example of Cu-Cu hybrid bonding between some chiplets before they are attached to the TSV interposer [168].

Figure 1.53 shows the 3.5D IC integration by Samsung [169]. It can be seen that the ASIC is vertically stacked with two chips which are partitioned by its functional blocks. Each chip is stacked face-to-face interconnected with Cu-Cu hybrid bonding. Samsung Foundry showed a roadmap for 3.5D configurations in 2024, with projections for stacking 2 nm and 4 nm chiplets in 2025, and 1.4 nm and 2 nm chiplets in 2027.

Figure 1.54 shows the AMD's compute and AI chiplets which are stacked on top of the I/O and cache chiplets in their MI300 shipped in 2024 [170]. All the compute chips are built using TSMC's N5 process. Neither the I/O functions nor the system's cache memory benefit from N5, so AMD chose a less-expensive technology (N6)

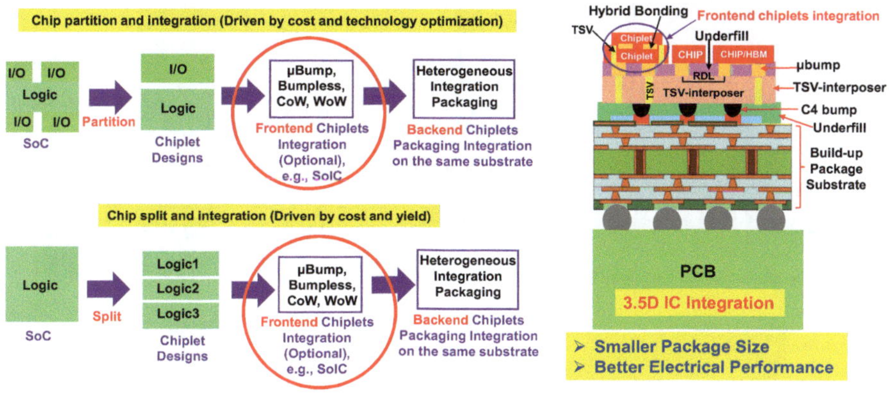


**Fig. 1.52** 3.5D IC heterogeneous integration

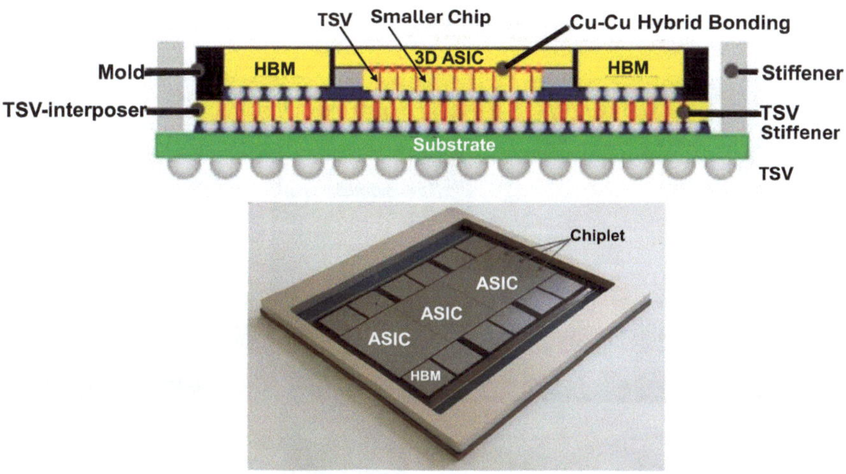


**Fig. 1.53** Samsung's 3.5D IC heterogeneous integration

for those. Therefore, those two functions could then be built together on the same chiplet. MI300 stacks (with hybrid bonding by TSMC's SoIC, the front-end optional part of Fig. 1.2 (a and b)) three CPU chiplets (called compute complex dies, or CCDs, in AMD's lingo) and six accelerator complex dies (XCDs) on top of four input-output dies (IODs), on top of a piece of TSV-interposer that links them together to eight stacks of high-bandwidth DRAM (HBM) that ring the superchip. AMD also called this 3.5D IC integration [176].

Broadcom has introduced its 3.5D eXtreme Dimension System in Package (3.5D XDSiP) platform for ultra-high performance processors driven by AI workloads [171]. This platform uses TSMC's CoWoS-L packaging technology that offers a maximum interposer size of approximately 5.5 times that of a reticle (about 858 mm$^2$), or 4719 mm$^2$ for compute chiplets, I/O chiplets, and up to 12 HBM3/HBM4

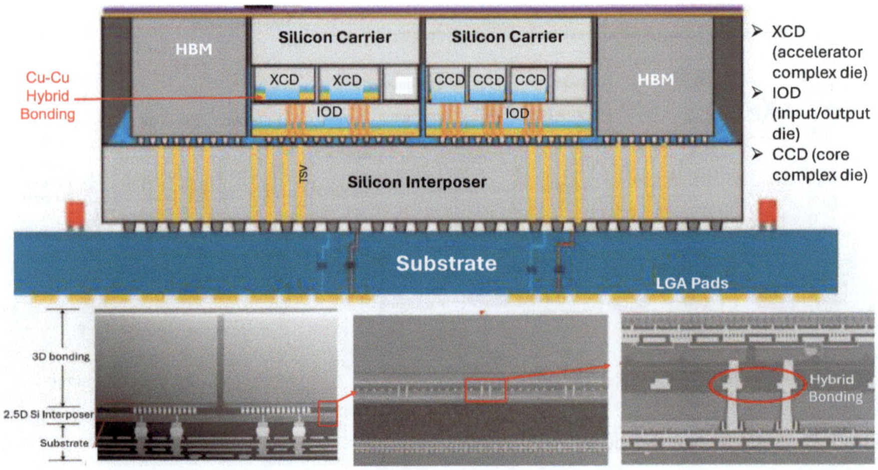


**Fig. 1.54** AMD's MI300 (3.5D IC integration)

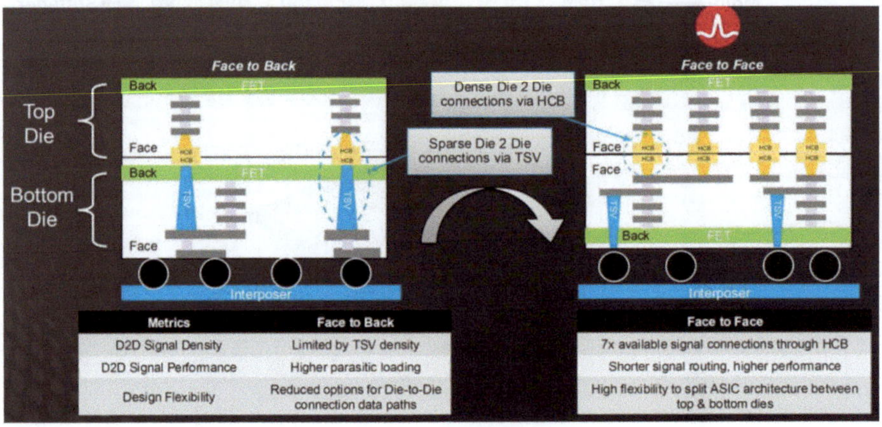


**Fig. 1.55** Broadcom's 3.5D IC heterogeneous integration (XDSiP)

packages. To maximize performance, Broadcom suggests disaggregating the design of compute chiplets and stacking one logic chiplet on top of another in a face-to-face (F2F) manner using Cu-Cu hybrid bonding, Fig. 1.55. Broadcom shipped this package to Fujitsu in February 2026.

## 1.12 Glass-core Packaging

Some of the glass packaging is shown in Fig. 1.56. It can be seen that the organic-core build-up package substrate, Fig. 1.56a, has been replaced by the glass-core build-up package substrate, Fig. 1.56b. Also, the traditional through-silicon via (TSV)-interposer has been replaced with the through-glass via (TGV)-interposer, Fig. 1.56c—a 2.5D IC integration. It can be seen that a TGV glass-core interposer that is supporting chips is also supported by an organic-core substrate as shown in Fig. 1.56c. Finally, a chip is embedded in the glass-core build-up package substrate—a 3D IC integration, Fig. 1.56d.

### *1.12.1 Advantages of Glass-core Packaging*

The advantages of glass-core packaging compared with the organic-core packaging are [169–204]: (a) ultra-high flatness for improved depth of focus for lithography, (b) dimensional stability needed for extremely tight layer to layer interconnect overlay, (c) higher interconnect density, (d) higher mechanical stability for ultra-large form-factor packages with high assembly yield, (e) improved flexibility in setting design rules for power delivery and signal routing, (f) better power delivery solutions while achieving high-speed signaling, (g) higher tolerance for higher temperatures, and (h) higher ability to seamlessly integrate optical interconnects.

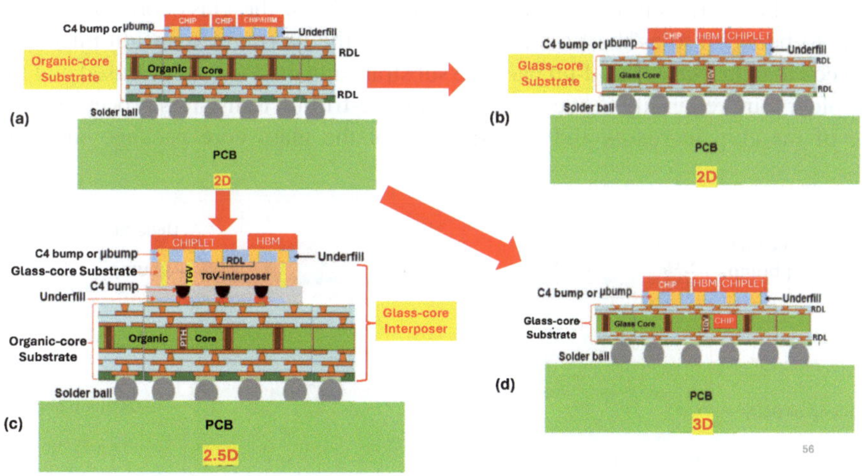


**Fig. 1.56** (**a**) Organic-core substrate. (**b**) Glass-core substrate. (**c**) Glass-core interposer. (**d**) Glass-core 3D IC integration

### 1.12.2 Disadvantages of Glass-core Packaging

The disadvantages of glass-core packaging compared with the organic-core packaging are [169–204]: (a) higher material cost, (b) higher processing (production) cost, (c) higher yield loss, (d) fragility (easy to break), (e) handling, (f) difficult in making TGVs, (g) glass (softens) at high temperatures, and (h) special and expensive equipment.

### 1.12.3 Glass-core Substrate vs. Glass-core Interposer

Figure 1.56b schematically shows a glass-core substrate supporting chips while Fig. 1.56c schematically shows a glass-core substrate supporting chips and then it is supported by an organic-core substrate (glass-core interposer).

### 1.12.4 Effects of Glass-core Substrate on Solder Joint Reliability

So far, solder joint reliability issues because of the glass core substrate are seldom discussed. In this study, the solder joint reliability because of the glass core substrate is investigated. Figure 1.57 shows the structures under consideration. In order to make the μbump solder joint reliable between the chip and the build-up package substrate the coefficient of thermal expansion (CTE) of the glass core substrate is made to be as close as the Si chip. In this case, the thermal expansion mismatch between the glass-core build-up package substrate and the PCB is increased and the C4 solder joint reliability is questionable! For one-trillion transistors processors, the size of the chip increases and thus the size of the glass core package substrate

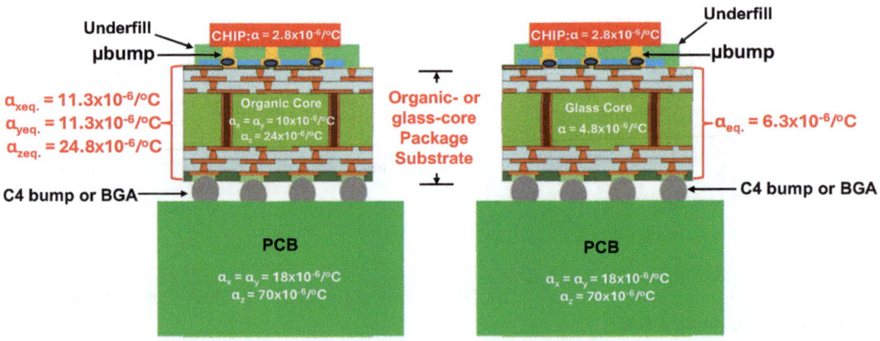


**Fig. 1.57** Coefficient of thermal expansion of key elements in the package assembly

increases. The thermal expansion mismatch between the glass core substrate and the PCB increases which may cause the C4 solder joint reliability issue.

In this study, the solder cap of the μbump is made of Sn0.7Cu (melting point = 227 °C) and the solder of the C4 bump is made of Sn3Ag0.5Cu (or SAC and the melting point = 217 °C). Both solders obey the Anand constitute model [205], i.e., the thermal-fatigue reliability of the solder joints is performed by a nonlinear temperature and time dependent finite element analysis.

(A) **The Structure**

Figure 1.58 schematically shows the structures under consideration. The silicon chip (10 mm × 10 mm × 350 μm) is supported by a build-up package substrate (20 mm × 20 mm × 832 μm), which is fabricated by the conventional material such as the organic core (800 μm-thick) and the new glass core (800 μm-thick) material. There are two build-up layers on the core's top and bottom sides which are fabricated by a PID (photoimageable dielectric) for the dielectric layer (5 μm-thick) and electrochemical deposition (ECD) Cu for the metal layer (3 μm-thick).

(B) **Finite Element Modelling**

Due to the double symmetries of the structure shown in Fig. 1.55, only a quarter of the structure is modeled. Figure 1.59 shows the chip, package substrate, PCB, μbumps and C4 bumps. Finner meshes are used for the critical locations (higher stress/strain areas) such as corner μbumps connecting the chip and the build-up package substrate and the corner C4 bumps connecting the build-up package substrate and the PCB.

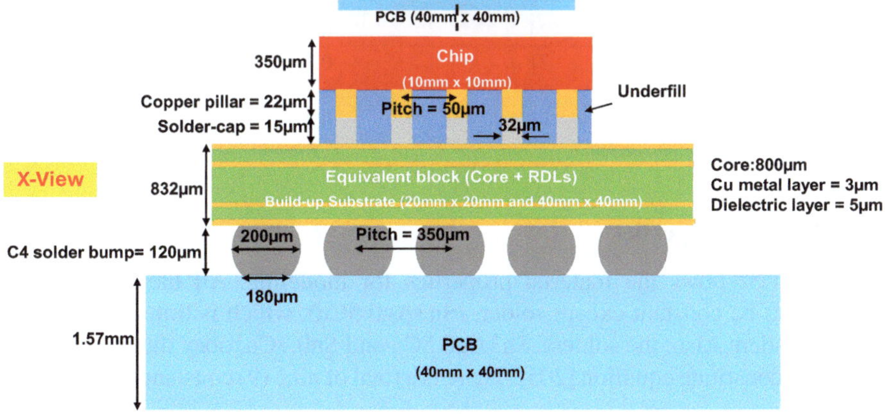


**Fig. 1.58** Top- and cross-section views of the structure

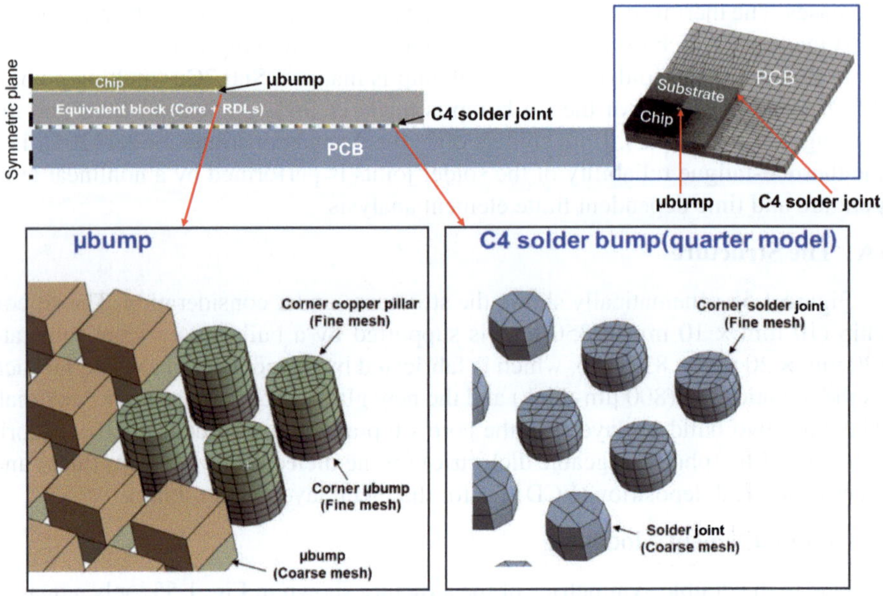


**Fig. 1.59** Finite element modeling of the structure

**Table 1.3** Material properties

| Materials | Young's modulus (GPa) | Poisson's ratio | CTE ($10 \times 10^{-6}$/°C) |
|---|---|---|---|
| Silicon | 131 | 0.278 | 2.8 |
| μbump (Sn0.7Cu) | 30 | 0.3 | 22 |
| C4 bump (SAC305) | 49 – 0.07T(°C) | 0.3 | 21 + 0.017T(°C) |
| PCB | $E_x = E_y = 22$; $E_z = 10$ | 0.28 | $\alpha_x = \alpha_y = 18$; $\alpha_z = 70$ |
| Underfill | 4.5 | 0.35 | 50 |
| RDL (Dielectric) | 2.0 | 0.3 | 60 |
| RDL (Copper) | 121 | 0.34 | 16.3 |
| Glass core | 72 | 0.29 | 4.8 |
| Organic core | 24 | 0.3 | $\alpha_x = \alpha_y = 10$; $\alpha_z = 24$ |
| Equivalent block (Glass core + RDLs) | 71 | 0.29 | 6.3 |
| Equivalent block (Organic core + RDLs) | 25 | 0.3 | $\alpha_x = \alpha_y = 11.3$; $\alpha_z = 24.8$ |

## (C) **Material Properties**

Table 1.3 shows the material properties for modeling. All the materials are assumed to be constant except solder, (Sn3Ag0.5Cu), which is time and temperature dependent Also, the solders, Sn3Ag0.5Cu and Sn0.7Cu, obey the Anand viscoplasticity constitute equation [205]. There is a total of nine (9) constants (parameters)

**Table 1.4** Parameters in Anand model. SAC305 and Sn0.7Cu

| Parameters | SAC305 | Sn0.7Cu |
|---|---|---|
| $s_0$ (Pa) | $18.07 \times 10^6$ | $33.8 \times 10^6$ |
| Q/R (K) | 9096 | 5276 |
| A ($s^{-1}$) | 3484 | 9875 |
| $\xi$ | 4 | 2 |
| m | 0.2 | 0.176 |
| $h_0$ (Pa) | $1.44 \times 10^{11}$ | $61987.3 \times 10^6$ |
| $\hat{s}$ (Pa) | $26.4 \times 10^6$ | $65.2 \times 10^6$ |
| n | 0.01 | 0.003 |
| a | 1.9 | 1.03 |

Thickness: Cu metal layer = 3μm; dielectric layer = 5μm, glass/organic core = 800μm

| | Young's modulus (GPa) | Poisson's ratio | CTE ($10^{-6}$/°C) |
|---|---|---|---|
| Dielectric (PID) | 2.0 | 0.3 | 60 |
| Metal (Cu) | 121 | 0.34 | 16.3 |
| Glass | 72 | 0.29 | 4.8 |
| Organic | 24 | 0.3 | $\alpha_x = \alpha_y = 10$; $\alpha_z = 24$ |

**For Glass Core**

☐ **For Equivalent Young's modulus:**

$$\frac{121\times(3*4)+2\times(5*4)+72*800}{(3*4)+(5*4)+800} = 71\text{GPa}$$

☐ **For Equivalent CTE:**

$$\frac{16.3\times(3*4)+60\times(5*4)+4.8\times800}{(3*4)+(5*4)+800} = 6.3\text{ppm/}^\circ\text{C}$$

☐ **For Equivalent Poisson's ratio:**

$$\frac{0.34\times(3*4)+0.3\times(5*4)+0.29\times800}{(3*4)+(5*4)+800} = 0.29$$

**For Organic Core**

☐ **For Equivalent Young's modulus:**

$$\frac{121\times(3*4)+2\times(5*4)+24*800}{(3*4)+(5*4)+800} = 25\text{GPa}$$

☐ **For Equivalent CTE in X and Y:**

$$\frac{16.3\times(3*4)+60\times(5*4)+10\times800}{(3*4)+(5*4)+800} = 11.3\text{ppm/}^\circ\text{C}$$

☐ **For Equivalent CTE in Z:**

$$\frac{16.3\times(3*4)+60\times(5*4)+24\times800}{(3*4)+(5*4)+800} = 24.8\text{ppm/}^\circ\text{C}$$

☐ **For Equivalent Poisson's ratio:**

$$\frac{0.34\times(3*4)+0.3\times(5*4)+0.3\times800}{(3*4)+(5*4)+800} = 0.3$$

**Fig. 1.60** Equivalent Young's modulus, CTE, and Poisson's ratio

of the Anand equation (model), namely $A$, $Q$, $\xi$, $m$, $s$,$n$, $h_0$, $\alpha$, and $s_0$. For Sn3Ag0.5Cu (SAC) and Sn0.7Cu, these constants are shown in Table 1.4.

The effective material properties of the organic build-up equivalent block can be calculated and are summarized in Fig. 1.60.

(D) **Thermal Boundary Condition**

The temperature boundary condition is shown in Fig. 1.61. It can be seen that the temperatures are −40 °C ↔ 85 °C, and the dwell-time at hot, dwell-time at cold, ramp-up time, and ramp-down time are 15 min each. Stress free is at room temperature (25 °C).

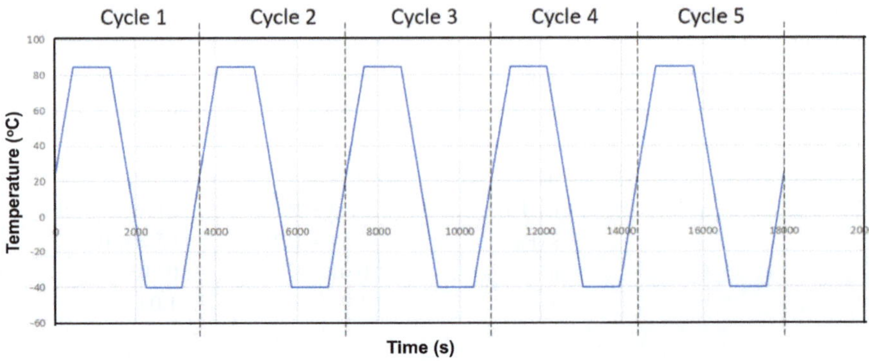


Fig. 1.61 Temperature boundary condition

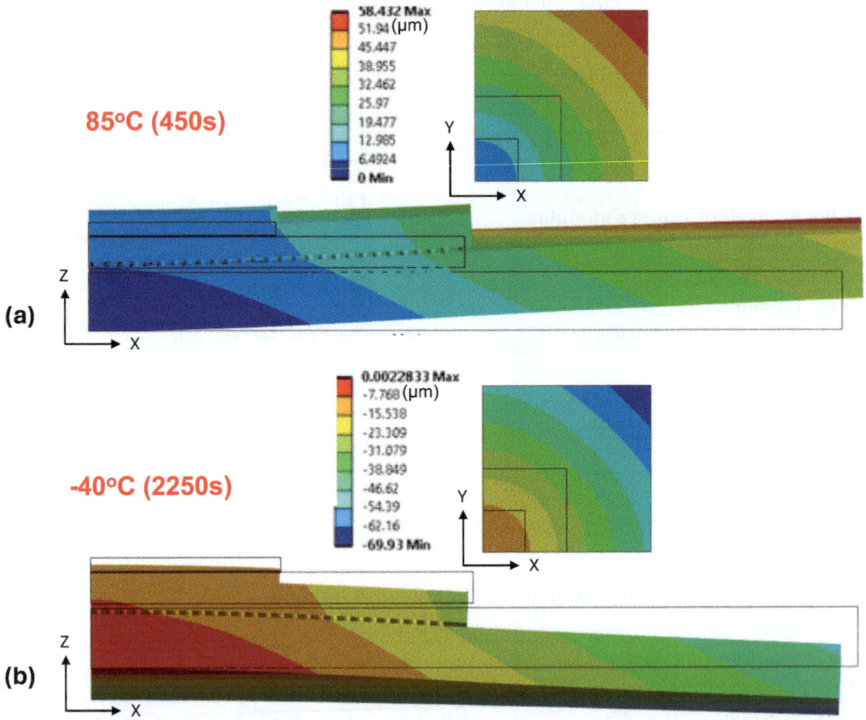


Fig. 1.62 (**a**) At 450s. (**b**) At 2250s. Deformation (organic core substrate)

(E) **Structural Deformation**

Deformed shape (color contours) and the undeformed shape (dark lines) of the flip chip PCB assembly with organic core substrate are shown in Fig. 1.62a at 450s (85 °C) and in Fig. 1.62b at 2250s (−40 °C). It can be seen that at 85 °C, the PCB

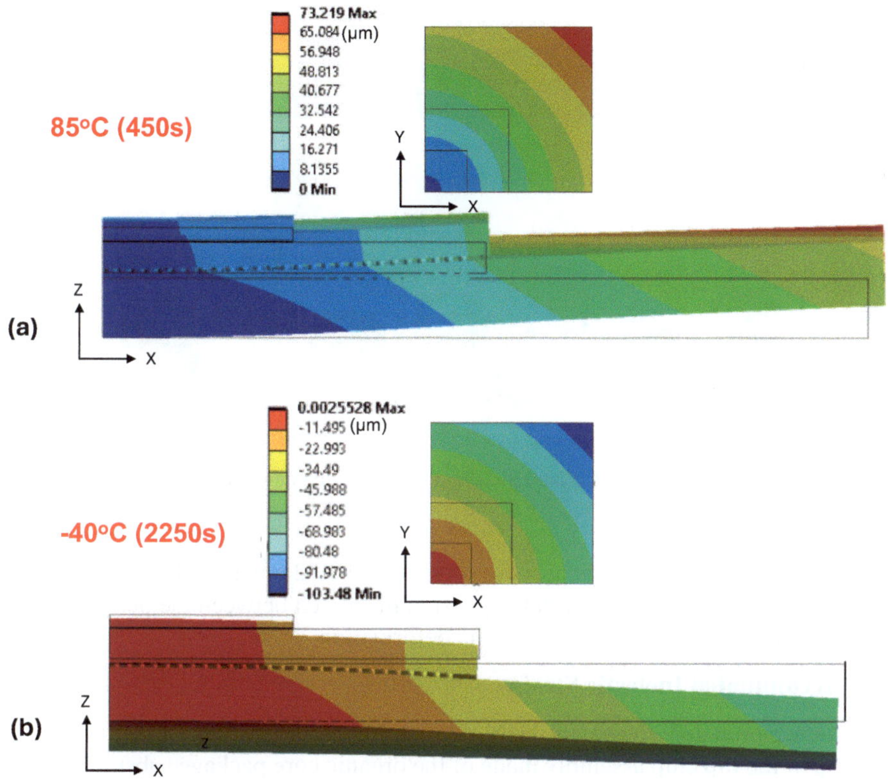


**Fig. 1.63** (**a**) At 450s. (**b**) At 2250s. Deformation (glass core substrate)

expands more than the organic package substrate and the silicon chip, and the structure is deformed in a convex shape (smiling face), and at −40 °C, the PCB shrinks more than the organic package substrate and the silicon chip, and the structure is deformed into a concave shape (crying face). These results show that the displacement, rotation, and temperature boundary conditions are prescribed correctly.

The deformed shape and the undeformed shape of the flip chip PCB assembly with glass core substrate are shown in Fig. 1.63a at 450s (85 °C) and in Fig. 1.63b at 2250s (−40 °C). It can be seen that, just like those with the organic package substrate, at 85 °C, the PCB expands more than the glass package substrate and the silicon chip, and the structure is deformed in a convex shape (smiling face), and at −40 °C, the PCB shrinks more than the glass package substrate and the silicon chip, the structure is deformed into a concave shape (crying face). However, because of the difference in thermal expansion coefficient between the organic core substrate ($\alpha_x = \alpha_y = 11.3 \times 10^{-6}/^{\circ}C$, $\alpha_z = 24.8 \times 10^{-6}/^{\circ}C$), the glass core substrate ($\alpha = 6.3 \times 10^{-6}/^{\circ}C$), and the PCB ($\alpha_x = \alpha_y = 18 \times 10^{-6}/^{\circ}C$, $\alpha_z = 70 \times 10^{-6}/^{\circ}C$) the magnitudes of deformation are different. The maximum deformations of the assembly with the glass core substrate (73.219 μm at 85 °C and −103.48 μm at −40 °C) are

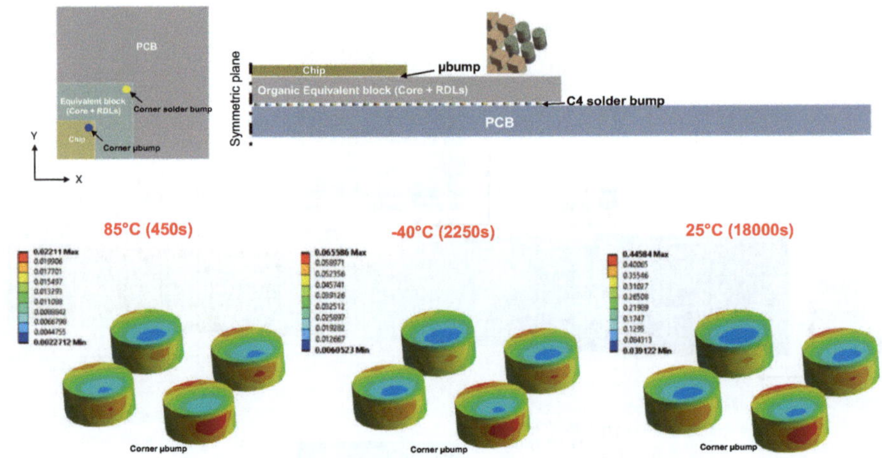


**Fig. 1.64** Maximum accumulated inelastic strain in corner µbump of structure with organic core substrate

larger than those with the organic core substrate (58.432 µm at 85 °C and −69.93 µm at −40 °C). The larger the thermal expansion mismatch between the build-up package substrate and the PCB the larger the deformation.

(F) **Accumulated Inelastic Strain**

The accumulated equivalent inelastic strain contour distributions at the corner µbump of the flip chip assembly made of the organic core package substrate at 450s (85 °C), at 2250s (−40 °C), and at 18000s (25 °C) are shown in Fig. 1.64. It can be seen that the maximum inelastic strain at the corner µbump solder joint occurs at a small local area, and the value is 2.2% at 450s (85 °C), 6.6% at 2250s (−40 °C), and 44.6% at 18000s (25 °C).

The accumulated equivalent inelastic strain contour distributions at the corner µbump of the flip chip assembly made of the glass core package substrate at 450s (85 °C), at 2250s (−40 °C), and at 18000s (25 °C) are shown in Fig. 1.65. It can be seen that the maximum inelastic strain at the corner µbump solder joint occurs at a small local area, and the value is 1.1% at 450s (85 °C), 3.2% at 2250s (−40 °C), and 21.8% at 18000s (25 °C). Comparing these values with those with organic core package substrate, the maximum equivalent inelastic strain at the corner µbump with glass core substrate shows the inelastic stain time-history in the corner µbump is smaller than that with organic core substrate. Figure 1.66 shows the inelastic stain time-history in the corner µbump with the organic and glass substrates. It can be seen that the strain is smaller in the glass structure. This is because of the thermal expansion mismatch between the glass core substrate, and the silicon chip is smaller than that between the organic core substrate and the silicon chip.

The accumulated equivalent inelastic strain contour distributions at the corner C4 solder joint of the flip chip assembly made of the organic core package substrate at 450s (85 °C), at 2250s (−40 °C), and at 18000s (25 °C) are shown in Fig. 1.67. It

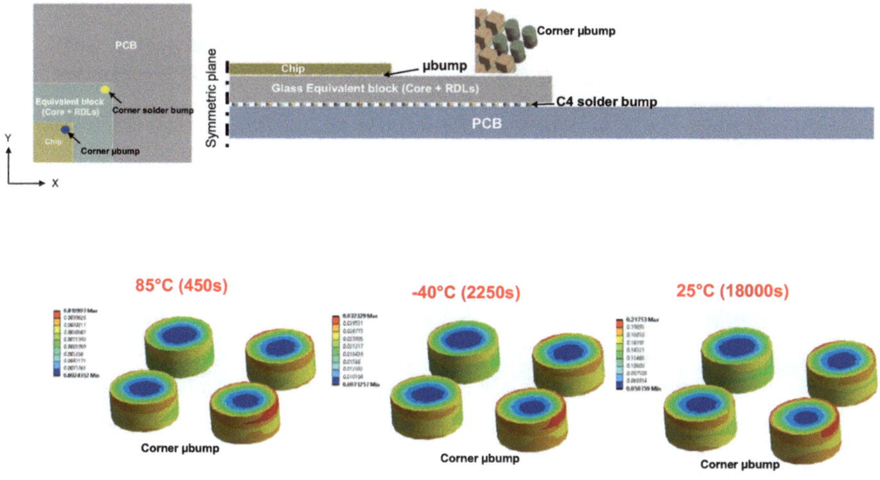


Fig. 1.65 Maximum accumulated inelastic strain in corner µbump of structure with glass core substrate

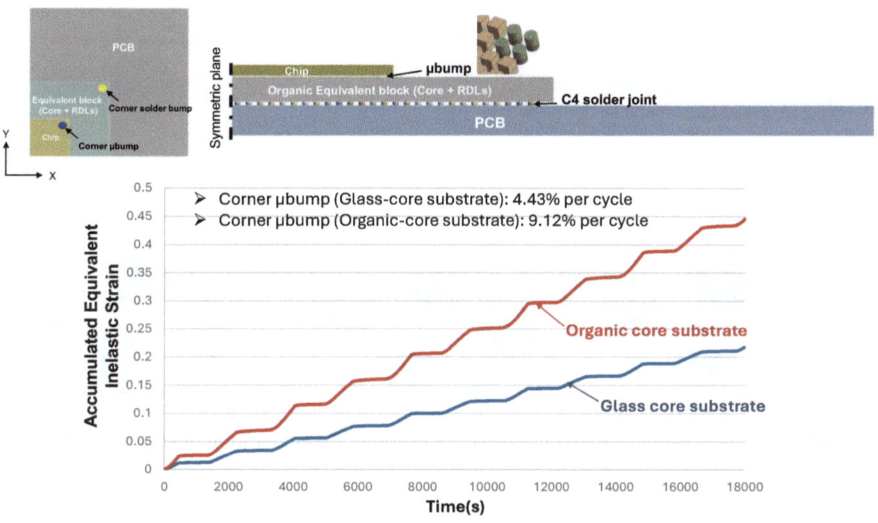


Fig. 1.66 Maximum accumulated inelastic strain time-history in corner µbump of structures with organic and glass substrates

can be seen that the maximum inelastic strain at the corner C4 solder joint occurs at a small local area, and the value is 2.0% at 450s (85 °C) and 6.4% at 2250s (−40 °C), and 41.4% at 18000s (25 °C).

The accumulated equivalent inelastic strain contour distributions at the corner C4 solder joint of the flip chip assembly made of the glass core package substrate at 450s (85 °C), at 2250s (−40 °C), and at 18000s (25 °C) are shown in Fig. 1.68. It can

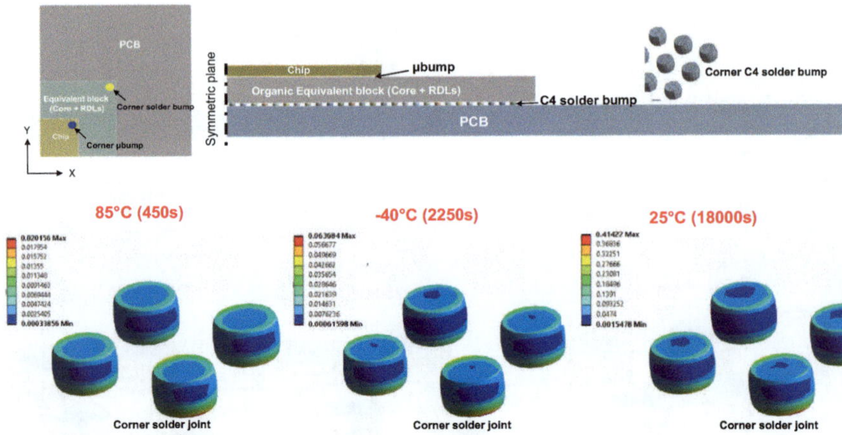


**Fig. 1.67** Maximum accumulated inelastic strain in corner C4 solder joint of structure with organic-core substrate

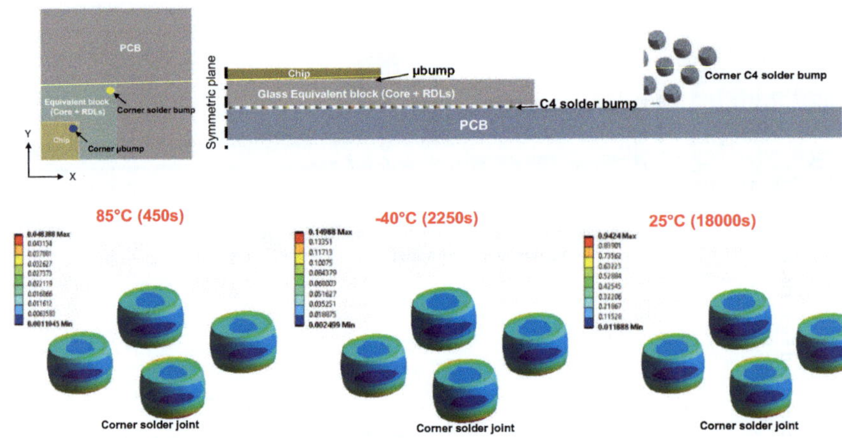


**Fig. 1.68** Maximum accumulated inelastic strain in corner C4 solder joint of structure with glass-core substrate

be seen that the maximum inelastic strain at the corner C4 solder joint occurs at a small local area, and the value is 4.8% at 450s (85 °C), 15.0% at 2250s (−40 °C), and 94.2% at 18000s (25 °C). Comparing these values with those with organic core package substrate, the maximum equivalent inelastic strain at the corner C4 solder joint with glass core substrate is larger than that with organic core substrate.

Figure 1.69 shows the maximum accumulated inelastic strain time-history in the corner C4 solder joint of structures made of the organic and glass substrates. It can be seen that the maximum accumulated inelastic strain in the corner C4 solder joint

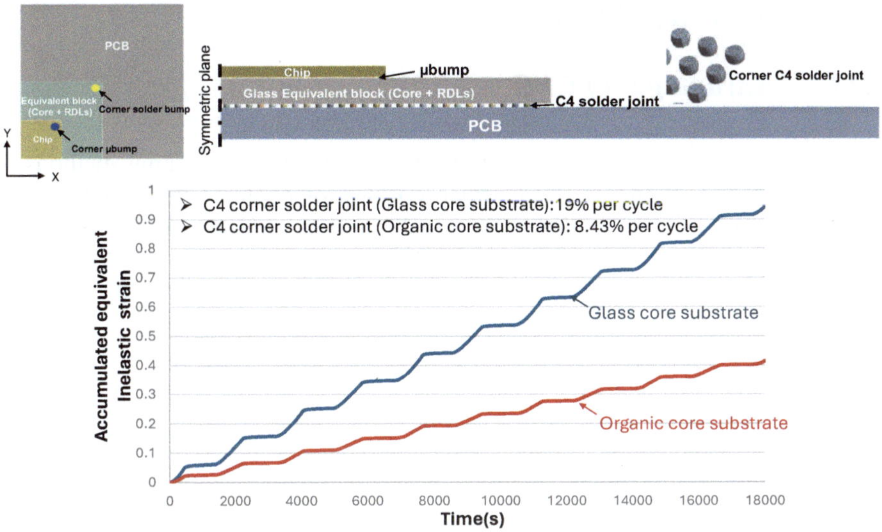


**Fig. 1.69** Maximum accumulated inelastic strain time-history in corner C4 solder joint of structures with organic and glass substrates

is larger with the glass core substrate. This is because of the thermal expansion mismatch between the glass core substrate ($\alpha = 6.3 \times 10^{-6}/^{o}C$) and the PCB ($\alpha_x = \alpha_y = 18 \times 10^{-6}/^{o}C$, $\alpha_z = 70 \times 10^{-6}/^{o}C$) is larger than that between the organic core substrate ($\alpha_x = \alpha_y = 11.3 \times 10^{-6}/^{o}C$, $\alpha_z = 24.8 \times 10^{-6}/^{o}C$) and the PCB.

(G) **Summary**

In packaging under thermal conditions, underfill is usually needed for silicon chips on package substrate to ensure the µbump solder joint reliability. However, underfill is seldom used for package substrate on PCB. Thus, in order to have a reliable C4 solder joint, the thermal expansion mismatch between the package substrate and PCB should be as small as possible especially for larger package substrates (because of larger chips). Unfortunately, the glass package substrate is heading in the other direction, i.e., the glass CTE is getting closer to the silicon CTE and farther to the PCB CTE. It is recommended when choosing the material properties such as the CTE of a glass substrate, care must be taken for the thermal expansion mismatch between the glass substrate and the PCB, especially for large glass substrates. As a matter of fact, the CTE of the glass substrate should be closer to the PCB CTE because there is underfill protection of the µbump solder joint on the glass substrate but there is no underfill protection of the C4 or BGA solder joints on the PCB especially for larger glass package substrates.

## 1.13 Co-Packaged Optics

### *1.13.1 Advantages of CPO*

Co-packaged optics (CPO) are heterogeneous integration packaging methods to integrate the optical engine (OE), which consists of photonic integrated circuits (PICs) and the electrical engine (EE), which consists of electronic ICs (EICs) as well as the switch application-specific IC (ASIC). The advantages of CPO are: (a) reducing the length of the electrical interface between the OE/EE (or PIC/ EIC) and the ASIC; (b) reducing the energy required to drive the signal; and (c) cutting the latency, which leads to better electrical performance [206–222].

### *1.13.2 Optical Transceivers*

A data center includes switches, routers, storage solutions, etc., such as those shown in Fig. 1.70a. Their functions include the network of computing and storage resources that enable the delivery of shared applications and data. Today, more than 90% of transceivers are pluggable transceivers (Fig. 1.70a). Figure 1.70b shows the future optical transceivers on a common package substrate (CPO) used in a data center.

Optical transceivers are the most important component in a data center. An optical transceiver is a key component of an optical transmission system that permits coupling of the transmission medium with the active components of the chain, such as switches, routers, optical interfaces or any optical transport equipment. An optical transceiver defines the process of converting electric signaling toward the

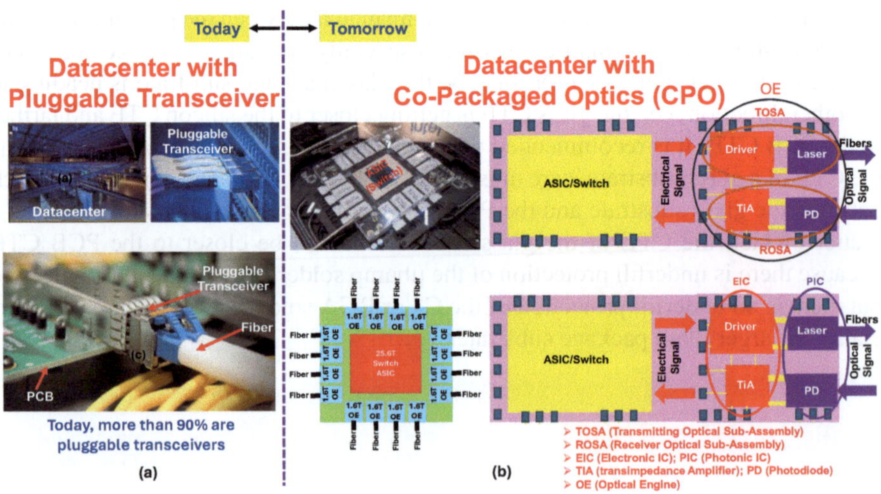


**Fig. 1.70** (**a**) Today. (**b**) Tomorrow. CPO for data center

optical transmission with the help of the transmission optical subassembly (TOSA), or Tx module— and performing inverse action through the receiver optical subassembly (ROSA), or Rx module (Fig. 1.70b). A TOSA contains a semiconductor laser diode (LD) and laser driver, while a ROSA contains a photodiode (PD), an optical interface such as a lens, a transimpedance amplifier (TIA), and a passive electrical interface. The TOSA module converts the electrical signal to the optical transmission light that lands on the fiber. The ROSA is used to receive an optical signal from a fiber and convert it back into an electrical signal (Fig. 1.70b).

The OE and EE are the key components of the optical transceiver for realizing photoelectric signal conversion. For example, the OE of an optical transceiver consists of the PIC, such as the PD and laser, etc., as shown in Fig. 1.70b. On the other hand, the EE of an optical transceiver consists of the EIC such as the laser driver and TIA, etc. The ASIC and the 16 OEs (PIC and EIC) are side by side on a CPO substrate such as that shown in Fig. 1.71 [223]. It can be seen that the CPO substrate is an organic interposer with fine linewidth and spacing RDLs.

### 1.13.3 3D Heterogeneous Integration of PIC and EIC

There are many different kinds of 3D heterogeneous integration (stacking) of the PIC and EIC as shown in Fig. 1.72. It can be seen from Fig. 1.72a that the light is coming from the fiber to the PIC and the PIC and EIC are stacked (face-to face) with the μbumps, and the signals are fanning out through the wire bonds from the PIC. Figure 1.72b is basically the same as Fig. 1.72a except the signals are fanning out through the C4 bumps, or μbumps, from the PIC. Figure 1.72c shows the EIC and the PIC are stacked (face-to back) with the μbumps, and the signals are running out through the TSVs and the C4 bumps or μbumps from the PIC. Figure 1.72d is basically the same as Fig. 1.72c except the μbumps between the EIC and PIC are

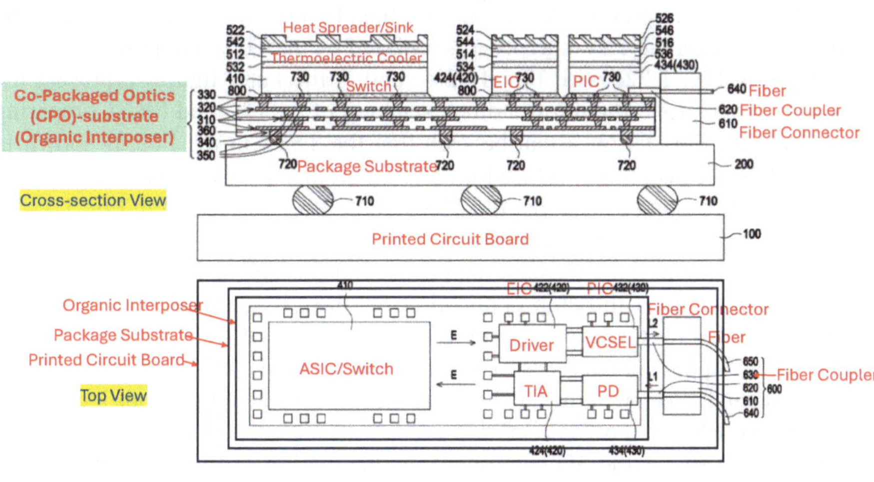


**Fig. 1.71** 2.3D heterogeneous integration of ASIC switch, PIC and EIC

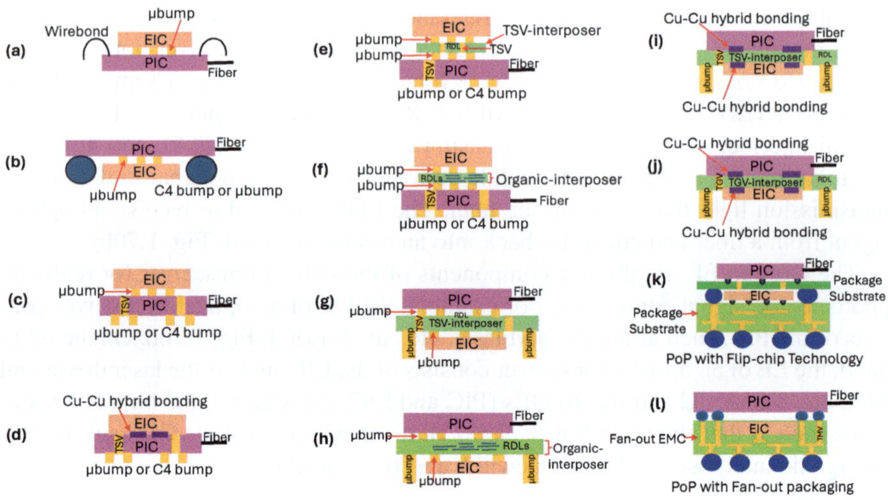


**Fig. 1.72** (**a**) to (**l**) (details are in the text) show various 3D heterogeneous integrations of PIC and EIC

eliminated by the bumpless Cu-Cu hybrid bonding. Figure 1.72e shows that the EIC and the PIC are stacked (connected) by a TSV-interposer with the μbumps, and the signals are fanning out through the RDLs and TSVs of the TSV-interposer and the TSVs and C4 bumps or μbumps from the PIC. Figure 1.72f is basically the same as Fig. 1.72€ except the TSV-interposer is replaced by the organic-interposer and the signals are fanning out through the RDLs of the organic-interposer and the TSVs and C4 bumps or μbumps from the PIC. Figure 1.72g shows that the PIC and EIC are stacked by a TSV interposer with μbumps, and the signals are running out through the μbumps of the TSV-interposer. Figure 1.72h is basically the same as Fig. 1.72g except the TSV-interposer is replaced by the organic-interposer. Figure 1.72i is basically the same as Fig. 1.72g except the μbumps on both sides of the TSV-interposer have been eliminated by the bumpless Cu-Cu hybrid bonding. Figure 1.72j is basically the same as Fig. 1.72i, except the TSV-interposer is replaced by the TGV-interposer.

Comparing Fig. 1.72c with Fig. 1.72 (e and f) the latter figures represent the cases where there are difficulties in direct stacking of the EIC and PIC. Comparing Fig. 1.72e with Fig. 1.72g, it can be seen that the TSVs in the PIC as shown in Fig. 1.72e have been eliminated (see Fig. 1.72g). Similarly, by comparing Fig. 1.72f with Fig. 1.72h, it can be seen that the TSVs in the PIC—as shown in Fig. 1.72f—have been eliminated (see Fig. 1.72h).

Figure 1.72 (k and l) are package-on package of the PIC and EIC. Figure 1.72k is for a solder-bumped flip-chip application for the EIC on a package substrate, while Fig. 1.72l is for the fan out of the EIC on a set of RDLs.

The patent application of case (h) has been granted [224] and is shown in Fig. 1.73. It can be seen that the CPO substrate for the 3D heterogeneous integration of the PIC and EIC is an organic interposer.

Figure 1.74a shows the 3D heterogeneous integration of PIC and EIC such as those shown in Fig. 1.72, and an ASIC with silicon bridges. It can be seen that the

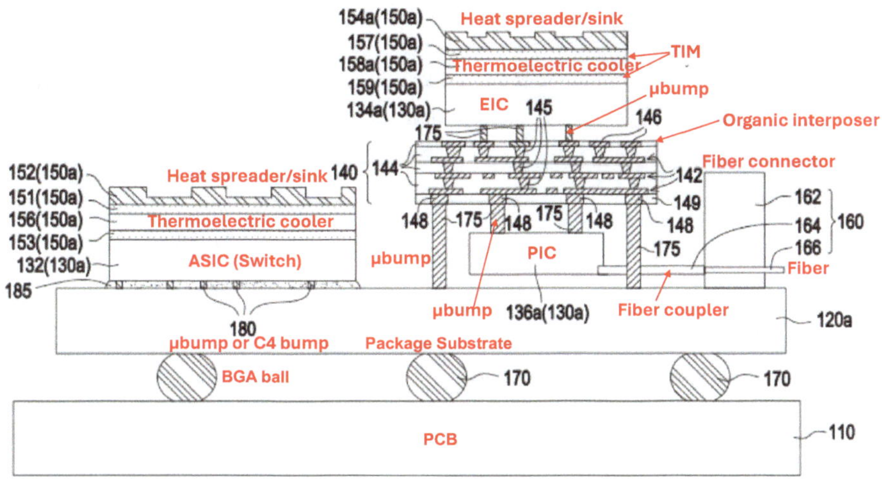


Fig. 1.73 3D heterogeneous integration of ASIC switch, PIC and EIC

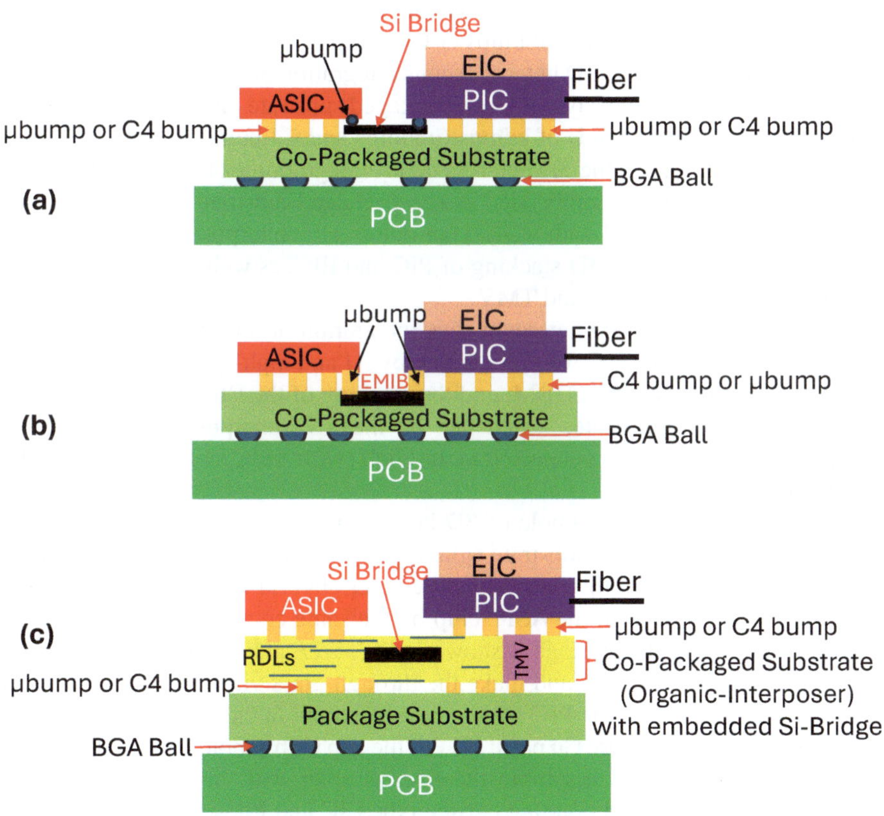


**Fig. 1.74** CPO with bridges. (**a**) Bridge connecting the switch and PIC. (**b**) EMIB connecting the switch and PIC. (**c**) Bridge embedded in fan-out EMC with RDLs connecting the switch and PIC

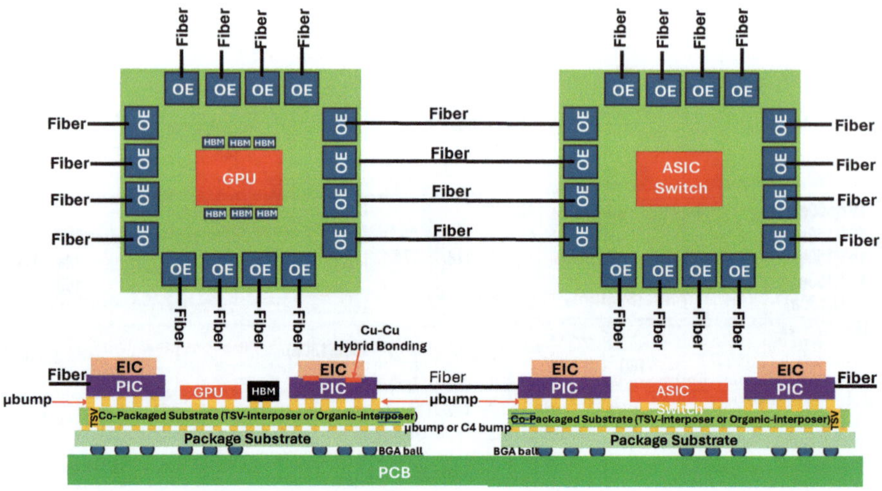


Fig. 1.75 3D CPO driven by AI and high-speed communication

ASIC and the 3D stacking of EIC and PIC are connected with a silicon bridge with μbumps. Then, the ASIC and the 3D stacking of EIC and PIC are attached to a co-packaged substrate with either μbumps or C4 bumps.

Figure 1.74b shows the 3D heterogeneous integration of PIC and EIC, and ASIC with Intel's (EMIB) in the cavity of a co packaged substrate (an organic interposer). There are two different kinds of bumps (μbumps and C4 bumps) on the ASIC and PIC. Figure 1.74c shows the 3D heterogeneous integration of PIC and EIC, and an ASIC with a silicon bridge embedded in a co-packaged substrate such as the fan-out epoxy molding compound with RDLs (fan-out organic interposer). The bridge will connect the ASIC and the 3D stacking of PIC and EIC, as well as the co-packaged substrate through the RDLs and TMV.

Figure 1.75 shows a 3D heterogeneous integration of a GPU, HBM or cHBM, ASIC switch, PIC, and EIC for HPC driven by AI and high-speed communication. It can be seen that the GPU, HBM, and the stacking of the EIC and PIC are on the same co-packaged substrate, which can be the TSV-interposer or the organic-interposer. This package is connected to the CPO (with ASIC switch, PIC, and EIC) through fibers.

Figure 1.76 shows an example of 3D heterogeneous integration of ASIC, PIC, and EIC on a co-packaged substrate (glass-core substrate). It can be seen that the EIC and PIC are stacked through Cu-Cu bumpless hybrid bonding. Then, they are arranged side-by-side with the ASIC chip on the same co-packaged substrate. The assembly process of the package is shown in Fig. 1.77 and can be described as follows: (1) First, etch a cavity and make the through-glass vias (TGVs) on the glass substrate; (2) Then place the PIC with a die attachment in the cavity with a holding device; (3) The PIC has bare Cu pads; (4) Fill the gap with a mold resin; (5) Fabricate the glass waveguide and assemble the fiber coupler with fibers; and lastly, (6) Perform the Cu-Cu hybrid bonding between the EIC and PIC.

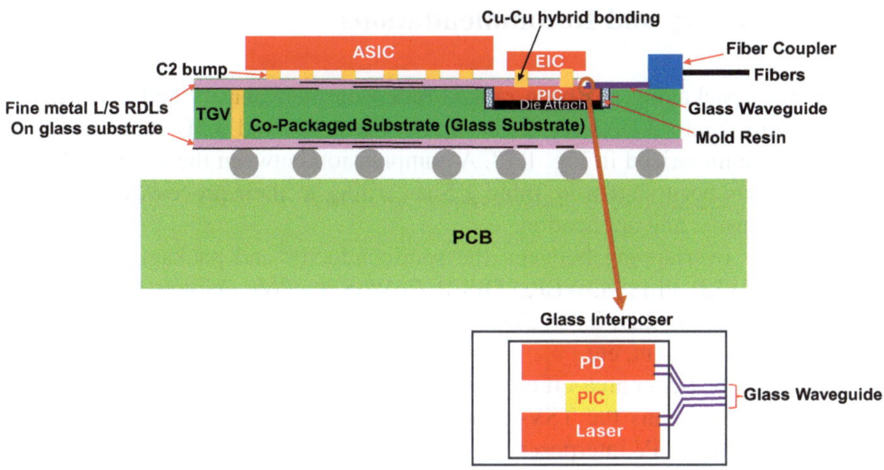


**Fig. 1.76** 3D heterogeneous integration of EIC and PIC and ASIC on a glass co-packaged substrate with Cu-Cu hybrid bonding

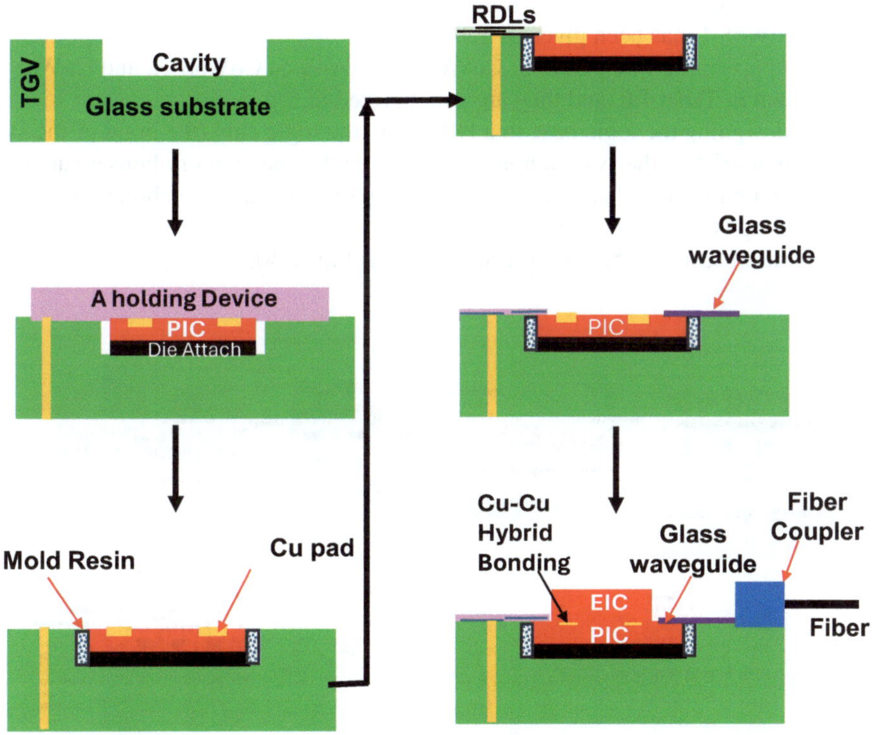


**Fig. 1.77** Processes used in the 3D heterogeneous integration of EIC and PIC and ASIC on a glass co-packaged substrate featuring Cu-Cu hybrid bonding

## 1.14 Summary and Recommendations

- Advanced packaging has been ranked based on their density and performance and grouped into 2.D, 2.1D, 2.3D, 2.5D, 3D, 3.3D, and 3.5D IC integrations. They are summarized in Fig. 1.78. A compression between these seven IC integrations has been shown in Table 1.5 according to their process, cost, performance, density, and application.
- Today, the overlapping between the semiconductor and packaging are wafer bumping, WLP, PLP, FOWLP, FOPLP, CoWoS, CoPoS, chiplets and heterogeneous integration, 2.D, 2.3D, 3D, 3.3D, 3.5D, RDLs, glass substrate/interposer, thermal management, etc.
- Because of the better area-efficiency of CoPoS, it is going to replace some of the CoWoS. This means the TSV-interpose fabricated on a silicon wafer will be replaced by the TGV-interposer fabricated on a glass panel.
- Because of the ever increase of the TSV-interposer size, the semiconductor yield loss of the large size TSV-interposer increases which means higher cost. The larger-size TSV-interposer in some of the CoWoS-S is replaced by the CoWoS-L or FO-EB-T.
- A comparison between 2.3D (chip-last) IC integration, CoWoS-S, CoWoS-R, and CoWoS-L has been given in Table 1.1.
- A comparison between CoPoS, CoWoS or CoWoS-S, CoWoS-R, and CoWoS-L is shown in Table 1.6, and they are summarized in Fig. 1.79.
- By redesigning the logic base in a HBM into the logic chip (the brain of the system) in a cHBM, the system houses (working with the memory houses) are able to stack up the cHBMs (area array) on the top of a SoC and then bond directly on a build-up (with either organic-core or glass-core) package substrate. A 3.3D IC integration structure has been recommended (Fig. 1.40).

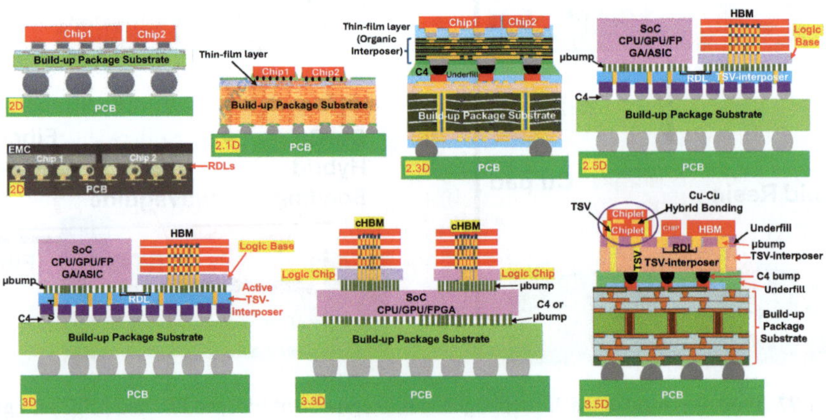


Fig. 1.78 2D, 2.1D, 2.3D, 2.5D, 3D, 3.3D, and 3.5D IC integrations

**Table 1.5** A comparison of 2D, 2.1D, 2.3D, 2.5D, 3D, 3.3D, and 3.5D IC integrations

| Advanced Substrate | Abbre. | Process | Cost | Perfor. | Density | Application |
|---|---|---|---|---|---|---|
| Build-up substrate or fan–out RDLs | 2D | Less | Low | Low | Low | Smartphones, tablets, notebooks, HPC, data center, etc. |
| Thin-film or bridge on build-up substrate or fan-out RDLs | 2.1D | More | Medium | Medium | Medium | Smartphones, tablets, notebooks, HPC, data center, etc. |
| Organic interposer on build-up substrate | 2.3D | More | Medium | Medium | High | HPC, data center, machine leaning, etc. |
| Passive TSV or TGV interposer on build-up substrate | 2.5D | More | High | Medium | High | HPC, data center, machine leaning, supercomputer, AI. etc. |
| Active TSV or TGV interposer on build-up substrate | 3D | Most | High | High | High | HPC, data center, machine leaning, supercomputer, AI. etc. |
| HBM on SoC (without TSV or TGV interposer) on build-up substrate | 3.3D | Most | Highest | Highest | Highest | HPC, data center, machine leaning, supercomputer, AI. etc. |
| Chiplets Cu-Cu hybrid bonding before heterogeneous integration on a common substrate | 3.5D | Most | Highest | Highest | Highest | HPC, data center, machine leaning, supercomputer, AI. etc. |

- In order to reduce the package size and enhance the performance, it is recommended to use Cu-Cu hybrid bonding to stack up some of the chiplets before heterogeneous integration of all the components on a common substrate. (3.5D IC integration)
- Hybrid bonding is one of the flip-chip technologies. The trends in flip-chip interconnects are shown in Fig. 1.80. Today, most C4 bump's diameters are 100 μm and are on 150 μm pitch. For some high-end applications, the diameter could be 50 μm and the pitch could be 100 μm. The 25 μm-diameter and 50 μm-pitch C4 bumps are in research and development (R&D). Today, most C2 bump's diameters are 50 μm and are on 100 μm pitch. For very high-end applications, the diameter is 25 μm and the pitch is 50 μm. The 10 μm-diameter and 20 μm-pitch are in R&D. Cu-Cu hybrid bonding is for $\leq$ 5 μm-diameter-pad and $\leq$ 10 μm-pitch.
- The assembly methods of Cu-Cu hybrid bonding are CoW, WoW, WoWoW, and reconstructed wafer. A comparison of them has been made in Table 1.2.
- As of today, there are very little papers published in the quality and reliability (such as the thermal-cycling, thermal shock, mechanical vibration, mechanical shock, high-temperature storage, highly accelerated temperature, highly accelerated moisture resistance etc.) of Cu-Cu hybrid bonding. In order for the hybrid

**Table 1.6** A comparison of CoPoS, CoWoS-S, CoWoS-R, and CoWoS-L

| Items | CoPoS | 2.5D (CoWoS-S) | 2.3D (CoWoS-R) | CoWoS-L |
|---|---|---|---|---|
| SoC Size | ≤ 25 × 25mm | ≤ 25 × 25mm | ≤ 25 × 25mm | ≤ 25 × 25mm |
| No. of HBMs | ≥ 4 | ≥ 4 | ≤ 8 | ≤ 8 |
| Interposer material | Glass | Silicon | Organic | Organic + Silicon bridges |
| RDL in interposer | Cu traces in $SiO_2$ | Cu traces in $SiO_2$ | Cu traces in $SiO_2$ | Cu traces in polymer or $SiO_2$ |
| RDL (L/S) interposer | < 1μm | < 1μm | ≥ 1μm | ≥1μm |
| RDL (Layer) interposer | ≤ 12 | ≤ 12 | ≤ 9 | ≤ 9 |
| Vertical interconnect | TGV in TGV interposer | TSV in TSV interposer | Via in RDL interposer | TSV and via in LSI and via in RDL interposer |
| Underfill (chip/interposer) | Yes | Yes | Yes | Yes |
| Underfill (interposer/sub.) | Yes | Yes | Yes | Yes |
| Total package profile | About the same | About the same | About the same | About the same |
| Package substrate size | Biggest | Biggest | Big | Bigger |
| Process steps | Most | Most | Less | More |
| Density | Highest | Highest | Higher | Higher |
| Performance | Highest | Highest | Higher | Higher |
| Decoupling capacitor | Embedded DTC | Embedded DTC | Discrete IPD | Discrete IPD |
| Cost | Highest | Highest | Higher | Higher |
| Applications | HPC, high bandwidth data center, etc. | HPC, high bandwidth data center, etc. | HPC, high bandwidth data center, etc. | HPC, high bandwidth data center, etc. |

bonding to be more popular, these quality and reliability data are desperately needed. Thermal management of Cu-Cu hybrid bonding will be an important topic [226].

- The advantages and disadvantages of glass-core packaging have been presented. Also, the difference between the glass-core substrate and the glass-core interposer has been provided. Glass-core build up package substrate is able to seamlessly integrate optical interconnects and will be used for niche applications, e.g., one-trillion transistors, e.g., [223]. Because of the on PCB solder joint reliability, the CTE of glass is recommended to be as close as that of the PCB, i.e., not as close as that of the silicon chip (there is underfill protection of the μbump between the chip and the glass-core substrate).
- The advantages of CPO have been given. More than ten different 3D heterogeneous integrations of PIC and EIC have been presented and discussed. A CPO

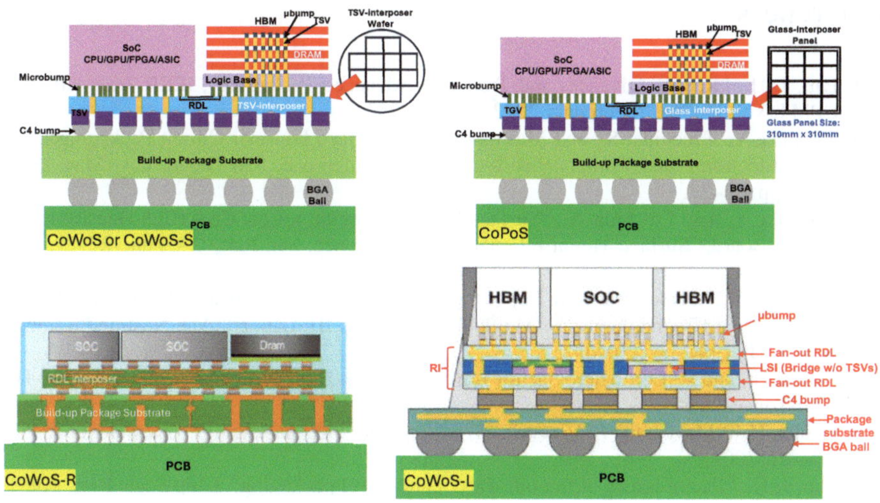


**Fig. 1.79** CoWoS, CoPoS, CoWoS-R, and CoWoS-L

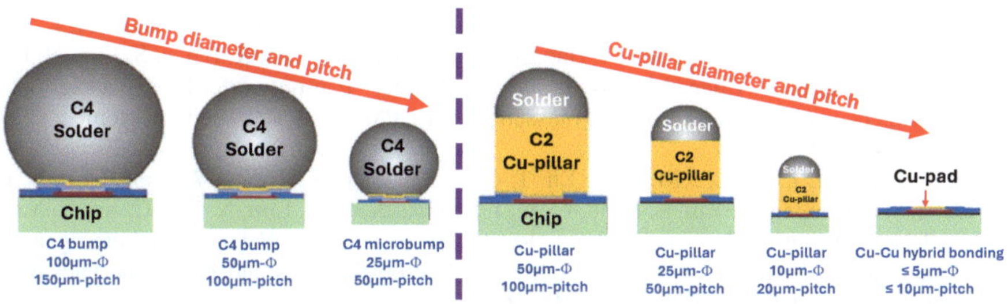


**Fig. 1.80** Trends in flip chip interconnects

structure has been recommended for HPC driven by AI and high-speed communication (Fig. 1.75). Also, another 3D CPO structure on glass substrate has been recommended (Fig. 1.76).

- As today, there are very small amount of papers published in the quality, reliability, and thermal management of CPO. In order for the CPO technology to be popular, the quality, reliability, and thermal management data of CPO are desperately needed.
- Today, the package of choice for HPC driven by AI is the 2.5D IC integration (CoWoS or CoWoS-S). 2.3D IC integration with an organic interposer is a lower cost package, but the performance is not as good as the 2.5D IC integration. Because of the higher area efficiency (lower cost) of the CoPoS with a glass interposer, which will take away some of the 2.5D IC integration market shares. 3.3D IC integration is another competitor of 2.5D IC integration with very large TSV interposers. cHBMs will be used more in the near future.

## References


1. Lau, J. H., "Recent Advances and Trends in Advanced Packaging", IEEE Transactions on CPMT, Vol. 12, No. 2, February 2022, pp. 228–252.
2. Tsukada, Y., Tsuchida, S., and Mashimoto, Y., "Surface Laminar Circuit Packaging," 42nd IEEE Electronic and Components Technology Conference, San Diego, CA, May 18–20, 1992, pp. 22–27.
3. Tsukada, Y., and Tsuchida, S., "Surface Laminar Circuit, a Low Cost High Density Printed Circuit Board," Surface Mount International Conference, San Jose, CA, Aug. 27–29, 1992, pp. 537–542.
4. Tsukada, Y., "Solder Bumped Flip Chip Attach on SLC Board and Multichip Module," Chip on Board Technologies for Multichip Modules, J. H. Lau, ed., Van Nostrand Reinhold, New York, 1994, pp. 410–443.
5. Lau, J. H., and R. Lee, Microvias for Low Cost, High Density Interconnects, McGraw-Hill, New York, 2001.
6. Hedler, H. T. Meyer, and B. Vasquez, "Transfer wafer level packaging," U.S. Patent 6727576, Apr. 27, 2001.
7. Lau, J. H., "Patent issues of fan-out wafer/panel-level packaging," Chip Scale Rev., vol. 19, pp. 42–46, Nov. 2015.
8. Brunnbauer, M., et al., "An embedded device technology based on a molded reconfigured wafer," in Proc. IEEE/ECTC, May 2006, pp. 547–551.
9. Meyer, T., G. Ofner, S. Bradl, M. Brunnbauer, and R. Hagen, "Embedded wafer level ball grid array (eWLB)," in Proc. IEEE/EPTC, Dec. 2006, pp. 1–5.
10. Lau, J. H., M. Li, M. Li, T. Chen, I. Xu, X. Qing, Z. Cheng, N. Fan, E. Kuah, Z. Li, K. Tan, Y. Cheung, E. Ng, P. Lo, K. Wu, J. Hao, S. Koh, R. Jiang, X. Cao, R. Beica, S. Lim, N. Lee, C. Ko, H. Yang, Y. Chen, M. Tao, J. Lo, and R. Lee, "Fan-out wafer-level packaging for heterogeneous integration," IEEE Transactions on CPMT, Vol. 8, Issue 9, September 2018, pp. 1544–1560.
11. Lau, J. H., M. Li, Y. Lei, M. Li, I. Xu, T. Chen, Q. Yong, Z. Cheng, W. Kai, P. Lo, Z. Li, K. Tan, Y. Cheung, N. Fan, E. Kuah, C. Xi, J. Ran, R. Beica, S. Lim, N. Lee, C. Ko, H. Yang, Y. Chen, M. Tao, J. Lo, and R. Lee, "Reliability of Fan-Out Wafer-Level Heterogeneous Integration", IMAPS Transactions, Journal of Microelectronics and Electronic Packaging, Vol. 15, Issue: 4, October 2018, pp. 148–162.
12. Ko, CT, H. Yang, Lau, J. H., M. Li, M. Li, C. Lin, J.W. Lin, T. Chen, I. Xu, C. Chang, J. Pan, H. Wu, Q. Yong, N. Fan, E. Kuah, Z. Li, K. Tan, Y. Cheung, E. Ng, K. Wu, J. Hao, R. Beica, M. Lin, Y. Chen, Z. Cheng, S. Koh, R. Jiang, X. Cao, S. Lim, N. Lee, M. Tao, J. Lo, and R. Lee, "Chip-First Fan-Out Panel Level Packaging for Heterogeneous Integration", IEEE Transactions on CPMT, 2018, Vol. 8, Issue 9, September 2018, pp. 1561–1572.
13. Ko, C. T., H. Yang, J. H. Lau, M. Li, M. Li, C. Lin, J.W. Lin, C. Chang, J. Pan, H. Wu, Y. Chen, T. Chen, I. Xu, P. Lo, N. Fan, E. Kuah, Z. Li, K. Tan, C. Lin, R. Beica, M. Lin, X. Cao, S. Lim, N. Lee, M. Tao, J. Lo, and R. Lee, "Design, Materials, Process, and Fabrication of Fan-Out Panel-Level Heterogeneous Integration", IMAPS Transactions, Journal of Microelectronics and Electronic Packaging, Vol. 15, Issue: 4, October 2018, pp. 141–147.
14. Shimizu, N., W. Kaneda, H. Arisaka, N. Koizumi, S. Sunohara, A. Rokugawa, and T. Koyama, "Development of Organic Multi Chip Package for High Performance Application", IMAPS Proceedings of International Symposium on Microelectronics, October 2013, pp. 414–419.
15. Oi, K., S. Otake, N. Shimizu, S. Watanabe, Y. Kunimoto, T. Kurihara, T. Koyama, M. Tanaka, L. Aryasomayajula, and Z. Kutlu, "Development of New 2.5D Package with Novel Integrated Organic Interposer Substrate with Ultra-fine Wiring and High Density Bumps", IEEE/ECTC Proceedings, May 2014, pp. 348–353.
16. Lau, J. H, R. Tain, C. Ko, T. J. Tseng, and C. Chien, US Patent 12,309,943, May 20, 2025, filed on Dec. 27, 2022.

17. Chiu, C., Z. Qian, and M. Manusharow, "Bridge interconnect with air gap in package assembly," US Patent No. 8,872,349, 2014.
18. Mahajan, R., R. Sankman, N. Patel, D. Kim, K. Aygun, Z. Qian, et al., "Embedded multi-die interconnect bridge (EMIB) – a high-density, high-bandwidth packaging interconnect," IEEE/ECTC Proceedings, May 2016, pp. 557–565.
19. Duan, G., et al., "EMIB-TSV Advanced Packaging Technology – EMIB's Next Evolution", IEEE/ECTC Proceedings, May 2025, pp. 269–273.
20. Pulido, C., et al., "Direct Bonded Heterogeneous Integration (DBHi): Surface bridge approach for die tiling", IEEE/ECTC Proceedings, May 2023, pp. 1053–1059.
21. Apple Inc., Apple unveils M1 Ultra, the world's most powerful chip for a personal computer, 2022-03-08.
22. Lau, J. H., C. Ko, P. Lin. T. J. Tseng, R. Tain, and K. Yang, US Paren 11,410,933, Aug. 9, 2022, Failed on May 7, 2021.
23. Yoon, S., P. Tang, R. Emigh, Y. Lin, P. Marimuthu, and R. Pendse, "Fanout Flipchip eWLB (Embedded Wafer Level Ball Grid Array) Technology as 2.5D Packaging Solutions", IEEE/ECTC Proceedings, 2013, pp. 1855–1860.
24. Chen, N. C., T. Hsieh, J. Jinn, P. Chang, F. Huang, J. Xiao, A. Chou, B. Lin, "A Novel System in Package with Fan-out WLP for high speed SERDES application", IEEE/ECTC Proceedings, May 2016, pp. 1496–1501.
25. Yip, L., R. Lin, C. Lai, and C. Peng, "Reliability Challenges of High-Density Fan-out Packaging for High-Performance Computing Applications", IEEE/ECTC Proceedings, May 2022, pp. 1454–1458.
26. Lin, Y., W. Lai, C. Kao, J. Lou, P. Yang, C. Wang, and C. Hseih, "Wafer warpage experiments and simulation for fan-out chip on substrate," IEEE/ECTC Proceedings, May 2016, pp. 13–18.
27. Yu, D., "Advanced system integration technology trends," SiP Global Summit, SEMICON Taiwan, Sept. 6, 2018.
28. Kurita, Y., T. Kimura, K. Shibuya, H. Kobayashi, F. Kawashiro, N. Motohashi, and M. Kawano, "Fan-out wafer level packaging with highly flexible design capabilities," IEEE/ESTC Proceedings, September 2010, pp. 1–6.
29. Motohashi, N., T. Kimura, K. Mineo, Y. Yamada, T. Nishiyama, K. Shibuya, H. Kobayashi, Y. Krita, and M. Kawano, "System in a Wafer-level package technology with RDL-first process," IEEE/ECTC Proceedings, May 2011, pp. 59–64.
30. Huemoeller, R., and C. Zwenger, "Silicon wafer integrated fan-out technology," Chip Scale Review, Mar/Apr 2015, pp. 34–37.
31. Lim, H., J. Yang, and R. Fuentes, "Practical Design Method to Reduce Crosstalk for Silicon Wafer Integration Fan-Out Technology (SWIFT) Packages", IEEE/ECTC Proceedings, May 2018, pp. 2205–2211.
32. Jayaraman. S., "Advanced Packaging: HDFO for Next Generation Devices", Proceedings of IWLPC, February 2022, pp. 1–28.
33. Suk, K., S. Lee, J. Kim, S. Lee, H. Kim, S. Lee, P. Kim, D. Kim, D. Oh, and J. Byun, "Low-Cost Si-less RDL Interposer Package for High Performance Computing Applications", IEEE/ECTC Proceedings, May 2018, pp. 64–69.
34. You, S., S. Jeon, D. Oh, K. Kim, J. Kim, S. Cha, and G. Kim, "Advanced Fan-Out Package SI/PI/Thermal Performance Analysis of Novel RDL Packages", IEEE/ECTC Proceedings, May 2018, pp. 1295–1301.
35. Lin, Y., M. Yew, M. Liu, S. Chen, T. Lai, P. Kavle, C. Lin, T. Fang, C. Chen, C. Yu, K. Lee, C. Hsu, P. Lin, F. Hsu, and S. Jeng, "Multilayer RDL Interposer for Heterogeneous Device and Module Integration", IEEE/ECTC Proceedings, May 2019, pp. 931–936.
36. Lin, P., M. Yew, S. Yeh, S. Chen, C. Lin, C. Chen, C. Hsieh, Y. Lu, P. Chuang, H. Cheng, and S. Jeng, "Reliability Performance of Advanced Organic Interposer (CoWoS-R) Packages", Proceedings of IEEE/ECTC, May 2021, pp. 723–728.

37. Lin, M., M. Liu, H. Chen, S. Chen, M. Yew, C. Chen, and S. Jeng, "Organic Interposer CoWoS-R (plus) Technology", Proceedings of IEEE/ECTC, May 2022, pp. 1–6.
38. Chang, K., C. Huang, H. Kuo, M. Jhong, T. Hsieh, M. Hung, and C. Wang, "Ultra High-Density IO Fan-Out Design Optimization with Signal Integrity and Power Integrity", IEEE/ECTC Proceedings, May 2019, pp. 41–46.
39. Lai, W., P. Yang, I. Hu, T. Liao, K. Chen, D. Tarng, and C. Hung, "A Comparative Study of 2.5D and Fan-out Chip on Substrate: Chip First and Chip Last", IEEE/ECTC Proceedings, May 2020, pp. 354–360.
40. Fang, J., M. Huang, H. Tu, W. Lu, and P. Yang, "A Production-worthy Fan-Out Solution – ASE FOCoS Chip Last", IEEE/ECTC Proceedings, May 2020, pp. 290–295.
41. Cao, L. "Advanced FOCOS (Fanout Chip on Substrate) Technology for Chiplets Heterogeneous Integration", Proceedings of IWLPC, February 2022, pp. 1–6.
42. Cao, L., T. Lee, R. Chen, Y. Chang, H. Lu, N. Chao, Y. Huang, C. Wang, C. Huang, H. Kuo, Y. Wu, and H. Cheng, "Advanced Fanout Packaging Technology for Hybrid Substrate Integration", Proceedings of IEEE/ECTC, May 2022, pp. 1362–1370.
43. Lee, T., S. Yang, H. Wu, and Y. Lin, "Chip Last Fanout Chip on Substrate (FOCoS) Solution for Chiplets Integration", Proceedings of IEEE/ECTC, May 2022, pp. 1970–1974.
44. Yin, W., W. Lai, Y. Lu, K. Chen, H. Huang, T. Chen, C. Kao, and C. Hung, "Mechanical and Thermal Characterization Analysis of Chip-last Fan-out Chip on Substrate", Proceedings of IEEE/ECTC, May 2022, pp. 1711–1719.
45. Li, J., F. Tsai, J. Li, G. Pan, M. Chan, L. Zheng, S. Chen, N. Kao, D. Lai, K. Wan, and Y. Wang, "Large Size Multilayered Fan-Out RDL Packaging for Heterogeneous Integration", IEEE/EPTC Proceedings, December 2021, pp. 239–243.
46. Kim, J., J. Choi, S. Kim, J. Choi, Y. Park, G. Kim, S. Kim, S. Park, H. Oh, S. Lee, T. Cho, and D. Kim, "Cost Effective 2.3D Packaging Solution by using Fanout Panel Level RDL", IEEE/ECTC Proceedings, June 2021, pp. 310–314.
47. Miki, S., H. Taneda, N. Kobayashi, K. Oi, K. Nagai, and T. Koyama, "Development of 2.3D High Density Organic Package using Low Temperature Bonding Process with Sn-Bi Solder", IEEE/ECTC Proceedings, May 2019, pp. 1599–1604.
48. Murayama, K., S. Miki, H. Sugahara, and K. Oi, "Electro-migration evaluation between organic interposer and build-up substrate on 2.3D organic package", IEEE/ECTC Proceedings, May 2020, pp. 716–722.
49. Lau, J. H., G. Chen, R. Chou, C. Yang, and T. Tseng, "Fan-Out (RDL-First) Panel-Level Hybrid Substrate for Heterogeneous Integration", IEEE/ECTC Proceedings, May 2021, pp. 148–156.
50. Lau, J. H., G. Chen, R. Chou, C. Yang, and T. Tseng, "Hybrid Substrate by Fan-Out RDL-First Panel-Level Packaging", IEEE Transactions on CPMT, Vol. 11, No. 8, August 2021, pp. 1301–1309.
51. Chou, R., J. H. Lau, G. Chen, J. Huang, C. Yang, N. Liu, and T. Tseng, "Heterogeneous Integration on 2.3D Hybrid Substrate Using Solder Joint and Underfill", IMAPS Transactions, Journal of Microelectronics and Electronic Packaging, Vol. 19, 2022, pp. 8–17.
52. Chen, G., J. H. Lau, R. Chou, C. Yang, J. Huang, N. Liu, and T. Tseng, "2.3D Hybrid Substrate with Ajinomoto Build-Up Film for Heterogeneous Integration", Proceedings of IEEE/ECTC, May 2022, pp. 30–37.
53. Lau, J. H., C. Lin, N. Liu, H. Yang, T. Xia, C. Ko, B. Lin, Y. Chuang, R. Chen, M. Ma, T. Tseng, M. Li, and K. Leung, "Hybrid Substrate with Ultra-Large Organic Interposer for Heterogeneous Integration", IEEE Transactions on CPMT, Vol. 13, No. 9, September 2023, pp. 1371–1379.
54. Chen, G., J. H. Lau, R. Chou, C. Yang, J. Huang, N. Liu, and T. Tseng, "2.3D Hybrid Substrate with Ajinomoto Build-Up Film for Heterogeneous Integration", Proceedings of IEEE/ECTC, May 2022, pp. 30–37.

55. Lau, J. H., G. Chen, C. Yang, V. Teng, A. Peng, J. Huang, N. Liu, Y. Chen, T. Tseng, and M. Li, "Hybrid Substrate for Heterogeneous Integration", ASME Transactions, Journal of Electronic Packaging, published online: August 11, 2023, Vol. 146, June 2024, pp. 021001-1–18.
56. Peng, P., J. H. Lau, C. Ko, P. Lee, E. Lin, K. Yang, P. Lin, T. Xia, L. Chang, N. Liu, C. Lin, T. Lee, J. Wang, M. Ma, and T. Tseng, "Development of High-Density Hybrid Substrate for Heterogeneous Integration," IEEE CPMT Symposium Japan, November 2021, pp. 5–8.
57. Peng, P., J. H. Lau, C. Ko, P. Lee, E. Lin, K. Yang, P. Lin, T. Xia, L. Chang, N. Liu, C. Lin, T. Lee, J. Wang, M. Ma, and T. Tseng, "High-Density Hybrid Substrate for Heterogeneous Integration", IEEE Transactions on CPMT, Vol. 12, No.3, March 2022, pp. 469–478.
58. Peng, C., J. H. Lau, K. Yang, P. Lin, C. Ko, and T. J. Tseng, US Patent 11,516,910, Nov. 29, 2011, Filed Jul. 9, 2021.
59. Lin, Y., M. Yew, M. Liu, S. Chen, T. Lai, P. Kavle, C. Lin, T. Fang, C. Chen, C. Yu, K. Lee, C. Hsu, P. Lin, F. Hsu, and S. Jeng, "Multilayer RDL Interposer for Heterogeneous Device and Module Integration", IEEE/ECTC Proceedings, May 2019, pp. 931–936.
60. Lin, P., Yew, M., Yeh, S., Chen, S., Lin, C., Chen, C., Hsieh, C., et al., 2021, "Reliability Performance of Advanced Organic Interposer (CoWoS-R) Packages," Proceedings of IEEE/ECTC, May 2021, pp. 723–728.
61. Lin, M., M. Liu, H. Chen, S. Chen, M. Yew, C. Chen, and S. Jeng, "Organic Interposer CoWoS-R (plus) Technology", Proceedings of IEEE/ECTC, May 2022, pp. 1–6.
62. Lee, C., Y. Hu, S. Chen, C. Lai, M. Liu, H. Chen, J. Lin, M. Yew, C. Hsu, M. Chiu, C. Chen, M. Chou, W. Chen, J. Chang, C. Hsieh, C. Chen, H. Chen, C. Wang, K. Yan, S. Jeng, and H. Jun, "Next Generation Large Size High Interconnect Density CoWoS-R Package", Proceedings of IEEE/ECTC, May 2024, pp. 259–263.
63. Yu, A., N. Khan, G. Archit, D. Pinjala, K. Toh, V. Kripesh, S. Yoon, and J. H. Lau, "Fabrication of Silicon Carriers With TSV Electrical Interconnections and Embedded Thermal Solutions for High Power 3-D Packages", IEEE Transactions on CPMT, Vol. 32, No. 3, September 2009, pp. 566–571.
64. Tang, G. Y., S. Tan, N. Khan, D. Pinjala, J. H. Lau, A. Yu, V. Kripesh, and K. Toh, "Integrated Liquid Cooling Systems for 3-D Stacked TSV Modules", IEEE Transactions on CPMT, Vol. 33, No. 1, March 2010, pp. 184–195.
65. Khan, N., H. Li, S. Tan, S. Ho, V. Kripesh, D. Pinjala, J. H. Lau, and T. Chuan, "3-D Packaging With Through-Silicon Via (TSV) for Electrical and Fluidic Interconnections", IEEE Transactions on CPMT, Vol. 3, No. 2, February 2013, pp. 221–228.
66. Khan, N., V. Rao, S. Lim, H. We, V. Lee, X. Zhang, E. Liao, R. Nagarajan, T. C. Chai, V. Kripesh, and J. H. Lau, "Development of 3-D Silicon Module With TSV for System in Packaging", IEEE Transactions on CPMT, Vol. 33, No. 1, March 2010, pp. 3–9.
67. Chai, T. C., X. Zhang, J. H. Lau, C. S. Selvanayagam, and D. Pinjala, "Development of Large Die Fine-Pitch Cu/low-k FCBGA Package with through Silicon via (TSV) Interposer", IEEE Transactions on CPMT, Vol. 1, No. 5, May 2011, pp. 660–672.
68. Lau, J. H., S. Lee, M. Yuen, J. Wu, C. Lo, H. Fan, and H. Chen, "Apparatus having thermal-enhanced and cost-effective 3D IC integration structure with through silicon via interposer". US Patent No: 8,604,603, Filed Date: February 19, 2010, Date of Patent: December 10, 2013.
69. Lau, J. H., M. S. Zhang, and S. W. R. Lee, "Embedded 3D Hybrid IC Integration System-in-Package (SiP) for Opto-Electronic Interconnects in Organic Substrates", ASME Transactions, Journal of Electronic Packaging, Vol. 133, September 2011, pp. 1–7.
70. Chien, H. C., J. H. Lau, Y. Chao, R. Tain, M. Dai, S. T. Wu, W. Lo, and M. J. Kao, "Thermal Performance of 3D IC Integration with Through-Silicon Via (TSV)", IMAPS Transactions, Journal of Microelectronic Packaging, Vol. 9, 2012, pp. 97–103.
71. Lau, J. H., "Overview and Outlook of TSV and 3D Integrations", Journal of Microelectronics International, V.28, No. 2, 2011, pp. 8–22.
72. Lau, J. H., "Critical Issues of 3D IC Integrations", IMAPS Transactions, Journal of Microelectronics and Electronic Packaging, First Quarter Issue, 2010, pp. 35–43.

73. Lau, J. H., "Design and Process of 3D MEMS Packaging", IMAPS Transactions, Journal of Microelectronics and Electronic Packaging, First Quarter Issue, 2010, pp. 10–15.
74. Lau, J. H., Lee, R., Yuen, M., and Chan, P., "3D LED and IC Wafer Level Packaging", Journal of Microelectronics International, Vol. 27, Issue 2, 2010, pp. 98–105.
75. Sheu, S., Z. Lin, J. Hung, J. H. Lau, P. Chen, S. Wu, K. Su, C. Lin, S. Lai, T. Ku, W. Lo, M. Kao, "An Electrical Testing Method for Blind Through Silicon Vias (TSVs) for 3D IC Integration", IMAPS Transactions, Journal of Microelectronic Packaging, Vol. 8, No. 4, Fourth Quarter 2011, pp. 140–145.
76. Chen, J. C., J. H. Lau, P. J. Tzeng, S. Chen, C. Wu, C. Chen, H. Yu, Y. Hsu, S. Shen, S. Liao, C. Ho, C. Lin, T. K. Ku, and M. J. Kao, "Effects of Slurry in Cu Chemical Mechanical Polishing (CMP) of TSVs for 3-D IC Integration", IEEE Transactions on CPMT, Vol. 2, No. 6, June 2012, pp. 956–963.
77. Lau, J. H., and G. Y. Tang, "Effects of TSVs (through-silicon vias) on thermal performances of 3D IC integration system-in-package (SiP)", Journal of Microelectronics Reliability, Vo. 52, Issue 11, November 2012, pp. 2660–2669.
78. Wu, C., S. Chen, P. Tzeng, J. H. Lau, Y. Hsu, J. Chen, Y. Hsin, C. Chen, S. Shen, C. Lin, T. Ku, and M. Kao, "Oxide Liner, Barrier and Seed Layers, and Cu-Plating of Blind Through Silicon Vias (TSVs) on 300mm Wafers for 3D IC Integration", IMAPS Transactions, Journal of Microelectronic Packaging, Vol. 9, No. 1, First Quarter 2012, pp. 31–36.
79. Lau, J. H., P. Tzeng, C. Lee, C. Zhan, M. Li, J. Cline, K. Saito, Y. Hsin, P. Chang, Y. Chang, J. Chen, S. Chen, C. Wu, H. Chang, C. Chien, C. Lin, T. Ku, R. Lo, and M. Kao, "Redistribution Layers (RDLs) for 2.5D/3D IC Integration", IMAPS Transactions, Journal of Microelectronic Packaging, Vol. 11, No. 1, First Quarter 2014, pp. 16–24.
80. Lau, J. H., C. Lee, C. Zhan, S. Wu, Y. Chao, M. Dai, R. Tain, H. Chien, J. Hung, C. Chien, R. Cheng, Y. Huang, Y. Lee, Z. Hsiao, W. Tsai, P. Chang, H. Fu, Y. Cheng, L. Liao, W. Lo, and M. Kao, "Low-Cost Through-Silicon Hole Interposers for 3D IC Integration", IEEE Transactions on CPMT, Vol. 4, No. 9, September 2014, pp. 1407–1419.
81. Hsieh, M. C., S. T. Wu, C. J. Wu, and J. H. Lau, "Energy Release Rate Estimation for Through Silicon Vias in 3-D Integration", IEEE Transactions on CPMT, Vol. 4, No. 1, January 2014, pp. 57–65.
82. Lee, C. C., C. S. Wu, K. S. Kao, C. W. Fang, C. J. Zhan, J. H. Lau, and T. H. Chen, "Impact of high density TSVs on the assembly of 3D-ICs packaging", Microelectronic Engineering, Vol. 107, July 2013, pp. 101–106.
83. Li, Y., J. Liu, et al., "Rapid estimation of anisotropic thermal conductivity in RDL for 2.5D chiplet design", IEEE ECTC Proceedings, Dallas, TX, May 2025, pp. 1541–1546.
84. http://press.xilinx.com/2013-10-20-Xilinx-and-TSMCReach-Volume-Production-on-all-28nm-CoWoS-based-All-Programmable-3D-IC-Families
85. Kim, M., S. Lee, K. Suk, J. Jang, G. Jeon, J. Choi, H. Yun, J. Hong, J. Choi, W. Lee, S. Jung, W. Choi, and D. Kim. "Novel 2.5D RDL Interposer Packaging: A Key Enabler for the New Era of Heterogenous Chip Integration", IEEE/ECTC Proceedings, May 2021, pp. 321–326.
86. Lee, D, S. Yoon, J. Jun, J. Park, T. Lee, U. Kang, J. Lee, "Improving Flip Chip Process for Large 2.5D Molded Interposer", IEEE/ECTC Proceedings, May 2021, pp. 1026–1030.
87. Lee, S., S. Han, J. Hong, S. O. D. Kwak, S. Nam, Y. Park, and J. Lee, "Novel Method of Wafer-Level and Package-Level Process Simulation for Warpage Optimization of 2.5D TSV", IEEE/ECTC Proceedings, May 2021, pp. 1527–1531.
88. Rajan, S., A. Kaul, T. Sarvey, G. May, and M. Bakir, "Design Considerations, Demonstration, and Benchmarking of Silicon Microcold Plate and Monolithic Microfluidic Cooling for 2.5D ICs", IEEE/ECTC Proceedings, May 2021, pp. 1418–1426.
89. Nam, S., Y. Kim, A. Jang, I. Hwang, S. Park, S. Lee, and D. Kim, "The Extremely Large 2.5D Molded Interposer on Substrate (MIoS) Package Integration - Warpage and Reliability", IEEE/ECTC Proceedings, May 2021, pp. 1998–2002.

90. Hayashi, T., P. R. Watanabe, and S. Ichikawa, "Development of Highly Reliable Crack Resistive Build-up Dielectric Material with Low Df Characteristic for Next-Gen 2.5D Packages", IEEE/ECTC Proceedings, May 2021, pp. 570–576.
91. Hayashi, T., P. Lin, R. Watanabe, and S. Ichikawa, "Development of Highly Reliable Crack Resistive Build-up Dielectric Material with Low Df Characteristic for Next-Gen 2.5D Packages", IEEE/ECTC Proceedings, May 2021, pp. 570–576.
92. Hwang, Y., S. Moon, S. Nam, J. Ahn, "Chiplet-based System PSI Optimization for 2.5D/3D Advanced Packaging Implementation", IEEE/ECTC Proceedings, May 2022, pp. 12–17.
93. Nam, S., J. Kang, I. Lee, Y. Kim, H. Yu, and D. Kim, "Investigation on Package Warpage and Reliability of the large size 2.5D Molded Interposer on Substrate (MIoS) Package", IEEE/ECTC Proceedings, May 2022, pp. 643–647.
94. Shao, A., S. Chen, K. Lee, H. Chen, S. Low, and I. Singh, "Investigation of Moisture-Induced Warpage of Chip-on-Wafer in 2.5D IC Package", IEEE/ECTC Proceedings, May 2022, pp. 1095–1102.
95. Zhao, G., Y. Yang, X. Peng, H. Wu, H. Lin, G. Li, P. Zhu, R. Sun, and C. Wong, "Effects of Temperature on the Adhesive Performance of High Tg Underfill in 2.5D Heterogeneous Integrated Packaging", IEEE/ECTC Proceedings, May 2022, pp. 2275–2280.
96. Razdan, S., P. Dobbelaere, J. Xue, and A. Prasad, "Advanced 2.5D and 3D Packaging Technologies for Next Generation Silicon Photonics in high Performance Networking Applications", IEEE/ECTC Proceedings, May 2022, pp. 428–435.
97. Satakel Y., T. Funaki, K. Kobinata, H. Matsuno, S. Hidaka, S. Abe, H. Ito, C. Hsiao, S. Li, Y. Km, and T. Ohba, "Functional Interposer Embedded with Multi-terminal Si Capacitor for 2.5D/3D Applications using Planarization and Bumpless Chip-on-Wafer (COW)", IEEE/ECTC Proceedings, May 2022, pp. 283–288.
98. Yip, L., J. Tsai, R. Lin, and I. Hsu, "2.5D MCM (Multi-Chip Module) Technology Development for Advanced Package", IEEE/ECTC Proceedings, May 2023, pp. 778–782.
99. Hu, Y., Y. Liang, H. Hu, C. Tan, C. Shen, C. Lee, and S. Hou, "CoWoS Architecture Evolution for Next Generation HPC on 2.5D System in Package", IEEE/ECTC Proceedings, May 2023, pp. 1022–1026.
100. Li, H., K. Chui, S. Goh, R. Budoyo, N. Long, Y. Ng, D. Lau, B. Jaafar, Y. Tan, and R. Dumke, "Design and Fabrication of 2.5D Cryogenic Interposer with Integrated Superconducting TSVs and Resonators", IEEE/ECTC Proceedings, May 2024, pp. 458–463.
101. Lee, B., J. Choi, S. Lee, J. Park, J. Kang, U. Kwon,, G. Lee, K. Lee, "Structural Characterization of 2.5D System in Package Combined with High Bandwidth Memory for Enhanced Quality and Reliability", IEEE/ECTC Proceedings, May 2024, pp. 643–646.
102. Hirano, K., D. Kang, M. Takahashi, T. Iseki, K. Murai, S. Katoh, "Study of Fabrication and Reliability for 120 mm x 120 mm Extremely Large Advanced 2.5D Package", IEEE/ECTC Proceedings, May 2024, pp. 955–962.
103. Sounart, T., T. Talukdar, H. Braunisch, A. Aleksov, K. Jun, A. Elsherbini, P. Nordeen, B. Rawlings, V. Strong, S. Liff, J. Swan, "IR Laser Debond from Silicon Carrier Wafers with Inorganic Thin Film Release Layers for High-Density 2.5D and 3D Integration", IEEE/ECTC Proceedings, May 2024, pp. 363–369.
104. Sun, H., B. Shi, T. Nguyen, J. Schutt-Ainé, "Modeling and Analysis of Heterogeneously Integrated Chiplet-to-Chiplet Communication Link in 2.5D Advanced Packaging", IEEE/ECTC Proceedings, May 2024, pp. 1078–1084.
105. Lin, Y., Y. Hsiao, C. Liu, F. Hsiao, C. Wang, C. Hung, "A Novel Detection Applied on Micro Defect in Bump Interface for 2.5DIC Package", IEEE/ECTC Proceedings, May 2024, pp. 1804–1807.
106. Li, Y., J. Liu, D. Lu, W. Zhang, R. Gao, E. Liu, M. Rotaru, D. Rahul, N. Sridhar, "Rapid Estimation of Anisotropic Thermal Conductivity in RDL for 2.5D Chiplet Design", IEEE/ECTC Proceedings, May 2025, pp. 1541–1546.

107. Chen, Z., Z. Xiong, K. Wang, A. Marconnet, T. Wei, "A Novel Thermal Isolation Method with Embedded 'Glass Bridge' Structures in Silicon-Based 2.5D Heterogeneous Integration Systems", IEEE/ECTC Proceedings, May 2025, pp. 895–901.
108. Lin, P., Y. Huang, Y. Lai, H. Lin, J. Paek, M. Wu, J. Zhang, T. Huber, "Optimized TIM1 Solution for Large 2.5D HPC Packages using Silicone Matrix Containing Liquid Metal Materials", IEEE/ECTC Proceedings, May 2025, pp. 1075–1081.
109. Hirano, K., D. Kang, M. Takahashi, M. Aoyama, Y. Inui, T. Iseki, S. Katoh, "Process Development and Reliability Investigation of 120 mm x 120 mm Large 2.5D Package with Low Melting Temperature Solder", IEEE/ECTC Proceedings, May 2025, pp. 122–128.
110. Wu, T., K. Chang, C. Wang, B. Wu, C. Wang, K. Yan, C. Lee, C. Hsieh, J. Lu, R. Song, S. Jiang, J. He, "Integrated Package-to-System Thermal Solution Evolution for High-Performance 2.5D CoWoS-R Advanced Packaging Technology Development", IEEE/ECTC Proceedings, May 2025, pp. 481–486.
111. https://www.prnewswire.com/news-releases/umc-enters-volume-production-for-tsv-process-used-to-enable-amds-high-performance-radeon-r9-fury-x-gpu-300115398.html
112. Hou, S., W. Chen, C. Hu, C. Chiu, K. Ting, T. Lin, W. Wei, W. Chiou, V. Lin, V. Chang, C. Wang, C. Wu, and D. Yu, "Wafer-Level Integration of an Advanced Logic-Memory System Through the Second-Generation CoWoS Technology", IEEE Transactions on Electron Devices, October 2017, pp. 4071–4077.
113. Moore, S., "Graphcore uses TSMC 3D Chip Tech to Speed AI by 40%", IEEE Spectrum, March 3, 2022, pp. 1–3.
114. Sato, M., "The Supercomputer "Fugaku"", Proceedings of IEEE/VLSI-DAT, April 2022, pp. 1–2.
115. https://news.samsung.com/global/samsung-announces-availability-of-its-leading-edge-2-5d-integration-h-cube-solution-for-high-performance-applications.
116. Nam, S., J. Kang, I. Lee, Y. Kim, H. Yu, and D. Kim, "Investigation on Package Warpage and Reliability of the large size 2.5D Molded Interposer on Substrate (MIoS) Package", Proceedings of IEEE/ECTC, May 2022, pp. 643–647.
117. TSMC North America Technology Symposium, Santa Clars, California, April 23, 2025.
118. Hou, S., C. Lee, T. Wang, and H. Hu, "Supercarrier redistribution layers to realize ultra-large 2.5D wafer-scale packaging by CoWoS," IEEE/ECTC Proceedings, June 2023, pp. 510–514.
119. Su, P., D. Ho, J. Pu, and Y. Wang, "FO-EB-T package solution for high-performance computing," IEEE/ECTC Proceedings, June 2023, pp. 1038–1042.
120. Lin, V., T. Shih, A. Jang, and Y. Wang, "Fan-out embedded bridge with TSV (FO-EB-T) package characterization and evaluation," IEEE/ECTC Proceedings, June 2023, pp. 1344–1347.
121. Agarwal, R., P. Cheng, P. Shah, B. Wilkerson, R. Swaminathan, J. Wun, and C. Mandalpu, "3D Packaging for Heterogeneous Integration", IEEE/ECTC Proceedings, May 2022, pp. 1103–1107.
122. Lau, J. H., M. Ma, T. Lan, T. Tseng, "3.3D IC integration for HPC Applications Driven by AI," U.S. patent application, Date: May 2025.
123. Zhou, W., M. Kwon, Y. Chiu, H. Guo, B. Bhushan, B. Street, K. Parekh, and A. Singh, "Critical Challenges with Copper Hybrid Bonding for Chip-to-Wafer Memory Stacking", Proceedings of IEEE/ECTC, May 2023, pp. 336–341.
124. Nakamura, Y., K. Tamura, Y. Kayaba, W. Okada, T. Shikama, and S. Inada, "A Study on Multi-Chip Stacking Process by Novel Dielectric Polymer Adhesive for Cu-Cu Hybrid Bonding", Proceedings of IEEE/ECTC, May 2023, pp. 1129–1134.
125. Kayaba, Y., Y. Nakamura, W. Okada, T. Shikama, K. Tamura, and S. Inada, "A New Adhesive for CoW Cu-Cu Hybrid Bonding with High Throughput and Room Temperature pre-Bonding", Proceedings of IEEE/ECTC, May 2023, pp. 571–578.
126. Kagawa, Y., T. Kamibayashi, N. Fujii, S. Furuse, T. Yamada, T. Hirano, and H. Iwamoto, "Cu-Cu Wiring: The Novel Structure of Cu-Cu Hybrid Bonding", Proceedings of IEEE/ECTC, May 2023, pp. 103–108.

127. Kamei, T., H. Yoshioka, T. Hiratsuka, A. Sakamoto, K. Shimizu, and H. Iwamoto, "Development of High Reliability 6 μm-Pitch Cu-Cu Connections Using over-400 $mm^2$-Large Chip on Wafer Bonding Process", Proceedings of IEEE/ECTC, May 2023, pp. 1285–1290.
128. Hirano; T., T. Yamada, S. Kobayashi, Y. Hagimoto, and H. Iwamoto, "Simulation of Device Structure Impacts on Bonding Wave and Strain in Wafer-to-Wafer Cu-Cu Hybrid Bonding", Proceedings of IEEE/ECTC, May 2023, pp. 1314–1318.
129. Yoshioka, H., N. Fujii, T. Shigetoshi, T. Kamei, K. Kotoo, N. Ogawa, S. Furuse, T. Horikiri, S. Saito, S. Saito, Y. Hagimoto, and H. Iwamoto, "Influence of H2O in Bonding Interfaces on Bonding Strength of Plasma-Activated Bonded Silicon Oxide", Proceedings of IEEE/ECTC, May 2023, pp. 1119–1123.
130. Berru, J., S. Nicolas, N. Bresson, M. Assous, and S. Borel, "Demonstration of a Wafer Level Face-To-Back (F2B) Fine Pitch Cu-Cu Hybrid Bonding with High Density TSV for 3D Integration Applications", Proceedings of IEEE/ECTC, May 2023, pp. 97–102.
131. Bourjot, E., A. Bond, N. Nadi, T. Enot, L. Sanchez, P. Montmeat, B. Martin, A. Campo, F. Fournel, F. Eid, and J. Swan, "Integration and Process Challenges of Self Assembly Applied to Die-to-Wafer Hybrid Bonding", Proceedings of IEEE/ECTC, May 2023, pp. 1397–1402.
132. Ryan, K., C. Netzband, A. Gildea, Y. Mimura, S. Hoshino, S. LeFevre, A. Tuchman, I. Son, H. Aizawa, and K. Maekawa, "Electrical Analysis of Wafer-to-Wafer Copper Hybrid Bonding at Sub-Micron Pitches", Proceedings of IEEE/ECTC, May 2023, pp. 114–117.
133. Netzband, C., K. Ryan, Y. Mimura, S. Ilseok, H. Aizawa, N. Ip, X. Chen, H. Fukushima, and S. Tan, "0.5μm Pitch Next Generation Hybrid Bonding with High Alignment Accuracy for 3D Integration", Proceedings of IEEE/ECTC, May 2023, pp. 1100–1104.
134. Saito, S., N. Fujii, S. Furuse, N. Ogawa, S. Saito, Y. Hagimoto, and H. Iwamoto, "Analysis of Vacancies in Wafer-Bonding Interface via Positron Annihilation Lifetime Spectroscopy", *IEEE/ECTC Proceedings*, May 2024, pp. 1916–1922.
135. Hirano, T., T. Hiratsuka, H. Yoshioka, N. Ogawa, S. Saito, S. Kobayashi, Y. Hagimoto, and H. Iwamoto, "Elucidating the Mechanism of four Corner Voids in Chip-on-Wafer Hybrid Bonding", *IEEE/ECTC Proceedings*, May 2024, pp. 1241–1245.
136. Maubert, M., M. Gottardi, P. Philip, E. Fragnaud, G. Romero, A. Cornelis, H. Hijazi, F. Fournel, and M. Calvo-Muñoz, "Copper Microstructure Optimization for Fine Pitch Low Temperature Cu/SiO_2 Hybrid Bonding", *IEEE/ECTC Proceedings*, May 2024, pp. 631–636.
137. Borel, S., M. Assous, R. Velard, J. Suarez-Berru, S. Nicolar, J. Derhamp, R. Bouis, L. Vignoud, P. Valentin, J. Marchand, A. Roman, and M. Vignoud, "Low Resistance and High Isolation HD TSV for 3-Layer CMOS Image Sensors", *IEEE/ECTC Proceedings*, May 2024, pp. 771–777.
138. Nicolas, S., J. Suarez-Berru, N. Bresson, C. Socquet-Clerc, M. Assous, S. Borel, R. VéLard, J. Dechamp, R. Bouis, A. Roman, K. Abadie, and D. Hebras, "3-Layer Fine Pitch Cu-Cu Hybrid Bonding Demonstrator With High Density TSV For Advanced CMOS Image Sensor Applications", *IEEE/ECTC Proceedings*, May 2024, pp. 305–311.
139. Renaud, P., C. Dubarry, N. Bresson, E. Deschaseaux, F. Fournel, C. Morales, K. Abadie, C. Thomas, and J. Charbonnier, "Fine Pitch Nb-Nb Direct Bonding for Quantum Applications", *IEEE/ECTC Proceedings*, May 2024, pp. 441–446.
140. Kim, D., "Development of 0.5 μm Pixel 3-Wafers Stacked CMOS Image Sensor With Through Silicon Deep Contact and In-Pixel Cu-Cu Bonding Process", *IEEE/ECTC Proceedings*, May 2024, pp. 394–398.
141. Lee, H., J. Kim, M. Kim, W. Lee, A. Jang, H. Lee, and D. Kim, "A Study on D2W Hybrid Cu Bonding Technology for HBM Multi-die Stacking", *IEEE/ECTC Proceedings*, May 2024, pp. 76–80.
142. Hahn, S., W. Kim, K. Lim, B. Moon, and M. Rhee, "Achieving Sub-nm Copper Recess Controllability for Advanced 3D Integration: An Experimental and Atomic-Scale Simulation Study on Wet Atomic Layer Etching Process", *IEEE/ECTC PROCEEDINGS*, MAY 2024, PP. 1322–1328.

143. Baek, K., M. Han, J. Lee, J. Lee, and K. Lim, "Finite Element Modeling for Wafer-to-Wafer Direct Bonding Behaviors and Alignment Prediction", *IEEE/ECTC Proceedings*, May 2024, pp. 1256–1259.
144. Lin, Y., P. Bex, K. Kennes, J. Derakhshandeh, P. Dhakras, S. Suhard, C. Gerets, S. Dewilde, V. Georgieva, A. Jourdain, G. Beyer, and E. Beyne, "Direct Die-to-Wafer Hybrid Bonding Using Plasma Diced Dies and Bond Pad Pitch Scaling Down to 2 μm", *IEEE/ECTC Proceedings*, May 2024, pp. 40–44.
145. Lianto, P., R Raphael, A. Tan, C. Chia, X. Li, H. Lee, and X. Chen, "SiCN CMP Integration for Hybrid Bonding Application", *IEEE/ECTC Proceedings*, May 2025, Dallas, Texas, pp. 347–350.
146. Chen, X., G. See, Y. Lim, P. Lianto, P. Suo, C.Andy, S. Rath, and X. Zhao, "Integration Solution for Thin D2W Hybrid Bonding for Yield and Reliability", *IEEE/ECTC Proceedings*, May 2025, Dallas, Texas, pp. 33–36.
147. Ryan, K., R. Sreenivasan, S. Ramamoorthy, R. Patlolla, J. Appell, J. Hebding, S. Krishnan, M. Chudzik, B. Weis, G. Probst, and B. Theile, "Integration, Materials and Equipment Innovations to Enable 100 nm Pitch W2W Bonding for Memory-to-Logic and Logic-to-Logic 3D Stacking", *IEEE/ECTC Proceedings*, May 2025, Dallas, Texas, pp. 542–546.
148. Kang, S., S. Iacovo, K. D. O. Okudur, A. Alexeev, T. Plach, T. Ding, and Markus Wimplinger, "Wafer-to-Wafer Bonding With Saddle-Shaped Wafers", *IEEE/ECTC Proceedings*, May 2025, Dallas, Texas, pp. 2028–2033.
149. Lin, Y., P. Bex, S. Suhard, B. Zhang, F. Okudur, A. Zerio, N. Reddy, E. Sanchez, Y. Li, A. Jourdain, G. Beyer, and E. Beyne, "2 μm Pitch Direct Die-to-Wafer Hybrid Bonding using Surface Protection During Wafer Thinning and Die Singluation", *IEEE/ECTC Proceedings*, May 2025, Dallas, Texas, pp. 287–291.
150. Okudur, O., S. Iacovo, S. Kang, D. Sharma, M. Gonzalez, and E. Beyne, "Mitigation of Wafer-to-Wafer Bonding Distortions Through Accelerated Simulations and Measurements", *IEEE/ECTC Proceedings*, May 2025, Dallas, Texas, pp. 1650–1655.
151. Huylenbroeck, S., S. Chew, B. Zhang, E. Chery, J. Messemaeker, P. Gupta, N. Jourdan, S. Dewilde, Z. El-Mekki, J. Vos, G. Beyer, and E. Beyne, "Self-formed Barrier using Cu-Mn Alloy Seed applied to a 400nm Pitch Wafer-to-Wafer Hybrid Bonding Technology", *IEEE/ECTC Proceedings*, May 2025, Dallas, Texas, pp. 342–346.
152. Huylenbroeck, S., S. Chew, B. Zhang, L. Bogaerts, C. Heyvaert, S. Dewilde, S. Iacovo, M. Stucchi, J. Vos, G. Beyer, and E. Beyne, "Wafer-to-Wafer Hybrid Bonding Technology with 300nm Interconnect Pitch", *IEEE/ECTC Proceedings*, May 2025, Dallas, Texas, pp. 565–569.
153. Kennes, K., P. Bex, Y. Lin, S. Suhard, D. Cuypers, A. Phommahaxay, G. Beyer, and E. Beyne, "Impact of Temporary Substrates and Adhesives on Die-to-Wafer Overlay", *IEEE/ECTC Proceedings*, May 2025, Dallas, Texas, pp. 1934–1038.
154. Li, W., F. Eid, A. Vyatskikh, R. Vreeland, W. Brezinski, M. Njuki, C. Jezewski, R. Mongia, K.Valavala, H. Mei, and C. Bakre, "Wafer-to-Wafer Bonding With Ultralow Thermal Resistance and High Bonding Energy", *IEEE/ECTC Proceedings*, May 2025, Dallas, Texas, pp. 83–86.
155. Widodo, T., P. Lianto, K. Prabhakar, J. Xu, A. Tan, P. Lim, G. See, and X. Brun, "Electrical Performance of CMP Process for Hybrid Bonding Application with Conventional / nt-Cu and Low Temperature of Si_xN_y / Si_xO_y Dielectrics", *IEEE/ECTC Proceedings*, May 2025, Dallas, Texas, pp. 1837–1842.
156. Lau, J. H., "Current Advances and Outlooks in Hybrid Bonding", *IEEE Transactions on CPMT*, Vol. 15, No. 3, April 2025, pp. 651–681.
157. Lau, J. H., "State of the art of Cu-Cu Hybrid Bonding", *IEEE Transactions on CPMT*, Vol. 14, No. 3, March 2024, pp. 376–396.
158. Lau, J. H., "Recent Advances and Trends in Cu-Cu Hybrid Bonding", *IEEE Transactions on CPMT*, Vol. 13, No. 3, March 2023, pp. 399–425.

159. Kagawa, Y., N. Fujii, K. Aoyagi, Y. Kobayashi, S. Nishi, and N. Todaka, "Novel stacked CMOS image sensor with advanced Cu2Cu hybrid bonding," *Proceedings of IEEE/IEDM*, Dec. 2016, pp. 8.4.1–4.
160. Haneda, M., Y. Ikegami, K. Kotoo, K. Shimizu, Y. Kagawa, and H. Iwamoto, "Wafer-on-wafer-on-wafer (WoWoW) integration having large-scale high reliability fine 1 μm pitch face-to-back (F2B) Cu-Cu connections and fine 6 μm pitch TSVs," in Proc. IEEE 74th Electron. Compon. Technol. Conf. (ECTC), May 2024, pp. 1342–1346.
161. AMD, "3D V-Cache Technology", Sunnyvale, CA January 2024.
162. Chen, M. F., C. S. Lin, E. B. Liao, W. C. Chiou, C. C. Kuo, C. C. Hu, C. H. Tsai, C. T. Wang and D. Yu, "SoIC for Low-Temperature, Multi-Layer 3D Memory Integration", *IEEE/ECTC Proceedings,* May 2020, pp. 855–860.
163. Chen, F., M. Chen, W. Chiou, D. Yu, "System on Integrated Chips (SoIC™) for 3D Heterogeneous Integration", *IEEE/ECTC Proceedings,* May 2019, pp. 594–599.
164. HPC, "Introducing Graphcore Bow IPU", Exxact Corporation, April 14, 2022.
165. M. Kim, H. Lee, A. Jang, W. Lee, S. Back, I. Kim, et al., "Multi-stack hybrid Cu bonding technology development using ultra-thin chips," *Proc. of IEEE/ECTC*, May 2023, pp. 1672–1676.
166. K. Taehwan, J. Lee, Y. Kim, H. Park, H. Hwang, J. Kim, et al., "Thermal improvement of HBM with joint thermal resistance reduction for scaling 12 stacks and beyond," *Proc. of IEEE/ECTC*, May 2023, pp. 767–771.
167. K. Kim, S. Lim, D. Jung, J. Choi, S. Na, J. Yeom, et al., "C2W hybrid bonding interconnect technology for higher density and better thermal dissipation of high bandwidth memory," *Proc. of IEEE/ECTC*, May 2023, pp. 1048–1052.
168. J. H. Lau, "Recent advances and trends in chiplet design and heterogeneous integration packaging," *ASME Trans., Journal of Electronic Packaging*, published online: June 2023, and hard-copy: March 2024, pp. 010801-1–31.
169. Lee, I., S. Nam, S. Kim, S. Shin, Y. Kim, S. Seo, H. Yu, and D. Kim, "Extremely Large 3.5D Heterogeneous Integration for the Next-Generation Packaging Technology", *Proceedings of IEEE/ECTC*, May 2023, pp. 893–898.
170. Mandalapu, C., C. Buch, P. Shah, R. Topacio, P. Cheng, L. Wang, L. Wang, R. Swaminathan, A. Smith, J. Wuu, K. Mysore, and A. Alam, "3.5D Advanced Packaging Enabling Heterogenous Integration of HPC and AI Accelerators," *IEEEECTC Proceedings*, May 2024, pp. 798–802.
171. https://www.tomshardware.com/tech-industry/artificial-intelligence/broadcom-unveils-gigantic-3-5d-xdsip-platform-for-ai-xpus-6000mm2-of-stacked-silicon-with-12-hbm-modules
172. Bhatt, D., K. Williams, D. A. Hutt, and P. P. Conway, "Process optimization and characterization of excimer laser drilling of microvias in glass," *IEEE EPTC Proceedings*, Singapore, 2007, pp. 196–201.
173. Cui, X., D. Bhatt, D. A. Hutt, K. Williams, and P. P. Conway, "Copper deposition and patterning for glass substrate manufacture," *IEEE EPTC Proceedings*, Singapore, 2007, pp. 37–42.
174. Brusberg, L., H. Schroder, M. Topper, N. Arndt-Staufenbiel, J. Roder, M. Lutz, and H. Reichl, "Thin glass based packaging technologies for optoelectronic modules," *IEEE ECTC Proceedings*, San Diego, CA, May 2009, pp. 207–212.
175. Brusberg, L., H. Schroder, M. Topper, and H. Reichl, "Photonic system in-package technologies using thin glass substrates," *IEEE EPTC Proceedings*, Singapore, December 2009, pp. 930–935.
176. Schroder, H., L. Brusberg, R. Erxleben, I. Ndip, M. Topper, N. F. Nissen, and H. Reichl, "GlassPack- A 3D glass based interposer concept for SiP with integrated optical interconnects," *IEEE ECTC Proceedings*, Las Vegas, NV, May 2010, pp. 1647–1652.
177. Topper, M., I. Ndip, R. Erxleben, L. Brusberg, N. Nissen, H. Schroder, H. Yamamoto, G. Todt, and H. Reichl, "3-D thin film interposer based on TGV (through glass vias): An alternative to Si-interposer," *IEEE ECTC Proceedings*, Las Vegas, NV, 2010, pp. 66–73.

178. Sukumaran, V., Q. Chen, F. Liu, N. Kumbhat, T. Bandyopadhyay, H. Chan, S. Min, C. Nopper, V. Sundaram, and R. Tummala, "Through package-via formation and metallization of glass interposers," *IEEE ECTC Proceedings*, Las Vegas, NV, 2010, pp. 557–563.
179. Tummala, R. R., V. Sundaram, T. Bandyopadhyay, V. Sridharan, and V. Sukumaran, "Low-cost silicon and glass interposers and packages," in Proc. Pan Pacific Microelectron. Symp., 2011, pp. 106–109.
180. Sukumaran, V., T. Bandyopadhyay, V. Sundaram and R. Tummala, "Low-Cost Thin Glass Interposers as a Superior Alternative to Silicon and Organic Interposers for Packaging of 3-D ICs," IEEE Transactions on Components, Packaging and Manufacturing Technology, vol. 2, no. 9, Sept. 2012, pp. 1426–1433.
181. McCann, S. R., "Experimental and theoretical assessment of thin glass panels as interposers for microelectronic packages", PhD thesis, Georga Institute of Technology, May 2014.
182. Ravichandran, S., M. Kathaperumal, M. Swaminathan, and R. Tummala, "Large-Body-Sized Glass-Based Active Interposer for High-Performance Computing", *IEEE Proceedings*, May 2020, pp. 879–884.
183. Wong, R., S. Ravichandran, H. Lee, and V. Smet, "Thermal Management of Glass Panel Embedded Packages: Package Architecture vs. Power Density", *IEEE Proceedings*, May 2020, pp. 2105–2111.
184. Ali, M., A. Watanabe, T. Kakutani, P. Raj, R. Tummala, and M. Swaminathan, "Heterogeneous Integration of 5G and Millimeter-Wave Diplexers with 3D Glass Substrates", *IEEE Proceedings*, May 2020, pp. 1376–1382.
185. Erdogan, S., S. Ravichandran, X. Jia, and M. Swaminathan, "Characterization of Chip-to-Package Interconnects for Glass Panel Embedding (GPE) for Sub-THz Wireless Communications", *IEEE ECTC Proceedings*, May 2021, pp. 2328–2333.
186. Sivapurapu, S., R. Chen, M. Rehman, K. Kanno, T. Kakutani, M. Letz, F. Liu, S. Sitaraman, and M. Swaminathan, "Flexible and Ultra-Thin Glass Substrates for RF Applications", *IEEE Proceedings*, May 2021, pp. 1638–1644..
187. Huang, K., and M. Swaminathan, "Antennas in Glass Interposer for sub-THz Applications", *IEEE Proceedings,* May 2021, pp. 1150–1155.
188. Jayaram, V., O. Gupte, and V. Smet, "Modeling and Design for System-Level Reliability and Warpage Mitigation of Large 2.5D Glass BGA Packages", *IEEE Proceedings*, May 2022, pp. 1060–1097.
189. Jia, X., X. Li, K. Moon, J. Kim, K. Huang, and M. Jordan, "Antenna-Integrated, Die-Embedded Glass Package for 6G Wireless Applications", *IEEE Proceedings*, May 2022, pp. 377–383.
190. Ravichandran, S., V. Smet, M. Swaminathan, and R. Tummala, "Demonstration of Glass-based 3D Package Architectures with Embedded Dies for High Performance Computing", *IEEE Proceedings*, May 2022, pp. 1114–1120.
191. Kim, J., X. Li, X. Jia, K. Moon, and M. Swaminathan, "Bottom Side Cooling for Glass Interposer with Chip Embedding Using Double-Sided Release Process for 6G Wireless Applications", *IEEE Proceedings,* May 2023, pp. 1609–1613.
192. Li, X., X. Jia, J. Kim, K. Moon, M. Jordan, M. Swaminathan, "Die Embedded Glass Interposer with Minimum Warpage for 5G/6G Applications", *IEEE Proceedings*, May 2023, pp. 2257–2264.
193. Kathaperumal, M., P. Bhaskar, A. Toro, P. Nimbalkar, L. Dahal, and M. Bakir, "Characterization of Warpage of Ultra-Low-K Dielectric Materials and Correlation with Modulus and Coefficient of Thermal Expansion", *IEEE Proceedings*, May 2024, pp. 1958–1962.
194. Molina-Mangual, C., E. Surillo, N. Nedumthakady, and V. Smet, "Mitigating Cracking from RDL Stress in Glass Substrates with Low-CTE Electrodeposited Copper-Graphene Composites", *IEEE Proceedings*, May 2024, pp. 1552–1557.
195. Bowrothu, R., Y. Yoon, and J. Zhang, "Through Glass Via (TGV) Based Band Pass Filter for 5G Communications", *IEEE Proceedings*, May 2018, pp. 1097–1102.

196. Hwangbo, S., Y. Yoon, and A. Shorey, "Millimeter-wave Wireless Chip-to-chip (C2C) Communications in 3D System-in Packaging (SiP) using Compact Through Glass Via (TGV)-integrated Antennas", *IEEE Proceedings*, May 2018, pp. 2068–2073.
197. Okoro, C., T. Allowatt, and S. Pollard, "Resolving Thermo-Mechanically Induced Circumferential Crack Formation in Copper Through Glass Vias", *IEEE Proceedings*, May 2021, pp. 954–958.
198. Pan, K., J. Xu, and S. Park, "Investigation of Copper and Glass Interaction in Through Glass Via (TGV) During Thermal Cycling", *IEEE Proceedings*, May 2021, pp. 1660–1666.
199. Pan, K., C. Okoro, Y. Lai, D. Joshi, S. Park, and S. Pollard, "A comparative study of the thermomechanical reliability of fully-filled and conformal through-glass via", *IEEE Proceedings*, May 2022, pp. 1211–1217.
200. Yeary, L., L. Brusberg, C. Kim, S. Seok, J. Noh, and A. Rozenvax, "Co-Packaged Optics on Glass Substrates for 102.4 Tb/s Data Center Switches", *IEEE Proceedings*, May 2023, pp. 224–227.
201. Brusberg, L., M. Dejneka, C. Okoro, D. McEnroe, A. Zakharian, and C. Terwillige, "Ultra Low-Loss Ion-Exchange Waveguides in Optimized Alkali Glass for Co-Packaged Optics", *IEEE Proceedings*, May 2024, pp. 85–89.
202. Yang, J., K. Pan, P. Yin, Y. Lai, S. Park, C. Okoro, D. Joshi, S. Pollard, "Adhesion Layer Influence on Thermomechanical Reliability of Electroplated Copper Through-Glass Via (TGV)", *IEEE Proceedings*, May 2024, pp. 1820–1825.
203. Lau, J. H., N. Liu, M. Ma, and T. Tseng, "Solder joint reliability - glass core substrate versus organic core substrate", *Journal of Microelectronic Reliability*, Vol. 168, May 2025, pp. 115701.
204. Lau, J. H., N. Liu, M. Ma, and T. Tseng, "Glass-Core Substrate versus Organic-Core Substrate", *IEEE/ECTC Proceedings*, May 2025, pp. 1014–1020.
205. Lau, J. H., and N. C. Lee, *Assembly and Reliability of Lead-Free Solder Joints*, (527 pages), Springer, New York, 2020.
206. Lau, J. H., "Co-Packaged Optics - Heterogeneous Integration of Photonic IC and Electronic IC", *ASME Transactions, J. of Electronic Packaging,* published online: June 1, 2024, hardcopy: Vol. 147, March 2025, pp. 1–12.
207. Takemura, K., D. Ohshima, A. Noriki, D. Okamoto, A. Ukita, J. Ushida, M. Tokushima, T. Shimizu, I. Ogura, D. Shimura, and T. Aoki, "Silicon-Photonics-Embedded Interposers as Co-Packaged Optics Platform", The *Transactions of The Japan Institute of Electronics Packaging*, Vol. 15, April 2022, pp. E21-012-1–E21-012-13.
208. Lee, B., N. Nedovic, T. Greer III, and C. Thomas Gray, "Beyond CPO: A Motivation and Approach for Bringing Optics Onto the Silicon Interposer", *Journal of Lightwave Technology*, Vol. 41, February 2023, pp. 1152–1162.
209. Noriki, A., A. Ukita, K. Takemura, S. Suda, T. Kurosu, Y. Ibusuki, I. Tamai, and D. Shimura, "Demonstration of Silicon-Photonics Hybrid Glass-Epoxy Substrate for Co-Packaged Optics", *Journal of Lightwave Technology*, Vol. 41, August 2023, pp. 5078–5083.
210. Tan, M., J. Xu, S. Liu, J. Feng, H. Zhang, C. Yao, et al. "Co-packaged optics (CPO): status, challenges, and solutions", *Frontiers of Optoelectronics*, Vol. 16, March 2023, pp. 1–40.
211. Grenier, J., L. Brusberg, K. Wieland, J. Matthies, and C. Terwilliger, "Ultrafast laser processing of glass waveguide substrates for multi-fiber connectivity in co-packaged optics", *Advanced* Optical Technology, Vol. 12, August 2023, pp. 1–21.
212. Yeary, L., L. Brusberg, C. Kim S. Seok, J. Noh, and A. Rozenvax, "Co-packaged Optics on Glass Substrates for 102.4 Tb/s Data Center Switches", *Proceedings of IEEE/ECTC*, May 2023, pp. 224–227.
213. Gupta, P., R. Bernson, N. Nudds, S. Collins, K. Gradkowski, P. Morrissey, and P. O'Brien, "Reference Thermal Chips for 2D and 3D Co-packaging Process Development", *Proceedings of IEEE/ECTC*, May 2023, pp. 2165–2170.
214. Bhuvanendran, S., N. Gourikutty, L. Long, S. Wei, M. Jong, D. Ho, J. Wu, T. Lim, R. Mandal, C. Choong, C. Yee, X. Li, J. Liow, and S. Bhattacharya, "A Heterogeneously Integrated Wafer-

level Processed Co-Packaged Optical Engine for Hyper-scale Data Centres", *Proceedings of IEEE/ECTC*, May 2023 pp. 207–211.
215. Li, J., T. Hsu, M. Zhuan, S. Lin, T. Shih, N. Kao, and Y. Eang, "Innovative Fan-Out Embedded Bridge Structure for Co-Packaged Optics", Proceedings of IEEE/ECTC, May 2023, pp. 228–232
216. McDonough, C., S. Kruger, T. Ngai, S. Baranowski, H. Yang, S. Deckoff-Jones, C. Poulton, M. Watts, and D. Harame, "AIM Photonics Demonstration of a 300 mm Si Photonics Interposer", *Proceedings of IEEE/ECTC*, May 2023, pp. 233–238.
217. Integrated Optical Interconnect Systems (iOIS) for Silicon Photonics Applications in HPC
218. Hsia, H., S. P. Tai, C. S. Liu, C. W. Tseng, S. W. Lu, Y. Wu, C. C. Chang, Jason Wu, K. C. Yee, C. Y. Wu, C. H. Tung, D. Yu, "Integrated Optical Interconnect Systems (iOIS) for Silicon Photonics Applications in HPC", *Proceedings of IEEE/ECTC*, May 2023, pp. 612–616.
219. Lu, M., S. Mu, C. Cheng, and J. Chen, "Advanced Packaging Technologies for Co-packaged Optics", *Proceedings of IEEE/ECTC*, May 2022, pp. 38–42.
220. Razdan, S., P. Dobbelaere, J. Xue, A. Prasad, and V. Patel, "Advanced 2.5D packaging technologies for next generation Silicon Photonics in high performance networking applications", *Proceedings of IEEE/ECTC*, May 2022, pp. 428–435.
221. Janta-Polczynski, A., and M. Robitaille, "Optical fiber pigtails integration in co-package", *Proceedings of IEEE/ECTC*, May 2022, pp. 410–421.
222. Maharry, A., L. A. Valenzuela, J. F. Buckwalter, and C. L. Schow, "A PCB Packaging Platform Enabling 100+ Gbaud Optoelectronic Device Testing", *Proceedings of IEEE/ECTC*, May 2021, pp. 1323–1328.
223. Lau, J. H., and T. J. Tseng, "2D heterogeneous integration of PIC and EIC", *US Patent No. 11,860,428*, Jan. 2, 2024, Filled on Jun. 9, 2022.
224. Lau, J. H., and T. J. Tseng, "3D heterogeneous integration of PIC and EIC", *US Patent No. 12,412,879*, Sep. 8, 2025, Filled on Jun. 9, 2023.
225. https://www.intel.com/content/www/us/en/newsroom/news/intel-unveils-industry-leading-glass substrates.html#gs.aub6ck, September 18, 2023.
226. Lau, J. H., and X. Fan, *Hybrid Bonging, Advanced Substrates, Failure Mechanisms, and Thermal Management for Chiplets and Heterogeneous Integration,* Springer, New York, April 2025.
227. Lau, J. H., *Flip Chip, Hybrid Bonding, Fan-In, and Fan-Out Technology*, Springer, New York, May 2024.

# Chapter 2
# Cu-Cu Hybrid Bonding

**Abstract** In this study, the state-of-the-art and outlooks of Cu-Cu hybrid bonding (HB) will be investigated. Emphasis is placed on the design, materials, process, fabrications, reliability, challenges, advantages and disadvantages, and examples of Cu–Cu HB. Specially, the works by Sony, Samsung, Applied Materials, EV Group, imec, Daicel Corporation, High Energy Accelerator Research Organization, Toray Industries, Entegris, National Yang Ming Chiao Tung University, Industrial Technology Research Institute, MediaTek, National Tsing Hua University, BASF, STMicroelectronics, CEA-Leti, Binghamton University, IBM, TAZMO, LINTEC, Institute of Science Tokyo, TSMC, Seoul National University of Science and Technology, YOKOHAMA National University, Tokyo Electron Kyushu Limited, SK Hynix, Tokyo Electron Limited, Toray, DISCO, University of Tsukuba, Kurita Water Industries, and Unimicron will be presented and discussed. Also, some recommendations will be provided.

## 2.1 Introduction

Hybrid bonding (HB) is one of the flip-chip assembly technologies. It can be applied to very fine-pitch pads and used for extremely high-density and high-performance applications. With HB, it leads to a low profile, low thermal resistance, and low insertion loss 3D package. HB is mostly suitable for silicon-to-silicon and silicon to-glass assemblies. The assembly methods can be die-to-die (D2D), die-to-wafer (D2W), wafer-on-wafer (WoW), chip-on-wafer-on-wafer (CoWoW), wafer-on-wafer-on wafer (WoWoW), reconstructed wafer, direct transfer bonding, etc.

HB is now being used for much of the complementary metal–oxide–semiconductor (CMOS) image sensor (CIS) market (e.g., Sony, OmniVision, and Samsung) in the world's smartphones and other image-based devices. For application processor examples, AMD has been shipping their 3D V-Cache and Graphcore have been shipping their intelligence processing unit (IPU) since 2022. AMD's compute and



J. Lau, K.-N. Chiang, *Cu-Interconnects, Glass, and AI-Assisted Simulation for Chiplets and Heterogeneous Integration*,
https://doi.org/10.1007/978-981-95-6891-8_2

artificial intelligence (AI) chiplets which are stacked on top of the I/O and cache chiplets in their MI300 have been shipped since 2024.

In 2022, Apple shipped their notebook computer with microbumps connecting the silicon bridge and the processors. In Q1 of 2026, Apple shipped their new notebook computer with Cu-Cu hybrid bonding connecting the silicon bridge and the GPU (graphics processing unit) and CUP (central processing unit).

For 3D NAND flash, Yangtze Memory Technologies Company Ltd. (YMTC) and Western Digital are using the HB technology for their products. It should be pointed out that both of them are using the HB with the monolithic manufacturing method within the chip (not the package).

The brief history and fundamentals, and the recent advances and trends in Cu–Cu HB from the IEEE/ECTC 2022, 2023, and 2024 papers have been reported, respectively, in [1–4]. The state-of-the-art and outlooks of Cu–Cu HB from the IEEE/ECTC 2025 papers will be discussed in this study. Some recommendations will be provided.

## 2.2 Sony's Hybrid Bonding

### *2.2.1 Development of a Novel WoWoW Process for 1/1.3-Inch 50 Megapixel Three-Wafer-Stacked CMOS Image Sensor with DNN Circuits*

During IEEE/ECTC 2025, Sony presented at least five papers on hybrid bonding [5–9]. In [5], they presented the development of a novel WoWoW process for 1/1.3-inch 50 megapixel three-wafer-Stacked CMOS image sensor with deep neural network (DNN) circuits. Figure 2.1 shows the 3-layer stacked image sensor of their study. It can be seen that the top wafer is for the pixel, the middle wafer is for the CIS logic (driver for pixel) and column analog-to-digital converter (ADC), and the third wafer is for the image signal processor (ISP), a DNN, and a static random-access memory (SRAM). The middle and bottom wafers are connected through through-silicon vias (TSVs).

The conventional WoW process is shown in Fig. 2.2a. While the WoWoW process is shown in Fig. 2.2b. For WoWoW process, first, the top and middle wafers are stacked by face-to-face (F2F) hybrid bonding to form a WoW stack. The thickness of the middle wafer is then reduced to 3 μm, and TSVs and bonding Cu pads are formed on the backside. In the second bonding step, the WoW stack and the bottom wafer are electrically connected by face-to-back (F2B) bonding. Finally, the top wafer substrate is thinned, and a color filter and on-chip lens are formed.

Figure 2.3 shows the scanning electron microscope (SEM) image of cross section of the 3-layer stacked image sensor. It can be seen that the Cu pads of the top and middle layers are connected by F2F hybrid bonding, ensuring precise alignment and a high degree of accuracy. The TSV is formed through a thinned middle wafer,

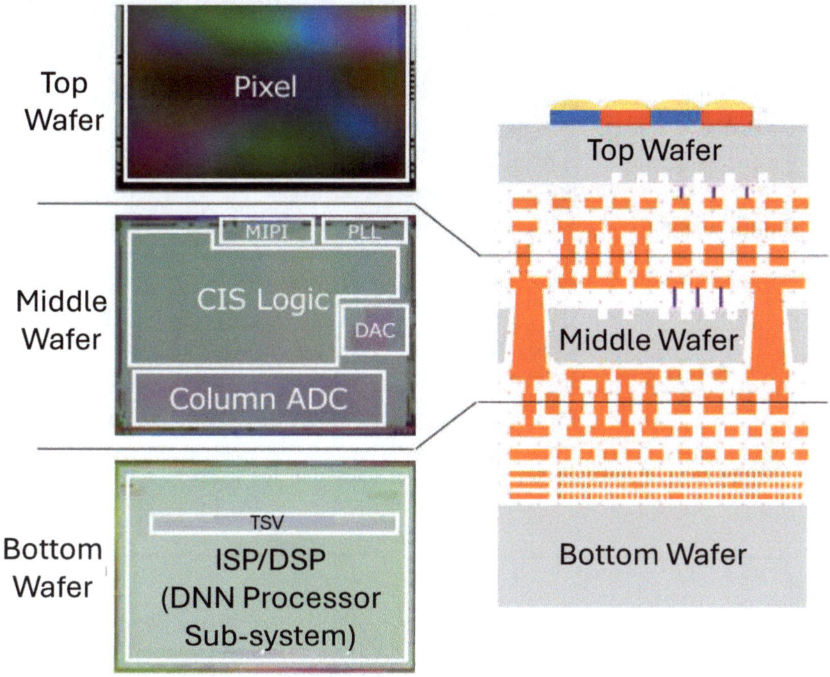


**Fig. 2.1** Sony's 3-layer stack sensor

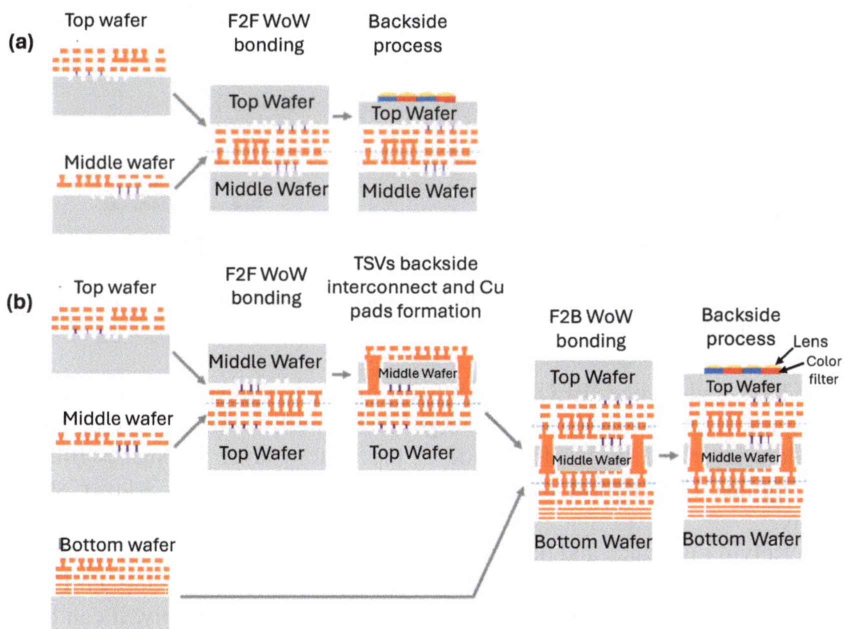


**Fig. 2.2** (**a**) WoW process. (**b**) WoWoW process

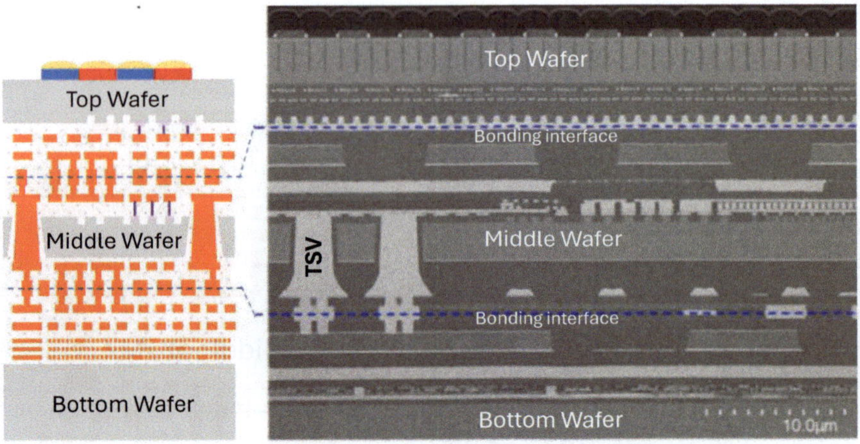


**Fig. 2.3** SEM image of cross-section of the 3-layer stacked image sensor

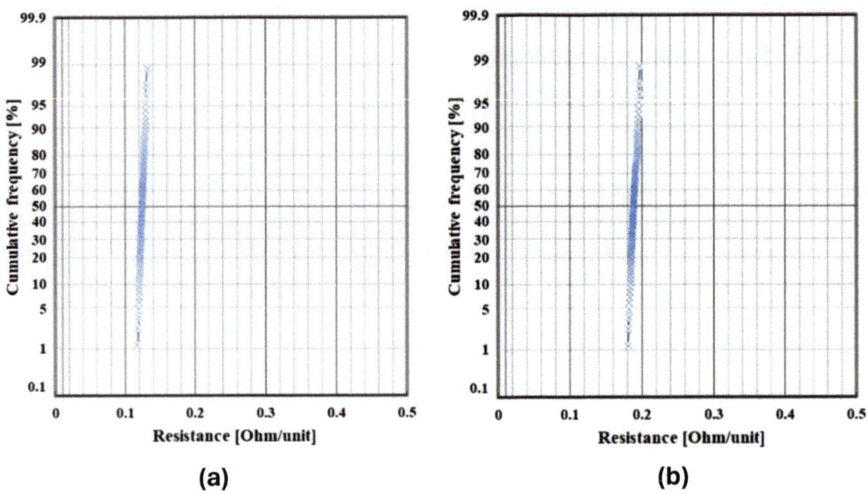


**Fig. 2.4** (**a**) Kelvin resistance of TSV-to-Cu interconnections. (**b**) Daisy chain resistance of TSV-to-Cu interconnections

and the back-end-of-line (BEOL) interconnections and Cu pads for the F2B hybrid bonding are formed with no abnormality. The Cu pads formed on the backside of the middle layer and on the front-side of the bottom layer are successfully connected by the F2B hybrid bonding, thereby constructing the 3-layer stacking structure.

The Kelvin resistance and daisy chain resistance of TSV-to-Cu interconnections (6 µm pitch F2F Cu-Cu hybrid bonding, 6 µm pitch TSV, and 6 µm pitch F2B Cu-Cu hybrid bonding) are shown in Figs. 2.4 (a) and (b), respectively. It can be seen

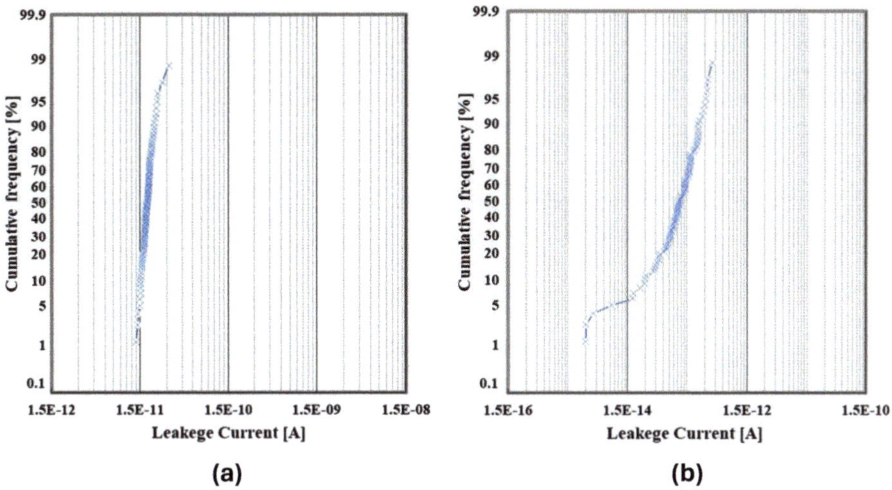


**Fig. 2.5** (**a**) Leakage current of the TSV-to-Si substrate. (**b**) Leakage current of the Cu-to-Cu interface during F2B bonding

that the small resistance distribution shows good electrical connectivity and the electrical path through bottom to top layer of three stacking is established by Cu-Cu connection and TSV (sufficiently low resistance of <0.2 Ω/unit and a 100% continuity confirmation.)

The leakage current of the TSV-to-Si substrate and leakage current of the Cu-to-Cu interface during F2B bonding are shown in Figs. 2.5 (a) and (b), respectively. It can be seen that the leakage current for TSV-to-Si substrate is suppressed below $1.0^{-10}$ A and for Cu-to-Cu interface during F2B bonding is suppressed below $1.0^{-12}$ A. These indicate that the TSV structure is well-isolated and there is no clear breakdown observed.

Electrical reliability is evaluated by the stress-induced voiding (SIV) test, which measured the electrical resistance shift during the 1000 h at 175 °C. Figure 2.6 shows the electrical resistance shift of the 6 μm pitch full coverage daisy chain of 50,000 links (6 μm pitch F2F Cu-Cu hybrid bonding, 6 μm pitch TSV, and 6 μm pitch F2B Cu-Cu hybrid bonding). No difference in the results is observed after stress-induced voiding (SIV) test.

### *2.2.2 Advanced Face-To-Back CoW 2.0-μm Pitch Cu-Cu Hybrid Bonding Process for Three Layer-Stacked 3D Heterogenous Integration*

Sony presented another paper on advanced F2B CoW 2.0 μm pitch Cu-Cu hybrid bonding process for three layer-stacked 3D heterogenous integration [6]. Figure 2.7 schematically shows a three-layer vertically stacked chip-on-wafer-on-wafer

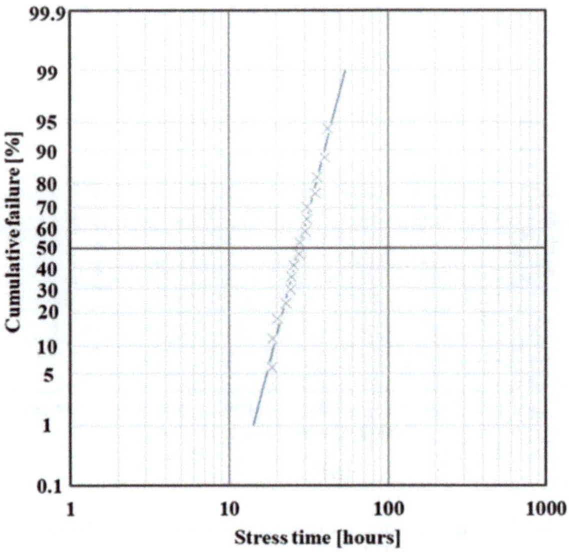


**Fig. 2.6** Electromigration test results on daisy chain resistance of TSV-to-Cu interconnections

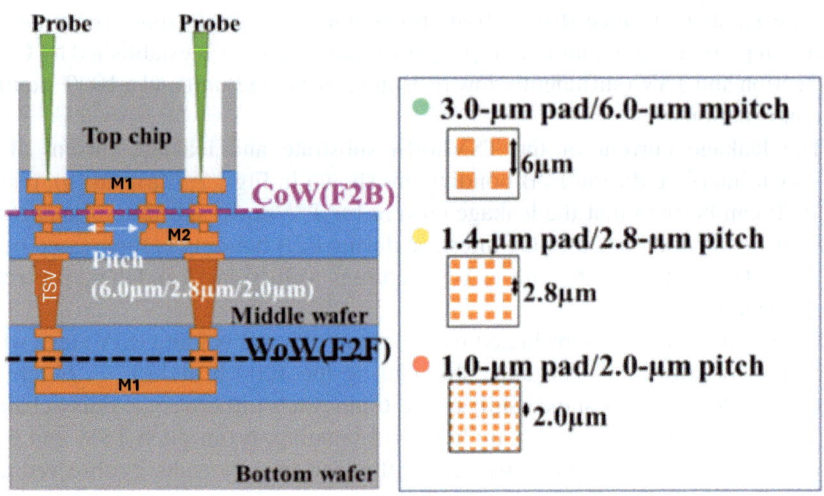


**Fig. 2.7** Schematic diagram of a three-layer vertically stacked structure

(CoWoW) structure. It can be seen that the middle and bottom wafers are connected via F2F WoW. The top chip is placed face-down on the backside Cu pads of the middle wafer (F2B). In this paper, Sony introduce a novel chemical mechanical polishing (CMP) process that incorporates oxidation inhibitors to optimize Cu recess control for 3.0 μm pad/6.0 μm pitch, 1.4 μm pad/2.8 μm pitch and 1.0 μm pad/2.0 μm pitch configurations in F2B CoW structures.

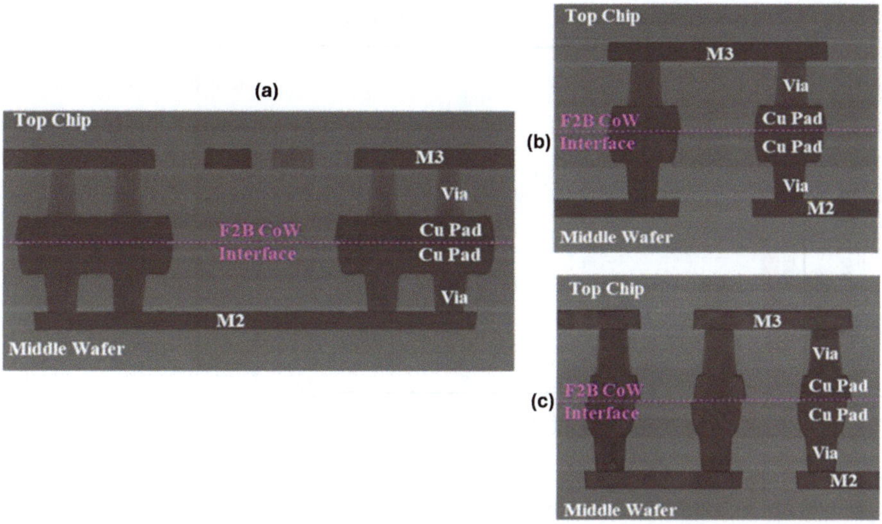


**Fig. 2.8** Cross-sectional SEM images of daisy chains for (**a**) 3.0 μm pad/6.0 μm pitch, (**b**) 1.4 μm pad/2.8 μm pitch, and (**c**) 1.0 μm pad/2.0 μm pitch

Successful Cu–Cu hybrid bonding implementation is demonstrated across multiple pad sizes (3.0, 1.4, and 1.0 μm) on a single chip. Process reliability is validated based on electromigration testing of Cu–Cu connections with 1.4 μm pad/2.8 μm pitch and 1.0 μm pad/2.0 μm pitch configurations in F2B CoW structures. Figure 2.8 shows the cross-sectional SEM images of daisy chains processed using the advanced CMP process for (a) 3.0 μm pad/6.0 μm pitch, (b) 1.4 μm pad/2.8 μm pitch, and (c) 1.0 μm pad/2.0-μm pitch. The results confirm both adequate Cu expansion and defect-free Cu–Cu hybrid bonding interfaces in the F2B CoW structures.

The cumulative probability distributions of contact resistance in Kelvin test structures for (a) conventional and (b) advanced CMP processes (enhanced Cu recess control) are shown in Figs. 2.9 (a) and (b), respectively. It can be seen that, for the conventional CMP process, the yield of successful Kelvin connection decreases from above 99% to approximately 90 and 50%, respectively. However, for the advanced CMP process, the successful Kelvin connection rate decreasing from above 99% to approximately 99% and 95% for decrease in pad size of 3.0 μm to 1.4 and 1.0 μm, respectively.

Figure 2.10 shows the cumulative probability distributions of electromigration lifetime measurements for F2B CoW structures with two distinct pad size and pitch configurations: 1.4 μm pad/2.8 μm pitch and 1.0 μm pad/2.0 μm pitch. The electromigration testing is conducted on daisy chains with Cu–Cu connections under accelerated stress conditions (constant current of 15 mA at 275 °C). It can be seen that the projected lifetimes exceeding 10 years, indicating excellent reliability interconnections.

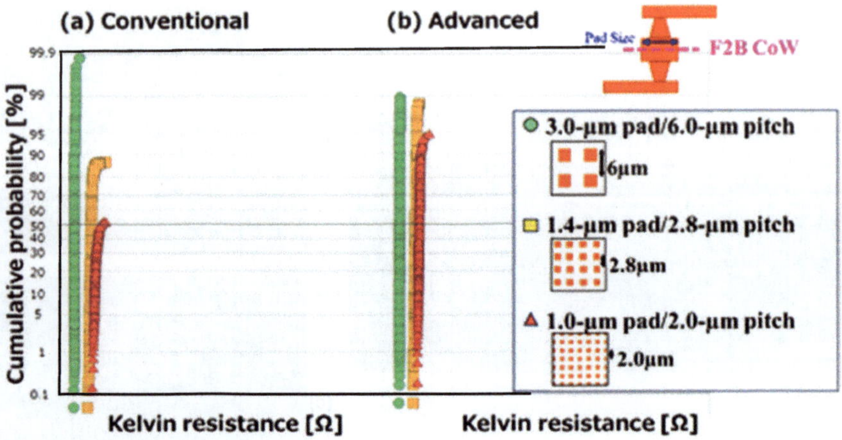


**Fig. 2.9** Comparison of cumulative probability distributions of contact resistances in Kelvin test structures for (**a**) conventional and (**b**) advanced CMP processes

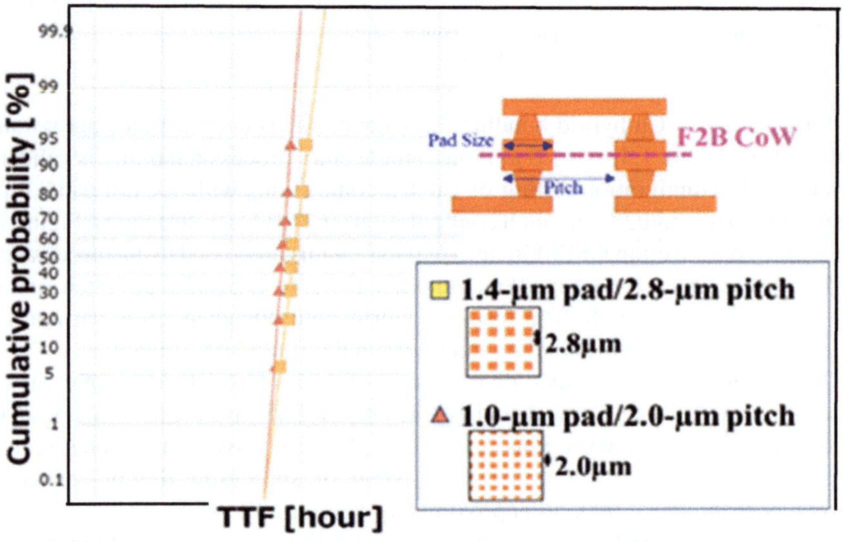


**Fig. 2.10** Cumulative probability distributions of the electromigration lifetime measurements in F2B CoW structures

### 2.2.3 *Influences of Chip Shape on Scaling in Chip-on-Wafer Hybrid Bonding*

Sony presented another paper on the influences of chip shape on scaling in chip-on-wafer hybrid bonding [7]. In this study, they investigated the influences of chip warpage and thickness on scaling in CoW hybrid bonding. The scaling values are

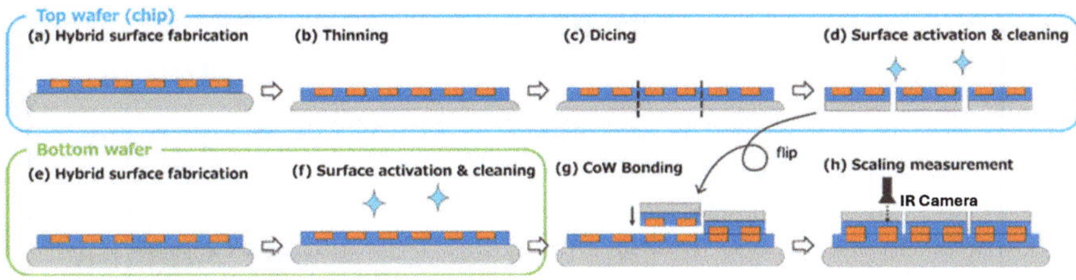


**Fig. 2.11** Process flow of the experiment

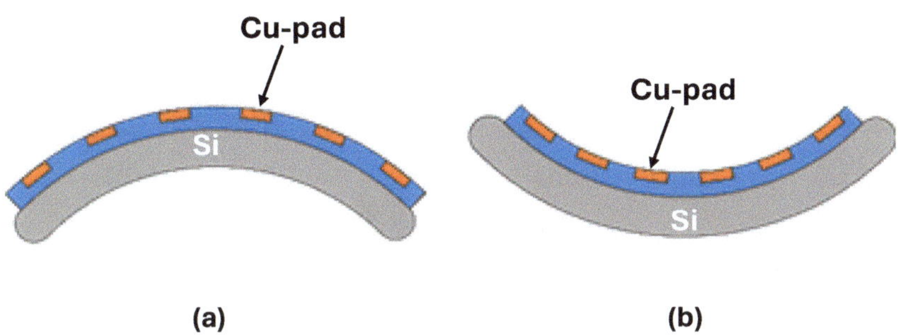


**Fig. 2.12** Illustrations of warpages: (**a**) negative and (**b**) positive

standardized and represented the ratio of the deformation of a chip relative to its original state before deformation. Figure 2.11 shows the process flow of their study. The positive and negative warpages are defined as concave and convex curvatures, respectively, as shown in Fig. 2.12. Figure 2.13 shows the standardized scaling values are functions of the wafer warpage prior to thinning. It can be seen that scaling values decrease as the warpage shifts from negative to positive. Also, the thicker a chip, the smaller the scaling value. Hence, a thinner chip is more susceptible to incoming scaling owing to the lower rigidity.

Figure 2.14 shows the general sequence of plasma activation bonding in ambient atmosphere. The top and bottom wafers are facing each other and secured with a vacuum chuck, Fig. 2.14a. Subsequently, to prevent void formation at the bonding interface owing to trapped air, an initial contact is induced by applying force to the center of the top wafer, Fig. 2.14b. The bonding front subsequently propagates spontaneously from the center to the edge, Fig. 2.14c, until bonding is complete, Fig. 2.14d. During this bonding sequence, the top wafer undergoes deformation, thereby influencing scaling and is referred to as bond wave scaling. Figure 2.15 shows the standardized scaling as a function of the wafer warpage before thinning. Figure 2.15a is for a chip bonded with a flat bonding head while Fig. 2.15b is for a chip bonded with a round bonding head.

By simulating bond wave propagation with varying chip thicknesses, Fig. 2.16 shows the induced maximum stress at the chip surface at the bonding front and the bond wave velocity. It can be seen as the chip thickness decreases, a high tensile

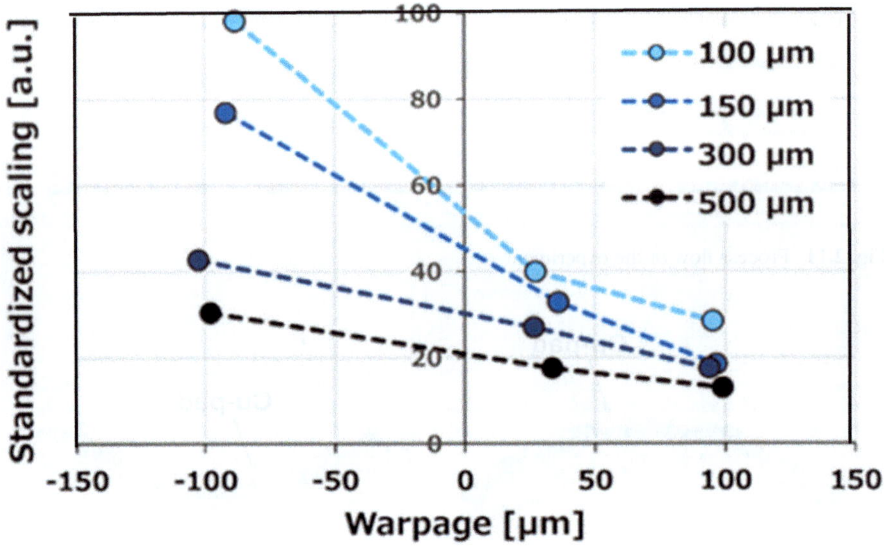


**Fig. 2.13** Standardized scaling as a function of the wafer warpage prior to thinning

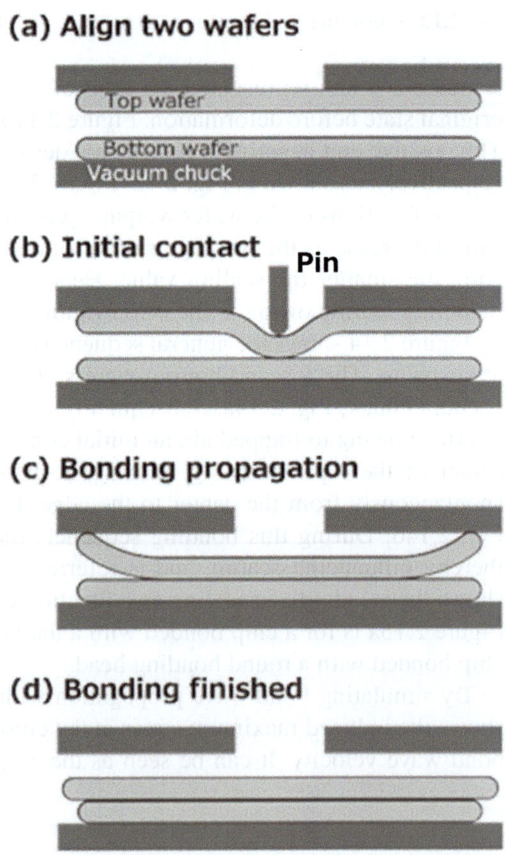


**Fig. 2.14** WoW bonding sequence. (**a**) Align two wafers. (**b**) Initial contact. (**c**) Bonding propagation. (**d**) Bonding finished

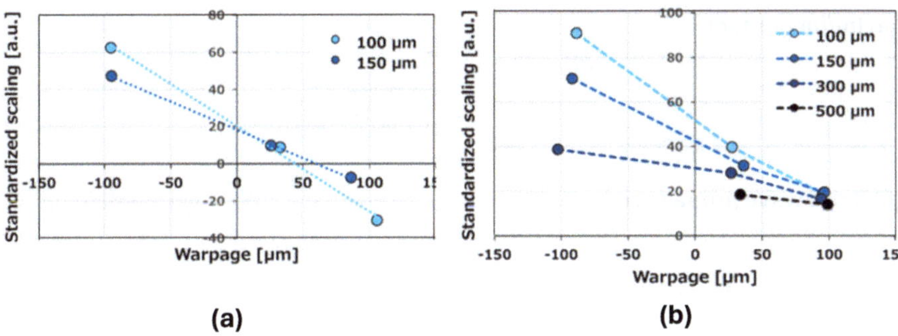


**Fig. 2.15** Standardized scaling as a function of the wafer warpage before thinning. (**a**) A chip is bonded with a flat bonding head. (**b**) A chip was bonded with a round bonding head

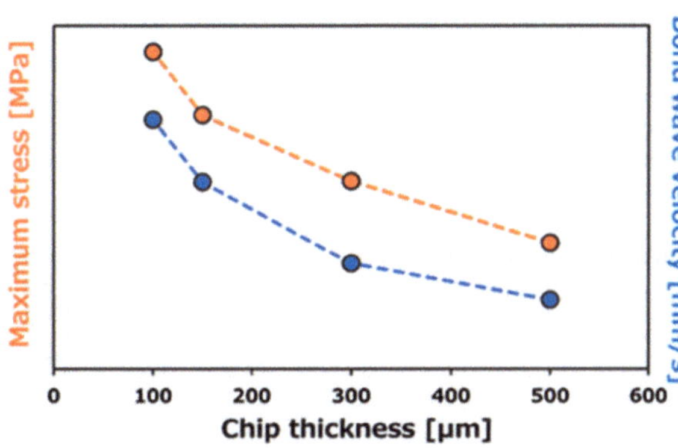


**Fig. 2.16** Simulated results of the bond wave velocity (blue) and maximum tensile stress at the bonding front (orange) as a function of the chip thickness

stress is induced and the bond wave velocity increases. These effects result from the chip deformation during bond wave propagation. The mechanism of bond wave scaling is shown in Fig. 2.17.

## 2.2.4 *Simulation of Mechanical Cu-Pad Expansion Mechanism and Measures to Increase Expansion*

Sony presented another paper on simulation of mechanical Cu-pad expansion mechanism and measures to increase expansion [8]. In this study, they investigated the mechanism underlying the mechanical expansion of Cu pads through simulations and measurements. Figure 2.18 shows the Cu-pad structure under consideration. The temperature-stress curve of single-layer Cu is shown in Fig. 2.19. It can be seen that the simulation and experiment results correlated very well. The elasto-plastic stress-strain curve of Cu is shown in Fig. 2.20 which will be used for all the simulations in this study. It can be seen that Cu becomes more plastic as the temperature increases.

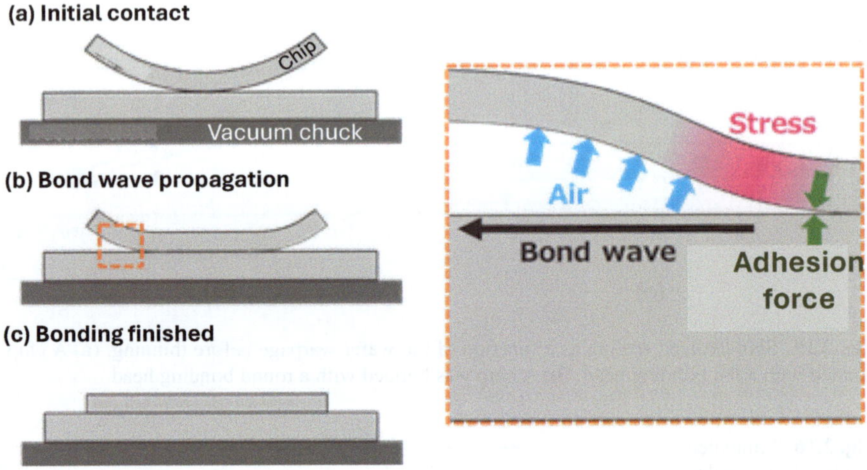


**Fig. 2.17** Schematic of the mechanism of bond wave scaling. (**a**) A chip is held by a bonding head and contacts the target wafer at the center. (**b**) The chip is released, and a bond wave spontaneously propagates. (**c**) Bonding finishes

**Fig. 2.18** Cu-pad structure

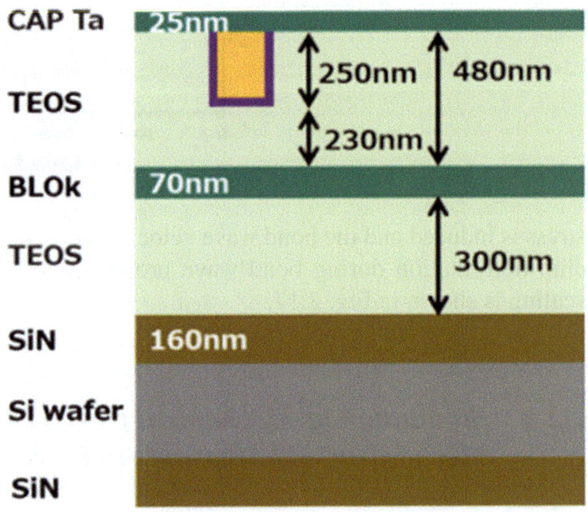


The Cu-pad size vs. expansion is shown in Fig. 2.21. It can be seen that the simulation and experiment results correlated very well, and the expansion amount of the Cu-pads decreases sharply when the size falls below a certain threshold. Figure 2.22 shows the deformation of Cu pads at 400 °C. Figure 2.22d shows the Cu pads are sufficiently large (2 μm). In this case, the deformation of Cu at the sidewalls is minimal, and the expansion gradually increases up to a certain distance from the sidewalls, after which it becomes saturated, reaching a maximum at the center of the pads.

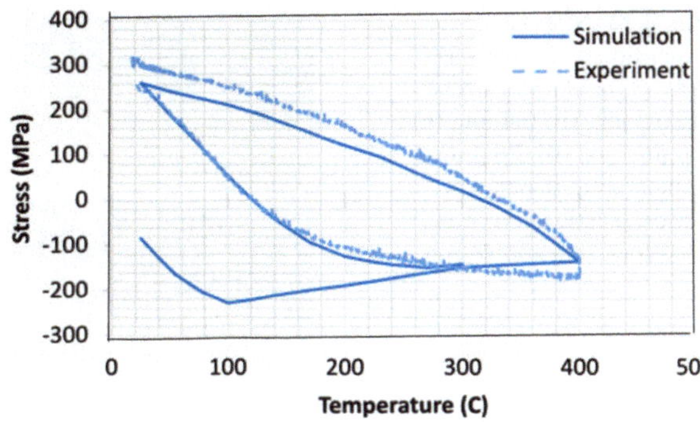


**Fig. 2.19** High-temperature stress curve of single-layer Cu

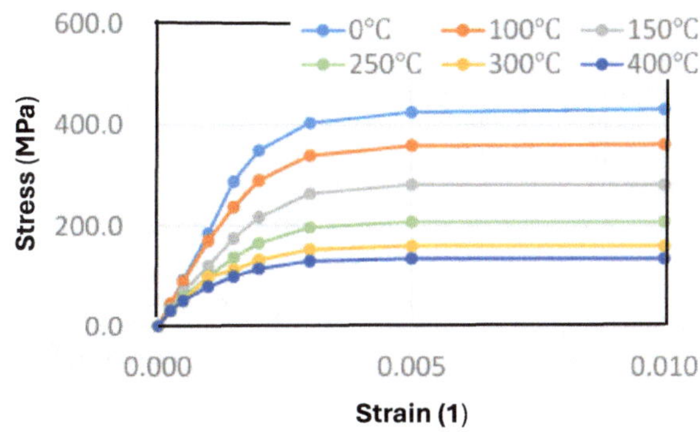


**Fig. 2.20** Identified plastic curve of Cu

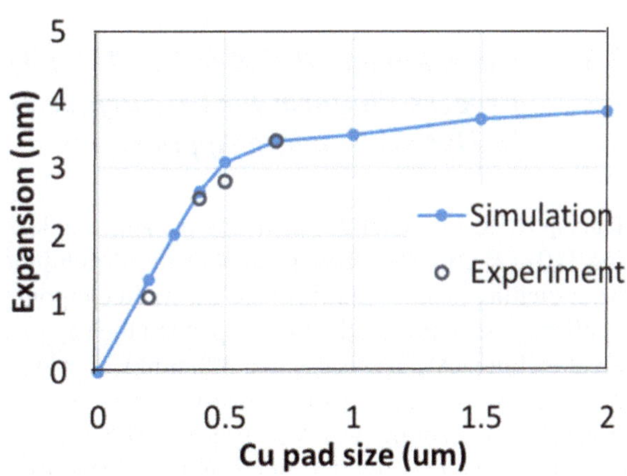


**Fig. 2.21** Dependence of Cu expansion on Cu-pad size

Figure 2.23 shows the deformation (expansion) of the Cu-pad surface at 400 °C with varying Cu-pad size. It can be seen that the distance affected by the sidewall constraints is approximately 0.25 μm from the spatial profile of Cu expansion in large Cu pads. Also, the size at which the expansion amount of the Cu pads begin to decrease sharply is approximately consistent with the Cu-pad size of 0.5 μm, that is, twice of the distance affected by the sidewall constraints.

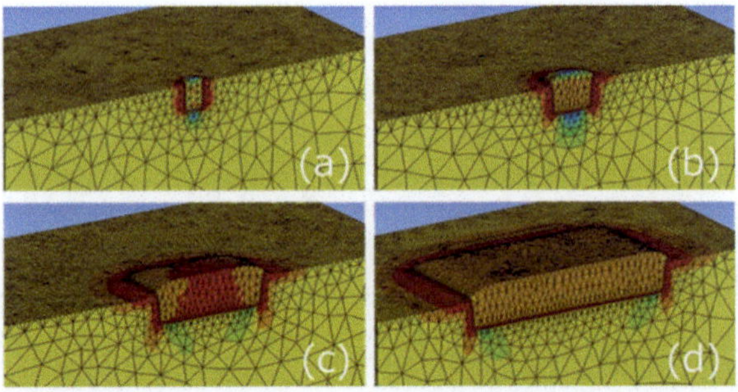


**Fig. 2.22** Stress and expansion of Cu pads at 400 °C for Cu-pad sizes: (**a**) 0.2 μm, (**b**) 0.4 μm, (**c**) 1.0 μm, and (**d**) 2.0 μm. Temperature-dependent elastic-plastic curve was used for Cu in simulation

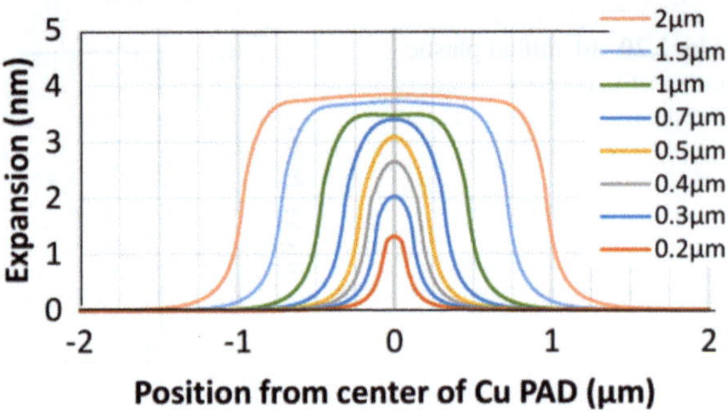


**Fig. 2.23** Deformation of the Cu-Pad surface at 400 °C with varying Cu-Pad size. Temperature-dependent elastic-plastic curve was used for Cu in simulation

## 2.3 Samsung's Hybrid Bonding

### *2.3.1 Development of Wafer-Level Wet Atomic Layer Etching Process Platform for Cu Surface Topography Control in Hybrid Bonding Applications*

During IEEE/ECTC 2025, Samsung presented at least three papers on hybrid bonding [10–12]. In [10], they presented the development of wafer-level wet atomic layer etching process platform for Cu surface topography control in hybrid bonding applications. In this study, they expanded the capability of the so-called wet atomic layer etching (ALE) process to a 300 mm wafer-level platform and reported results of targeting Cu recess within a nanometer range. Figure 2.24 shows the configuration schematics of the wet ALE process. The key components of the ALE prototype system consist of a revolution per minute (RPM) variable 300 mm vacuum chuck, auto dispensing module with flow rate controller for each chemical varieties and motion-enabled dispensing arm.

As shown in Fig. 2.25, bulk Cu CMP step is first applied to remove electrodeposited Cu and then, barrier metal removal CMP process is performed to reveal SiCN surface. This process flows in Cu protrusion of around ~20 Å across the test vehicle wafers and these wafers are employed to evaluate the wet ALE process as well as the dishing control CMP-recessed specimen.

**Fig. 2.24** Hardware configuration schematics of the wet ALE process platform

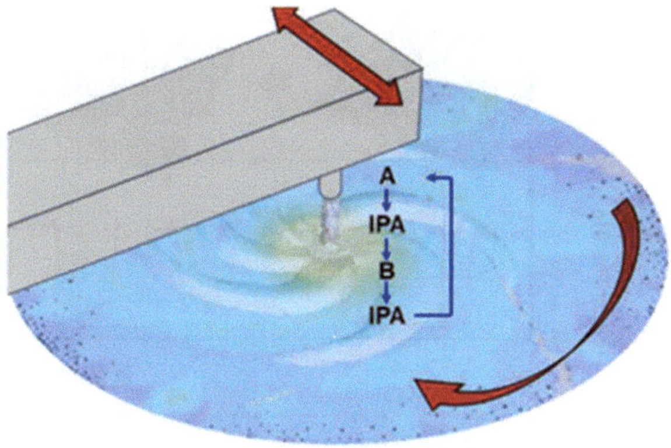


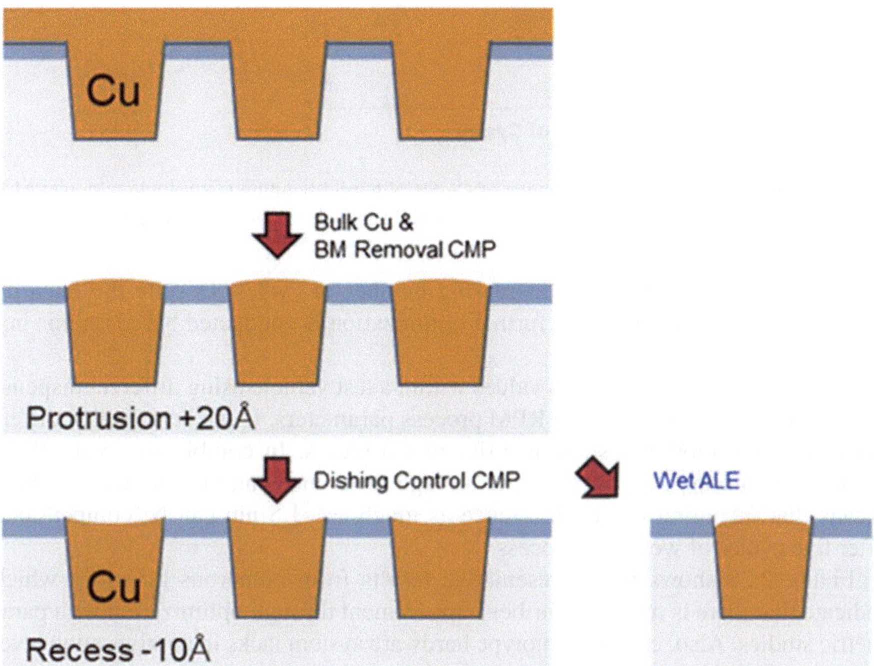


**Fig. 2.25** CMP process flow for test vehicle preparation

Figure 2.26a shows surface height profiles after 1, 2, 3, 5, and 10 application runs of wet ALE process cycle on 20 Å Cu protrusion test vehicle wafers. It is evident from the result that protrusion state of the incoming wafer exhibits dishing profile afterwards due to cyclical treatment of wet chemicals. It should also be noted that dielectric surface is intact from chemicals as line profiles on dielectric surface region overlap regardless of number of cycles applied to the test wafer. Figure 2.26b shows in-wafer variation results which indicate the Cu etching trend on center,

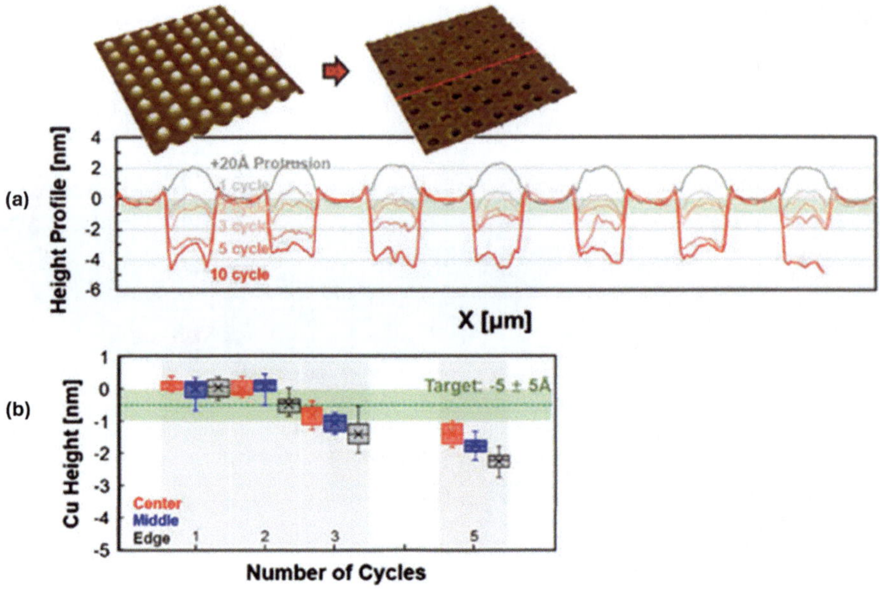


**Fig. 2.26** (**a**) AFM height profile change of Cu/SiCN hybrid bonding test vehicle with wet ALE process. (**b**) In-wafer variation results at center, middle and edge region for each process cycle

middle, and edge region with increasing number of cycles. Despite the linearity observed for wet ALE process, further optimization is conducted based on this initial result.

Figure 2.27 shows Cu recess values within a test vehicle using different dispensing methodologies and varying RPM process parameters. It can be seen that different dispensing motion results in different Cu recess. In combination with RPM parameters, around 3 Å/cycle center-to-edge difference can be established which means that incoming height difference as much as ~1.5 nm can be compensated after five cycles of wet ALE process.

Figure 2.28 shows the representative results from numerous iterations which indicate that there is room for further improvement through optimization with parametric studies. Also, current prototype hardware system lacks in configuration level of maturity and seeks better performance through revisiting components such as nozzle type, flow rate and dispensing amount controller.

Figure 2.29 shows the comparison of CMP (Cu pad recess −10 Å) and wet ALE (10 cycles) process on larger scale features within the test vehicle. The incoming wafer for both cases is +20 Å. It can be seen that, for CMP Cu process, during the process of recess control, dielectric surface is also affected by CMP and exhibits surface topography change around bonding key area. On the other hand, for the ALE process, the dielectric surface remains without notable change even after application of 10 cycles, only affecting and eliminating the Cu region.

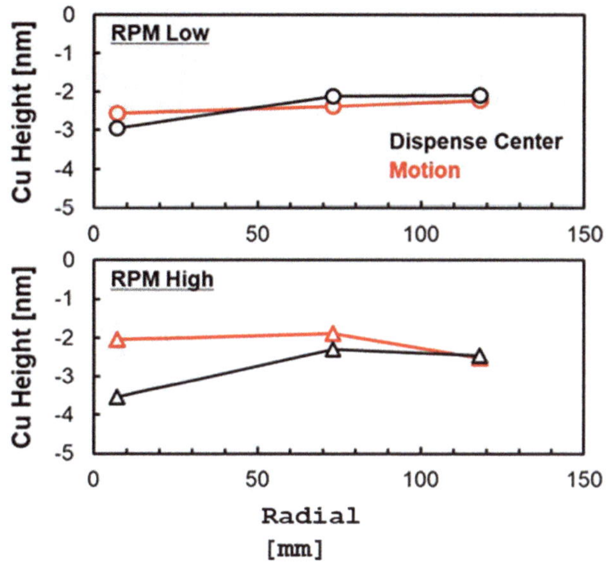


**Fig. 2.27** Cu recesses within a test vehicle using different dispensing methodologies and varying RPM process parameters

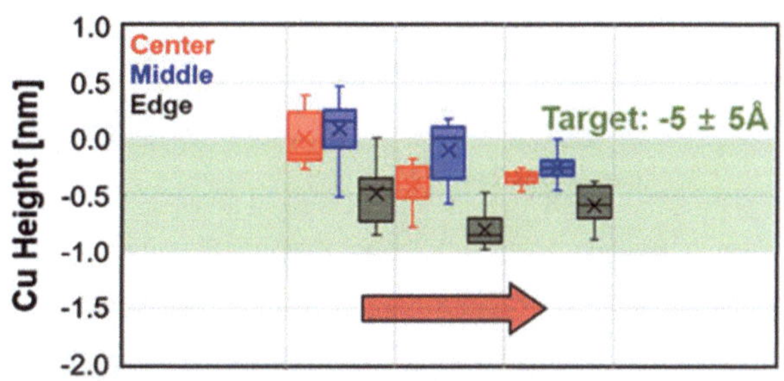


**Fig. 2.28** Optimization of 2-cycle operations of wet ALE process to achieve target Cu recess of −10 ~ 0 Å over the entire wafer surface

### *2.3.2 A Novel 3D Heterogeneous Integration using 2 μm Bond Pitch Die-to-Wafer Hybrid Cu Bonding and Wafer Reconstruction Process*

Samsung presented another paper on a novel 3D heterogeneous integration using 2 μm bond pitch D2W hybrid Cu bonding and wafer reconstruction process [11]. In this study, they explored die preparation using stealth laser sawing, < 2 μm bond pitch direct D2W HB, island die grinding (IDG), gap-fill with thick oxide and CMP planarization. Figure 2.30 shows the process flow for verification of key processes: D2W bonding, island die grinding, inter-die gap-fill and planarization.

Figure 2.31 shows the optimized 5-step CMP process for Cu pad dishing and global erosion control. It can be seen that the Cu pad dishing is controlled by the number of barrier metal (BM) and buffing CMP. Figure 2.31 also shows that the larger the number of buffing CMP the smaller the Cu dishing.

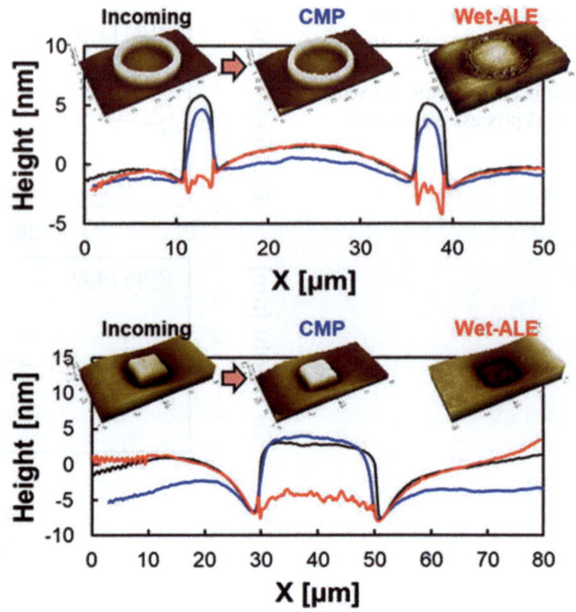


**Fig. 2.29** Comparison of CMP (Cu pad recess −10 Å) and wet ALE (10 cycles) process on larger scale features within the test vehicle

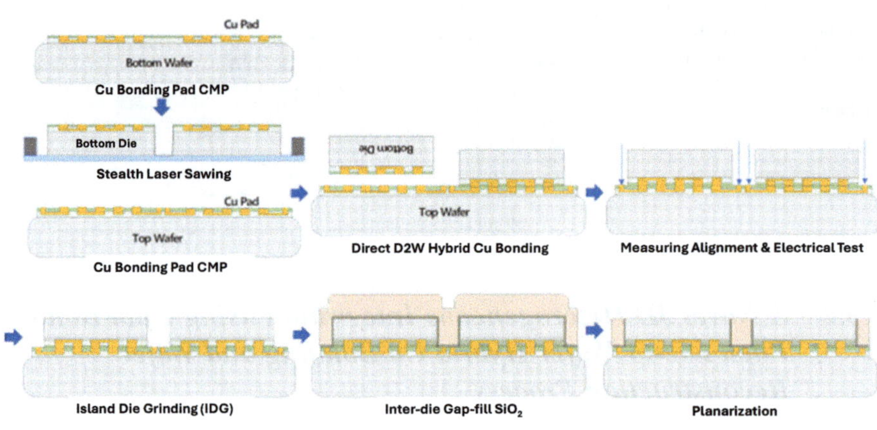


**Fig. 2.30** 3D heterogeneous integration manufacturing process flow using D2W HCB wafer reconstruction

The D2W overlay error consists of translational, rotational, run out, and residual. Among them, run-out error can be improved through chip warpage changing toward tensile through evaluating three types of chip warpage, and the best overlay for 2 μm bond pitch achieved is shown in Fig. 2.32. Figure 2.33 shows the cross section of HB interface with transmission electron microscope (TEM) images. It can be seen that there are no void and delamination observed in Cu-Cu and SiCN-SiCN dielectric bonding interfaces.

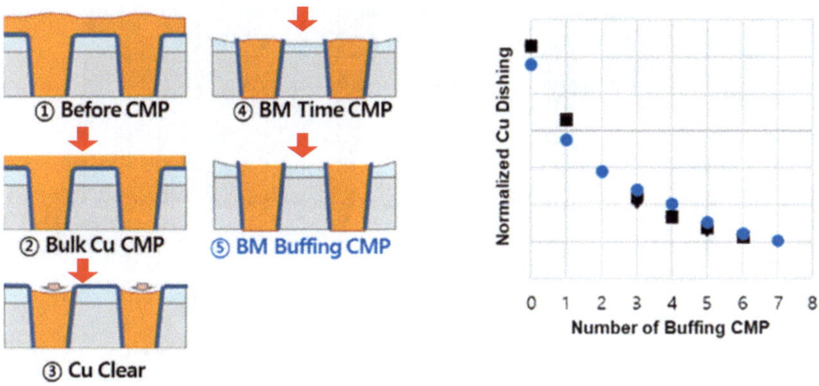


Fig. 2.31 Cu pad CMP process flow and dishing control

Fig. 2.32 Bonding overlay measurement results

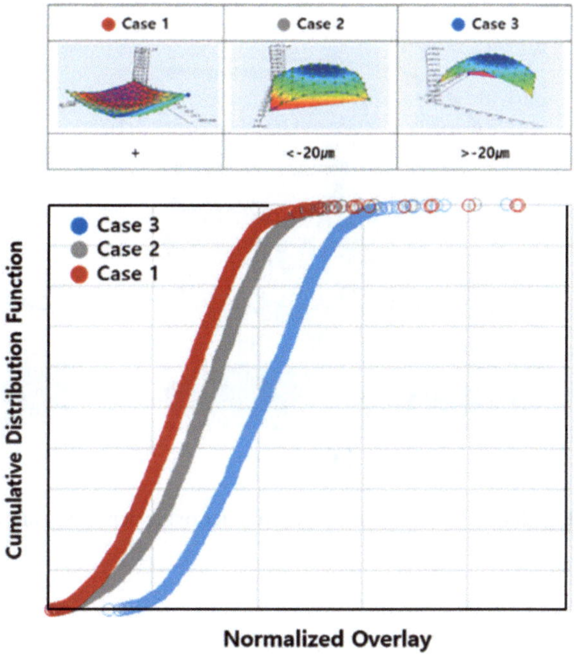


Figure 2.34a shows the deposition of low temperature thick oxide over 30 μm (measured form the top wafer) without any oxide cracking. Figure 2.34b shows the cross section SEM image after planarization using high removal rate CMP process with non-selective slurry. Figure 2.35 shows good topography, measuring dishing between dies after planarization.

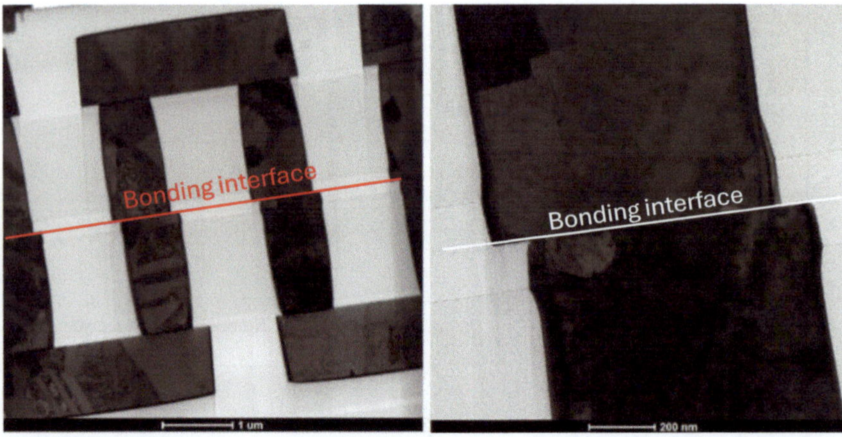


Fig. 2.33 TEM images of cross section of HCB interface

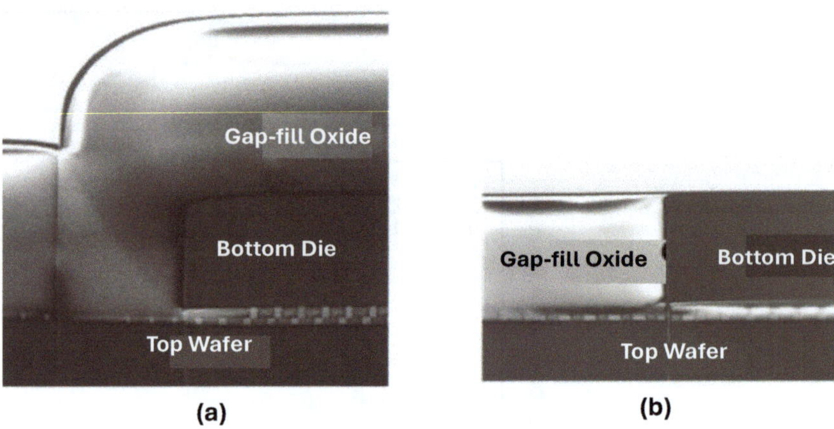


**Fig. 2.34** Cross section images (**a**) after oxide deposition and (**b**) after planarization

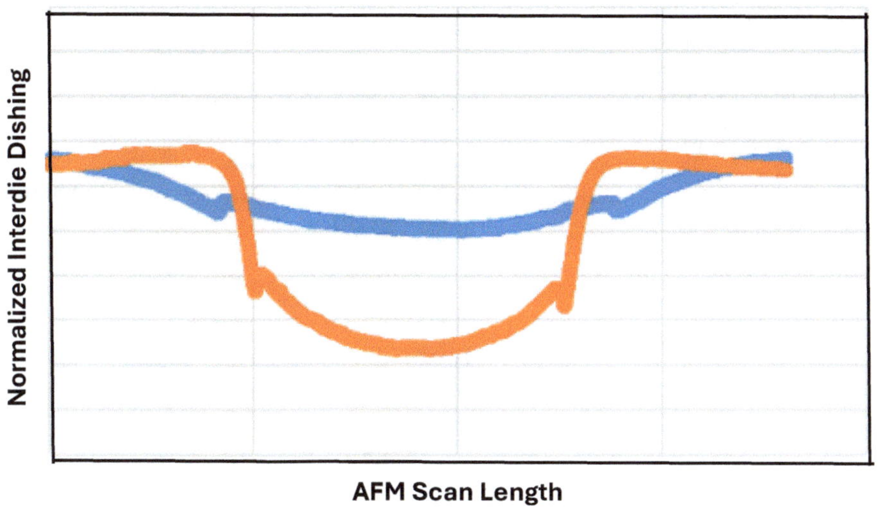


**Fig. 2.35** AFM results of oxide dishing between dies after planarization

## 2.4 Applied Materials' Hybrid Bonding

### *2.4.1 Process Technology Enablers for Aggressive Scaling of Hybrid Bonding Temperature and Pitch*

During IEEE/ECTC 2025, Applied Materials presented at least four papers on hybrid bonding [13–16]. In [13], they presented the recess technology enablers for aggressive scaling of hybrid bonding temperature and pitch. In this study, they attempt to use a W2W hybrid bonding test vehicle to evaluate two process technology solutions for enabling the scaling of hybrid bonding temperatures down to 200 °C regime and at deeply scaled pitches.

Figure 2.36 schematically shows the 6-layer W2W hybrid bonding test vehicle. The bonding layer used a SiCN film with optimized composition for high bonding energy. The top and bottom via sizes may be the same or different. Figure 2.37 shows some examples of atomic force microscopy (AFM) line scans of post- CMP Cu/SiCN bonding layer. Figure 2.37a shows the optimized CMP process by adopting hardware settings, state-of-the-art multi-zone control, and software and metrology knobs for topography control. It can be seen that flat SiCN surface is achieved with minimized erosion. Figure 2.37b shows the non-optimal CMP process when hardware or process control parameters were not optimized. It can be seen that the SiCN top rounding and the larger dielectric roll-off slope are not desired for good bonding quality.

Figure 2.38 (a) and (b) show the TEM image and Figs. 2.38c shows the electron energy loss spectroscopy (EELS) images of a continuous and uniform (~2.5 nm)

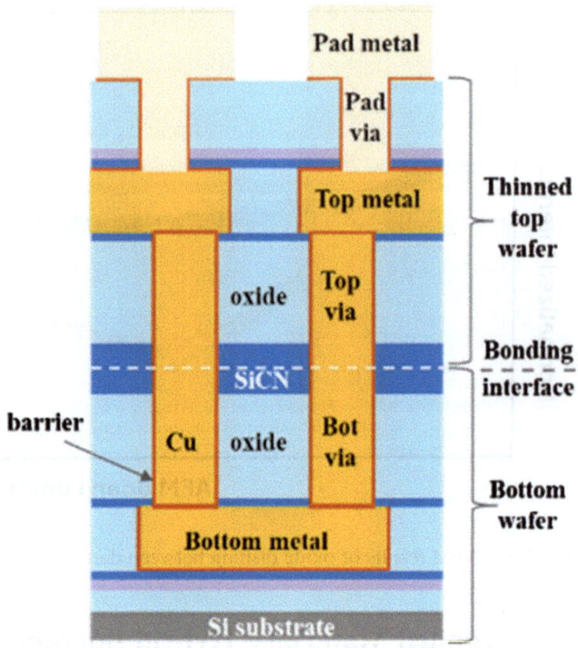


**Fig. 2.36** Illustration of the 6-layer W2W hybrid bonding test vehicle. The bonding layer used a SiCN film with optimized composition for high bonding energy

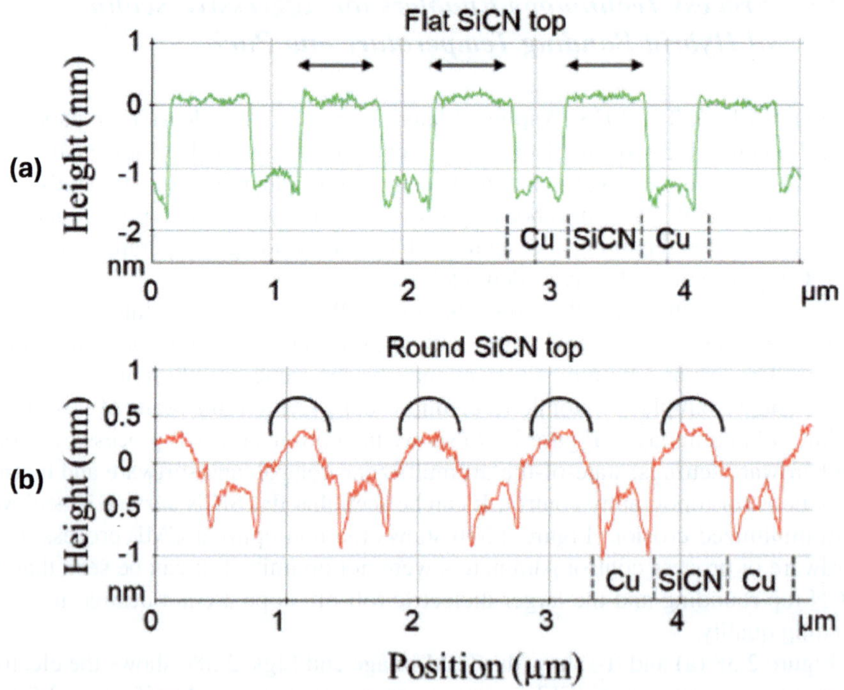


**Fig. 2.37** Example AFM line scans of post-CMP Cu/SiCN bonding layer. (**a**) Optimized CMP process. (**b**) Non-optimal CMP process

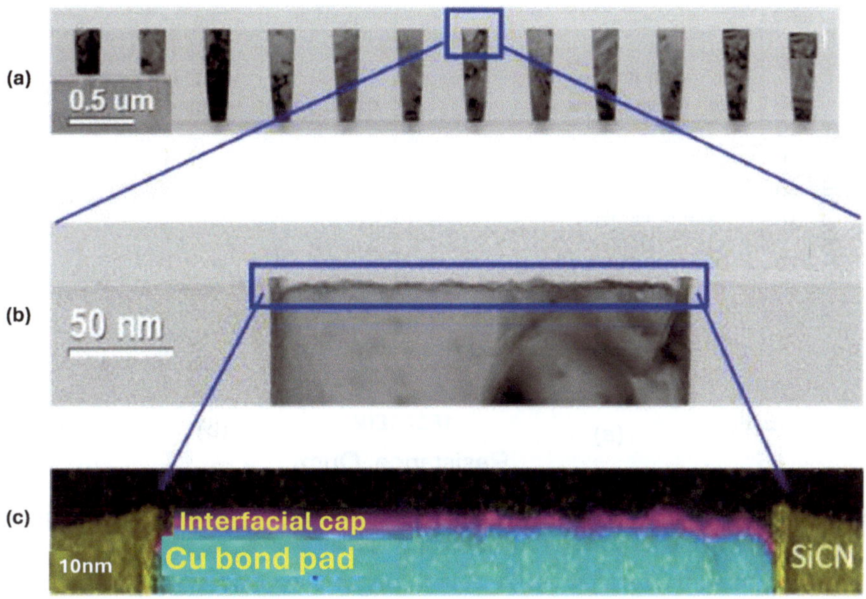


**Fig. 2.38** TEM (a, b) and EELS (**c**) images

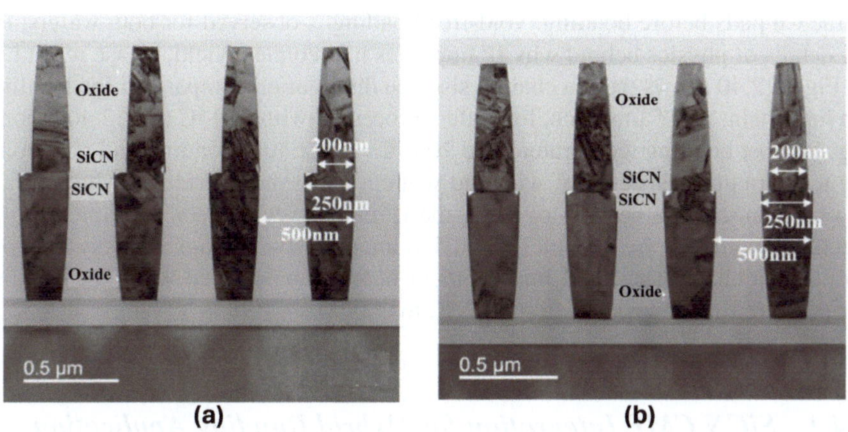


**Fig. 2.39** TEM images of two 500 nm pitch W2W SiCN/Cu bonded via chains. (**a**) No IF metal film is used. (**b**) a ~ 2.5nm thick IF metal layer is used

thick interfacial metal diffusion layer selectively deposited only on top of the Cu pads but not on SiCN.

Figure 2.39 shows the TEM images of two 500 nm pitch W2W SiCN/Cu bonded via chains. In both bonded wafer pairs, the top wafer via CD is 200 nm and the bottom wafer via CD is 250 nm. Both wafers are bonded at room temperature and then annealed to 200 °C for 2 h. Water is cut right after the post-bonding anneals. No

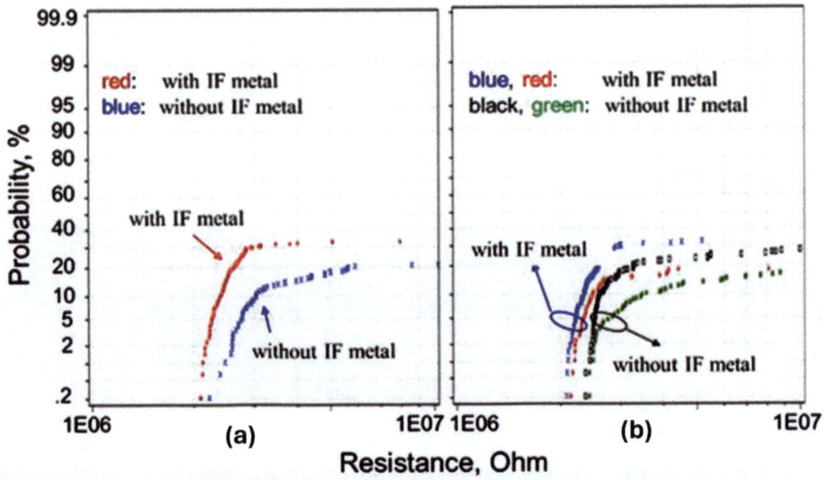


**Fig. 2.40** Via chain resistance distribution comparison. (**a**) 280 °C bonding and post bonding temperature. (**b**) 200 °C bonding and post-bonding temperature

further thermal process is done after the anneals. Figure 2.39a shows the one with no interfacial (IF) metal film is used. While Fig. 2.39b shows the one with a ~ 2.5 nm thick IF metal diffusion layer which is electroless selective deposited on top of the Cu pads before bonding. Void-free bonding is observed for both wafers. For the details of physics behind why IF improves interconnect yield, please read [13].

Figure 2.40 shows the via chain resistance distribution comparison for 1 million via link chains at 0.5 μm pitch, for wafers processed with 280 °C (Fig. 2.40a) bonding and post bonding temperature and 200 °C (Fig. 2.40b) bonding and post bonding temperature, respectively, with and without using the interfacial metal layer. It can be seen that the benefits of using the IF metal are still seen in improving the electrical yields for the extremely tight 500 nm pitch and the aggressively long one million via chain. The yield improvement benefits maintained when the bonding temperature further lowered from 280 °C to 200 °C.

### 2.4.2 SiCN CMP Integration for Hybrid Bonding Application

Applied Materials presented another paper on SiCN CMP integration for hybrid bonding application [14]. In this paper, they explored the integration of SiCN dielectric into hybrid bonding. First, they characterize the CMP selectivity on blanket level to understand the operating regime with which bonding pad profile can be tuned reversibly. In particular, they built on previous learnings that Cu dishing must be maintained to be <5 nm to achieve successful hybrid bonding. Then they develop a CMP process to achieve that Cu dishing requirement on their D2W hybrid bonding test vehicle. Subsequently, they demonstrate hybrid bonding integration of the

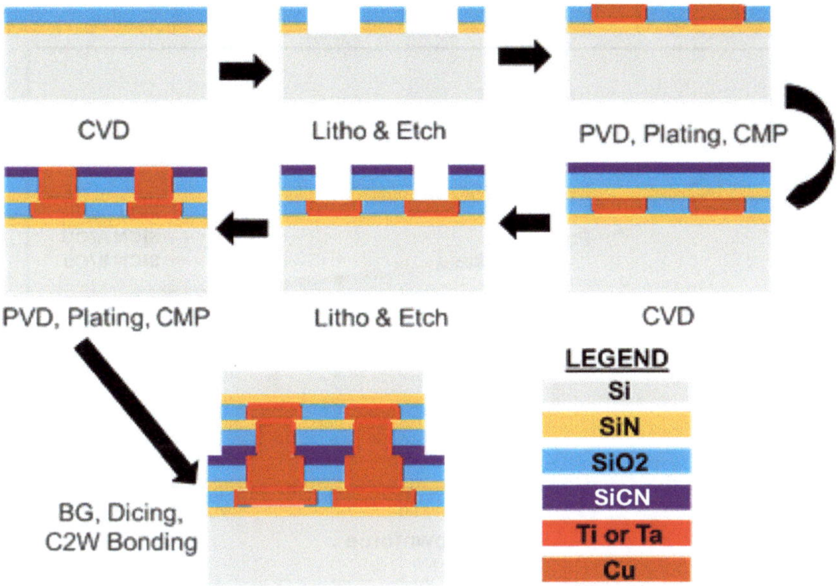


**Fig. 2.41** Process flow of D2W HB TV

metal-dielectric pair with clear cross-boundary Cu grain growth and no dielectric delamination. Finally, they conclude the qualification by demonstrating post-bonding key electrical performance metrics, namely daisy chain resistance < 300 Ω, Kelvin resistance <30 mΩ, and leakage current <1 pA. The process flow of their D2W test vehicle is shown in Fig. 2.41.

Figure 2.42 shows the SiCN/Cu removal rate for two types of SiCN, namely SiCN A and SiCN B. SiCN A and B differ in their deposition temperature, namely <200 °C and <300 °C and SiCN B is the film used for subsequent hybrid bonding integration. It can be seen that for both types of SiCN, the removal rate selectivity is consistently >1 over downforce range up to 2 psi.

Figure 2.43 shows the validation of the CMP performance on the component and substrate wafers. It can be seen that, it is possible to achieve dishing the thinned and singulated component chips are bonded to the substrate wafer, and the cross-section after 350 °C post-bond anneals (PBA). Figure 2.44 shows the cross section SEM images after 350 °C PBA. It can be seen that the Cu-Cu interface is fully fused with clear cross-boundary Cu grain growth and no dielectric delamination, demonstrating compatibility of the metal-dielectric pair to D2W HB integration.

Figure 2.45 shows bonding electrical performance of the D2W hybrid bonding at 300–350 °C in terms of Kelvin—Fig. 2.45a, 2.5k-via direct current resistance—Fig. 2.45b, and leakage current—Fig. 2.45c. It can be seen that: (a) Cu-Cu connectivity with yield >90% at 300 °C (resistance <30 Ω), (b) the electrical daisy chain connection involving numerous vias can produce lower yield, and (c) the leakage current on average stays below 1 pA for both 300 °C and 350 °C, demonstrating low electrical leakage property of the SiCN film.

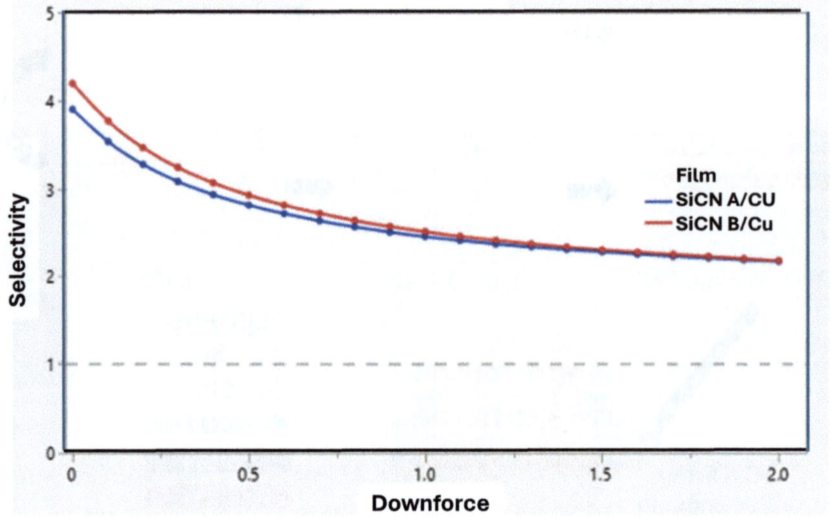


**Fig. 2.42** SiCN/Cu CMP removal rate selectivity for 2 types of SiCN

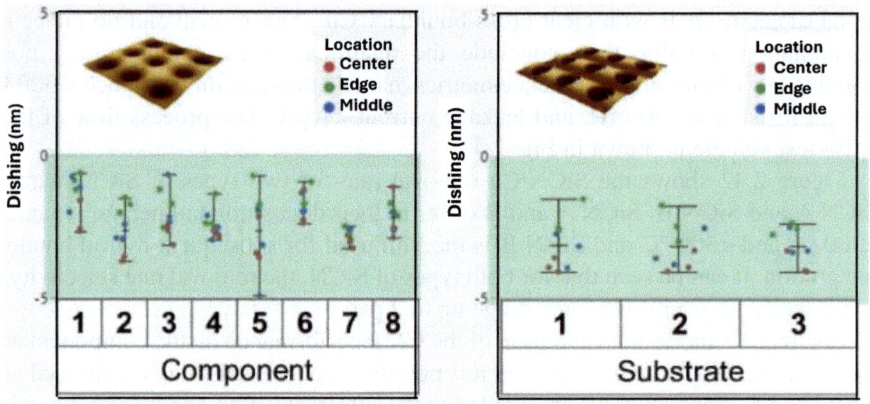


**Fig. 2.43** Post-CMP bond pad dishing performance for Component and Substrate wafers

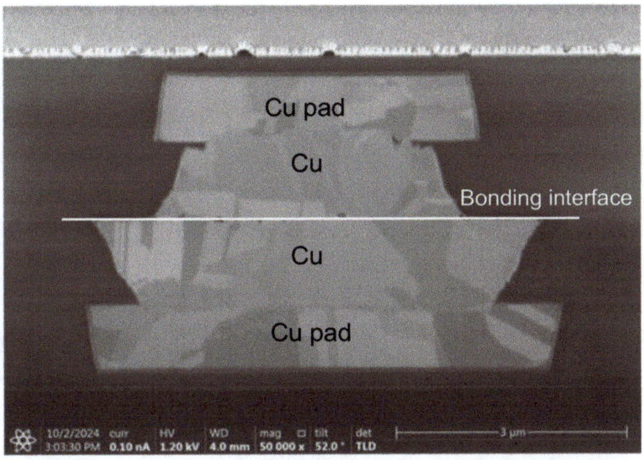


**Fig. 2.44** Cross-section of the bonded interface of the D2W HB TV after 350 °C PBA

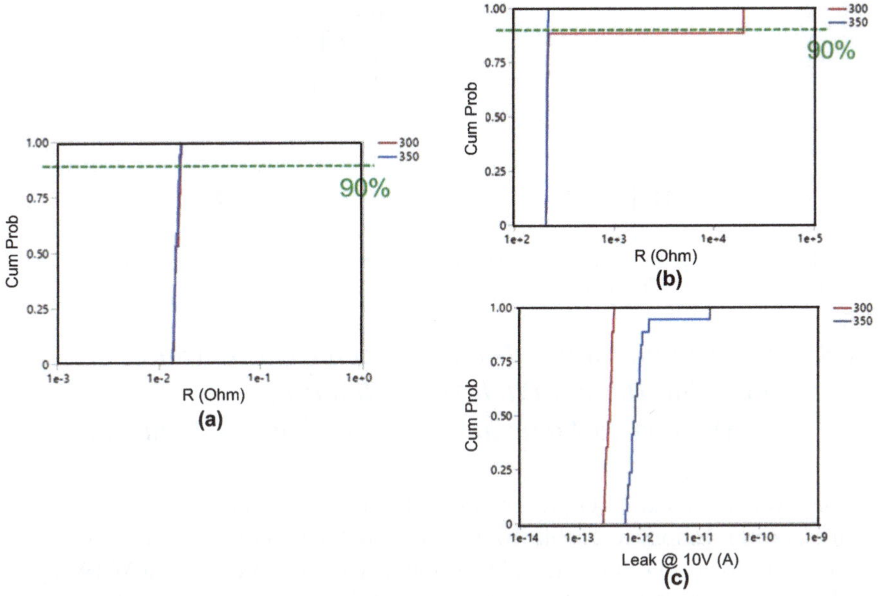


**Fig. 2.45** Bonding electrical performance of the D2W HB TV at 300–350 °C, in terms of (**a**) Kelvin resistance; (**b**) 2.5k-via DC resistance; (**c**) leakage current

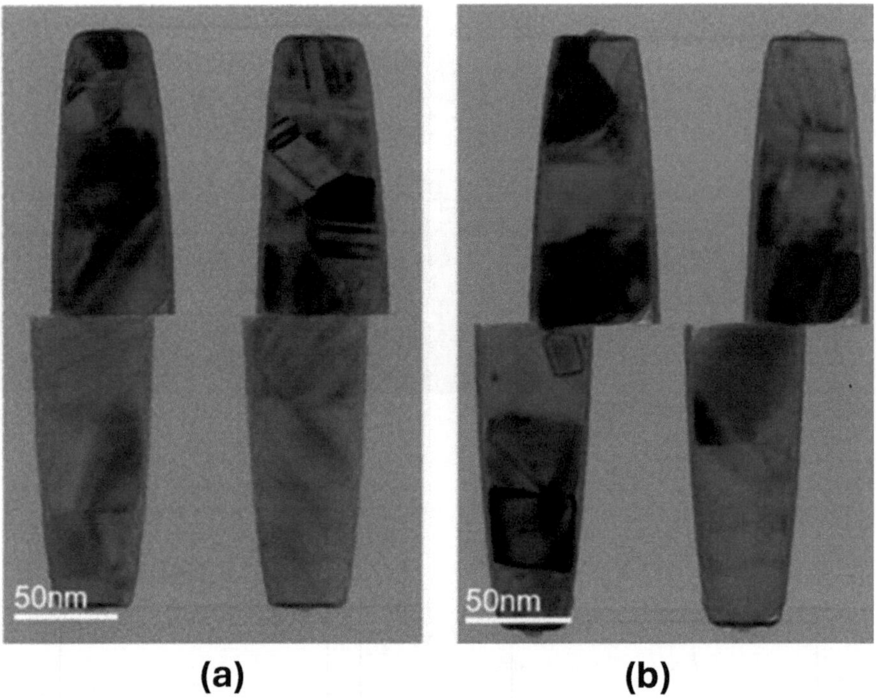


**Fig. 2.46** TEM image of (**a**) an excellent aligned 100nm pitch hybrid bonded Cu via and (**b**) ~50% misaligned 100nm pitch hybrid bonded Cu via

### *2.4.3 Integration, Materials and Equipment Innovations to Enable 100 nm Pitch W2W Bonding for Memory-to-Logic and Logic-to-Logic 3D Stacking*

Applied Materials and EVG presented another paper on integration, materials and equipment innovations to enable 100nm pitch W2W bonding for memory-to-logic and logic-to-logic 3D Stacking [15]. In the study, they presented a W2W hybrid bonding at a bond pitch of 100nm through careful design, materials and module co-optimization. Their results are shown in Figs. 2.46 and 47. Figure 2.46a shows the TEM image of an excellent aligned 100 nm hybrid bonded Cu vias, while Fig. 2.46b shows the TEM image of ~50% misaligned 100 nm hybrid bonded Cu via. Figure 2.47 shows the TEM image of (a) 100 nm pitch hybrid bonded Cu via chain link, and (b) zoomed in image of bonding interface showing excellent Cu contact with a smooth transition to the bonding dielectric with minimal key-hole voids.

(a)

(b)

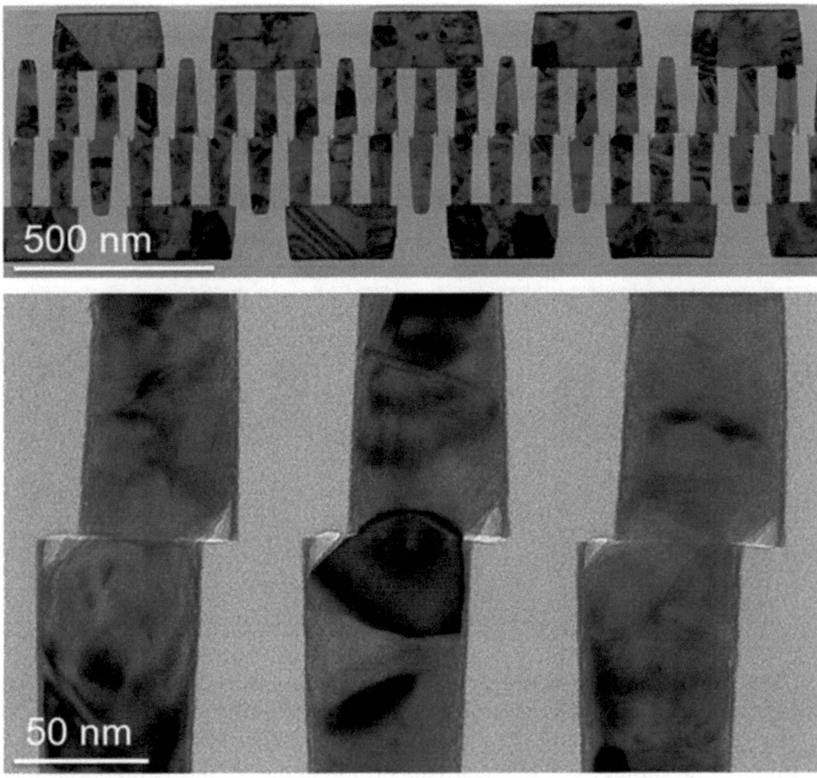


**Fig. 2.47** TEM image of (**a**) 100nm pitch hybrid bonded Cu via chain link, and (**b**) zoomed in image of bonding interface

## 2.5 imec's Hybrid Bonding

### *2.5.1 Wafer-to-Wafer Bonding With Saddle-Shaped Wafers*

During IEEE/ECTC 2025, imec and EVG published at least six papers on hybrid bonding [17–22]. In [17], they presented the wafer-to-wafer bonding with saddle-shaped wafers. In this study, they investigate the bonding of wafers with saddle shape, a deformation commonly observed in 3D NAND wafers for 3D NAND-to-logic CMOS bonding. Figure 2.48a schematically shows the design for patterning the stress layer strips, while Fig. 2.48b shows the cross-sections of wafers with layer stacks for saddle-shaped wafers.

The saddle-shaped wafers, including all layer stacks, show a warpage of 230 μm (90 to 100 μm in the X-direction and −130 to −140 μm in the Y-direction), Fig. 2.49. Two sets of wafers are considered: one is with warpage of approximately 20 μm and the other is umbrella-shaped with a warpage of around 220 μm.

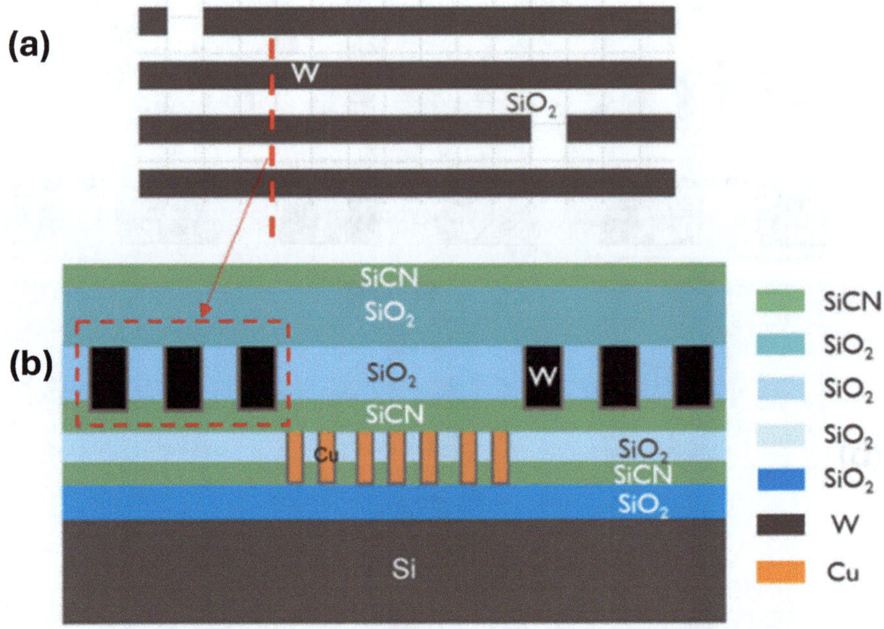


**Fig. 2.48** (**a**) Schematic illustrating the design for patterning the stress layer strips, (**b**) Schematic cross-sections of wafers with layer stacks for saddle-shaped wafers

Figure 2.50 shows the wafer warpage profile in the X- and Y-directions measured by wafer shape and warpage profile in the X and Y directions with a patterned wafer geometry metrology tool on a relatively flat wafer (a) and umbrella-shaped wafer (b).

After the detailed saddle-shaped wafer preparation, wafer bonding, thinning and metrology, overlay model, bonding with flat bottom chuck of bonder, bonding with adjusted bottom chuck bow, asymmetric scaling compensation, total wafer bonding overlay comparison, asymmetric scaling compensation for high precision hybrid bonding, and post-thinning backside lithography overlay correction studies (please read [17] for more information), they conclude that (a) placing the saddle-shaped wafer on the bonder's bottom chuck, rather than the top chuck, yields better overlay results when applying asymmetric top wafer release, (b) while asymmetric top wafer control significantly reduces the delta of scaling components in the X and Y directions, the remaining uncompensated scaling components remains high, (c) for flat and umbrella wafers exhibiting slight asymmetry, a scaling component difference of 0.2 ppm is observed between the X and Y directions, corresponding to a scaling error of 30 nm at the wafer edge, (d) the backside lithography overlay on the thinned wafers is significantly improved with higher order corrections per exposure, and (e) compared to bonded umbrella shaped wafers, bonded saddle-shaped wafers required significantly larger non-linear corrections.

(a)

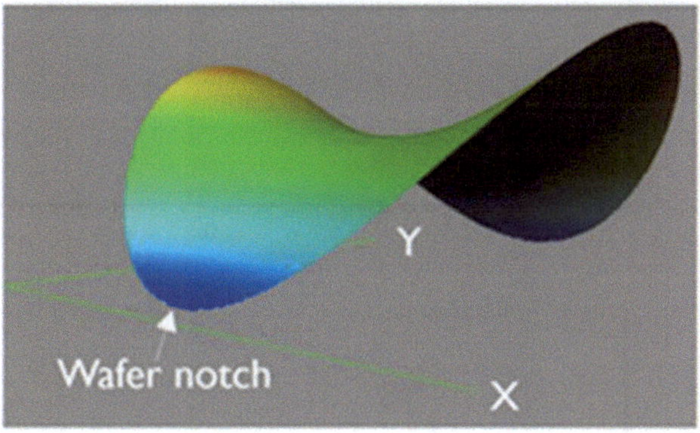


(b)

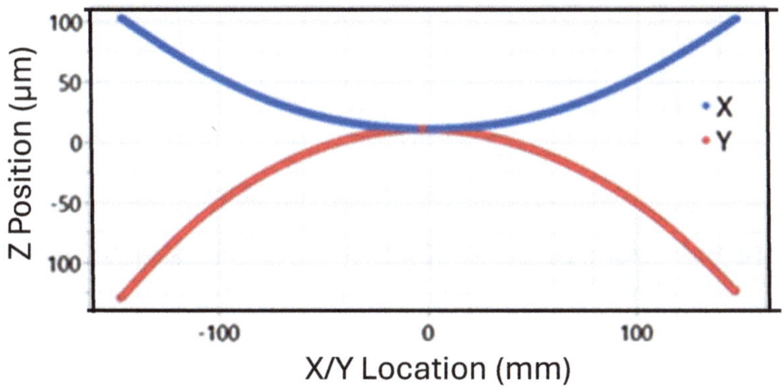


**Fig. 2.49** Saddle-shaped wafer deformation measurement result: (**a**) wafer shape, (**b**) warpage profile in the X and Y directions

Figure 2.51 shows the cumulative distribution of total bonding overlay vector length for bonding with varied wafer shape and placement configurations. It can be seen that for an umbrella-shaped wafer with a warp of −220 µm, when placed on the bottom chuck of the bonder and bonded to a flat wafer, the bonding overlay performance is comparable to the baseline with flat wafers.

Figure 2.52 shows the backside lithography overlay results for four scenarios: one pair of flat wafers, one pair of an umbrella shaped wafer and a flat wafer (with the umbrella wafer on the bonder's bottom chuck during bonding and later thinned down), and two pairs of saddle-shaped wafers and flat wafers. In the first saddle-flat pair, the saddle wafer is placed on the bonder's top chuck, while in the second pair, it is positioned on the bottom chuck. It can be seen that the bonded saddle-shaped wafers require much larger non-linear corrections but their final correctability is only slightly worse compared to using an umbrella wafer.

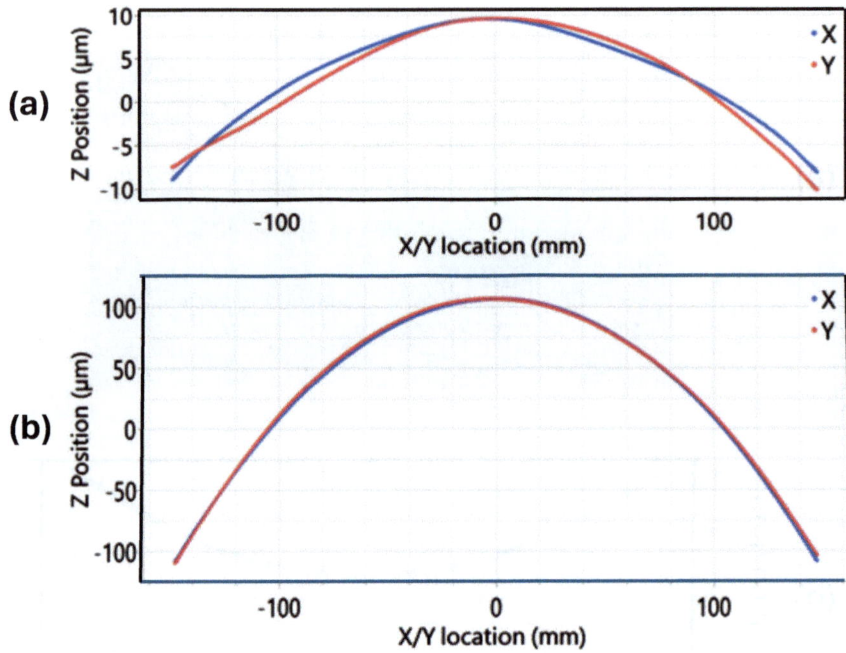


**Fig. 2.50** Wafer warpage profile in the X and Y directions measured by PWG on a relatively flat wafer (**a**) and umbrella-shaped wafer (**b**)

**Fig. 2.51** Cumulative distribution of total bonding overlay vector length for bonding with varied wafer shape and placement configurations

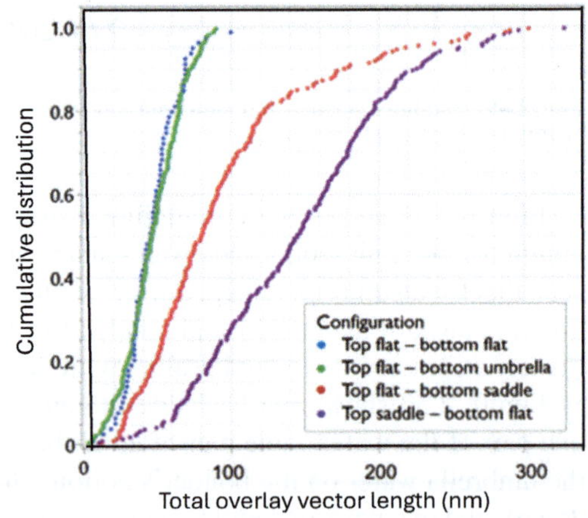

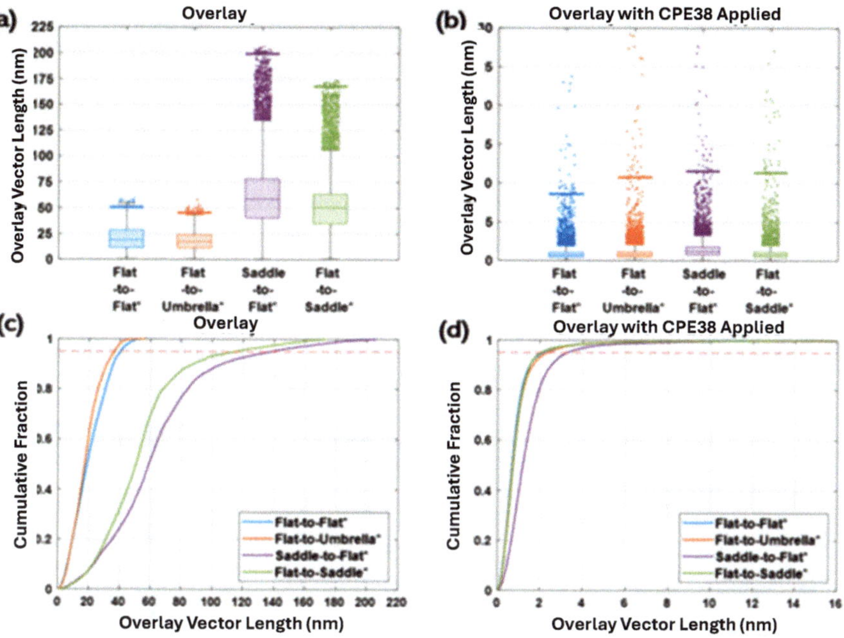


**Fig. 2.52** Backside post-thinning lithography overlay box plots (**a**) and cumulative distributions (**c**) without corrections and box plots (**b**) and cumulative distributions (**d**) with higher order corrections per exposure

### *2.5.2 2 μm Pitch Direct Die-to-Wafer Hybrid Bonding using Surface Protection During Wafer Thinning and Die Singulation*

imec presented another paper on 2 μm pitch direct D2W hybrid bonding using surface protection during wafer thinning and die singulation [18]. In this study, they proposed and demonstrated the feasibility of incorporating an inorganic protective layer in the direct D2W hybrid bonding flow. Figure 2.53 shows an example of the AFM comparison results of the protrusion of Cu pads after wafer thinning. Figure 2.53a is for the wafer after hybrid CMP and Fig. 2.53b is for the wafer after hybrid CMP and wafer thinning. It can be seen that the Cu has been oxidized, and it is because of the spin-coating and curing of adhesives on the die surface for temporary bonding.

Figure 2.54 shows the proposed process flow of D2W bonding with an inorganic protective layer incorporated to preserve the bonding surface of dies. Figure 2.55a shows the SEM image after the etching of protective layer and dielectric stack (landing on Si), while Fig. 2.55b shows the image of after complete plasma dicing as shown in Fig. 2.54g.

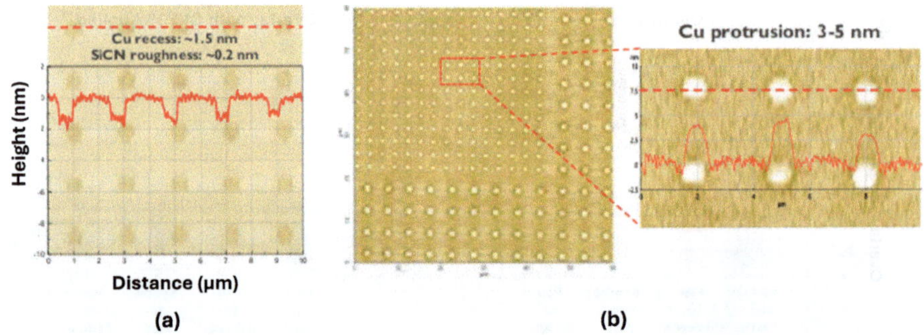


**Fig. 2.53** Comparison of AFM results show the protrusion of Cu pads after wafer thinning

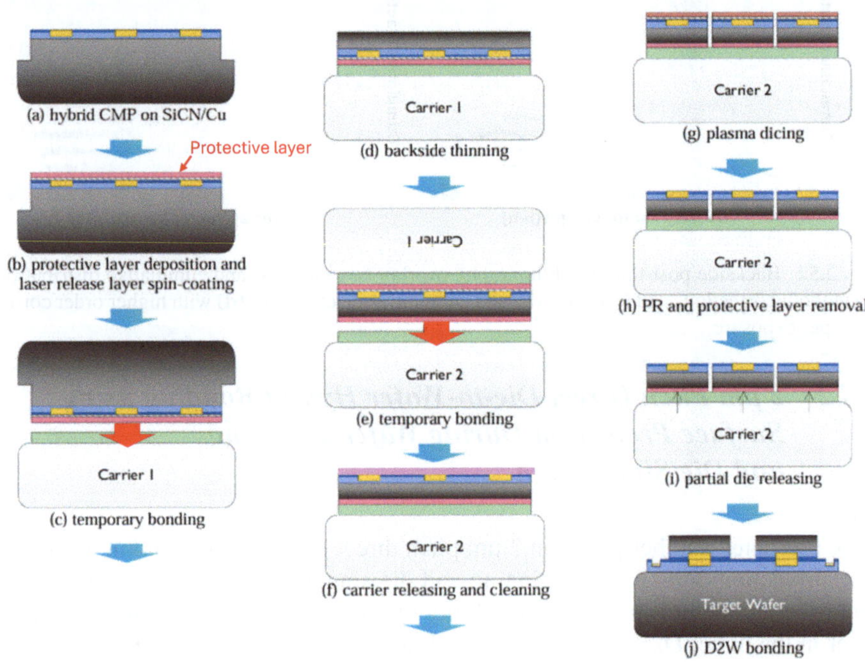


**Fig. 2.54** Process flow of D2W hybrid bonding with protective layer incorporated to preserve the bonding surface of dies

Figure 2.56a shows the TEM images of cross section of bonding interface in a 2 μm pitch daisy chain area. It can be seen that both SiCN-SiCN and Cu-Cu bonding are seamless. Also, the 2 μm pitch bond pads are well aligned between top die and bottom wafer. Figure 2.56b shows the energy-dispersive x-ray (EDX) mapping between Cu-Cu bonding. It can be seen that there is no oxidation between Cu-Cu bonding. This is also confirmed by the high angle annular dark field (HAADF) image, Fig. 2.56c.

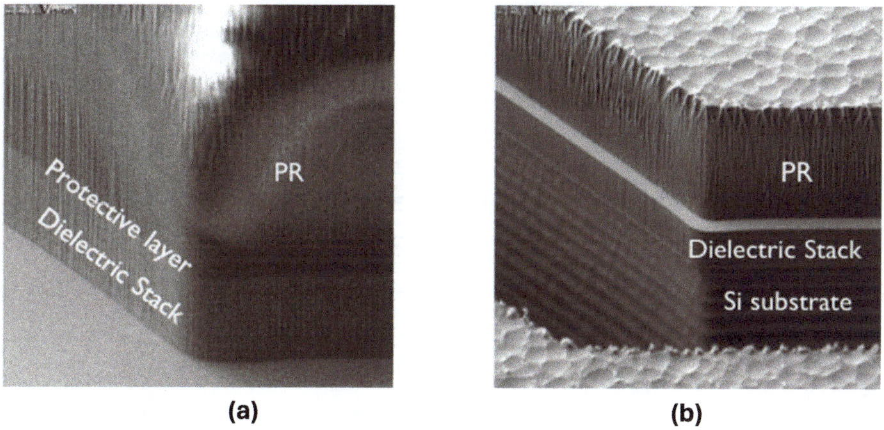


**Fig. 2.55** SEM images of plasma dicing with protective layer incorporated: (**a**) after etching and (**b**) after complete plasma dicing

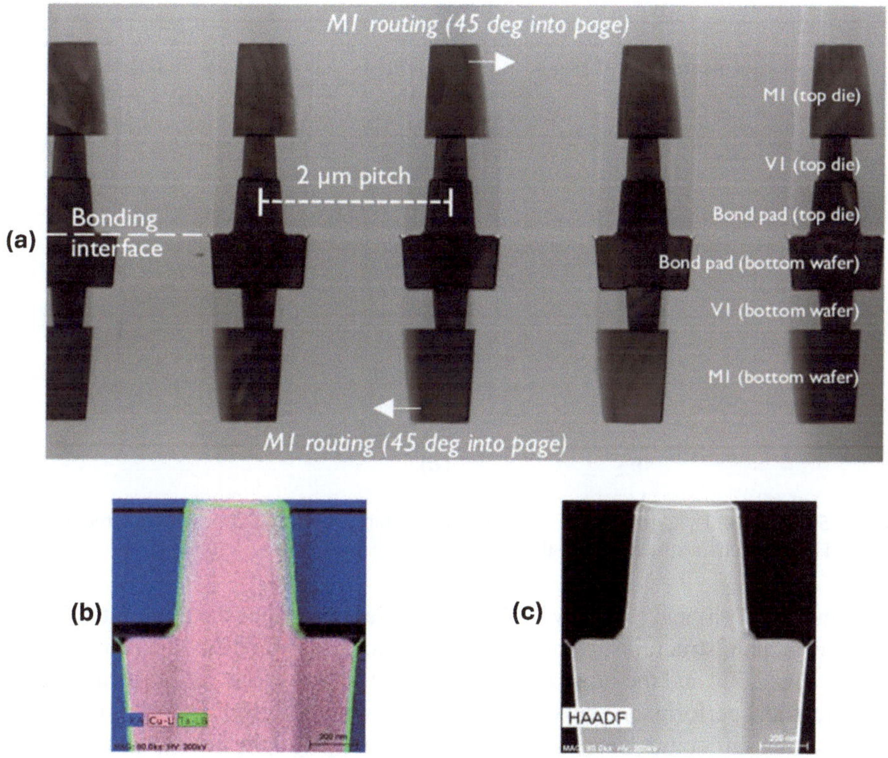


**Fig. 2.56** Bonding interface with 2 μm pitch. (**a**) TEM image. (**b**) EDX image. (**c**) HAADF image

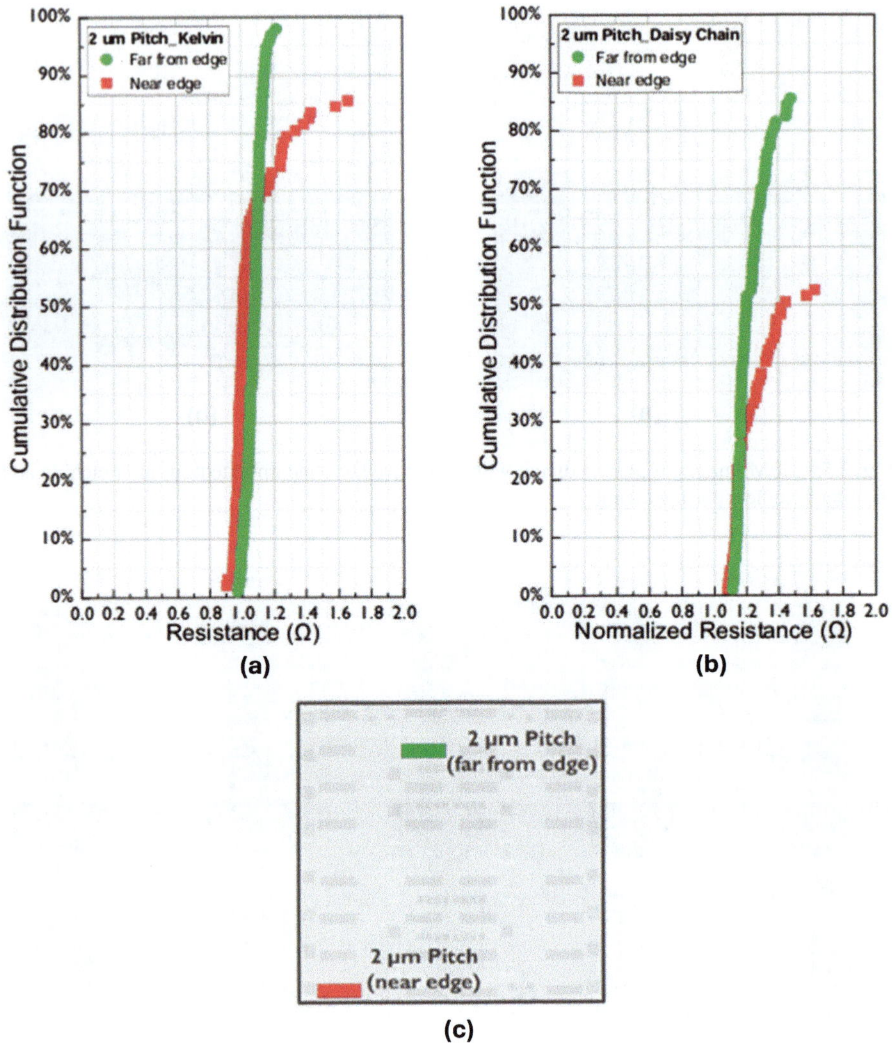


**Fig. 2.57** Electrical yield results of 2 μm pitch structures. (**a**) E-yield of Kelvin structures. (**b**) E-yield of daisy chain structures. (**c**) the locations on the die layout

Figure 2.57 (a) and (b) show the electrical yield results of the 2 μm pitch Kelvin and daisy chain structures, respectively at two locations on dies: one near the edge in red, Fig. 2.57c and the other far from the edge in green, Fig. 2.57c. It can be seen that electrical performance depends on the location of the chip. This could be due to the cleaning difficulties near the die edges due to topographical variations.

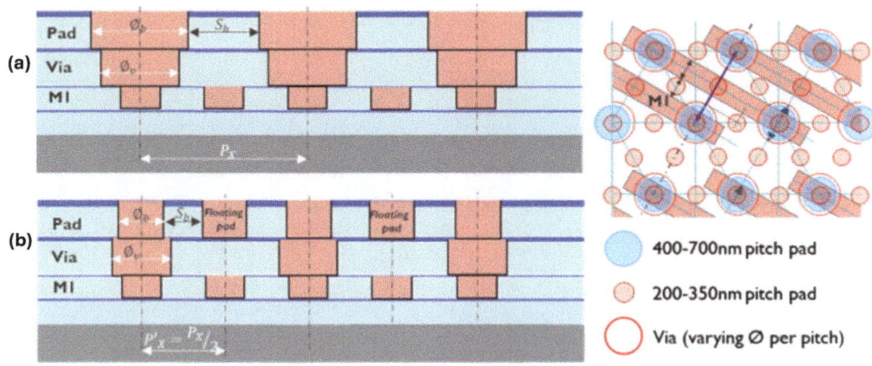


**Fig. 2.58** Daisy chain test structures: (**a**) original test vehicle and (**b**) small pitch test vehicle

### *2.5.3 Wafer-to-Wafer Hybrid Bonding Technology with 300 nm Interconnect Pitch*

imec presented another paper on wafer-to-wafer hybrid bonding technology with 300 nm interconnect pitch [19]. In this study, they scale imec's WoW hybrid bonding technology to a pad pitch of 300 nm. Figure 2.58a shows the daisy chain test structures for the investigation of hybrid pad pitch scaling, with a hexagonal grid arrangement of the circular hybrid pads. Fig. 2.58a shows the larger (400–700 nm) pitch test vehicle while Fig. 2.58b shows the smaller (200–350 nm) pitch test vehicle.

Figure 2.59 shows the TEM of Kelvin structure on 300 nm hexagonal pad grid (30° cut through hexagonal array) with unequal hybrid pad size. Fig. 2.59a is for 15% designed Cu density for top wafer and 25% designed Cu density for bottom wafer, while Fig. 2.59b is for 25% designed Cu density for both top wafer and bottom wafer. It can be seen that the hybrid top to bottom pads on the unequal pad structure are well connected. However, micro voids are present at the corners of the pads, as the copper diffuses from the pad edges to the center, forming the pad-to-pad hybrid interconnections. Figure 2.60 shows the end of line TEM of daisy chain structure on 300 nm hexagonal pad grid (horizontal cut) with top and bottom designed pad density of 20%. It can be seen that good void free bonding has been obtained.

Figure 2.61a shows the electrical yield on hybrid bonded Kelvin structures as a function of pad pitch for unequal pad size. Top wafer hybrid pad designed Cu density = 15%, bottom wafer hybrid pad designed Cu density = 25%, the number of links > 100k. Fig. 2.61b shows electrical yield on hybrid bonded Kelvin structures as a function of pad pitch for equal pad size. The solid line is for hybrid pad designed Cu density = 25%, while dashed line is for hybrid pad designed Cu density = 20%. The number of links > 100k.

(a)

(b)

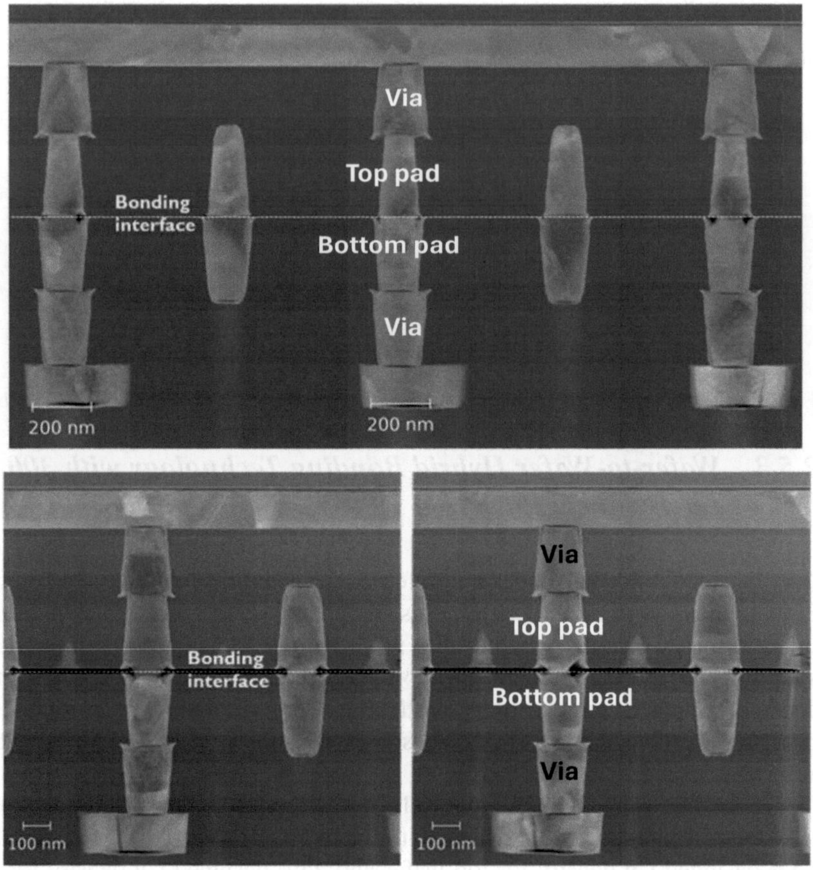


**Fig. 2.59** TEM of Kelvin structure on 300nm hexagonal pad grid (30° cut through hexagonal array) with (**a**) unequal hybrid pad size, 15% designed Cu density for top wafer and 25% designed Cu density for top wafer, and (**b**) equal hybrid pad size, 25% designed Cu density for both top wafer and bottom wafer

## *2.5.4 Impact of Temporary Substrates and Adhesives on Die-to-Wafer Overlay*

imec presented another paper on impact of temporary substrates & adhesives on die-to wafer overlay [20]. In this study, they presented a critical review of the alignment factors contributing to the final die-to-target wafer overlay in a collective die to wafer bonding flow and discussed in terms of minimal residual bonding errors. Figure 2.62 shows the used collective-die-to-wafer bonding flow within this study, showing the die preparation, carrier preparation, cleaning and finally, collective bonding and debonding steps.

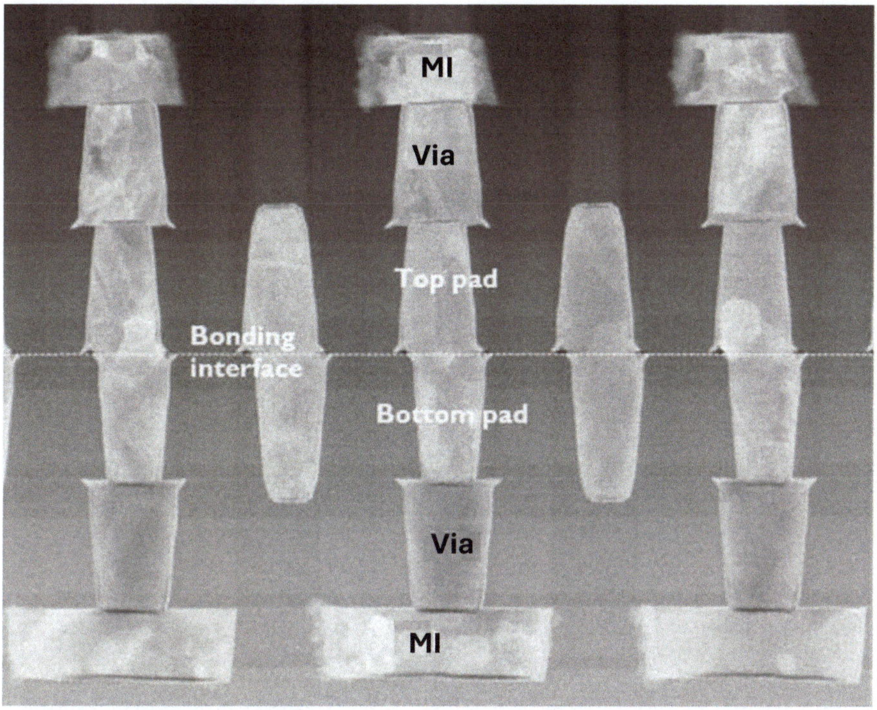


**Fig. 2.60** TEM of Daisy Chain structure on 300nm hexagonal pad grid (horizontal cut) with equal hybrid pad size, 20% designed Cu density

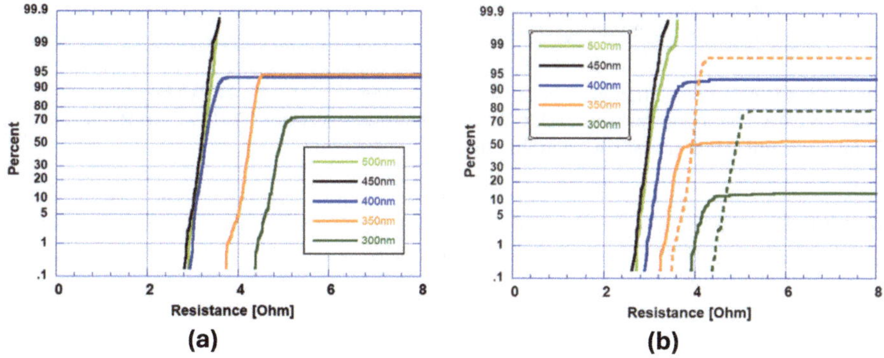


**Fig. 2.61** (**a**) Electrical yield on hybrid bonded Kelvin structures as a function of pad pitch for (**a**) unequal pad size, and (**b**) for equal pad size

Figure 2.63a shows the alignment strategies during a collective die-to-wafer (CoD2W) bonding. Figure 2.63b shows the four systems, A(Si/Si), B(Si/TBM/Si), C(Glass/TBM/Si), and D(Glass/TBM/Glass) under study. The results are shown in Fig. 2.64, where the bar diagram of the four different systems that are evaluated for

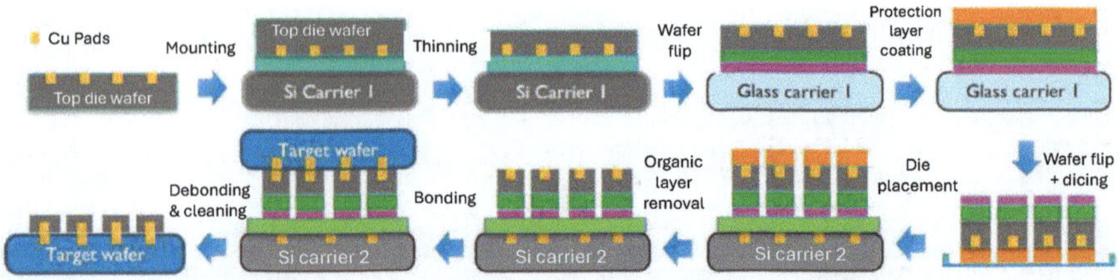


**Fig. 2.62** The used collective-die-to-wafer bonding flow showing the die preparation, carrier preparation, cleaning and finally, collective bonding and debonding steps

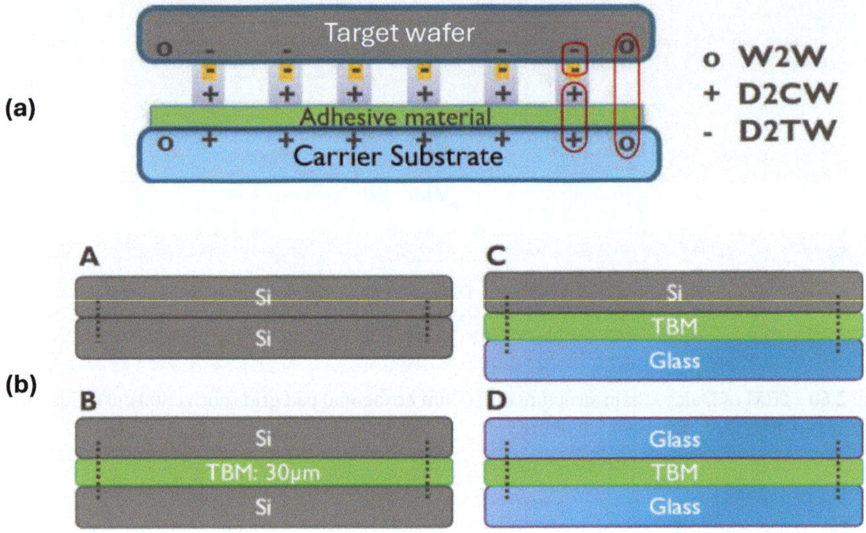


**Fig. 2.63** (**a**) Alignment strategies during a collective die-to-wafer (CoD2W) bonding. Wafer-to-wafer (W2W), Die-to-carrier wafer (D2CW) and Die-to-target wafer (D2TW). (**b**) Representation of the 4 systems under investigation in this study

**Fig. 2.64** Bar diagram of the four different systems that were evaluated for their scaling (S) performance and residuals ($\mu \pm 3\sigma$ and max values)

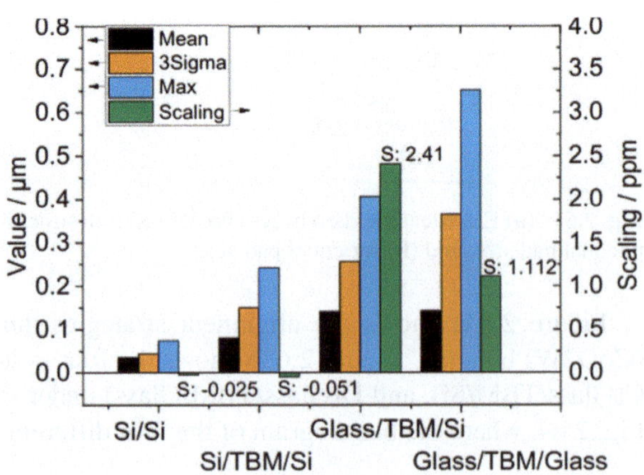

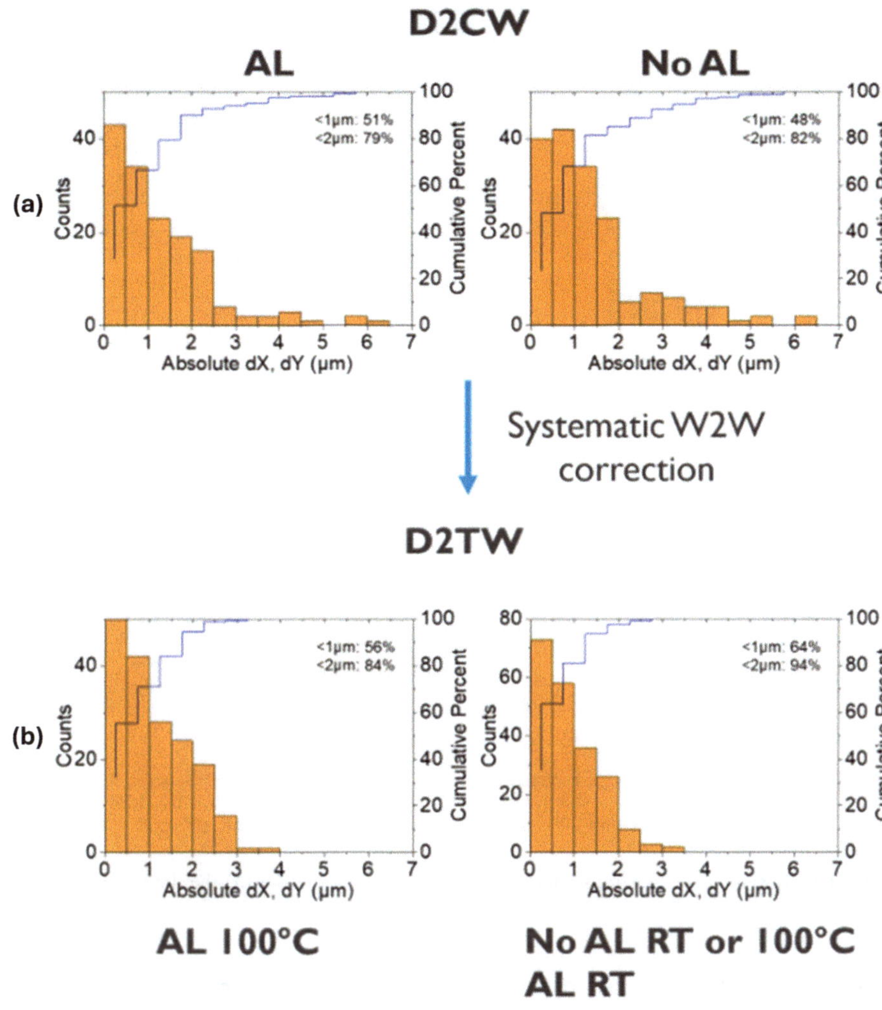


**Fig. 2.65** (**a**) Raw die-to-wafer translational alignment distributions for the die-to-carrier wafer. (**b**) Die-to-target wafer with or without acoustic layer and at room temperature or 100 °C as indicated

their scaling (S) performance and residuals ($\mu \pm 3\sigma$ and max values). It can be seen that for the Si-Si pair (system A) a mean ± 3 sigma ($\mu \pm 3\sigma$) of 36 ± 45nm residuals is obtained, with a maximum of 75nm, showing the base performance of the used wafer bonder tool for these types of wafers. Also, when adding a 30 µm temporary bond material (TBM) on one of the silicon wafers (System B), the $\mu \pm 3\sigma$ residual vector increases to 81 ± 150 nm.

Figure 2.65 shows the raw die-to-wafer translational alignment distributions for the die-to-carrier, Fig. 2.65a, and die-to-target wafer, Fig. 2.65b, with or without

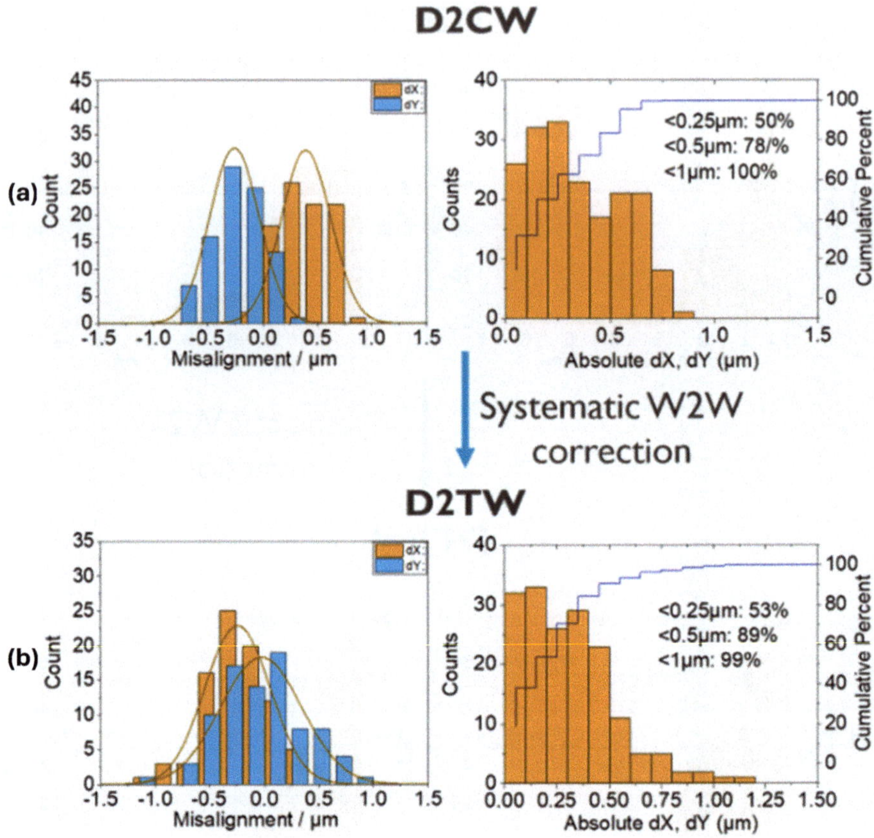


**Fig. 2.66** Raw die-to-wafer translational alignment distributions for (**a**) the die to-carrier wafer and (**b**) die-to-target wafer

acoustic layer (AL) and at room temperature or 100 °C. It can be seen that: (a) there is no significant difference observed between dies with or without acoustic layer on the die back side, (b) a significant effect could be observed for dies with acoustic layer bonded at 100 °C compared to the dies with acoustic layer bonded at RT, and (c) the D2CW distributions are nearly identical the D2TW distributions show a difference of about 10% in dies that have a better alignment than 2 µm.

Figure 2.66 shows the raw die-to-wafer translational alignment distributions for the die to-carrier wafer (D2CW), Fig. 2.66a, and die-to-target wafer (D2TW), Fig. 2.66b. It can be seen that: (a) below 1 µm, there is no significant difference observed between dies with or without acoustic layer on the die back side, (b) a significant effect could be observed for dies with acoustic layer bonded at 100 °C compared to the dies with acoustic layer bonded at RT, and (c) the D2CW distributions are nearly identical the D2TW distributions show a difference of about 10% in dies that have a better alignment than 2 µm.

## 2.6 Intel's Hybrid Bonding

### *2.6.1 Electrical Performance of CMP Process for Hybrid Bonding Application with Conventional/nt-Cu and Low Temperature of Si_xN_y/Si_xO_y Dielectrics*

During IEEE/ECTC 2025, Intel and Applied Materials presented at least five papers [23–27]. In [23], they presented electrical performance of CMP process for hybrid bonding application with conventional/nano-twinned (nt)-Cu and low temperature SixNy/SixOy dielectrics. In their study, they demonstrate that an optimal post-hybrid bonding anneal temperature, in conjunction with their CMP process, can achieve device yields exceeding 99%. Their samples are subjected to progressive thermal cycling (TC) in accordance with JEDEC-standard JESD22-A104D condition G.

Figure 2.67a shows the yield counts for daisy chain (10k vias) direct current (DC) resistance after different post-bonding anneal (PBA) temperatures. It can be seen that at 250 °C, nt-Cu with SixOy achieves a yield greater than 90% with resistance below 1 kΩ, in contrast to standard copper (std-Cu) with the same SixOy dielectric, which exhibits a yield of less than 25%. Upon increasing the PBA temperature to 300 °C, the nt-Cu DC yield improves to over 99%, whereas the std-Cu yield remains below 40%.

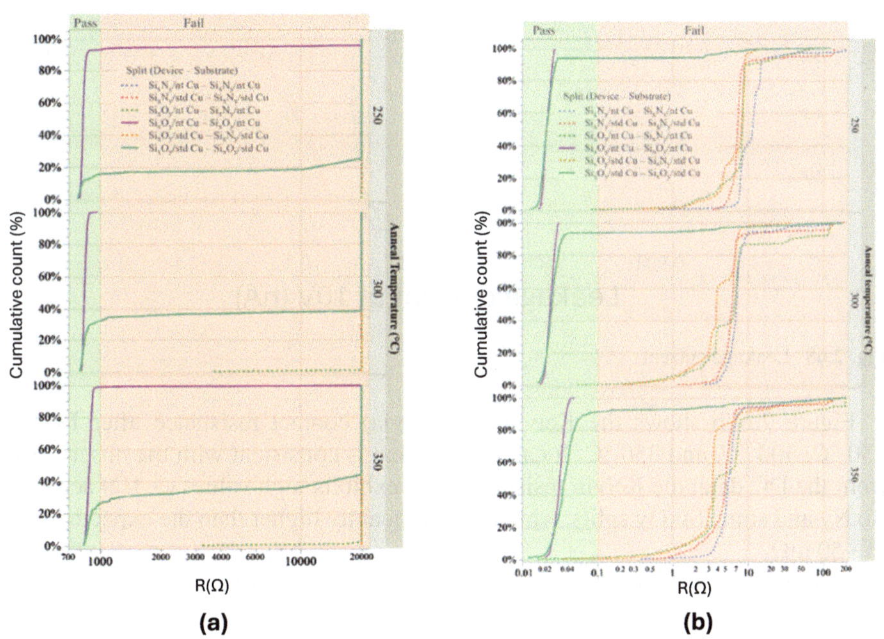


**Fig. 2.67** (**a**) Yield counts for daisy chain (10k via) resistance after different PBA temperatures. (**b**) Kelvin (single via) contact resistance after PBA of 250 °C, 300 °C, and 350 °C

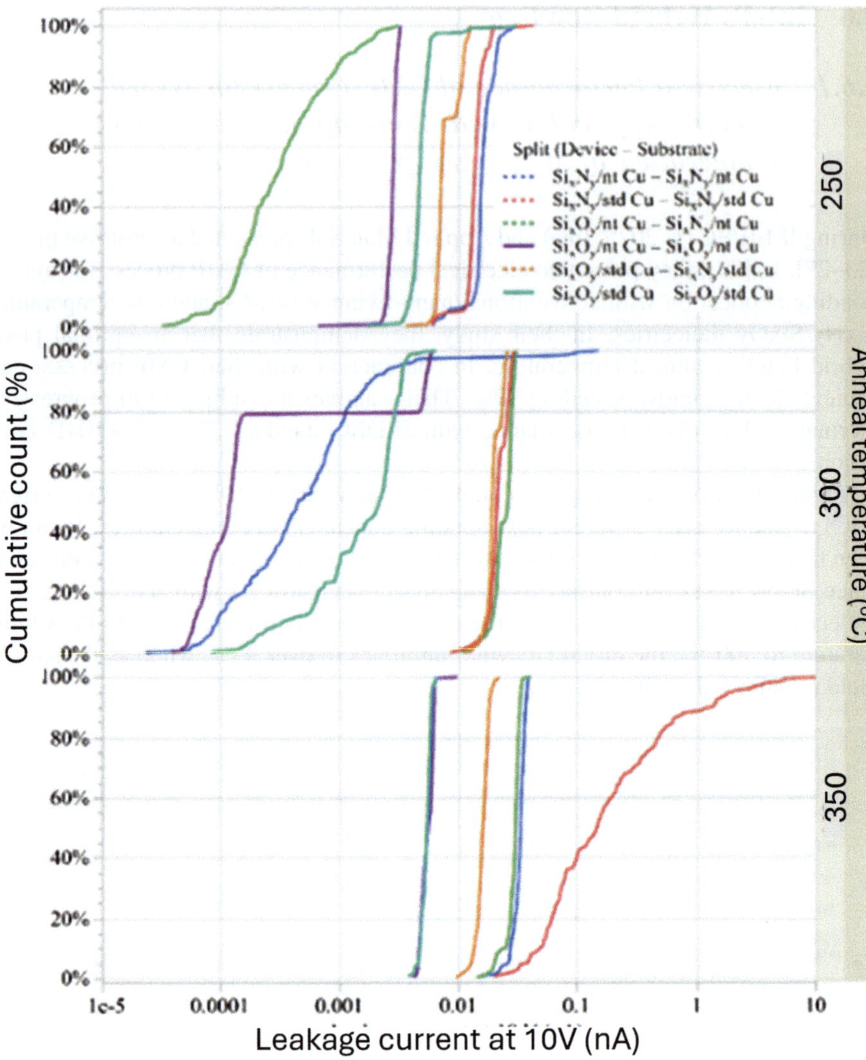


**Fig. 2.68** Leakage current

Figure 2.67b shows the Kelvin (single via) contact resistance after PBA of 250 °C, 300 °C, and 350 °C. It can be seen that is consistent with the observations from the DC data, the Kelvin resistance also exhibits high values (> 1Ω) across all SixNy and some SixOy splits, which is significantly higher than the expected value of ~50 mΩ.

Figure 2.68 shows the leakage current which is measured to be <10 nA at an applied potential of 10 V for all combinations of metallization and dielectric. It can be seen that, by focusing on the two yielding pairs of wafers with SixOy dielectrics

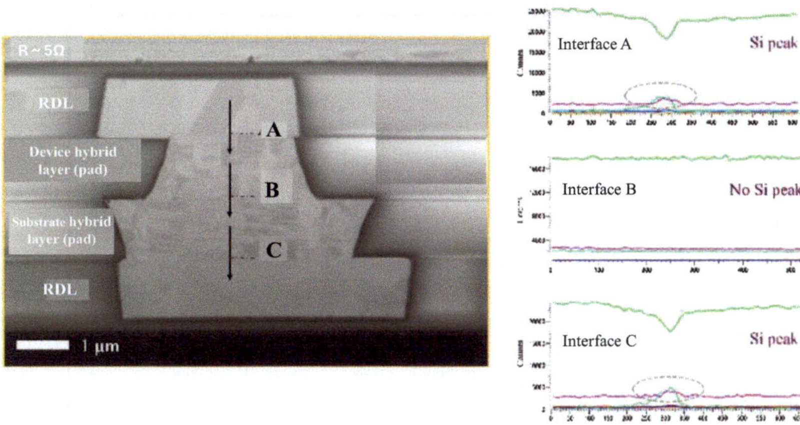


**Fig. 2.69** Cross-sectional of the Kelvin structure and EDX line-scan sampling location

on both the device die and substrate wafer, the leakage current partially improved at 300 °C before returning to baseline values after annealing at 350 °C.

Multiple yielding samples with Kelvin resistance values ranging from 30 mΩ to 163 Ω are examined using EDX. Figure 2.69 shows the cross-section of the Kelvin structure and EDX line-scan sampling location. Line-scan data is collected at three locations across the interface region of the device: the interface of the device redistribution layer (RDL) and the device side of the bonding pad (A), the interface of the device bonding pad and the substrate bonding pad (B), and the interface of the substrate bonding pad and the substrate RDL (C). It can be seen that there is a distinct elemental silicon peak at interfaces A and C, but not at the bonding interface B. Also, from the table in Fig. 2.69, it is evident that for all moderate and high resistance values, there is a presence of the silicon peak at interface A and/or C. Conversely, interface B consistently lacks the silicon peak across all splits. This is the phenomenon to the presence of silicon-containing etch residue.

Figure 2.70 shows the thermal cycling reliability performance of 10-k via DC structure for (a) SixOy/ nt-Cu – SixOy/nt-Cu (top) and (b) SixOy/Std-Cu – SixOy/Std-Cu (bottom) splits. It can be seen that (a) for both SixOy/nt-Cu – SixOy/nt-Cu and SixOy/Std-Cu – SixOy/Std Cu splits their daisy chain resistance degraded, (b) there is about 50% rate of degradation for every thermal cycling increment for SixOy/nt-Cu – SixOy/nt-Cu pair, while the rate of degradation of SixOy/Std-Cu – SixOy/Std-Cu pair is slower (about 5–10%), and (c) the degradation distribution of TC250 and TC500 look similar for both pairs.

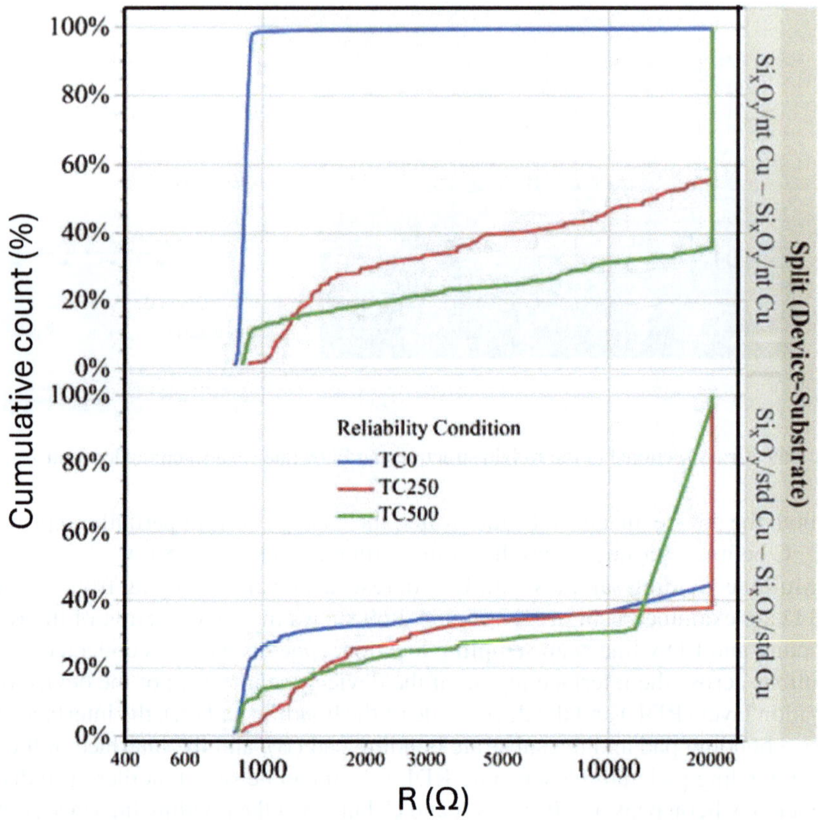


**Fig. 2.70** Thermal cycling reliability performance of 10-k via DC structure for SixOy/nt-Cu – SixOy/nt-Cu and SixOy/Std-Cu – SixOy/Std-Cu splits

### *2.6.2 Enabling Chip-to-Wafer Hybrid Bonding Scaling to 1 um Pitch With Optimal Power Delivery Using New Bond Via Architectures*

Intel presented another hybrid bonding paper on enabling C2W hybrid bonding scaling to 1 μm pitch with optimal power delivery using new bond via architectures [24]. They presented a new bond via architecture, multi-level via (MLVs), for the bonding layer by splitting the bond layer vias into two types of vias, namely (a) direct vias that land on the power delivery planes to minimize the power delivery losses, and (b) multi-level vias that land on the lower metal layers to enable continued hybrid bonding pitch scaling.

Figure 2.71a shows the cross section of a D2D interconnect and (b) the standard die backend for hybrid bonding showing the power delivery section and the die to die signaling. Figure 2.71c shows the thinned top metals to avoid exceeding a

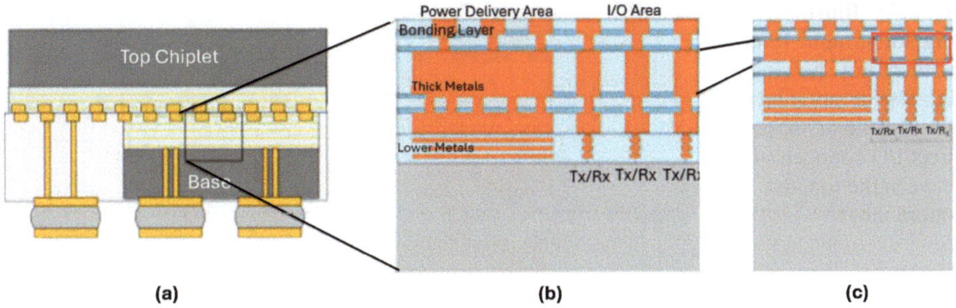


**Fig. 2.71** (**a**) A section of a die-to-die interconnect, (**b**) standard die backend for hybrid bonding showing the power delivery section, and (**c**) the die to die signaling

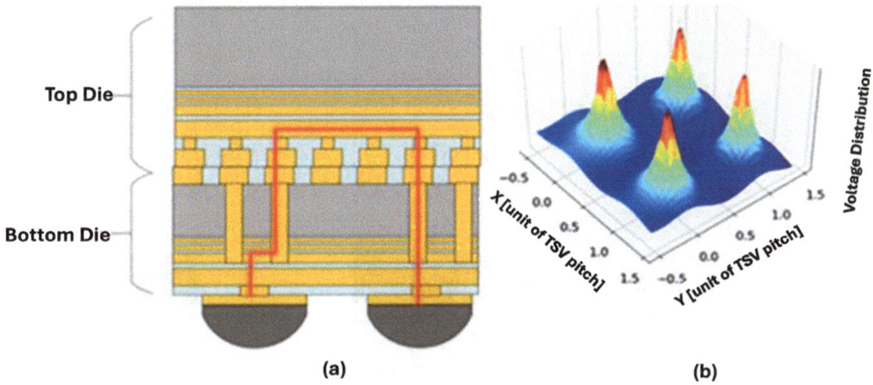


**Fig. 2.72** (**a**) Current path for power delivery in a 3D die stack. (**b**) The voltage distribution on the top die power plane between 4 TSV clusters

damascene fill processing aspect ratio. Figure 2.72a shows the current path for power delivery in a 3D die stack, going through the base die metals and the base die TSVs to reach the top die metals. Figure 2.72b shows the voltage distribution on the top die power plane between four TSV clusters (peaks at the 4 corners). Please note how the voltage drops in the area between the TSVs.

Figure 2.73 shows the illustrative diagram of MLV architectures showing the high aspect ratio and deep vias capable of supporting direct I/O connections between the top and bottom dies while still enabling thick metals for high efficiency power distribution. By utilizing a via that bypasses the thick planes and wires of the power grid, it is possible to decouple the interconnect density and the power delivery and radio frequency (RF) passives requirements and optimize each feature separately. Figure 2.74 shows the SEM images of MLV landing on Cu pads. Figure 2.74a is for aspect ratio (AR) ~ 10 post ash while Fig. 2.74b is for AR ~ 20 but pre ash. Figure 2.75 shows the barrier and seed layers. It can be seen that they are continuous coverage of over the entire structure including sidewall under hang region. Figure 2.76 shows the MLV landing on Cu pads. In both figures, the Cu filled vias are ~1 × 9 μm with 3 μm pitch.

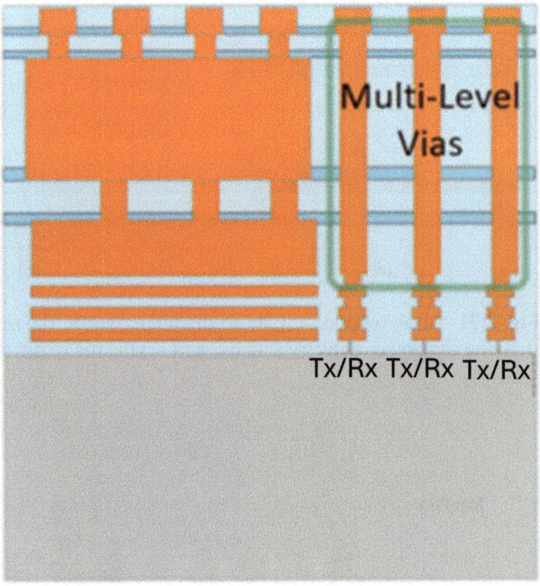


**Fig. 2.73** Illustrative diagram of multi-level via architectures showing the high aspect ratio and deep vias capable of supporting direct I/O connections between the top and bottom dies

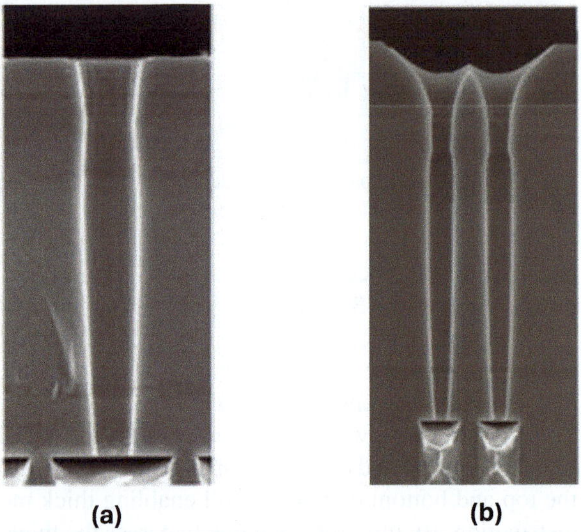


**Fig. 2.74** Multi-layer vias landing on Cu pads. (**a**) AR ~ 10 post ash, and (**b**) AR ~ 20 but pre ash

## 2.6.3 *Hybrid Bonding with Fluidic Self Alignment: Process Optimization and Electrical Test Vehicle Fabrication*

Intel presented another hybrid bonding paper on hybrid bonding with fluidic self-alignment: process optimization and electrical test vehicle fabrication [25]. In this study, they reported the progress on fluidic self- alignment to increase the throughput of D2W hybrid bonding. Their self-alignment (SA) hybrid bonding process flow

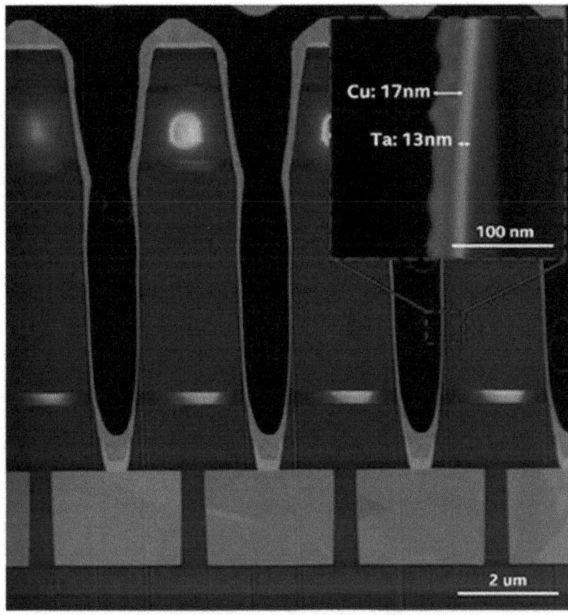


**Fig. 2.75** The barrier and seed layers show continuous coverage of over the entire structure including sidewall under hang region

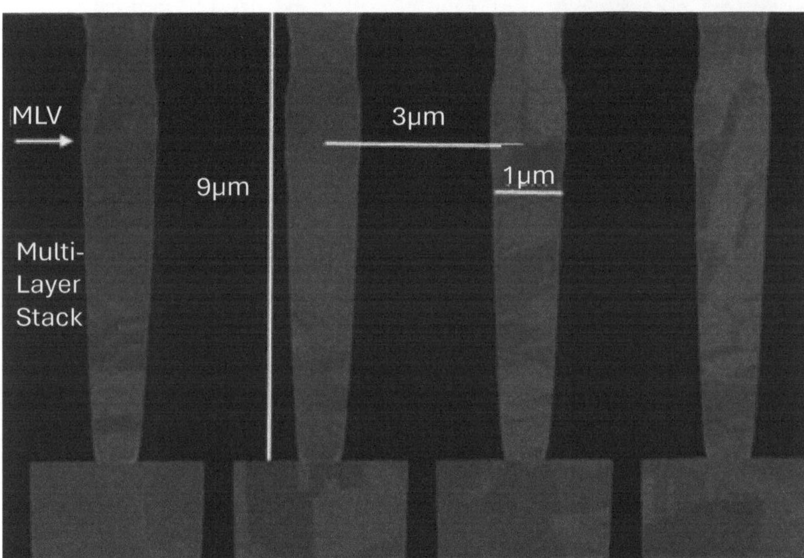


**Fig. 2.76** Multi-layer vias landing on Cu pads

is shown in Fig. 2.77. It can be seen that by leveraging the capillary action of fluid to enable fast, precise alignment of a top die to a bottom die or wafer. Figure 2.78 shows the side view schematics of test vehicles TV1, Fig. 2.78a, and TV2, Fig. 2.78b. TV1 is designed for fusion bonding, with the top layer containing oxide only. TV2 is designed for hybrid bonding, with the top layer containing both oxide and 3 µm

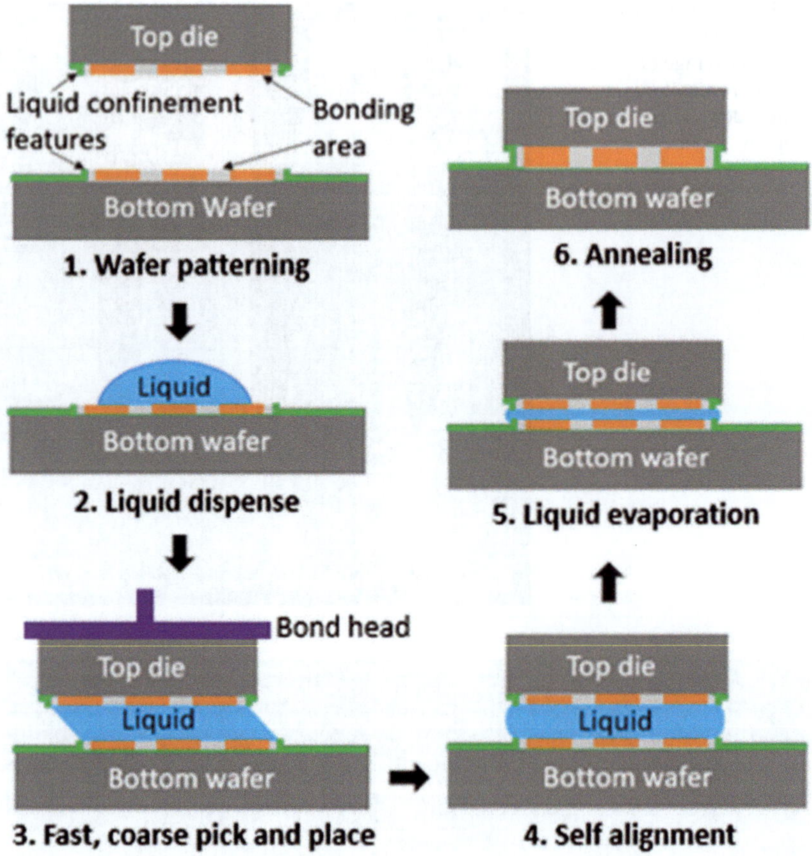


**Fig. 2.77** SA HB process steps: (1) wafer patterning, (2) liquid dispense, (3) die pick and place, (4) self-alignment, (5) evaporation, and (6) annealing

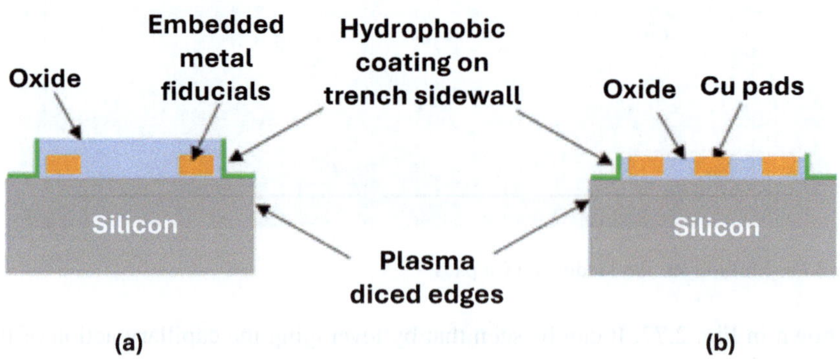


**Fig. 2.78** Side view schematics of TV1 (**a**) and TV2 (**b**)

**Fig. 2.79** Schematic of the multi-die SA bonding process: (1) incoming dies that may be misaligned on multi-die bond head, (2) transfer, and (3) capillary alignment and liquid evaporation

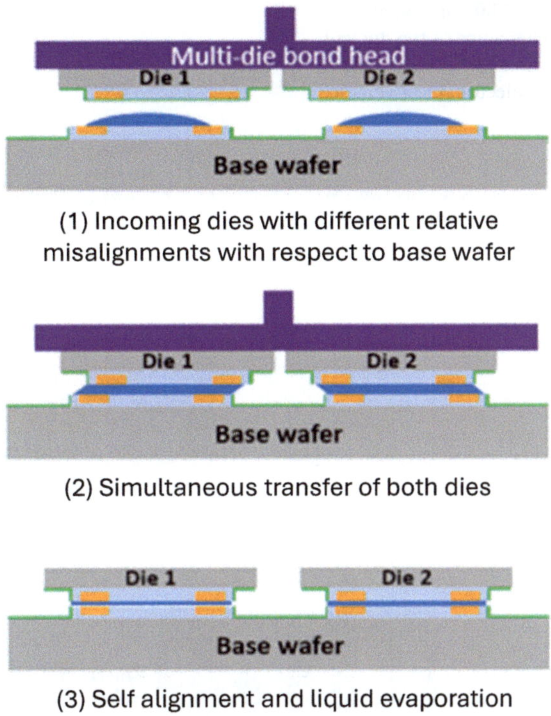


pitch Cu pads. Both test vehicles have self-alignment features in their respective top layers and are singulated to an identical chiplet size of 6.2mm × 7.7mm.

Figure 2.79 shows the schematic of the multi-die self-alignment bonding process: (1) incoming dies that may be misaligned on multi-die bond head, (2) transfer, and (3) capillary alignment and liquid evaporation. Figure 2.80 shows the cross section view of the top side and base wafer components of self-alignment electrical test vehicle. Figure 2.80a shows a cross-section of one of those assemblies, demonstrating successful Cu-Cu bonding with a misalignment of 13nm in the x-direction between the top die and base wafer. Figure 2.81 shows the electrical testing results (cumulative probability distributions) for D2W assemblies that are hybrid bonded with and without self-alignment. It can be seen that there are no major differences in electrical resistance magnitudes or distributions, indicating that the hydrophobic patterning and use of a fluid for self-alignment do not negatively impact the electrical connections across the interface.

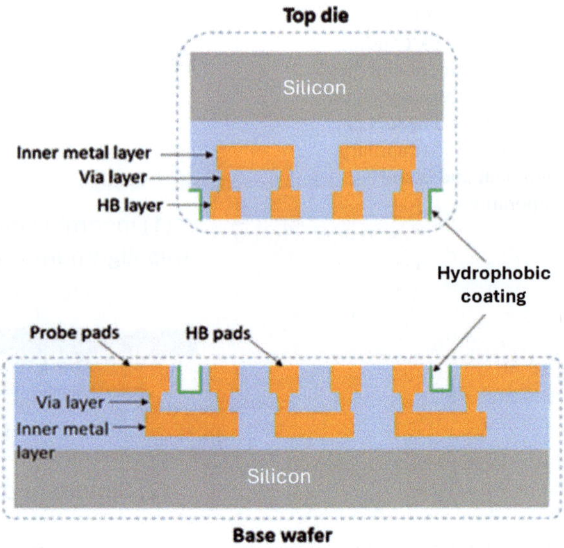


**Fig. 2.80** Side view schematics of top die and base wafer components of SA electrical test vehicle

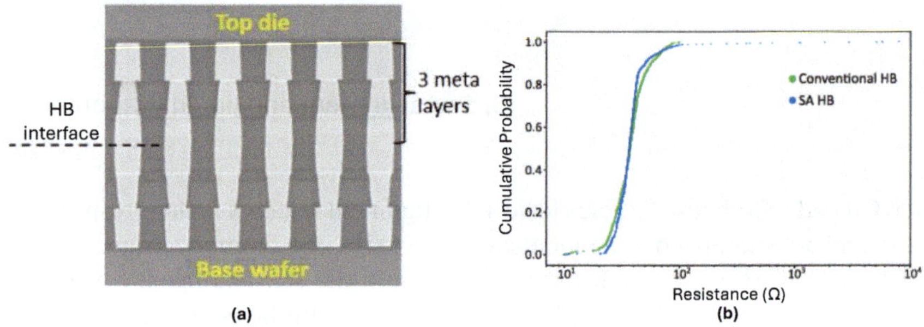


**Fig. 2.81** (**a**) Cross-section of an electrical test vehicle D2W assembly using SA. (**b**) Electrical testing results

## 2.7 Tohoku University's Hybrid Bonding

### *2.7.1 Characterization of Self-Nanoparticulated Cu-Cu Interconnection for Low-temperature Hybrid Bonding*

During IEEE/ECTC 2025, Tohoku University, Daicel Corporation, High Energy Accelerator Research Organization, Applied Physics Research Center, and Meiden Nanoprocess Innovations published at least four papers on hybrid bonding [28–31]. In [28], they presented characterization of self-nanoparticulated (SNP) Cu-Cu interconnection for low-temperature hybrid bonding. In this study, they propose and evaluate a novel bonding process with SNP by Cu-Cu direct bonding.

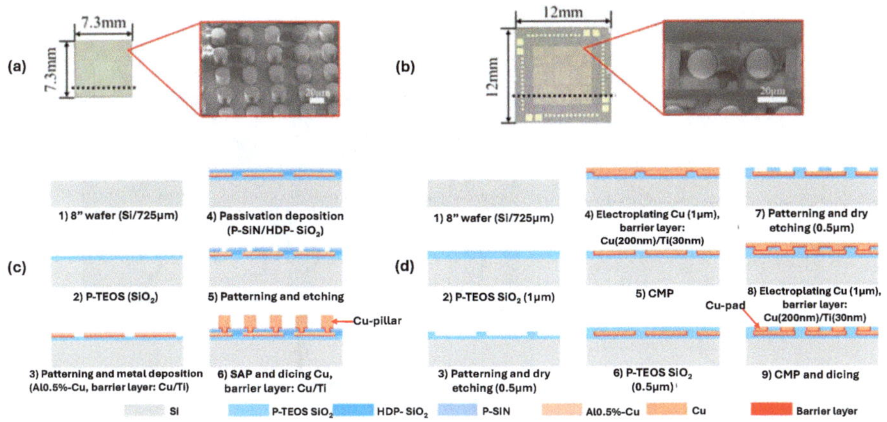


**Fig. 2.82** Top views and SEM images of top (**a**) and bottom (**b**) chip's pads, the chip fabrication process flow of top chip (**c**) and bottom chip (**d**) as observed from a cross-sectional view of the dotted line

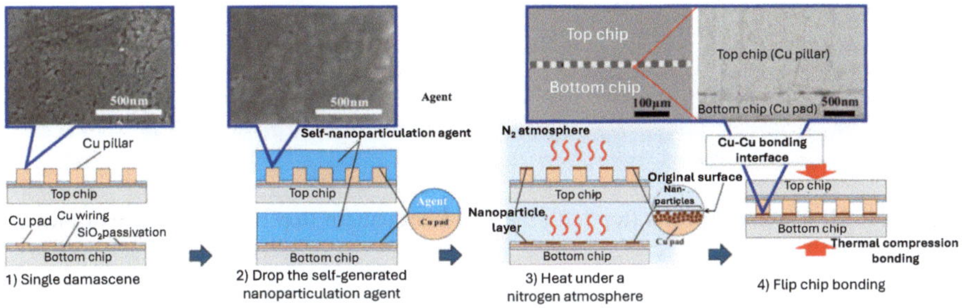


**Fig. 2.83** Bonding process flow with SNP

Figures 2.82 (a) and (b) show the SEM images of two kinds of chips for the SNP hybrid bonding evaluation. The top chip with a size of 7.3 mm square, and the bottom chip with a size of 12.0 mm square, are fabricated. The top chip has Cu pillars by semi-additive process (SAP) with a diameter of 20 μm, a height of 20 μm, and a pitch of 40 μm, Fig. 2.82c. The bottom chip has Cu pads with a diameter of 20 μm and a pitch of 40 μm, Fig. 2.82d. In total, a pair of chips has 28,224 pairs of Cu pillars and pads. Both chips are made from 725 μm-thick 8-inch Si wafers. Please note the structural difference between the top chip, Fig. 2.82c-6 and the bottom chip, Fig. 2.82d-9.

Figure 2.83 shows the bonding process with SNP. It can be seen that after SNP, the surface of the chips no longer shows the scratches and is covered with nanoparticles ranging from several tens to several hundreds of nanometers in size, Fig. 2.83 (2). The SNP process involves complex reactions between the Cu pads and the chemical agents upon heating, leading to the volatilization of components in the

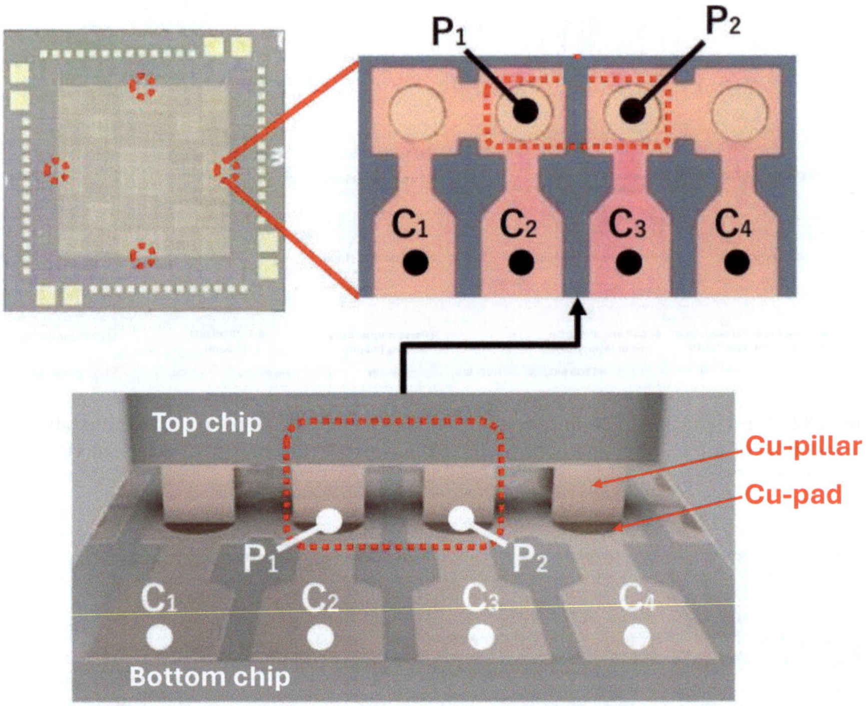


**Fig. 2.84** Measurement points (top left), measurement pattern (top right), 3D view of measurement pattern (bottom)

agents, the subsequent generation of Cu nanoparticles, and finally their growth to the desired size. The SNP is heated (180 °C/5 min or 150 °C/5 min → 200 °C/5 min) under a nitrogen atmosphere. Then, the chips are thermal compression bonded at 200 °C, 250 °C, or 350 °C. This SNP Cu-Cu hybrid bonding is different from the conventional Cu-Cu hybrid bonding. Figure 2.84 shows a 3D view of a SNP Cu-Cu hybrid bonding D2D assembly.

### *2.7.2 Are Voids Restricted to Cu-Cu Bonding Interface? Truth Revealed by Conversion Electron Yield-Scanning Transmission X-ray Microscopy*

Tohoku University and High Energy Accelerator Research Organization also published another paper on are voids restricted to Cu-Cu bonding interface? Truth revealed by conversion electron yield scanning transmission x-ray microscopy [29]. In this study, they proposed and demonstrated the cutting-edge conversion electron yield – scanning transmission x-ray microscopy (CEY-STXM), an authentic method

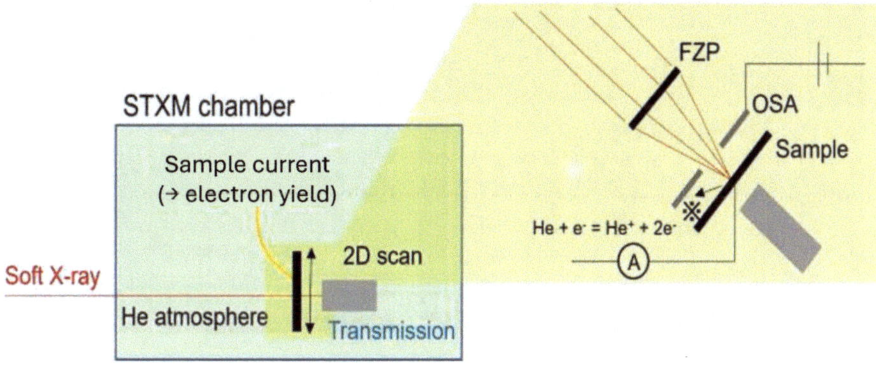


**Fig. 2.85** Schematic diagram illustrates the measurement principle

**Fig. 2.86** SAT image of C-t-C hybrid bonded chip used for CEY STXM

to precisely locate the voids/delamination present in the Cu-Cu μ-joints formed by C2W hybrid bonding using synchrotron-radiation has been proposed and demonstrated. Figure 2.85 Schematically illustrates the measurement principle. It can be seen that the electrons are efficiently collected by applying +100V voltage between the order selecting aperture (OSA) and the sample.

Figure 2.86 shows the scanning acoustic tomography (SAT) image of D2D hybrid bonded chips used for CEY STXM. Despite the whole chip area is electrically connected, one can clearly see the presence of voids in the SAT image. Upon carefully looking into these voids, it can be observed that all the voids are with different contrasts (as marked as number 1–4 in black circles). This appearance of different contrasts may be due to the fact that all these voids are at different layers of the Cu-joint.

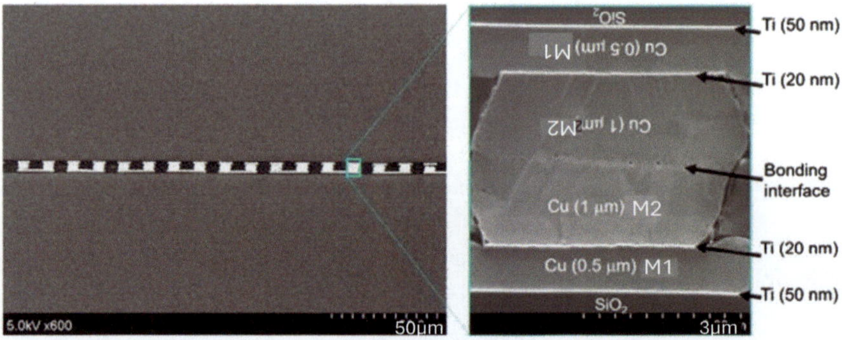


Fig. 2.87 X-sec. SEM image of C-t-C hybrid bonded chip used for CEY-STXM

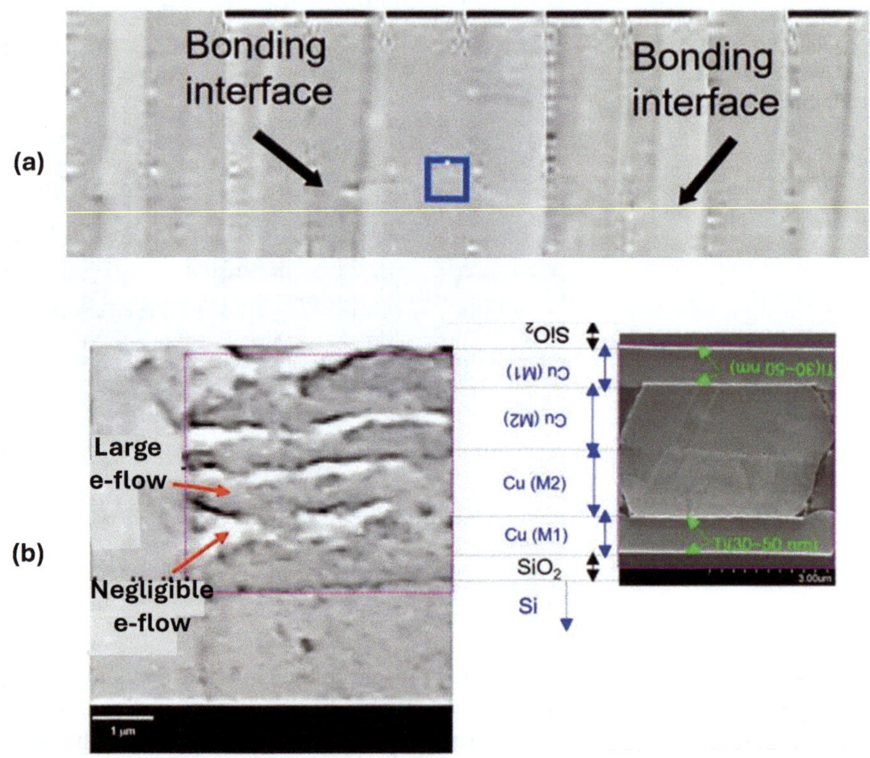


Fig. 2.88 (a) A scanned image of 350 μm × 350 μm area. (b) A scanned image of 8 μm × 8 μm area, and equivalent SEM image

Figure 2.87 shows the cross SEM image of D2D hybrid bonded chips used for CEY-STXM. The right side image in Fig. 2.87 is the magnified view of single Cu μ-joint. It is inferred from the SEM images that bonding between Cu bond-pads are perfect, and the bonding interface is free from delamination. Figure 2.88a shows the scanned STXM image obtained under CEY mode of 350 μm × 350 μm area using the x-ray energy equivalent to Cu-L(III) edge absorption energy 932eV under the conversion electron yield mode.

Figure 2.88b shows the enlarge view of the blue square in Fig. 2.88a. It can be seen that (a) the white regions correspond to the insulating region where there is negligible election flow which is attributed to the void/delamination, (b) the delamination occurs from above the M1/M2 interface of the bottom Cu bond-pad to the M2/M1 interface of the top Cu bond-pad, and (c) the substrate/P-TEOS oxide interface and oxide/barrier (Ti)/M1 interfaces are intact.

## 2.8 IME's Hybrid Bonding

### *2.8.1 Hybrid Bonding With Particle Accommodation Using Polymer Dielectric: Design, Process and Yield Study*

During IEEE/ECTC 2025, IME, Toray, Applied Materials, Aurora Technology Center, and Entegris, published at least ten papers on hybrid bonding [32–41]. In [32], IME and Toray presented a hybrid bonding with particle accommodation using polymer dielectric: design, process and yield study. In this study, they introduce a low temperature-curable polymer dielectric designed for hybrid bonding using mechanically diced chips. Figure 2.89 shows the conventional oxide-oxide C2W bonding. Figure 2.89a shows the schematic drawing of particle contamination causing void in bonding interface, and Fig. 2.89 (b) and (c) show the C-mode scanning

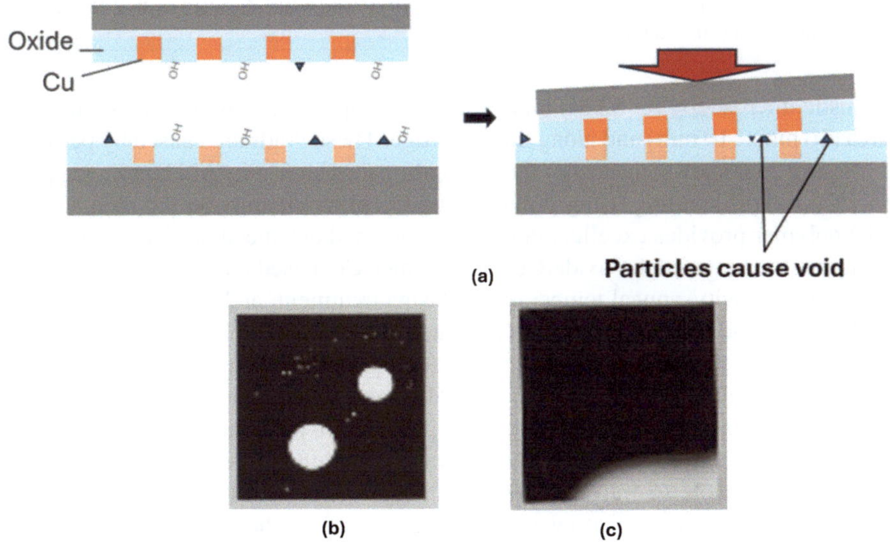


**Fig. 2.89** C2W bonding, (**a**) schematic drawing of particle contamination causing void in bonding interface, and C-SAM image shows voids (**b**) within the chip area and (**c**) along the chip edge

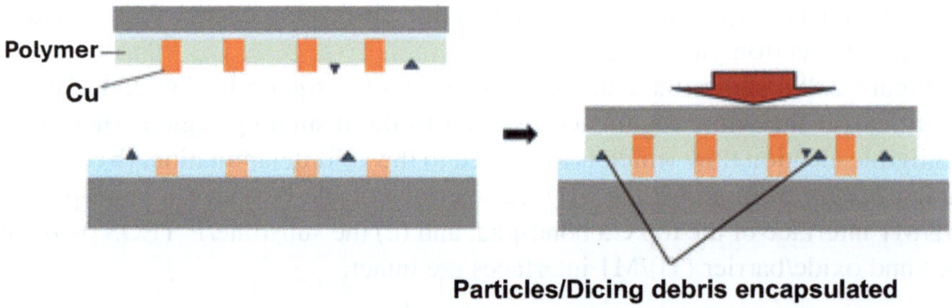


**Fig. 2.90** Illustration of polymer dielectric used in assembly

**Table 2.1** Comparison between organic and inorganic material hybrid bonding

| | Hybrid Bonding Methods | |
|---|---|---|
| | Oxide—Copper | Polymer—Copper |
| Bond-pad pitch | 1 μm or sub-micron Cu pad alignment | 3–10 μm |
| Particle sensitivity | Highly sensitive to particles | Accommodating particles |
| Anneal temperature | ≥300 °C for annealing and Cu expansion (The Cu expansion distance should be sufficient to bridge the gap in the bonding interface cause by Cu dishing during CMP an establish contact.) | 230–250 °C for polymer cure, polymer covalent bond formation and Cu diffusion (The protruding Cu bump is already in contact with the UBM on the substrate during tacking, so no additional Cu expansion is needed.) |
| Plasma treatment | To activate the oxide surface before hybrid bonding | To remove Cu oxide on the Cu bump before bonding |
| Suggest application | Logic/SRAM with 1 μm or sub-micron pitch | HBM and multi-die stack with 3–20 μm |

acoustic microscopy (C-SAM) image of voids (appearing in white) within the chip area (10 mm × 10 mm) and along the chip edge. These voids are caused by particles.

Figure 2.90 shows, instead of the conventional oxide/Cu hybrid bonding, a polymer/Cu hybrid bonding using polymer and Cu pillars/bumps on the chip surface. The polymer provides excellent defect coverage and enhanced particle tolerance. A compression between the oxide-Cu and polymer-Cu, based on their bond pad pitch, particle sensitivity, anneal temperature, plasma treatment, and application is shown in Table 2.1 and the one for the present study is shown in Table 2.2.

Figure 2.91 shows the process flow in identifying the potential polymer material, followed by design of the test vehicle, wafer fabrication, mechanical dicing, automated optical inspection (AOI) to scan the particles on the bonding surfaces, surface treatment by plasma and deionized (DI) water cleaning before the tacking process.

The test vehicle (6 mm × 13 mm memory) is shown in Fig. 2.92a, and it is mounted on a 300 mm silicon wafer by the process flow shown in Fig. 2.92b. It can be seen that the memory chip is coated with a polymer and then CMP. Figure 2.93 (a), (b), and (c) show the Cu-Cu interconnect after annealing. It can be seen that

**Table 2.2** Developed polymer dielectric material properties

| | Design Target | Actual Parameters |
|---|---|---|
| Curing Condition | <250 °C | 230 °C for 60 mins |
| Tg (°C) | >200 | 150, 280 |
| $Td_5$ (°C) | High | 350 |
| CTE (ppm/K) | Low | 55 |
| Elongation (%) | – | 100 |
| Modulus (GPa) | – | 1.7 |
| Design | – | Low modulus type |

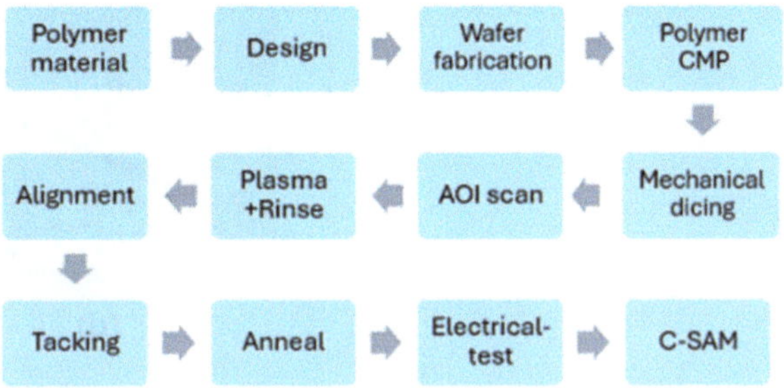


**Fig. 2.91** Evaluation process flow

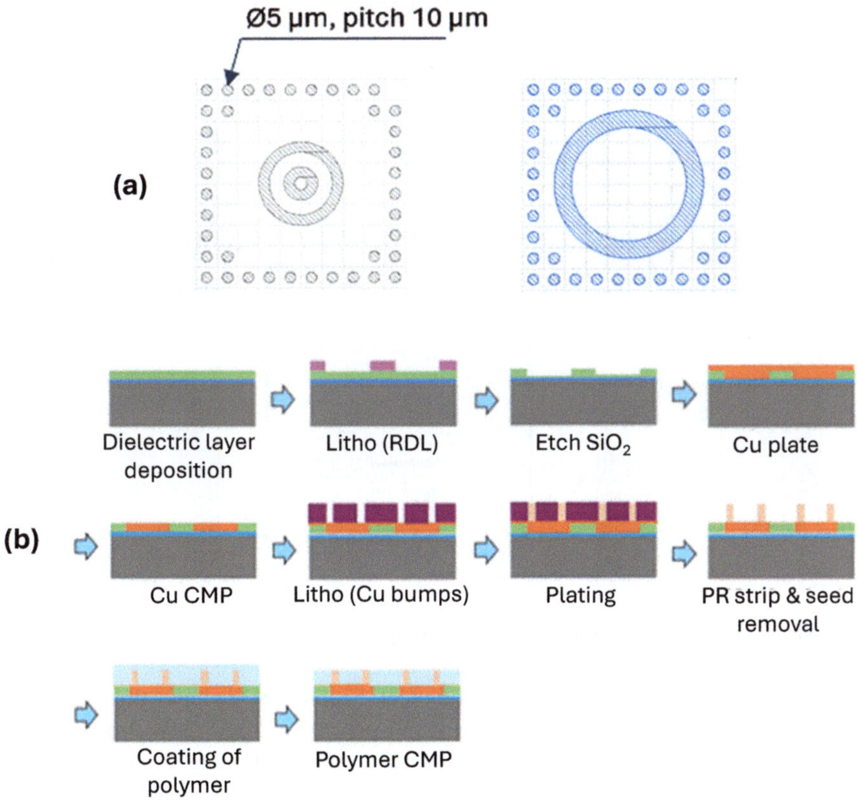


**Fig. 2.92** (**a**) Fiducial mark for bonding on (L) chip (R) interposer. (**b**) Memory wafer fabrication process flow

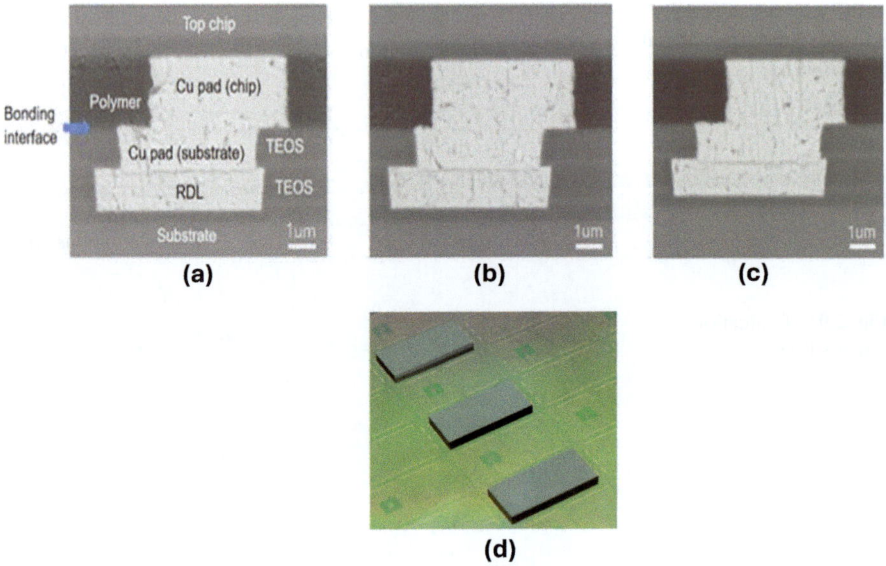


**Fig. 2.93** DOE of key bonding parameters result, (**a**) Bonding Case 1. (**b**) Bonding Case 2. (**c**) Boding Case 3. (**d**) Memory chips bonded on interposer

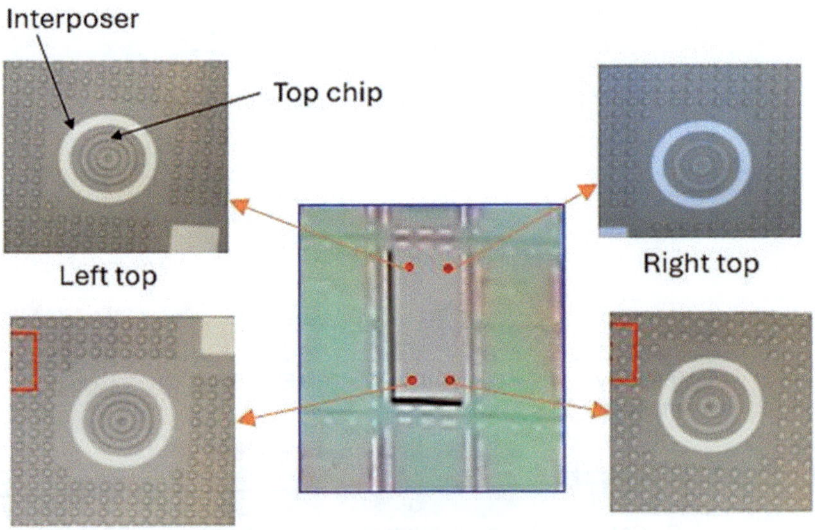


**Fig. 2.94** IR image of chip corners after HB bonding

there is no void within the Cu-Cu interconnect. Figure 2.93d shows the daisy-chain memory chips are bonded to the silicon wafer properly. Figure 2.94 shows the very well alignment accuracy of C2W hybrid bonding with polymer.

### 2.8.2 *Novel Selective Metal Capping on Cu pad Enabling Void-Free Hybrid Bonding at Low Thermal Budgets, Irrespective of Cu Topography and Microstructure*

IME published another paper on hybrid bonding, novel selective metal capping on Cu pad enabling void-free hybrid bonding at low thermal budgets, irrespective of Cu topography and microstructure [33]. They introduced a novel approach to achieving hybrid bonding below 250 °C by implementing an innovative metal passivation engineering technique for Cu bond pads.

Figure 2.95 shows challenges in conventional Cu/dielectric hybrid bonding with potential void formation sites: (i) stress-induced crack propagation at the dielectric and dielectric-to-barrier interface, (ii) overlapping region between the barrier and Cu, (iii) voids at the Cu-Cu interface caused by surface roughness, (iv) voids resulting from galvanic corrosion, and (v) overlapping region between Cu and the dielectric. Figure 2.96 shows the test vehicle fabrication process using the damascene process flow for an ultrathin layer metal passivation selective to the Cu pad as highlines in Fig. 2.96 (h) and (i).

Figure 2.97 shows the AFM employed to analyze surface topography and morphological variations. Figure 2.97 (a) and (b) are for Cu surface without passivation while Fig. 2.97 (c) and (d) are for Cu surface with passivation. It can be seen that (a) the un-passivated Cu surface exhibits a non-uniform topography with a dishing depth of approximately 5–7 nm, as shown in Fig. 2.97 (a) and (b), (b) a pronounced increase in dishing is observed at the Cu-barrier layer interface, likely resulting from galvanic corrosion during the CMP process, (c) the metal-passivated Cu pad exhibits a significantly smoother surface profile with dishing reduced to ≤ 1nm, with no evidence of excessive dishing at the Cu-barrier interface, as shown in Fig. 2.97 (c) and (d).

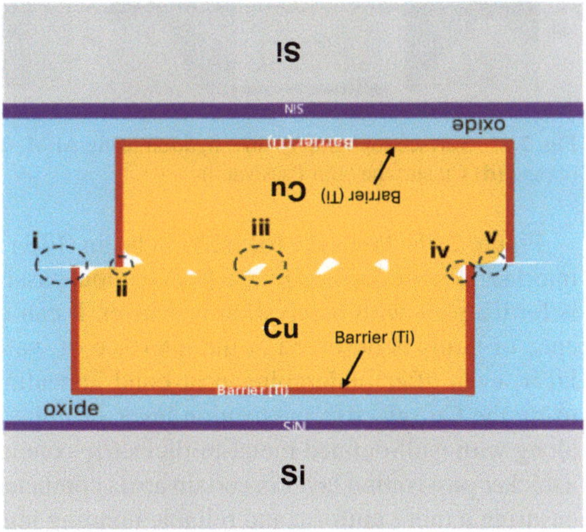


**Fig. 2.95** Challenges in conventional Cu/dielectric hybrid bonding with potential void formation sites

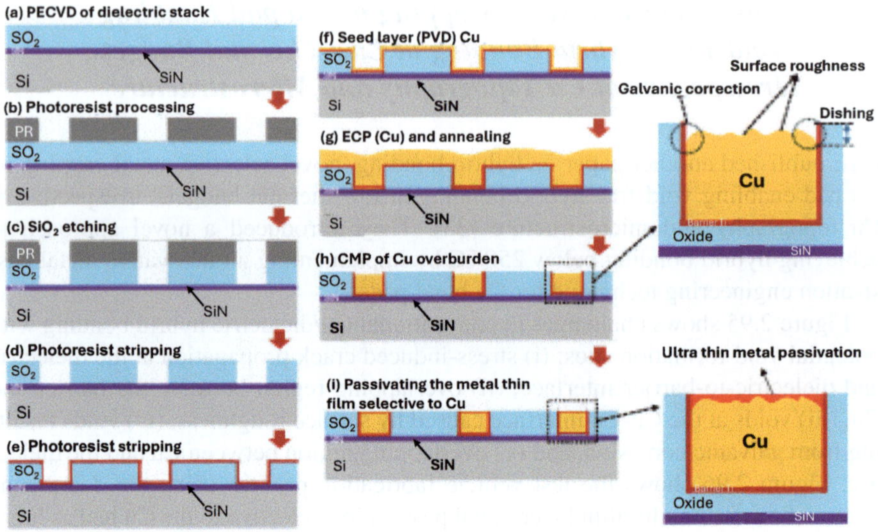


**Fig. 2.96** Test vehicle fabrication process using the damascene process flow

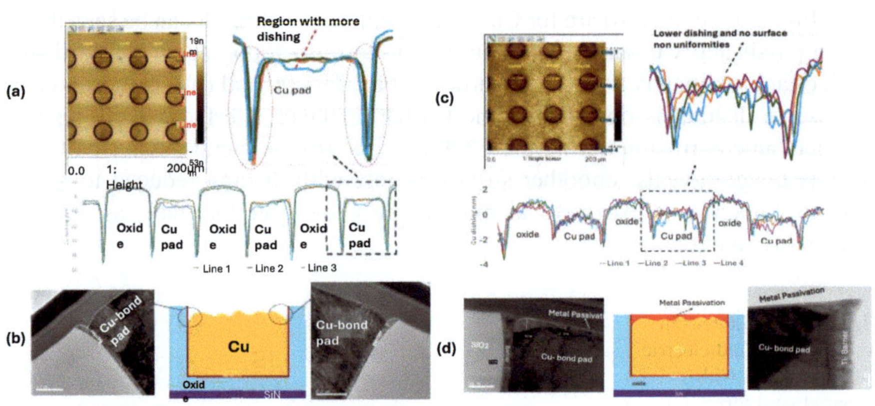


**Fig. 2.97** Surface topography investigation using AFM. (**a**) and (**b**) Surface without passivation. (**c**) and (**d**) Cu surface with Passivation

Figure 2.98 shows the focused ion beam (FIB) and TEM investigation of bonded interface. Figure 2.98a is for Cu pads without passivation interlayers while Fig. 2.98b is for Cu pads with passivation interlayer. It can be seen that (a) a significant presence of voids is observed at the interface of wafers bonded without a passivation layer, even after undergoing post-bond annealing at 350 °C, (b) wafers bonded using the Cu selective passivation layer exhibited a promising void-free interface, along with well-defined metal-to-dielectric contact regions, and (c) the presence of a thicker passivation layer in certain areas compensates for surface non-uniformities, ensuring a more uniform and reliable bonding interface.

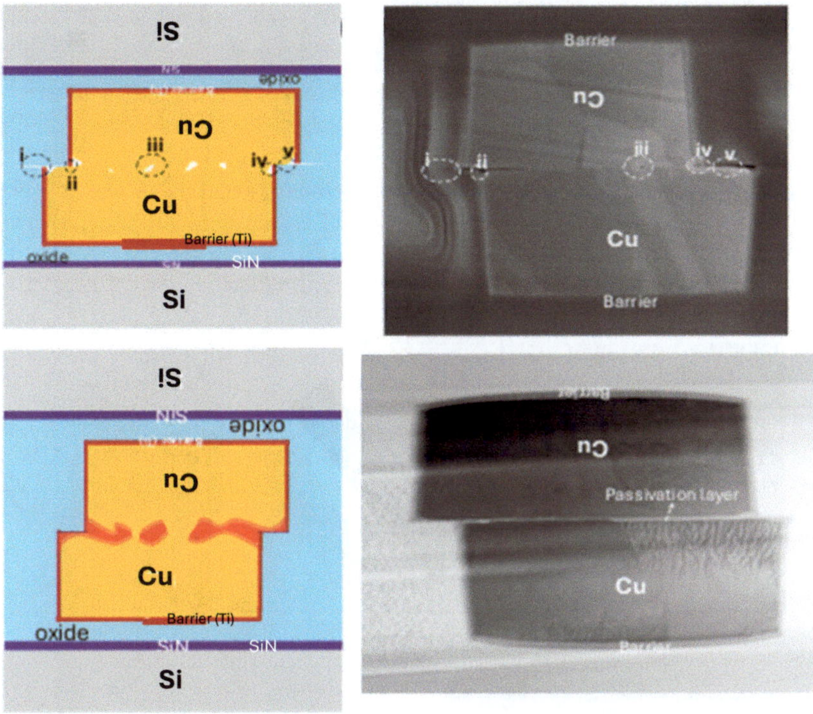


**Fig. 2.98** FIB and TEM investigation of bonded interface. (**a**) Without passivation interlayer. (**b**) With passivation interlayer

Figure 2.99 shows the interface inspection using TEM. Figure 2.99a is for high-resolution transmission electron microscopy (HRTEM) interface investigation and energy dispersive x-ray spectroscopy (EDS) line scan elemental analysis while Fig. 2.99b is for bonding mechanism with ultrathin passivation interlayer. It can be seen that (a) the five critical void- nucleation sites shown in Fig. 2.95, are completely mitigated through the implementation of an ultrathin metal passivation layer, (b) the bonded dielectric-to- metal overlap region exhibited no evidence of crack propagation voids, indicating a robust interfacial adhesion between $SiO_2$ and the passivation metal, and (c) a thinner passivation layer facilitates immediate copper diffusion, promoting direct Cu-Cu bonding. Figure 2.100 shows the interface inspection using TEM for sub-0.5 μm pitch. Figure 2.100a is for HRTEM interface investigation while Fig. 2.100b is for EDS elemental mapping. It can be seen that the process is not only suitable for larger pitch dimensions but is also extendable to fine-pitch interconnects below 0.5 μm.

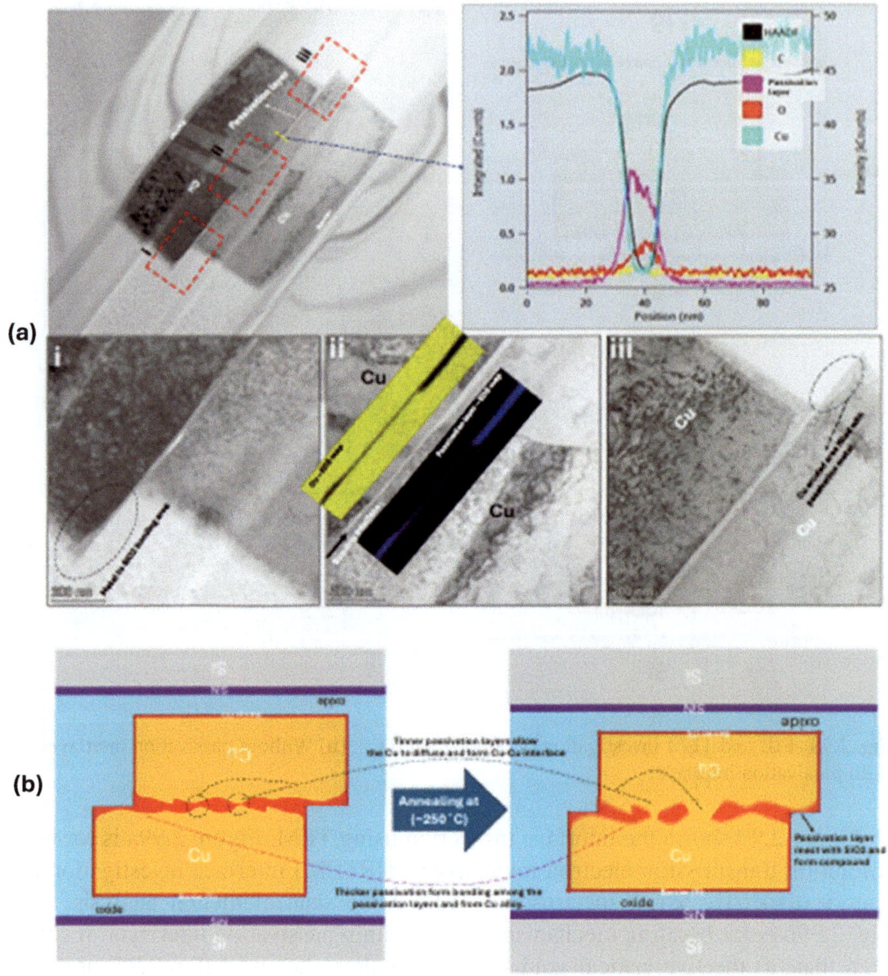


**Fig. 2.99** Interface inspection using TEM. (**a**). X-TEM interface investigation and EDS elemental analysis. (**b**) Bonding mechanism with and ultrathin passivation interlayer

## *2.8.3 Scalability and Process Optimization in Hybrid Bonding with Fine-Pitch Interconnections*

IME and Applied Materials presented another paper on hybrid bonding, scalability and process optimization in hybrid bonding with fine-pitch interconnections [34]. They used tetraethyl orthosilicate (TEOS) oxide and SiCN interface dielectric films to achieve void-free hybrid bonding. Figure 2.101 shows the schematic damascene

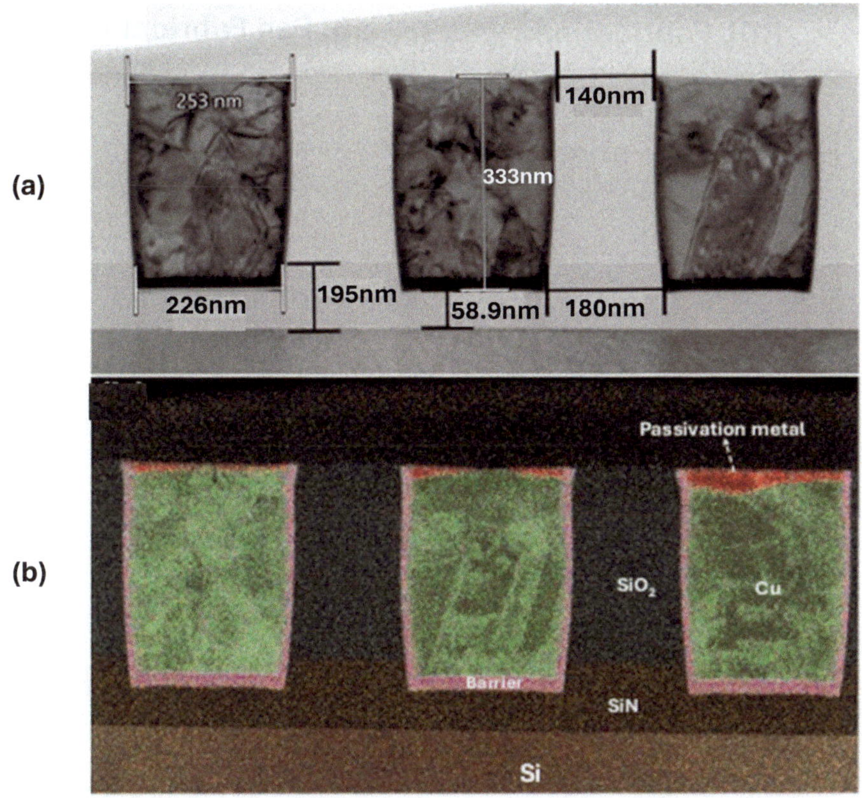


**Fig. 2.100** Interface inspection using TEM for sub-0.5 μm pitch (**a**) X TEM interface investigation. (**b**) EDS elemental mapping

process flow for fabricating the redistribution layer (RDL) and hybrid bonding interface (HBI) pad for both component and substrate wafers. Figure 2.102 shows the FIB SEM interface inspection. It can be seen that (a) HBI pads with micro trenching at Ti/Cu interface, Fig. 2.102a, and (b) hybrid bonding of HBI pads with Ti/Cu for oxide dielectric as bond layer. Figure 2.103 shows the FIB SEM interface inspection for 3 μm HBI pad with SiCN dielectric layer for CMP condition of 0 to 1 nm Cu protrusion. It indicates a perfect hybrid bonding with good alignment for Cu-Cu hybrid bonding.

Figure 2.104 shows the FIB SEM analysis for a C2W hybrid bonding for the optimized CMP condition. It can be seen that the C2W hybrid bonding without edge voids, no barrier spike and perfect alignment of HBI pads with Ta barrier layer. Figure 2.105 shows the FIB-SEM images of C2W hybrid bonding, (a) 6 μm bond pitch, (b) 10 μm bond pitch, and (c) 1 μm HBI pad with 2.4 μm bond pitch. It can be seen that perfect bonding alignment has been achieved.

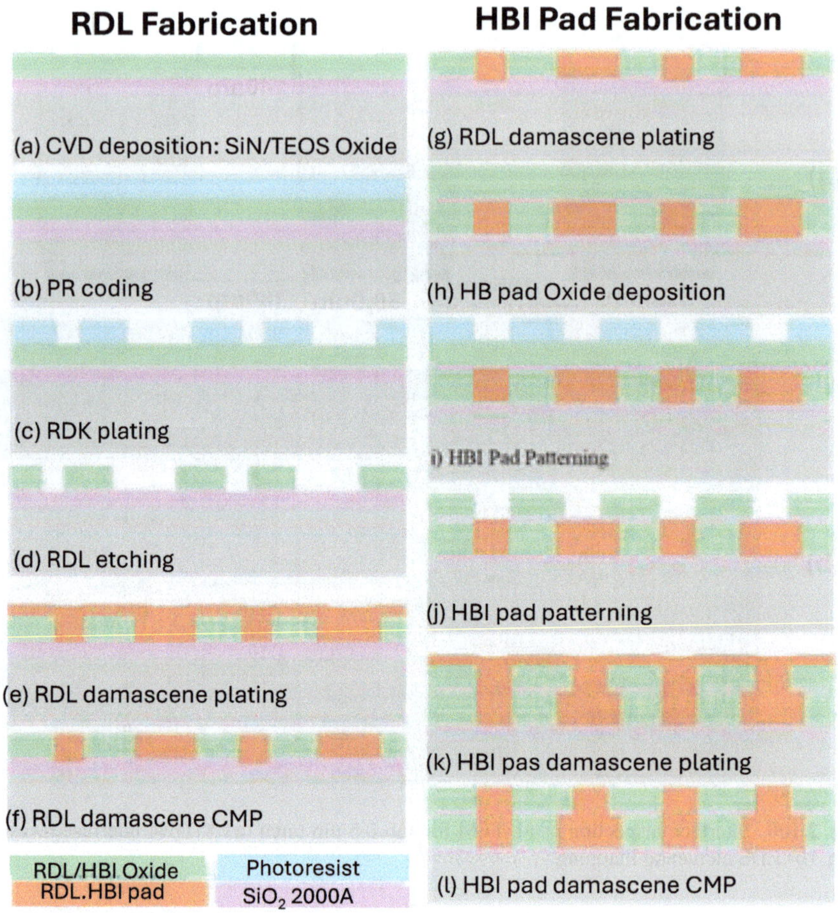


**Fig. 2.101** Schematic process flow of pre-bond component and substrate fabrication

## *2.8.4 Thermal Characterization of HBMs Integrated via Hybrid Bonding*

IME presented another paper on thermal characterization of HBMs integrated via hybrid bonding [35]. They reported two hybrid bonding process flows, double side fabrication (DSF) and inter-die gap filling (IDGF). Also, they analyze and compared the thermal performance of HBMs integrated through these two process flows. Figure 2.106 shows the (a) DSF process flow and (b) IDGF process flow. In the DSF flow, the TSVs, RDL, and HB pads are fabricated on both sides of chips first, then the chips are stacked via hybrid bonding, and then followed by over molding after stacking. In the IDGF flow, HB pads are fabricated on one-side of chips, bonded to the substrate first, then followed by dielectric film deposition and CMP of the

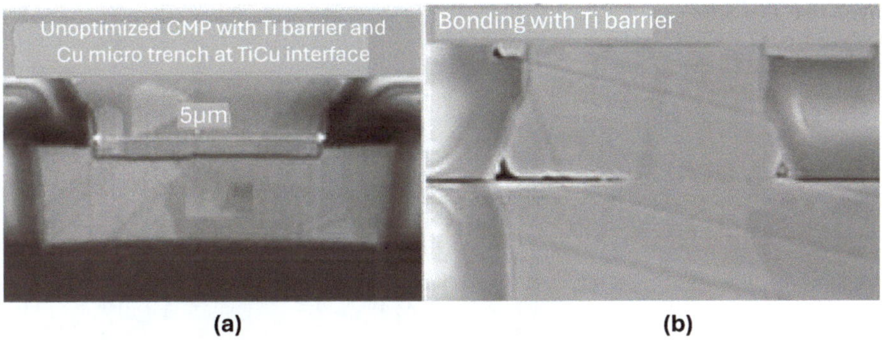


**Fig. 2.102** FIB SEM interface inspection. (**a**) HBI pads with micro trenching at Ti/Cu interface. (**b**) Hybrid bonding of HBI pads with Ti/Cu for oxide dielectric as bond layer

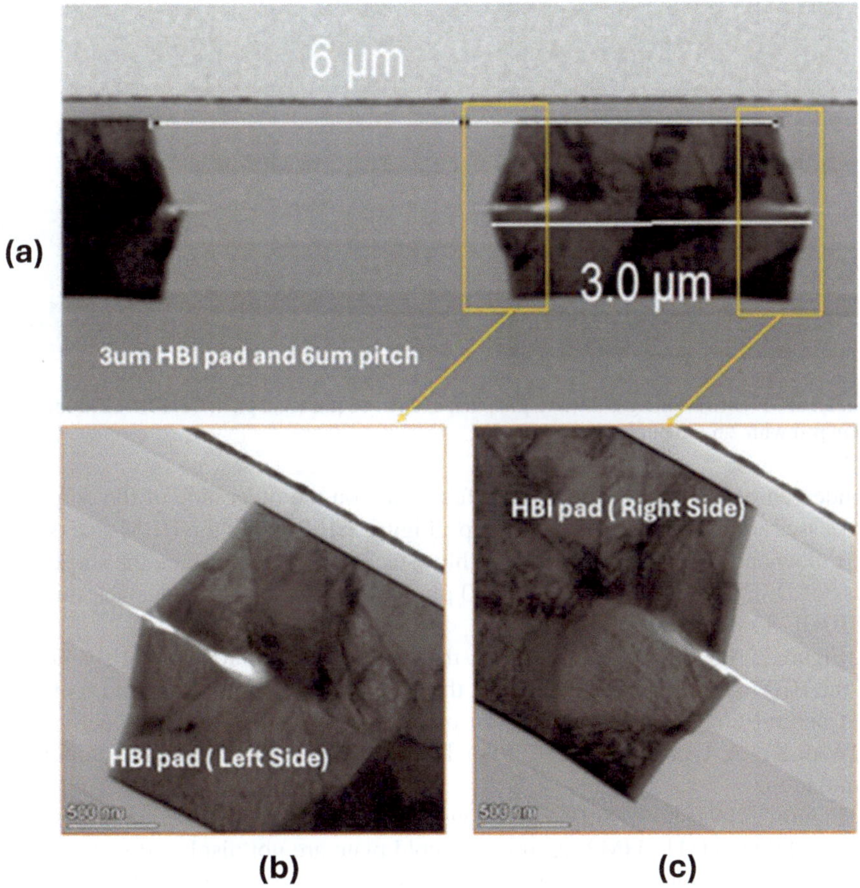


**Fig. 2.103** (**a**) FIB SEM interface inspection for 3 μm HBI pad. (**b**) Left-side HBI pad. (**c**) Right side HBI pad

**Fig. 2.104** FIB SEM analysis for mTV C2W hybrid bonding for the optimized CMP condition

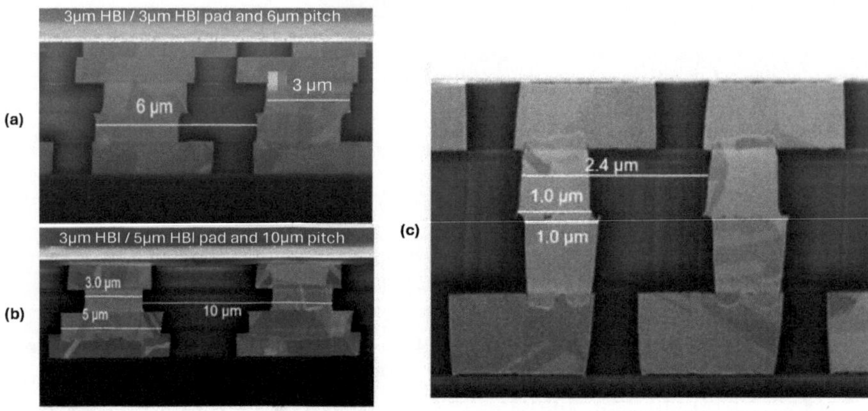


**Fig. 2.105** FIB-SEM images of (**a**) 6 μm pitch and (**b**) 10 μm bond pitch. (**c**) FIB-SEM images of 1 μm pad with 2.5 μm pitch

bonded chips, and then HB pads are fabricated on the other side of the chips. The processes are repeated to stack the chips. Figure 2.107 shows the HBM stacks of (a) DSF flow and (b) IDGF flow. For 12 chips stacked on the substrate, the stack thickness of DSF flow is 899 μm (with 100 μm top molding), while the stack thickness of IDGF flow is 418 μm.

Figure 2.108 shows (a) HBM stack thermal simulation model and (b) a unit cell of the HBM stack. In the stack model, the cold plate is kept at temperature of 25 °C and natural convection is applied on other sides (the heat transfer coefficient is 10W/m.K). A typical power (2.5W) is uniformly applied to each chip in the simulation.

Figure 2.109a shows the isometric view of temperature distribution of DSF-100. Top molding, TIM1, TIM2, casing, and cold plate are not displayed so as to show the chip temperature. Figure 2.109b shows the cross-sectional temperature

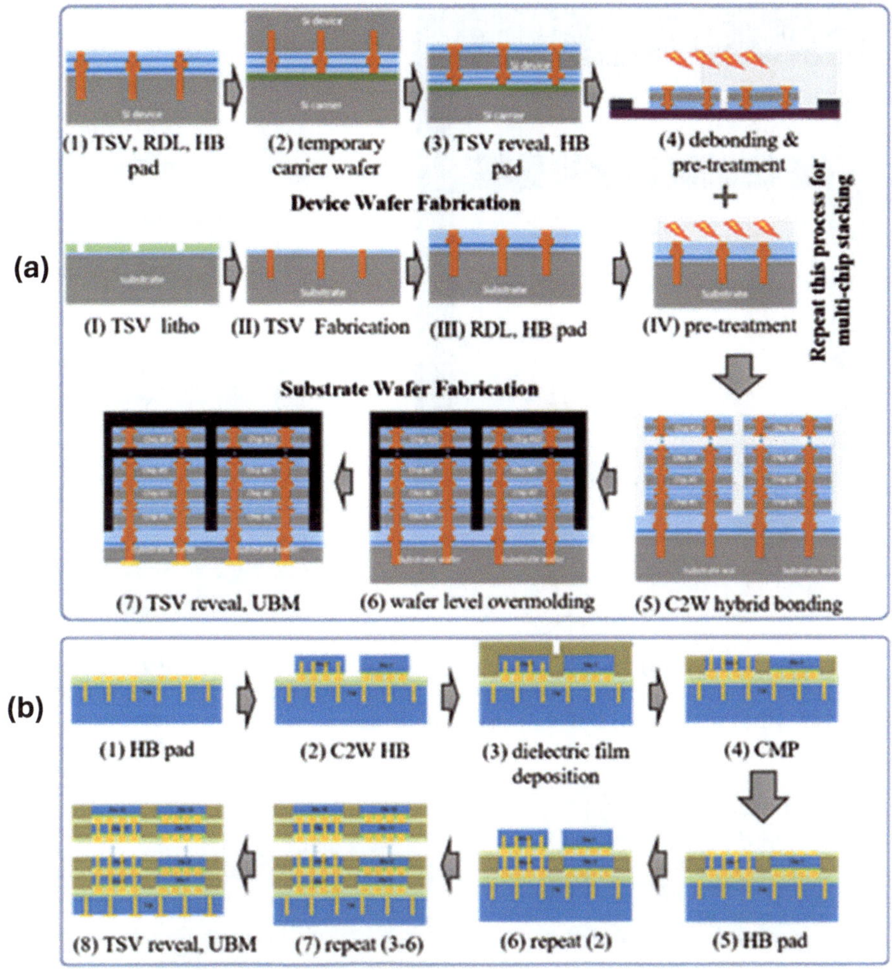


**Fig. 2.106** (**a**) DSF process flow and (**b**) IDGF process flow

distribution of DSF-100 while Fig. 2.109c shows the temperature rise across the layers of DSF-100. Figure 2.110a shows isometric view of temperature distribution of IDGF-12. TIM1, TIM2, casing, and cold plate are not shown here. Figure 2.110b shows the cross-sectional temperature distribution of IDGF-12 while Fig. 2.110c shows the temperature rise across the layers of IDGF-12. It can be seen that, comparing with the DSF flow, the IDGF flow achieves exceptionally lower thermal resistance.

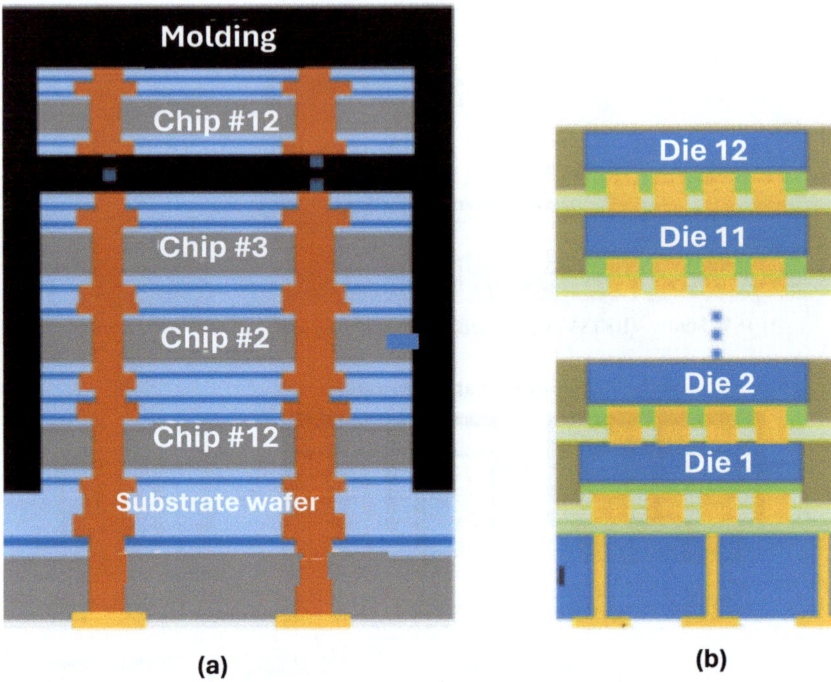


**Fig. 2.107** HBM stacks of (**a**) DSF flow and (**b**) IDGF flow

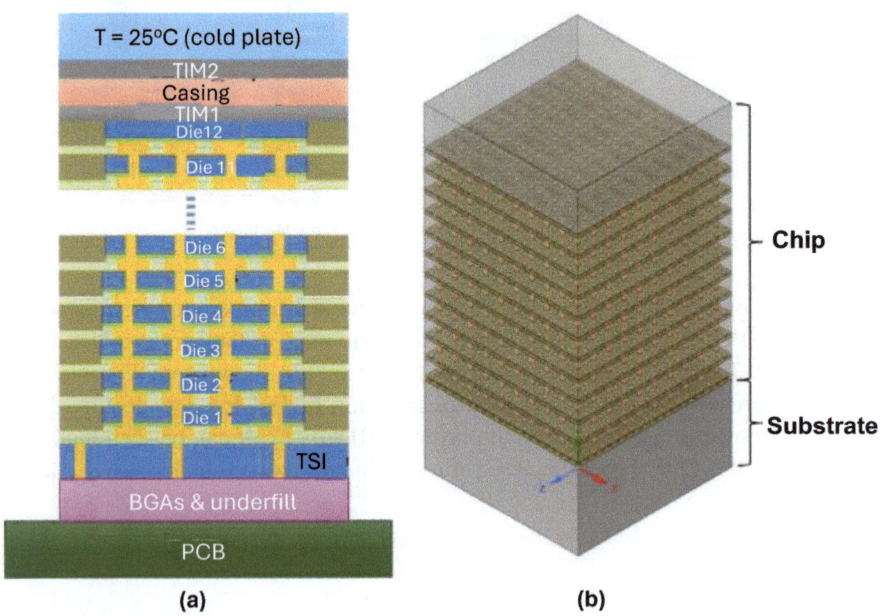


**Fig. 2.108** (**a**) HBM stack thermal simulation model and (**b**) a unit cell of the HBM stack

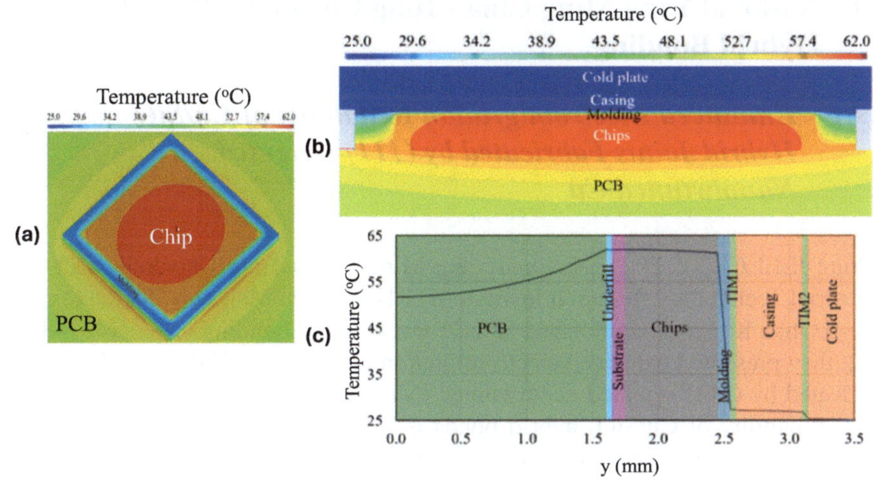


**Fig. 2.109** (**a**) Isometric view of temperature distribution of DSF-100. (**b**) Cross-sectional temperature distribution of DSF-100. (**b**) Temperature rise across the layers of DSF-100

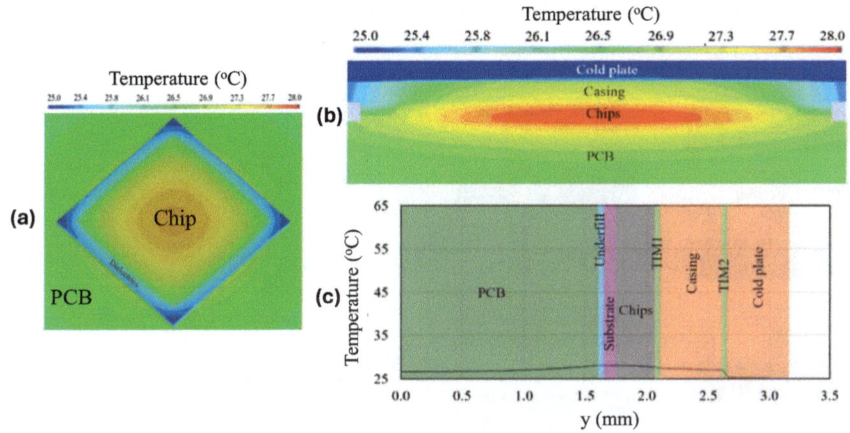


**Fig. 2.110** (**a**) Isometric view of temperature distribution of IDGF-12. (**b**) Cross-sectional temperature distribution of IDGF-12. (**c**) Temperature rise across the layers of IDGF-12

## 2.9 National Yang Ming Chiao Tung University (NYCU)'s Hybrid Bonding

### *2.9.1 Enhanced Electromigration Reliability of Cu/SiO$_2$ Hybrid Joints Fabricated by (111)-Oriented Nanotwinned Cu*

During IEEE/ECTC 2025, National Yang Ming Chiao Tung University (NYCU), Industrial Technology Research Institute (ITRI), Applied Materials, MediaTek, and Tokyo Ohka Kogyo Co. published at least seven hybrid bonding papers [42–48]. In [42], they presented enhanced electromigration reliability of Cu/SiO$_2$ hybrid joints fabricated by (111)-oriented nanotwinned (NT) Cu. In this study, electromigration (EM) reliability of Cu/SiO$_2$ hybrid bonds is investigated. Figure 2.111 schematically shows the sequences of NT-Cu/SiO$_2$ damascene process on a 300 mm wafer. It can be seen that the NT-CuSiO$_2$ is fabricated in Fig. 2.111 (h)–(k). Figure 2.112 shows the cross-sectional inspections on as-bonded NT-Cu/SiO$_2$ hybrid joints. Figure 2.112a is for crossbar Kelvin resistor (CKR) while Fig. 2.112 (b) and (c) are for dummy joints. Figure 2.113 schematic illustrating the CKR test key subjected to current stressing, where the electron flow is denoted.

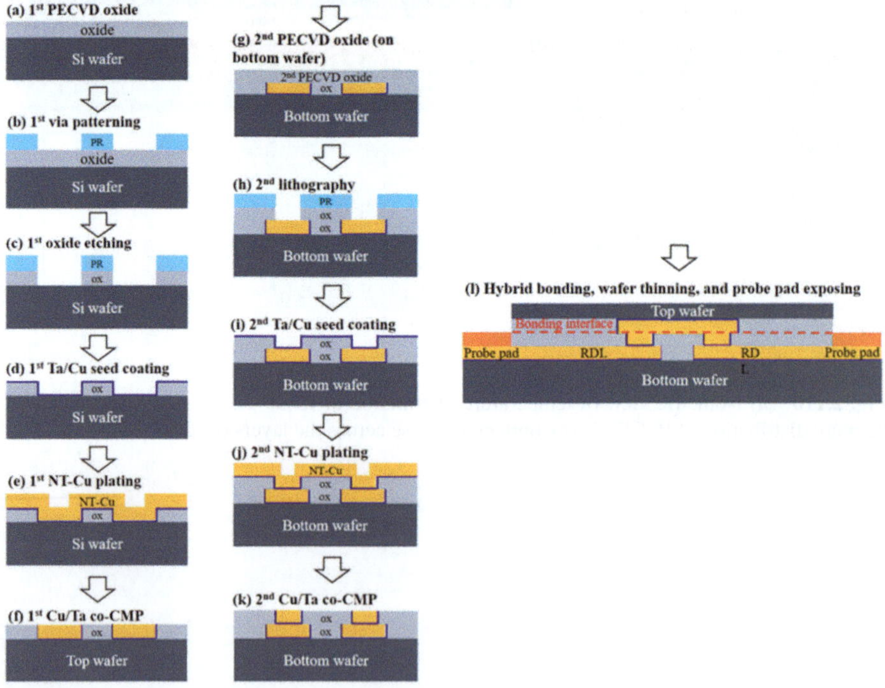


**Fig. 2.111** Schematic illustrations of the sequences of NT-Cu/SiO$_2$ damascene process on 12-inch Si wafers

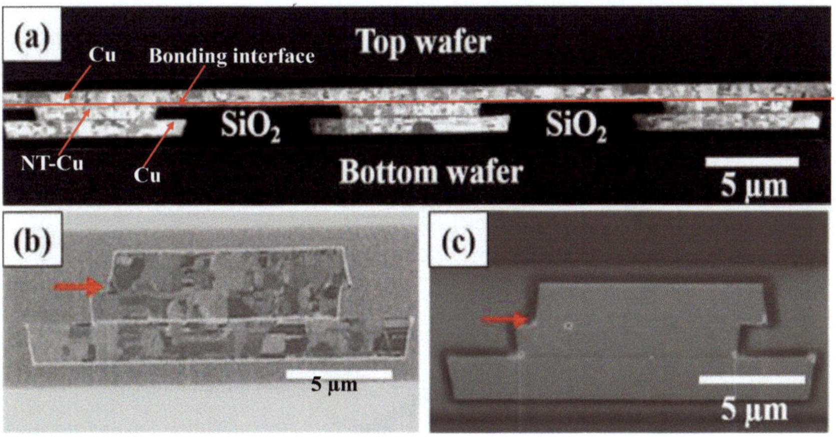


**Fig. 2.112** Cross-sectional inspections on as-bonded NT-Cu/$SiO_2$ hybrid joints. (**a**) CKR test key. (**b**)–(**c**) Dummy joint

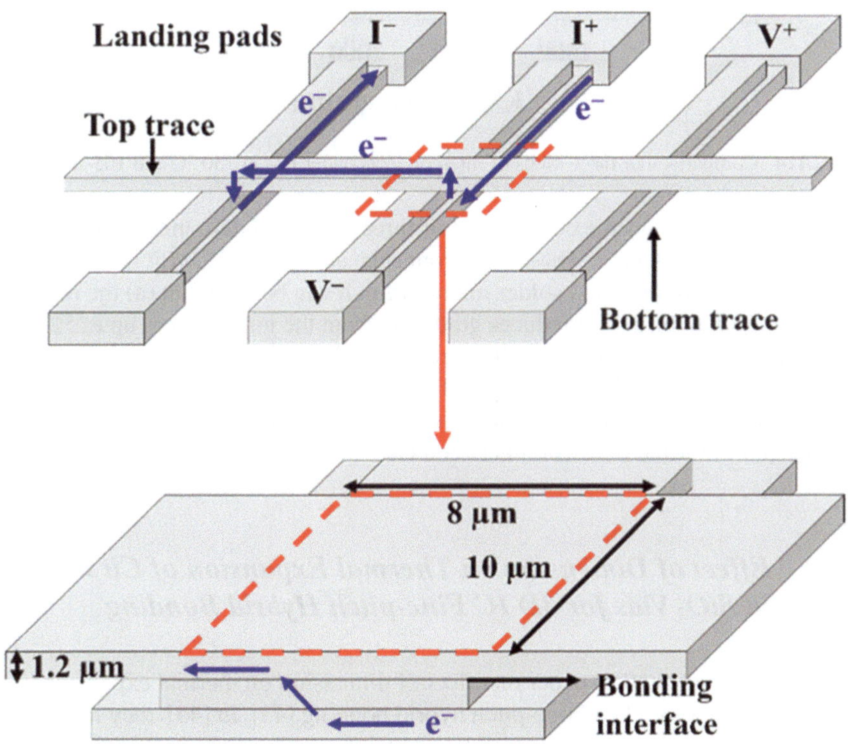


**Fig. 2.113** Schematic illustrating the CKR test key subjected to current stress. The electron flow is denoted

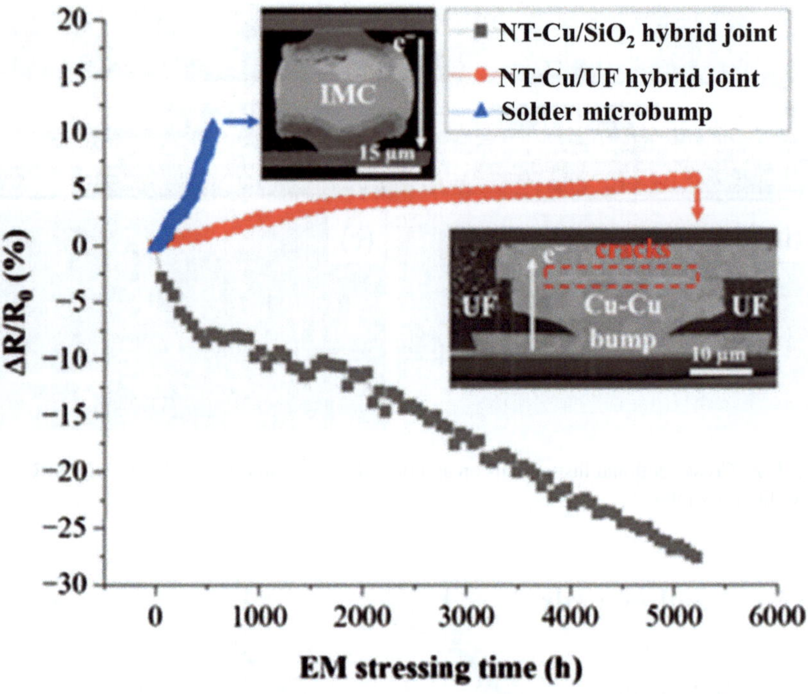


**Fig. 2.114** Comparison of three kinds of interconnects under EM at 150 °C and 105 A/cm$^2$

Figure 2.114 shows the comparison of three kinds of interconnects under EM at 150 °C and 105 A/cm$^2$. These interconnects are the NT-Cu/$SiO_2$ hybrid joint, NT-Cu/UF hybrid joint, and solder microbump. It can be seen that (a) the resistance of NT-Cu/$SiO_2$ hybrid joint reduces gradually from the initial stages up to 5200 h of stressful durations and (b) a steadily increase profile and an abrupt resistance increment is observed in NT-Cu/UF hybrid joint and solder microbump, respectively. Such discrepancy is attributed to the reverse stress states caused by mismatches in coefficient of thermal expansion between Cu and dielectrics.

## *2.9.2 Effect of Dimension on Thermal Expansion of Cu Pads in $SiO_2$ Vias for 3D IC Fine-pitch Hybrid Bonding*

NYCU published another paper on effect of dimension on thermal expansion of Cu pads in $SiO_2$ vias for 3D IC fine-pitch hybrid bonding [43]. In [43], they investigated the effect of dimension on thermal expansion of Cu pads in hybrid bonding. Figure 2.115 shows the schematic process flow for the fabrication of Cu/$SiO_2$ vias. Figure 2.116 shows the schematic drawing of the setup for in-situ heating atomic force microscopy (AFM).

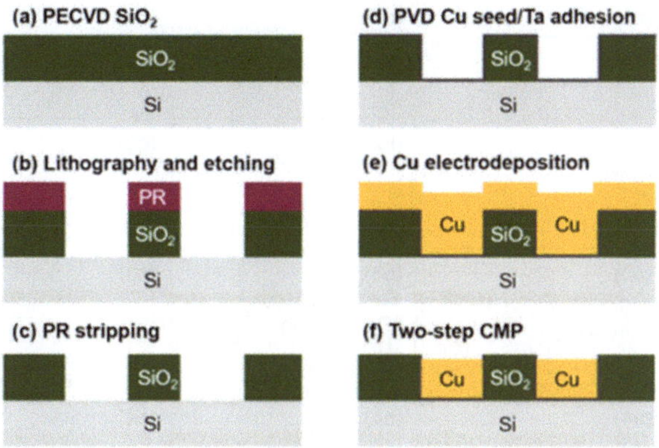


**Fig. 2.115** Schematic process flow for the fabrication of $Cu/SiO_2$ vias

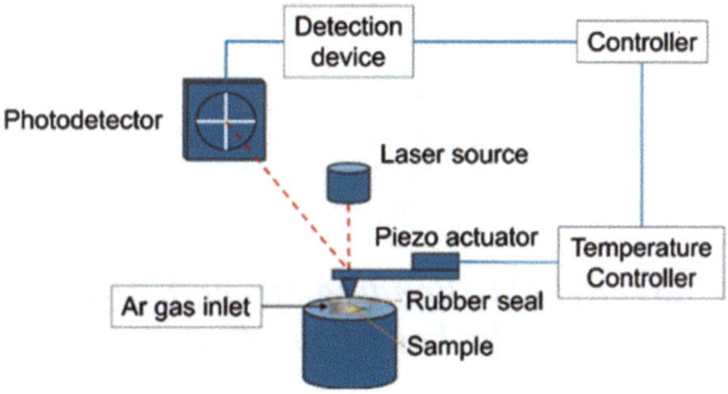


**Fig. 2.116** Schematic drawing showing the setup for in-situ heating AFM

Figure 2.117 shows the cross-sectional AFM surface profiles at room temperature and 200 °C for 1 μm, Fig. 2.117 (a1), 2.5 μm, Fig. 2.117 (b1), 4 μm, Fig. 2.117 (c1), and 5 μm, Fig. 2.117 (d1). Expansion values at 200 °C are shown in Fig. 2.117 (a2) for 1 μm, Fig. 2.117 (b2) for 2.5 μm, Fig. 2.117 (c2) for 4 μm, and Fig. 2.117 (d2) for 5 μm. Surface topographies of $Cu/SiO_2$ vias at room temperature and 200 °C with diameters of 1 μm, are shown respectively in Figs. 2.117 (a3) and (a4), 2.5 μm in Figs. 2.117 (b3) and (b4), 4 μm in Figs. 2.117 (c3) and (c4), and 5 μm in Figs. 2.117 (d3) and (d4). The colors in the AFM 2D profiles indicate the surface heights of the samples, as defined by the color scale at the bottom right—darker colors correspond to lower heights, while lighter colors indicate higher regions. The darker areas in the images represent copper pads, whereas the lighter regions correspond to the $SiO_2$ surface. Random white particles scattered across the samples are identified as slurry residues from the CMP process. It can be seen that (a) at 200 °C, the central regions of the Cu pads shift to lighter colors, indicating an increase in height with rising temperature, (b) this color change becomes less pronounced as the via diameter decreases from 5 μm to 1 μm, (c) the red line at the center of the Cu bump is higher than the green line, Fig. 2.117 (a1)–(d1), indicating thermal expansion during heat treatment, (d) the surface profiles of the $SiO_2$ regions remain, and (e) the expansion values of the Cu pads decrease as the pad size decreases, with an average difference of approximately 3 nm between the 5 μm and 1 μm vias.

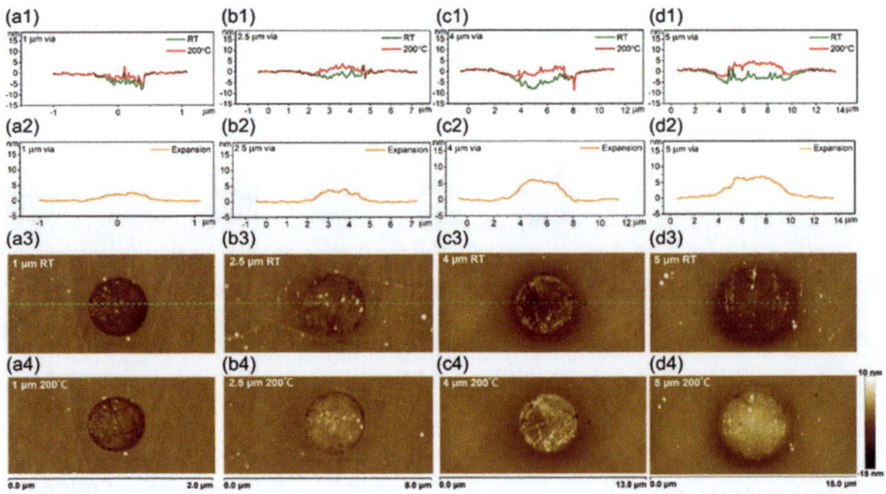


**Fig. 2.117** Cross-sectional AFM surface profiles at room temperature and 200 °C for (a1) 1 μm, (b1) 2.5 μm, (c1) 4 μm, and (d1) 5 μm. Expansion values at 200 °C for (a2) 1 μm, (b2) 2.5 μm, (c2) 4 μm, and (d2) 5 μm. Surface topographies of Cu/$SiO_2$ vias with diameters of (a3, a4) 1 μm, (b3, b4) 2.5 μm, (c3, c4) 4 μm, and (d3, d4) 5 μm at room temperature and 200 °C

### *2.9.3 Reliable Bonding Strength Measurement of SiCN/SiCN Films by Four-Point Bending Methodology*

NYCU published another paper on reliable bonding strength measurement of SiCN/SiCN films by four-point bending methodology [44]. In [44], they investigate bonded SiCN films through four-point bending tests, evaluating energy release rates and fracture behavior under various bending rates, sample sizes, and pin pitches. Figure 2.118 shows the (a) process flow of test samples (b) CSAM images of the bonded samples. Figure 2.119 shows the set up for four-point bending test. The interface of interest, which in this case would be SiCN/SiCN bonding interface, is sandwiched between two silicon substrates. A notch is created on the top plate using the diamond blade dicing machine. The interfacial adhesion energy, often expressed in terms of the critical energy release rate Gc can be determined by the equation in Fig. 2.119, where P is the steady-state force, L is the pin spacing, and b is the width of the specimen. E, $v$, and h are Young's modulus, Poisson's ratio, and the thickness of the substrate, respectively.

Figure 2.120 shows a typical load-displacement curve of four-point bending tests. It can be seen that (a) the load-displacement relationship is linear until the notch breaks, resulting in the first drop of the load and (b) in the later stage, a steady-state delamination develops and breaks along the weak interface between SiCN/Si.

Figure 2.121 shows the experimental results of different bending conditions (a) fixed sample width and pin pitch, with varying bending rate between 0.05 μm/s and

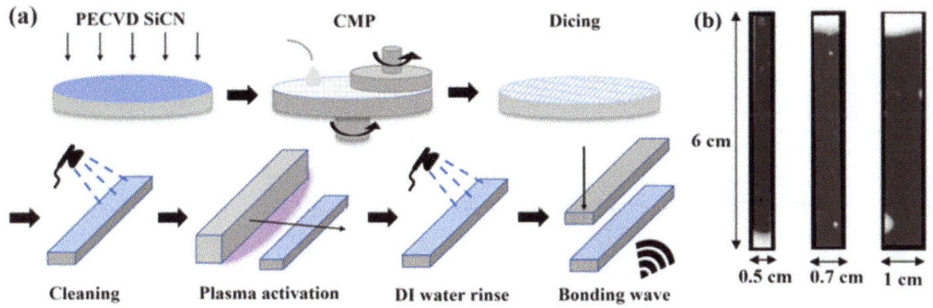


**Fig. 2.118** (**a**) Process flow of test samples. (**b**) CSAM images of the bonded samples

$$G_c = \frac{21(1- {}^2)P^2L^2}{16Eh^3b^2}$$

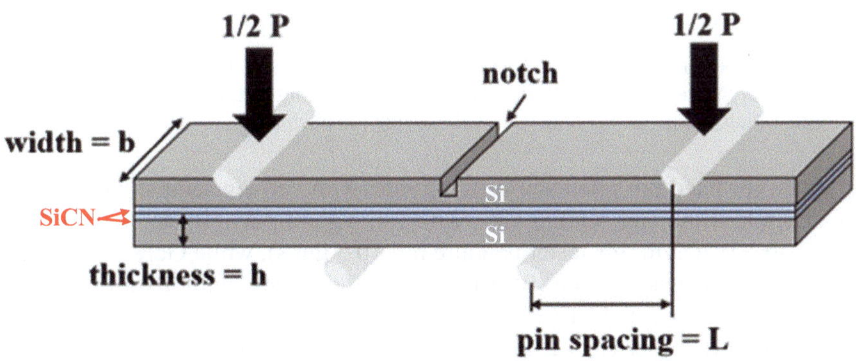


**Fig. 2.119** Set up for four-point bending test

**Fig. 2.120** Load-displacement curve of four-point bending tests

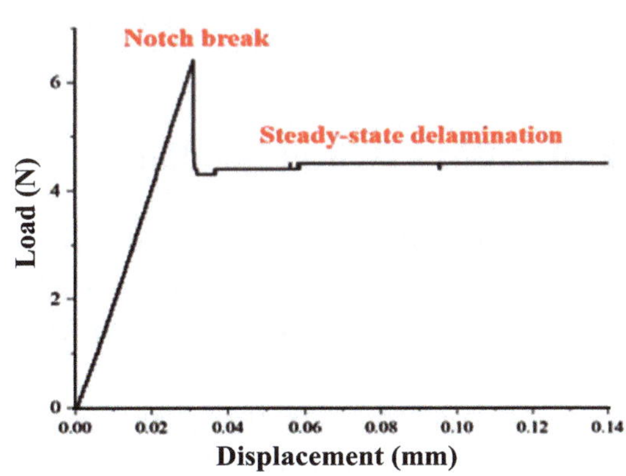

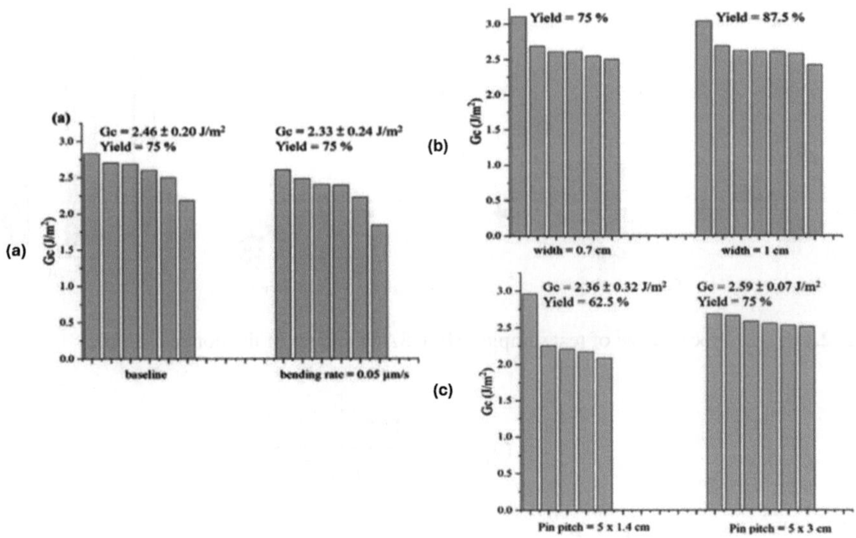


**Fig. 2.121** Experimental results of different bending conditions. (**a**) Fixed sample width and pin pitch, with varying bending rate. (**b**) Fixed pin pitch and bending rate, with varying sample width. (**c**) Fixed sample width and bending rate, with varying pin pitch

0.5 μm/s, (b) fixed pin pitch and bending rate, with varying sample width, and (c) fixed sample width and bending rate, with varying pin pitch. It can be seen that (a) Gc = 2.46 ± 0.20 J/m$^2$ for faster bending rate (0.5 μm/s), while Gc = 2.33 ± 0.24 J/m$^2$ for a slower bending rate (0.05 μm/s), (b) the sample with a width of 0.7 cm exhibits Gc = 2.51 ± 0.44 J/m$^2$ and a yield of 75%, while the 1 cm width sample shows a slightly higher Gc = 2.66 ± 0.18 J/m$^2$ with an improved yield of 87.5%, and (c) Gc is 2.36 ± 0.32 J/m$^2$ with a yield of 62.5% for a pin pitch of 5 × 1.4 cm, while Gc = 2.59 ± 0.07 J/m$^2$ and a yield of 75% for a pin pitch of 5 × 3 cm. It is recommended (a) the bending rate = 0.5 μm/s, (b) 1 cm sample width size is a better choice, and (c) 5 × 3 cm pitch is suitable.

## 2.10 IBM's Hybrid Bonding

### *2.10.1 Characterization of Interfacial Fracture Strength in Hybrid Bonded Wafers: A Novel Approach for High-Resolution Spatial Profiling*

During IEEE/ECTC 2025, IBM published at least one paper on characterization of interfacial fracture strength in hybrid bonded wafers: a novel approach for high-resolution spatial profiling [49]. They proposed a novel approach for characterizing the interfacial fracture strength across a 300 mm hybrid bonded wafer. Figure 2.122

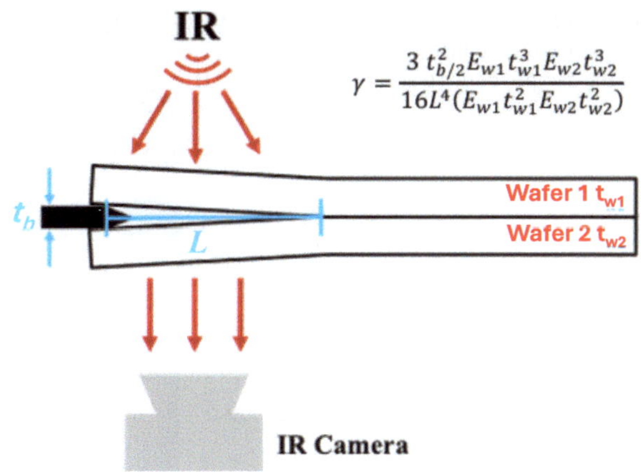


**Fig. 2.122** Schematic of razor blade test

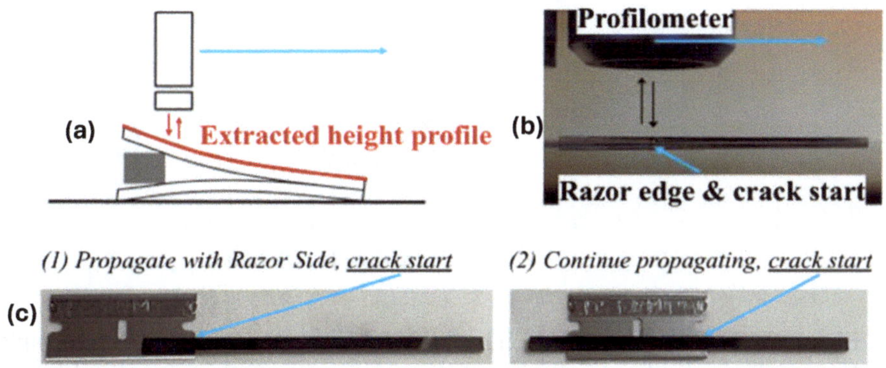


**Fig. 2.123** Extended Razor Blade Test setup. (**a**) Schematic of E-RBT. (**b**) Physical setup of E-RBT, (**c**) Top down view of two insertions

shows the schematic of razor blade test, in which γ is the interfacial bond strength, $t_w$ is the wafer thickness, $t_{b/2}$ is the half the thickness of the blade, $L$ is the crack distance measured between blade and crack tip, $w_1$ and $w_2$ are the thickness of wafer one and wafer two, and $E$ is the Young's modulus of the wafers. Both test wafers are 300 mm silicon wafer 775 μm thick with 2 μm TEOS layer. The total thickness of the bonded wafers is 1558 μm.

Figure 2.123 shows the extended razor blade test setup (E-RBT). Figure 2.123a is the schematic of E-RBT, while Fig. 2.123b is the physical setup of E-RBT, and Fig. 2.123c is the top down view of two insertions. It can be seen that (a) a razor blade cuts into the sample edge for crack initiation, (b) an optical profilometer is aligned above a bonded sample with a path sweep programmed along the entire razor-sample length, (c) the optical profilometer then sweeps along the path of the sample, collecting height information for crack length extraction based on beam deflection.

Fig. 2.124 XY height maps from three insertions of the sample under test

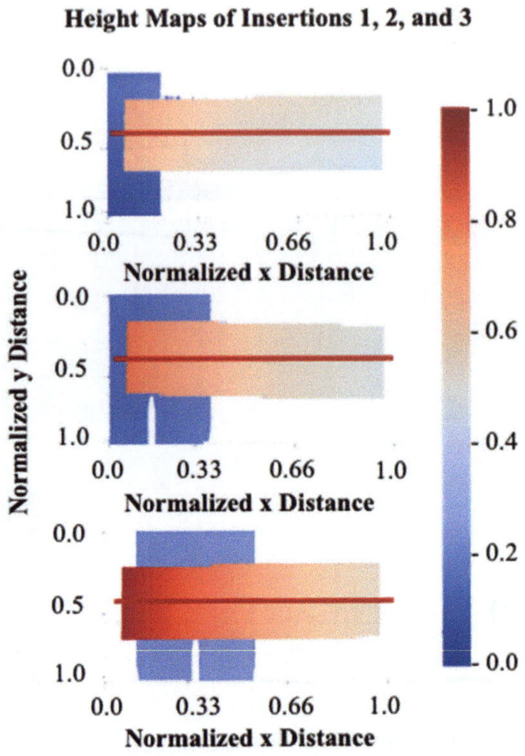


Figure 2.124 shows the XY height maps from three insertions of the sample under test. The color bar on the right hand side represents the normalized height with red being at a higher level and blue being at a lower level. The razor inserted into the sample is blue as it is below the deflecting sample. As the test continues, the samples can be seen deflecting higher on the left edge. Furthermore, the razor is also lifted higher by the deflected sample, seen by the lighter blue.

Figure 2.125 shows the test results from three insertions of the sample under test and the axes are normalized to the sample maximum. Figure 2.125a is for the transformed profiles (sample deflection vs. distance from wafer edge) while Fig. 2.125b is for the gradient profiles (deflection gradient vs. distance from wafer edge). It can be seen that (a) the gradient is constant in the beginning, increases, and then returns to being constant, (b) the initial constant value corresponds to linear deflection in the sample, (c) the second region with increasing gradient represents the curvature of the sample due to the razor blade insertion, and (d) the final region represents the flat undeformed bonded sample region.

Figure 2.126 shows the interfacial bond strength results from E-RBT. Axes are normalized to sample set's global maximum and minimum. Colors represent different samples taken from the same wafer. It can be seen that interfacial bond strength gradually grows with insertion distance.

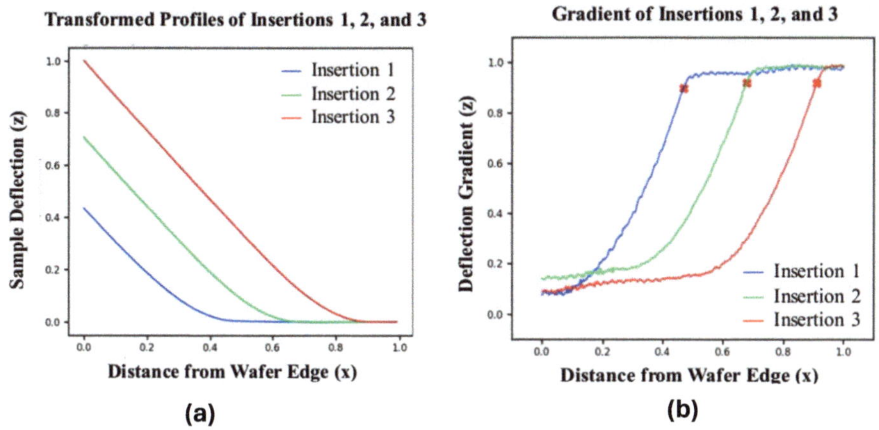


**Fig. 2.125** (**a**) Transformed profiles from three insertions of the sample under test are pictured. (**b**) Gradient profiles from three insertions of the sample under test are pictured

**Fig. 2.126** Interfacial bond strength results from extended razor blade test

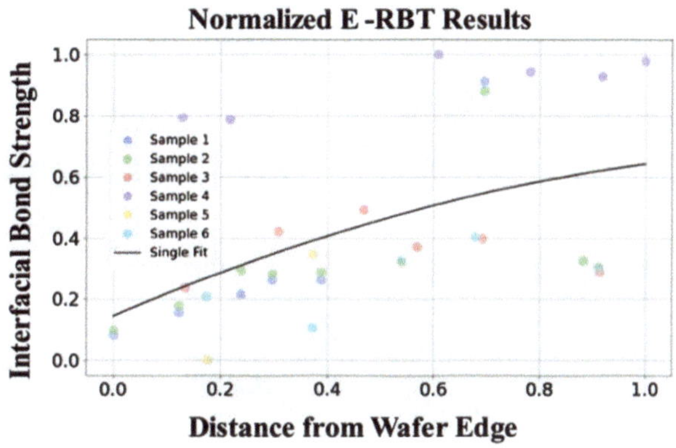


## 2.11 Sungkyunkwan University's Hybrid Bonding

### *2.11.1 Crystal Plasticity-Based Modeling and Experimental Validation of the Influence of Microstructures and Grain Boundary Junction Types on the Cu-Cu Bonding Interface*

During IEEE/ECTC 2025, Sungkyunkwan University and BASF published at least one paper on crystal plasticity-based modeling and experimental validation of the influence of microstructures and grain boundary junction types on the Cu-Cu bonding interface [50]. They investigated void closure and interfacing mechanisms at the bonding interface with experimental observations and crystal plasticity finite element method (CPFEM) simulations.

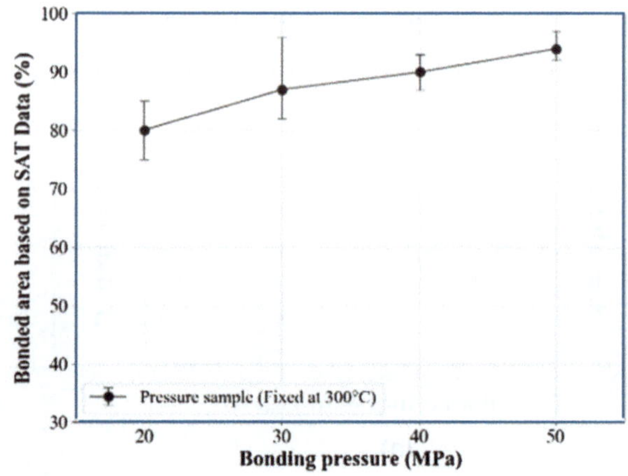


**Fig. 2.127** Pressure effect on bonded area based on SAT data

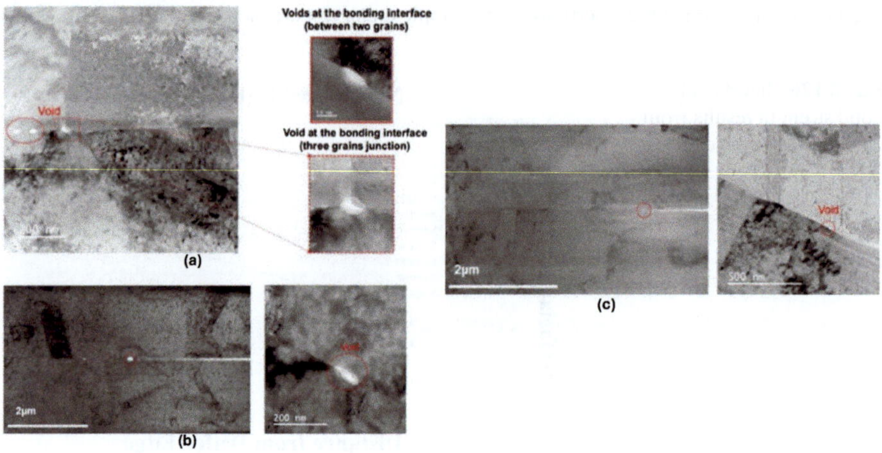


**Fig. 2.128** TEM data at bonding interface of various conditions (**a**) 300 °C and 30 MPa, (**b**) 300 °C and 40 MPa, (**c**) 300 °C and 50 MPa

Figure 2.127 shows the pressure effect on bonded areas based on scanning acoustic tomography (SAT) analysis data. It can be seen that (a) the larger the bonding pressure the larger the bonding area, and (b) the best bonding quality is confirmed under 50 MPa conditions.

Figure 2.128 shows TEM data at bonding interface of various conditions (a) 300 °C and 30 MPa, (b) 300 °C and 40 MPa, (c) 300 °C and 50 MPa. It can be seen that (a) under 30 MPa conditions, distinct voids are observed at grain boundary and additional voids are identified at triple junctions, Fig. 2.127a, (b) under 40 MPa and 50 MPa, voids are still present but with significantly lower density, (c) at 50 MPa, void size is further reduced while achieving a more homogeneous bonding interface.

Figure 2.129 shows the (111) Cu simulation results during void closure under 300 °C and 50 MPa condition. Figure 2.128a is for normalized pressure field results according to time progression, while Fig. 2.128b is for different grains. It can be seen from Fig. 2.128a that (a) voids are formed due to the initial roughness, (b) due to

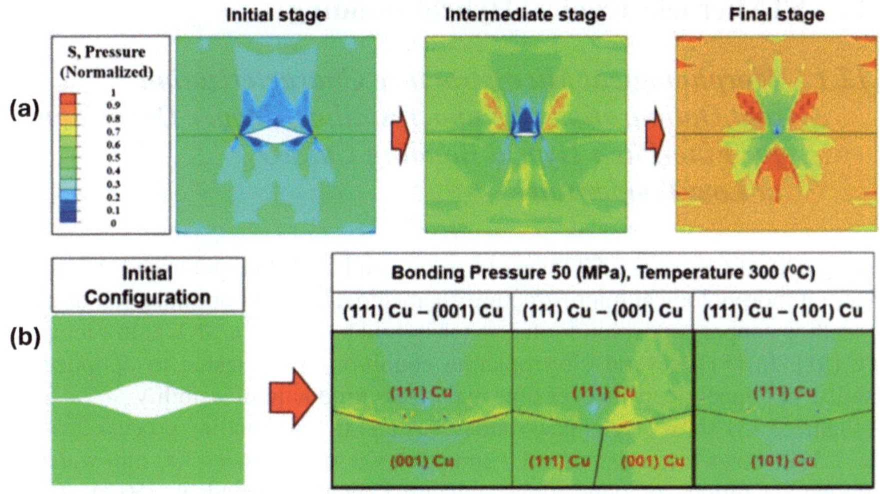


**Fig. 2.129** (**a**) Cu-(111) simulation results during void closure under 300 °C and 50 MPa condition. (**b**) Simulation results during void closure under 300 °C and 50 MPa condition with different grains

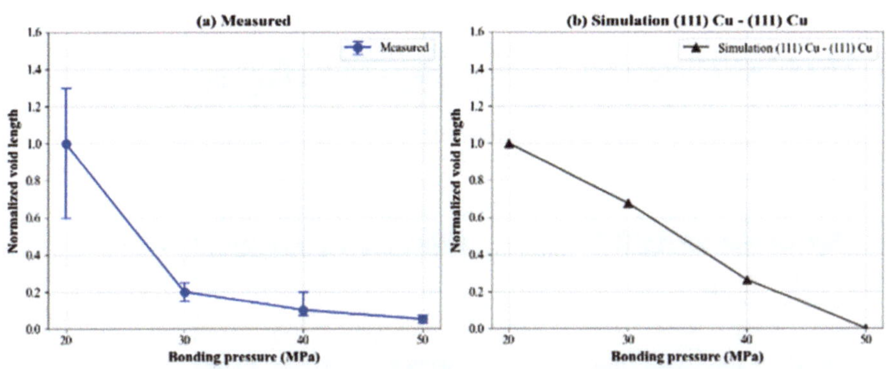


**Fig. 2.130** Comparison of measured and simulation results in normalized void length with respect to bonding pressure: (**a**) Measured case, (**b**) Simulation (111) Cu- (111) Cu case

pressure gradient, the voids begin to close, and (c) once the voids begin to close, the pressure gradient decreases. The impact of the microstructure is shown in Fig. 2.128b with the same initial configuration but different microstructures (the image on the left shows (111) grain at the top and (001) grain at the bottom, while the right image shows the result of a (111)-(101) grain combination). It can be seen that (a) the (111) grain penetrates the other two grains, creating an uneven interface and (b) in the intermediate image, where three grains meet at a junction, the different slip behaviors between grains result in a tripod shaped interface, as shown in Fig. 2.128a.

Figure 2.130 shows the comparison of measured and simulation results in normalized void length with respect to bonding pressure: (a) measured case, (b) simulation (111) Cu-(111) Cu. It can be seen that (a) the void size decreases as pressure increases and (b) the simulation closely follows the experimental trend.

## 2.12 STMicroelectronics' Hybrid Bonding

### 2.12.1 *Morphological Microstructure Characterization and Optimization of Nanocrystalline Copper Deposition for Fine-Pitch Hybrid Bonding Cu/SiO$_2$ at Low Temperature*

During IEEE/ECTC 2025, STMicroelectronics and CEA-Leti published at least one paper on morphological microstructure characterization and optimization of nanocrystalline copper deposition for fine-pitch hybrid bonding Cu/SiO$_2$ at low temperature [51]. In [51], several electroplating conditions were tested to optimize the stability of copper deposits over time and assess production feasibility.

Figure 2.131 shows the copper damascene and hybrid bonding process flow and Fig. 2.132 shows the resistivity of 1 μm thick Cu on full-sheet 300 mm wafer for different conditions of deposition; standard copper annealed at 250 °C (green square), standard copper self-annealed (dark green triangle), nanocrystalline copper (NC-Cu, also known as fine grain copper) condition A (red triangle), NC-Cu condition B (orange diamond), NC-Cu condition C (violet hexagon), NC-Cu condition D

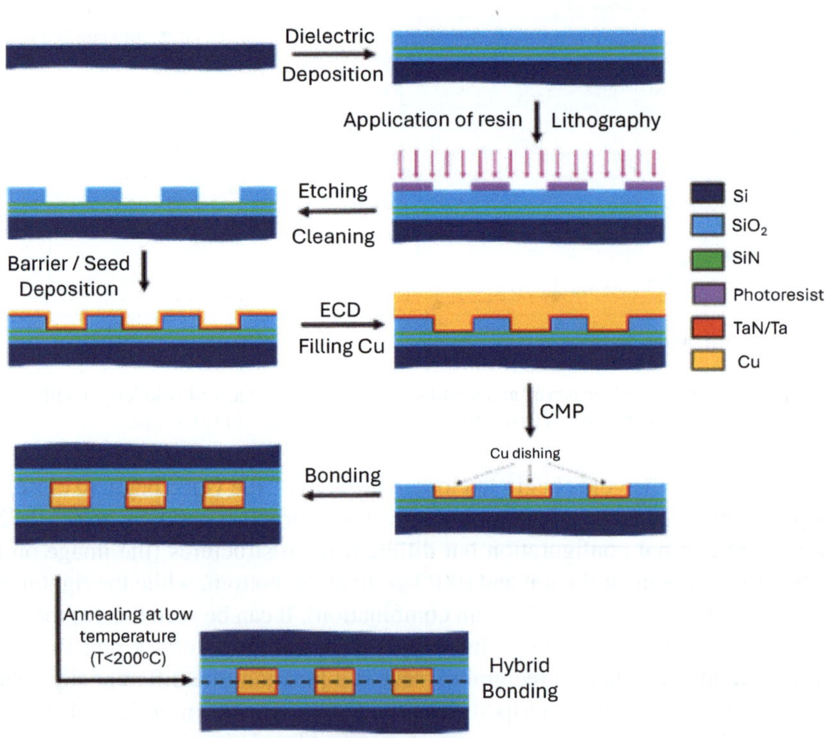


**Fig. 2.131** Copper damascene and hybrid bonding process flow

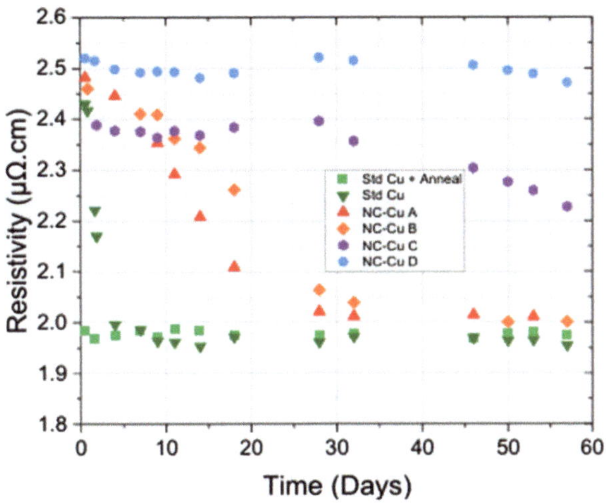


**Fig. 2.132** Resistivity of 1 μm thick Cu on fullsheet 300 mm wafer for different conditions of deposition

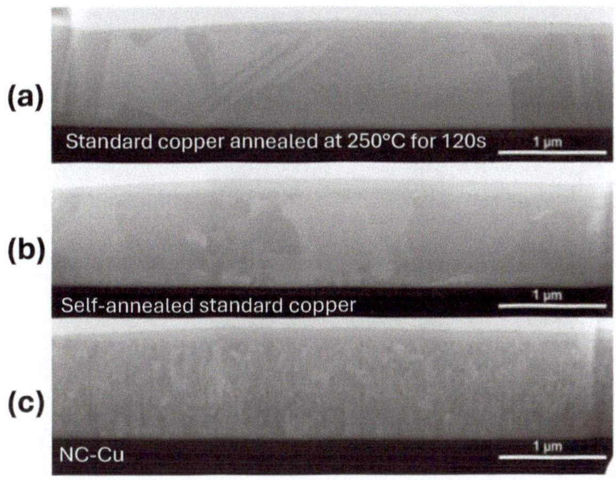


**Fig. 2.133** FIB-SEM cross-sections of 1 μm copper film on full-sheet wafer after 8 days in an ambient environment. (**a**) standard copper annealed at 250 °C for 120s, (**b**) self-annealed standard copper, (**c**) nanocrystalline copper

(blue circle). It can be seen that (a) the resistivity of standard copper decreases after a few days reaching a similar resistivity as annealed copper and (b) the NC-Cu remains stable for a longer period, maintaining a higher resistivity than standard copper.

Figure 2.133 shows the FIB-SEM cross-sections of 1 μm copper film on full-sheet wafer after 8 days in an ambient environment. Figure 2.133a is for standard copper annealed at 250 °C for 120s, while Fig. 2.133b is for self-annealed standard copper, and Fig. 2.133c is for NC-Cu. It can be seen that (a) standard copper crystallizes by "self-annealing" into a large-grain structure within a few days, (b) with the predominantly present (111) peak for fresh deposited copper becoming sharper with time, and (c) with the optimized electroplating process, NC-Cu remains stable for 2–3 months depending on electrochemical deposition (ECD) process conditions.

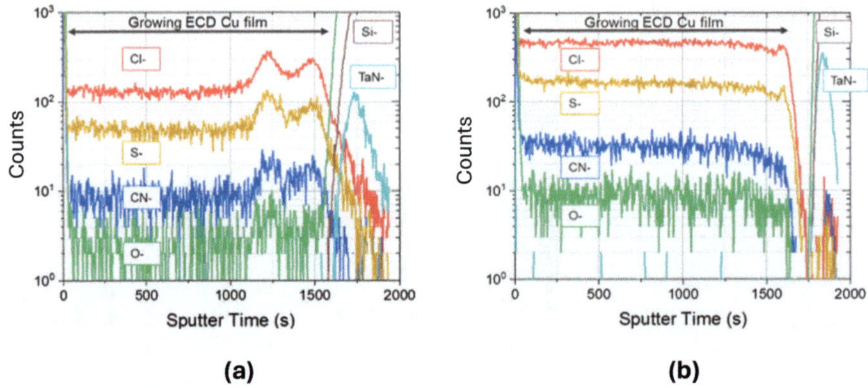


**Fig. 2.134** (**a**) TOF-SIMS analysis of self-annealed standard copper. (**b**) TOF-SIMS analysis of nanocrystalline copper condition D

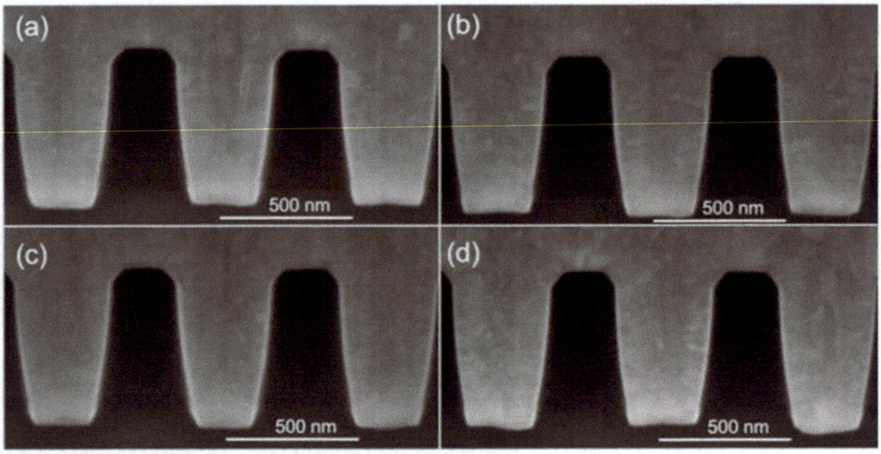


**Fig. 2.135** FIB SEM images of features with 0.6 μm pitch and CD 0.3 μm, filled with copper after electroplating NC-Cu condition (**a**) A, (**b**) B, (**c**) C, and (**d**) D

Figure 2.134 shows the time-of-flight secondary ion mass spectrometry (TOF-SIMS) analysis of (a) self-annealed standard copper and (b) TOF-SIMS analysis of nanocrystalline copper condition D. It can be seen that (a) more contaminants such as chlorine, sulfur, and CN species are observed in the NC-Cu compared to the standard copper and (b) these species are also present in the standard copper but in lower quantities, with a higher concentration of contaminants near the seed and barrier layers, (c).

Figure 2.135 shows the FIB SEM images of features with 0.6 μm pitch and CD 0.3 μm, filled with copper after electroplating (a) NC-Cu condition A, (b) NC-Cu condition B, (c) NC-Cu condition C, (d) NC-Cu condition D. While Fig. 2.136 shows the same thing except for critical dimension (CD) 3 μm. It can be seen that

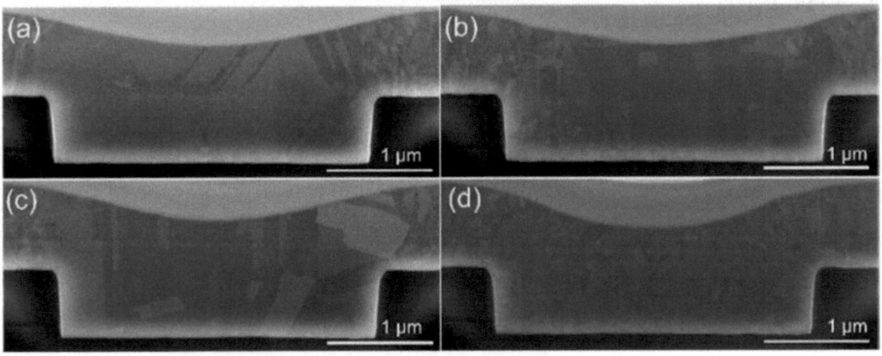


**Fig. 2.136** FIB SEM images of features with 6 μm pitch and CD 3 μm filled with copper after electroplating NC-Cu condition (**a**) A, (**b**) B, (**c**) C, and (**d**) D

(a) the selected conditions enable the filling of all cavity's sizes, with CD ranging from 0.3 to 5 μm, (b) for the smallest cavities, the 0.3 μm round pads exhibit uniform filling and all process conditions sets successfully lead to NC-Cu, and (c) for 3 μm cavities only condition D maintains a stable nanocrystalline state after deposition and after few days.

## 2.13 Binghamton University's Hybrid Bonding

### *2.13.1 Electrical Performance of Hybrid Bonding with Sub Micron Cu-Cu Bonding Contacts: Effects of Scaling, Microstructure, and Surface Morphology*

During IEEE/ECTC 2025, Binghamton University and IBM published at least one paper on electrical performance of hybrid bonding with sub-micron Cu-Cu bonding contacts: effects of scaling, microstructure, and surface morphology [52]. They demonstrated surface preparation methods involving acetone and dilute potassium hydroxide (KOH) solution, which effectively removed the surface protection layer, as verified by Fourier transform infrared (FTIR) and Raman spectroscopy techniques.

Figure 2.137 shows the Kelvin structures in hybrid bonding. Figure 2.138 shows the FTIR spectra of die surface with a protection layer and after its removal using acetone and 1% KOH in an ultrasonic bath. Figure 2.139 shows the Raman spectra of: (a) Cu pads and (b) $SiO_2$ dielectric surfaces after protection layer removal with 1% KOH and acetone. It can be seen that (a) Cu-Cu bonding connects a Cu pad on top of a via(s) to the corresponding Cu pad on the other die, (b) for the die surface with the protection layer, the FTIR spectra showed the distinctive spectral fingerprint of a phenolic-resin-based material, (c) after the protection layer removal, FTIR spectra showed peaks at 790, 960, and 1123 $cm^{-1}$, corresponding to Si–O–Si symmetric stretching, Si–OH stretching (silanol group), and Si–O–Si asymmetric

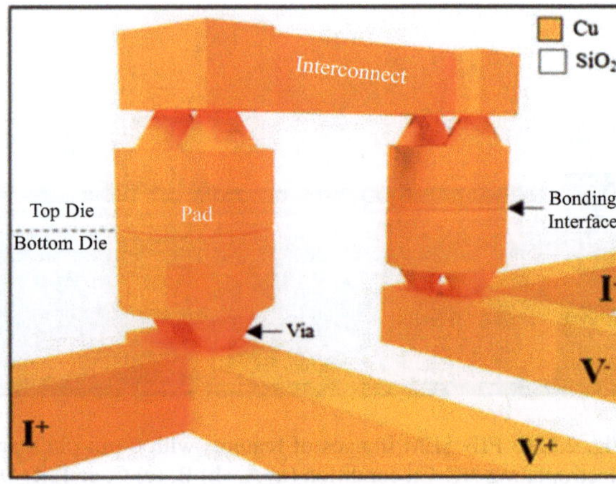


**Fig. 2.137** Kelvin structures in hybrid bonding

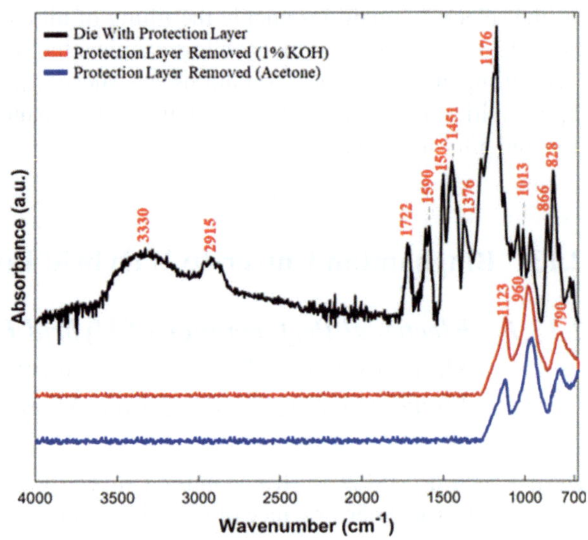


**Fig. 2.138** FTIR spectra of die surface with a protection layer and after its removal using acetone and 1% KOH in an ultrasonic bath

stretching vibrations, respectively, (d) for dies treated with 1% KOH, a peak that corresponds to Si appears on the Cu pads, suggesting some contamination, and (e) the strong presence of $Cu(OH)_2$ (at 488 $cm^{-1}$) on the $SiO_2$ surface after KOH treatment indicates the dissolution of the Cu pad surface and subsequent redeposition onto the dielectric layer.

Figure 2.140 shows the four-point resistance of Kelvin structures for the five pad/pitch configurations (measured at room temperature). It can be seen that (a) an increase in resistance as the Cu pad size decreases and (b) the increase in resistance spread and the decrease in yield as the pad size and pitch distance decrease.

Figure 2.141 shows the electron backscatter diffraction (EBSD) maps before (top-down view) and after (cross section) annealing bonding, showing the

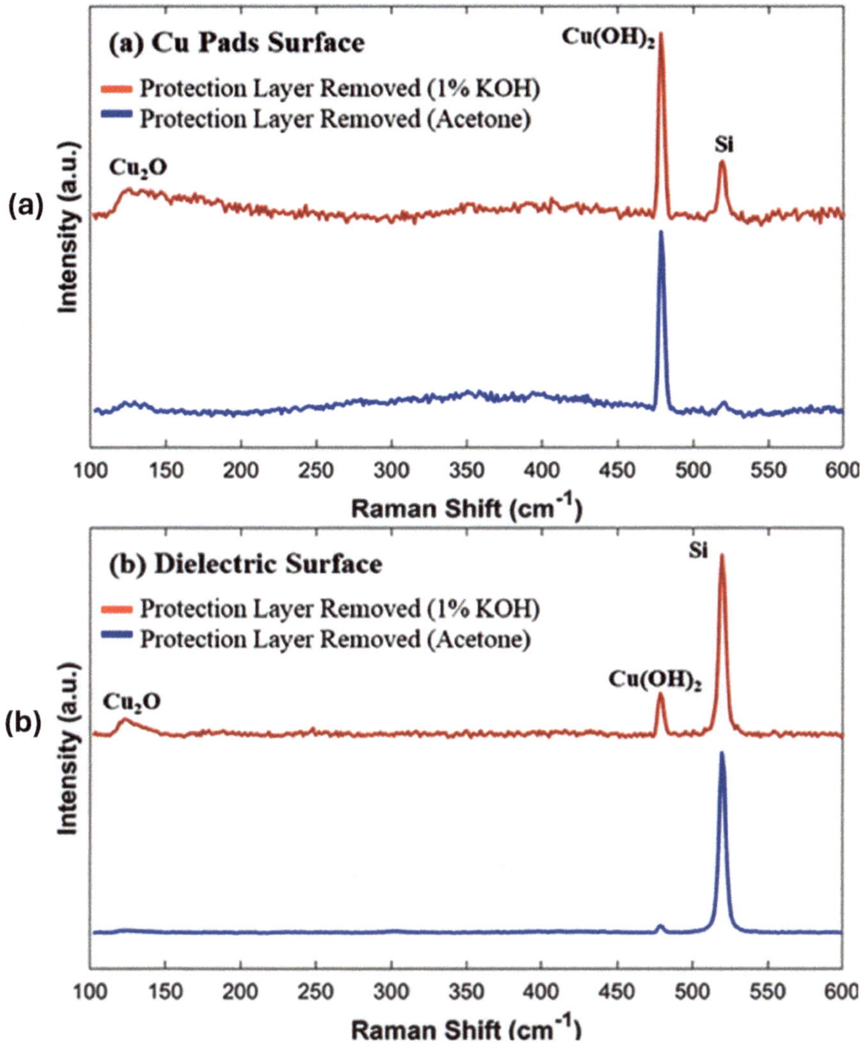


**Fig. 2.139** Raman spectra of: (**a**) Cu pads and (**b**) $SiO_2$ dielectric surfaces after protection layer removal with 1% KOH and acetone

microstructure evolution up on bonding of (a) 4 μm pre bonding, (b) 4 μm post-bonding, (c) 0.8 μm pre-bonding, (d) 0.8 μm post-bonding, and (e) inverse pole Figure (IPF) color map. It can be seen that (a) there is grain growth in both Cu pads during bonding, (b) there are still multiple grains within the 4 μm pads, and (c) the 0.8 μm pads contained only 1–2 grains along the interface line, and (d) the abnormal grain growth in the top 4 μm Cu pad. Figure 2.142 shows the current stressing of Kelvin structures at 20 MA/cm$^2$ and 300 °C. It can be seen that a slight drop in the in-situ resistance values is observed for the 0.8 μm Cu pad.

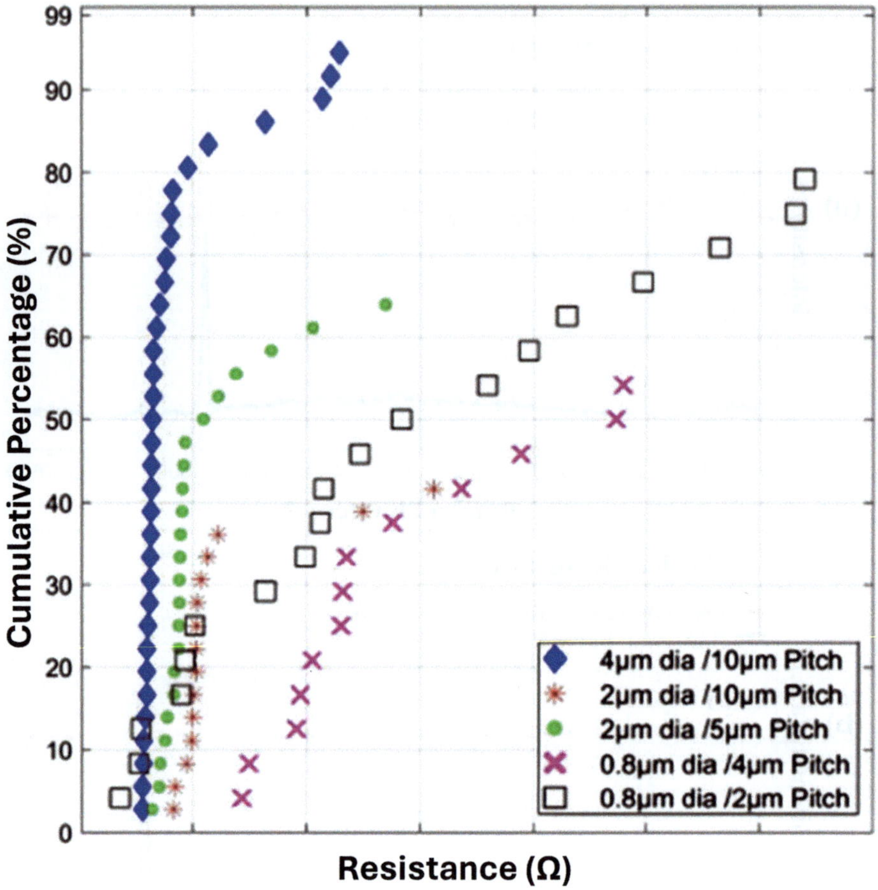


**Fig. 2.140** Four-point resistance of Kelvin structures for the five pad/pitch configurations (measured at room temperature)

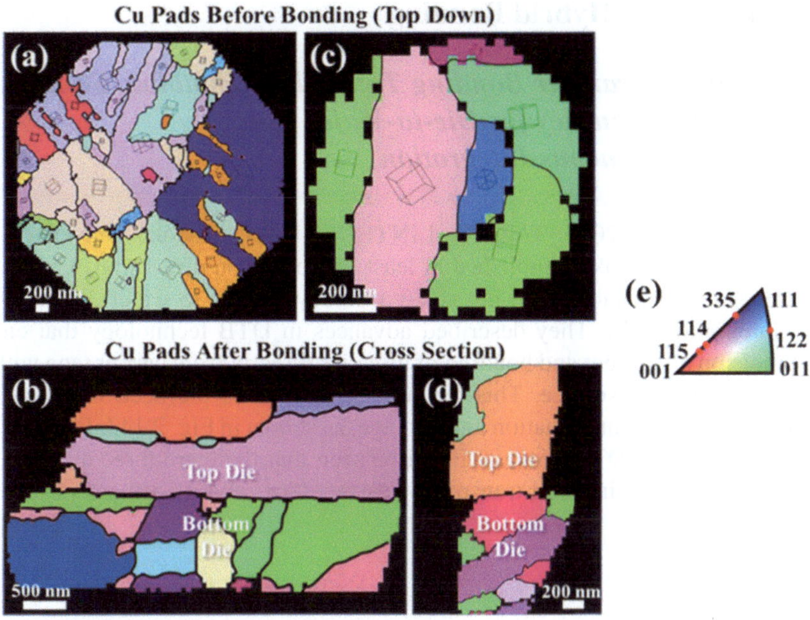


**Fig. 2.141** EBSD maps of: (**a**) 4 μm pre bonding, (**b**) 4 μm post-bonding, (**c**) 0.8 μm pre-bonding, (**d**) 0.8 μm post-bonding, (e) IPF color map

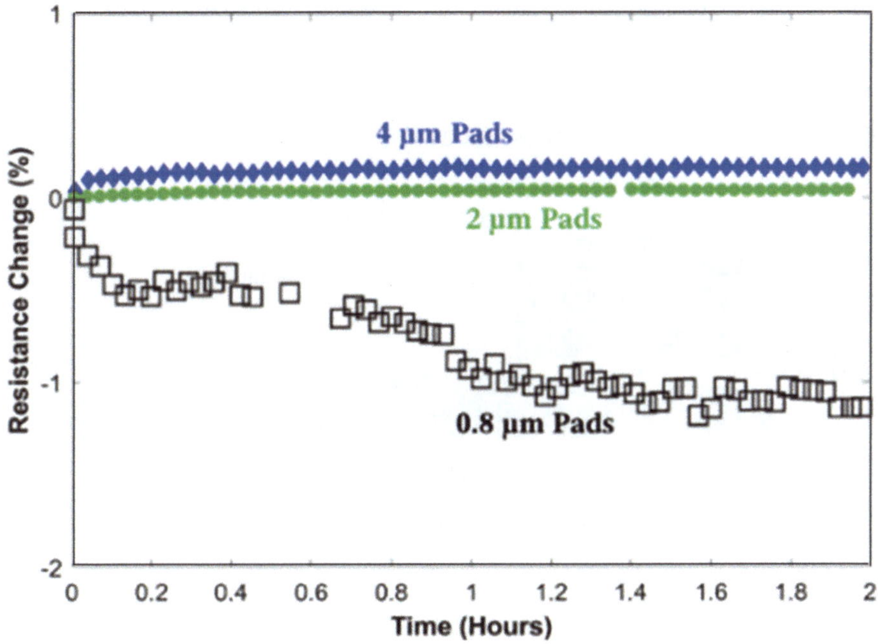


**Fig. 2.142** Current stressing of Kelvin structures at 20 MA/cm² and 300 °C

## 2.14 TAZMO's Hybrid Bonding

### *2.14.1 Direct Transfer Bonding Technology Enabling 50-nm Scale Accuracy for Die-to-Wafer 3D/ Heterogeneous Integration*

During IEEE/ECTC 2025, TAZMO, LINTEC, ULVAC, Tohoku University, and Institute of Science Tokyo published at least one paper on direct transfer bonding (DTB) technology enabling 50-nm scale accuracy for die-to-wafer 3D/heterogeneous Integration [53]. They described advances in DTB technology that enable high-precision alignment and bonding of thin dies directly from carrier tape without contact with the die surface. This technology eliminates the need for chip-level handling, reducing contamination and damage, as shown in Fig. 2.143. Figure 2.144 shows the advanced DTB process. It can be seen that there are three technologies (recognition system, infrared-transparent carrier tape, and alignment just before

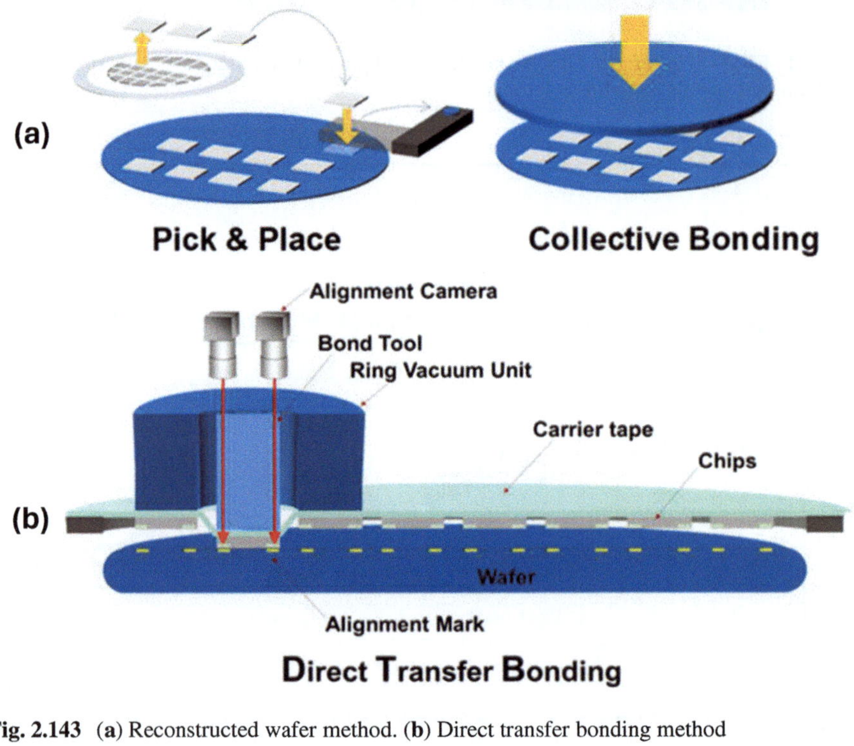


**Fig. 2.143** (**a**) Reconstructed wafer method. (**b**) Direct transfer bonding method

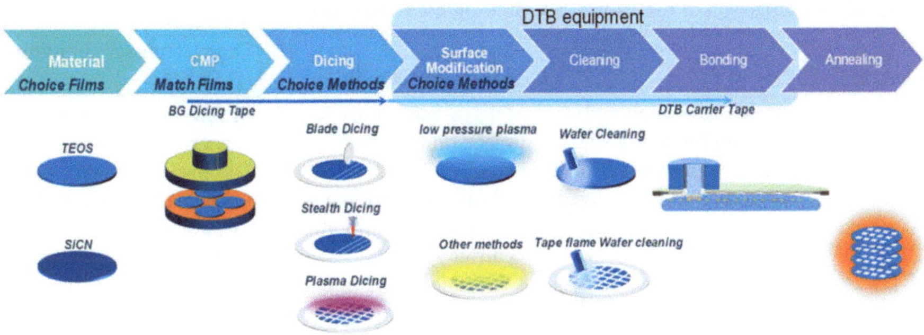


**Fig. 2.144** DTB process flow

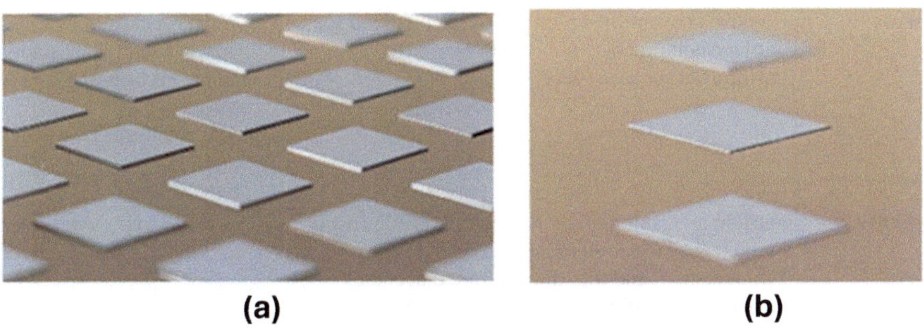


**Fig. 2.145** (**a**) External view of bonded dies on wafer. (**b**) Close-up view

bonding) which are required to achieve high-precision bonding. To realize these technologies, they have developed a new DTB equipment. This equipment enables high-speed, high precision bonding by combining a high-magnification optical system and image processing, a high speed stage and a precision stage, and an optimized carrier tape. Figure 2.145 shows the DTB results of a 6 mm × 6 mm chip with 150 μm thick and 50 μm thick after 200 °C/2h annealing.

A comparison of D2W with pick & place, conventional collective bonding, and DTB is shown in Table 2.3. The advantages of DTB over the pick & place and conventional collective bonding are obvious.

**Table 2.3** Comparison of die-to-wafer technologies: CB vs. DTB

| Method | Pick & Place | Collective Bonding | Direct Transfer Bonding (DTB) |
|---|---|---|---|
| Process | Plasma activation and cleaning on dicing sheet<br>Die-to-wafer pick & place<br>Annealing at wafer-level | Die-to-wafer pick & place on temporary carrier<br>Plasma activation and cleaning on temporally carrier<br>Wafer-level die transfer<br>Annealing at wafer-level | Plasma activation on carrier sheet<br>Die-to-wafer direct transfer from carrier sheet to wafer<br>Annealing at wafer-level<br>(No pick & place process) |
| Die-to-Wafer Equipment | Modified pick & place | Pick & place to temporally carrier<br>Wafer and wafer-level transfer | Direct transfer from sheet to wafer |
| Pros | Can use conventional pick & place equipment customized for clean-level | Commercialized for HVM | Low risks caused by particle and contamination<br>Simple process and low cost<br>Ultra small and thin die handling<br>High accuracy |
| Cons | Particle and contamination by pick & place equipment<br>Limited small and thin die handling capability<br>Limited accuracy by ball screw mechanism<br>(~0.1 μm) | Complicated process and expensive equipment<br>Limited accuracy<br>(Alignment × 2 times) | No HVM experience |

## 2.15 Purdue University's Hybrid Bonding

### *2.15.1 Atomistic Modeling of Interfacial Cracking in Copper-to-Copper Direct Bonding*

During IEEE/ECTC 2025, Purdue University published at least one paper on atomistic modeling of interfacial cracking in copper-to-copper direct bonding [54]. They employed large-scale molecular dynamics (MD) simulations to investigate the atomic-scale mechanisms of void evolution in Cu-Cu direct bonding under cyclic mechanical loading. Their atomistic simulations are performed using LAMMPS, while atomic configurations are visualized using OVITO, and the embedded atom method (EAM) potential is employed to accurately model atomic interactions in copper.

Figure 2.146 shows the cross-sectional view of the atomistic model for Cu-Cu bonding. Grain colors represent crystallographic orientations. The voids are designed to mimic experimental observations, with an elliptical shape and a width-to-height ratio of 2:1. To introduce realistic size variability, the void diameters were randomly assigned, ranging from 5 Å (minimum) to 15 Å (maximum).

Figures 2.147 and 2.148 show atomic structure of the Cu-Cu bonding interface for five samples with a 70% (Fig. 2.147) and 80% (Fig. 2.148) interfacial bonding

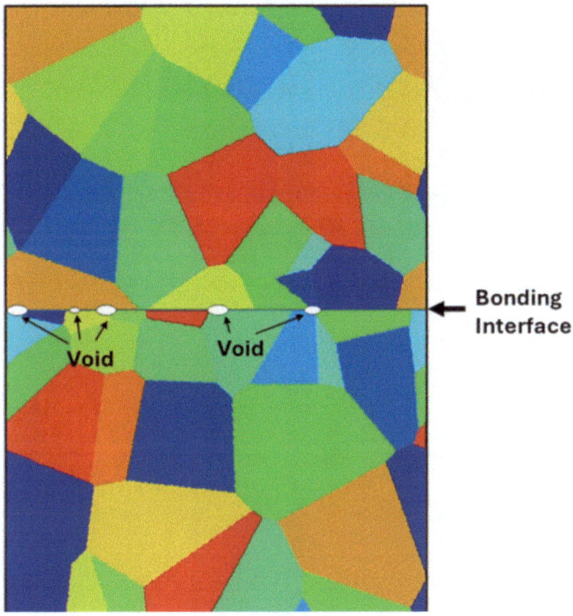


**Fig. 2.146** Cross-sectional view of the atomistic model for Cu-Cu bonding

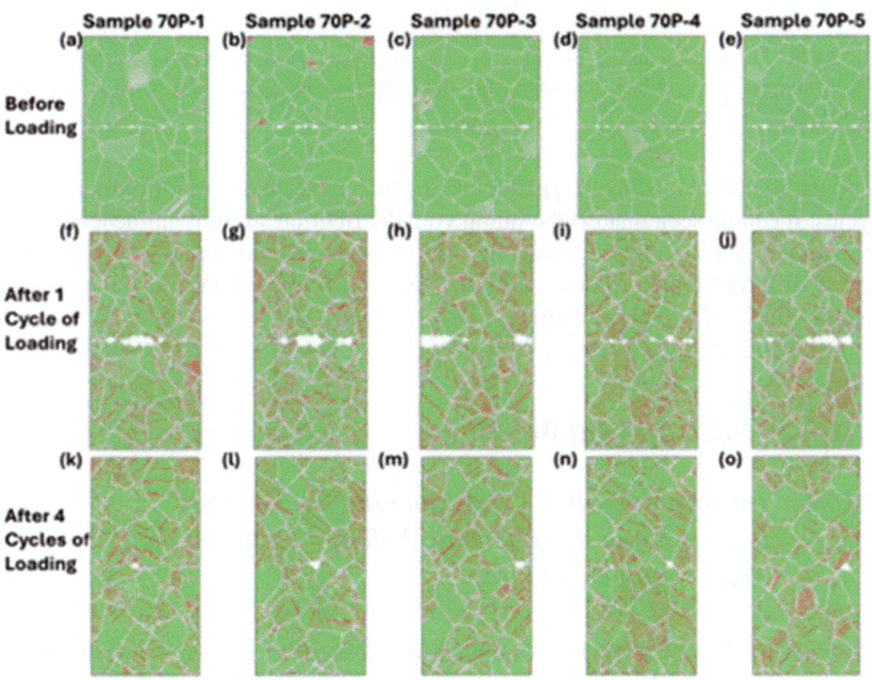


**Fig. 2.147** Atomic structure of the Cu-Cu bonding interface for five samples with a 70% interfacial bonding ratio: (a–e) before loading, (f–j) after one loading cycle, and (k–o) after four loading cycles

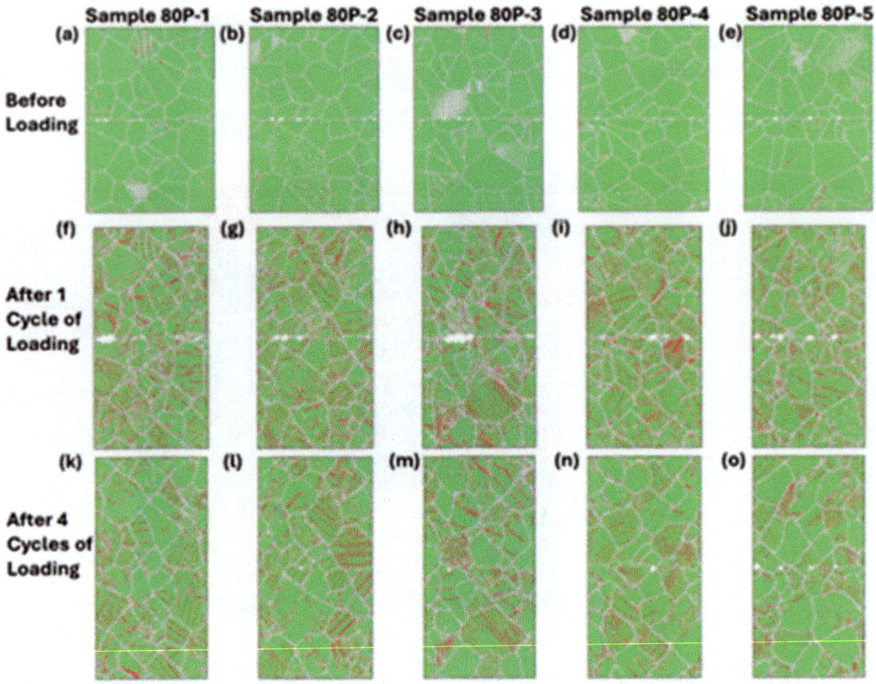


**Fig. 2.148** Atomic structure of the Cu-Cu bonding interface for five samples with an 80% interfacial bonding ratio: (a–e) before loading, (f–j) after one loading cycle, and (k–o) after four loading cycles

ratio: (a–e) before loading, (f–j) after one loading cycle, and (k–o) after four loading cycles. It can be seen that (a) the initial void distribution strongly influences microstructural evolution, (b) extend cyclic loading promotes void reduction (shrinkage) across different bonding ratios, and (c) mechanical cycling can enhance bonding integrity in Cu-Cu hybrid bonding.

## 2.16 TSMC's Hybrid Bonding

### *2.16.1 Innovative SoIC® Cool-Stacking Technology to Overcome the Thermal Wall of Future High Performance Compute*

During IEEE/ECTC 2025, TSMC published at least one paper on innovative SoIC® cool-stacking technology to overcome the thermal wall of future high performance compute [55]. They demonstrated the feasibility of their next-generation system on integrated chip (SoIC®) cooling stacking (SoIC®-CS) technology, a 3D IC

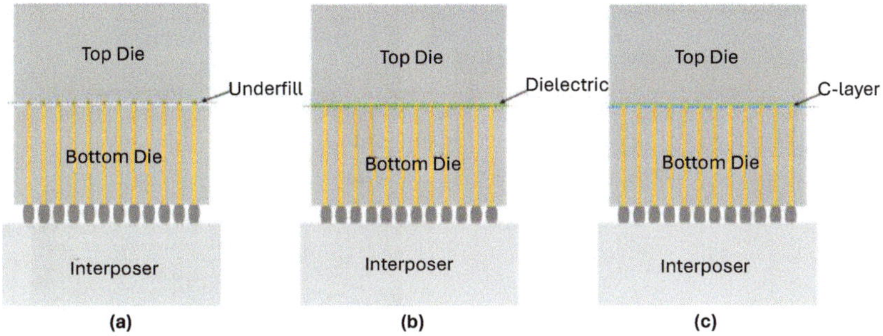


**Fig. 2.149** Not-to-scale Face-to-Back bonding schemes: (**a**) micro-bump (**b**) standard SoIC® (**c**) SoIC®-CS

**Table 2.4** Comparison of thermal resistance for bonding scheme

| Scheme | Microbump | SoIC® | SoIC®-CS |
|---|---|---|---|
| Face-to-Back bonding | 1.58X | 1X | 0.36X |
| Face-to-Fack bonding | 1.16X | 1X | 0.40X |

**Table 2.5** Comparison of electrical performance and energy efficiecy for F2B Bonding

| Face-to-Back Bonding | Microbump | SoIC® | SoIC®-CS |
|---|---|---|---|
| Bonding Density | 0.06X | 1X | 1X |
| Parasitic Resistance, R | 0.47X | 1X | 0.76X |
| Parasitic Capacitance, C | 6.44X | 1X | 1.06X |
| Bandwidth Density, 1/(RC)xBond-Density | 0.02X | 1X | 1.23X |
| Energy Efficiency, EE | 0.16X | 1X | 0.94X |
| Energy Efficient Performance, EEP | < 0.01X | 1X | 1.16X |

consisting of a high-performance compute base die and multilayered high-capacity memory top dies. Figure 2.149 shows the stacked structures with microbump and underfill (a), with standard SoIC (b), and with SoIC-CS (c). The difference between the SoIC and SoIC-CS is the dielectric layer of the chips for hybrid bonding. For SoIC-CS, the dielectric material is with high thermally conductive C-Layer to some degree, and the bonding structure is carefully optimized to ensure the quality of interconnection. The comparison of thermal resistance for bonding scheme is shown in Table 2.4, while the comparison of electrical performance and energy efficiency for face-to-back hybrid bonding is shown in Table 2.5. It can be seen that among these three structures, SoIC-CS performs the best.

The schematic of the two CoWoS structures for high performance computing driven by artificial intelligence with direct liquid cooling cold plate is shown in Fig. 2.150a for lidded CoWoS, and Fig. 2.150b for lidless CoWoS. There are 2 GPUs and 8 HBMs. The area of GPU, TSV interposer, substrate, and the 2-mm-thick lid are 800 mm$^2$, 3000 mm$^2$, 5000 mm$^2$ and 5000 mm$^2$, respectively. The area of HBM base die and DRAM core are 143 mm$^2$ and 131 mm$^2$, respectively.

(a)

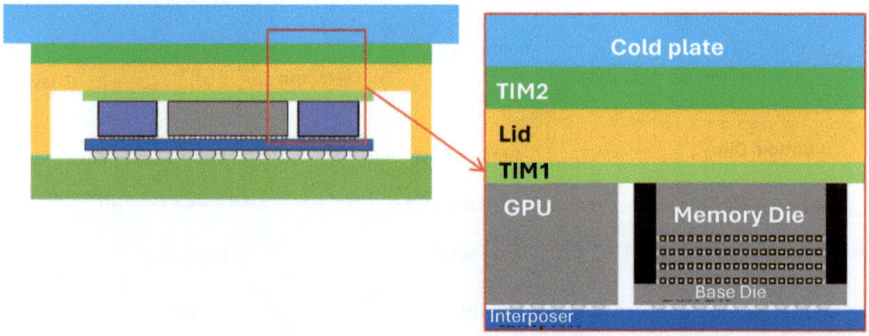


(b)

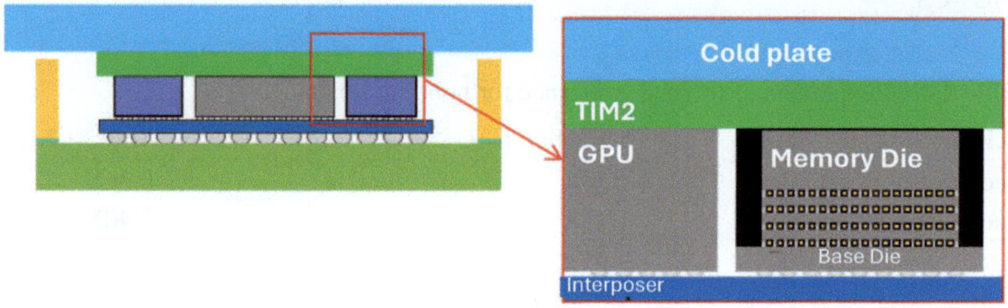


**Fig. 2.150** Schematic of two CoWoS chips with liquid-cooled cold plate: (**a**) Lidded CoWoS (**b**) Lidless CoWoS

The DRAM core is constantly heated at 0.015W/mm$^2$ while the GPU is 500W each. The micro-channeled cold plate is slightly larger than the substrate of CoWoS and fed with 40 °C deionized water running at 4 L per minute. There are six CoWoS structures in consideration, i.e., 8-hi, 12-hi and 16-hi HBMs multiplied by lidded and lidless. All the structures are subject to, on top of the aforementioned TDP of DRAM core and GPU, four heating scenarios, namely "PHY" being heated at respectively 0.4W/mm$^2$, 0.6W/mm$^2$, 0.8W/mm$^2$ and 1W/mm$^2$, and "TSV" always heated at 90% of "PHY". The results are shown in Fig. 2.151. It can be seen that (a) in all the six structures, SoIC-CS performs the best, and (b) the performance of SoIC is better than that of microbump and underfill.

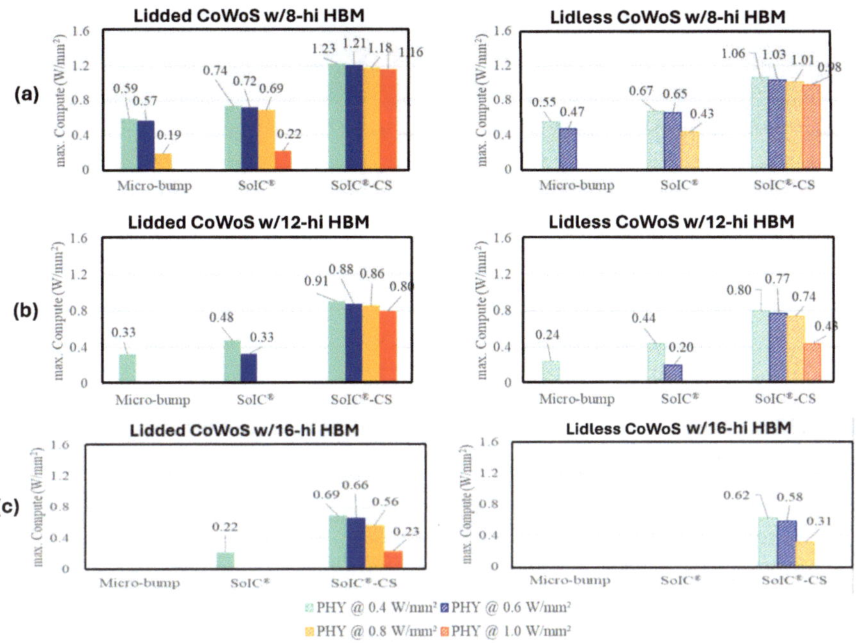


**Fig. 2.151** The derived maximal power density of "Compute": (**a**) Lidded CoWoS and 8-hi HBM (**b**) Lidless CoWoS and 8-hi HBM (**c**) Lidded CoWoS and 12-hi HBM (**d**) Lidless CoWoS and 12-hi HBM (e) Lidded CoWoS and 16-hi HBM (f) Lidless CoWoS and 16-hi HBM

## 2.17 Seoul National University of Science and Technology's Hybrid Bonding

### *2.17.1 The Role of Preparation and Bonding Parameters in Improving Hybrid Bonding Quality*

During IEEE/ECTC 2025, Seoul National University of Science and Technology published at least one paper on the role of preparation and bonding parameters in improving hybrid Bonding quality [56]. In this study, $O_2$, $N_2$, and forming gas (95% Ar, 5% $H_2$) plasma treatments were applied to improve $SiO_2$ bonding quality while reducing Cu surface damage, and bonding was performed under different bonding conditions.

After the room temperature bonding of $SiO_2$, Fig. 2.152 shows the comparison of bonding areas (by SAT observation) for different gas types and bonding pressures: (a) as bonded and (b) after annealed. It can be seen that (a) at zero pressure, the as bonded bonding area for $O_2$ is the largest, while $H_2$ is the smallest, (b) after 4 MPa, the as bonded bonding area increases for all the cases, (c) at zero pressure, after annealed bonding area of $O_2$ increases while that of $N_2$ and $H_2$ reduces, and (d) after 4 MPa, the after annealed bonding area of $O_2$ and $H_2$ decreases while that of $N_2$ increases.

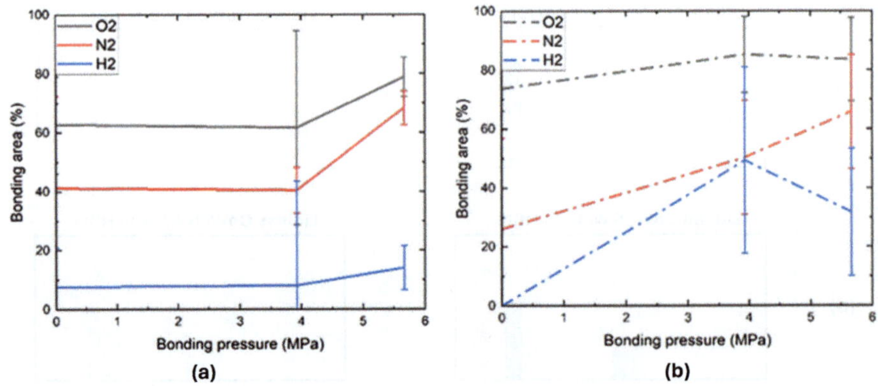


**Fig. 2.152** Comparison of bonding areas for different gas types and bonding pressures: (**a**) as bonded and (**b**) after annealed

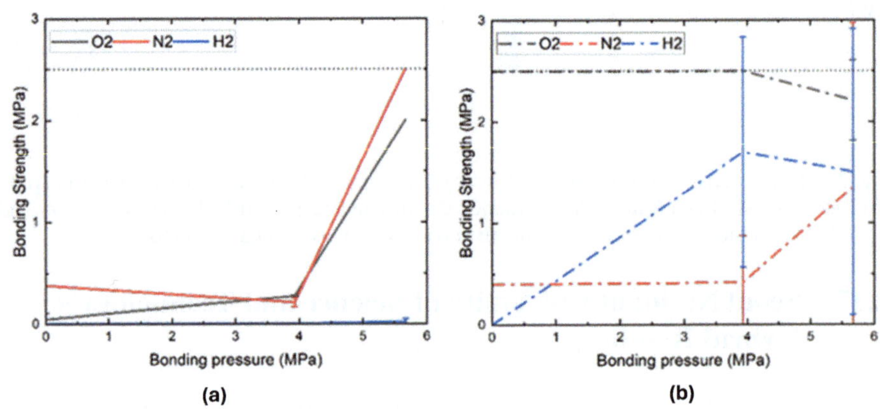


**Fig. 2.153** Pull-test results for different gas types and bonding pressures: (**a**) as bonded and (**b**) after annealed

After the room temperature bonding of $SiO_2$, Fig. 2.153 shows the comparison of pull test results for different gas types and bonding pressures: (a) as bonded and (b) after annealed. It can be seen that (a) at zero pressure, the as bonded bonding strength of $N_2$ is the largest while that of $H_2$ is the smallest, (b) after 4 MPa, the as bonded bonding strength of $N_2$ and $O_2$ increases sharply, while there is no change in that of $H_2$, (c) at lower pressure, the after annealed bonding strength of $O_2$ is the largest while of $N_2$ is the smallest, (d) after 4 MPa, the after annealed bonding strength of $O_2$ and $H_2$ decreases while that of $N_2$ increases.

Figure 2.154 shows the double cantilever beam (DCB) test results of $O_2$ plasma (a), $N_2$ plasma (b), and $H_2$ plasma (c) treated $SiO_2$. It can be seen that, for $O_2$, $N_2$, and $H_2$ plasmas, the adhesive energy of the annealed at 250 °C w/o pressure is larger than that of the as bonded at room temperature with pressure. Except in one case (3.93 MPa) of the $H_2$ plasma, which is very small anyway.

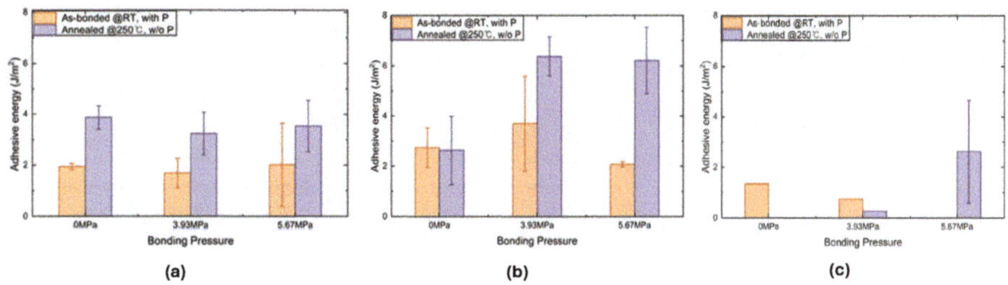


**Fig. 2.154** SAT image and DCB test results of (**a**) $O_2$ plasma treated $SiO_2$, (**b**) $N_2$ plasma treated $SiO_2$, and (**c**) $H_2$ plasma treated $SiO_2$

## 2.18 YOKOHAMA National University's Hybrid Bonding

### *2.18.1 Bond Wave Analysis of SiCN for Fine Pitch Hybrid Bonding*

During IEEE/ECTC 2025, YOKOHAMA National University, Tokyo Electron Limited, SK Hynix, Toray, DISCO, Tohoku University, University of Tsukuba, imec, and Kurita Water Industries published at least three papers on hybrid bonding [57–59]. In [57], YOKOHAMA National University, Tokyo Electron Limited, and SK Hynix investigated the bonding dynamics with plasma enhanced chemical vapor deposition (PECVD)-SiCN as a superior dielectric material. Figure 2.155 shows the image of the bond wave propagation, which is used to study the relationship between the pre-bonding dielectric surface condition and the bond wave speed.

Figure 2.156 shows the water contact angle (WCA) and surface free energy for SiCN films with different plasma ($N_2$ and $O_2$) conditions. In Fig. 2.156, $\gamma^d$, $\gamma^p$, and $\gamma^h$ represent the dispersive force, the orientation force, and the hydrogen bonding force, respectively. It can be seen that (a) the total surface free energy ($\gamma^d + \gamma^p + \gamma^h$) with both $N_2$ and $O_2$ plasmas at 100 watts (W) and 400W is larger than that of the post CMP SiCN, and (b) the total surface free energy with both $N_2$ and $O_2$ plasmas at 100W and 400W is about the same.

Figure 2.157 shows the vibrational sum-frequency generation (VSFG) spectra of a modulus square of second-order susceptibility $|\chi^{(2)}|^2$ spectra specifically focused (a) on amine (NH) group for SiCN and $SiO_2$ films with different plasma conditions, and (b) on hydroxyl ion (OH) groups for SiCN and $SiO_2$ films with different plasma conditions. It can be seen that (a) SiCN exhibits a prominent peak around 3400 $cm^{-1}$ compared with $SiO_2$, and this peak can be attributed to NH stretching vibrations, indicating the presence of amine (–NH) groups on the SiCN surface, (b) for plasma-activated SiCN, an increase in intensity on the lower wavenumber side of the peak is observed compared with the pre-activation state, (c) a peak near 3750 $cm^{-1}$ corresponds to free, non-hydrogen bonded (isolated) OH groups, while a shift towards approximately 3700 $cm^{-1}$ indicates the formation of hydrogen-bonded OH networks, (d) further shifts to lower wavenumbers suggest the presence of

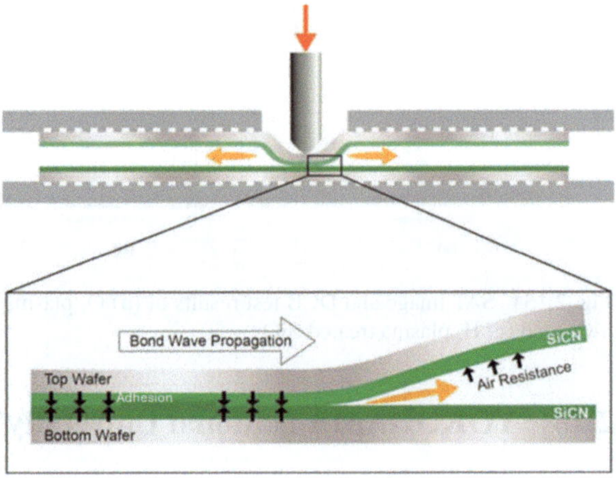


**Fig. 2.155** A schematic image of the bond wave propagation

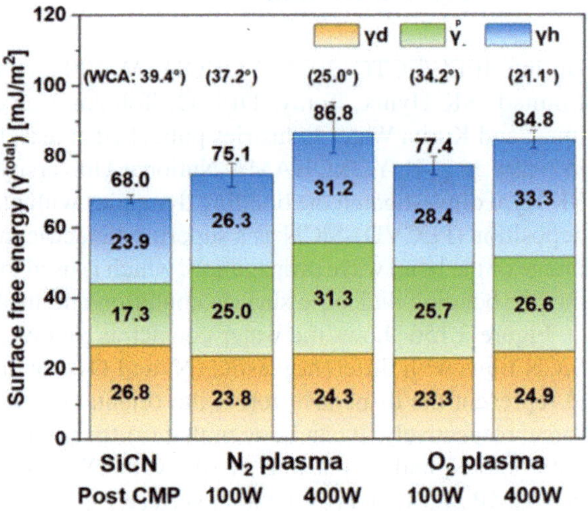


**Fig. 2.156** Water contact angle and surface free energy for SiCN films with different plasma conditions

surface-adsorbed wafer molecules, (e) in the untreated SiCN film, OH-related peak, which is distinctly observed in $SiO_2$, is significantly weaker, (f) after the plasma treatment, the peak associated with OH groups and water became more distinct, and (g) the background signal in SiCN is higher than in $SiO_2$.

Figure 2.158 shows (a) the bond wave speed for SiCN-SiCN direct bonding with different plasma conditions and (b) the adhesion energy for SiCN-SiCN direct bonding with different plasma conditions. It can be seen that (a) the bond wave speed for SiCN-SiCN direct bonding with both $N_2$ and $O_2$ plasmas at 50W, 100W, and 400W is almost the same, and (b) the adhesion energy for SiCN-SiCN direct bonding with $N_2$ and $O_2$ plasmas is about the same for different powers.

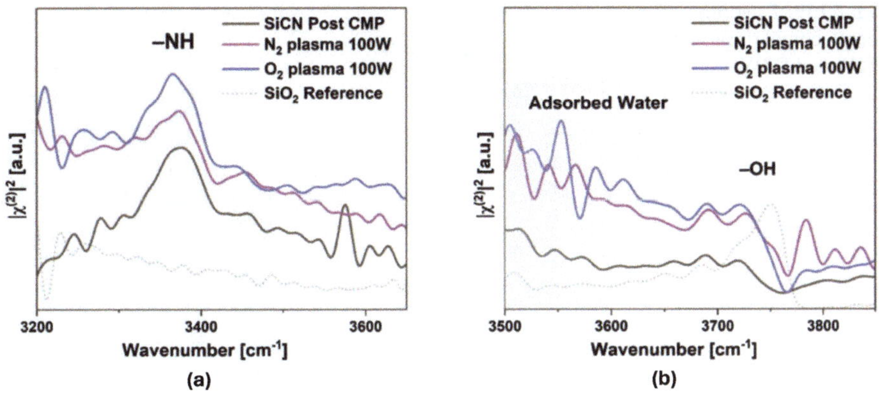


**Fig. 2.157** VSFG spectra of a modulus square of second-order susceptibility specifically focused on (**a**) NH groups for SiCN and $SiO_2$ films with different plasma conditions and (**b**) OH groups for SiCN and $SiO_2$ films with different plasma conditions

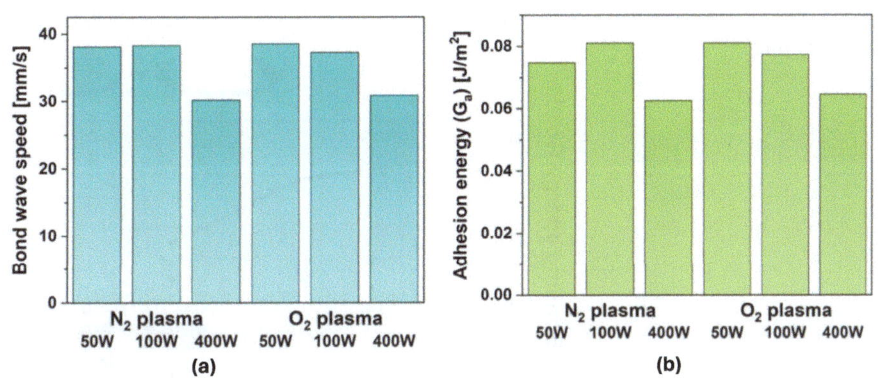


**Fig. 2.158** (**a**) Bond wave speed for SiCN-SiCN direct bonding with different plasma conditions. (**b**) Adhesion energy for SiCN-SiCN direct bonding with different plasma conditions

## *2.18.2 Assessing Queue Time in D2W Hybrid Bonding Through Precise Bond Strength Measurements*

During IEEE/ECTC 2025, YOKOHAMA National University, Toray, DISCO, Tohoku University, University of Tsukuba published another paper on assessing queue time in D2W hybrid bonding through precise bond strength measurements [58]. In this study, the queue time (Q-time) in D2W hybrid bonding has been comprehensively investigated using a novel method for accurately measuring bond strength. It is well known that the Q-time for D2W is longer than W2W due to activation/deionized water rinse and bonding. Figure 2.159 shows the bonding energy comparison with W2W and D2W by the double cantilever beam (DCB) test.

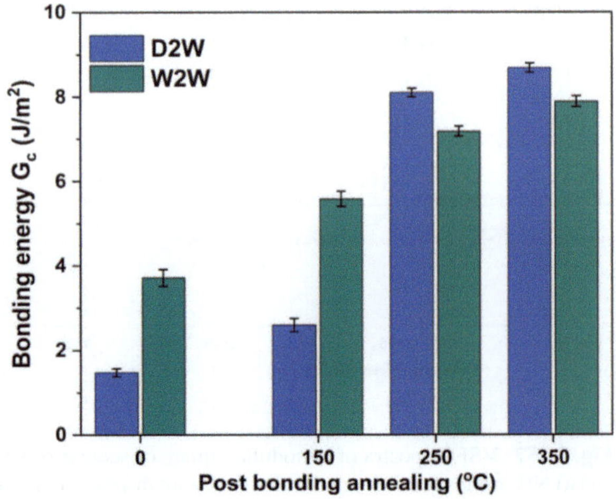


**Fig. 2.159** Bonding energy comparison with W2W and D2W

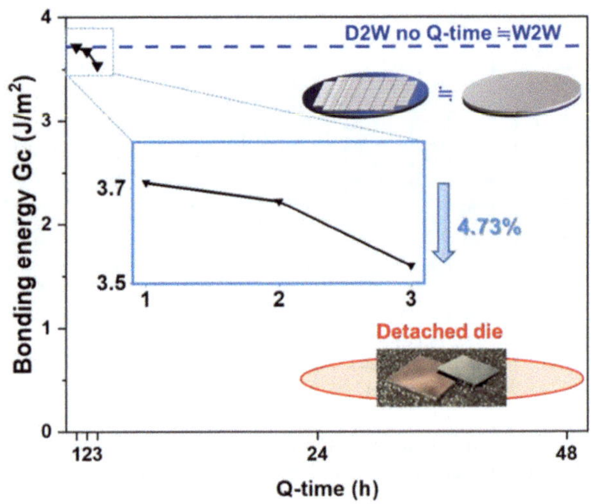


**Fig. 2.160** Changing bonding energy with D2W varying Q-time

Both samples undergo the same bonding film formation and plasma treatment. It can be seen that (a) a significant discrepancy in bonding energy between W2W and D2W bonding in the absence of post bonding annealing or at an annealing temperature of 150 °C, and (b) at higher annealing temperatures, both bonding processes exhibit nearly identical bonding energies.

Figure 2.160 shows the changing of bonding energy in D2W with varying Q-time by the nanoindentation test. It can be seen that there is a decrease in bonding energy with increasing Q-time. The bonding energy is estimated using the W2W bonding energy value for samples with a Q-time of less than 1 h as a reference.

Figure 2.161 shows the zeta potential transition with Q-time of the surface after plasma activation/DIW rinse of hydroxyl (OH) group. It can be seen that (a) the

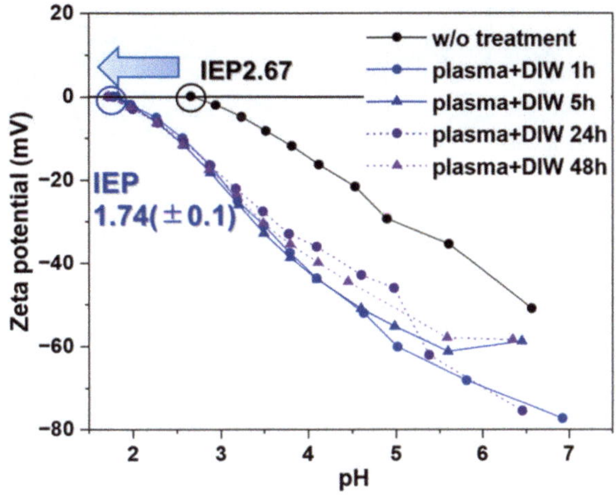


**Fig. 2.161** Zeta potential transition with Q-time of the surface after plasma activation/DIW rinse

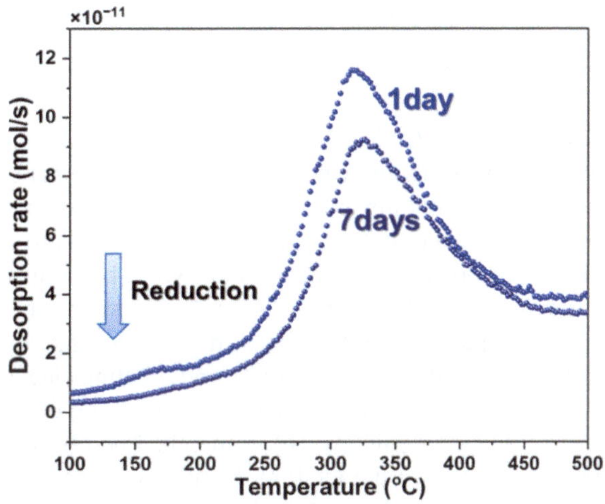


**Fig. 2.162** TDS spectra of mass number of 18M/e (water) for $SiO_2$

isoelectric point (IEP) shifts significantly toward the acidic region after surface treatment, (b) the ionization of these OH groups result in a negatively charged $SiO_2$ surface, and (c) there is no change in the IEP over time.

Figure 2.162 shows the thermal desorption spectroscopy (TDS) spectra of mass number of 18 M/e (water) for $SiO_2$. It can be seen that in the temperature range of 150–250 °C, where surface-adsorbed water typically desorbs, samples with longer Q-time exhibits a lower amount of desorbed water.

Figure 2.163 shows the water content at the bonding interface (a) low (b) middle (c) high. I can be seen that the reduction of surface water due to Q-time (presumably through evaporation) weakens the bond strength.

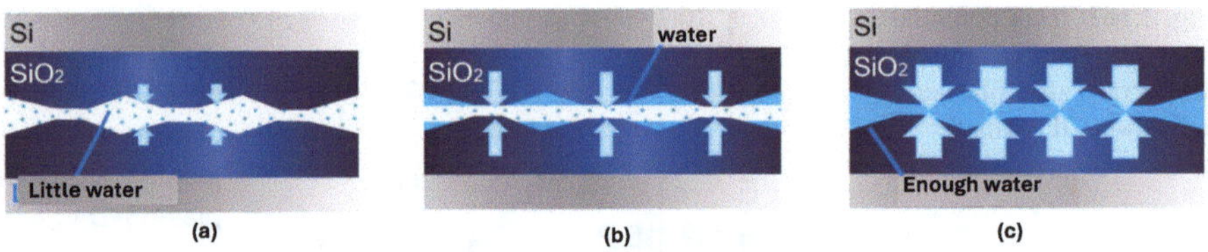


**Fig. 2.163** Water content at the bonding interface, (**a**) low (**b**) middle, and (**c**) high

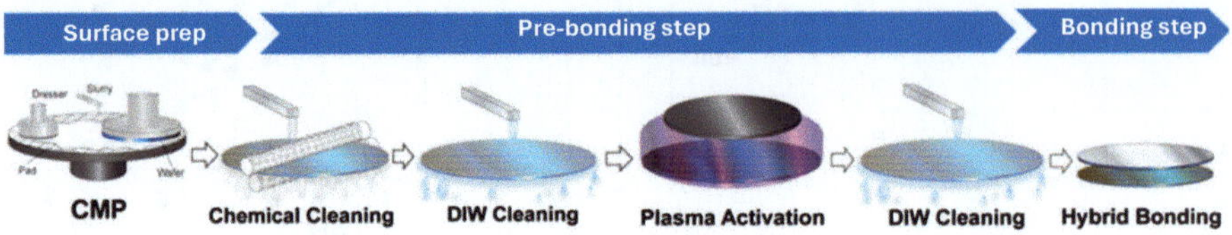


**Fig. 2.164** Process flow diagram of hybrid bonding process

## 2.18.3 *Wet-Chemical Cu Cleaning for Fine-Pitch Hybrid Bonding*

During IEEE/ECTC 2025, YOKOHAMA National University, imec, and Kurita Water Industries published another paper on wet-chemical Cu cleaning for fine-pitch hybrid bonding [59]. In [59], they focused on the post-CMP cleaning process, evaluating functional water compared to formulate chemical cleaners. Figure 2.164 shows the process flow diagram of hybrid bonding such as the surface preparation, pre-bonding, and bonding. Figure 2.165a shows the electrochemical quartz crystal microbalance (EQCM) mass loss monitoring and Fig. 2.165b shows the open-circuit potential (OCP) on Cu for acidic and alkaline cleaners, DIW, and functional water. It can be seen that (a) the mass loss in acidic cleaner during 10 min of exposure decreases monotonically, (b) in the alkaline cleaner, the mass rapidly decreases in the first 100 s, (c) the OCP in the alkaline cleaner also decreases especially in the first 100s, and (d) in the functional water, the mass loss is substantially lower than the other three solutions.

Figure 2.166 shows the x-ray photoelectron spectroscopy (XPS) $Cu^{2p}$ spectrum (a) and XPS $N^{1s}$ spectrum (b) before and after cleaning. It can be seen from Fig. 2.165a that (a) on the bare Cu sample, the strong peaks at 934eV and 932.6eV corresponding to CuO, and Cu and Cu2O are detected. (b) the satellite peak at 940—945eV corresponding to CuO is also detected, (c) for the before cleaning (after CMP) samples, Cu, and Cu2O peaks are detected, (c) after cleaning with acidic cleaner, alkaline cleaner, and functional water, the Cu, and Cu2O peaks are still detected.

From Fig. 2.166b, it can be seen that (a) prior to cleaning (after CMP), a strong peak at 399.7eV, corresponding to Cu-benzotriazole (Cu-BTA), is observed. (b) the Cu-BTA peak remains after cleaning with both the acidic cleaner and functional

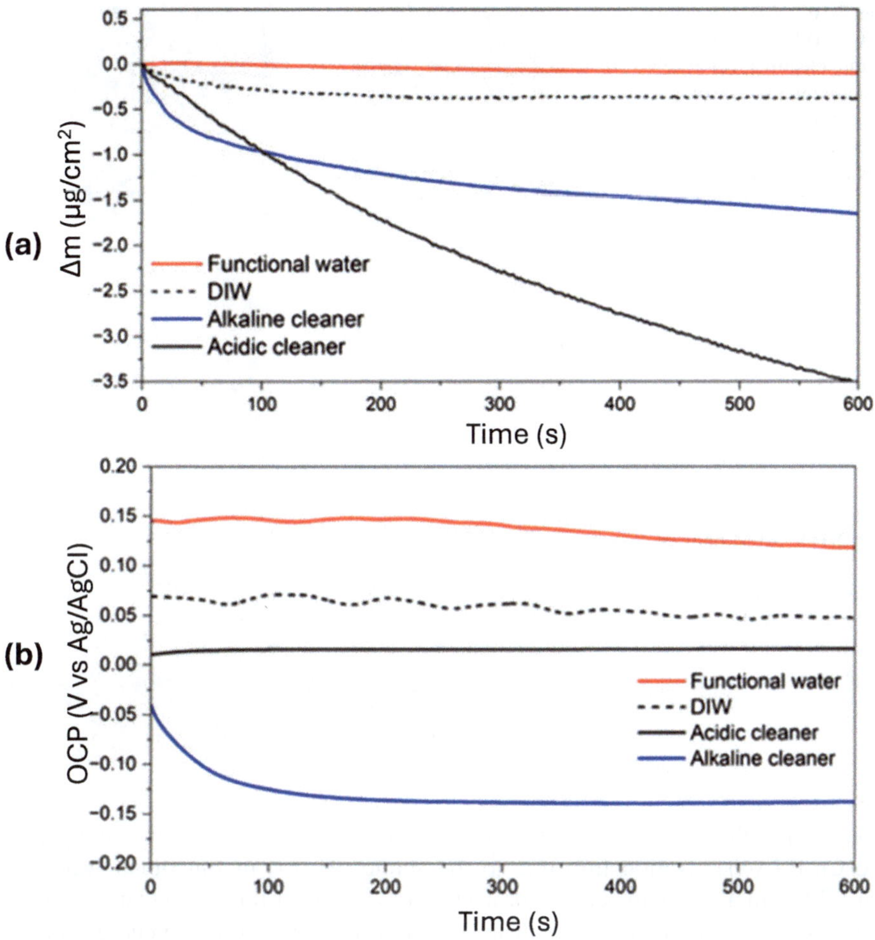


**Fig. 2.165** (**a**) EQCM mass loss monitoring, (**b**) OCP on Cu for acidic and alkaline cleaners, DIW, and functional water

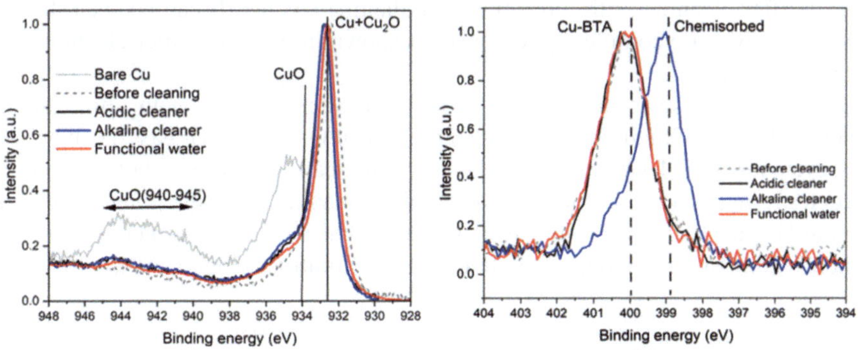


**Fig. 2.166** (**a**) XPS Cu2p spectrum before and after cleaning. (**b**) XPS N1s spectrum before and after cleaning

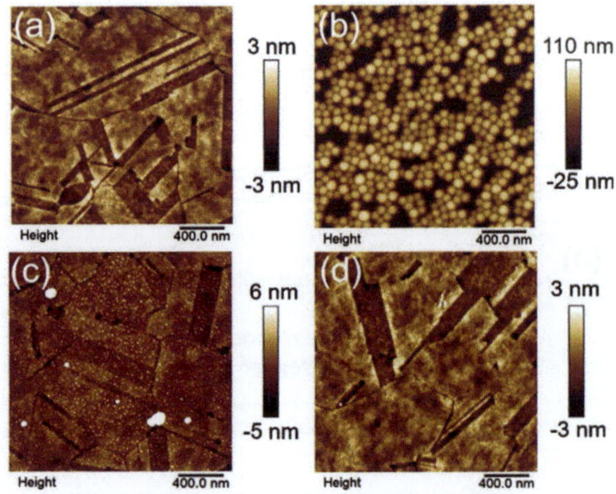


**Fig. 2.167** AFM images of Cu (**a**) bare Cu, (**b**) before cleaning, (**c**) after cleaning with ultrasonic and acidic cleaner, (**d**) after cleaning with ultrasonic and alkaline cleaner

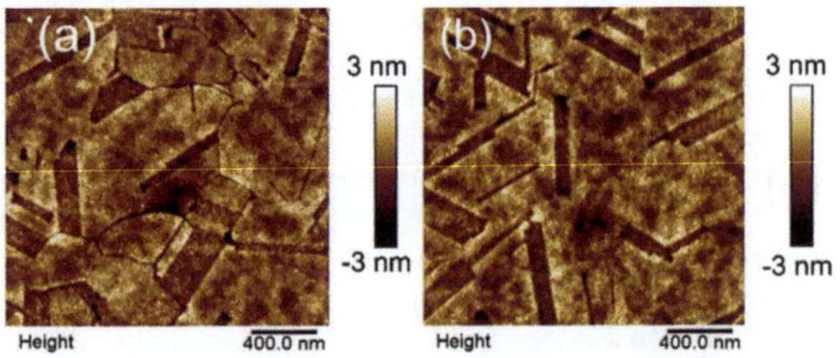


**Fig. 2.168** AFM images of Cu after cleaning with ultrasonic (**a**) functional water without H2 gas, (**b**) functional water with H2 gas

water, (c) in the case of the acidic cleaner, Cu is readily dissolved through gaps in the Cu-BTA passivation film due to the acidic pH, (d) in functional water, the Cu-BTA film appears to act as a protective layer against further oxidation, and (e) after alkaline cleaning, the main peak shifts toward a lower binding energy, and the Cu-BTA peak disappears, and a new strong peak at 398.7eV is detected.

Figure 2.167 shows the AFM images of Cu after being dipped in the slurry: (a) bare Cu, (b) before cleaning, (c) after cleaning with ultrasonic and acidic cleaner, and (d) after cleaning with ultrasonic and alkaline cleaner. It can be seen that the slurry abrasives are completely removed, as shown in Fig. 2.6d, after cleaning with the alkaline cleaner.

Figure 2.168 shows the AFM images of Cu after cleaning with ultrasonic (a) functional water without $H_2$ gas, and (b) functional water with $H_2$ gas. It can be seen that even functional water without $H_2$ gas can clean the surface, but some small particles remain on the surface. However, after cleaning with functional water containing $H_2$ gas, both slurry abrasives and small particles are completely removed, which is comparable to that of alkaline cleaning solutions (Fig. 2.169).

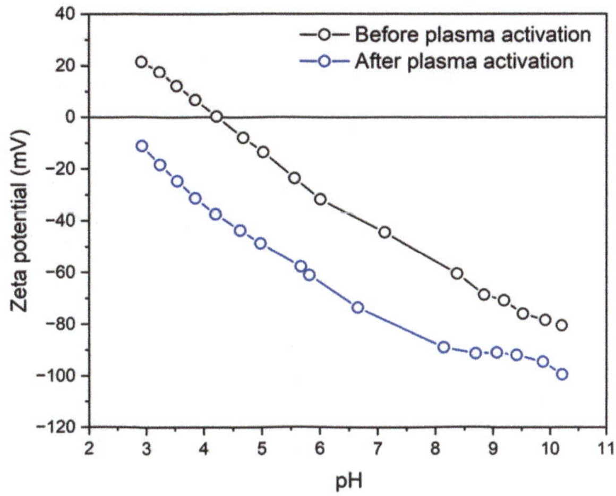


**Fig. 2.169** Zeta potential of SiCN surfaces before and after plasma activation

## 2.19 Unimicron's Hybrid Bonding

### *2.19.1 Flip Chip on Glass-Core Substrate with Cu-Cu Hybrid Bonding*

During IEEE/ECTC 2025, Unimicron published one paper on flip chip on glass-core substrate with hybrid bonding [60]. They studied the solder joint reliability of the flip chip on glass-core package substrate with Cu-Cu hybrid bonding. Figure 2.170 shows the cross section of a flip chip on glass-core substrate with Cu-Cu hybrid bonding and then solder balled on a printed circuit board (PCB). The silicon chip (10 mm × 10 mm × 350 μm) with Cu-pad (5 μm × 5 μm × 1.5 μm) is bonded to a glass-core build-up package substrate (20 mm × 20 mm × 815 μm) with Cu-Cu hybrid bonding. There are two build-up layers on the glass-core's top and bottom sides which are fabricated by PECVD (plasma-enhanced chemical vapor deposition) for the $SiO_2$ dielectric layer (1.5 μm-thick) and Cu dual-damascene for the metal layer (1.5 μm-thick). The Cu-pad pitch is 10 μm.

The temperature boundary condition is shown in Fig. 2.171a. Stress free is at room temperature (25 °C). The maximum deflection of the flip chip bonded on the glass-core package substrate with Cu-Cu hybrid bonding and then solder balled on a PCB is shown in Fig. 2.71b. It can be seen that the deformation shape of time history is following that of the temperature boundary condition.

The accumulated equivalent (von Mises) stress contours at the corner Cu-pads at 450s and 2250s are shown in Fig. 2.172. The maximum stress location is at the interface between the Si chip and the Cu-pad. This is due to the large thermal expansion mismatch between the silicon ($2.8 \times 10^{-6}$/°C) and the Cu ($16.3 \times 10^{-6}$/°C). However, the magnitudes of this stress are acceptable for most applications.

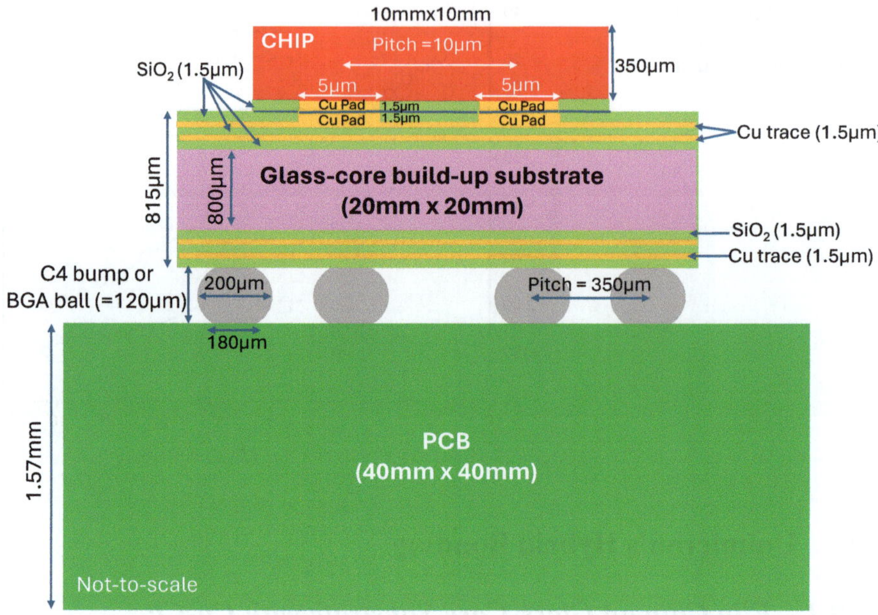


**Fig. 2.170** Cross section of flip chip on glass-core substrate with bumpless Cu-Cu hybrid bonding

The accumulated equivalent inelastic strain contours at the corner solder ball at 450s and 2250s are shown in Fig. 2.173. It can be seen that the maximum inelastic strain occurs near the interface between the solder ball and the PCB. This is due to the large thermal expansion mismatch between the glass-core ($4.8 \times 10^{-6}$/°C) and the PCB ($18 \times 10^{-6}$/°C). The time-history of accumulated equivalent inelastic strain is shown in Fig. 2.174a. It can be seen that the accumulated equivalent inelastic strain per cycle is 16.02%, which is quite high for the thermal fatigue life of the corner solder ball.

If the CTE of the glass core is increased from $4.8 \times 10^{-6}$/°C to $10 \times 10^{-6}$/°C (which is closer to the PCB) and subjected to the same temperature boundary condition (Fig. 2.171a), then the time-history of accumulated equivalent inelastic strain is shown in Fig. 2.174b. It can be seen that the be seen that the accumulated equivalent inelastic strain per cycle is reduced from 16% per cycle to 8% per cycle. This is due to the reduction of the CTE of the glass core material, i.e., smaller thermal expansion mismatch between the glass core and the PCB.

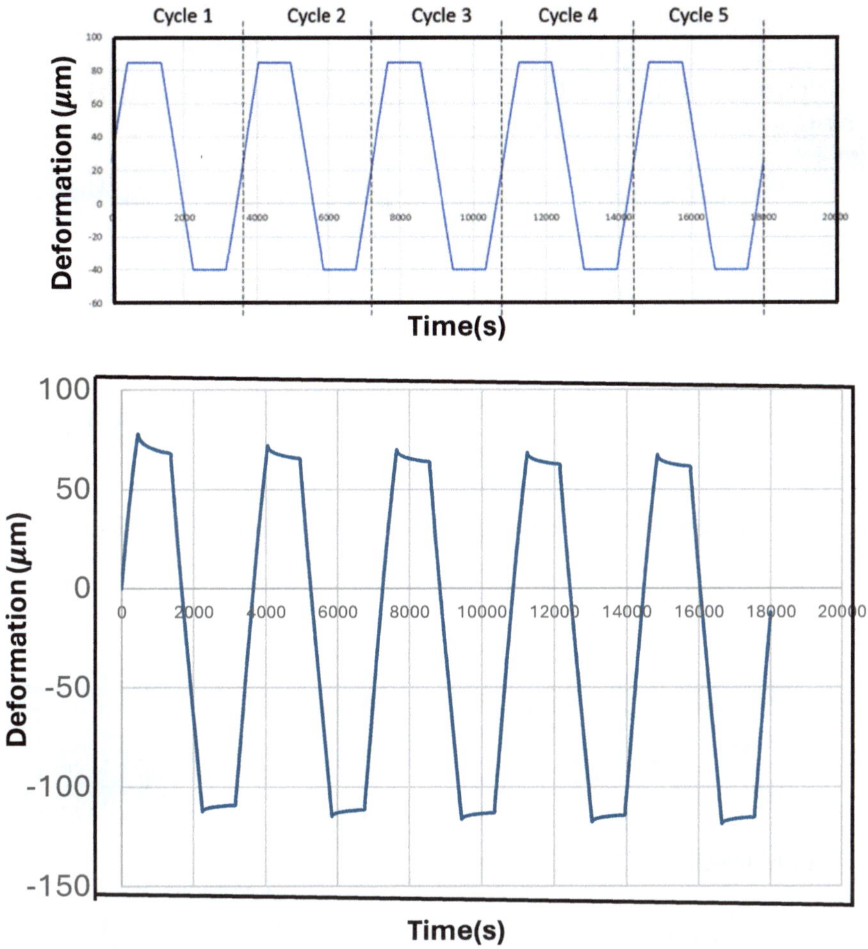


**Fig. 2.171** Top: Thermal cycling boundary condition. Bottom: Time-history deformation of the structure

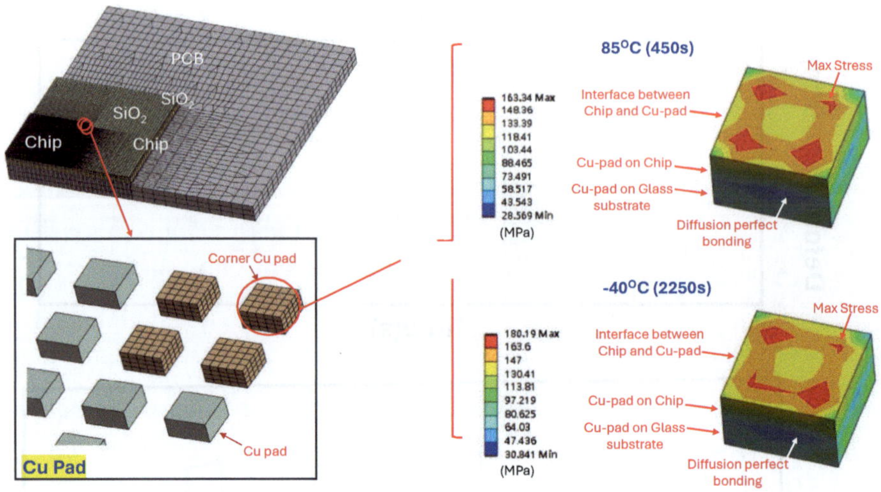


**Fig. 2.172** Accumulated equivalent (von Mises) stress contours at the corner Cu-pads at 450s and 2250s

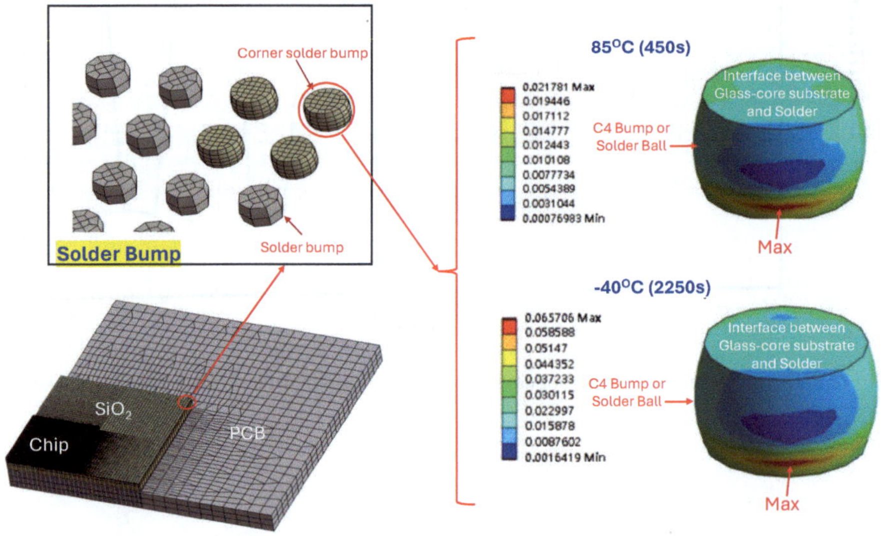


**Fig. 2.173** Accumulated equivalent inelastic strain contours at the corner solder ball at 450s and 2250s

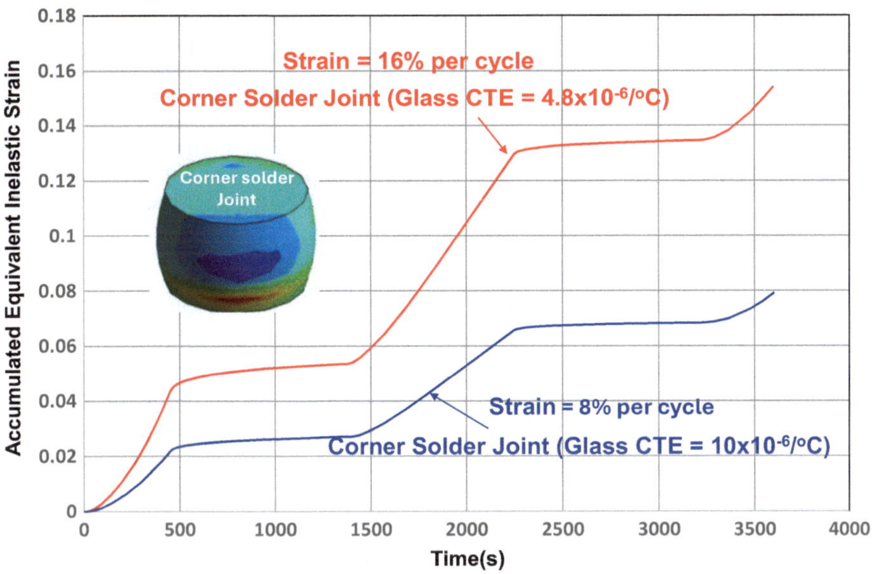


**Fig. 2.174** Accumulated equivalent inelastic strain at the corner solder joint (CTE = $4.8 \times 10^{-6}$/°C and $10 \times 10^{-6}$/°C)

## 2.20 Summary and Recommendations

Some important results and recommendations are summarized as follows.

- Hybrid bonding is an old flip chip assembly (packaging) technology. It has been in high volume manufacturing (HVM) since 2016. For examples, from the CIS for smartphones and cars by Sony, Samsung, OmniVision, etc., memory-on-logic and logic-on-I/O for HPC by Samsung, AMD, Graphcore, Broadcom (shipped in February 2026), etc., HBM by Samsung (2026), SK Hynix (2026), Micron (2027), etc., to the GPU-bridge-CPU for notebooks by Apple (shipped in Q1 of 2026).
- Because of the great demand for CIS in self-drive cars and mobile products, and HPC driven by data center, 5G and AI, the use of Cu-Cu hybrid bonding will be increased substantially.
- Three-dimensional NAND flash memory is a 3D monolithic hybrid bonding manufacturing method within the chip (not a packaging technology). YMTC and Western Digital have been manufacturing their 3D NAND flash by monolithic hybrid bonding within a chip.
- Just like the 3D NAND, 3D DRAM is a 3D monolithic hybrid bonding within a chip manufacturing method, which is a very hot topic today. Products are planned to be shipped in 2030.
- The works by Sony, Samsung, Applied Materials, EV Group, imec, Daicel Corporation, High Energy Accelerator Research Organization, Toray Industries,

Entegris, National Yang Ming Chiao Tung University, Industrial Technology Research Institute, MediaTek, National Tsing Hua University, BASF, STMicroelectronics, CEA-Leti, Binghamton University, IBM, TAZMO, LINTEC, Institute of Science Tokyo, TSMC, Seoul National University of Science and Technology, YOKOHAMA National University, Tokyo Electron Kyushu Limited, SK Hynix, Tokyo Electron Limited, Toray, DISCO, University of Tsukuba, Kurita Water Industries, and Unimicron have been briefly mentioned in this report. The Cu-Cu bonding pad pitches, as small as 100 nm [15] and 300 nm [19], have been studied.

- As of today, there are not many papers published in the quality and reliability (such as the thermal-cycling, thermal shock, mechanical vibration, mechanical shock, high-temperature storage, highly accelerated temperature, highly accelerated moisture resistance etc.) of Cu-Cu hybrid bonding. In order for the hybrid bonding to be more popular, these quality and reliability data are desperately needed [61]. Thermal management of Cu-Cu hybrid bonding will be an important topic.
- Also, we would like to know the answers to the following questions, (a) What is the effect of Cu-Cu bonding misalignment on the quality and reliability of Cu-Cu hybrid bonding interconnect? (b) What is the effect of bonding interface void (size, location, and propagation) on the quality and reliability of Cu-Cu hybrid bonding interconnect?
- The most important equipment and process steps in hybrid bonding are CMP for Cu recess (~3 nm) and surface planarization, and high precision pick and place equipment for alignment accuracy ($\leq \pm 0.5$ μm). Advanced cleaning techniques and plasma activation to improve bond strength are very important research topics for hybrid bonding.
- Unlike the organic-core substrate, glass-core substrate is flat enough to perform Cu-Cu hybrid bonding. The current research/practice is to use glasses with thermal coefficient of expansion (CTE) as close as to the silicon chip ($2.8 \times 10^{-6}$/°C). Thus, the thermal expansion mismatch between the glass substrate and PCB ($18.5 \times 10^{-6}$/°C) is very larger and the solder joint on PCB is subjected to very large strain and stress. A larger CTE of glass substrate such as $10 \times 10^{-6}$/°C is recommended.
- The hybrid bonding process can be CoW, WoW, WoWoW, CoWoW, reconstructed wafer, direct transfer bonding (DTB), etc. A comparison between CoW, WoW, WoWoW, reconstructed wafer, and DTB processes, based on their advantages, disadvantages, and applications is shown in Table 2.6.

**Table 2.6** Comparison of CoW, WoW, WoWoW, reconstructed wafer, and direct transfer bonding

| Hybrid Bonding | Advantages | Disadvantages (Challenges) | Applications |
|---|---|---|---|
| CoW | High assembly yield with KGD<br>Design flexibility | The edge effects due to singulation<br>Contaminants due to singulation<br>Particles due to singulation<br>Higher accuracy pick & place (P&P) machines<br>Slightly larger pads to compensate the P&P tolerance<br>CMP for metal recess, clean, and flat surface<br>Topography control of the Cu/dielectric surface | AI, HPC, Machine leaning<br>Logic, SoC, HBM |
| WoW | High throughput<br>Mutual technology<br>High Volume manufacturing | High assembly yield loss<br>Similar die sizes<br>Topography control of the Cu/dielectric surface<br>CMP for metal recess, clean, and flat surface | AI, HPC, Machine leaning<br>Logic, SoC<br>CIS |
| WoWoW | High throughput | More process steps<br>Alignment issues | CIS |
| Reconstructed Wafer | High throughput | More process steps<br>Alignment issues | AI, HPC, Machine leaning<br>Logic, SoC, HBM |
| Direct Transfer Bonding | High throughput & low cost<br>High accuracy | No HVM experience | AI, HPC, Machine leaning<br>Logic, SoC, HBM |

**Acknowledgements** The author would like to thank all the authors of the papers cited in this chapter for their significant contributions to Cu-Cu hybrid bonding.

## References

1. J. H. Lau, "Recent Advances and Trends in Advanced Packaging", *IEEE Transactions on CPMT*, Vol. 12, No. 2, February 2022, pp. 228–252.
2. J. H. Lau, "Recent Advances and Trends in Cu-Cu Hybrid Bonding", *IEEE Transactions on CPMT*, Vol. 13, No. 3, March 2023, pp. 399–425.
3. J. H. Lau, "State of the art of Cu-Cu Hybrid Bonding", *IEEE Transactions on CPMT*, Vol. 14, No. 3, March 2024, pp. 376–396.
4. J. H. Lau, "Current Advances and Outlooks in Hybrid Bonding", *IEEE Transactions on CPMT*, Vol. 15, No. 3, April 2025, pp. 651–681.
5. K. Shimizu, T. Kamibayashi, K. Saito, N. Araki, R. Nakamura, W. Otsuka, Y. Kagawa, and H. Iwamoto, "Development of a Novel WoWoW Process for 1/1.3-Inch 50 Megapixel Three-Wafer-Stacked CMOS Image Sensor with DNN Circuits", *IEEE/ECTC Proceedings*, May 2025, Dallas, Texas, pp. 559–564.

6. A. Urata, T. Kaei, I. Imanishi, M. Chiyozono, T. Osako, K. Shimizu, Y. Kagawa, and H. Iwamoto, "Advanced Face-To-Back CoW 2.0-μm Pitch Cu-Cu Hybrid Bonding Process for Three Layer-Stacked 3D Heterogenous Integration", *IEEE/ECTC Proceedings*, May 2025, Dallas, Texas, pp. 1896–1900.
7. K. Onishi, S. Saito, T. Osako, T. Hirano, T. Amei, N. Ogawa, S. Saito, Y. Hagimoto, and H. Iwamoto, "Influences of Chip Shape on Scaling in Chip-on-Wafer Hybrid Bonding", *IEEE/ECTC Proceedings*, May 2025, Dallas, Texas, pp. 297–301.
8. T. Hirano, T. Yamada, Y. Yamano, N. Komai, T. Onodera, Y. Ikegami, S. Kobayashi, Y. Hagimoto, and H. Iwamoto, "Simulation of Mechanical Cu-Pad Expansion Mechanism and Measures to Increase Expansion", *IEEE/ECTC Proceedings*, May 2025, Dallas, Texas, pp. 1488–1492.
9. M. Haneda, T. Miyawaki, Y. Fukumizu, K. Kubota, K. Ono, T. Hiratsuka, K. Shimizu, Y. Kagawa, and H. Iwamoto, "Study of 2 × 4 μm Pitch Capsule Shaped TSVs and 2 μm Pitch TSVs in WoWoW Integration", *IEEE/ECTC Proceedings*, May 2025, Dallas, Texas, pp. 1347–1352.
10. S. Hahn, W. Kim, S. Son, K. Lee, J. Lee J. Jang, K. Lim, B. Moon, M. Rhee, and K. Moon, "Development of Wafer-Level Wet Atomic Layer Etching Process Platform for Cu Surface Topography Control in Hybrid Bonding Applications", *IEEE/ECTC Proceedings*, May 2025, Dallas, Texas, pp. 554–558.
11. P. Kang, J. Lee, M. Kim, E. Kim, K. Lee, C. Lee, J. Park, M. Rhee, K. Moon, and S. Hyun, "A Novel 3D Heterogeneous Integration using 2 μm Bond Pitch Die-to-Wafer Hybrid Cu Bonding and Wafer Reconstruction Process", *IEEE/ECTC Proceedings*, May 2025, Dallas, Texas, pp. 16–19.
12. S. Kim, Y. Kim, C. Shin, G. Kim, M. Je, S. Han, K. Lee, H. Lee, H. Shin, H. Jeong, S. Son, H. Bae, S. Kim, D. Lim, Y. Kim, H. Lee, S. Lee, S. Lee, K. Moon, and D. Kim, "Simulation Analysis of Edge Dot-Void Formation and Edge Roll-Off Effects in Wafer-to-Wafer Bonding", *IEEE/ECTC Proceedings*, May 2025, Dallas, Texas, pp. 1242–1246.
13. K. Ma, L. Xue, E. Bergman, N. Materials, T. Sekine, C. Hsu, S. Rath, T. Ding, B. Weis, G. Probst, T. Uhrmann, and M. Wimplinger, "Process Technology Enablers for Aggressive Scaling of Hybrid Bonding Temperature and Pitch", *IEEE/ECTC Proceedings*, May 2025, Dallas, Texas, pp. 1901–1907.
14. P. Lianto, R Raphael, A. Tan, C. Chia, X. Li, H. Lee, and X. Chen, "SiCN CMP Integration for Hybrid Bonding Application", *IEEE/ECTC Proceedings*, May 2025, Dallas, Texas, pp. 347–350.
15. K. Ryan, R. Sreenivasan, S. Ramamoorthy, R. Patlolla, J. Appell, J. Hebding, S. Krishnan, M. Chudzik, B. Weis, G. Probst, and B. Theile, "Integration, Materials and Equipment Innovations to Enable 100 nm Pitch W2W Bonding for Memory-to-Logic and Logic-to-Logic 3D Stacking", *IEEE/ECTC Proceedings*, May 2025, Dallas, Texas, pp. 542–546.
16. X. Chen, G. See, Y. Lim, P. Lianto, P. Suo, C. Andy, S. Rath, and X. Zhao, "Integration Solution for Thin D2W Hybrid Bonding for Yield and Reliability", *IEEE/ECTC Proceedings*, May 2025, Dallas, Texas, pp. 33–36.
17. S. Kang, S. Iacovo, K. D, O. Okudur, A. Alexeev, T. Plach, T. Ding, and Markus Wimplinger, "Wafer-to-Wafer Bonding With Saddle-Shaped Wafers", *IEEE/ECTC Proceedings*, May 2025, Dallas, Texas, pp. 2028–2033.
18. Y. Lin, P. Bex, S. Suhard, B. Zhang, F. Okudur, A. Zerio, N. Reddy, E. Sanchez, Y. Li, A. Jourdain, G. Beyer, and E. Beyne, "2 μm Pitch Direct Die-to-Wafer Hybrid Bonding using Surface Protection During Wafer Thinning and Die Singluation", *IEEE/ECTC Proceedings*, May 2025, Dallas, Texas, pp. 287–291.
19. S. Huylenbroeck, S. Chew, B. Zhang, L. Bogaerts, C. Heyvaert, S. Dewilde, S. Iacovo, M. Stucchi, J. Vos, G. Beyer, and E. Beyne, "Wafer-to-Wafer Hybrid Bonding Technology with 300nm Interconnect Pitch", *IEEE/ECTC Proceedings*, May 2025, Dallas, Texas, pp. 565–569.

20. K. Kennes, P. Bex, Y. Lin, S. Suhard, D. Cuypers, A. Phommahaxay, G. Beyer, and E. Beyne, "Impact of Temporary Substrates and Adhesives on Die-to-Wafer Overlay", *IEEE/ECTC Proceedings*, May 2025, Dallas, Texas, pp. 1934–1038.
21. O. Okudur, S. Iacovo, S. Kang, D. Sharma, M. Gonzalez, and E. Beyne, "Mitigation of Wafer-to-Wafer Bonding Distortions Through Accelerated Simulations and Measurements", *IEEE/ECTC Proceedings*, May 2025, Dallas, Texas, pp. 1650–1655.
22. S. Huylenbroeck, S. Chew, B. Zhang, E. Chery, J. Messemaeker, P. Gupta, N. Jourdan, S. Dewilde, Z. El-Mekki, J. Vos, G. Beyer, and E. Beyne, "Self-formed Barrier using Cu-Mn Alloy Seed applied to a 400nm Pitch Wafer-to-Wafer Hybrid Bonding Technology", *IEEE/ECTC Proceedings*, May 2025, Dallas, Texas, pp. 342–346.
23. T. Widodo, P. Lianto, K. Prabhakar, J. Xu, A. Tan, P. Lim, G. See, and X. Brun, "Electrical Performance of CMP Process for Hybrid Bonding Application with Conventional / nt-Cu and Low Temperature of SixNy / SixOy Dielectrics", *IEEE/ECTC Proceedings*, May 2025, Dallas, Texas, pp. 1837–1842.
24. G. Naderi, A. Elsherbini, B. Rawlings, S. Liff, S. Chauhan, Q. Yu, A. Rahimi, C. Cooper, T. Bejitual, P. Shivapooja, M. Stolt, and M. Hossain, and C. Atay, "Enabling Chip-to-Wafer Hybrid Bonding Scaling to 1um Pitch With Optimal Power Delivery Using New Bond Via Architectures", *IEEE/ECTC Proceedings*, May 2025, Dallas, Texas, pp. 20–24.
25. F. Eid, C. Helou, S. Sinha, G. Naderi, F. Mahmoudabadi, B. Rawlings, B. Barley, T. Mahdi, S. Benson, S. Dong, Y. Shi, W. Brezinski, F. Bedoya, W. Li, and F. Bard, "Hybrid Bonding with Fluidic Self Alignment: Process Optimization and Electrical Test Vehicle Fabrication", *IEEE/ECTC Proceedings*, May 2025, Dallas, Texas, pp. 868–872.
26. W. Li, F. Eid, A. Vyatskikh, R. Vreeland, W. Brezinski, M. Njuki, C. Jezewski, R. Mongia, K.Valavala, H. Mei, and C. Bakre, "Wafer-to-Wafer Bonding With Ultralow Thermal Resistance and High Bonding Energy", *IEEE/ECTC Proceedings*, May 2025, Dallas, Texas, pp. 83–86.
27. A. Elsherbini, T. Talukdar, T. Sounart, S. Chauhan, K. Radhakrishnan, Y. Shi, H. Niazi, P. Nordeen, A. Vyatskikh, B. Rawlings, W. Brezinski, R. Vreeland, S. Chang, F. Eid, S. Liff, and J. Swan, "Mid-BEOL Heterogeneous Integration Through Sub-1um Pitch Hybrid Bonding & Advanced Silicon Carrier Technologies for AI & Compute Applications", *IEEE/ECTC Proceedings*, May 2025, Dallas, Texas, pp. 753–760.
28. R. Kawashima, H. Sakamoto, B. Tanaka, T. Tanaka, and T. Fukushima, "Characterization of Self-Nanoparticulated Cu-Cu Interconnection for Low-temperature Hybrid Bonding", *IEEE/ECTC Proceedings*, May 2025, Dallas, Texas, pp. 99–105.
29. M. Mariappan, S. Yamashita, and T. Fukushima, "Are Voids Restricted to Cu-Cu Bonding Interface? Truth Revealed by Conversion Electron Yield-Scanning Transmission X-ray Microscopy", *IEEE/ECTC Proceedings*, May 2025, Dallas, Texas, pp. 1296–1301.
30. B. Tanaka, M. Mariappan, S. Motoda, T. Nishiguchi, Y. Okadome, T. Tanaka, and T. Fukushima, "Ozone-Ethylene Radical Activation of SiCN/Cu without Water Rinsing for Hybrid Bonding", *IEEE/ECTC Proceedings*, May 2025, Dallas, Texas, pp. 1948–1951.
31. M. Mariappan, M. Ochi, and T. Fukushima, "High-Quality Cu μ-Joints by High-Throughput Contactless Hot-Isostatic-Pressure (HIP) Annealing for Chip-to-Wafer and Wafer-to-Wafer Hybrid Bonding", *IEEE/ECTC Proceedings*, May 2025, Dallas, Texas, pp. 1908–1914.
32. L. Xie, Y. Shoji, M. Jukei, K. Nomura, T. Fujiwara, H. Araki, S. Chong, and N. Vasarla, "Hybrid Bonding With Particle Accommodation Using Polymer Dielectric: Design, Process and Yield Study", *IEEE/ECTC Proceedings*, May 2025, Dallas, Texas, pp. 92–98.
33. H. Cheemalamarri, M. Fujino, C. Bhesetti, L. Long, M. Dileep, S. Vempati, and N. Singh, "Novel Selective Metal Capping on Cu pad Enabling Void-Free Hybrid Bonding at Low Thermal Budgets, Irrespective of Cu Topography and Microstructure", *IEEE/ECTC Proceedings*, May 2025, Dallas, Texas, pp. 576–581.
34. B. Rao, P. Lim, A. Sundaram, S. Rath, M. Kumar, C. Kumar, R. Umralkar, V. Sekhar, Y. Liang, H. Tupaen, G. See, and V. Rao, "Scalability and Process Optimization in Hybrid Bonding

with Fine-Pitch Interconnections", *IEEE/ECTC Proceedings*, May 2025, Dallas, Texas, pp. 318–323.
35. H. Feng, Y. Han, G. Tang, L. Xie, and V. Sekhar, "Thermal Characterization of HBMs Integrated via Hybrid Bonding", *IEEE/ECTC Proceedings*, May 2025, Dallas, Texas, pp. 1621–1627.
36. S. Chong, L. Xie, V. Sekhar, M. Kumar, C. Rao. and V. Rao, "Process Development, Challenges, and Strategies for Void-Free Multi-Chip Stacking in Hybrid Bonding Applications", *IEEE/ECTC Proceedings*, May 2025, Dallas, Texas, pp. 1943–1947.
37. M. Kumar, V. Sekhar, X. Ling, C. Kumar, S. Tippabhotla, B. Rao, H. Tupaen, S. Chong, and V. Rao, "Warpage Engineering in C2W Hybrid Bonding using Inter-Die Gap Fill Dielectrics for 2.5D/3D Integration", *IEEE/ECTC Proceedings*, May 2025, Dallas, Texas, pp. 302–307.
38. Y. Yeo, S. Bhuvanendran, N. Gourikutty, H. Cheemalamarri, S. Chong, E. Wai, and A. Fong, "Die-to-Die Bonding Development Employing Atomic Layer Deposition to Enable Heterogeneous Integration of Optoelectronic Integrated Circuits", *IEEE/ECTC Proceedings*, May 2025, Dallas, Texas, pp. 1805–1810.
39. S. Tippabhotla, M. Kumar, V. Sekhar, C. Kumar, B. Rao, and V. Rao, "Digital Design of Inter-Die Gap Fill Dielectric Film Processing for C2W Hybrid Bonding using Finite Element Modelling", *IEEE/ECTC Proceedings*, May 2025, Dallas, Texas, pp. 1644–1649.
40. V. Sekhar, R. Ivanov, J. Zhang, F. Acquaye, T. Masuda, S. Guddati, C. Ganguly, M. Kumar, B. Rao, and V. Rao, "Innovative Silicon Die Thinning and Edge Protection of Chip-to-Wafer Hybrid Bonded Wafer for High-Density Multi-Chip Stacking", *IEEE/ECTC Proceedings*, May 2025, Dallas, Texas, pp. 862–867.
41. A. Sundaram, C. Kumar, and B. Rao, "Patterning of Sub 1 μm CD TSV and Bond Pad for Fine-pitch 2.5D/3D Hybrid Bonding Applications", *IEEE/ECTC Proceedings*, May 2025, Dallas, Texas, pp. 1939–1942.
42. S. Yang, Y. Lin, W. Chiu, H. Chang, and C. Chen, "Enhanced Electromigration Reliability of Cu/$SiO_2$ Hybrid Joints Fabricated by (111)-Oriented Nanotwinned Cu", *IEEE/ECTC Proceedings*, May 2025, Dallas, Texas, pp. 653–658.
43. P. Chen, and C. Chen, "Effect of Dimension on Thermal Expansion of Cu Pads in $SiO_2$ Vias for 3D IC Fine-pitch Hybrid Bonding", *IEEE/ECTC Proceedings*, May 2025, Dallas, Texas, pp. 1929–1933.
44. Y. Chen, G. You, P. He, Y. Chung, R. Lee, C. Liu, C. Lee, and C. Chen, "Reliable Bonding Strength Measurement of SiCN/SiCN Films by Four-Point Bending Methodology", *IEEE/ECTC Proceedings*, May 2025, Dallas, Texas, pp. 2313–2317.
45. C. Li, Y. Liu, T. Sun, C. Hsiung, Y. Yang, W. Tsai, M. Hsu, Y. Huang, and K. Chen, "Fabrication Optimization and Reliability Study of Hyper RDL (HRDL) Interposer for Advanced Packaging and Heterogeneous Integration", *IEEE/ECTC Proceedings*, May 2025, Dallas, Texas, pp. 1134–1139.
46. Y. Liu, T. Chen, K. Ebisawa, M. Irie, C. Li, J. Lin, Y. Chen, Y. Chuang, H. Yeh, S. Fujimura, and K. Chen, "Ultra-Fast Cu/Polymer Hybrid Bonding with Electroless Passivation Layer for Cost-Effective High I/O Interconnection Stacking", *IEEE/ECTC Proceedings*, May 2025, Dallas, Texas, pp. 1302–1307.
47. L. Cheng, C. Wang, K. Lai, M. Hsu, and K. Chen, "First Demonstration of Superior Characteristics of Co-Co Bonding with Passivation Structure at Low Thermal Budget for Advanced Packaging and Ultra-Fine Pitch Applications", *IEEE/ECTC Proceedings*, May 2025, Dallas, Texas, pp. 357–361.
48. P. Lee, M. Hsu, Y. Tsai, A. Hou, W. Wu, and K. Chen, "Void Formation Elimination in Ultra-Thin Au-Sn Solid-Liquid Interdiffusion Bonding using Rapid Cooling Process for Advanced Packaging and MEMS Applications", *IEEE/ECTC Proceedings*, May 2025, Dallas, Texas, pp. 2243–2247.
49. S. Raghavan, B. Yang, N. Polomoff, and K. Sakuma, "Characterization of Interfacial Fracture Strength in Hybrid Bonded Wafers: A Novel Approach for High-Resolution Spatial Profiling", *IEEE/ECTC Proceedings*, May 2025, Dallas, Texas, pp. 2054–2058.

50. J. Lee, H. Lee, S. Oh, S. Park, W. Cho, Y. Park, A. Haag, S. Watanabe, M. Arnold, H. Lee, and E. Lee, "Crystal Plasticity-Based Modeling and Experimental Validation of the Influence of Microstructures and Grain Boundary Junction Types on the Cu-Cu Bonding Interface", *IEEE/ECTC Proceedings*, May 2025, Dallas, Texas, pp. 1211–1215.
51. M. Loyer, M. Maubert, M. Gottardi, P. Philip, G. Romero, E. Deloffre, and M. Calvo-Muñoz, "Morphological Microstructure Characterization and Optimisation of Nanocrystalline Copper Deposition for Fine-Pitch Hybrid Bonding Cu/$SiO_2$ at Low Temperature", *IEEE/ECTC Proceedings*, May 2025, Dallas, Texas, pp. 76–82.
52. S. Zerey, A. Bennett, M. Patel, J. Cho, R. Yu, N. Polomoff, L. Darling, and K. Sakuma, "Electrical Performance of Hybrid Bonding with Sub Micron Cu-Cu Bonding Contacts: Effects of Scaling, Microstructure, and Surface Morphology", *IEEE/ECTC Proceedings*, May 2025, Dallas, Texas, pp. 570–575.
53. I. Sano, M. Yamagishi, S. Takyu, T. Kirihata, R. Kitazawa, T. Fukushima, and Y. Kurita, "Direct Transfer Bonding Technology Enabling 50-nm Scale Accuracy for Die-to-Wafer 3D/ Heterogeneous Integration", *IEEE/ECTC Proceedings*, May 2025, Dallas, Texas, pp. 308–312.
54. S. Yang, and J. Lu, "Atomistic Modeling of Interfacial Cracking in Copper-to-Copper Direct Bonding", *IEEE/ECTC Proceedings*, May 2025, Dallas, Texas, pp. 1206–1210.
55. T. Ku, C. Tsai, R. Tsui, J. Twu, S. Lu, C. Hsieh, C. Liu, and D. Yu, "Innovative SoIC® Cool-Stacking Technology to Overcome the Thermal Wall of Future High Performance Compute", *IEEE/ECTC Proceedings*, May 2025, Dallas, Texas, pp. 292–296.
56. I. Kim, S. Lee, M. Kang, J. Jang, H. Jin, and S. Kim, "The Role of Preparation and Bonding Parameters in Improving Hybrid Bonding Quality", *IEEE/ECTC Proceedings*, May 2025, Dallas, Texas, pp. 1827–1831.
57. J. Sato, A. Nagata, H. Kitagawa, Y. Kondo, K. Saito, J. Park, C. Ahn, Y. Kim, and F. Inoue, "Bond Wave Analysis of SiCN for Fine Pitch Hybrid Bonding", *IEEE/ECTC Proceedings*, May 2025, Dallas, Texas, pp. 2046–2053.
58. Y. Yuki, J. Fuse, S. Aoki, T. Hre, K. Mihara, T. Miyoshi, N. Yamamoto, S. Teranishi, T. Fukushima, A. Uedono, and F. Inone, "Assessing Queue Time in D2W Hybrid Bonding Through Precise Bond Strength Measurements", *IEEE/ECTC Proceedings*, May 2025, Dallas, Texas, pp. 878–884.
59. K. Nakayama, K. Hayama, F. Tanaka, S. Dewilde, S. Deckers, N. Heylen, Y. Tanaka, Y. Okazaki, N. Gan, H. Iino, F. Inoue, and H. Philipsen, "Wet-chemical Cu Cleaning for Fine-Pitch Hybrid Bonding", *IEEE/ECTC Proceedings*, May 2025, Dallas, Texas, pp. 1859–1863.
60. J. H. Lau, N. Liu, M. Ma, and T. Tseng, "Glass-Core Substrate versus Organic-Core Substrate", *IEEE/ECTC Proceedings*, May 2025, pp. 1014–1020.
61. J. H. Lau and X. Fan, *Hybrid Bonding, Advanced Substrate, Failure Mechanisms, and Thermal Management for Chiplets and Heterogeneous Integration*, Springer, New York, May 2025.

# Chapter 3
# Glass Packaging

**Abstract** In this chapter, two problems of flip chip on glass-core package substrate will be investigated. The first problem deals with the flip chip on glass-core package substrate with microbumps and the other deals with the flip chip on glass-core package substrate with Cu-Cu hybrid bonding. Emphasis is placed on the solder joint reliability due to the glass-core substrate. Some recommendations will be provided. The advantages and disadvantages, glass-core substrate vs. glass-core interposer, through-glass via (TGV), and redistribution-layers (RDLs) will be briefly mentioned first.

## 3.1 Introduction

Some of the early research of glass in semiconductor packaging dated back almost 20 years ago. For example, the flip chip on glass substrate has been studied by Loughborough University [1, 2]. At Fraunhofer IZM, they have studied: (a) a thin glass based packaging technology for optoelectronic modules [3], (b) a glassPack SiP uses thin glass foils as substrate material [4], (c) a photonic SiP technology using thin glass substrates [5], and (d) 3D glass interposer for system-in-package (SiP) with integrated optical interconnects [6]. Georgia Institute of Technology (GIT) have done the most research in applying glass to semiconductor packaging, e.g., [7–24]. Another well-known company which has been working on applying glass to semiconductor packaging is Corning Research & Development Corporation, e.g., [25–32].

Absolics Inc., a US subsidiary of South Korean Chaebol SK Group (SKC) in Georgia, which is in tight cooperation with GIT, will be in high volume manufacturing of the glass-core package substrate and the TGV-interposer in 2026 [33].

On September 18, 2023, Intel announced they will use the glass-core substrate to replace the traditional organic-core substrate for their one-trillion transistors processors to be shipped before 2030 [34].


J. Lau, K.-N. Chiang, *Cu-Interconnects, Glass, and AI-Assisted Simulation for Chiplets and Heterogeneous Integration*,
https://doi.org/10.1007/978-981-95-6891-8_3

On April 23, 2025, during the TSMC 2025 North America Technology Symposium (April 23), the company announced its chip-on-panel-on substrate (CoPoS) to replace some of their chip-on-wafer-on substrate (CoWoS) for high-performance computing (HPC) driven by artificial intelligence (AI). The CoPoS products are planned to be shipped in Q1 of 2029 and the glass panel size is 310 mm × 310 mm. During February 2026, while ASE originally planned for a 600 mm × 600 mm panel line, it scaled back to 310 mm × 310 mm to align with TSMC's CoPoS. ASE aims to have a fully automated 310 mm × 310 mm production line operational for pilot production by the end of 2026. The 310 mm × 310 mm panel format strikes a critical balance. It is large enough to capture the area efficiency and small enough to maintain the processing yield and control.

## 3.2 Advantages and Disadvantages of Glass Packaging

The advantages of glass-core packaging compared with the organic-core packaging are: (a) ultra-high flatness for improved depth of focus for lithography, (b) dimensional stability needed for extremely tight layer to layer interconnect overlay, (c) higher interconnect density, (d) higher mechanical stability for ultra-large form-factor packages with high assembly yield, (e) improved flexibility in setting design rules for power delivery and signal routing, (f) better power delivery solutions while achieving high-speed signaling, (g) higher tolerance for higher temperatures, and (h) higher ability to seamlessly integrate optical interconnects.

The disadvantages of glass-core packaging compared with the organic-core packaging are: (a) higher material cost, (b) higher processing (production) cost, (c) higher yield loss, (d) fragility (easy to break), (e) handling, (f) difficult in making TGVs, (g) glass (softens) at high temperatures, and (h) special and expensive equipment,

## 3.3 Glass-Core Substrate vs. Glass-Core Interposer

Some of the glass packaging is shown in Fig. 3.1. It can be seen that the organic-core build-up package substrate, Fig. 3.1a, has been replaced by the glass-core build-up package substrate, Fig. 3.1b. Also, the traditional through-silicon via (TSV)-interposer [35] has been replaced with the TGV-interposer, Fig. 3.1c—a 2.5D IC integration. It can be seen that a TGV glass-core interposer that is supporting chips is also supported by an organic-core substrate as shown in Fig. 3.1c. Finally, a chip is embedded in the glass-core build-up package substrate—a 3D IC integration, Fig. 3.1d.

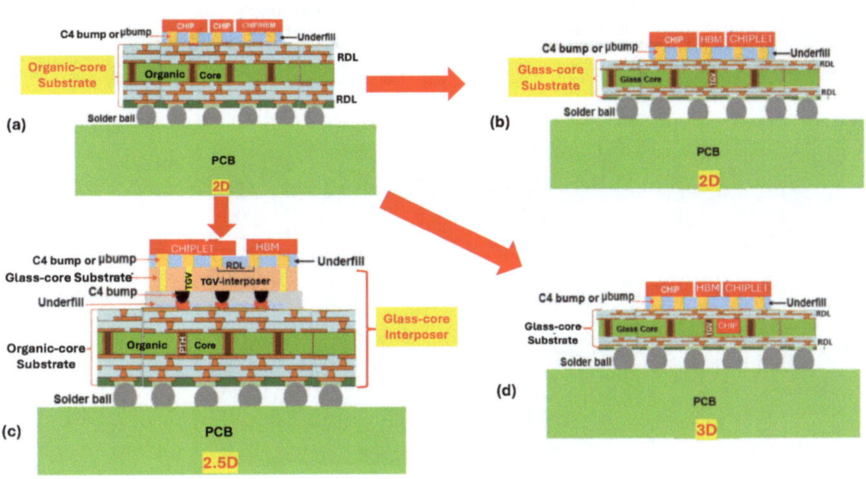


**Fig. 3.1** (**a**) Organic-core package substrate. (**b**) Glass-core package substrate. (**c**) TGV-interposer (2.5D IC integration). (**d**) Embedded chip in glass-core package substrate (3D IC integration)

## 3.4 Through-Glass Vias and RDLs

The fabrication process of TGV is very different from that of TSV. Most of the TSVs are fabricated by the deep reactive-ion etching (DRIE) (Sect. 1.5.2). However, today most of the TGVs are fabricated by laser drilling such as the laser induced deep etching (LIDE) developed by LPKF Laser & Electronics AG in 2017 [36–39]. The most common used glass materials are the SCHOTT AF 32 alkali-free flat glass and the CORNING HPFS 7980 high purity non crystalline fused silica glass. Both materials have low coefficients of thermal expansion (CTE), which are close to that of silicon.

The process flow for fabricating the TGV is shown in Fig. 3.2a. It can be seen that the vias are formed by high-speed laser and the modified area of the glass are removed by anisotropic wet chemical etching, e.g., hydrofluoric acid (HF) or sodium hydroxide (NaOH). Figure 3.2b shows a typical TGV. It can be seen that there is a taper angle of the TGV. This is due to the rate of the circulation and temperature of the etching solution, and the concentration difference of the etching solution. This process results in making the outer ionizer move inward and be replenished as shown in Fig. 3.2c. The higher the etching rate the larger the taper angle. It is followed by the metallization of the seed layer, which can be materials such as Ti/Cu, electroless Cu, etc. as shown in Fig. 3.3a. Then, electroplate the Cu to fill the via as shown in Fig. 3.3(b). Figure 3.3(c) shows a SEM image of the Cu-filled TGV. The process flow for fabricating the RDLs (build-up layers) of the glass core is very similar to the RDLs of the TSV interposer (Sect. 1.5.2): for

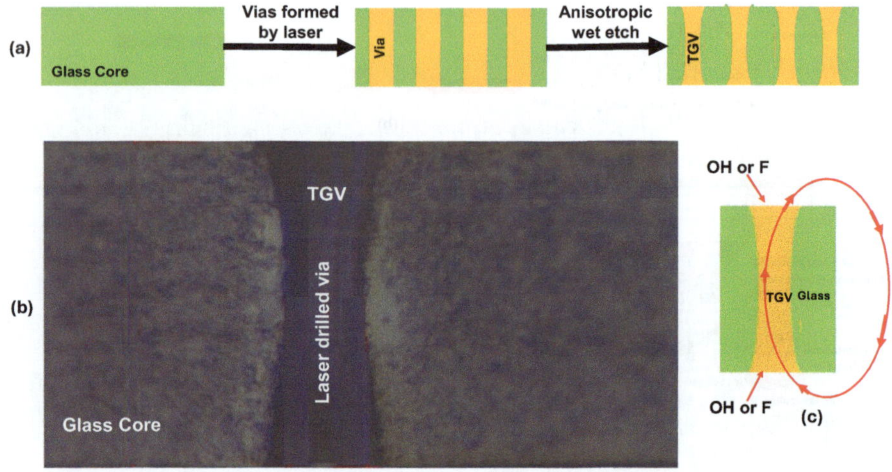


Fig. 3.2 (a) TGV process flow. (b) SEM image of a TGV. (c) Etch solution mechanism

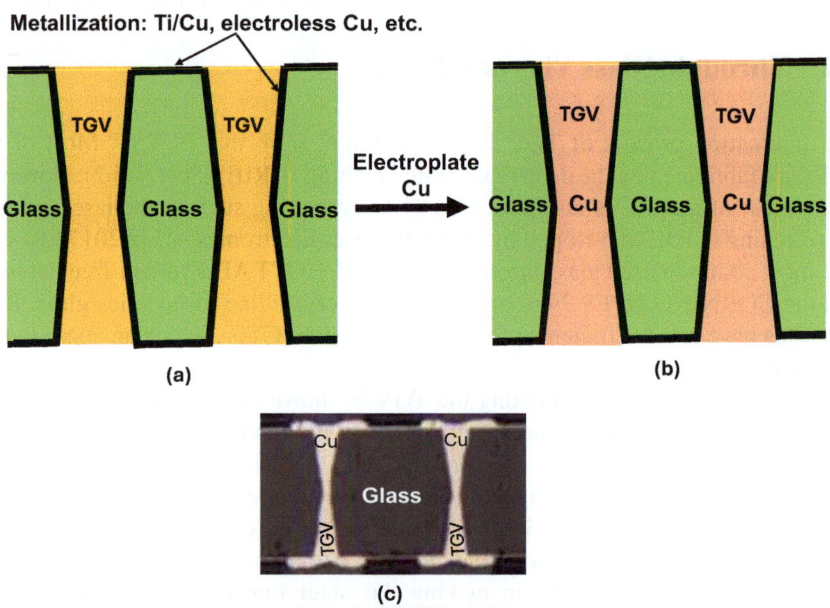


Fig. 3.3 (a) TGV sidewall metallization. (b) Cu plating to fill the via. (c) SEM image of a Cu-filled TGV

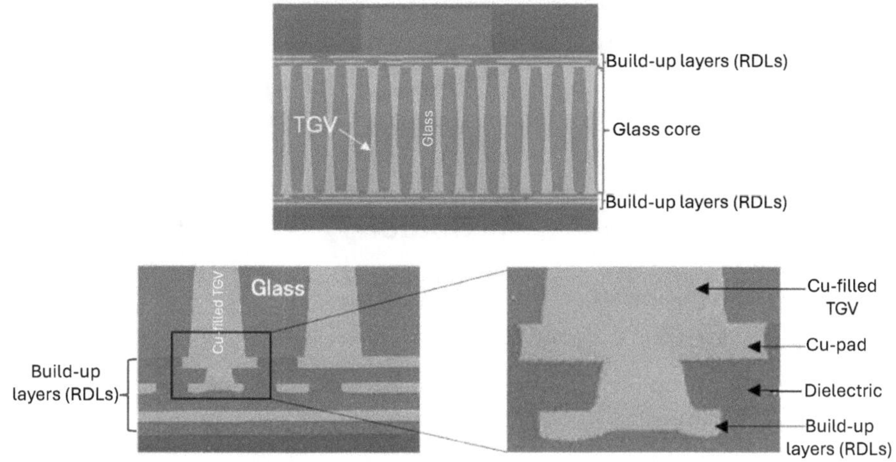


**Fig. 3.4** SEM image of a Cu-filled TGV with build-up layers (RDLs)

L/S $\geq$ 2 μm, the dielectric materials is either photoimageable dielectric (PID) or Ajinomoto Build-Up Film (ABF), and for L/S < 2 μm, the dielectric material is $SiO_2$. For L/S $\geq$ 10/15 μm, the dielectric material can be resin. Figure 3.4 shows the SEM images of the Cu-filled TGV with RDLs (build-up layers).

## 3.5 Effects of Glass-Core Substrate on Solder Joint Reliability

By now, reliability issues because of the glass-core substrate are seldom discussed, especially the semiconductor packaging interconnects such as the solder joint reliability. The trends of flip chip interconnects are shown in Fig. 3.5 [40–45]. Today, most C4 bump's diameters are 100 μm and are on 150 μm pitch. For some high-end applications, the diameter could be 50 μm and the pitch could be 100 μm. The 25 μm-diameter and 50 μm-pitch C4 bumps are in research and development (R&D). Today, most C2 bump's diameters are 50 μm and are on 100 μm pitch. For very high-end applications, the diameter is 25 μm and the pitch is 50 μm. The 10 μm-diameter and 20 μm-pitch are in R&D. Cu-Cu hybrid bonding is for $\leq$5 μm-diameter pads and $\leq$ 10 μm-pitch.

In this study, the solder joint reliability of semiconductor packaging, because of the glass-core substrate, is investigated and there are two problems. One is shown in Fig. 3.6 for high-performance and high-density products, and the other is shown in Fig. 3.7 for extremely fine-pitch, high-performance and high-density niche products such as the one-trillion transistors processors.

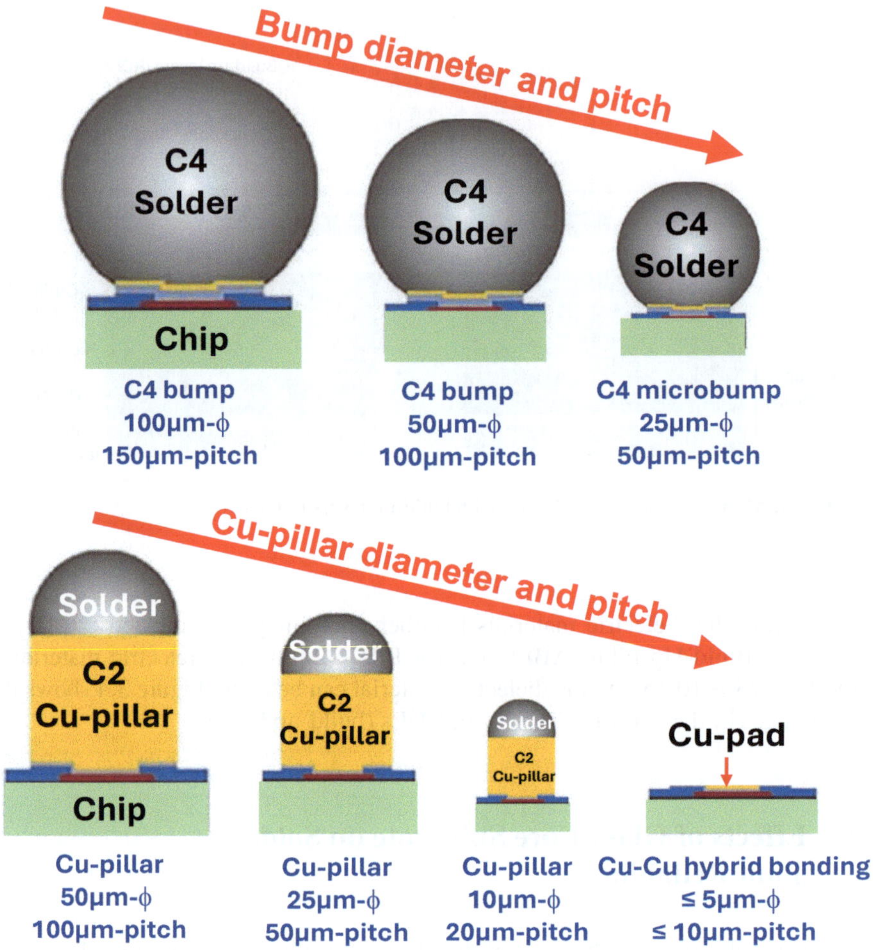


**Fig. 3.5** Trends in solder bump and Cu-pillar with solder cap in flip chip technology

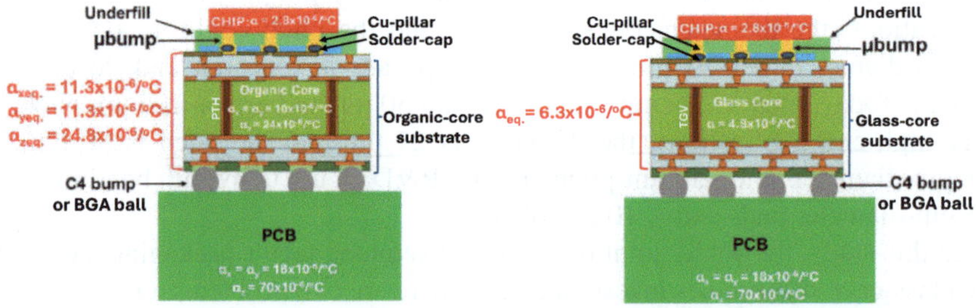


**Fig. 3.6** Cross section of the structures. (**a**) Organic-core substrate. (**b**) Glass-core substrate

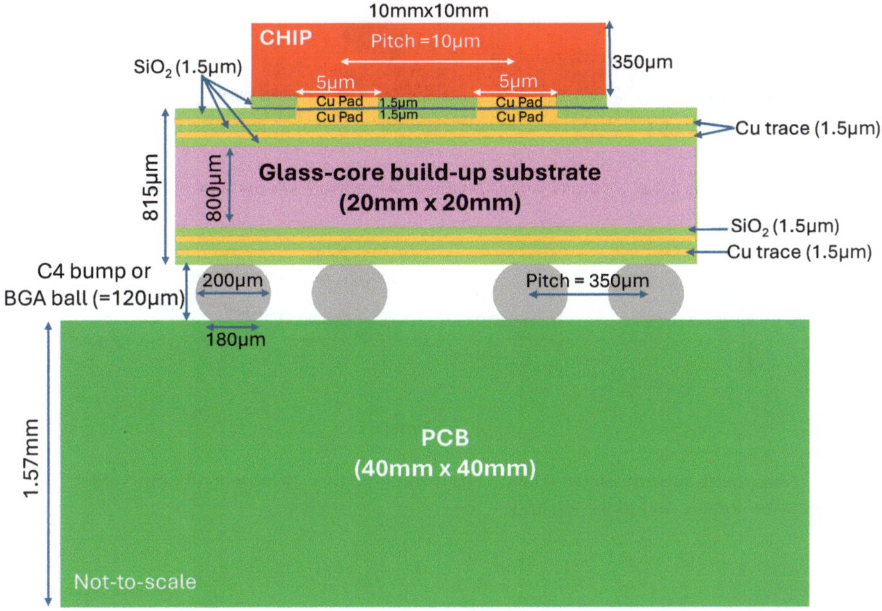


**Fig. 3.7** Cross section of flip chip on glass-core substrate with bumpless Cu-Cu hybrid bonding

## 3.6 Flip Chip on Glass-Core Substrates with μbumps

Figure 3.6 schematically shows the structures of Problem-One under consideration. From [46], we had shown that for the glass-core substrate, the accumulated equivalent inelastic strain in the μbump is 4.43% per cycle and in the solder joint is 19% per cycle (Figs. 3.8 and 3.9). For the organic-core substrate, the accumulated equivalent inelastic strain in the μbump is 9.12 per cycle and in the solder joint is 8.43% per cycle. The temperature boundary condition is −45 ↔ 125 °C. For more details, please read [46].

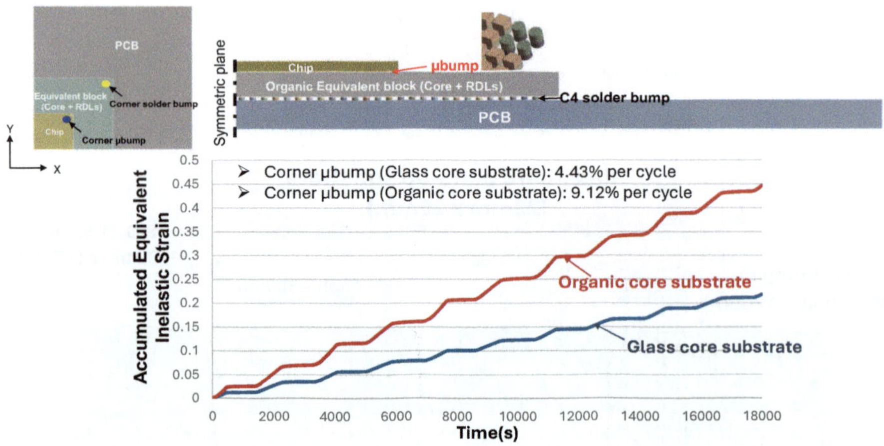


**Fig. 3.8** Maximum accumulated inelastic strain time-history in corner μbump solder joint of structures with organic and glass substrates

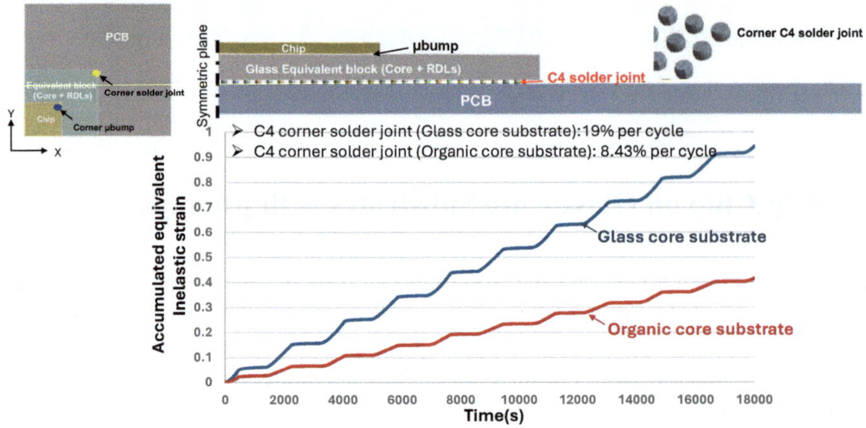


**Fig. 3.9** Maximum accumulated inelastic strain time-history in corner BGA solder joint of structures with organic and glass substrates

## 3.7 Flip Chip on Glass-Core Substrates with Cu-Cu Hybrid Bonding [47]

### *3.7.1 The Structure*

Figures 3.7 and 3.10 schematically show the structure of Problem-Two under consideration. The silicon chip (10 mm × 10 mm × 350 μm) with Cu-pad (5 μm × 5 μm × 1.5 μm) is bonded to a glass-core build-up package substrate

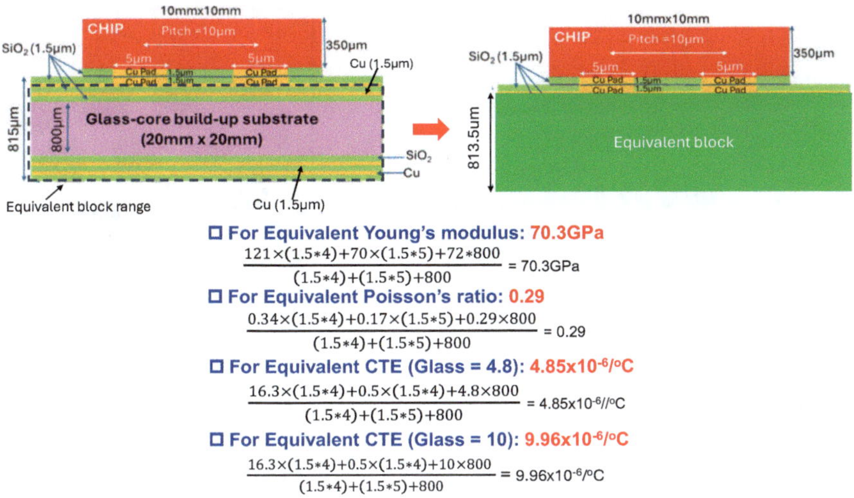


**Fig. 3.10** Flip chip on glass-core substrate with Cu-Cu hybrid bonding—equivalent block

(20 mm × 20 mm × 815μm) with Cu-Cu hybrid bonding. There are two build-up layers on the glass-core's top and bottom sides which are fabricated by PECVD (plasma-enhanced chemical vapor deposition) for the $SiO_2$ dielectric layer (1.5 μm-thick) and Cu dual-damascene for the metal layer (1.5 μm-thick) [47]. The Cu-pad pitch is 10 μm. The geometry and material properties of C4 bump (ball-grid array (BGA) solder joint) between the glass-core substrate and the PCB are the same as [46]. It should be emphasized that unlike the organic-core substrate, the glass-core substrate is flat enough to perform Cu-Cu hybrid bonding.

### *3.7.2 Finite Element Modeling*

Figures 3.10 and 3.11 show the finite element models for the chip, package substrate, PCB (printed circuit board), Cu-Cu bonding interconnect, and C4 bumps. Finner meshes are used for the higher stress/strain areas such as corner Cu-Cu pads connecting the chip and the glass-core build-up package substrate and the corner C4 bumps (solder balls) connecting the build-up glass-core package substrate and the PCB. The glass-core substrate is modeled as an equivalent block as shown in Fig. 3.10. The effective material properties of the equivalent block are shown in Table 3.1.

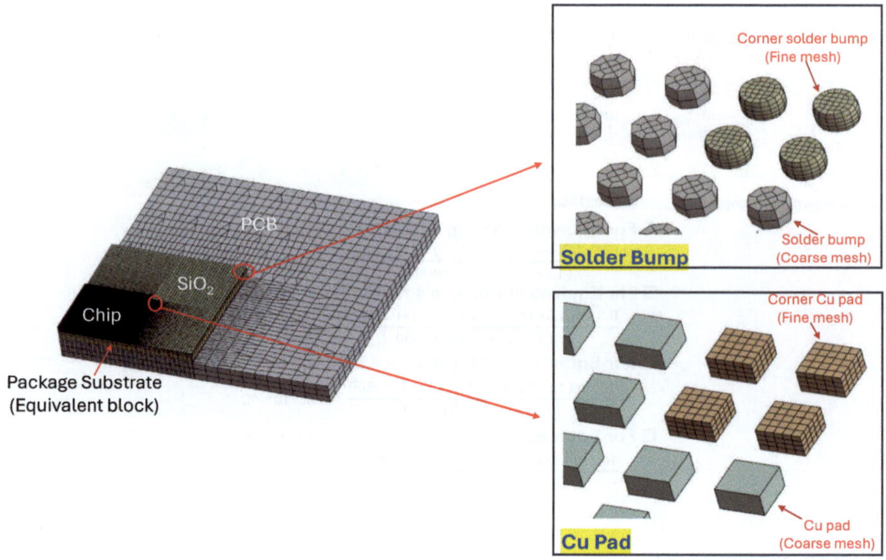


**Fig. 3.11** Finite element modeling of flip chip on glass-core substrate with Cu-Cu hybrid bonding

## 3.7.3 Materials Properties

The effective material properties of the glass-core build-up equivalent block can be calculated by those shown in Fig. 3.7 and are summarized in Table 3.1. The solder, Sn3Ag0.5Cu, obeys the Anand viscoplasticity constitutive equation [48] and the parameters of the Anand equation are shown in Table 3.2. The dual-damascene Cu is assumed to be an elastoplastic (bilinear kinematic hardening) material with the stress-strain relation shown in Fig. 3.12 (first Young's modulus = 121 Gpa, second Young's modulus = 1.2 Gpa, and yield strength = 173 MPa).

## 3.7.4 Temperature Boundary Condition

The temperature boundary condition is shown in Fig. 3.13. It can be seen that the temperatures are −40 °C↔85 °C, and the dwell-time at hot, dwell-time at cold, ramp-up time, and ramp-down time are 15 min each. Stress free is at room temperature (25 °C), just like the real thermal cycling test condition in a thermal cycling chamber.

**Table 3.1** Material properties for simulations

| Materials | Young's modulus (GPa) | Poisson's ratio | CTE (10 × $10^{-6}$/°C) |
|---|---|---|---|
| Silicon | 131 | 0.278 | 2.8 |
| C4 bump (SAC305) | 49–0.07 T(°C) | 0.3 | 21 + 0.017 T(°C) |
| PCB | $E_x = E_y = 22$; $E_z = 10$ | 0.28 | $\alpha_x = \alpha_y = 18$; $\alpha_z = 70$ |
| RDL (copper) cu | 121 | 0.34 | 16.3 |
| RDL (dielectric) $SiO_2$ | 70 | 0.17 | 0.5 |
| Glass core (1) | 72 | 0.29 | 4.8 |
| Glass core (2) | 72 | 0.29 | 10 |
| Equivalent block (glass core (1) + cu + $SiO_2$) | 70.3 | 0.29 | 4.85 |
| Equivalent block (glass core (2) + cu + $SiO_2$) | 70.3 | 0.29 | 9.96 |

**Table 3.2** Anand parameters for Sn3Ag0.5Cu (SAC305)

| Anand Model | |
|---|---|
| Parameters | SAC305 |
| $s_0$ (Pa) | 18.07 × $10^6$ |
| Q/R (K) | 9096 |
| A ($s^{-1}$) | 3484 |
| $\xi$ | 4 |
| m | 0.2 |
| $h_0$ (Pa) | 1.44 × $10^{11}$ |
| $\hat{s}$ (Pa) | 26.4 × $10^6$ |
| n | 0.01 |
| a | 1.9 |

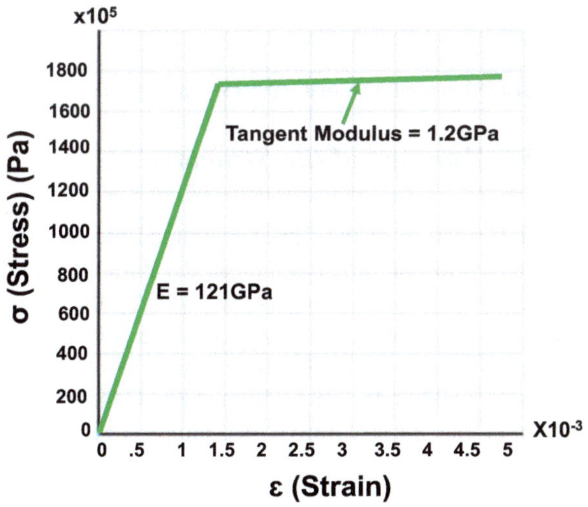


**Fig. 3.12** Elastoplastic of dual-damascene copper

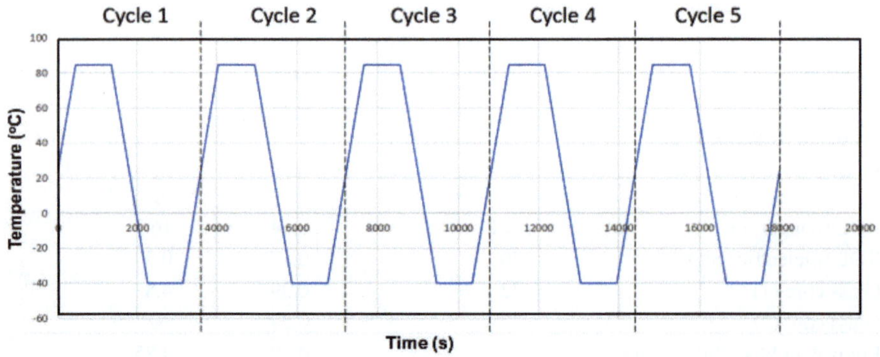


**Fig. 3.13** Thermal cycling boundary condition

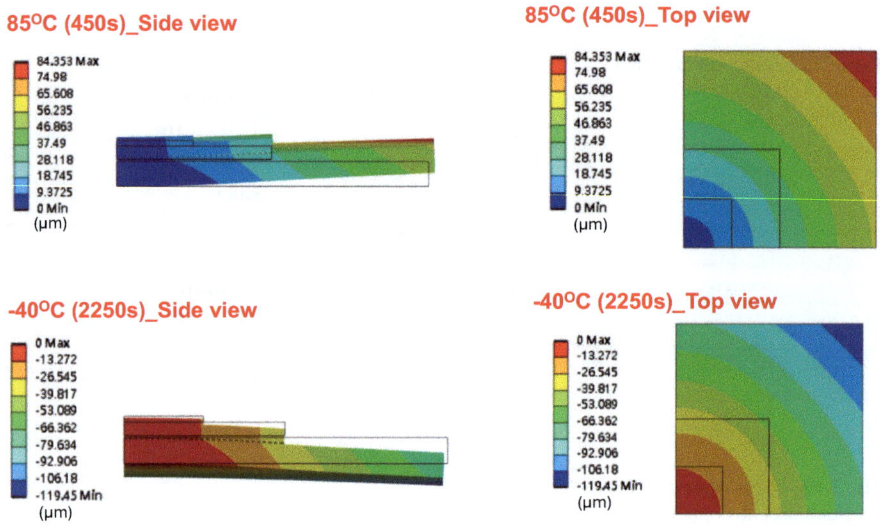


**Fig. 3.14** Structural deformation at 450 s and 2250 s

### 3.7.5 *Warpage of the Structure*

The deformation of the flip chip bonded on the glass-core package substrate with Cu-Cu hybrid bonding and then solder balled on a PCB is shown in Fig. 3.14. First of all, the deformation shapes are basically the same as those shown in Figure 6 of [46] (flip chip bonded on the glass-core package substrate with μbumps and then solder balled on a PCB). However, the deformation magnitudes of the structure with Cu-Cu hybrid bonding (84.35 μm at 85 °C (450 s) and −119.45 μm at −40 (2250 s)) are larger than the structure with μbumps (73.22 μm at 85 °C (450 s) and −103.48 μm

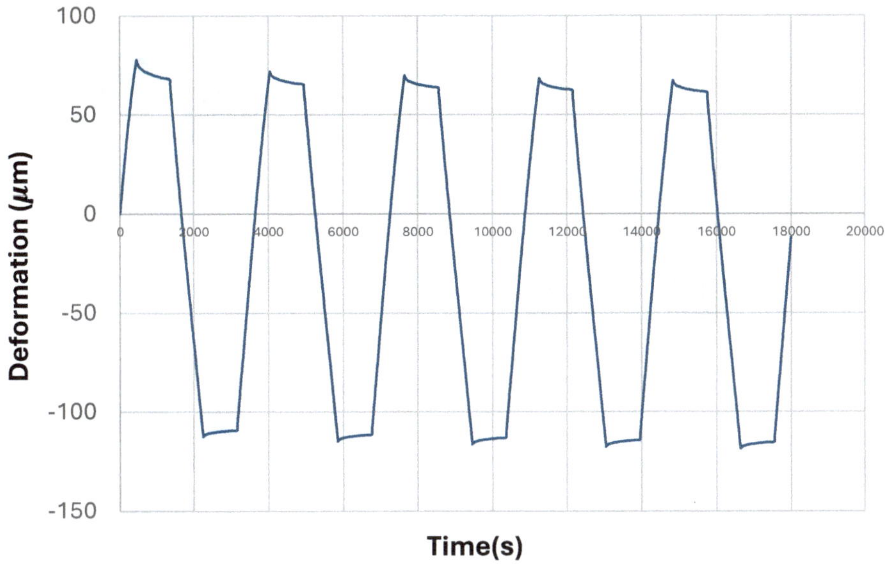


**Fig. 3.15** Time-history deformation of the structure

at −40 (2250 s)). This could be due to the relaxation of the μbumps. (In this study, perfect diffusion Cu-Cu hybrid bonding has been assumed). The time-history deformation of the structure with Cu-Cu hybrid bonding is shown in Fig. 3.15. It can be seen that the deformation patterns are following the applied temperature boundary condition, Fig. 3.13. The deformation of this size of chip is acceptable in most applications.

### *3.7.6 Accumulated Equivalent (von Mises) Stress*

For flip chip solder interconnects, the focus is on the inelastic strain [47] while for flip chip Cu-Cu interconnects, the focus is on the stress. The strains in Cu-Cu perfect bonding are very small (in the $10^{-6}$ ranges) as shown in Fig. 3.16.

The Accumulated equivalent (von Mises) stress contours at the corner Cu-pads at 450 s and 2250 s are shown in Fig. 3.17. The maximum stress location is at the interface between the Si chip and the Cu-pad. This is due to the large thermal expansion mismatch between the silicon ($2.8 \times 10^{-6}$/°C) and the Cu ($16.3 \times 10^{-6}$/°C) and stress concentration.

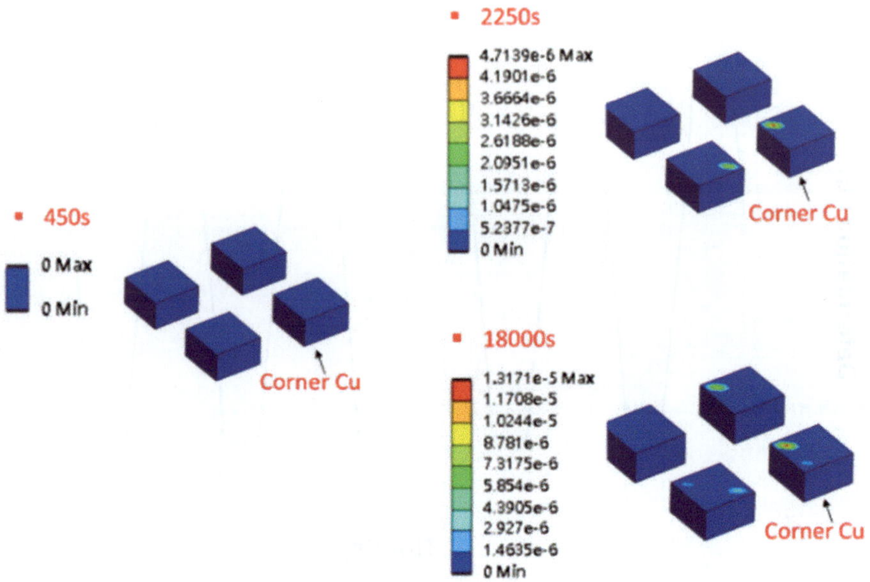


**Fig. 3.16** Accumulated equivalent inelastic strain contours at the corner Cu-pads at 450 s, 2250 s, and 18,000 s

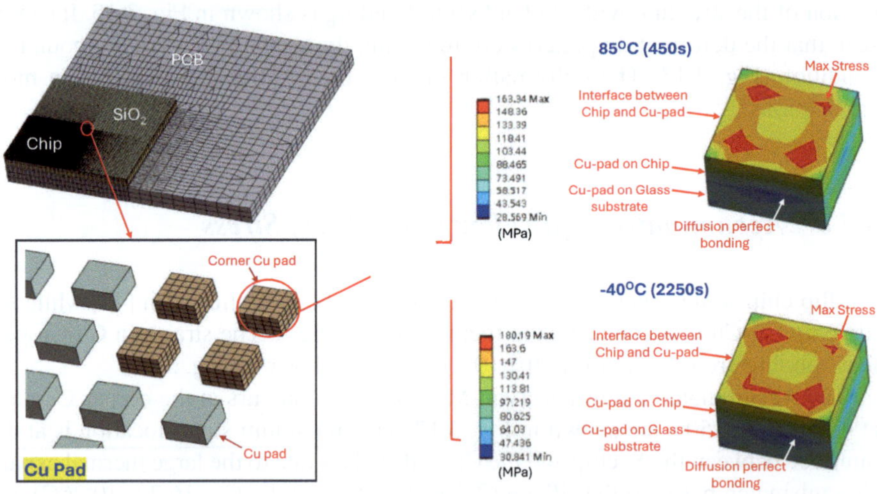


**Fig. 3.17** Accumulated equivalent (von Mises) stress contours at the corner Cu-pads at 450 s and 2250 s

### 3.7.7 Accumulated Equivalent Inelastic Strain

As shown in [48], a simple thermal-fatigue life equation for the lead-free solder joints is given by

$$N_f = \sum_j \alpha_j \left[ \frac{\sum_i \Delta W_i x V_i}{\sum_i V_i} \right]^{\beta_j}$$

where $N_f$ is the thermal fatigue life of the solder joint, and $\alpha_j$ and $\beta_j$ (which is a negative value) are constant to be determined by experiments such as the isothermal fatigue tests for a specific component/package and solder joint. $\Delta W_i$ is the accumulated inelastic strain energy density per cycle in the element determined from finite element simulations and $V_i$ is the volume of that *i*th element. It can be seen that the larger the accumulated inelastic strain energy density per cycle the smaller (shorter) the thermal fatigue life of the solder joint.

The accumulated equivalent inelastic strain contours at the corner solder ball at 450 s and 2250 s are shown in Fig. 3.18. It can be seen that the maximum inelastic strain occurs near the interface between the solder ball and the PCB. This is due to the large thermal expansion mismatch between the glass-core ($4.8 \times 10^{-6}/^{o}C$) and the PCB ($18 \times 10^{-6}/^{o}C$). The time-history of accumulated equivalent inelastic strain is shown in Fig. 3.19. It can be seen that the accumulated equivalent inelastic strain per cycle is 16.02%, which is quite high for the thermal fatigue life of the corner solder ball.

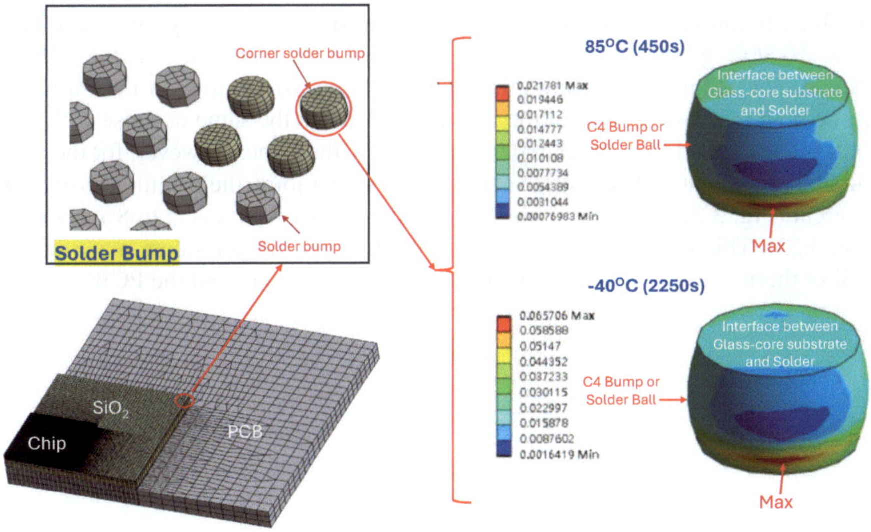


**Fig. 3.18** Accumulated equivalent inelastic strain contours at the corner solder ball at 450 s and 2250 s

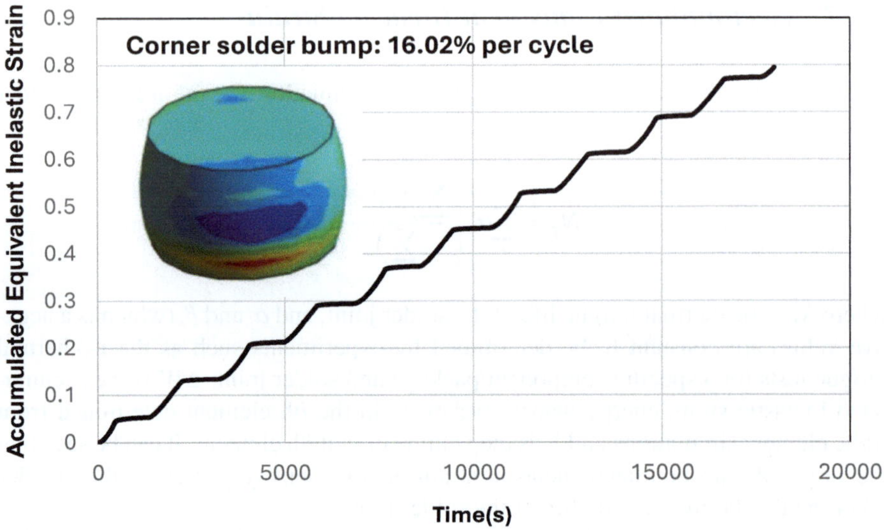


**Fig. 3.19** Time-history accumulated equivalent inelastic strain at the corner solder ball

### 3.7.8 *The CTE of Glass Increased from 4.8 × 10⁻⁶/oC to 10 × 10⁻⁶/oC*

If the CTE of the glass core is increased from $4.8 \times 10^{-6}/^{o}C$ to $10 \times 10^{-6}/^{o}C$ (which is closer to the PCB) and subjected to the same temperature boundary condition (Fig. 3.13), then the warpage, stress, and strain are shown in Figs. 3.20, 3.21, 3.22 and 3.23. It can be seen from Fig. 3.20 that the structure warpage for the CTE = $10 \times 10^{-6}/^{o}C$ glass-core substrate is slightly less than that for the CTE = $4.8 \times 10^{-6}/^{o}C$. Figure 3.21 shows the equivalent von Mises stress at the Cu-pad. It can be seen that the location of the maximum stress is the same as those with CTE = $4.8 \times 10^{-6}/^{o}C$ and their magnitudes are basically the same. However, for the accumulated equivalent inelastic strain at the corner solder joint, the magnitudes are very different (Fig. 3.22), and the strain is reduced from 16% per cycle to 8% per cycle (Fig. 3.23). This is due to the increase in the CTE of the glass core material, i.e., smaller thermal expansion mismatch between the glass core and the PCB.

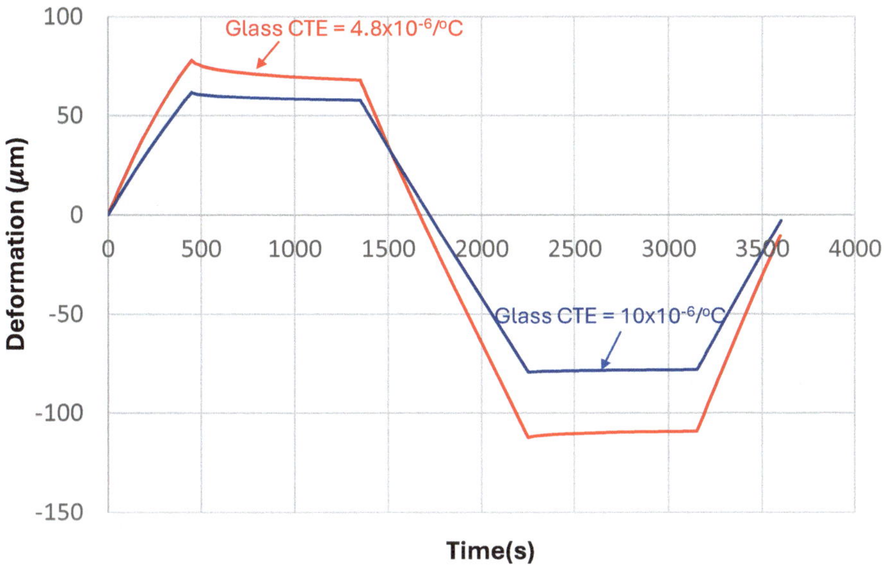


Fig. 3.20 Warpage for glass-core substrate with glass CTE = 4.8 × 10⁻⁶/°C and 10 × 10⁻⁶/°C

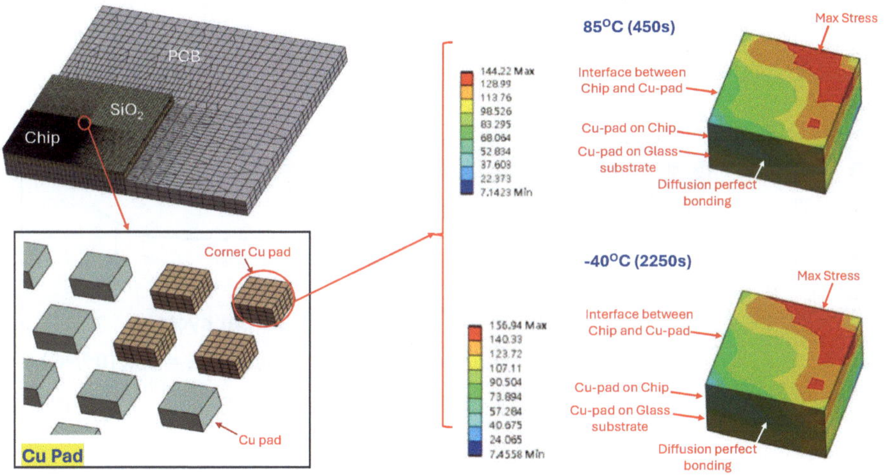


Fig. 3.21 The von Mises stress at the corner Cu-pad (glass CTE = 10 × 10⁻⁶/°C)

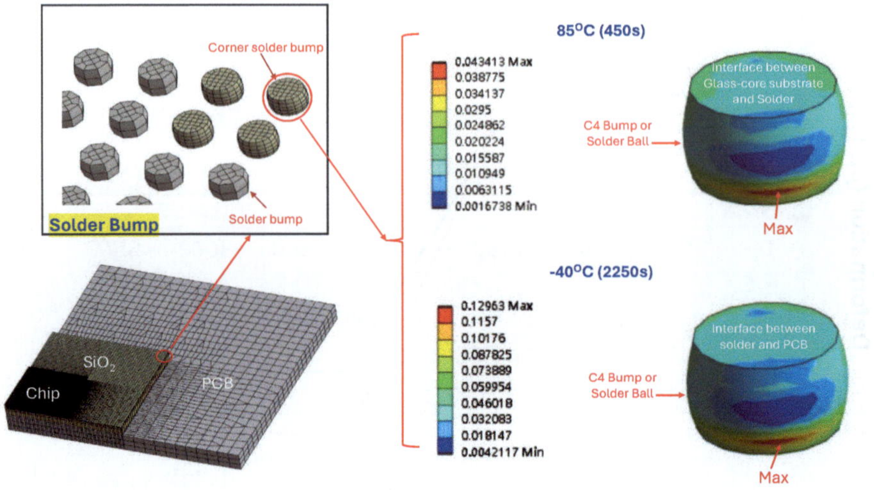


**Fig. 3.22** Accumulated equivalent inelastic strain at the corner solder joint (glass CTE = $10 \times 10^{-6}$/°C)

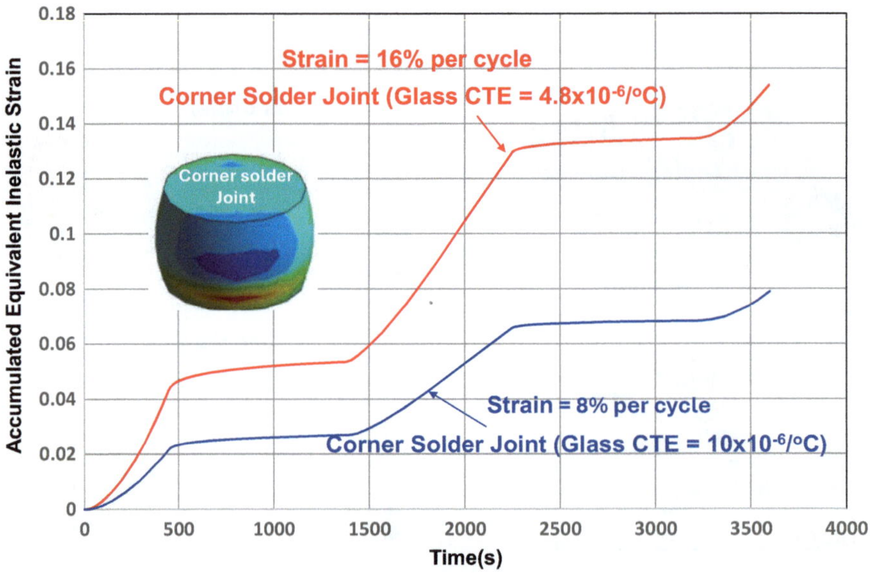


**Fig. 3.23** Accumulated equivalent inelastic strain at the corner solder joint (glass CTE = $4.8 \times 10^{-6}$/°C and $10 \times 10^{-6}$/°C)

## 3.8 Summary and Recommendations

Some important results and recommendations are summarized as follows.

- Just like the PCB, the organic-core build-up package substrate will be here to stay and used at huge volume for a long time. Glass-core build-up package substrate is able to seamlessly integrate optical interconnects, e.g., [49] and will be used for niche applications, one-trillion transistors, e.g., [34] and very high-performance and high-density.
- For flip chips bonded to glass-core and organic-core substrates with μbumps.
  - For the μbump solder joint, the maximum accumulated equivalent inelastic strain is smaller in the structure with glass-core substrate than with organic-core substrate.
  - For the C4 solder joint, the maximum accumulated equivalent inelastic strain is more than two times larger in the structure with glass-core substrate than with organic-core substrate.
  - It is recommended that the CTE of the glass-core substrate should be closer to the CTE of PCB (not to the CTE of Si chip) because there is no underfill protection of the C4 or BGA solder joints on the PCB, especially for larger glass-core substrates.
- For flip chips bonded to glass-core substrate with Cu-Cu hybrid bonding.
  - Unlike organic-core substrates, the glass-core substrates are flat enough to perform Cu-Cu hybrid bonding.
  - For extremely fine-pitch, high-density, and high-performance applications, the L/S of the RDLs of the glass-core substrate are usually less than 2 μm. The dielectric layer ($SiO_2$) can be fabricated by PECVD and the metal layer by dual-damascene Cu + CMP (chemical-mechanical polishing).
  - The warpage of the flip chip bonded on glass-core substrate with Cu-Cu hybrid bonding are very similar to that of the flip chip bonded on glass-core substrate with μbumps, except the magnitudes. The magnitudes of the Cu-Cu hybrid bonding are larger than those of the μbumps. This is because of the relaxation of the μbumps.
  - The maximum accumulated equivalent inelastic strains in Cu-Cu hybrid bonding are very small.
  - The maximum accumulated equivalent (von Mises) stress occurs at the interface between the Si chip and the Cu-pad.
  - The maximum accumulated equivalent inelastic strain occurs near at the interface between the solder ball and the PCB.
  - The accumulated equivalent inelastic strain per cycle is 16.02%, which is smaller than that with μbump (19%). However, it is still quite high for the solder joint reliability. A larger CTE for the glass-core substrate is recommended.

- By increasing the CTE of glass core from $4.8 \times 10^{-6}/^{o}C$ to $10 \times 10^{-6}/^{o}C$, the warpage of the structure reduced by 48% and the accumulated maximum equivalent inelastic strain in the corner BGA solder joint reduced by 50%.

## References


1. Bhatt, D., K. Williams, D. A. Hutt, and P. P. Conway, "Process optimization and characterization of excimer laser drilling of microvias in glass," *IEEE EPTC Proceedings*, Singapore, 2007, pp. 196–201.
2. Cui, X., D. Bhatt, D. A. Hutt, K. Williams, and P. P. Conway, "Copper deposition and patterning for glass substrate manufacture," *IEEE EPTC Proceedings*, Singapore, 2007, pp. 37–42.
3. Brusberg, L., H. Schroder, M. Topper, N. Arndt-Staufenbiel, J. Roder, M. Lutz, and H. Reichl, "Thin glass based packaging technologies for optoelectronic modules," *IEEE ECTC Proceedings*, San Diego, CA, May 2009, pp. 207–212.
4. Brusberg, L., H. Schroder, M. Topper, and H. Reichl, "Photonic system in-package technologies using thin glass substrates," *IEEE EPTC Proceedings*, Singapore, December 2009, pp. 930–935.
5. Schroder, H., L. Brusberg, R. Erxleben, I. Ndip, M. Topper, N. F. Nissen, and H. Reichl, "GlassPack- A 3D glass based interposer concept for SiP with integrated optical interconnects," *IEEE ECTC Proceedings*, Las Vegas, NV, May 2010, pp. 1647–1652.
6. Topper, M., I. Ndip, R. Erxleben, L. Brusberg, N. Nissen, H. Schroder, H. Yamamoto, G. Todt, and H. Reichl, "3-D thin film interposer based on TGV (through glass vias): An alternative to Si-interposer," *IEEE ECTC Proceedings*, Las Vegas, NV, 2010, pp. 66–73.
7. Kim, S., V. Govindarajulu, C. Tao, A. Brar, and Z. Karim, "High Aspect Ratio (AR) Through Glass Via (TGV) Etch Performance on Glass Core Substrates for High Density 3D Advanced Packaging Applications", *IEEE Proceedings*, May 2024, pp. 551–555.
8. Sukumaran, V., Q. Chen, F. Liu, N. Kumbhat, T. Bandyopadhyay, H. Chan, S. Min, C. Nopper, V. Sundaram, and R. Tummala, "Through package-via formation and metallization of glass interposers," *IEEE ECTC Proceedings*, Las Vegas, NV, 2010, pp. 557–563.
9. Tummala, R. R., V. Sundaram, T. Bandyopadhyay, V. Sridharan, and V. Sukumaran, "Low-cost silicon and glass interposers and packages," in Proc. Pan Pacific Microelectron. Symp., 2011, pp. 106–109.
10. Sukumaran, V., T. Bandyopadhyay, V. Sundaram and R. Tummala, "Low-Cost Thin Glass Interposers as a Superior Alternative to Silicon and Organic Interposers for Packaging of 3-D ICs," IEEE Transactions on Components, Packaging and Manufacturing Technology, vol. 2, no. 9, Sept. 2012, pp. 1426–1433.
11. McCann, S. R., "Experimental and theoretical assessment of thin glass panels as interposers for microelectronic packages", PhD thesis, Georga Institute of Technology, May 2014.
12. Ravichandran, S., M. Kathaperumal, M. Swaminathan, and R. Tummala, "Large-Body-Sized Glass-Based Active Interposer for High-Performance Computing", *IEEE Proceedings*, May 2020, pp. 879–884.
13. Wong, R., S. Ravichandran, H. Lee, and V. Smet, "Thermal Management of Glass Panel Embedded Packages: Package Architecture vs. Power Density", *IEEE Proceedings*, May 2020, pp. 2105–2111.
14. Ali, M., A. Watanabe, T. Kakutani, P. Raj, R. Tummala, and M. Swaminathan, "Heterogeneous Integration of 5G and Millimeter-Wave Diplexers with 3D Glass Substrates", *IEEE Proceedings*, May 2020, pp. 1376–1382.
15. Erdogan, S., S. Ravichandran, X. Jia, and M. Swaminathan, "Characterization of Chip-to-Package Interconnects for Glass Panel Embedding (GPE) for Sub-THz Wireless Communications", *IEEE ECTC Proceedings*, May 2021, pp. 2328–2333.

16. Sivapurapu, S., R. Chen, M. Rehman, K. Kanno, T. Kakutani, M. Letz, F. Liu, S. Sitaraman, and M. Swaminathan, "Flexible and Ultra-Thin Glass Substrates for RF Applications", *IEEE Proceedings*, May 2021, pp. 1638–1644.
17. Huang, K., and M. Swaminathan, "Antennas in Glass Interposer for sub-THz Applications", *IEEE Proceedings,* May 2021, pp. 1150–1155.
18. Jayaram, V., O. Gupte, and V. Smet, "Modeling and Design for System-Level Reliability and Warpage Mitigation of Large 2.5D Glass BGA Packages", *IEEE Proceedings*, May 2022, pp. 1060–1097.
19. Jia, X., X. Li, K. Moon, J. Kim, K. Huang, and M. Jordan, "Antenna-Integrated, Die-Embedded Glass Package for 6G Wireless Applications", *IEEE Proceedings*, May 2022, pp. 377–383.
20. Ravichandran, S., V. Smet, M. Swaminathan, and R. Tummala, "Demonstration of Glass-based 3D Package Architectures with Embedded Dies for High Performance Computing", *IEEE Proceedings*, May 2022, pp. 1114–1120.
21. Kim, J., X. Li, X. Jia, K. Moon, and M. Swaminathan, "Bottom Side Cooling for Glass Interposer with Chip Embedding Using Double-Sided Release Process for 6G Wireless Applications", *IEEE Proceedings,* May 2023, pp. 1609–1613.
22. Li, X., X. Jia, J. Kim, K. Moon, M. Jordan, M. Swaminathan, "Die Embedded Glass Interposer with Minimum Warpage for 5G/6G Applications", *IEEE Proceedings*, May 2023, pp. 2257–2264.
23. Kathaperumal, M., P. Bhaskar, A. Toro, P. Nimbalkar, L. Dahal, and M. Bakir, "Characterization of Warpage of Ultra-Low-K Dielectric Materials and Correlation with Modulus and Coefficient of Thermal Expansion", *IEEE Proceedings*, May 2024, pp. 1958–1962.
24. Molina-Mangual, C., E. Surillo, N. Nedumthakady, and V. Smet, "Mitigating Cracking from RDL Stress in Glass Substrates with Low-CTE Electrodeposited Copper-Graphene Composites", *IEEE Proceedings*, May 2024, pp. 1552–1557.
25. Bowrothu, R., Y. Yoon, and J. Zhang, "Through Glass Via (TGV) Based Band Pass Filter for 5G Communications", *IEEE Proceedings*, May 2018, pp. 1097–1102.
26. Hwangbo, S., Y. Yoon, and A. Shorey, "Millimeter-wave Wireless Chip-to-chip (C2C) Communications in 3D System-in Packaging (SiP) using Compact Through Glass Via (TGV)-integrated Antennas", *IEEE Proceedings*, May 2018, pp. 2068–2073.
27. Okoro, C., T. Allowatt, and S. Pollard, "Resolving Thermo-Mechanically Induced Circumferential Crack Formation in Copper Through Glass Vias", *IEEE Proceedings*, May 2021, pp. 954–958.
28. Pan, K., J. Xu, and S. Park, "Investigation of Copper and Glass Interaction in Through Glass Via (TGV) During Thermal Cycling", *IEEE Proceedings*, May 2021, pp. 1660–1666.
29. Pan, K., C. Okoro, Y. Lai, D. Joshi, S. Park, and S. Pollard, "A comparative study of the thermomechanical reliability of fully-filled and conformal through-glass via", *IEEE Proceedings*, May 2022, pp. 1211–1217.
30. Yeary, L., L. Brusberg, C. Kim, S. Seok, J. Noh, and A. Rozenvax, "Co-Packaged Optics on Glass Substrates for 102.4 Tb/s Data Center Switches", *IEEE Proceedings*, May 2023, pp. 224–227.
31. Brusberg, L., M. Dejneka, C. Okoro, D. McEnroe, A. Zakharian, and C. Terwillige, "Ultra Low-Loss Ion-Exchange Waveguides in Optimized Alkali Glass for Co-Packaged Optics", *IEEE Proceedings*, May 2024, pp. 85–89.
32. Yang, J., K. Pan, P. Yin, Y. Lai, S. Park, C. Okoro, D. Joshi, S. Pollard, "Adhesion Layer Influence on Thermomechanical Reliability of Electroplated Copper Through-Glass Via (TGV)", *IEEE Proceedings*, May 2024, pp. 1820–1825.
33. https://www.absolicsinc.com/about. 2024.
34. https://www.intel.com/c.ontent/www/us/en/newsroom/news/intel-unveils-industry-leading-glass-substrates.html#gs.aub6ck, September 18, 2023
35. Lau, J. H., *Through-Silicon Via (TSV) for 3D Integration*, McGraw-Hill, New York, 2013.

36. Ostholt, R., N. Ambrosius, and R. Kruger, "High speed through glass via manufacturing technology for interposer", IEEE Proceedings of Electronics System-integration Technology Conference, September 2014, pp. 1–3.
37. Xing, D., C. Che, X. Hou, X. Li, Y. Ning, H. Lian, Q. Wang, and H. Zhao, "Preparation Process of Through Glass Via Based on Laser Induced Deep Etching", SID, V. 56, Issue 1, June 2025, pp. 1105–1108.
38. Chen L., H. Wu, M. Ahang, F. Jiang, T. Yu, and D. Yu, "Development of Laser-Induced Deep Etching Process for Through Glass Via", Proceedings of IEEE ICEPT, August 2025, pp. 1–4.
39. Sunohara, M., J. Yoshike, H. Taneda, Y. Nakabayashi, and N. Shimizu, "Development of Glass Core Build-up Substate with TGV", *IEEE ECTC Proceedings*, May 2025, pp. 1021–1026.
40. Lau, J. H., and X. Fan, *Hybrid Bonding, Advanced Substrates, Failure Mechanisms, and Thermal Management for Chiplets and Heterogeneous Integration*, Springer, New York, 2025.
41. Lau, J. H., *Flip Chip, Hybrid Bonding, Fan-In, and Fan-Out Technology*, Springer, New York, 2024.
42. Lau, J. H., "Current Advances and Outlooks in Hybrid Bonding", *IEEE Transactions on CPMT*, Vol. 15, No. 3, March 2025.
43. *Lau, J. H.*, "State of the art of Cu-Cu Hybrid Bonding", *IEEE Transactions on CPMT*, Vol. 14, No. 3, March 2024, pp. 376–396.
44. Lau, J. H., "Recent Advances and Trends in Cu-Cu Hybrid Bonding", *IEEE Transactions on CPMT*, Vol. 13, No. 3, March 2023, pp. 399–425.
45. Lau, J. H., "Recent Advances and Trends in Advanced Packaging", *IEEE Transactions on CPMT*, Vol. 12, No. 2, February 2022, pp. 228–252.
46. Lau, J. H., N. Liu, M. Ma, and T. J. Tseng, "Solder Joint Reliability - Glass Core Substrate versus Organic Core Substrate", *Microelectronic Reliability*, Vol. 168, May 2025, pp. 115701.
47. Lau, J. H., N. Liu, M. Ma, and T. Tseng, "Flip Chip on Glass-Core Substrates with Microbump and Cu-Cu Hybrid Bonding", *Journal of Microelectronic Packaging*, Vol. 22, Issue 3, September 2025, pp. 44–60.
48. Lau, J. H., and Lee, N. C., Assembly and Reliability of Lead-Free Solder Joints, Springer, New York, 2020.
49. Lau, J. H., P. Lin, K. Yang, C. Lin, C. Ko, and T. J. Tseng, "3D Heterogeneous Integration of PIC and EIC with PIC embedded in a Glass Substrate," US Patent application filed on March 1, 2024.

# Chapter 4
# Heterogeneous Integration of PIC, EIC, and Switch (CPO)

**Abstract** There have been strong demands for lower power consumption and higher bandwidth in optical/electrical interconnects used for artificial intelligence (AI) and networks in a data center. The adoption of co-packaged optics (CPO) has been expected for both high-performance computing (HPC) driven by AI and high-bandwidth and high-speed communications networks in a data center. In this chapter, on-board optics (OBO), near package optics (NPO), and CPO will be discussed. Emphasis is placed on the trends of 3D heterogeneous integration of photonic integrated circuits (PIC), electronic integrated circuits (EIC), and application specific IC (ASIC) switch with or without bridges on various CPO substrates such as organic, silicon, and glass. Some recommendations will be provided.

## 4.1 Introduction

Co-packaged optics (CPO) are heterogeneous integration packaging methods to integrate the chiplets such as the optical engine (OE) and the switch application-specific IC (ASIC). An optical engine consists of PIC such as photodiode and laser, and EIC such as transimpedance, microcontroller, and laser driver. The EICs are complementary metal-oxide semiconductor (CMOS) device and are sued to receive and transmit light (photon) and convert it to electrical signal (electron) and vice versa. The advantages of CPO are (a) Reducing the length of the electrical interface between the OE/EE (or PIC/ EIC) and the ASIC; (b) Reducing the energy required to drive the signal; and (c) Cutting the latency, which leads to better electrical performance [1–44].

Twenty years ago, silicon integration was very popular. Companies like Intel have been promoting the silicon photonics which integrate some of the PIC and EIC on a chip from a silicon wafer with CMOS technology. The holy grail of silicon photonic is to integrate all the PIC and EIC into a single chip fabricated from a silicon wafer with CMOS technology. However, in the past few years, driven by cost,



J. Lau, K.-N. Chiang, *Cu-Interconnects, Glass, and AI-Assisted Simulation for Chiplets and Heterogeneous Integration*,
https://doi.org/10.1007/978-981-95-6891-8_4

the PIC and EIC are partitioned and spit into small chiplets [45–49]. The key advantages of silicon photonics are better performance and easier to test. The key disadvantages of silicon photonics are higher costs and more rigid infrastructure. The key advantages of chiplets are lower costs and more flexible. The key disadvantages of chiplets are not as good performance as silicon photonics and larger package size.

## 4.2 Photonic Devices and Silicon Photonics

Photonic devices are components that generate, manipulate, and detect light, e.g., LED (light emitting diode), lasers, photodiode, etc.

Silicon photonics is a technology that integrates optical components onto a silicon chip, enabling the transmission of data using light instead of traditional electrical signals. In another words, silicon photonics are the semiconductor integration of PIC and EIC on a silicon substrate (wafer) with CMOS technology. As shown in Fig. 4.1, after the IC is fabricated on the silicon wafer, the indium phosphide (InP; a light-emitting material) is bonded to the top of the silicon wafer with a thin layer of glass glue. This is what Intel called the hybrid laser technology, in which the IC and laser are manufactured by the CMOS technology on the same silicon wafer. Some of the advantages include the automation of on-wafer optical, electrical, and high speed test and wafer-level burn-in. This book is focused on the heterogeneous integration packaging methods to integrate the chiplets (including the silicon photonics of the silicon integration of some but not all) of the PICs and EICs, as well as the ASIC switch.

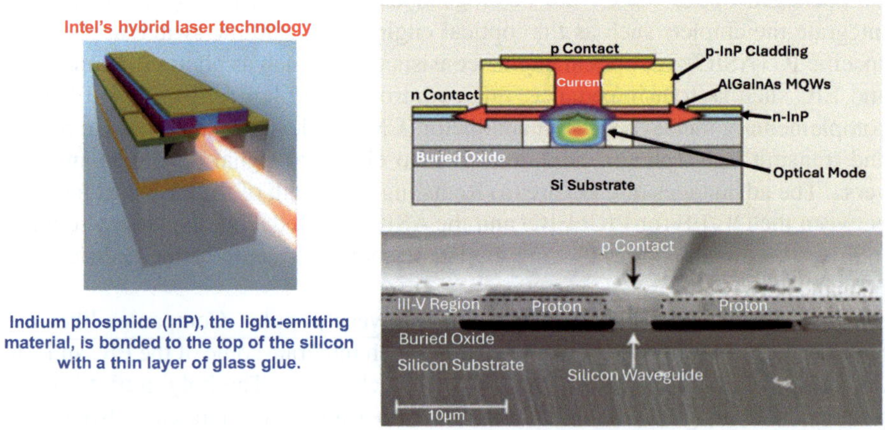


Fig. 4.1 A hybrid silicon laser

**Fig. 4.2** Data center with pluggable transceiver

## 4.3 Data Centers

A data center includes switches, routers, storage solutions, etc., such as those shown in Fig. 4.2. Their functions include the network of computing and storage resources that enable the delivery of shared applications and data. Today, more than 90% of transceivers are pluggable transceivers [50–53]. Figure 4.3 shows the future optical transceivers and switch on a common package substrate used in a data center.

## 4.4 Optical Transceivers

Optical transceivers are the most important component in a data center. An optical transceiver is a key component of an optical transmission system that permits coupling of the transmission medium with the active components of the chain, such as switches, routers, optical interfaces or any optical transport equipment. An optical transceiver defines the process of converting electric signaling toward the optical transmission with the help of the transmission optical subassembly (TOSA), or Tx module and performing inverse action through the receiver optical subassembly

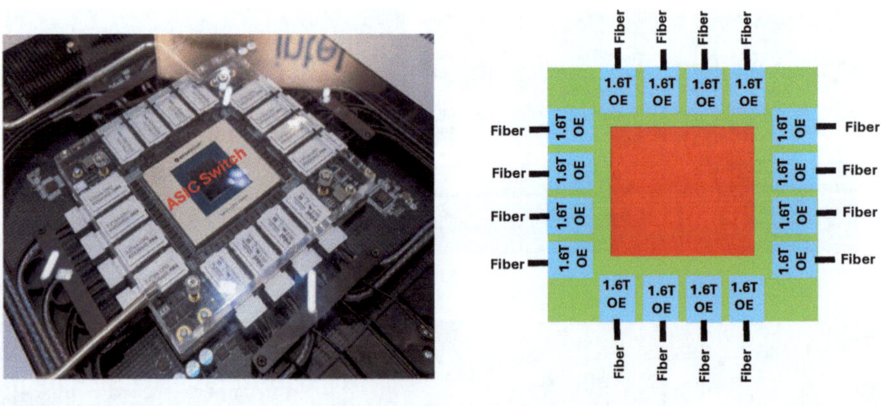


Fig. 4.3 CPO in a data center

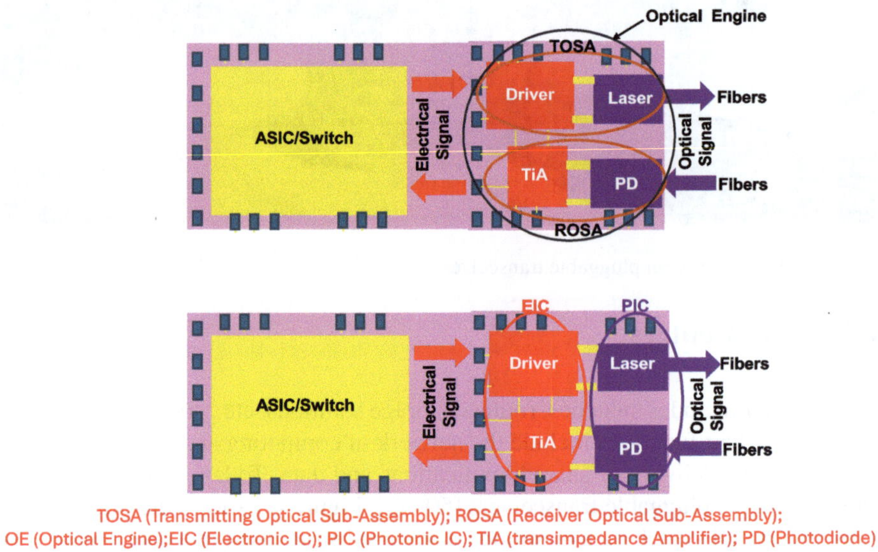


Fig. 4.4 Switch, TOSA, ROSA, OE, PIC, and EIC

(ROSA), or Rx module (Fig. 4.4). A TOSA contains a semiconductor laser diode (LD), laser driver, etc., while a ROSA contains a photodiode (PD), an optical interface such as a lens, a transimpedance amplifier (TIA), a passive electrical interface, etc. The TOSA module converts the electrical signal to the optical transmission light that lands on the fiber. The ROSA is used to receive an optical signal from a fiber and convert it back into an electrical signal.

There are many forms of transceivers in CPO as shown in Fig. 4.5. The most common form is the CPO with the basic chiplets such as switch, laser driver, TIA, and the photonic devices such as the laser, and PD. The advantages are lower cost

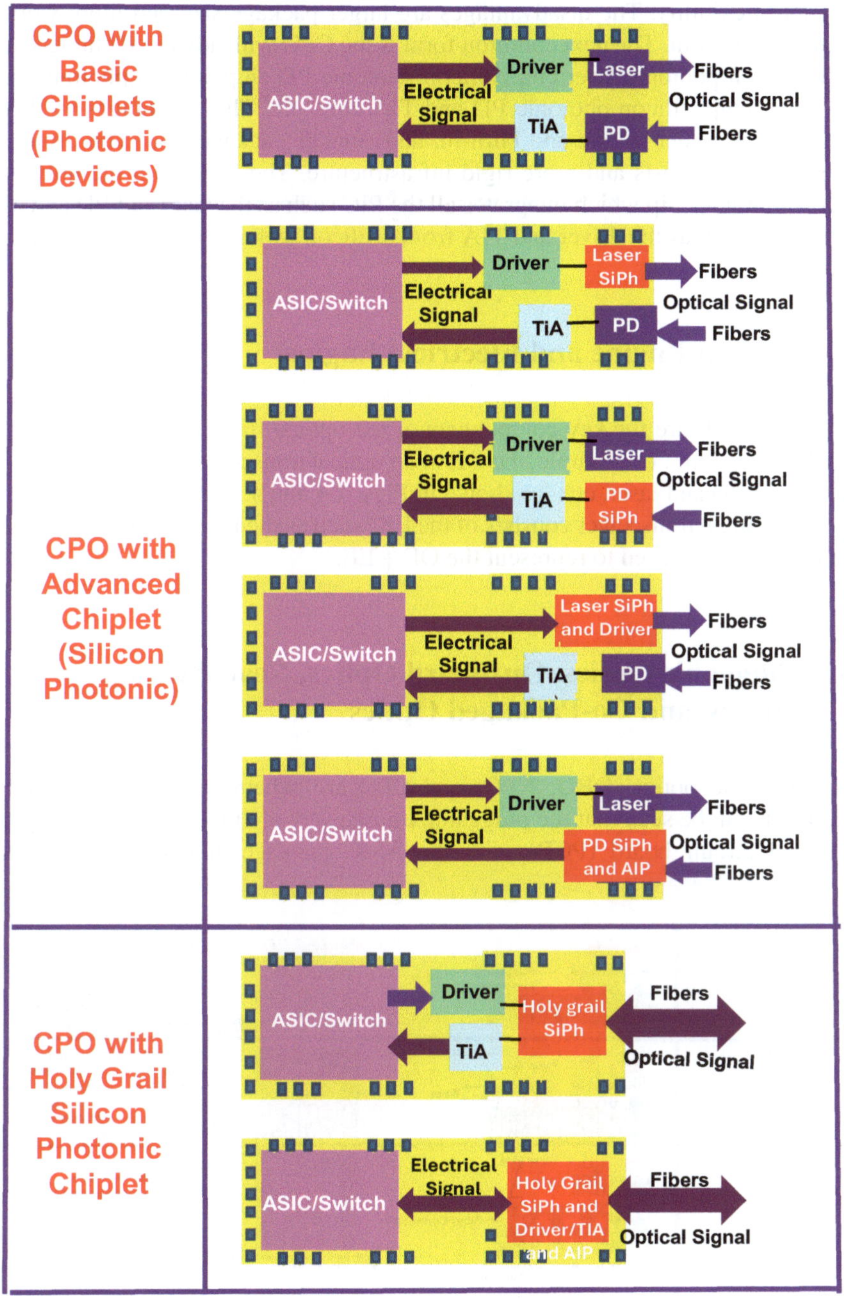


Fig. 4.5 Various forms of CPO

and more flexibility. The disadvantages are larger package size and more work in testing and burn-in. The next common form is the CPO with advanced chiplets such as the silicon photonic laser chip, silicon photonic PD chip, silicon photonic laser and driver chip, silicon photonic PD and TIA chip, etc. The advantages are better performance, easier to test and burn-in, and smaller package size. The disadvantages are higher costs and more rigid infrastructure. The CPO with the holy grail silicon photonic chip which integrates all the PIC such as the laser and PD and even all the EIC such as the driver and TIA from a silicon wafer is nowhere in sight.

## 4.5 Optical Engine and Electrical Engine

The OE and EE are the key components of the optical transceiver for realizing in signal conversion. For example, the OE of an optical transceiver consists of the PIC, such as the PD and laser, etc., as shown in Figs. 4.4 and 4.5. On the other hand, the EE of an optical transceiver consists of the EIC such as the laser driver and TIA, etc. Sometimes, OE is used to represent the OE + EE.

## 4.6 Pluggable Optics, On-Board Optics, Near-Package Optics, and Co-Packaged Optics

The pluggable optics, OBO, NPO, and CPO are schematically summarized in Fig. 4.6. It can be seen that the pluggable transceivers are mounted on the edges of a printed circuit board (PCB) with the ASIC package attached to a package

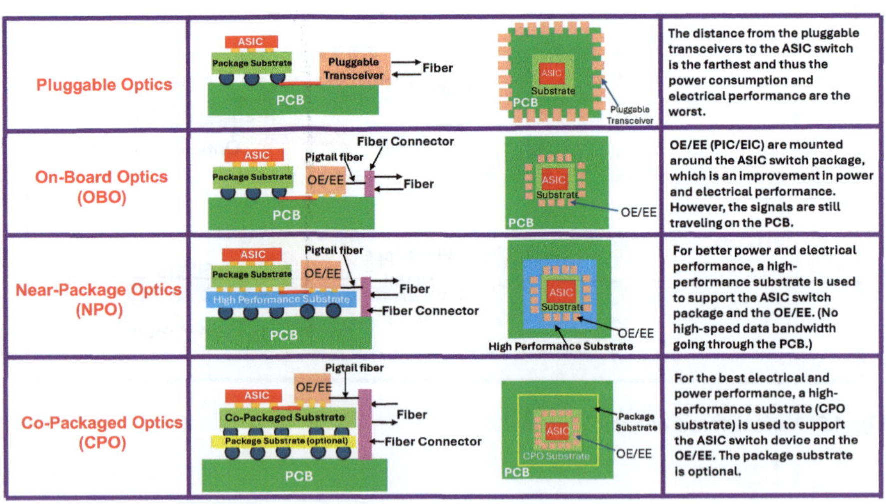


**Fig. 4.6** Pluggable optics, OBO, NPO, and CPO

substrate. The distance from the PIC/EIC of the pluggable transceivers to the ASIC chip is the farthest and therefore, the power consumption and electrical performance are the worst.

Since 2018, the key components such as OE and EE (PIC/EIC) of the transceivers have been mounted on the same PCB as the packaged ASIC and arranged on the four sides of the packaged ASIC. This scheme is called OBO (on-board optics), which uses the PCB to perform the connection between the packaged ASIC and the OE/EE. The distance between the PIC/EIC and the ASIC is closer than the pluggable optics and therefore, there is an improvement in power and electrical performance.

Since 2020, some people placed the OE/EE side-by-side with the packaged ASIC on the same high-performance substrate. This is called NPO (near-package optics) and uses the high-performance substrate to perform the connection between the packaged ASIC and the OE/EE. There is no high speed data bandwidth going through the PCB, so better power and electrical performance can be achieved. According to the Optical Internetworking Forum (OIF)—for NPO applications—lengths between the ASIC and optics can be up to 150 mm with the channel loss constrained to 13 dB. This is why the OIF is developing the extra short reach (XSR) + electrical interface, to expand the XSR's reach for NPO.

Since 2023, after the shipment of CPO by Intel and Broadcom, people are very interested in CPO. In CPO, the OE/EE is placed side-by-side along the four edges of the ASIC chip on the same co-packaged substrate. In this case, the distance between the ASIC and the PIC/EIC is the shortest and therefore has the best electrical performance. In short, NPO brings the PIC/EIC closer to the ASIC package on a high-performance substrate, while CPO brings the PIC/ EIC and the ASIC chip side-by-side on the same co-packaged substrate. CPO offers superior formfactor, power consumption, and latency benefits compared to NPO. According to OIF, CPO confines the modules OE/EE and host ASIC chip to be within 50 mm of each other. The channel loss here is limited to 10 dB. CPO has a lower power consumption, Fig. 4.7 [38], because of the shorter spans and 3 dB saving.

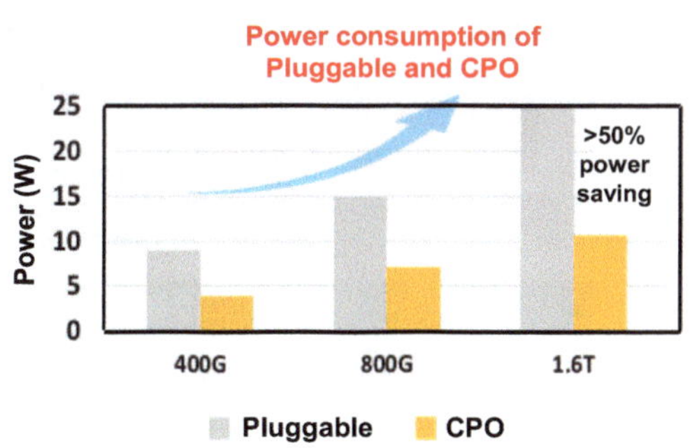


**Fig. 4.7** Power consumption of pluggable and CPO

## 4.7 CPO Substrates

There are many different types of CPO substrates such as silicon, Fig. 4.8 [39, 40], glass, Fig. 4.9 [40], and organic, Fig. 4.10 [42], etc. Silicon substrates such as the through-silicon via (TSV)-interposer have been presented in great detail in, e.g [54, 55]. Glass substrates such as the laser-induced deep etching (LIDE) process developed by LPKF Laser & Electronics AG in 2017 has been reported in [56–59]. Figure 4.10 shows a 2,3D heterogeneous integration of ASIC, EIC and PIC on a CPO organic interposer [42]. It can be seen that the ASIC switch chip, EIC chips and PIC chips are side-by-side attached to the CPO organic interposer with μbumps. In this chapter the fabrication of CPO organic interposer (Figs. 4.10 and 4.11) will be briefly mentioned. The flow chart fabricating the CPO organic hybrid substrate is shown in Fig. 4.12. It can be seen that the organic interposer with a temporary glass panel carrier and the build-up package substrate are fabricated separately and then are combined into a hybrid CPO organic substrate.

The CPO organic interposer (20 × 15 × 0.053 mm) is shown in Fig. 4.13. It can be seen that there are three RDLs, each consisting of a dielectric layer (DL) and a Cu metal layer (ML). The DL materials are photoimageable dielectric (PID) and Ajinomoto build-up film (ABF), and their physical and mechanical material properties are shown in Table 4.1. For both materials the L/S of ML1 (metal layer 1) are 2 μm/2 μm, of ML2 are 5 μm/5 μm, and of ML3 are 10 μm/10 μm. The thickness (H) of ML1 is the same (3 μm) for both PID and ABF and of ML2 and ML3 is 5 μm

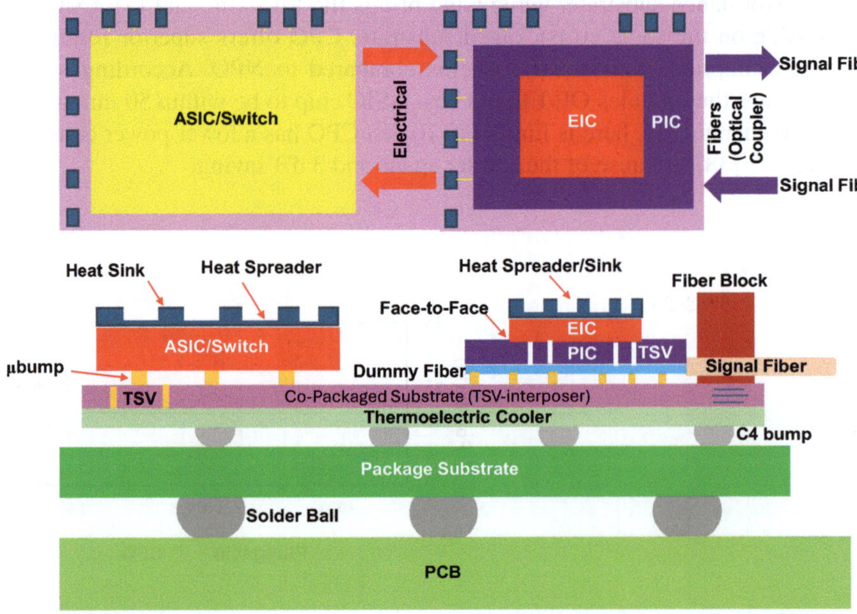


**Fig. 4.8** CPO on TSV-interposer

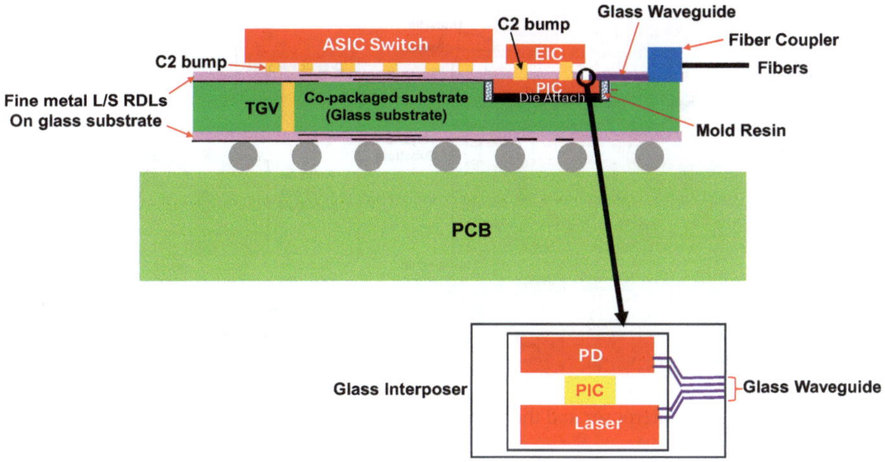


**Fig. 4.9** CPO on glass-core build-up substrate

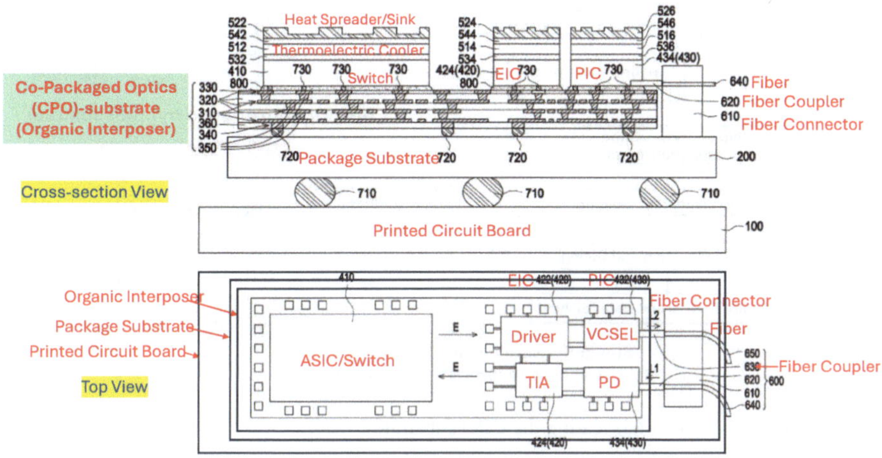


**Fig. 4.10** CPO on organic-interposer

for PID and 8 μm for ABF. The dielectric layer between the contact pad and ML1 (DL01) is 3 μm for PID and 3.5 μm for ABF, the DL12 (dielectric layer between ML1 and ML2) is 6 μm for PID and 10 μm for ABF, and the DL23 (dielectric layer between the ML2 and ML3) is 4 μm for PID and 7.5 μm for ABF. The via opening between contact pad and ML1 (V01) is 10 μm for PID and 20 μm for ABF, the via opening between ML1 and ML2 (V12) is the same (20 μm) for both PID and ABF, and the via opening between ML2 and ML3 (V23) is 18 μm for PID and 20 μm for ABF.

The fabrication processes of the fine metal L/S/H RDL substrate are shown in Figs. 4.14 and 4.15, respectively for PID and ABF. For the one with PID, a released

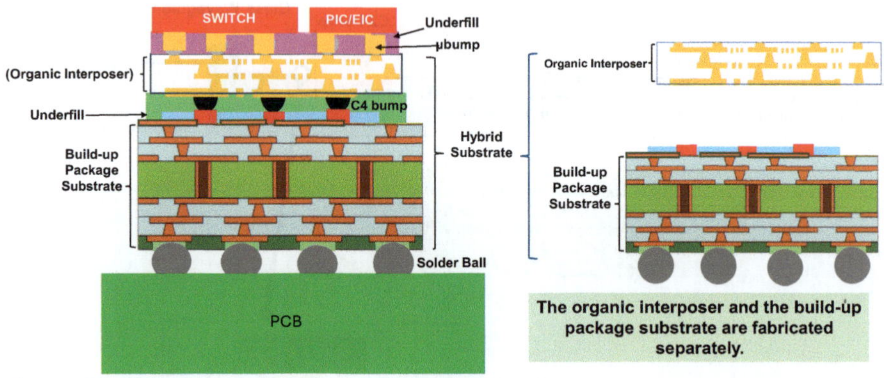


**Fig. 4.11** The organic interposer and the package substrate are fabricated separately

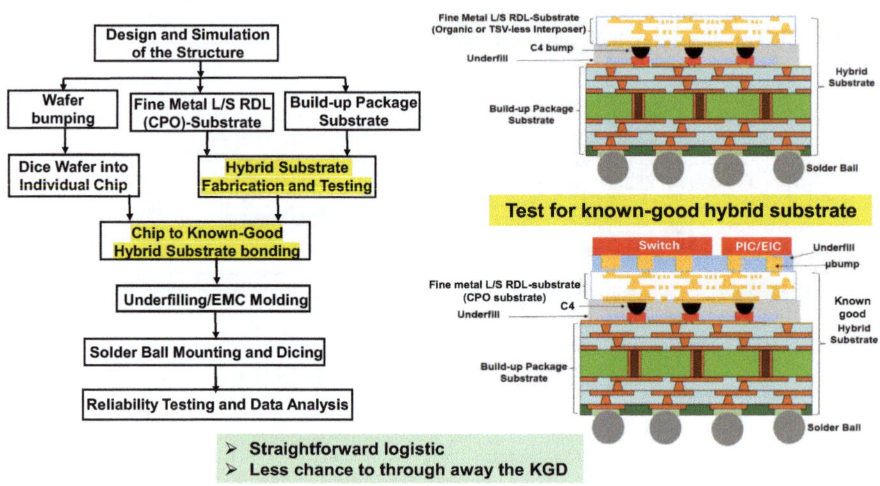


**Fig. 4.12** Process flow in fabricating the CPO substrate

film (sacrificial layer) is slit coating on a (515 × 510 × 1.1 mm) glass temporary panel and then a Ti/Cu seed layer is deposited on the top, as shown in Fig. 4.13. The thickness of the sacrificial layer is 1 μm and it is a light-to-heat conversion film. The contact pad can be obtained by photoresist, laser direct imaging (LDI), development, electrochemical deposit (ECD) Cu, and stripping off the photoresist. Then, a PID is slit coating and the first DL (DL01) of RDL1 is obtained by LDI. It is followed by physical vapor deposition (PVD) the Ti/Cu, photoresist, LDI and development, ECD the Cu, strip off the photoresist, and etch off the TiCu to obtain the first ML (ML1) of RDL1. DL12 and ML2 of RDL2, and DL23 and ML3 of RDL3 can be obtained by repeating the same process steps. The solder mask (passivation) can be obtained by slit coating a PID and LDI. Then, the surface is finishing with electroless nickel electroless palladium immersion gold (ENEPIG).

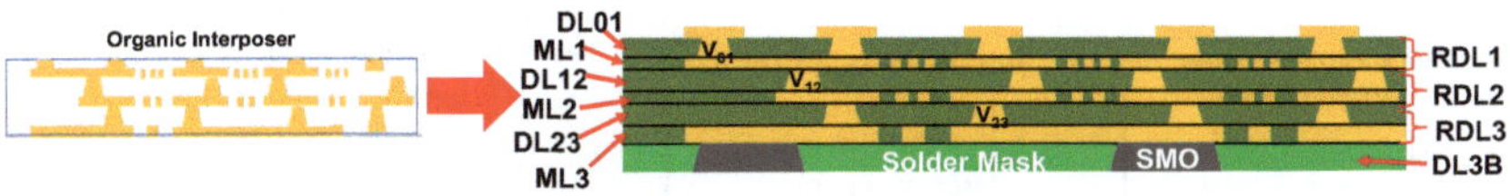


| RDLs | Key Elements of RDL | Line width (L) / Spacing (S) | Thickness (H) | |
|---|---|---|---|---|
| | | | PID | ABF |
| RDL1 | ML1 (Metal layer 1) | 2/2μm | 3μm | 3μm |
| | DL01 (Dielectric layer between the contact pad and ML1) | NA | 3μm | 3.5μm |
| | $V_{01}$ (Via opening between contact pad and ML1) | NA | 10μm | 20μm |
| RDL2 | ML2 (Metal layer 2) | 5/5μm | 5μm | 8μm |
| | DL12 (Dielectric layer between ML1 and ML2) | NA | 6μm | 10μm |
| | $V_{12}$ (Via opening between ML1 and ML2) | NA | 20μm | 20μm |
| RDL3 | ML3 (Metal layer 3) | 10/10μm | 5μm | 8μm |
| | DL23 (Dielectric layer between ML2 and ML3) | NA | 4μm | 7.5μm |
| | $V_{23}$ (Via opening between ML2 and ML3) | NA | 18μm | 20μm |

**Fig. 4.13** Fine metal L/S/H RDL-substrate (organic interposer)

**Table 4.1** PID and ABF material properties

| Items | PID | ABF (Organic Interposer) | ABF (Build-up Substrate) |
|---|---|---|---|
| Curing condition, °C for min | 200 for 60 | 200 for 90 | 190 for 90 |
| CTE (25–150 °C), $10^{-6}$/°C | 60 | 37 | 39 |
| CTE (150–240 °C), $10^{-6}$/°C | – | 98 | 117 |
| Tg, °C | 300 | 156 | 153 |
| Dielectric constant (Dk) | 3 (1 MHz) | 3.2 (5.8 GHz) | 3.2 (5.8 GHz) |
| Loss tangent (Df) | – | 0.011 (5.8 GHz) | 0.017 (5.8 GHz) |
| Young' modulus (23 °C), GPa | 2.1 | 7.5 | 5.0 |
| Tensile strength (23 °C), GPa | 0.125 | 125 | 98 |
| Elongation (23 °C), % | 60 | 5.4 | 5.6 |

Figure 4.16a shows the scanning electron microscopy (SEM) images of the PID RDL substrate and Table 4.2 shows the measured L, S, and H. For ML1, L = 2.07 and 2.08 μm (not exactly 2 μm), S = 1.74 and 1.81 μm (not 2 μm), and H = 2.06 and 2.17 μm (not 3 μm). For ML2, L = 5.6 and 6.03 μm (not 5 μm), S = 3.91 and 4.5 μm (not 5 μm), and H = 4.35 and 4.48 μm (not 5 μm). For ML3, L = 10.24 and 10.69 μm (not 10 μm), S = 9.46 and 10.18 μm (not 10 μm), and H = 3.39 and 5.11 μm (not 5 μm). Thus, there are rooms for improvements, e.g., a better estimation of the compensation of photoresists, LDI, ECD Cu, Cu etching, selection of dielectric material, etc. The metal lines are not flat.

The process flow (Fig. 4.15) for the ABF case is very similar to that with PID, except instead of slit costing a PID, laminate a 12.5 μm-thick raw ABF with nano-filler on the whole panel. There are two operating stages of the ABF: (1) at the first stage, the temperature is 120oC for 30 s at vacuum condition and then press (0.68 MPa) for 30 s with the temperature and vacuum on, and (2) at the second

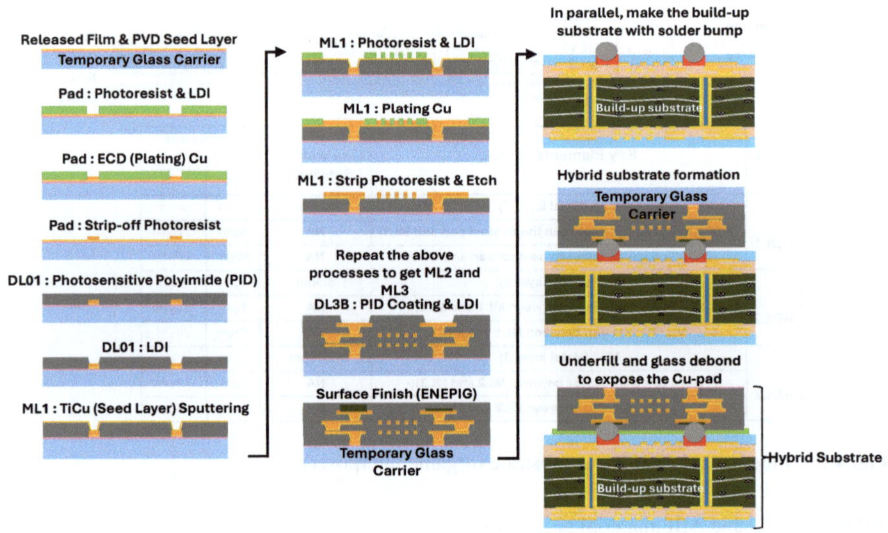


**Fig. 4.14** Process flow of fine metal L/S/H RDL organic interposer with PID

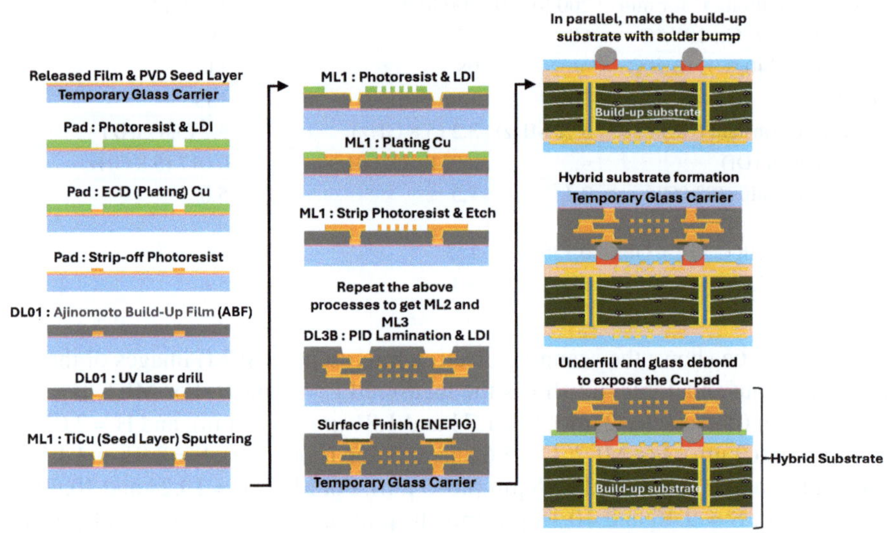


**Fig. 4.15** Process flow of fine metal L/S/H RDL organic interposer with ABF

stage, the temperature is 100oC and press (0.58 MPa) for 60 s. The first DL (DL01) (3.5 μm-thick) of RDL1 is drilled by a UV laser to obtain the blind via. It is followed by PVD the Ti/Cu, photoresist, LDI and development, ECD the Cu, strip off the photoresist, and etch off the TiCu to obtain the first ML (ML1) of RDL1. DL12 and ML2 of RDL2, and DL23 and ML3 of RDL3 can be obtained by repeating the

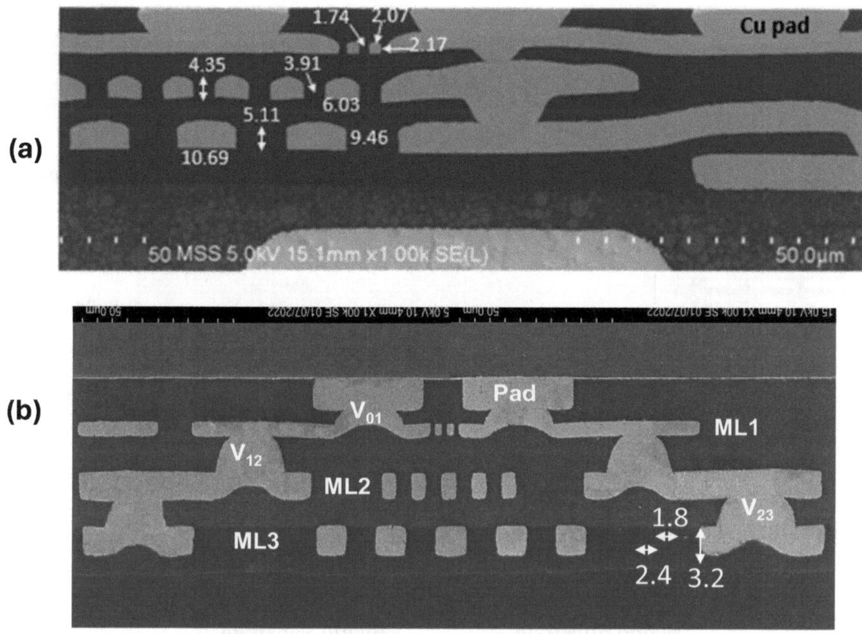


**Fig. 4.16** SEM images of RDLs with (**a**) PID (**b**) ABF

**Table 4.2** Dimensions of the organic interposer

| RDLs | Items | Design (μm) | | Measured (μm) | |
|---|---|---|---|---|---|
| | | PID | ABF | PID | ABF |
| Metal of RDL1 (ML1) | Linewidth (L) | 2 | 2 | 2.07, 2.08, … | 2.4, 2.4, … |
| | Spacing (S) | 2 | 2 | 1.74, 1.81, … | 1.8, 2.0, … |
| | Thickness (H) | 3 | 3 | 2.06, 2.17, … | 3.2, 3.5, … |
| Metal of RDL2 (ML2) | Linewidth (L) | 5 | 5 | 5.6, 6.03, … | 5.1, 5.0, … |
| | Spacing (S) | 5 | 5 | 3.91, 4.5, … | 4.8, 4.4, … |
| | Thickness (H) | 5 | 8 | 4.4, 4.5, … | 7.2, 7.4, … |
| Metal of RDL3 (ML3) | Linewidth (L) | 10 | 10 | 10.2, 10.7, … | 9.9, 10.3, … |
| | Spacing (S) | 10 | 10 | 9.5, 10.2, … | 9.4, 9.2, … |
| | Thickness (H) | 5 | 8 | 3.4, 5.1, … | 8.2, 8.6, … |

same process steps. The SEM images of the RDL-substrate with ABF are shown in Fig. 4.16b. It can be seen that the metal lines are very flat.

The cross section of the 2-2-2 build-up package substrate (23 × 23 × 1.3 mm) is shown in Fig. 4.17. The truncated spherical 30 μm diameter C4 bumps are fabricated by stencil printing a Sn3Ag0.5Cu solder paste with a 29-μm-thick stainless-steel stencil on the build-up package substrate.

The hybrid substrate is fabricated by combining (assembling) the fine metal L/S/H RDL-substrate (with the temporary glass carrier) and the build-up package substrate with C4 solder bumps and enhanced with underfill. Figure 4.18 shows the

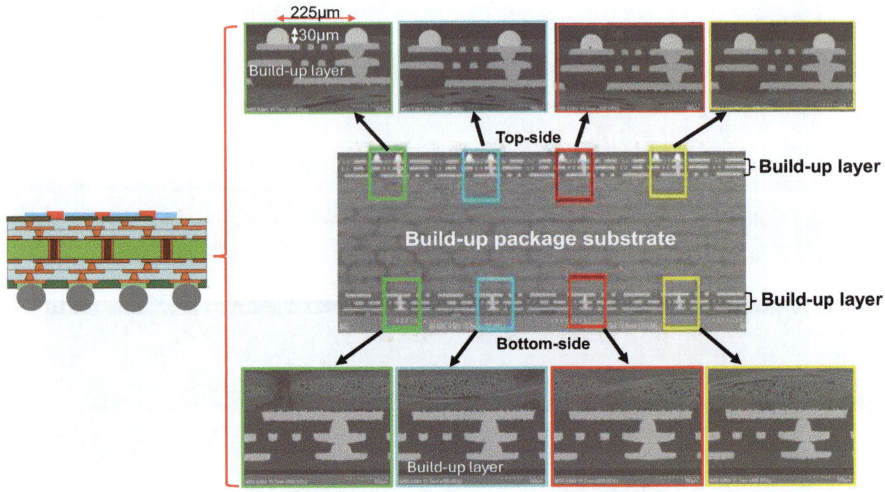


**Fig. 4.17** Build-up package substrate with C4 bumps

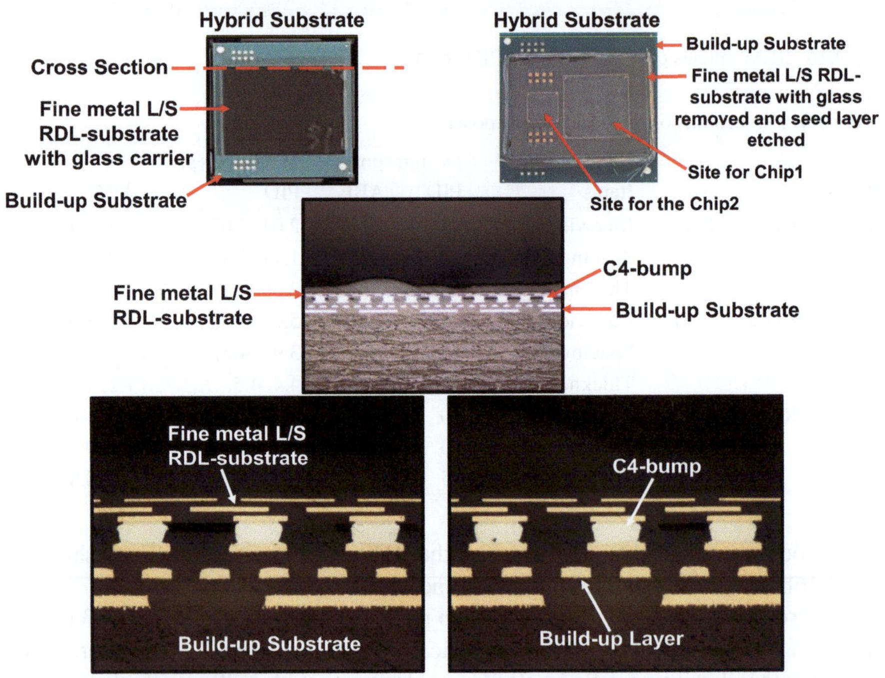


**Fig. 4.18** Hybrid substrate (fine metal L/S/H RDL organic interposer + package substrate)

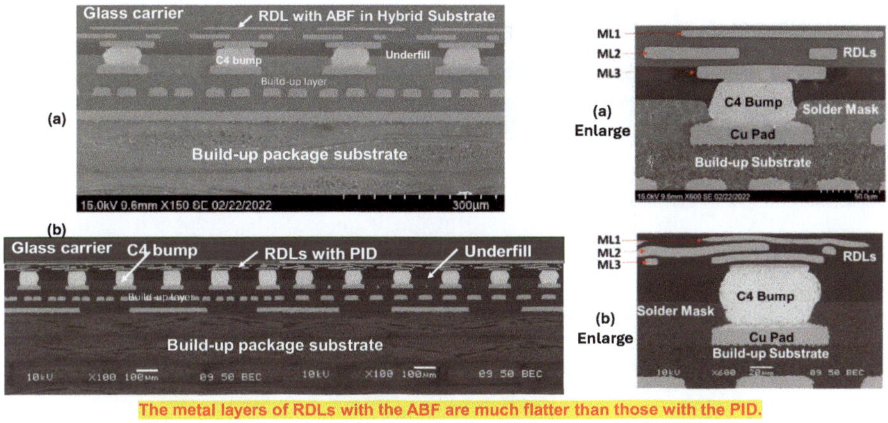


**Fig. 4.19** (**a**) SEM images with ABF. (**b**) SEM images with PID

top view and cross section view. Figure 4.19 shows the metal lines (ML1, ML2, and ML3) in the hybrid substrate. Figure 4.19 (a and b) show the close-up images of the hybrid substrate with PID and ABF, respectively. It can be seen that the metal lines with ABF are much flatter than those with PID. However, the hybrid substrate fabricated with ABF is thicker than that with PID.

## 4.8 Heterogeneous Integration of Switch, PIC and EIC

### *4.8.1 2D Heterogeneous Integration of ASIC Switch, PIC and EIC*

Figure 4.20a shows a 2D COP system. It can be seen that the ASIC switch, PIC, and EIC are side-by-side attached on a CPO substrate, which can be an organic-core build-up package substrate or a glass-core build-up package substrate. Figure 4.19b shows another CPO system. It can be seen that the switch, PIC, and EIC are side-by-side attached to a CPO substrate which can be either a TSV-interposer, a TGV-interposer, or an organic-interposer. The interposer is then attached to an organic build-up package substrate. The CPO system shown in Fig. 4.20b is for higher performance applications than that of Fig. 4.19a, but with higher cost.

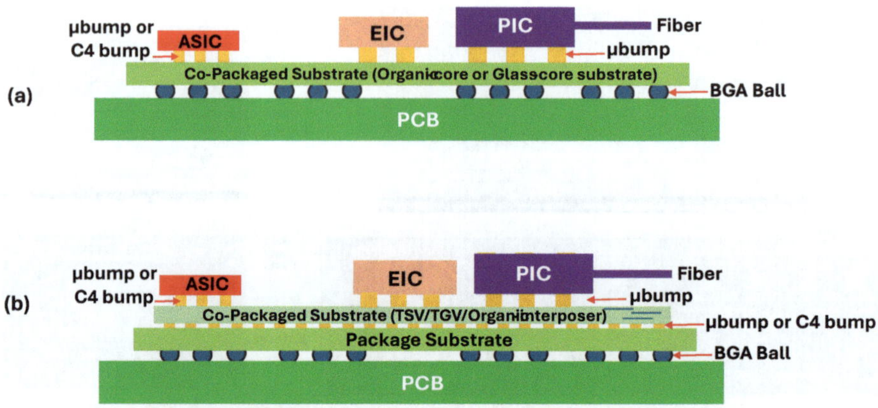


**Fig. 4.20** (**a**) CPO on an organic-core or a glass-core package substrate. (**b**) CPO on a TSV, a TGV, or an organic interposer

### *4.8.2 2D Heterogeneous Integration of ASIC Switch, PIC and EIC with Si Bridges*

Figure 4.21a shows a CPO system. It can be seen that the chiplets such as the switch, PIC, and EIC are side-by-side attached to an organic build-up package substrate with embedded multi-die interconnect bridge (EMIB) [60] connecting the switch and EIC and the EIC and PIC. Figure 4.21b shows another CPO system. It can be seen that the switch, PIC, and EIC are side-by-side attached to either an organic-core substrate or a glass-core substrate. The Si bridge is connected between the switch and the EIC and the EIC and the PIC with microbumps. It is noted that the Si bridges are not embedded in the CPO organic-core or glass-core build-up package substrate. Figure 4.21c shows another CPO structure. It can be seen that the switch, PIC, and EIC are side-by-side attached to either a glass-core substrate or an organic-core substrate. The Si bridge is connected between the switch and the EIC and the EIC and the PIC with Cu-Cu hybrid bonding. A comparison of these three CPO structures is shown in Fig. 4.22.

### *4.8.3 3D Heterogeneous Integration of ASIC Switch, PIC and EIC*

With CPO, the switch chip is typically surrounded by 16 OEs, all placed on a package substrate. Currently, the 25.6-terabit Ethernet switch chip (Fig. 4.23a) requires 16@1.6 terabits-per-second (1.6Tbps) OEs, while the upcoming 51.2-terabit switch chips (Fig. 4.23b) will use 3.2Tbps OEs (the size of the EIC and PIC of the OE will be larger). The issue is that the multi-chip module can only be so large. It is challenging with today's packaging technology to surround the 51.2Tbps ASIC with

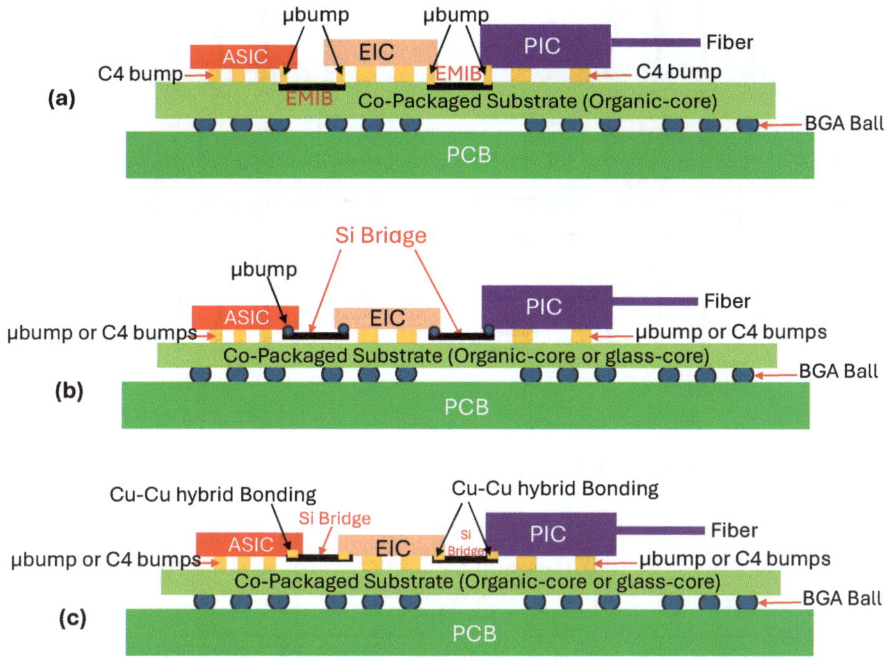


**Fig. 4.21** (**a**) CPO with EMIB embedded in an organic build-up package substrate. (**b**) CPO with Si bridges (micro bumped) between chips on a glass-core or an organic-core build-up package substrate. (**c**) CPO with Si bridges (hybrid bonding) between chips on a package substrate

| CPO Structures | Advantages | Disadvantages |
|---|---|---|
| 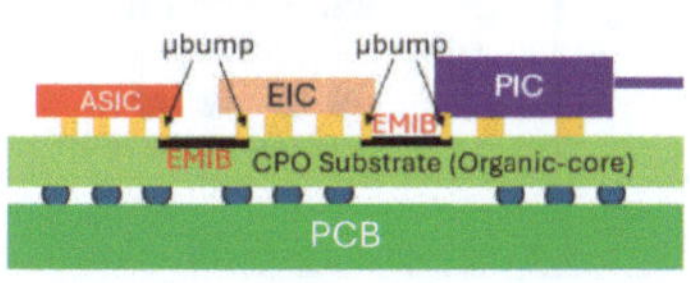 | ➢ High performance<br>➢ Fine pitch<br>➢ High density | ➢ Higher cost<br>➢ EMIB involves complicated CPO substrate fabrication |
| 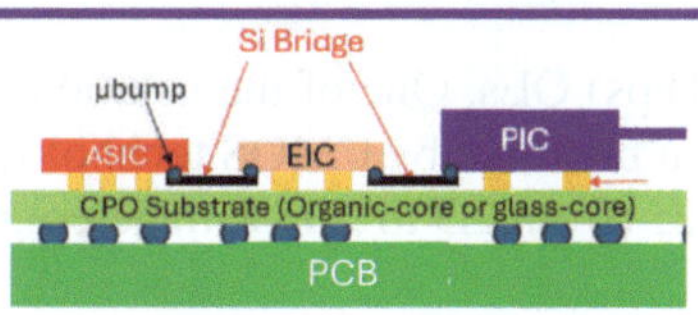 | ➢ Higher performance<br>➢ Finer pitch<br>➢ Higher density<br>➢ Simple CPO substrate | ➢ High cost |
| 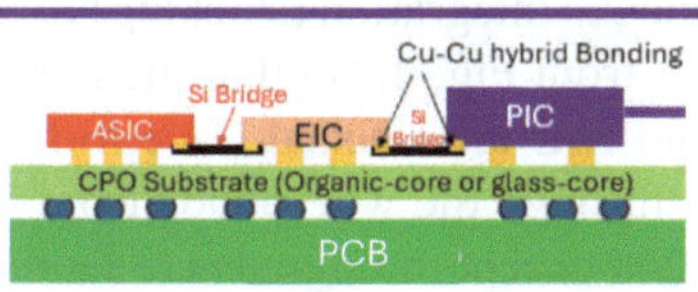 | ➢ Highest performance<br>➢ Finest pitch<br>➢ Highest density<br>➢ Simple CPO substrate | ➢ Highest cost<br>➢ Hybrid bonding involves complicated chip bonding processes |

**Fig. 4.22** Comparison between various CPO systems

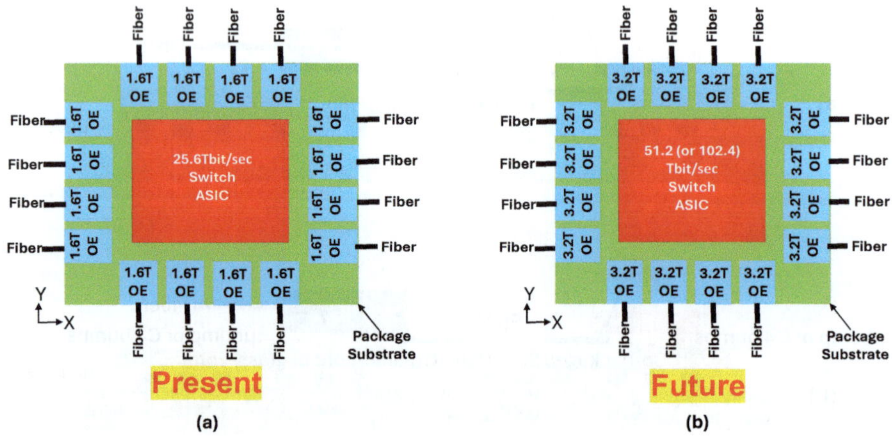


**Fig. 4.23** Present and future CPO systems

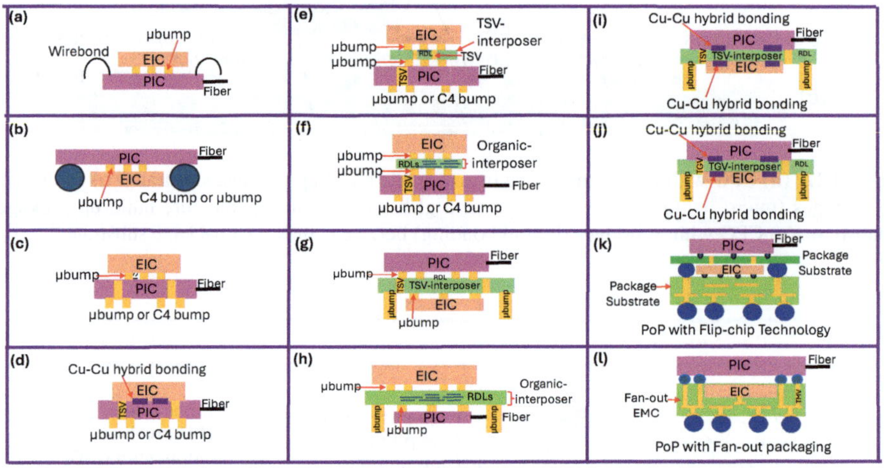


**Fig. 4.24** Various 3D stacks of PIC and EIC

16@3.2Tbps (or 102.4Tbps ASIC with 16@6.2Tbps) OEs. One of the solutions is to integrate some of the PIC and EIC into a silicon photonic by CMOS technology. Another solution is to stack up the PIC and the EIC (from 2D to 3D) with heterogeneous integration packaging method.

There are many different kinds of 3D heterogeneous integration (stacking) of the PIC and EIC as shown in Fig. 4.24. It can be seen from Fig. 4.24a that the light is coming from the fiber to the PIC and the PIC and EIC are stacked (face-to-face) with the μbumps, and the signals are fanning out through the wire bonds from the PIC. Figure 4.24b is basically the same as Fig. 4.24a except the signals are fanning out through the C4 bumps, or μbumps, from the PIC. Figure 4.24c shows the EIC and the PIC are stacked (face-to back) with the μbumps, and the signals are running out through the TSVs and the C4 bumps or μbumps from the PIC. Figure 4.24d is basically the same as Fig. 4.24c except the μbumps between the EIC and PIC are eliminated by the bumpless Cu-Cu hybrid bonding. Figure 4.24e shows that the EIC and the PIC are stacked (connected) by a TSV-interposer with the μbumps, and the signals are fanning out through the RDLs and TSVs of the TSV-interposer and the TSVs and C4 bumps or μbumps from the PIC. Figure 4.24f is basically the same as Fig. 4.24e except the TSV-interposer is replaced by the organic-interposer and the signals are fanning out through the RDLs of the organic-interposer and the TSVs and C4 bumps or μbumps from the PIC. Figure 4.24g shows that the PIC and EIC are stacked by a TSV interposer with μbumps, and the signals are running out through the μbumps of the TSV-interposer. Figure 4.24h is basically the same as Fig. 4.24g except the TSV-interposer is replaced by the organic-interposer. Figure 4.24i is basically the same as Fig. 4.24g except the μbumps on both sides of the TSV-interposer have been eliminated by the bumpless Cu-Cu hybrid bonding. Figure 4.24j is basically the same as Fig. 4.24i, except the TSV-interposer is replaced by the TGV-interposer.

Comparing Fig. 4.24c with Fig. 4.24 (e and f) the latter figure represents the case for extremely high-density situations and the case where there are difficulties in direct stacking of the EIC and PIC. Comparing Fig. 4.24e with Fig. 4.24g, it can be seen that the TSVs in the PIC as shown in Fig. 4.24e have been eliminated (see Fig. 4.24g). Similarly, by comparing Fig. 4.24f with Fig. 4.24h, it can be seen that the TSVs in the PIC—as shown in Fig. 4.24f—have been eliminated (see Fig. 4.24h).

Figure 4.24 (k and l) are package-on package (PoP) of the PIC and EIC. Figure 4.24k is for a solder-bumped flip-chip application for the EIC on a package substrate, while Fig. 4.24l is for the fan out of the EIC on a set of RDLs.

The patent application of case (h) has been granted [42] and is shown in Fig. 4.25. It can be seen that the EIC and PIC are stacked with an organic interposer and the CPO substrate for the 3D heterogeneous integration of the PIC and EIC as well as

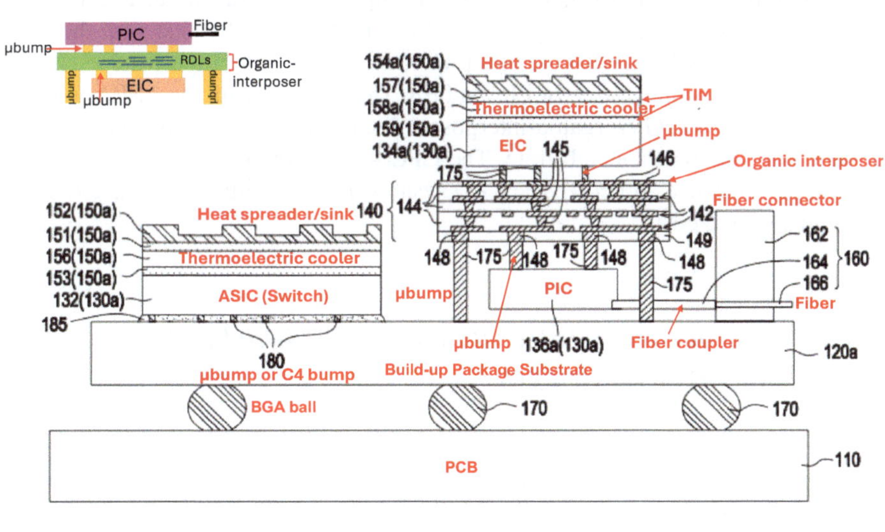


**Fig. 4.25** PIC and EIC are stacked with an organic interposer

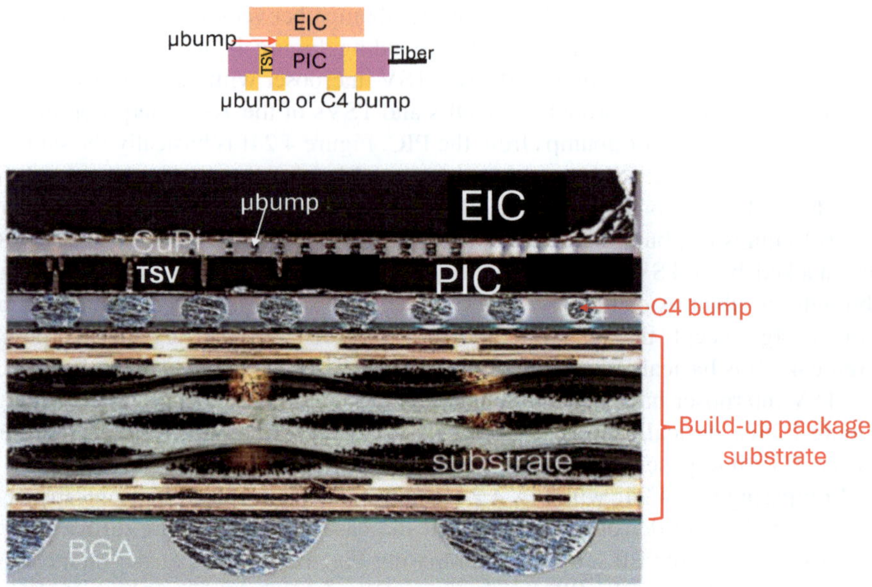


**Fig. 4.26** Cisco's 3D stacking of EIC and PIC with TSVs

the switch is a build-up package substrate. Figure 4.26 shows a CPO system with EIC stacked on top of PIC with TSVs [34], which is case (c) of Fig. 4.24. Figure 4.27 shows a CPO system with POP structure [17]. It can be seen that the PIC is stacked on top of the EIC which is embedded in the fan-out EMC, case (l) of Fig. 4.24.

Figure 4.28a shows a 3D heterogeneous integration (stacking) of PIC and EIC (such as those shown in Fig. 4.24) side-by-side with the ASIC switch with μbumps or C4 bumps on a CPO organic-core build-up package substrate. Figure 4.28b shows a 3D heterogeneous integration of PIC and EIC side-by side with the ASIC switch on a same CPO glass-core build-up package substrate. Figure 4.27c shows a 3D heterogeneous integration of PIC and EIC side-by side with the ASIC switch on a same CPO TSV/TGV/organic-interposer. Then, the interposer is attached to an organic build-up package substrate. The CPO system shown in Fig. 4.28c has the highest performance and cost. The CPO system shown in Fig. 4.27a has the lowest performance and cost.

### *4.8.4 Heterogeneous Integration of ASIC Switch, PIC and EIC with Bridges*

Figure 4.29a shows a 3D heterogeneous integration of PIC and EIC (such as those shown in Fig. 4.24), as well as the ASIC switch with Intel's embedded multi-die interconnect bridge (EMIB) in the cavity of a CPO organic-core package substrate.

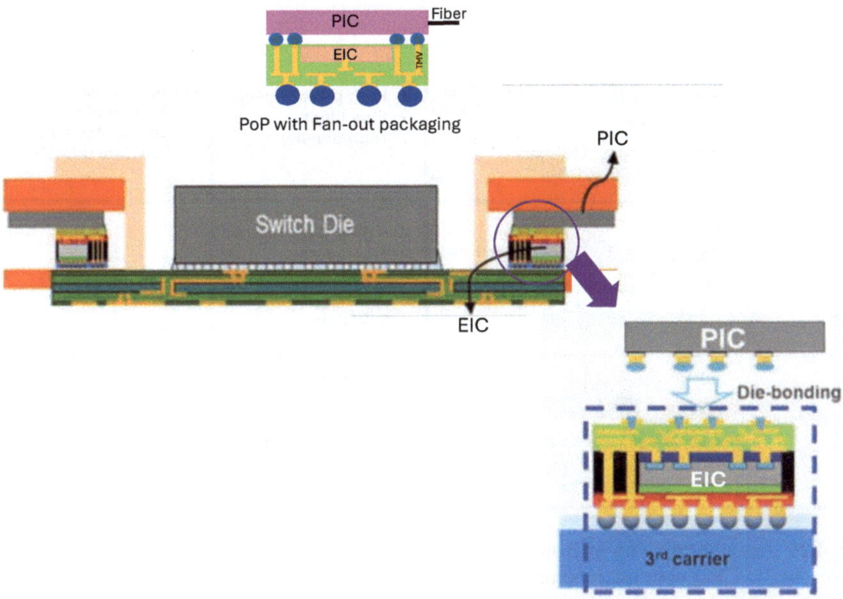


Fig. 4.27 Spill's 3D stacking of PIC and EIC in the fan-out EMC

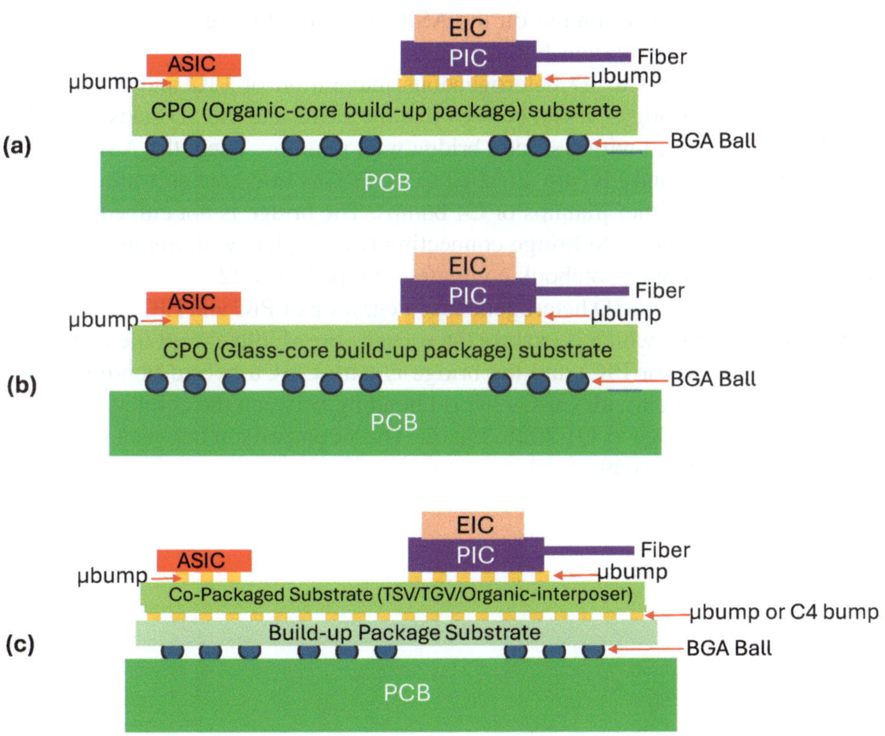


**Fig. 4.28** (**a**) 3D CPO on organic-core package substrate. (**b**) 3D CPO on glass-core package substrate. (**c**) 3D CPO on a TSV, a TGV, or an organic interposer

| | CPO Structures | Advantages | Disadvantages |
|---|---|---|---|
| (a) | 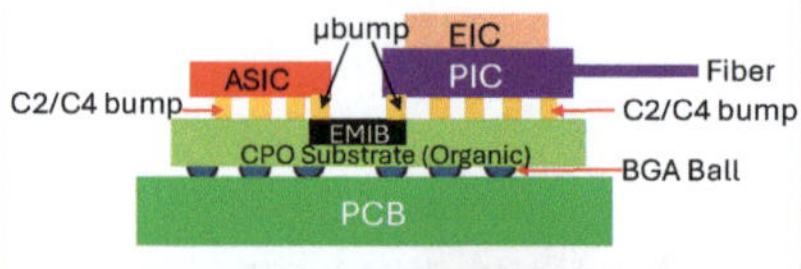  | ➢ High performance<br>➢ Fine pitch<br>➢ High density | ➢ Higher cost<br>➢ EMIB involves complicated CPO substrate fabrication |
| (b) | 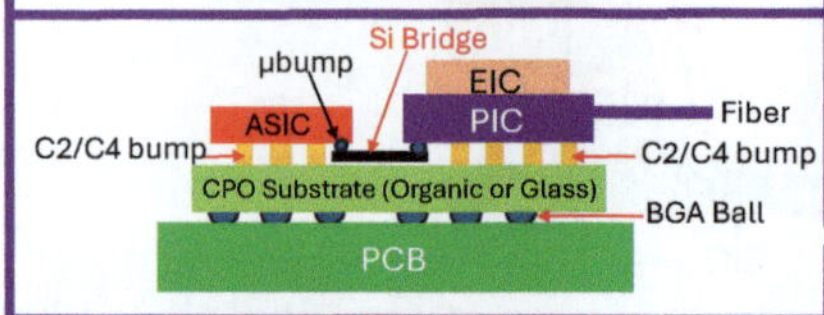  | ➢ Higher performance<br>➢ Finer pitch<br>➢ Higher density<br>➢ Simple CPO substrate | ➢ High cost |
| (c) | 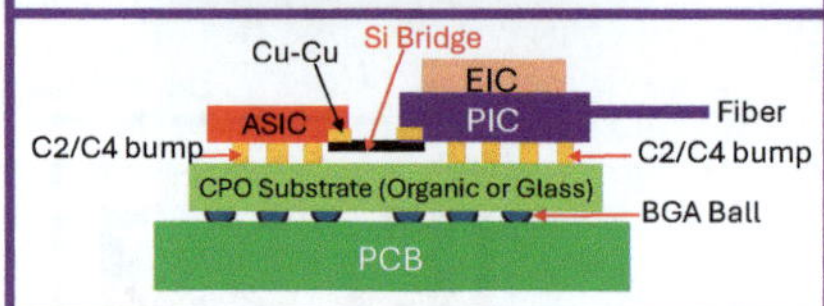  | ➢ Highest performance<br>➢ Finest pitch<br>➢ Highest density<br>➢ Simple CPO substrate | ➢ Highest cost<br>➢ Hybrid bonding involves complicated chip bonding processes |

**Fig. 4.29** (**a**) 3D CPO with microbumped EMIB embedded in an organic build-up package substrate. (**b**) 3D CPO with microbumped Si bridge on an organic or a glass build-up package substrate. (**c**) 3D CPO with Cu-Cu Si bridge

The bridge is connecting the switch and the PIC. There are two different kinds of bumps (μbumps and C4 bumps) on the ASIC and PIC. Figure 4.30 shows one of Intel's CPO systems with their EMIB [13].

Figure 4.29b shows a 3D heterogeneous integration of PIC and EIC, and an ASIC with silicon bridges. It can be seen that the ASIC and the 3D stacking of EIC and PIC are connected with a silicon bridge with μbumps. Then, the ASIC and the 3D stacking of EIC and PIC are attached to a CPO organic-core substrate or glass-core substrate with either μbumps or C4 bumps. The bridge is not embedded in the substrate. An example of Si bridge connecting two chiplets with μbumps is shown in Fig. 4.31a for Apple's notebook computers shipped in 2022.

Figure 4.29c shows a 3D heterogeneous integration of PIC and EIC, and an ASIC with a silicon bridge, which is exactly the same as Fig. 4.29b except the connection between the switch (or PIC) and the bridge is with Cu-Cu hybrid bonding. Apple shipped their notebook with Cu-Cu hybrid bonding between the chiplets (CPU and GPU) and the Si bridge in Q1 2026, Fig. 4.31b. A comparison between these three cases is shown in Fig. 4.29.

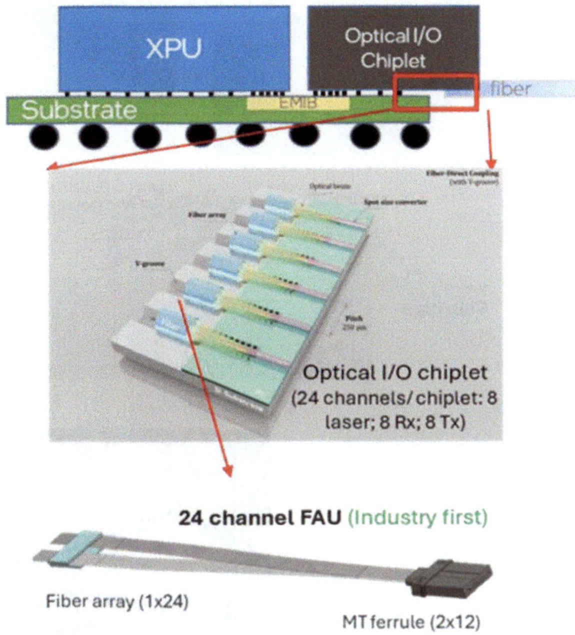


**Fig. 4.30** Intel's CPO with EMIB

## 4.8.5 3D Heterogeneous Integration of PIC, EIC, and ASIC Switch on Fan-out RDL Substrate

Figure 4.32a shows a 3D heterogeneous integration of PIC and EIC, and an ASIC with a silicon bridge embedded in a co-packaged substrate such as the fan-out epoxy molding compound with RDLs (fan-out organic-interposer). The bridge will connect the ASIC and the 3D stacking of PIC and EIC, as well as the co-packaged substrate through the RDLs and through-mold via (TMV). An example of Si bridge embedded in fan-out EMC with RDLs (organic interposer) is shown in Fig. 4.32b [61]. The interposer is then attached to the PCB. The structure is very simple with high performance and reasonable cost.

## 4.8.6 3D Heterogeneous Integration of PIC, EIC, and ASIC Switch on Fan-out RDLs and Polymer Waveguides

Figure 4.33 shows a CPO system. It can be seen that the PIC and EIC are 3D stacked with hybrid bonding and then they are side-by-side with a switch attached to a fan-out RDLs CPO substrate. Figure 4.34 shows another CPO system. It can be seen that the PIC and EIC are 3D stacked with hybrid bonding and then they are side-by-side with a switch attached to a fan-out RDLs CPO substrate with holes. The holes

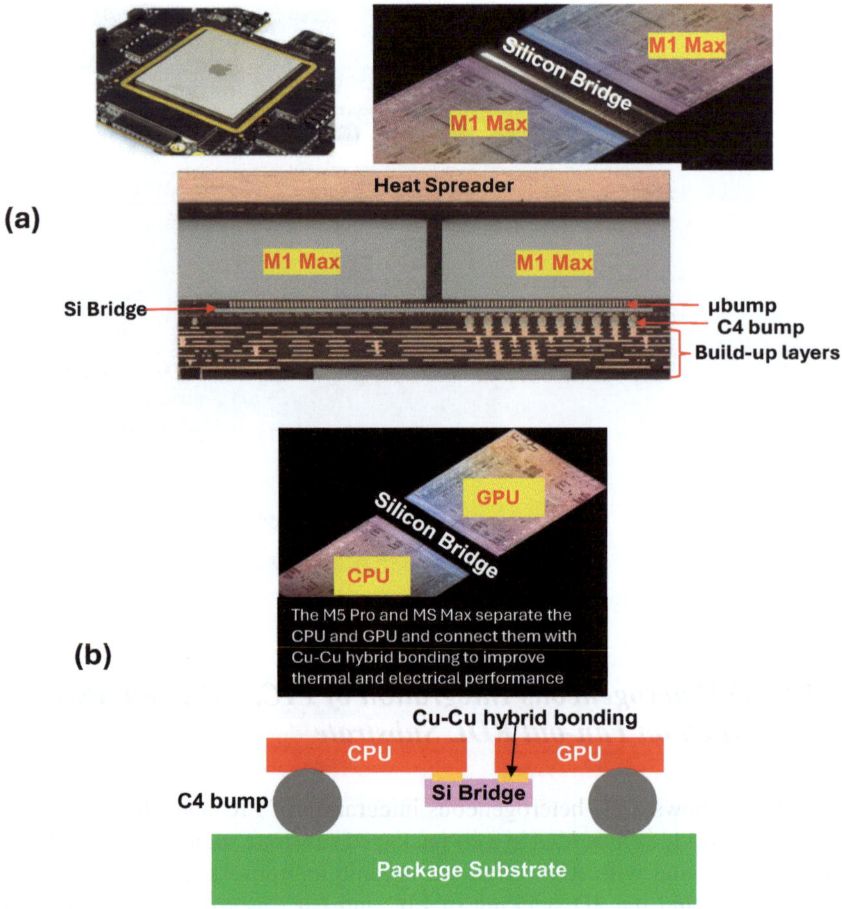


**Fig. 4.31** Apple's notebook (**a**) with microbumped Si bridge between chiplets and (**b**) with Cu-Cu hybrid bonding

are used for lights to pass through. There are polymer waveguides with 45° mirrors at their both ends semi-embedded in the PCB. The light from the external fiber through the V-groove hits the 45° physical mirror and then reflects and passes through the hole and hits one end of the mirror of the polymer waveguide. The light travels along the polymer waveguide, hits the other end of the 45° mirror, and then reflects on the PIC. Finally, the PD of the PIC receives the light. This system has better performance than that shown in Fig. 4.32 but with higher cost.

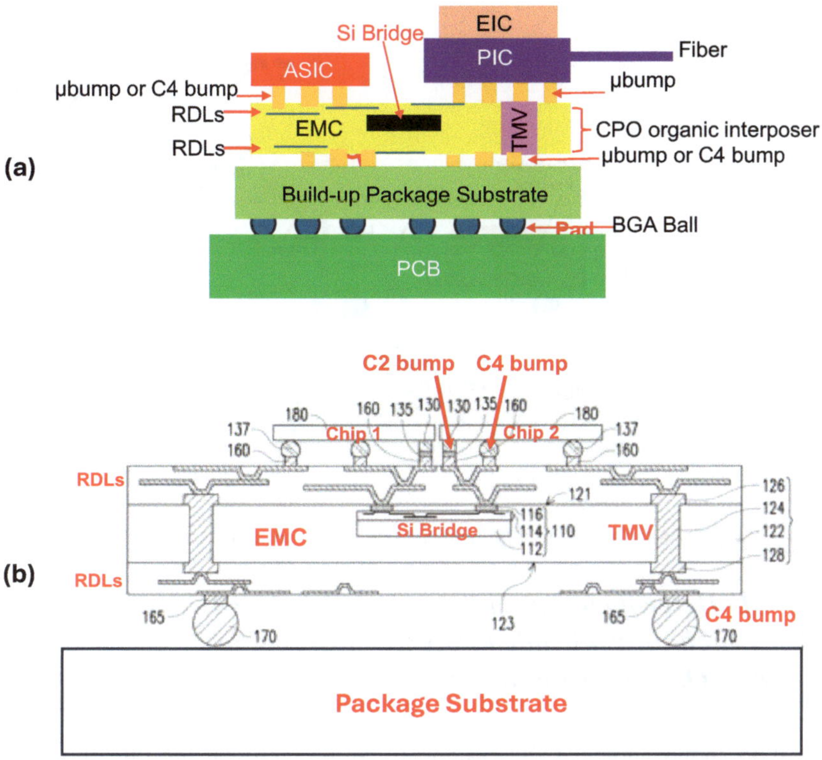


**Fig. 4.32** (**a**) COP on fan-out EMC with embedded Si bridge. (**b**) Si bridge embedded in fan-out EMC with RDLs

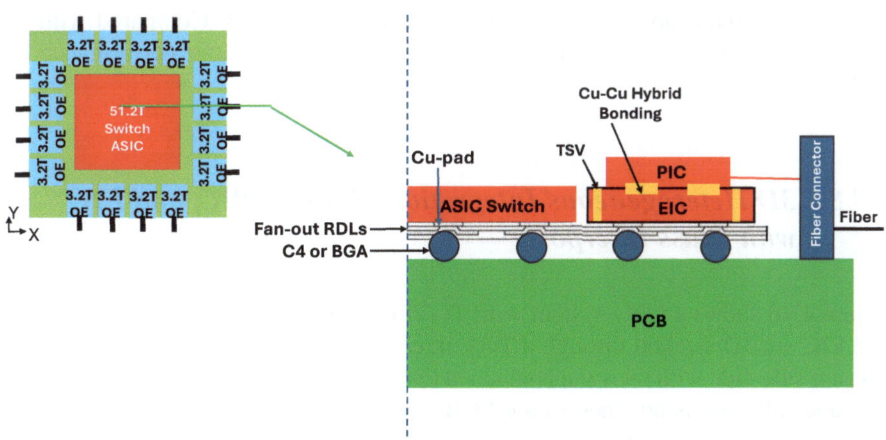


**Fig. 4.33** 3D CPO on fan-out RDL

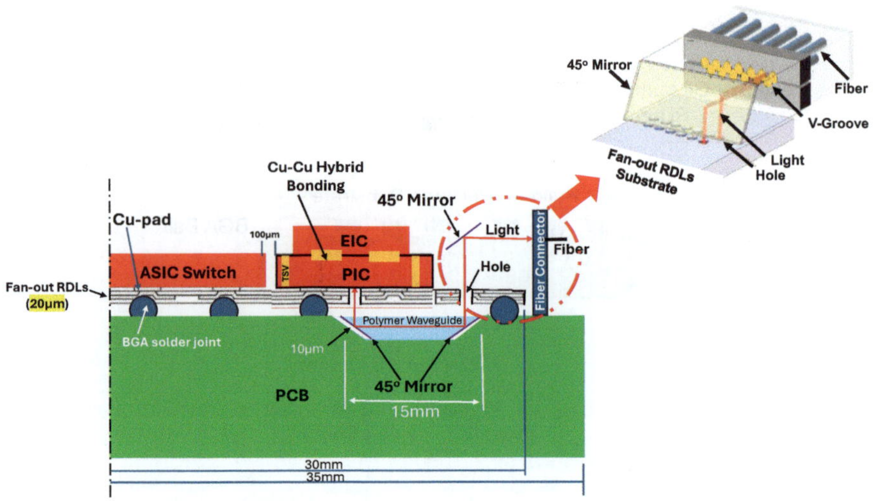


Fig. 4.34 3D CPO on fan-out RDL with holes and polymer waveguides on PCB

### 4.8.7 3D Heterogeneous Integration of EIC, PIC, and ASIC Switch on Embedded Polymer Waveguide and Organic Interposer

Figure 4.35 shows another CPO system. It can be seen that the PIC and EIC are stacked with hybrid bonding and then they are side-by-side with the switch attached to an organic interposer with an embedded polymer waveguide with 45° mirrors at its both ends. There are holes in the organic interposer. The CPO interposer is attached to a build-up package substrate and then on a PCB. Compared with the CPO system shown in Figs. 4.34 and 4.35, the latter has better performance but higher cost.

### 4.8.8 3D Heterogeneous Integration of ASIC, PIC, and EIC with Glass Interposer

Figure 4.36 shows a CPO system with TGV-interposer. It can be seen that the PIC and EIC are 3D stacked through a TGV-interposer with microbumps. The PIC and EIC are side-side with the switch attached to a build-up package substrate. Then, the package substrate is attached to the PCB. The glass waveguide on the glass interposer receives the light from the external fiber and sends the light to the PD of the PIC. Figure 4.37 shows exactly the same structure except the microbumps are replaced by Cu-Cu hybrid bonding, which yields higher density, fine pitch, and better performance, but with a higher cost.

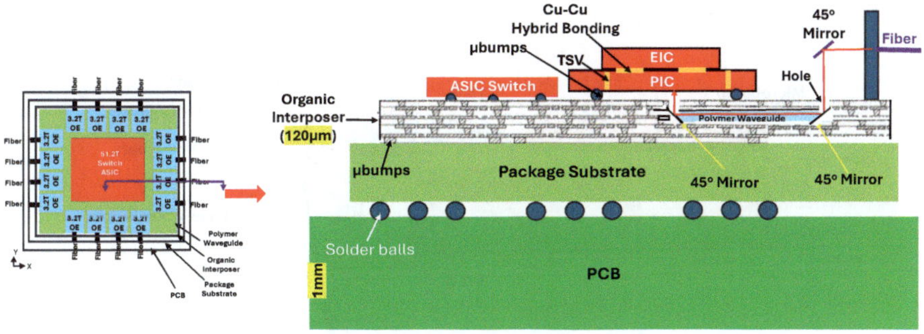


Fig. 4.35 3D CPO on organic interposer with holes and embedded polymer waveguides

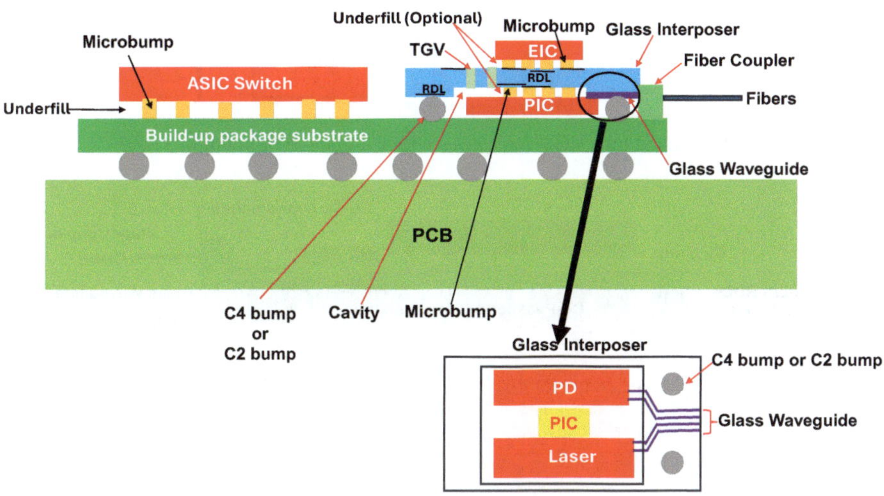


Fig. 4.36 3D CPO stacked by a TGV interposer with microbumps

### 4.8.9 3D Heterogeneous Integration of ASIC, PIC and EIC with Glass Substrate

Figure 4.38 shows a CPO structure with a glass-core build-up package substrate. It can be seen that there is a cavity on the CPO glass-core substrate. The PIC and EIC are 3D stacked with hybrid bonding and the PIC is semi-embedded in the cavity of the substrate. The switch and the PIC/EIC stack are side-by-side attached to the CPO substrate. Compared to the CPO structure shown in Fig. 4.35, this CPO structure has better performance but has higher cost.

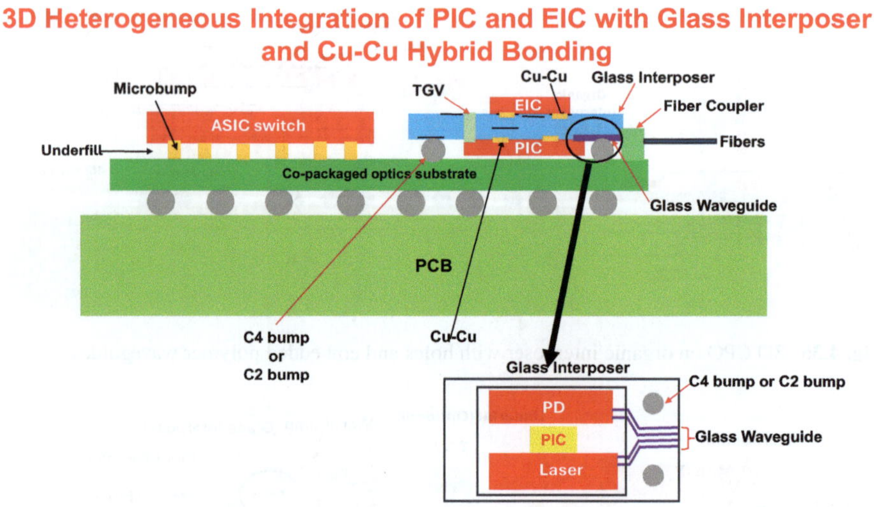


Fig. 4.37 3D CPO stacked by a TGV interposer with Cu-Cu hybrid bonding

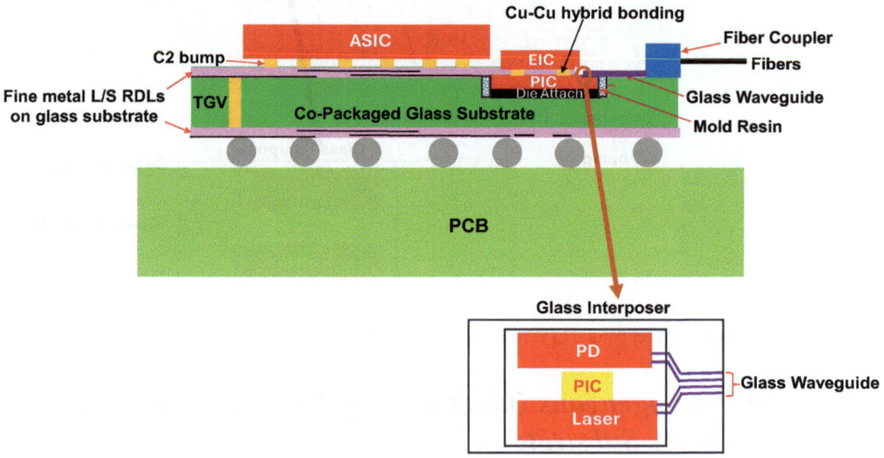


Fig. 4.38 3D CPO semi-embedded in a glass-core substrate

## 4.9 3D CPO for HPC Driven by AI and High-Speed and High-Bandwidth Communications

Figure 4.39 shows a CPO system for HPC driven by AI and high-speed and high bandwidth communication applications. It can be seen that the graphics processing unit (GPU) is surrounded by the high-bandwidth memory (HBM) and then surrounded by the OE such as the PIC and EIC. The ASIC switch is surrounded by the PIC and EIC. For both structures, the PIC and EIC are 3D stacked by Cu-Cu hybrid bonding. The GPU, HBM, and the stacking of the EIC and PIC are side-by-side on

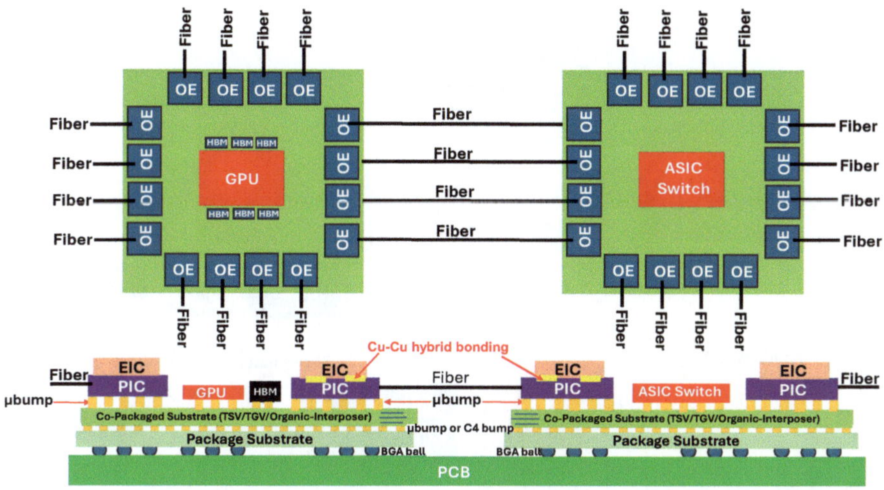


Fig. 4.39 3D heterogeneous integration of GPU, HBM, switch, PIC, and EIC for HPC driven by AI and high-speed and high-bandwidth communications

the same co-packaged substrate, which can be a TSV-interposer, TGV-interposer, or organic-interposer. Similar to the ASIC switch and PIC/EIC. These two structures are then attached to a PCB.

## 4.10 CPO with Silicon Photonics

Recently, CPO with silicon photonics such as silicon InP laser, silicon Germanium photodetector, silicon waveguides and silicon modulators/resonators have been getting lots of tractions for future AI and data center applications. These components are fabricated on a silicon wafer with CMOS technology. These components enable high-speed, high-bandwidth, energy-efficient optical interconnects that overcome the physical limitations of traditional copper wiring in high-performance computing driven by artificial intelligence (AI) and data centers. High volume manufacturing of CPO with silicon photonics are expected in 2027.

For ROSA in CPO with silicon photonics, the light from a fiber coupled to the silicon waveguide which confines the light. Then the silicon modulator changes light to optical signal by altering the light's properties. Finaly, the photodetector (usually made by Germanium) of the PIC changes the optical signals to electrical signals, which are then amplified by the transimpedance amplifier (TIA) of the EIC before connecting to the switch, Fig. 4.40.

For TOSA in CPO with silicon photonics, the silicon modulator injects or depletes the electrons and holes (charge carriers) of the electrical signal (voltage) in the silicon waveguide. While the silicon modulator processes the data, the laser driver of the EIC is the power and control engine for the light source itself. It provides stable power supply and signal amplification before the light is shot out from the laser (usually made by InP) of the PIC to the fiber, Fig. 4.41.

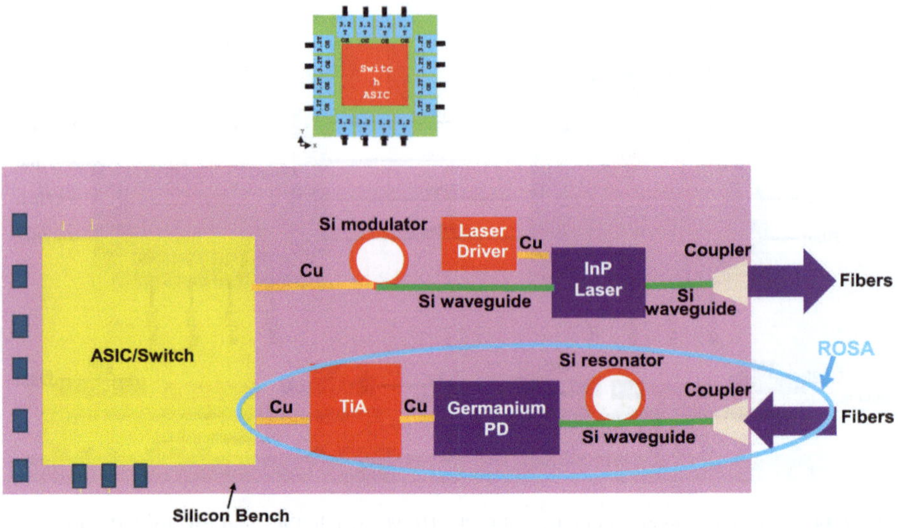


**Fig. 4.40** Receiver optical sub-assembly (ROSA) in CPO with silicon photonics (silicon waveguides and silicon modulators)

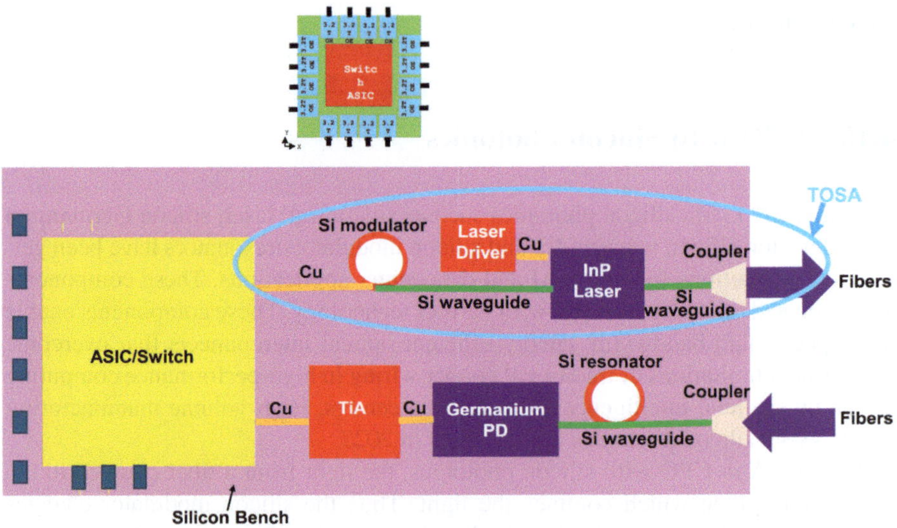


**Fig. 4.41** Transmitting optical sub-assembly (TOSA) in CPO with silicon photonics (silicon waveguides and silicon modulators)

## 4.11 Summary and Recommendations

Some important results and recommendations are summarized in the follows.

- Photonic devices are components that generate, manipulate, and detect light, e.g., LED (light emitting diode), lasers, photodiode, etc.
- Silicon photonics is the integration of some or all of the PIC and EIC, silicon waveguides, and silicon modulators into a 2D chip from a silicon wafer with CMOS technology.
- Twenty years ago, silicon integration was very popular. Companies like Intel have been promoting the silicon photonics. The holy grail of silicon photonic is to integrate all of the PIC and EIC into a single 2D chip fabricated from a silicon wafer with CMOS technology.
- However, in the past few years, driven by cost, the PIC and EIC are partitioned and spit into small chipets.
- CPO is the heterogeneous integration packaging method to integrate all the chiplets such as the switch, photonic devices, EIC, and PIC, silicon waveguides, and silicon modulators on a CPO substrate or interposer.
- Various forms of CPO have been shown in Fig. 4.5, where their advantages and disadvantages have been given. CPO with basic chiplets will be used the most. CPO with advanced chiplet (silicon photonic) will be used more. CPO with holy grail silicon photonic is nowhere in sight.
- Various systems (structures) of CPO with Si bridges have been shown in Fig. 4.22, where their advantages and disadvantages have been given.
- Twelve different 3D stacks of PIC and EIC have been given in Fig. 4.24 and some of their advantages and disadvantages have also been briefly mentioned.
- Various new structures (systems) of CPO have been proposed. Their substrate can be fan-out RDLs, fan-out RDLs with holes and polymer waveguides in the PCB, an organic interposer with holes and embedded polymer waveguides, a TGV interposer, a glass-core build-up package substrate, etc.
- Figure 4.39 shows a future CPO system for HPC driven by AI and high-speed and high-bandwidth communications.
- As of today, there are not many papers published in the quality and reliability (such as the thermal-cycling, thermal shock, mechanical vibration, mechanical shock, high-temperature storage, highly accelerated temperature, highly accelerated moisture resistance etc.) of CPO. In order for CPO to be more popular, these quality and reliability data are desperately needed. Thermal management of 3D CPO will be an important topic.
- Just like electrical leakage, light leakage due to coupling loss, waveguide cracks, scattering, and absorption can all become invisible killers and should also be noted.
- Because glass has a higher ability to seamlessly integrate optical interconnects, more use of the glass substrate or glass interposer for CPO is recommended.
- Due to the requirement of higher interconnect density, finer pitch, more compact 3D integration, lower power consumption, lower latency, higher bandwidth, and better signal integrity of a CPO system—the Cu Cu hybrid bonding [9–13] of some of the PICs and EICs are strongly recommended.

## References


1. Takemura, K., D. Ohshima, A. Noriki, D. Okamoto, A. Ukita, J. Ushida, M. Tokushima, T. Shimizu, I. Ogura, D. Shimura, and T. Aoki, "Silicon-Photonics-Embedded Interposers as Co-Packaged Optics Platform", The *Transactions of The Japan Institute of Electronics Packaging*, Vol. 15, April 2022, pp. E21-012-1–E21-012-13.
2. Ackermann, M., B. Shen, F. Merget, M. Wolz, and J. Witzens, "Glass-molded optical interposers for wafer scale photonic integrated circuit packaging in 800G modules and co-packaged optics," *Proceedings of SPIE*, March 2022, pp. M1–11.
3. Lee, B., N. Nedovic, T. Greer III, and C. Thomas Gray, "Beyond CPO: A Motivation and Approach for Bringing Optics Onto the Silicon Interposer", *Journal of Lightwave Technology*, Vol. 41, February 2023, pp. 1152–1162.
4. Lars Brusberg, Aramais R. Zakharian, Jason R. Grenier, Barry J. Paddock, Daniel W. Levesque, Cliff G. Sutton, Robin M. Force, Lucas W. Yeary, Robert A. Bellman, Chad C. Terwilliger, Betsy J. Johnson, "Glass Substrate for Co-Packaged Optics", *Proceedings of IMAPS 55th International Symposium on Microelectronics*, October 2022, pp. 236–241.
5. Noriki, A., A. Ukita, K. Takemura, S. Suda, T. Kurosu, Y. Ibusuki, I. Tamai, and D. Shimura, "Demonstration of Silicon-Photonics Hybrid Glass-Epoxy Substrate for Co-Packaged Optics", *Journal of Lightwave Technology*, Vol. 41, August 2023, pp. 5078–5083.
6. Brusberg, L., J. Grenier, Ş. Kocabaş, A. Zakharian, L. Yeary, D. Levesque, B. Paddock, R. Bellman, R. Force, C. Terwilliger, C. Sutton, J. Clark, and K. Rousseva, "Glass Interposer for High-Density Photonic Packaging", *Proceedings of OFC (Optical Fiber Communication)*, March 2022, pp. Tu3A.3 1–3.
7. Tan, M., J. Xu, S. Liu, J. Feng, H. Zhang, C. Yao, et al. "Co-packaged optics (CPO): status, challenges, and solutions", *Frontiers of Optoelectronics*, Vol. 16, March 2023, pp. 1–40.
8. Grenier, J., L. Brusberg, K. Wieland, J. Matthies, and C. Terwilliger, "Ultrafast laser processing of glass waveguide substrates for multi-fiber connectivity in co-packaged optics", *Advanced* Optical Technology, Vol. 12, August 2023, pp. 1–21.
9. Yoshida, W., K. Nagashima, K. Nishizaki, S. Yoneyama, and H. Nasu, "An Ultra-Compact 8-Channel Linear-Drive VCSEL-Based CPO Transceiver for Networks and AI/ML Applications in a Data Center", *IEEE/ECTC Proceedings*, May 2025, pp. 590–595.
10. Brusberg, L., B. J. Johnson, C. Herbst, J. R. Grenier, M. J. Dejneka, C. C. Terwilliger, and J. Schwietering, "Large-Scale Glass Waveguide Circuit for Board-Level Optical Interconnects Between Faceplate and Co-Packaged Optical Transceivers", *IEEE/ECTC Proceedings*, May 2025, pp. 59–63.
11. Knickerbocker, J., J. Herou, A. Ramos, H. Hsu, N. Liu, Y. Taira, C. Marushima, S. Kohara, H. Mori, M. Schultz, A. Horibe, and H. Numata, "Co-Packaged Optics (CPO) Technology Full Module Test Vehicle Demonstrations", *IEEE/ECTC Proceedings*, May 2025, pp. 274–281.
12. Lo, K.H., Y.R. Liang, T.W. Chen, E. Lin, S.W. Liang, J.Y. Wu, C.J. Wang, S.W. Lu, H.K. Chiu, H. Hsia, K.C. Yee, and D. Yu, "Broadband Optical Engine System Integration by Wafer Level Process in HPC/AI Era", *IEEE/ECTC Proceedings*, May 2025, pp. 1672–1675.
13. Singh, K., Z. Lin, P. Williams, D. A. Vance, J. Wright, B. Chowdhury, T. R. Coons, S. Mani, J. Decker, M. Fritz, P. Mohapatra, V. Meesala, S. Syal, J. Oh, and S. Jayaraman, "Heterogeneous Integration of Fiber-Based Co-Packaged Optics with EMIB Technology: Assembly, Performance, and Reliability:, *IEEE/ECTC Proceedings*, May 2025, pp. 37–42.
14. Chen, M. F., C. H. Tsou, S. Y. Hou, T. H. Yu, S. Liu, R. Lu, and K. C. Hsu, "Optical and Electrical Characterization of a Compact Universal Photonic Engine", *IEEE/ECTC Proceedings*, May 2025, pp. 54–58.
15. Mehta, N, S. Zarkovsky, B. G. Lee, R. Chen, S. Cohen, L. Gantz, and C. Gray, "All-SMF Arrays for Co-Packaged Optics: Optimizing Cost, Complexity and Performance", *IEEE/ECTC Proceedings*, May 2025, pp. 64–69.
16. Noriki, A., F. Nakamura, S. Suda, T. Murao, H. Kuwatsuka, and T. Amano, "Study of High-Density Optical Redistribution Layer Enabling Advanced Chiplet Edge Bandwidth Density on Active Optical Package Substrate", *IEEE/ECTC Proceedings*. May 2025, pp. 1313–1317.

17. Lin, V., T. Shih, A. Kang, and Y. Wang, "Fan-Out Embedded Bridge Structure for Co-packaged Optics Characterization and Evaluation", *IEEE/ECTC Proceedings*, May 2024, pp. 1246–1250.
18. Deo, K., Y. Lai, J. Yang, J. Ha, and S. Park, "Influence of Stiffener Design on Co-Packaged Optics (CPO) 2.5D Heterogeneous Packages", *IEEE/ECTC Proceedings*, May 2024, pp. 1815–1819.
19. Brusberg, L., M. Dejneka, C. A. Okoro, D. J. McEnroe, A. R. Zakharian, and C. C. Terwilliger, "Ultra Low-Loss Ion-Exchange Waveguides in Optimized Alkali Glass for Co-Packaged Optics", *IEEE/ECTC Proceedings*, May 2024, pp. 85–89.
20. Mizuno, Y., H. Uemura, T. Saeki, K. Tanaka, and K. Uesaka, "3D-Printed Beam Expanding Lens for Chip to Fiber Vertical Coupling", *IEEE/ECTC Proceedings*, May 2024, pp. 107–111.
21. Umeta, K., T. Sawamura, Y. Shiroishi, and H. Nasu, "Characterization of QSFP and OSFP CPO ELS Modules Employing an 8-Channel CWDM TOSA in Practical air-Cooling Conditions", *IEEE/ECTC Proceedings*, May 2024, pp. 112–117.
22. Steenberge, G., J. Van Asch, V. Geudens, T. De Baere, N. De Moerlooze, J. Missinne, A. Podpod, G. Lepage, N. Golshani, R. Magdziak, H. Sar, S. Bipul, D. Bode, Y. Ban, F. Ferraro, and J. Van Campenhout, "Interfacing Silicon Photonics for CPO", *IEEE/ECTC Proceedings*, May 2024, pp. 1161–1165.
23. Nakamura, F., K. Suzuki, A. Noriki, S. Suda, T. Kurosu, and T. Amano, "Scalable Fabrication of 3D Optical Re-distribution for Co-Packaged Optics using an Developed Nanoimprint Stepper", *IEEE/ECTC Proceedings*, May 2024, pp. 1404–1408.
24. Noriki, A., and T. Amano, "Self-aligned optical connector assembly on polymer waveguide integrated package substrate for co-packaged optics", *Proceedings of IEEE/ECTC*, May 2023, pp. 1892–1895.
25. Yeary, L., L. Brusberg, C. Kim S. Seok, J. Noh, and A. Rozenvax, "Co-packaged Optics on Glass Substrates for 102.4 Tb/s Data Center Switches", *Proceedings of IEEE/ECTC*, May 2023, pp. 224–227.
26. Gupta, P., R. Bernson, N. Nudds, S. Collins, K. Gradkowski, P. Morrissey, and P. O'Brien, "Reference Thermal Chips for 2D and 3D Co-packaging Process Development", *Proceedings of IEEE/ECTC*, May 2023, pp. 2165–2170.
27. Bhuvanendran, S., N. Gourikutty, L. Long, S. Wei, M. Jong, D. Ho, J. Wu, T. Lim, R. Mandal, C. Choong, C. Yee, X. Li, J. Liow, and S. Bhattacharya, "A Heterogeneously Integrated Wafer-level Processed Co-Packaged Optical Engine for Hyper-scale Data Centres", *Proceedings of IEEE/ECTC*, May 2023 pp. 207–211.
28. Li, J., T. Hsu, M. Zhuan, S. Lin, T. Shih, N. Kao, and Y. Eang, "Innovative Fan-Out Embedded Bridge Structure for Co-Packaged Optics", Proceedings of IEEE/ECTC, May 2023, pp. 228–232
29. McDonough, C., S. Kruger, T. Ngai, S. Baranowski, H. Yang, S. Deckoff-Jones, C. Poulton, M. Watts, and D. Harame, "AIM Photonics Demonstration of a 300 mm Si Photonics Interposer", *Proceedings of IEEE/ECTC*, May 2023, pp. 233–238.
30. Integrated Optical Interconnect Systems (iOIS) for Silicon Photonics Applications in HPC
31. Hsia, H., S. P. Tai, C. S. Liu, C. W. Tseng, S. W. Lu, Y. Wu, C. C. Chang, Jason Wu, K. C. Yee, C. Y. Wu, C. H. Tung, D. Yu, "Integrated Optical Interconnect Systems (iOIS) for Silicon Photonics Applications in HPC", *Proceedings of IEEE/ECTC*, May 2023, pp. 612–616.
32. Lu, M., S. Mu, C. Cheng, and J. Chen, "Advanced Packaging Technologies for Co-packaged Optics", *Proceedings of IEEE/ECTC*, May 2022, pp. 38–42.
33. Application Chou, B., B. M. Sawyer, G. Lyu, E. Timurdugan, C. Minkenberg, A. J. Zilkie, and D. McCann, "Demonstration of Fan-out Silicon Photonics Module for Next Generation Co-Packaged Optics (CPO)", *Proceedings of IEEE/ECTC*, May 2022, pp. 394–402.
34. Razdan, S., P. Dobbelaere, J. Xue, A. Prasad, and V. Patel, "Advanced 2.5D packaging technologies for next generation Silicon Photonics in high performance networking applications", *Proceedings of IEEE/ECTC*, May 2022, pp. 428–435.
35. Janta-Polczynski, A., and M. Robitaille, "Optical fiber pigtails integration in co-package", *Proceedings of IEEE/ECTC*, May 2022, pp. 410–421.
36. Maharry, A., L. A. Valenzuela, J. F. Buckwalter, and C. L. Schow, "A PCB Packaging Platform Enabling 100+ Gbaud Optoelectronic Device Testing", *Proceedings of IEEE/ECTC*, May 2021, pp. 1323–1328.

37. Nieweglowski, K., L. Lorenz, K. Bock, M. Catuneanu, K. Jamshidi, "Electro-optical co-integration of chip-components in optical transceivers for optical inter-chip communication", *Proceedings of IEEE/ECTC*, May 2020, pp. 139–147.
38. Johnson, J. E., K. Bacher, R. Schaevitz, and V. Raghunathan, "Performance and Reliability of Advanced CW Lasers for Silicon Photonics Applications", Optical Fiber Communications Conference and Exhibition (OFC), 2022, Tu2D.1.
39. Lau, J. H., "Recent Advances and Trends in Advanced Packaging", *IEEE Transactions on CPMT*, Vol. 12, No. 2, February 2022, pp. 228–252.
40. Lau, J. H., "Co-Packaged Optics - Heterogeneous Integration of Photonic IC and Electronic IC", *ASME Transactions, J. of Electronic Packaging,* published online: June 1, 2024, hard-copy: Vol. 147, March 2025, pp. 1–12.
41. Lau, J. H., "Recent Advances and Trends in Multiple System and Heterogeneous Integration with TSV-Interposers", *IEEE Transactions on CPMT*, Vol. 13, No. 1, January 2023, pp. 3–25.
42. Lau, J. H., and T. J. Tseng, "2D Package Structure and Optical Signal Transmitter", US Patent No. 11,860,428, Date of Patent: Jan. 2, 2024, Filing Date: Jun. 9, 2022.
43. Lau, J. H., and T. J. Tseng, "3D Integration of Electronic IC and Photonic IC", US Patent No. 12,412,879, Date of Patent: Sep. 9, 2025, Filing Date: Jun. 8, 2023.
44. Lau, J. H., P. Lin, K. Yang, C. Lin, C. Ko, and T. J. Tseng, "3D Heterogeneous Integration of Electronic IC and Photonic IC embedded in a glass interposer", US Patent application filed on March 1, 2024.
45. Lau, J. H., and X. Fan, *Hybrid Bonging, Advanced Substrates, Failure Mechanisms, and Thermal Management for Chiplets and Heterogeneous Integration,* Springer, New York, April 2025.
46. Lau, J. H., *Flip Chip, Hybrid Bonding, Fan-In, and Fan-Out Technology*, Springer, New York, 2024.
47. Lau, J. H., *Chiplet Design and Heterogeneous Integration Packaging*, Springer, New York, 2023.
48. Lau, J. H., *Semiconductor Advanced Packaging*, Springer, New York, 2021.
49. Lau, J. H., *Heterogeneous Integrations*, Springer, New York, 2019.
50. Lau, J. H., S. Erasmus, F. Sporon-Fiedler, S. Murphy, R. Herms, and S. Pan, "Creep Analysis of a Fiber-Optic Transceiver Housing", *ASME Paper No. 15868, ASME Inter-PACK'01*, Kauai, Hawaii, July 8–13, 2001.
51. Lau, J. H., S. Erasmus, F. Sporon-Fiedler, and C. Chang, "DSC, TMA, DMA, and TGA of Epoxies for Fiber-Optic Transceiver Applications", *ASME Paper No. IMECE 2001/ EPP-24720, ASME IMEC*, November 2001.
52. Lau, J. H., C. L. Jiaa, and S. Erasmus, "Corrosion Analysis of a Fiber-Optic Transceiver Housing", *ASME Paper No. IMECE 2001/EPP-24721, ASME IMEC*, November 2001.
53. Lau, J. H., et. al., "Corrosion Analysis of a Fiber-Optic Transceivers Housing", *ASME Transactions, Journal of Electronic Packaging*, Vol. 124, December 2002, pp. 379–384.
54. Lau, J. H., *3D IC Integration and Packaging*, McGraw-Hill, New York, 2016.
55. Lau, J. H., *Through-Silicon Via (TSV) for 3D Integration*, McGraw-Hill, New York, 2013.
56. Lau, J. H., "Effect of a through-glass via (TGV) substrate on solder joint reliability", *Chip Scale Review*, September/October Issue, 2025.
57. Xing, D., C. Che, X. Hou, X. Li, Y. Ning, H. Lian, et al., "Preparation process of through glass via based on laser induced deep etching," *SID*, V. 56, Issue 1, June 2025, pp. 1105–1108.
58. Chen, L., H. Wu, M. Ahang, F. Jiang, T. Yu, and D. Yu, "Development of laser induced deep etching process for through glass via," *Proc. of IEEE ICEPT*, Aug. 2025, pp. 1–4.
59. Sunohara, M., J. Yoshike, H. Taneda, Y. Nakabayashi, N. Shimizu, "Development of glass core build up substate with TGV," *IEEE ECTC Proc.*, May 2025, pp. 1021–1026.
60. Mahajan, R., R. Sankman, N. Patel, D. Kim, K. Aygun, Z. Qian, et al., "Embedded multi-die interconnect bridge (EMIB)—a high-density, high-bandwidth packaging interconnect," *IEEE/ ECTC Proc.*, May 2016, pp. 557–565.
61. Lau, J. H., C. Ko, P. Lin, T. Tseng, R. Tain, and H. Yang, "Package structure and manufacturing method thereof," US Patent No. 11,410,933, Patent Date: Aug. 9, 2022, filed on May 7, 2021.

# Chapter 5
# Chiplets Connections—Bridges

**Abstract** In this chapter, the connections (bridges) between chiplets are presented. Emphasis is placed on the bridges embedded in organic build-up package substrate, bridges embedded in fan-out epoxy molding compound (EMC) and redistribution-layers (RDLs), and hybrid bonding bridges. Finally, thermal-mechanical simulations of a silicon bridge connecting two chiplets with microbumps w/o underfill are presented. Also, some recommendations are provided.

## 5.1 Introduction

The original idea of chiplets is from the most famous paper in semiconductor industry "Cramming more components onto integrated circuits" authored by Gordon E. Moore in *Electronics*, Volume 38, Number 8, April 19, 1965. In the paper, besides the Moore Law "… the complexity for minimum component costs has increased at a rate of roughly a factor of two per year.", Moore also mention that: "… it may prove to be more economical to build large systems out of smaller functions, which are separately packaged and interconnected." The last sentence is the basic idea of chiplets [1].

There are many chiplet designs such as the chip partitioning and chip splitting [2]. Their key advantage is to lower the semiconductor manufacturing yield loss (which leads to lower cost) because of smaller chip sizes. However, the key disadvantage is to increase the package size, complexity, and cost. Usually, the semiconductor cost is a few times the packaging cost, thus it is worthwhile to go for chiplets [3].

The most important element in chiplet design is the connections between chiplets, also called bridges. There are many different kinds of bridges [4–64]. In this book, three different kinds of bridges will be briefly discussed: (a) bridge embedded in package substrate, (b) bridge embedded in fan-out EMC with RDLs, and (c) hybrid bonding bridge.



J. Lau, K.-N. Chiang, *Cu-Interconnects, Glass, and AI-Assisted Simulation for Chiplets and Heterogeneous Integration*,
https://doi.org/10.1007/978-981-95-6891-8_5

The thermal-mechanical simulations of a silicon chiplets packaging system with silicon bridge by microbumps w/o underfill are presented. It is demonstrated that even if there is not any coefficient of thermal expansion (CTE) mismatch between the Si chiplets and the Si bridge, the maximum accumulative equivalent inelastic strain at the corner solder cap of the bridge is very large without underfill.

## 5.2 Bridges Embedded in Build-up Package Substrate

Figure 5.1 shows one of Intel's embedded multidie interconnect bridge (EMIB) patents [50]. It can be seen that the EMIB die is embedded in the cavity of a build-up package substrate, which is supporting the chiplets. For EMIB, there are at least three important tasks, namely: (a) wafer bumping of two different kinds of bumps (C2 and C4) on the chiplets wafer; (b) embedding the bridge in the cavity of a build-up substrate and then laminating the top surface of the substrate; and (c) bonding the chiplets on the substrate with the embedded bridge.

The silicon bridge is very small, just like a Basmati Rice, Fig. 5.1. Today, the minimum metal L/S/H is 2 μm/2 μm/2 μm and the bridge size is from 2 mm × 2 mm to 8 mm × 8 mm [51], but most are less than 5 mm × 5 mm [52]. The dielectric layer is 2 μm. Usually, there are 4 RDLs. One of the challenges of the EMIB technology is to fabricate the organic build-up package substrate with cavities for the silicon bridges and then laminate (with pressure and temperature) another build-up layer on top (to meet the substrate surface flatness requirement) for chiplets (with both C2 and C4 bumps) bonding. Figure 5.2 shows the Intel spaceship of GPU (Pont Vecchio), which has 11 EMIBs.

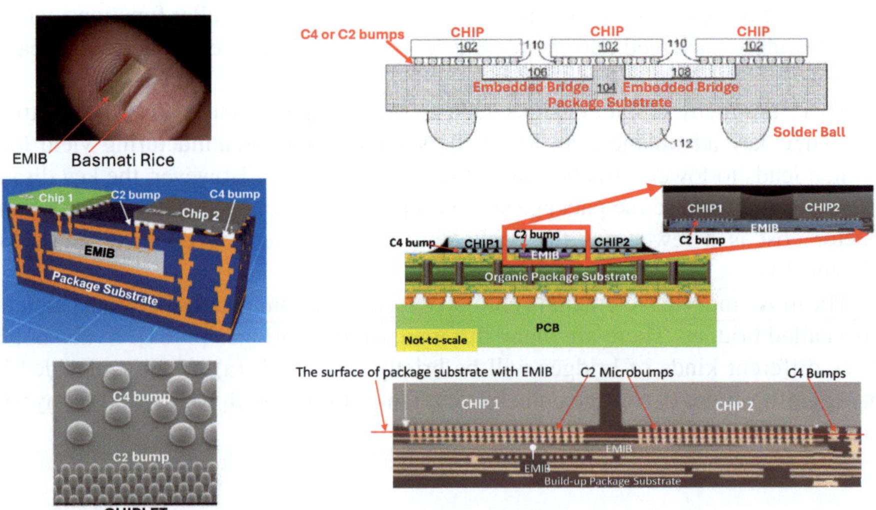


**Fig. 5.1** Intel's EMIB

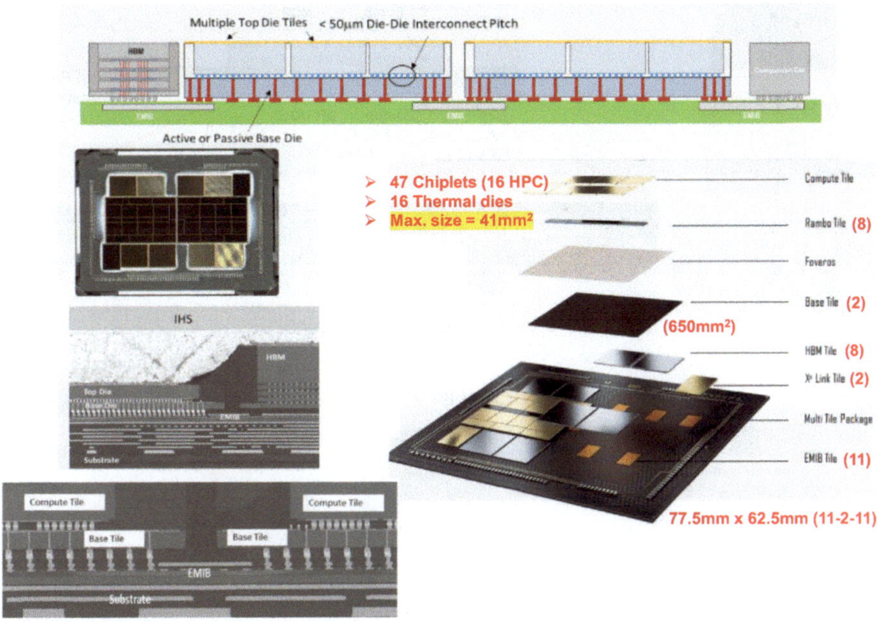


**Fig. 5.2** Intel's Pont Vecchio

In the past few years, IBM presented many papers on "Direct Bonded Heterogeneous Integration (DBHi) Si Bridge" [38–42, 44–46], Fig. 5.3. The major differences between Intel's EMIB and IBM's DBHi are as follows. (a) For Intel's EMIB, there are two different (C4 and C2) bumps on the chiplets, Fig. 5.1. However, for IBM's DBHi the C4 bump is on the chiplet and the C2 bump is on the silicon bridge, Fig. 5.3. (b) For Intel's EMIB, the bridge is embedded in the cavity of a build-up substrate with a die-attach material and then laminated with another build-up layer on top. For IBM's DBHi, the substrate is just a regular build up substrate with a cavity on top as shown in Fig. 5.3.

The challenges in IBM's DBHi are as the follows. (a) Handling and bonding of a portion of the tiny rigid bridge on a portion of the large chiplet with very fine-pitch pads. (b) Dealing with a situation in which there is more than one rigid bridge on a chiplet. (c) Dealing with a situation in which there are more than two chiplets on a package substrate. If the bridge is very thin, e.g., 50 μm and the C2 bump is very short, e.g., 30 μm, then the cavity of the package substrate is not needed (Fig. 5.4) if the C4 solder bump height is >85 μm [28].

UltraFusion is Apple's innovative packaging architecture that interconnects the die of two M1 Max chips to create a SoC with unprecedented levels of stunningly compact design, extensive connectivity, performance and capabilities. This architecture does not combine two M1 Max dies into a single chip package—it also makes the two dies present themselves as a single chip. The interconnection between the two M1 Max dies is by a "silicon bridge" as shown in Fig. 5.5. The chiplets and

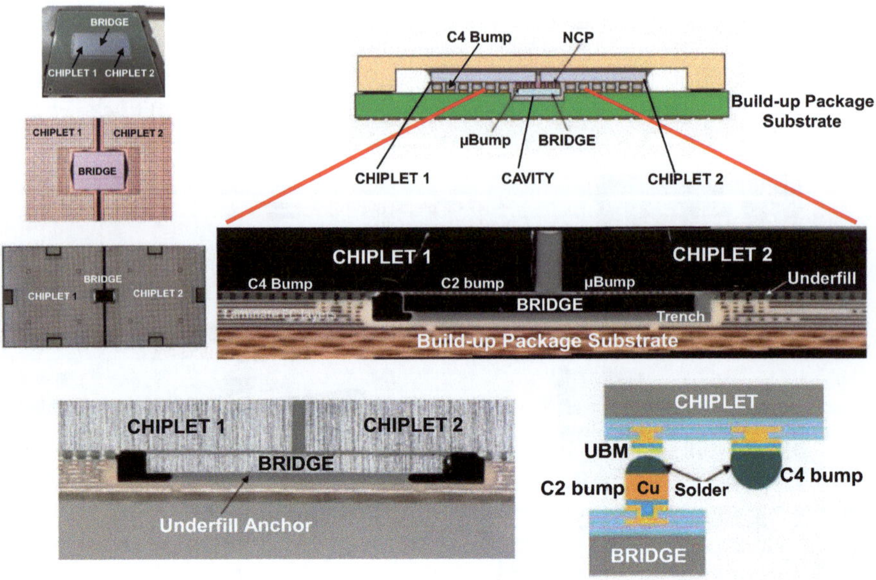


**Fig. 5.3** IBM's DBHi

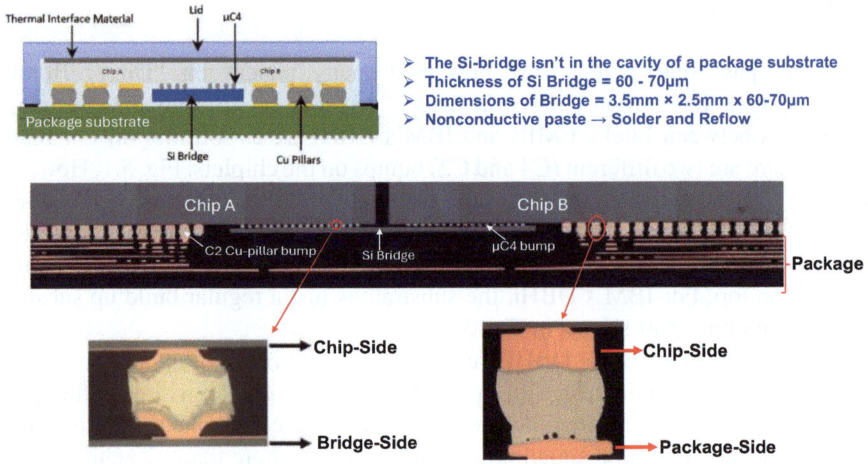


**Fig. 5.4** IBM's Si bridge above the package substrate

the Si-bridge are supported by an ordinary build-up package substrate without any cavity. Apple shipped this product in March 2022.

The AMD Instinct MI250X compute accelerator is shown in Fig. 5.6. It can be seen that there are two GPUs (the second-generation matrix cores for HPC driven by AI) with each having a size of 790mm$^2$. Each GPU is connected to four HBM2e with Si bridges on microbumps. The GPU and the HBM2E are supported by a build-up package substrate without any cavity. AMD shipped this product in Q1 of 2023.

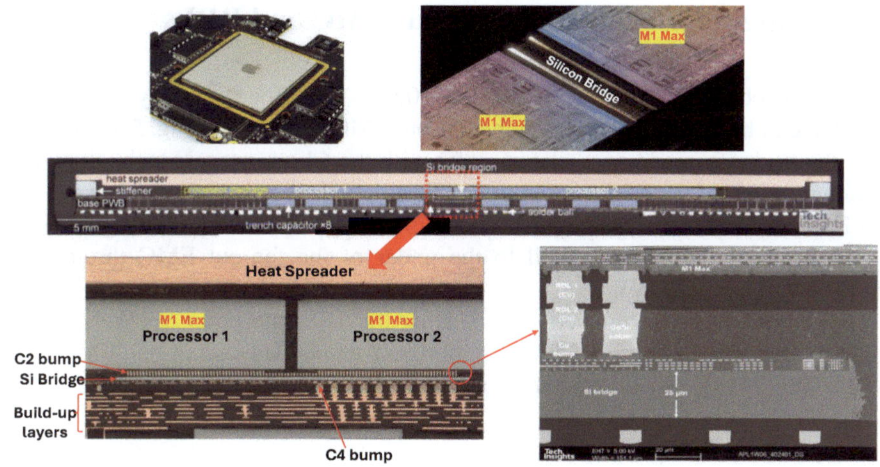


**Fig. 5.5** Apple's UltraFusion

**Fig. 5.6** AMD's Instinct MI250X

## 5.3 Bridges Embedded in Fan-Out EMC and RDLs

Unimicron obtained the U.S. patent 11,410,933 [54] (filed on May 7, 2021) in which the Si bridge is embedded in the fan-out EMC with RDLs (Fig. 5.7). It can be seen the bridge is made on a silicon wafer with the 64 nm process technology (the minimum pitch of the RDLs is 0.4 μm.) The silicon bridge is embedded in the fan-out EMC and connects the chiplet 1 and chiplet 2 through RDLs. The signals from above the bridge are transmitted to the under of the fan-out EMC through the through-mold via (TMV).

TSMC published two papers on replacing the TSV-interposer (CoWoS) by the local silicon interconnects (LSIs) (Si bridges) embedded in fan-out EMC with RDLs [26, 27] (CoWoS-L) as shown in Fig. 5.8. The new interposer consists of

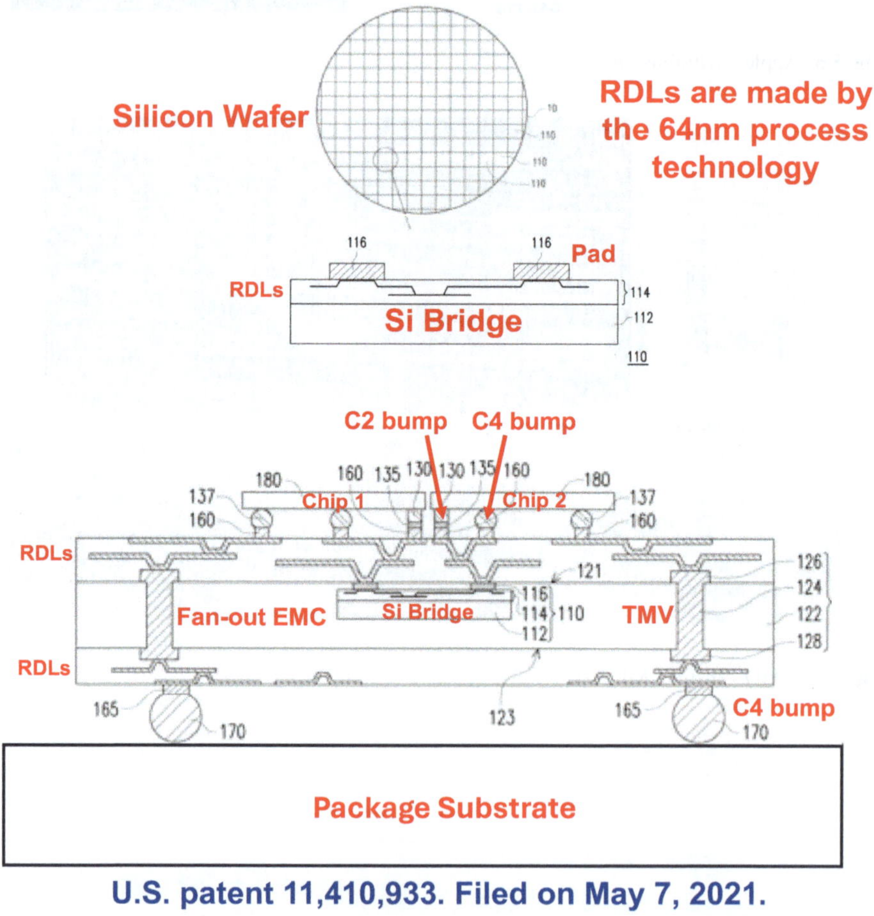


**Fig. 5.7** Unimicron's US Patent 11,410,933 (filed on May 7, 2021). Si bridge embedded in fan-out EMC with RDLs

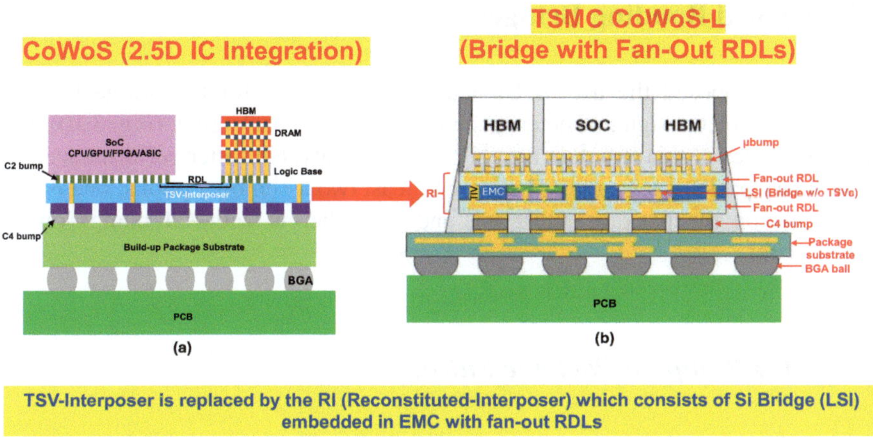


**Fig. 5.8** (**a**) CoWoS. (**b**) TSMC's CoWoS-L

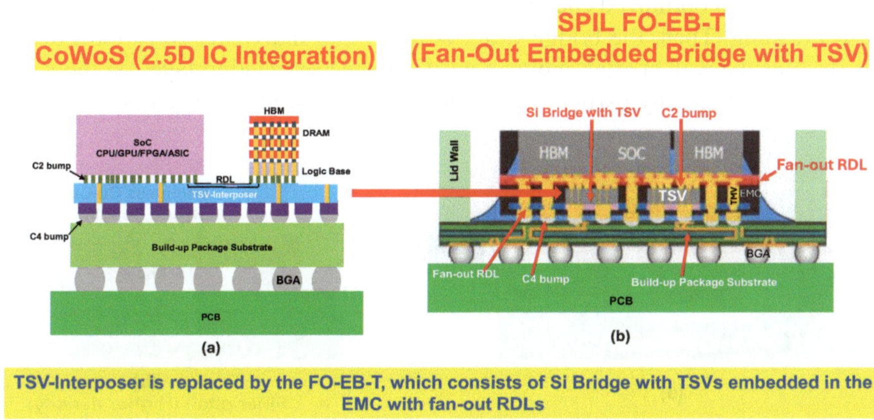


**Fig. 5.9** (**a**) CoWoS. (**b**) FO-EB-T

multiple LSIs or Si bridges w/o TSV and global integrated fan-out (InFO) EMC with RDLs to form a reconstituted interposer (RI). The small size LSI (Bridge) inherits all the attractive features of TSV interposers by retaining sub-micron Cu interconnects, through silicon vias (TSV), and embedded deep trench capacitors (eDTCs) to ensure good system performance, while avoiding the issues associated with one large TSV-interposer, such as yield loss.

In Ref. [20], Siliconware Precision Industries Co., Ltd. (SPIL) called its similar technology fan-out embedded bridge (FO-EB-T) (Fig. 5.9).

## 5.4 Hybrid Bonding Bridges

Unimicron proposed the use of Cu–Cu hybrid bonding for the bridge between chiplets in chiplet design and heterogeneous integration packaging, (Fig. 5.10) [64]. The advantages of this structure are: (1) higher density, (2) better performance, and (3) ordinary package substrate. There are at least two assembly options: one is with C4 bumps on the package substrate, and the other is with C4 bumps on the chiplet wafer.

### *5.4.1 C4 Bump on Package Substrate*

Figure 5.11 shows the process flow of a hybrid bonding bridge with C4 bumps on the package substrate. For the bridge wafer, the processing starts off with chemical vapor deposition (CVD) to make a dielectric material such as $SiO_2$ and then it is planarized by an optimized chemical mechanical polishing (CMP) process to make the Cu dishing. Then, the bridge wafer is diced into individual chips (still on the blue tape of the wafer) after application of a protective coating layer on the wafer surfaces to prevent any particles and contaminants that may cause interface voids

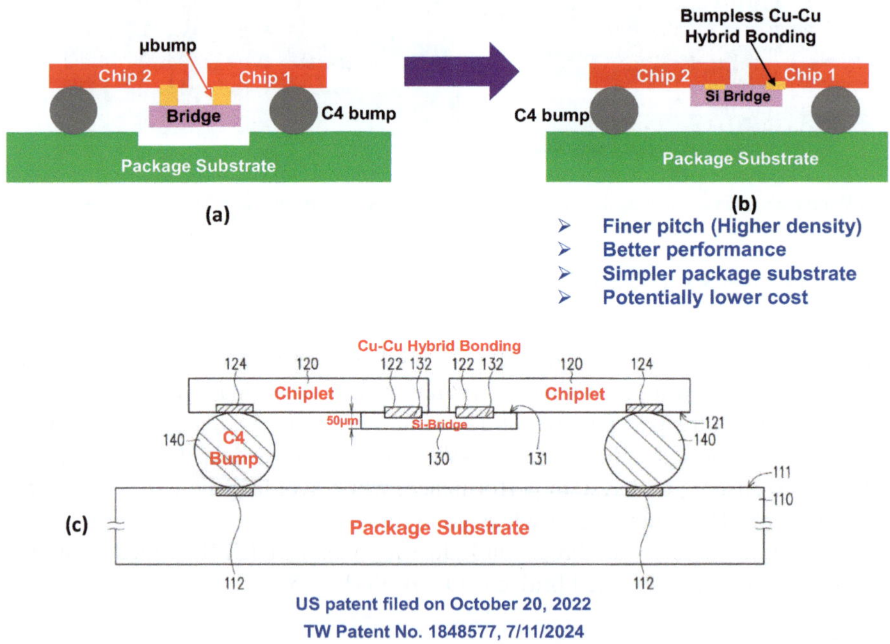


**Fig. 5.10** (**a**) Bridge with microbumps. (**b**) Bridge with hybrid bonding. (**c**) Unimicron's TW Patent 1,848,577 (filed on Oct. 20, 2022). Chiplets with Si-bridge by Cu-Cu hybrid bonding

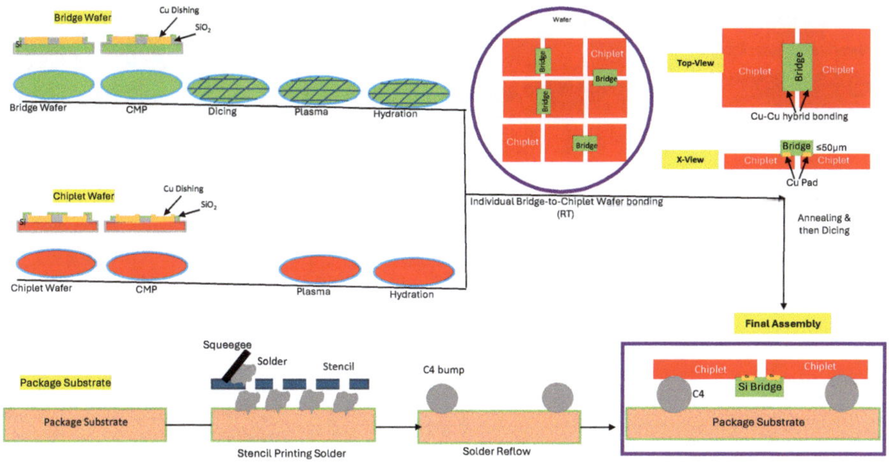


**Fig. 5.11** Process flow of the hybrid bonding bridge with C4 bumps on the package substrate

during the subsequent bonding process. These steps are followed by activating the bonding surface by using plasma and hydration processes for better hydrophilicity and a higher density of a hydroxyl group on the bonding surface. To process the chiplet wafer, repeat the CVD process for the $SiO_2$, CMP for the Cu dishing, and plasma and hydration of the activation of the bonding surface. Then, pick and place the individual bridge chip on the chiplet wafer and perform the $SiO_2$-to-$SiO_2$ bonding at room temperature. These steps are followed by annealing to achieve covalent bonding between oxide layers and metallic bonding between Cu–Cu contacts and the diffusion of Cu atoms. For the package substrate, the process is to stencil print the solder paste on the substrate and then reflow into C4 solder bumps. For the final assembly, the bridge + chiplets module is picked and placed on the package substrate, then the C4 bumps are reflowed.

### 5.4.2 *C4 Bump on Chiplet Wafer*

Figure 5.12 shows the process flow of the hybrid bonding bridge with C4 bumps on the chiplet wafer. It can be seen that, compared with the C4 bumps for the package substrate case, the process steps for the bridge wafer and the chiplet wafer are the same up to the bridge-to-chiplet wafer bonding step. After that, the C4 bumps are fabricated by wafer bumping on the chiplet wafer. Then, the chiplet wafer is diced into individual modules (bridge + chiplets with C4 bumps). The final assembly is accomplished by picking and placing the individual module on the package substrate and reflowing the C4 solder bumps.

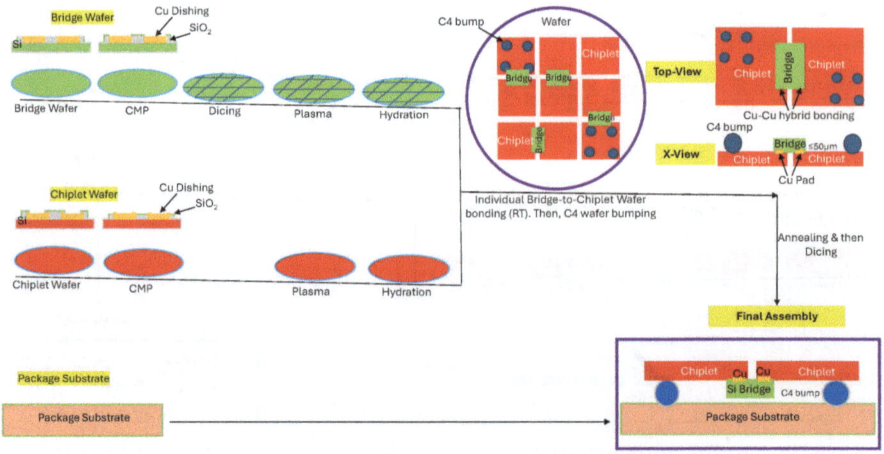


**Fig. 5.12** Process flow of the hybrid bonding bridge with C4 bumps on the chiplet wafer

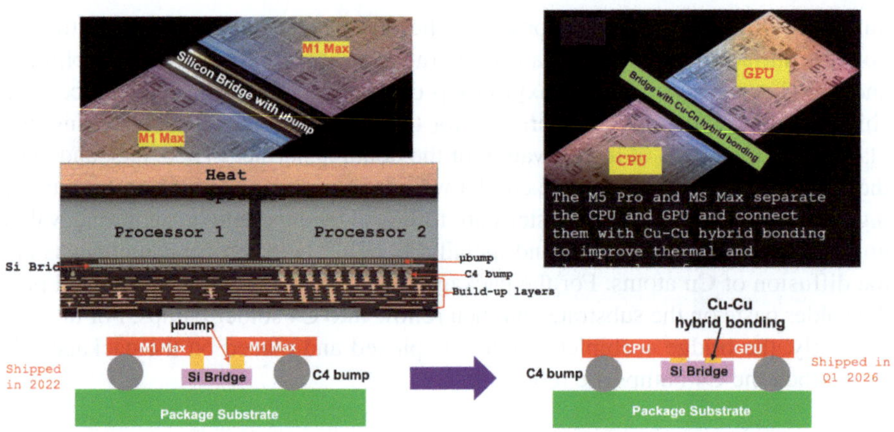


**Fig. 5.13** CPU and GPU are hybrid bonded with the Si bridge (Apple's notebook shipped in the end of 2025)

## 5.5 New Applications of Si Bridge in Packaging

Apple is using Cu-Cu hybrid bonding packaging technology for the Si bridge in its high-end M5 Pro, M5 Max, and M5 Ultra chiplets (CPU and GPU) as shown in Fig. 5.13. Figure 5.14 shows a package-on-package, where the application processor is split into small chiplets (in order to get higher semiconductor manufacturing yield). The chiplets are connected with Si bridge by Cu-Cu hybrid bonding. Figure 5.14 shows a Si bridge connecting (on top of) two chiplets with microbumps. The signals from the bridge to the chiplets and then wraparound the edge of the chiplets to the next level of interconnect (Fig. 5.15).

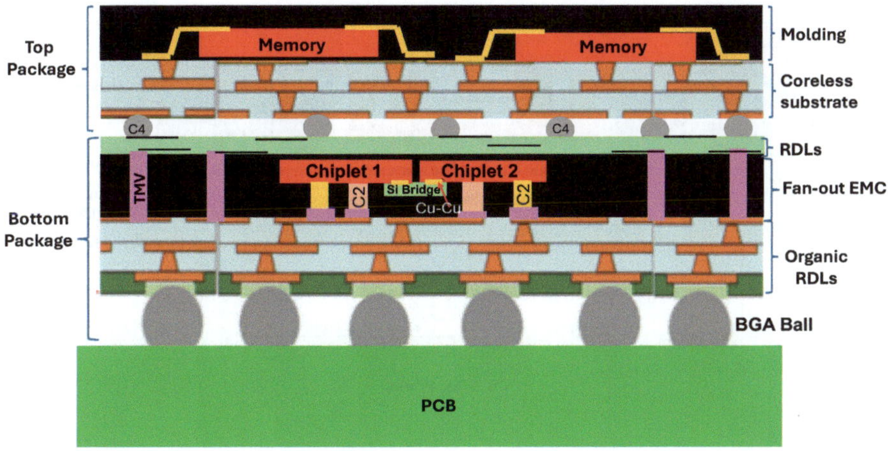


**Fig. 5.14** PoP with chiplets application processors hybrid bonded with the Si bridge

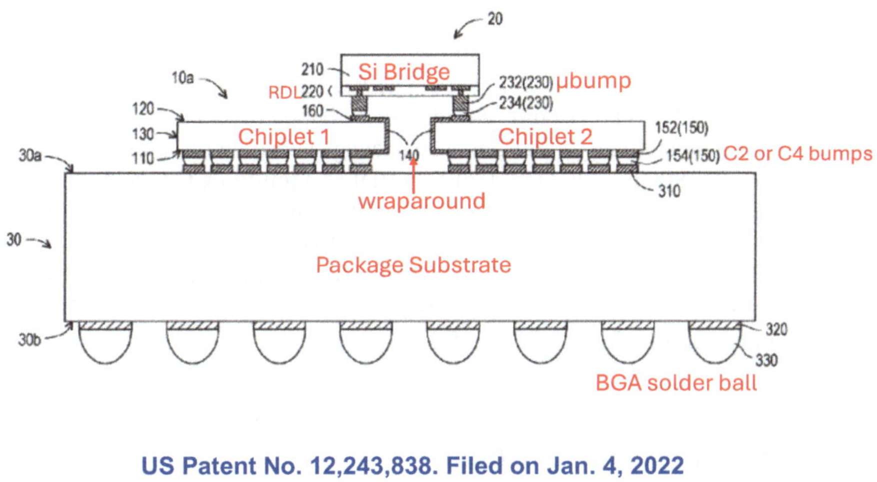


**Fig. 5.15** Unimicron's US Patent 12,243,838 (filed on Jan. 4, 2022). Chiplets are bonded with Si-bridge by microbumps

## 5.6 Chiplets Interconnect with Bridges by Microbumps without Underfill

### *5.6.1 The Structure*

The structure under consideration is shown in Fig. 5.16. There are two chiplets which are connected with a Si bridge with μbump (Cu-pillar with solder cap or C2 bump). The chiplets are attached to a package substrate with C4 bump, and the

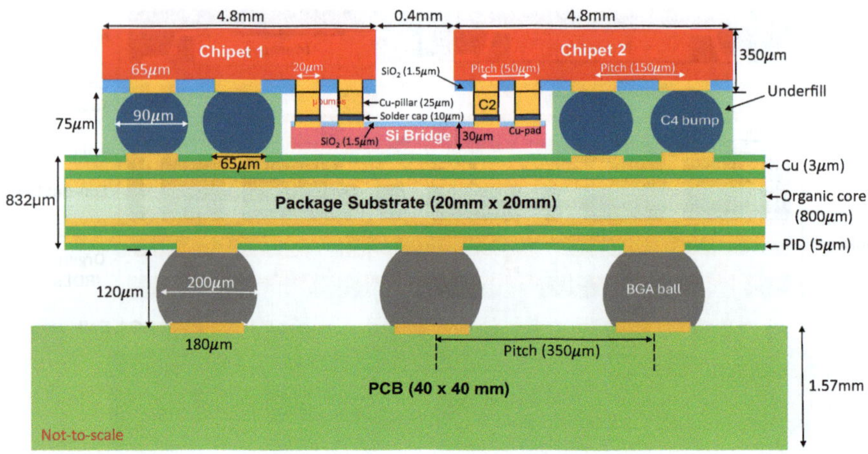


**Fig. 5.16** Coress section of chiplets connected by a Si-bridge with microbumps

package substrate is then attached to a printed circuit board (PCB) with ball grid array (BGA) solder ball. The chiplets dimensions are 4.8 mm × 4.8 mm × 350 μm. The passivation of the chiplets is 1.5 μm. The C2 bump consists of the Cu-pillar (25 μm-tall) and solder cap (10 μm-tall) and their diameter is 20 μm. The C4 solder joint is 90 μm-diameter and 75 μm-tall. The package substrate (20 mm × 20 mm) is 2-2-2 with a core (800 μm) and two build-up RDLs (3 μm-thick Cu and 5 μm-thick photoimageable dielectric (PID)). The BGA solder ball has a diameter of 200 μm and tall of 120 μm. The dimensions of the PCB are 40 mm × 40 mm × 1,57 mm. The solder cap material for the C2 bump is Sn0.7Cu, while the material for the C4 solder joint and the BGA solder joint is Sn3Ag0.5Cu (SAC). The Si bridge dimensions are 3 mm × 2 mm × 30 μm, Figs. 5.16 and 5.17.

## 5.6.2 *Finite Element Modelling*

Due to double symmetries of the structure shown in Figs. 5.16, 5.17 and 5.18, only a quarter of the structure is modeled (Figs. 5.19 and 5.20). Figures 5.19 and 5.20 show the chiplets, Si bridge, package substrate, PCB, C2 μbumps, C4 bumps, and BGA solder ball. Finer meshes are used for the critical locations (higher stress/strain areas) such as the corner C2 μbumps connecting the chiplets and the Si bridge (Fig. 5.21), the corner C4 bumps connecting the chiplets and the build-up package substrate (Fig. 5.22), and the BGA solder ball connecting the package substrate and the PCB (Fig. 5.23).

The displacement/rotation boundary conditions along the x-axis are that there is no displacement in the y-direction (UY = 0), and there is no rotation about the x-axis and the z-axis. The boundary conditions along the y-axis are that there is no

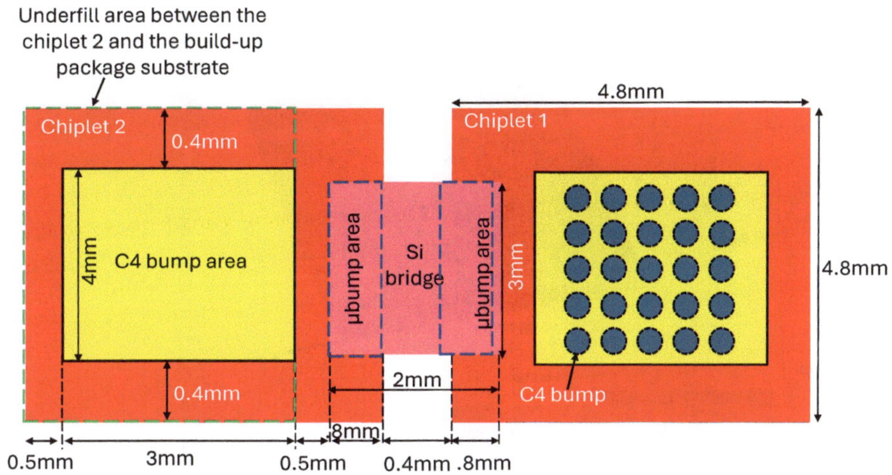


**Fig. 5.17** Bottom-view of the interconnection between the chiplets and the Si-bridge with microbumps

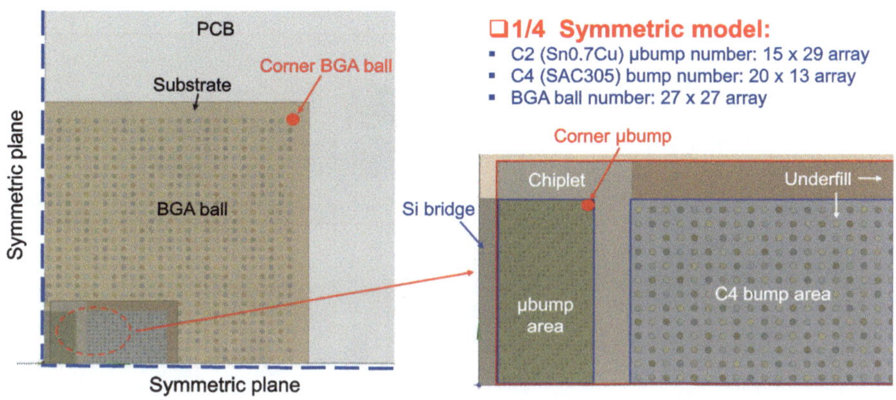


**Fig. 5.18** Quarter of the structure

displacement in the x-direction (UX = 0), and there is no rotation about the y-axis and the z-axis. At the center, the boundary conditions are that there are not any displacement and rotation. Since the focus of the present study is on the solder joint reliability, in order to simplify the modeling and save the computing time, the organic build-up package substrate is modeled as an equivalent block which will be discussed in the next section.

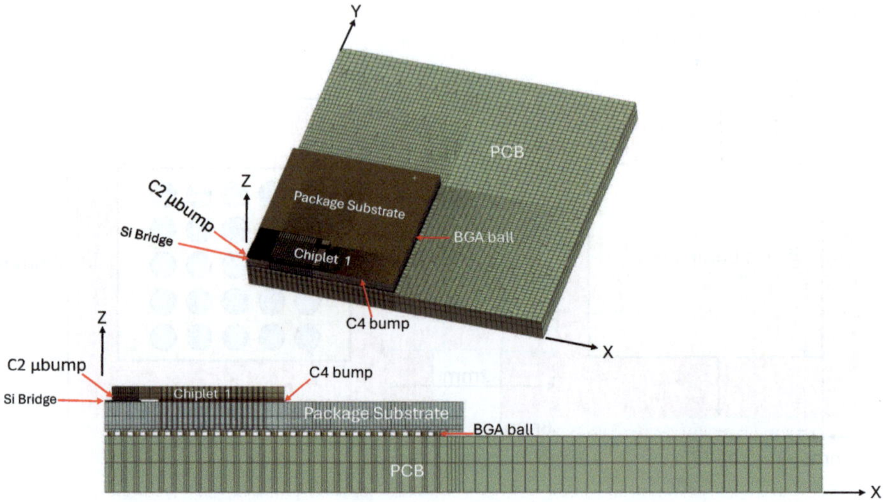


**Fig. 5.19** Quarter of finite element model

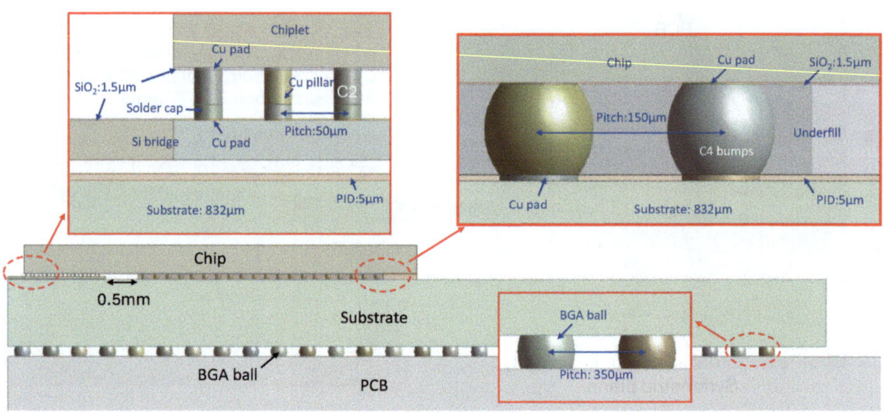


**Fig. 5.20** Cross section of the quarter model

## 5.6.3 *Material Properties*

Table 5.1 shows the material properties for modeling. All the materials are assumed to be constant except solder (Sn3Ag0.5Cu), which is time and temperature dependent. Also, the solders, Sn3Ag0.5Cu and Sn0.7Cu, obey the Anand viscoplasticity constitute equation [65–70]. There is a total of nine (9) constants (parameters) of the Anand equation (model), namely A, Q, $\xi$, m, n, $\hat{s}$, $h_0$, $\alpha$, and $s_0$. For Sn3Ag0.5Cu (SAC) [66–70] and Sn0.7Cu [66, 70], these constants are shown in Table 5.2. The Cu is assumed as an elastic-plastic (bilinear kinematic hardening) material with the

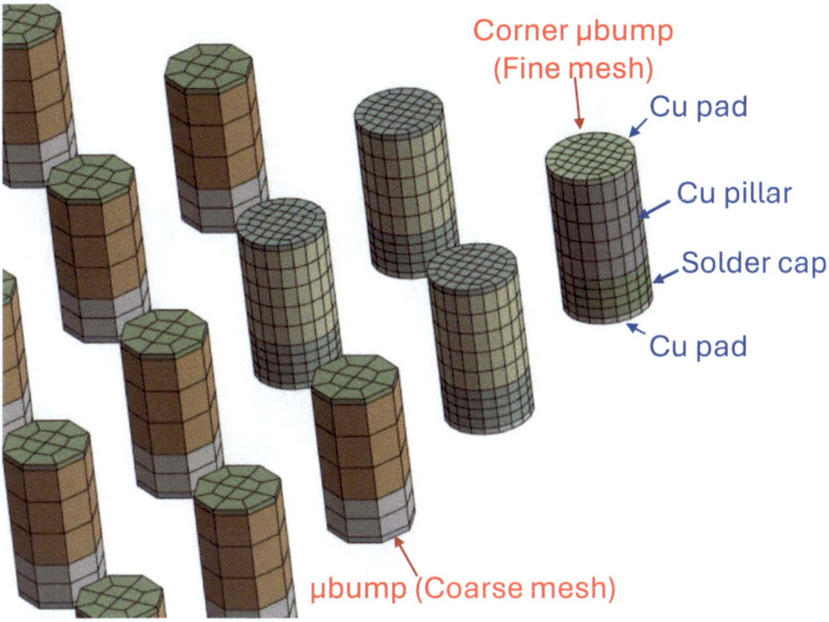


**Fig. 5.21** Finer meshes for the C2-bump solder joint

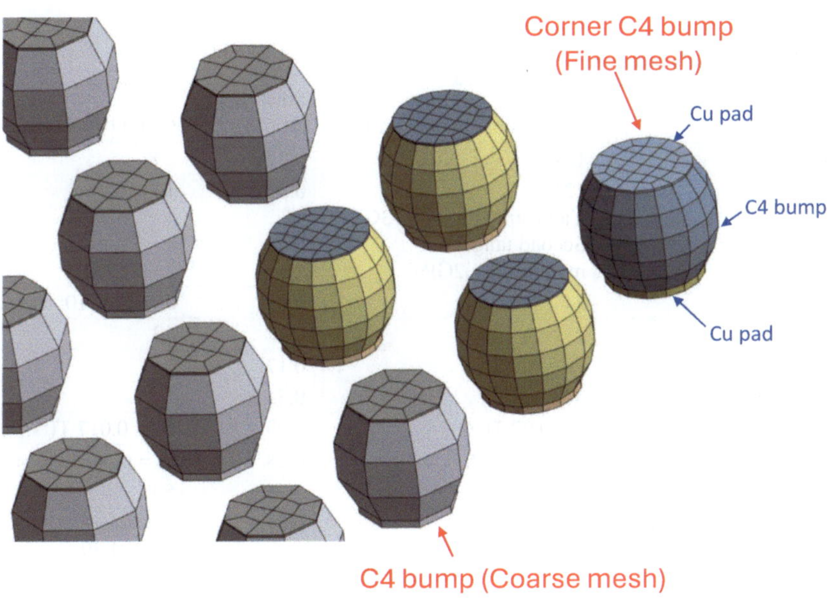


**Fig. 5.22** Finer meshes for the C4-bump solder joints

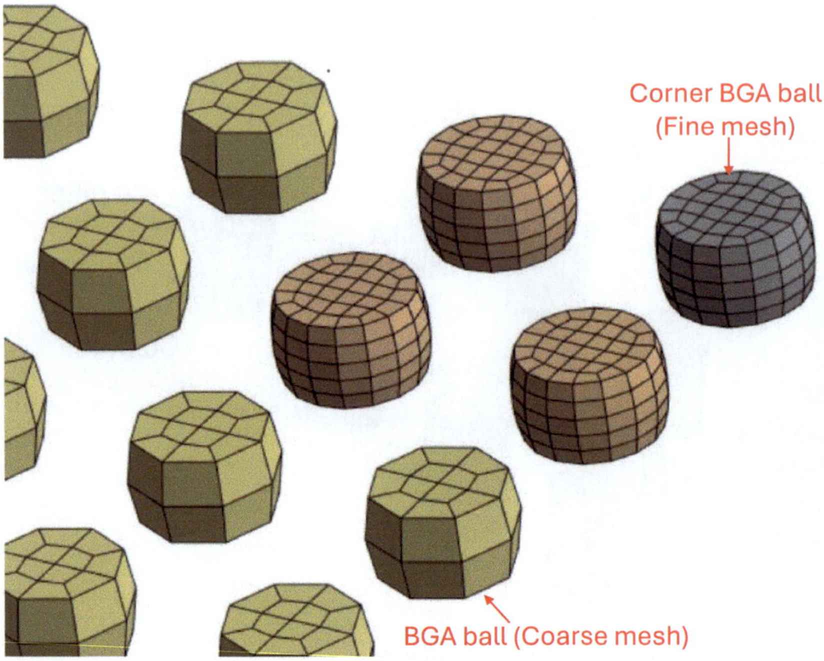


**Fig. 5.23** Finer meshes for the BGA solder joints

**Table 5.1** Materials properties

| Material | Young's modulus (GPa) | Poisson's ratio | CTE ($10 \times 10^{-6}$/°C) |
|---|---|---|---|
| Silicon | 131 | 0.278 | 2.8 |
| Copper | 121<br>Yield strength = 0.173GPa<br>Second tangent modulus = 1.2GPa | 0.34 | 16.3 |
| Organic core | 24 | 0.3 | $\alpha_x = \alpha_y = 10$; $\alpha_z = 24$ |
| Dielectric | 2 | 0.3 | 60 |
| $SiO_2$ | 70 | 0.17 | 0.5 |
| C2 or μbump (Sn0.7Cu) | 30 | 0.3 | 22 |
| C4 bump (SAC305) | 49–0.07 T(°C) | 0.29 | 21 + 0.017 T(°C) |
| PCB | $E_x = E_y = 22$; $E_z = 10$ | 0.28 | $\alpha_x = \alpha_y = 18$; $\alpha_z = 70$ |
| Underfill | 4.5 | 0.35 | 50 |
| Equivalent block (organic core + RDL) | 24.87 | 0.3 | $\alpha_x = \alpha_y = 11.29$; $\alpha_z = 24.39$ |

**Table 5.2** Anand parameters for SAC and Sn0.7Cu

| Anand Parameters | SAC305 | Sn0.7Cu |
|---|---|---|
| $s_0$ (Pa) | $18.07 \times 10^6$ | $33.8 \times 10^6$ |
| Q/R (K) | 9096 | 5276 |
| A ($s^{-1}$) | 3484 | 9875 |
| $\xi$ | 4 | 2 |
| m | 0.2 | 0.176 |
| $h_0$ (Pa) | $1.44 \times 10^{11}$ | $61987.3 \times 10^6$ |
| $\hat{s}$ (Pa) | $26.4 \times 10^6$ | $65.2 \times 10^6$ |
| n | 0.01 | 0.003 |
| a | 1.9 | 1.03 |

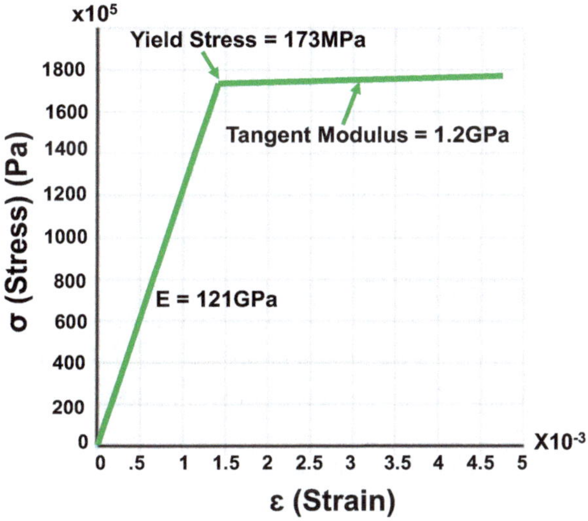


**Fig. 5.24** Elastoplastic stress-strain of the electroplated Cu

stress–strain relation shown in Fig. 5.24 (first Young's modulus = 121 GPa, second Young's modulus = 1.2 GPa, and yield strength = 173 MPa). The effective material properties of the organic build-up equivalent block can be calculated by Fig. 5.25 and the following and are summarized in Table 5.1. For example, the equivalent Young's modulus = [Young's modulus of metal layer) (thickness of metal layers) + (Young's modulus of dielectric layer) (thickness of dielectric layer)]/ (thickness of metal layers + thickness of dielectric layer + thickness of organic core layer).

For equivalent Young's modulus:

$$\frac{121\times(3\times4)+2\times(5\times4)+24\times800}{(3\times4)+(5\times4)+800}=24.87\,\text{GPa}$$

| | Young's modulus (GPa) | Poisson's ratio | CTE ($10\times^{-6}/^{o}C$) |
|---|---|---|---|
| Organic core | 24 | 0.3 | $\alpha_x=\alpha_y=10$; $\alpha_z=24$ |
| RDL | 121 | 0.34 | 16.3 |
| Dielectric | 2 | 0.3 | 60 |

□ **For Equivalent Young's modulus**

$$\frac{121\times(3*4)+2\times(5*4)+24\times800}{(3*4)+(5*4)+800} = 24.87\text{GPa}$$

□ **For Equivalent CTE-X/Y**

$$\frac{16.3\times(3*4)+60\times(5*4)+10\times800}{(3*4)+(5*4)+800} = 11.29\times10^{-6}/^{o}C$$

□ **For Equivalent Poisson's ratio**

$$\frac{0.34\times(3*4)+0.3\times(5*4)+0.3\times800}{(3*4)+(5*4)+800} = 0.3006$$

□ **For Equivalent CTE-Z**

$$\frac{16.3\times(3*4)+60\times(5*3)+24\times800}{(3*4)+(5*4)+800} = 24.39\times10^{-6}/^{o}C$$

**Fig. 5.25** Equivalent block for the package substrate

For equivalent CTE in X and Y:

$$\frac{16.3\times(3\times4)+60\times(5\times4)+10\times800}{(3\times4)+(5\times4)+800} = 11.29\times10^{-6}\ /\,C$$

For equivalent CTE in Z:

$$\frac{16.3\times(3\times4)+60\times(5\times4)+24\times800}{(3\times4)+(5\times4)+800} = 24.39\times10^{-6}\ /\,C$$

For equivalent Poisson's ratio:

$$\frac{0.34\times(3\times4)+0.3\times(5\times4)+0.3\times800}{(3\times4)+(5\times4)+800} = 0.3006$$

## *5.6.4 Thermal Boundary Condition*

The temperature boundary condition is shown in Fig. 5.26a. It can be seen that the temperatures are −40 °C ↔ 85 °C, and the dwell-time at hot, dwell-time at cold, ramp-up time, and ramp-down time are15 min each. Stress free is at room temperature (25 °C).

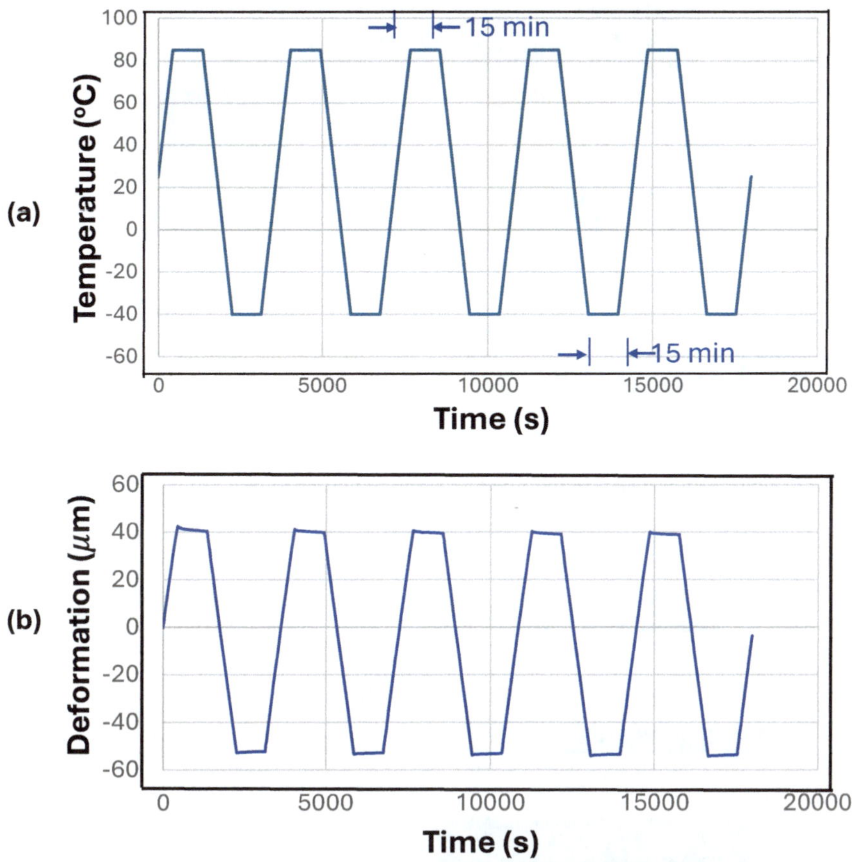


**Fig. 5.26** (**a**) Temperature cycling boundary conditions. (**b**) Deformation time-history

### *5.6.5 Structural Deformation*

The deformed shape (color contours) and the undeformed shape (dark lines) of the whole flip chip PCB assembly with Si bridge without underfill between the chiplet and the bridge are shown in Fig. 5.27 at 450 s (85 °C) and 2250 s (−40 °C). It can be seen that at 85 °C (450 s), the PCB expands more than the chiplets and Si bridge, and the structure is deformed (maximum deformation = 49 μm) in a convex shape (smiling face), and at −40 °C (2250 s), the PCB shrinks more than the chiplets and Si bridge, and the structure is deformed (maximum deformation = −59.9 μm) into a concave shape (crying face). These results show that the displacement, rotation, and temperature boundary conditions have been properly prescribed.

A closer look at the deformation of the Chiplets with Si bridge is shown in Fig. 5.28. It can be seen that even though there is not any CTE mismatch between the Si bridge and the Si chiplets, the solder cap of the C2 μbump is subjected to very

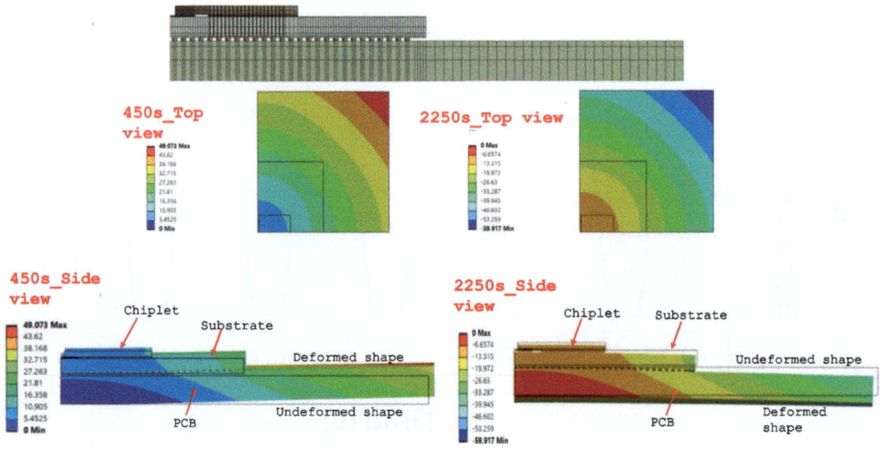


**Fig. 5.27** Deformation shapes of the structure at different time steps

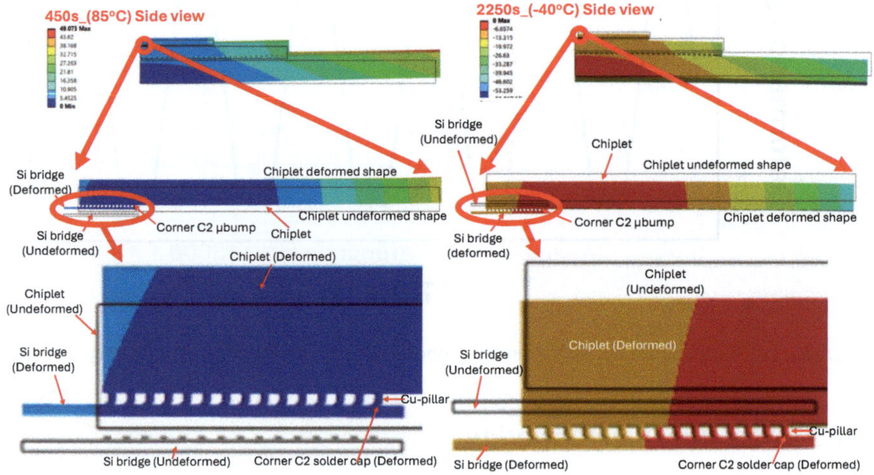


**Fig. 5.28** Local deformation of the Si bridge connecting the chiplets without underfill

large shearing strain. When at 450 s (85 °C), the structure is in a convex shape, and the corner μbump solder joint is subjected to tension besides the shear (with the chiplet moving outward more than the bridge). On the other hand, at 2250 s (−40 °C), the corner μbump solder joint is subjected to compression in addition to the shear (with the chiplet moving inward more than the bridge).

The deformation time-history of the flip chip with Si bridge assembly is shown in Fig. 5.26b. It can be seen that the deformation pattern of the structure is following the temperature boundary condition prescribed in Fig. 5.26a.

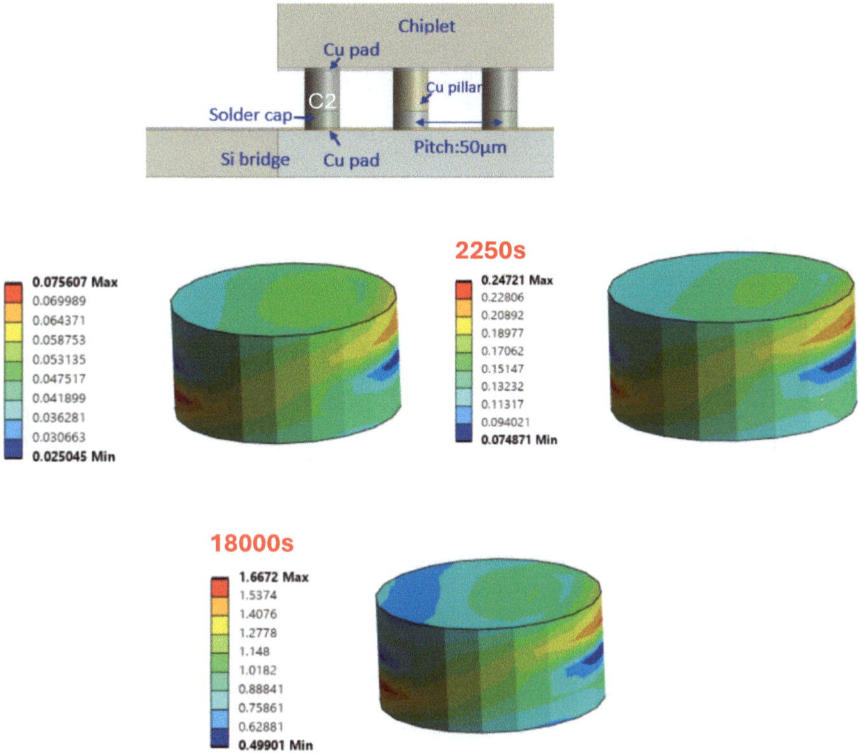


**Fig. 5.29** Accumulative equivalent inelastic strain at different time steps (Corner C2 solder joint)

### *5.6.6 Accumulated Equivalent Inelastic Strain and Equivalent Stress*

The accumulated equivalent inelastic strain contour distributions at the corner C2 μbump of the flip chip assembly with Si-bridge without underfill at 450 s (85 °C), 2250 s (−40 °C), and at 18000 s (25 °C) are shown in Fig. 5.29. It can be seen that the maximum inelastic strain at the corner C2 μbump solder cap occurs at a small local area and the value is 7.6% at 450 s (85 °C), 24.7% at 2250 s (−40 °C), and 166% at 18000 s (25 °C). Figure 5.30 shows the accumulated equivalent inelastic stain time-history in the corner C2 μbump without underfill. It can be seen that the maximum accumulated equivalent inelastic stain is 34% per cycle.

The accumulated equivalent inelastic strain contour distributions at the corner C4 μbump of the flip chip assembly with Si-bridge at 450 s (85 °C), 2250 s (−40 °C), and at 18000 s (25 °C) are shown in Fig. 5.31. It can be seen that the maximum inelastic strain at the corner C4 solder joint occurs at a small local area and the value is 0.6% at 450 s (85 °C), 1.8% at 2250 s (−40 °C), and 12.2% at 18000 s (25 °C). Figure 5.32 shows the accumulated equivalent inelastic stain time-history in the

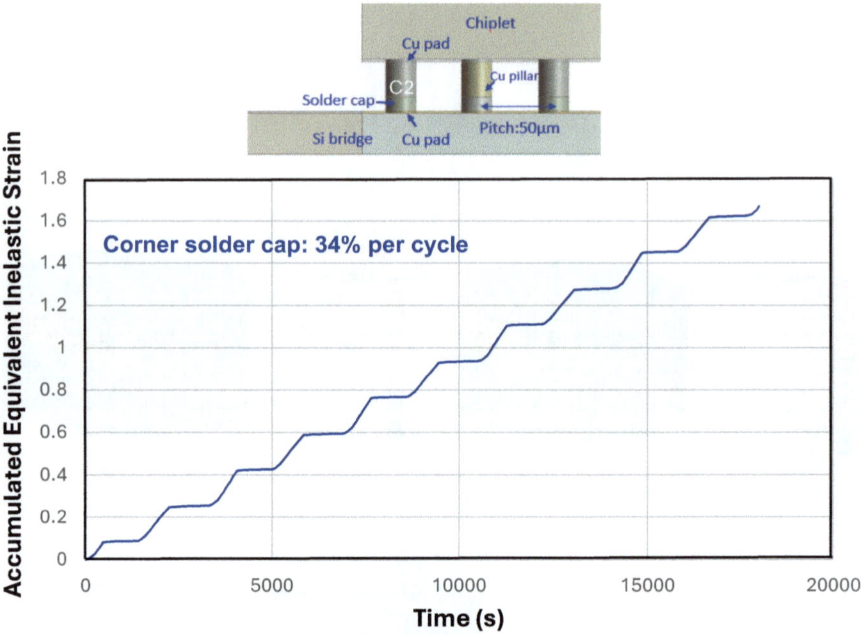


**Fig. 5.30** Accumulative equivalent inelastic strain time-history at corner C2 solder joint

corner C4 solder joint. It can be seen that the maximum accumulated equivalent inelastic stain is 2.48% per cycle.

The accumulated equivalent inelastic strain contour distributions at the corner BGA solder joint of the flip chip assembly with Si-bridge at 450 s (85 °C), 2250 s (−40 °C), and at 18000 s (25 °C) are shown in Fig. 5.33. It can be seen that the maximum inelastic strain at the corner BGA solder joint occurs at a small local area and the value is 1.4% at 450 s (85 °C), 4.2% at 2250 s (−40 °C), and 25.8% at 18000 s (25 °C). Figure 5.34 shows the accumulated equivalent inelastic stain time-history in the corner BGA solder joint. It can be seen that the maximum accumulated equivalent inelastic stain is 5.22% per cycle.

The equivalent stress contour distributions at the Cu-pad of the corner C2 μbump with Si bridge without underfill are shown in Fig. 5.35. It can be seen that at 450 s (85 °C) the maximum equivalent stress at the Cu-pad between the chiplet and the Cu-pillar is 197.7 MPa, at the Cu-pillar is 170 MPa, and at the Cu-pad between the solder cap and the Si bridge is 185 MPa. At 2250 s (-40 °C) the maximum equivalent stress at the Cu-pad between the chiplet and the Cu-pillar is 185 MPa, at the Cu-pillar is 175 MPa, and at the Cu-pad between the solder cap and the Si bridge is 199 MPa.

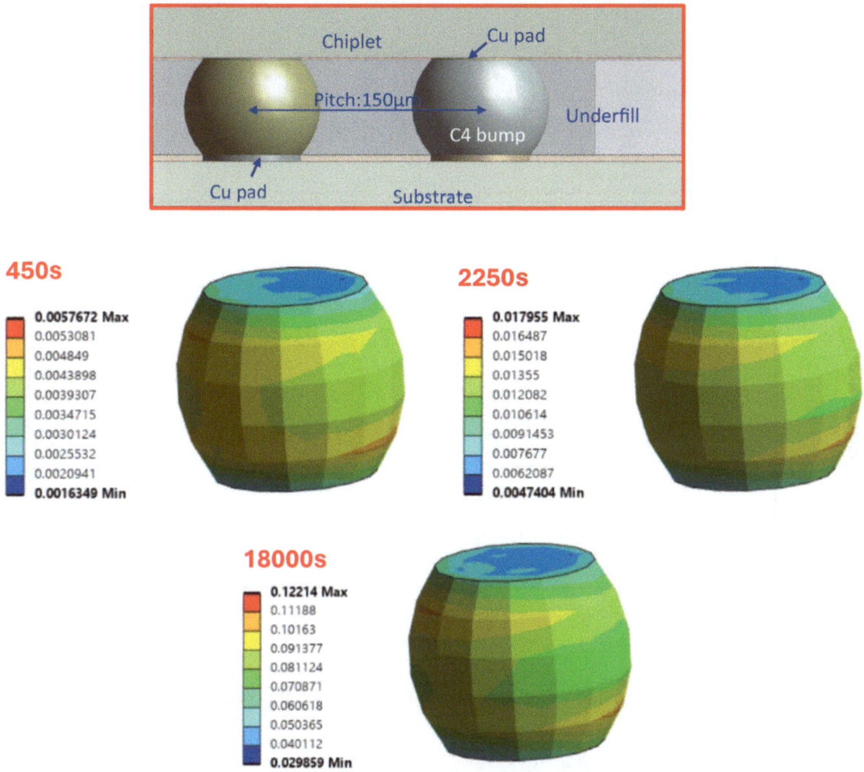


**Fig. 5.31** Accumulative equivalent inelastic strain at different time steps (Corner C4 solder joint)

The equivalent stress contour distributions at the Cu-pad of the corner C4 solder joint with Si bridge without underfill are shown in Fig. 5.36. It can be seen that at 450 s (85 °C) the maximum equivalent stress at the Cu-pad between the chiplet and the C4 solder joint is 199.7 MPa, and at the Cu-pad between the C4 solder joint and package substrate is 155.35 MPa. At 2250 s (-40 °C) the maximum equivalent stress at the Cu-pad between the chiplet and the C4 solder joint is 189.4 MPa, and at the Cu-pad between the C4 solder joint and the substrate is 180.41 MPa.

It can be seen that the accumulated equivalent inelastic strain in the solder cap of the C2 bump for the interconnect between the chiplets and the bridge without underfill is very large (34% per cycle). The solder joint reliability could be an issue.

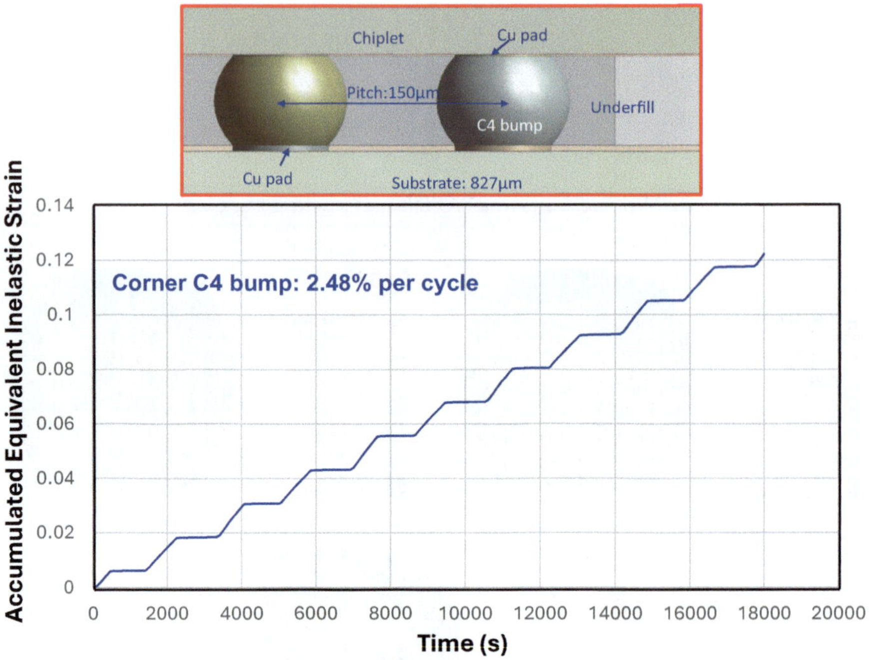


**Fig. 5.32** Accumulative equivalent inelastic strain time-history at corner C4 solder joint

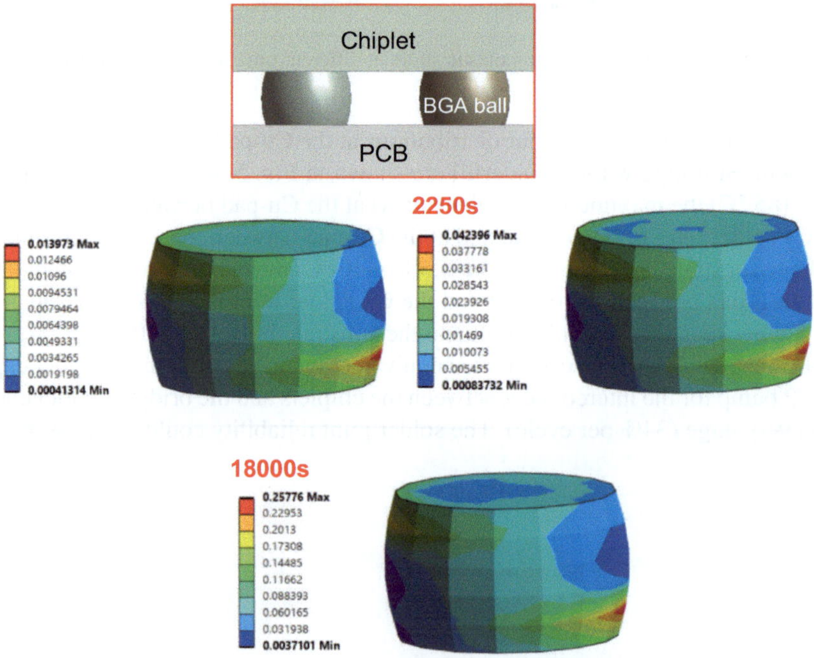


**Fig. 5.33** Accumulative equivalent inelastic strain at different time steps (Corner BGA solder joint)

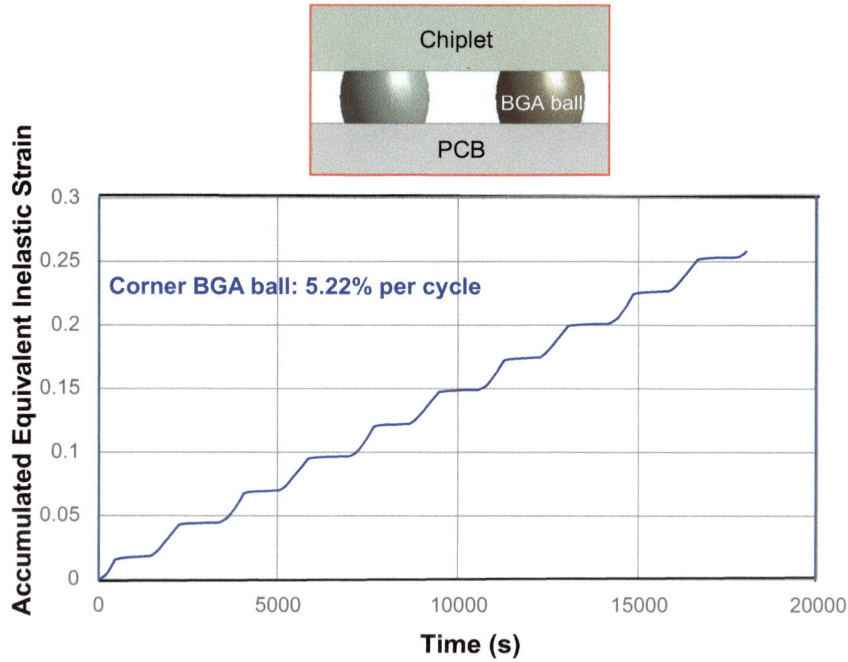


**Fig. 5.34** Accumulative equivalent inelastic strain time-history at corner BGA solder joint

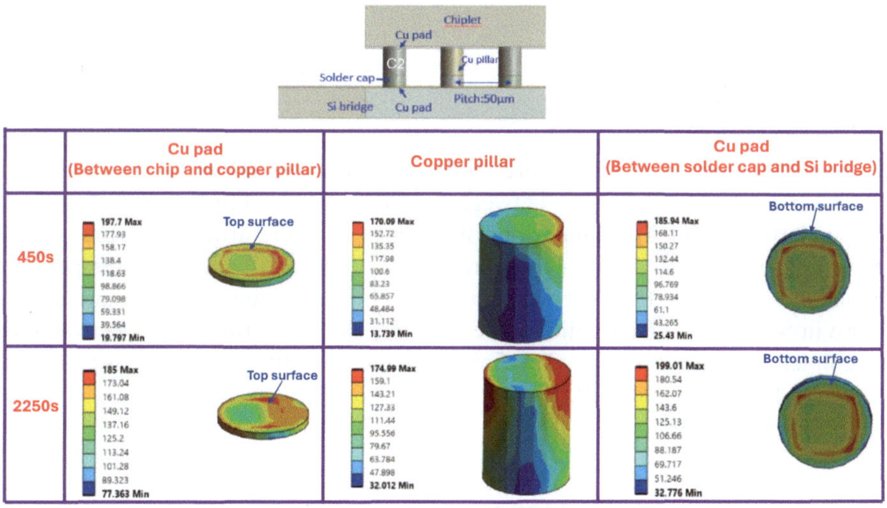


**Fig. 5.35** Accumulative equivalent stress at Cu-pad and Cu-pillar at different time steps (corner C2 joint)

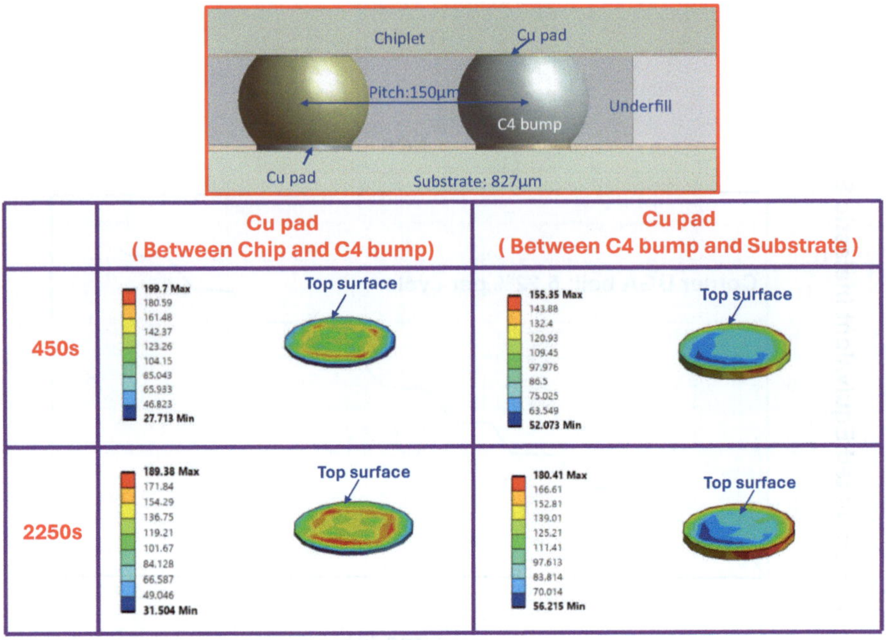


**Fig. 5.36** Accumulative equivalent stress at Cu-pads at different time steps (corner C4 joint)

## 5.7 Chiplets Interconnect with Bridges by Microbumps with Underfill

### *5.7.1 The Structure*

Figure 5.37 shows a new structure under consideration. It can be seen that even there are not CTE mismatches between the Si chiplets and Si bridge, unlike Sect. 5.6, underfill is used. The dimensions of all the parts of the structure are the same as pervious section. All the material properties (Tables 5.1 and 5.2), temperature boundary conditions, Fig. 5.26a, and kinematic boundary conditions are the same as those in Sect. 5.6. The only difference is in this case, underfill is used between the Si chiplets and the Si bridge. The finite element modeling of the new structure is shown in Figs. 5.38 and 5.39.

### *5.7.2 Structural Deformation*

The deformation time-history of the chiplets with Si bridge by microbumps with underfill is shown in Fig. 5.40b. It can be seen that the deformation time-history is following the temperature boundary condition prescribed in Fig. 5.26 or Fig. 5.40a.

The deformed shape (color contours) and the undeformed shape (dark lines) of the flip chip PCB assembly with Si bridge with underfill between the chiplets and

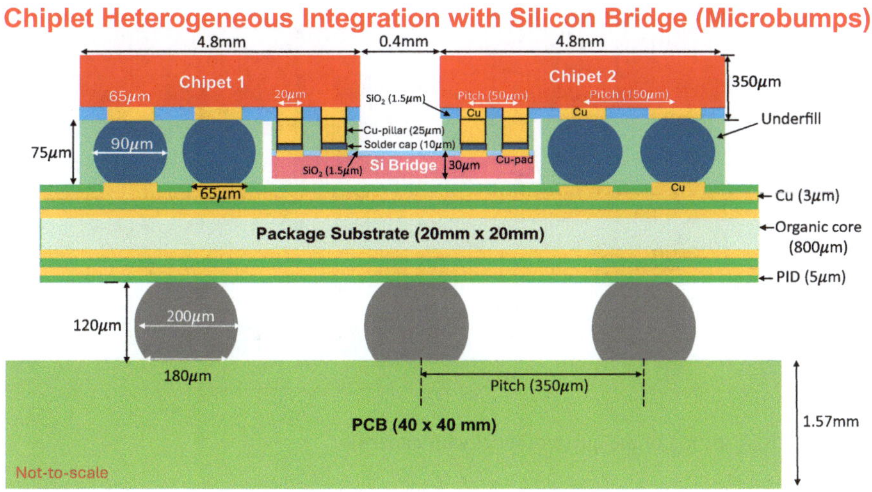


**Fig. 5.37** Coress section of chiplets connected by a Si-bridge with microbumps and underfill

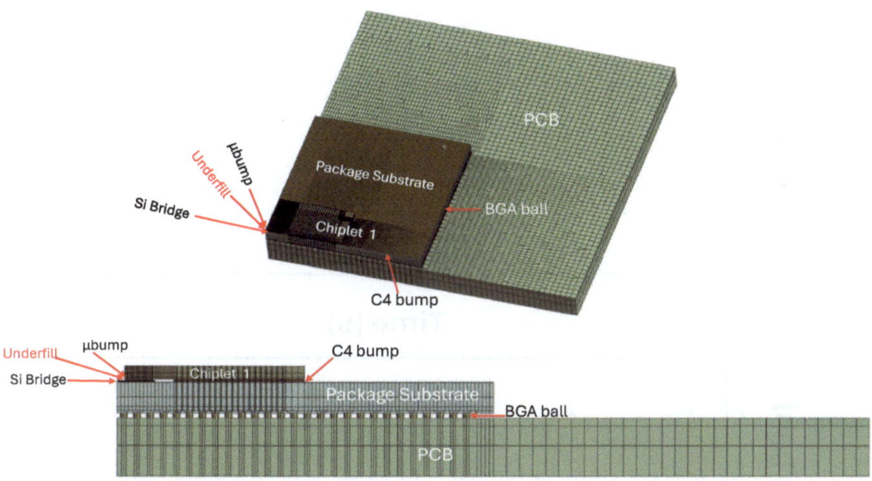


**Fig. 5.38** Quarter of finite element model with underfill

the bridge are shown in Fig. 5.41 at 450 s (85 °C) and 2250 s (−40 °C). It can be seen that at 85 °C (450 s), the PCB expands more than the chiplets and Si bridge, and the structure is deformed (maximum deformation = 49.8 μm) in a convex shape (smiling face), and at −40 °C (2250 s), the PCB shrinks more than the chiplets and Si bridge, and the structure is deformed (maximum deformation = – 60.4 μm) into a concave shape (crying face). These global deformation results are very similar to those reported in Sect. 5.6.

A closer look at the deformation (with the underfill removed) of the Chiplets with Si bridge is shown in Fig. 5.42. It can be seen that the shear as well as the

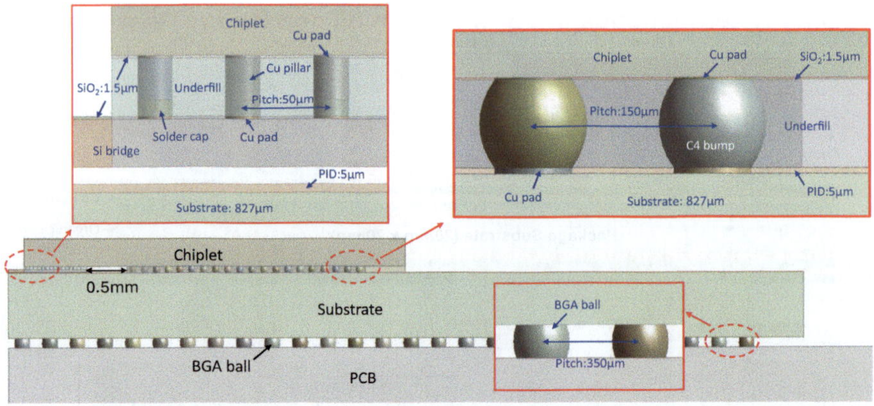


**Fig. 5.39** Cross section of the quarter model with underfill

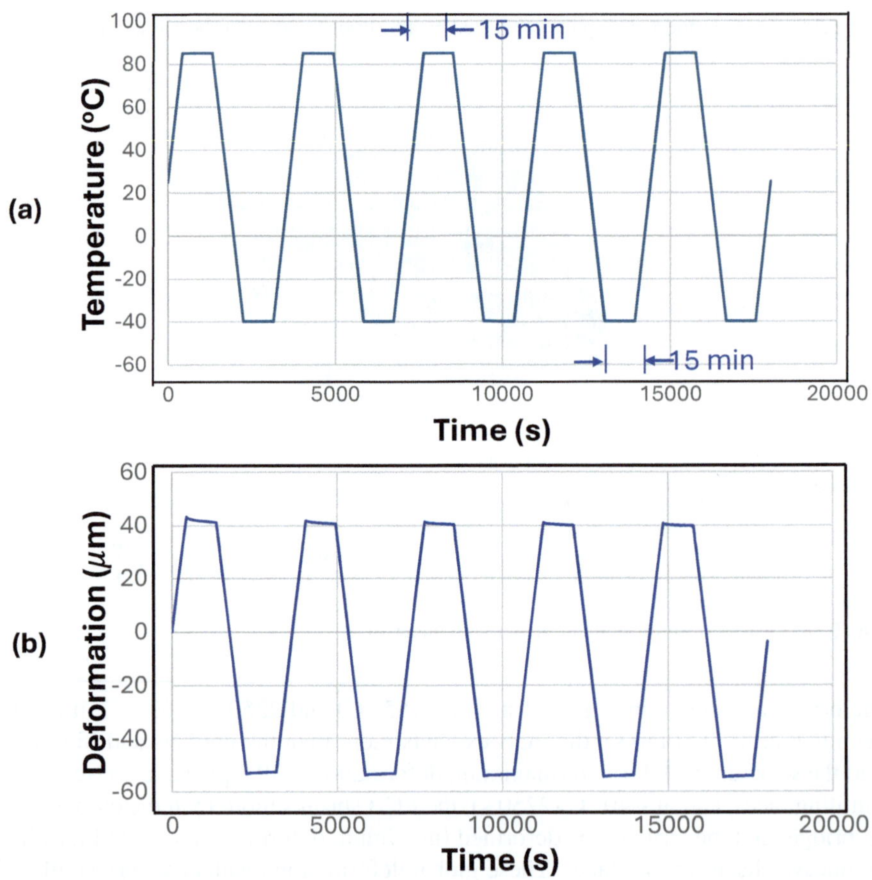


**Fig. 5.40** (**a**) Temperature boundary condition. (**b**) Deformation time-history

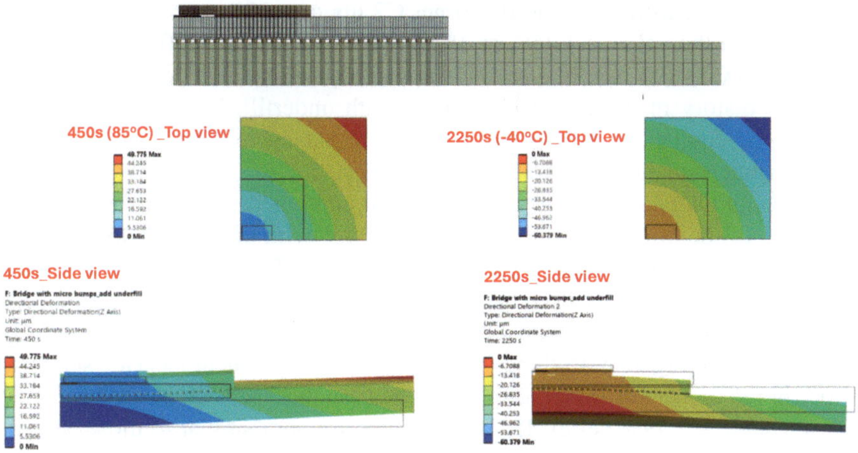


**Fig. 5.41** Deformation shapes of the structure at different time steps

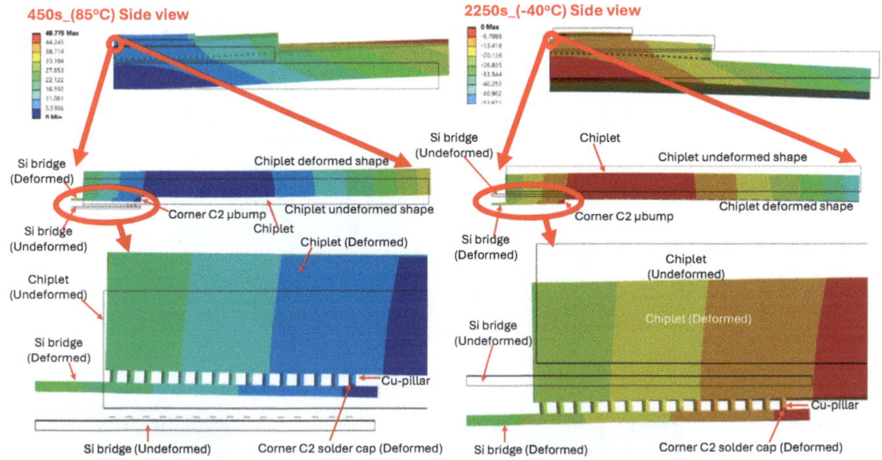


**Fig. 5.42** Local deformation of the Si bridge connecting the chiplets with underfill

tension and compression of the C2 solder cap are very small compared to those of the case without underfill, Fig. 5.28. Both Figs. 5.28 and 5.42 are plotted on the same scale.

### *5.7.3 Accumulated Equivalent Inelastic Strain and Equivalent Stress*

The accumulated equivalent inelastic strain contour distributions at the corner C2 μbump of the flip chip assembly with Si-bridge with underfill at 450 s (85 °C), 2250 s (−40 °C), and at 18000 s (25 °C) are shown in Fig. 5.43. It can be seen that the maximum inelastic strain at the corner C2 μbump solder cap occurs at a small local area and the value is 1.46% at 450 s (85 °C), 4.3% at 2250 s (−40 °C), and 28.6% at 18000 s (25 °C). Figure 5.44 shows the accumulated equivalent inelastic stain time-history in the corner C2 μbump with underfill. It can be seen that the accumulated equivalent inelastic stain is 5.81% per cycle. These values are very different (much smaller) than those shown in Sect. 5.6.

The accumulated equivalent inelastic strain contour distributions at the corner C4 μbump of the flip chip assembly with Si-bridge with underfill at 450 s (85 °C), 2250 s (−40 °C), and at 18000 s (25 °C) are shown in Fig. 5.45. It can be seen that the maximum inelastic strain at the corner C4 solder joint occurs at a small local area and the value is 0.6% at 450 s (85 °C), 1.8% at 2250 s (−40 °C), and 12.2% at 18000 s (25 °C). Figure 5.46 shows the accumulated equivalent inelastic stain time-history in the corner C4 solder joint. It can be seen that the accumulated equivalent inelastic stain is 2.49% per cycle. These values are very similar to those shown in Sect. 5.6 for the case without underfill.

The accumulated equivalent inelastic strain contour distributions at the corner BGA solder joint of the flip chip assembly with Si-bridge with underfill at 450 s (85 °C), 2250 s (−40 °C), and at 18000 s (25 °C) are shown in Fig. 5.47. It can be seen that the maximum inelastic strain at the corner BGA solder joint occurs at a

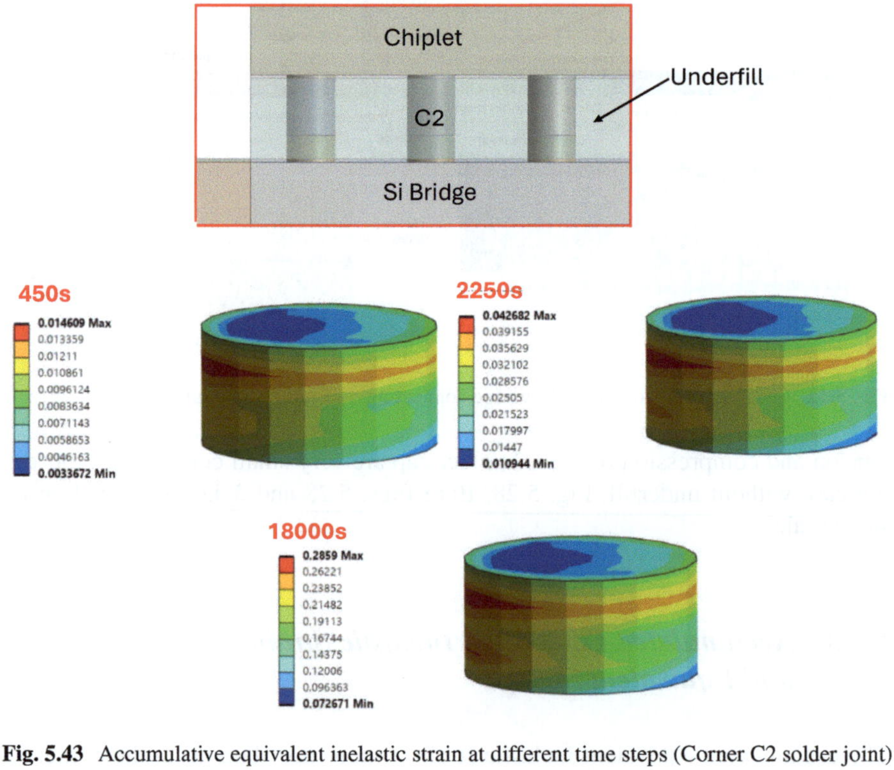


**Fig. 5.43** Accumulative equivalent inelastic strain at different time steps (Corner C2 solder joint)

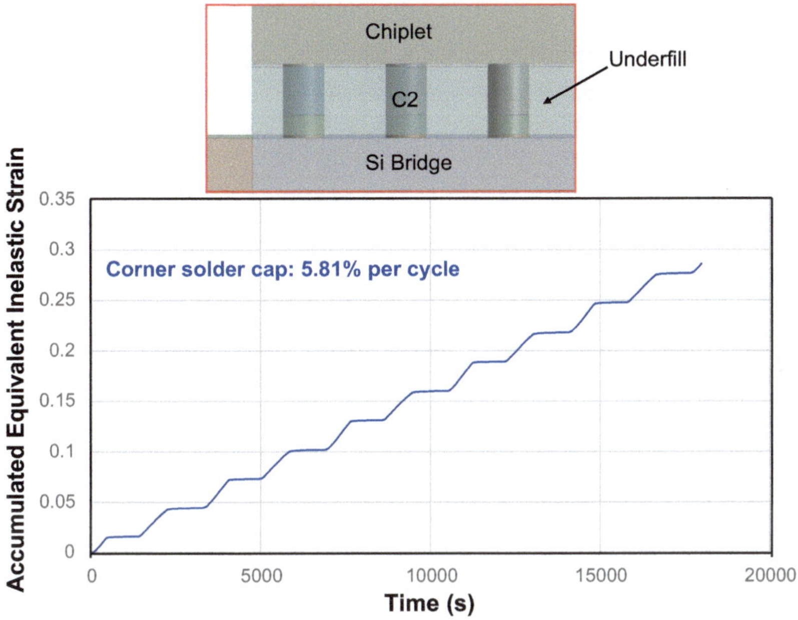


Fig. 5.44 Accumulative equivalent inelastic strain time-history at corner C2 solder joint

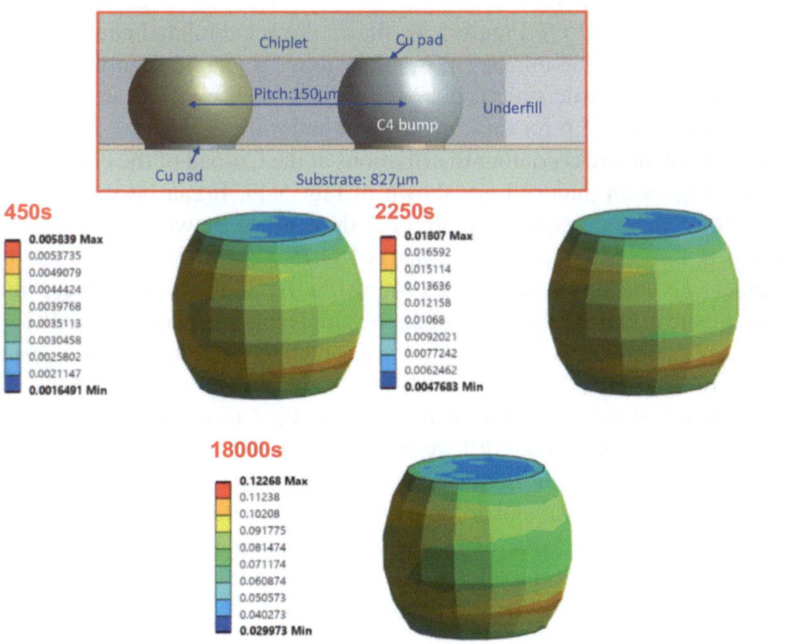


Fig. 5.45 Accumulative equivalent inelastic strain at different time steps (Corner C4 solder joint)

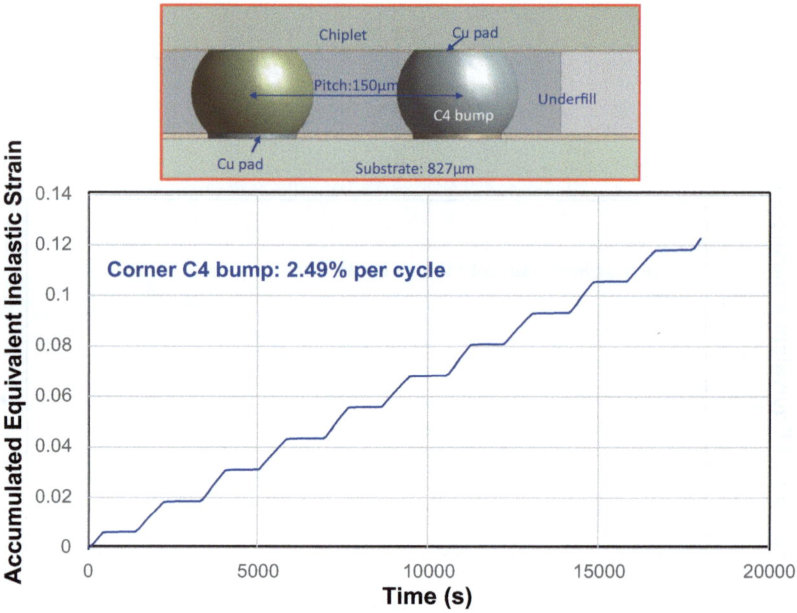


**Fig. 5.46** Accumulative equivalent inelastic strain time-history at corner C4 solder joint

small local area and the value is 1.4% at 450 s (85 °C), 4.2% at 2250 s (−40 °C), and 25.5% at 18000 s (25 °C). Figure 5.48 shows the accumulated equivalent inelastic strain time-history in the corner BGA solder joint. It can be seen that the accumulated equivalent inelastic stain is 5.17% per cycle. These values are very similar to those shown in Sect. 5.6 for the case without underfill.

The equivalent stress contour distributions at the Cu-pad of the corner C2 μbump with Si bridge with underfill are shown in Fig. 5.49. It can be seen that at 450 s (85 °C) the maximum equivalent stress at the Cu-pad between the chiplet and the Cu-pillar is 174.8 MPa, at the Cu-pillar is 184.15 MPa, and at the Cu-pad between the solder cap and the Si bridge is 174.9 MPa. At 2250 s (-40 °C) the maximum equivalent stress at the Cu-pad between the chiplet and the Cu-pillar is 181.46 MPa, at the Cu-pillar is 178.88 MPa, and at the Cu-pad between the solder cap and the Si bridge is 183.74 MPa.

The equivalent stress contour distributions at the Cu-pad of the corner C4 solder joint with Si bridge with underfill are shown in Fig. 5.50. It can be seen that at 450 s (85 °C) the maximum equivalent stress at the Cu-pad between the chiplet and the C4 solder joint is 199.2 MPa, and at the Cu-pad between the C4 solder joint and package substrate is 156.56 MPa. At 2250 s (-40 °C) the maximum equivalent stress at the Cu-pad between the chiplet and the C4 solder joint is 188.83 MPa, and at the Cu-pad between the C4 solder joint and the substrate is 181.61 MPa.

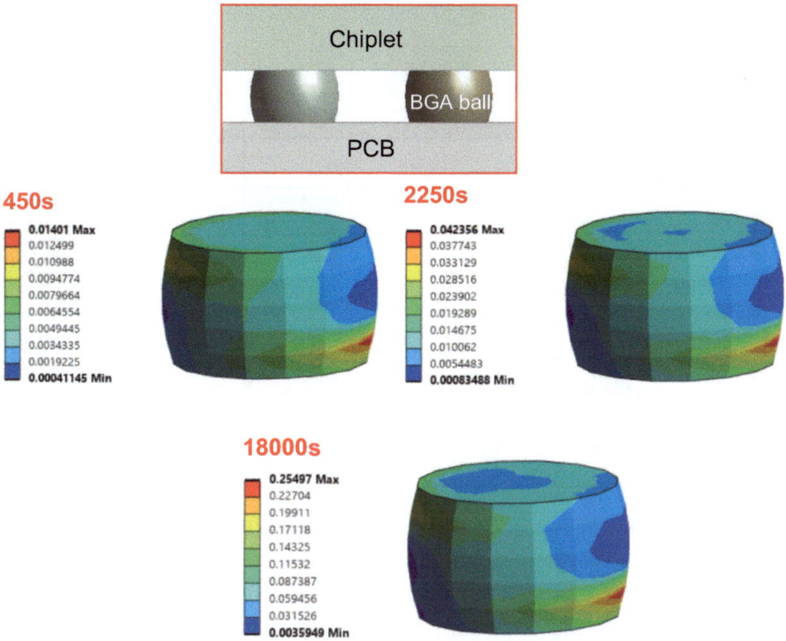


**Fig. 5.47** Accumulative equivalent inelastic strain at different time steps (Corner BGA solder joint)

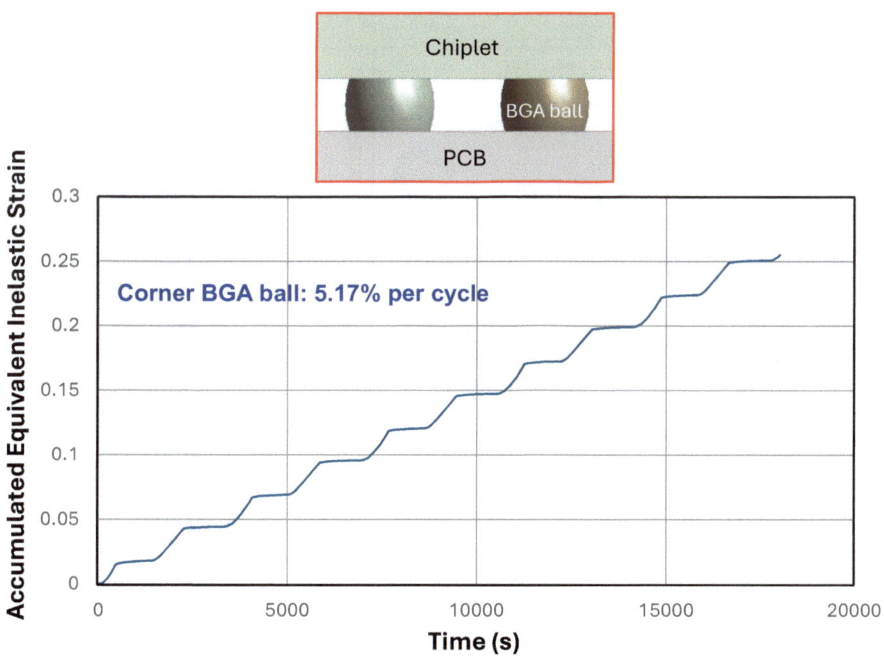


**Fig. 5.48** Accumulative equivalent inelastic strain time-history at corner BGA solder joint

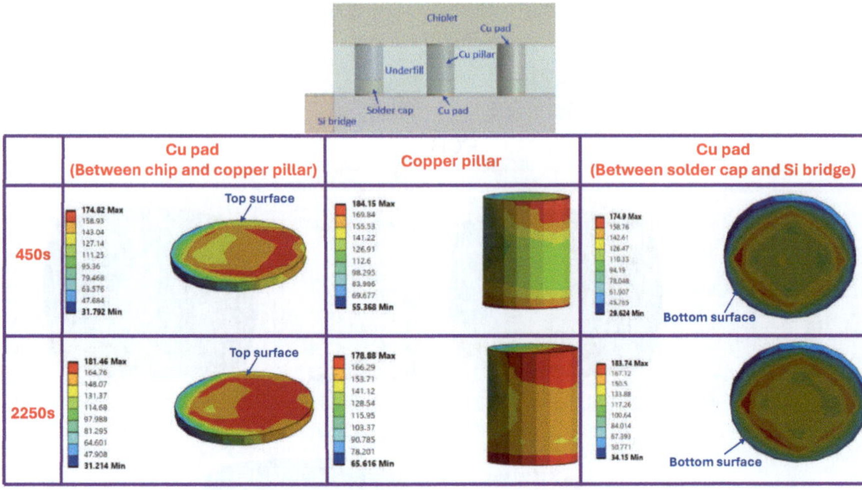


**Fig. 5.49** Accumulative equivalent stress at Cu-pad and Cu-pillar at different time steps (corner C2 joint)

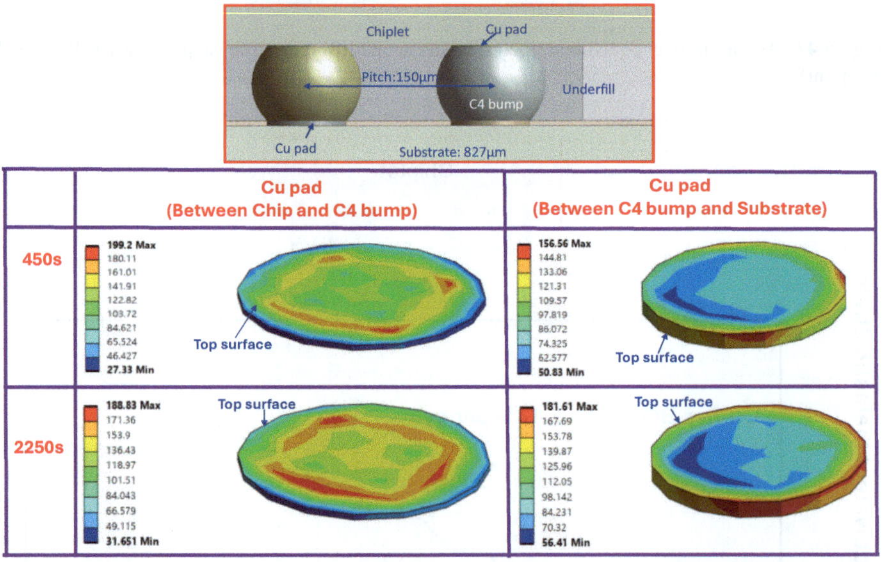


**Fig. 5.50** Accumulative equivalent stress at Cu-pads at different time steps (corner C4 joint)

| | Max. Global Deformation (µm) | |
|---|---|---|
| | Without Underfill for C2 µbump | With Underfill for C2 µbump |
| 450s (85°C) | 49.07 | 49.78 |
| 2250s (- 40°C) | - 59.92 | - 60.38 |

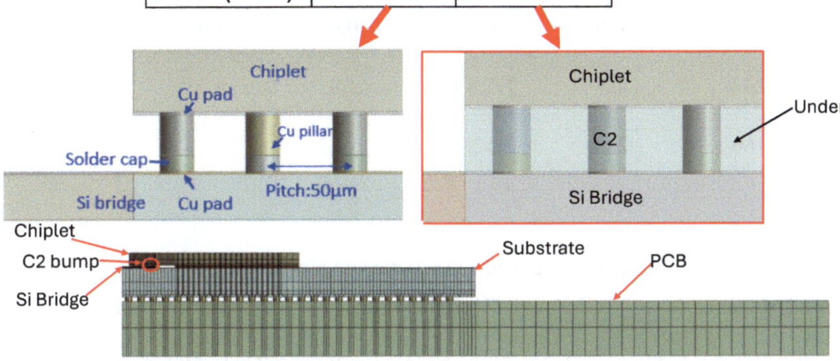


**Fig. 5.51** Maximum global deformation

## *5.7.4 Comparison of Si Bridge Connecting Chiplets w/o Underfill*

Figure 5.51 shows the comparison of the maximum global deflection of a system consisting of chiplets connected with Si bridge by C2 µbump w/o underfill. It can be seen that the effect of the underfill for the C2 µbump is negligible for the global deflection. This is reasonable because the size of the Si bridge is very small in the whole structure.

Table 5.3 shows the maximum accumulated equivalent inelastic strain at the corner C2 solder cap, corner C4 solder joint, and corner BGA solder joint w/o underfill for the C2 µbump at various time steps. It can be seen that (a) the effect of the underfill for the C2 µbump is very significant for the maximum accumulated equivalent inelastic strain at the corner C2 solder cap at all time steps (1.46% vs. 7.6% at 450 s, 4.3% vs. 24.7% at 2250 s, and 28.6% vs. 166% at 18000 s), and (b) the effect of the underfill for the C2 µbump is negligible for the corner C4 solder joint and the corner BGA solder joint.

Table 5.4 shows the maximum accumulated equivalent inelastic strain per cycle at the corner C2 solder cap, corner C4 solder joint, and corner BGA solder joint w/o underfill for the C2 µbump. Again, the effect of the underfill for the C2 µbump is very significant for the maximum accumulated equivalent inelastic strain per cycle at the corner C2 solder cap (5.81% vs. 34%) and the effect of the underfill for the C2 µbump is negligible for the corner C4 solder joint and the corner BGA solder joint.

The simulation results show that even there is not any CTE mismatch between the Si chiplets and the Si bridge, however due to the global CTE mismatch of the

**Table 5.3** Maximum accumulated equivalent inelastic strain (%)

| | Corner C2 solder cap | | Corner C4 solder joint | | Corner BGA solder joint | |
|---|---|---|---|---|---|---|
| | Without Underfill for C2 bump | With Underfill for C2 bump | Without Underfill for C2 bump | With Underfill for C2 bump | Without Underfill for C2 bump | With Underfill for C2 bump |
| 450 s (85 °C) | 7.6 | 1.46 | 0.6 | 0.6 | 1.4 | 1.4 |
| 2250 s (−40 °C) | 24.7 | 4.3 | 1.8 | 1.8 | 4.2 | 4.2 |
| 18,000 s (25 °C) | 166 | 28.6 | 12.2 | 12.2 | 25.8 | 25.5 |

**Table 5.4** Maximum accumulated equivalent inelastic strain per cycle (%)

| Corner C2 solder cap | | Corner C4 solder joint | | Corner BGA solder joint | |
|---|---|---|---|---|---|
| Without Underfill for C2 bump | With Underfill for C2 bump | Without Underfill for C2 bump | With Underfill for C2 bump | Without Underfill for C2 bump | With Underfill for C2 bump |
| 34 | 5.81 | 2.48 | 2.49 | 5.22 | 5.17 |

whole structure, which affects the local deformation, stress, and strain of the C2 μbump solder joint. Thus, underfill is needed between the chiplets and the Si bridge to protect the C2 μbump solder cap from failure.

## 5.8 Summary

Some important results and recommendations are summarized as follows.

- The silicon bridges embedded in (or above) the build-up package substrate, silicon bridges embedded in fan-out EMC with RDLs, and Cu-Cu hybrid bonding bridges have been briefly mentioned and discussed. Hybrid bonding bridges will lead to highest density, finest pitch, best electrical and thermal performance, and lowest package profile.
- Besides the horizontal communications by the RDLs in the Si bridge, there could be TSVs in the Si bridge for vertical communications.
- The dimensions of the Si bridge are very small.
- The effects of the underfill between the chiplets and the Si bridge (for the C2 μbump) on the global deformation of the whole package assembly are negligible.
- The effects of the underfill between the chiplets and the Si bridge on the C4 solder joint and the BGA solder joint are negligible.
- However, the effects of the underfill between the Si chiplets and the Si bridge on the C2 μbump solder cap are very significant. Even though there is not any CTE mismatch between the Si chiplets and the Si bridge, due to the global CTE mismatch of all the parts and the geometry of the structure, the deformation, strain,

and stress of the solder cap of the C2 μbump without underfill is very large which could lead to solder joint failure.

- Thus, it is recommended to add underfill between the Si chiplets and the Si bridge even if there is not any CTE mismatch.

## References


1. Lau, J. H., and X. Fan, *Hybrid Bonging, Advanced Substrates, Failure Mechanisms, and Thermal Management for Chiplets and Heterogeneous Integration*, Springer, New York, 2025.
2. Lau, J. H., *Flip Chip, Hybrid Bonding, Fan-In, and Fan-Out Technology*, Springer, New York, 2024.
3. Lau, J, H., *Chiplet Design and Heterogeneous Integration Packaging*, Springer, New York, 2023.
4. Lesueur, T., D. Danovitch, D. Drouin, I. de Sousa, D. Taneja, A. Horibe, and S. Kohara, "Self-Alignment of Active Si Bridge using Solder Joints Capillary Forces", *IEEE/ECTC Proceedings*, June 2025, pp. 256–263.
5. Lim, J., S. Jang, Y. Jang, Y. Koh, J. Jang, and J. Kim, "Novel Packaging Platform Based on Bridge Dies with Top and Bottom I/O Connections on Standard Substrates", *IEEE/ECTC Proceedings*, June 2025, pp. 2034–2038.
6. Arioka, S., Y. Naka, K. Meiten, F. Ootake, Y. Morikawa, T. Srichanthamit, O. Okada, and Y. Kurita, "Pillar-Suspended Bridge (PSB); Transmission Simulation and Fabrication Process of 2-Micron Diameter / 5-Micron Pitch Dry Etched Stacked Via in Low-k Polymer for High Performance Redistribution Layer Bridge (RDL Bridge)", *IEEE/ECTC Proceedings*, June 2025, pp. 785–792.
7. Ahn, S., S. Song, H. Kang, S. Yong, D. Kim, D. Kim, S. Lee, J. Kim, J. Yun, and E. Lee, "A Novel Approach of Multi-Chip FOPKG Chiplet Interconnection using a Land-Side Si Bridge", *IEEE/ECTC Proceedings*, June 2025, pp. 748–752.
8. Kuo, K., J. Hung, T. Lin, and F. Chien, "The Study and Challenges of Backside Metal Interface Material to Enhance Fan-Out Embedded Bridge Die Package (FO-EB) Thermal Performance", *IEEE/ECTC Proceedings*, June 2025, pp. 761–766.
9. Seo, H., J. Lee, and I. Han, "Study of the Optimal Cu Core Size Based on Surface Evolver for Solder Bridge and HiP Free of Large PBGA Package", *IEEE/ECTC Proceedings*, June 2025, pp. 1264–1268.
10. Na, Y., S. Kim, Y. Shin, H. Kim, S. Yang, S. Im, S. Ha, W. Lee, S. Yeum, and J. Lim, "Signal and Power Integrity Optimization using Novel Bridge Die and Copper Post Interconnect Design in 2.xD Packaging for Wide I/O Applications", *IEEE/ECTC Proceedings*, June 2025, pp. 1780–1685.
11. Aggarwal, A., S. Pothukuchi, A. Rathi, S. Sahni, P. Winterbottom, and D. Lazovsky, "Optical Multi-Chip Interconnect Bridge (OMIB™) Interposer Assembly Process to Enable High-Density Photonic Interconnects for High-Performance Computing Applications", *IEEE/ECTC Proceedings*, June 2025, pp. 331–336.
12. Chen, Z., Z. Xiong, K. Wang, A. Marconnet, and T. Wei, "A Novel Thermal Isolation Method with Embedded 'Glass Bridge' Structures in Silicon-Based 2.5D Heterogeneous Integration Systems", *IEEE/ECTC Proceedings*, June 2025, pp. 895–901.
13. Miao, W., C. Tan, W. Goh, M. Raju, and M. Rotaru, "AI-Driven Automatic Routing for UCIe Bridge with Heterogeneous Configurations in Advanced System-in-Package", *IEEE/ECTC Proceedings*, June 2025, pp. 1757–1764.
14. Han, H., J. Park, M. Park, J. Lee, W. Choi, and D. Kim, "Development of Embedded Multi Si Bridge Package in Panel Level Process for HPC/AI Applications", *IEEE/ECTC Proceedings*, June 2025, pp. 251–255.
15. Duan, G., Y. Zhang, A. Gunawan, Y. Fang, J. Mousavi, A. Apte, N. Ahmed, S. Sharma, S. Alur, A. Chandolu, X. Li, M. Mo, J. Jones, R. McRee, R. Limvorapitux, C. Subramani, M. Rahman,

T. Ibrahim, R. Eluri, S. Kumar, S. Agraharam, and R. Manepalli, "EMIB-TSV Advanced Packaging Technology - EMIB™s Next Evolution", *IEEE/ECTC Proceedings*, June 2025, pp. 269–273.
16. Lee, T., C. Chen, L. Cao, Y. Su, C. Zhuang, H. Kuo, P. Pan, and C. Hung, "Technology Enhancements on FOCoS-Bridge for Emerging Trends in High-Performance Computing (HPC) and Artificial Intelligence (AI)", *IEEE/ECTC Proceedings*, June 2025, pp. 902–905.
17. Cheng, J., L. Zhu, B. Wu, C. Yang, and H. Che, "Warpage Simulation and Experimental Validation of The X-Dimension Fan-Out Integration-Bridge Wafer Level Packaging Process", *IEEE/ECTC Proceedings*, June 2025, pp. 1454–1459.
18. Jang, Y., J. Lee, Y. Jang, S. Jang, N. Cho, and J. Jang, "Microbump Interconnections Using Vertical Wire Technology for the Power Connections between SoC and Substrate in the Bridge Die-Based Packaging Platform", *IEEE/ECTC Proceedings*, June 2025, pp. 1964–1969.
19. Wudjud, W., S. Lin, Y. Chang, J. Yen, R. Liao, L. Cheng, Y. Wu, S. Huang, J. Chen, C. Lee, J. Huang, L. Cao, "Advanced Thermocompression Bonding on High Density Fan-Out Embedded Bridge Technology for HPC/AI/ML Applications", *IEEE/ECTC Proceedings*, June 2024, pp. 929–935.
20. Kuo, K., A. Li, S. Wu, and F. Chien, "Challenges and Analysis for Pitch 25 μm -100 μm Mixed Micro Bumps And Interconnection in Fan-Out Embedded Bridge Die with TSV Package (FO-EB-TSV)", *IEEE/ECTC Proceedings*, June 2024, pp. 23–28.
21. Kang, L., J. Kim, B. Lee, J. Park, H. Yu, S. Lee, J. Lim, M. Lee, S. Han, J. Lee, and H. Hwang, "Single and Multi NPU Chiplet Heterogeneous Integration Packaging Based on Fanout RDL Interposer with Silicon Bridge Technology", *IEEE/ECTC Proceedings*, June 2024, pp. 1603–1608.
22. Kono, I., Y. Kometani, A. Kuroha, and Y. Kurita, "High Precision and Productivity Bridge-Die-Last Bonding Process and Its Reliability for Pillar-Suspended Bridge (PSB) Architecture", *IEEE/ECTC Proceedings*, June 2024, pp. 29–33.
23. Horibe, A., T. Watanabe, C. Marushima, S. Kohara, Y. Yamada, H. Mori, Di Taneja, K. Pilger, A. Jacques-Fortin, M. Godard, Q. Chen, E. Perfecto A. Jain, J. Ross, T. Wassick, and I. Sousa, "Scalable Advanced DBHi Chiplet Package Using Silicon Bridge with 30 μm-Pitch Solder Joints", *IEEE/ECTC Proceedings*, June 2024, pp. 269–273.
24. Lin, V., T Shih, A. Kang, and Y. Wang, "Fan-Out Embedded Bridge Structure for Co-packaged Optics Characterization and Evaluation", *IEEE/ECTC Proceedings*, June 2024, pp. 1246–1250.
25. Brun, X., S. Wong, M. Mota, and E. Udofia, "World's First UCIe Interoperability Silicon Enabling Open Standards Heterogeneous Integration", *IEEE/ECTC Proceedings*, June 2024, pp. 264–268.
26. Hu, Y., Y. Liang, H. Hu, C. Tan, C. Shen, C. Lee, et al., "CoWoS architecture evolution for next-generation HPC on 2.5D system in package," *IEEE/ECTC Proceedings*, June 2023, pp. 1022–1026.
27. Hou, S., C. Lee, T. Wang, and H. Hu, "Supercarrier redistribution layers to realize ultra-large 2.5D wafer-scale packaging by CoWoS," *IEEE/ECTC Proceedings*, June 2023, pp. 510–514.
28. Paulido, C., S. Kohara, A. Jain, T. Wassick, D. Taneja, P. McInnes, A. Horibe, and I. Sousa, "Direct Bonded Heterogeneous Integration (DBHi): Surface Bridge Approach for die Tiling", *IEEE/ECTC Proceedings*, June 2023, pp. 1053–1059.
29. Jang, H., K. Ahn, H. Park, G. Yong, J. Kim, Y. Jeong, E. Sohn, D. Hiner, M. Kelly, T. Hwang, W. Do, and J. Khim, "Reliability Performance of S-Connect Module (Bridge Technology) for Heterogeneous Integration Packaging", *IEEE/ECTC Proceedings*, June 2023, pp. 1027–1031.
30. Morikawa, Y., M. Koh H. Hashimoto, C. Chen, Y. Wang, I. Kono, S. Wakisaka, K. Ukawa, T. Fukushima, K. Suganuma, and Y. Kurita, "A Novel Chiplet Integration Architecture Employing Pillar-Suspended Bridge with Polymer Fine-Via Interconnect", *IEEE/ECTC Proceedings*, June 2023, pp. 34–39.
31. Lin, V., T. Shih, A. Kang, and Y. Wang, "Fan-Out Embedded Bridge with TSV(FO-EB-T) Package Characterization and Evaluation", *IEEE/ECTC Proceedings*, June 2023, pp. 1344–1347.

32. He, L., Z. Xie, S. Deshpande, A. Jimenez, J. Han, G. Duan, and R. Manepalli, "Warpage Modulation Study on Panel-Level Compression Molding Technology for Heterogenous Integration Applications", *IEEE/ECTC Proceedings*, June 2023, pp. 526–530.
33. Li, J., T. Hsu, M. Zhuan, S. Lin, T. Shih, N. Kao, and Y. Wang, "Innovative Fan-Out Embedded Bridge Structure for Co-Packaged Optics", *IEEE/ECTC Proceedings*, June 2023, pp. 228–232.
34. Al-Haidari, R., M. Alhendi, D. Richmond, E. Abbara, A. Obeidat, U. Somarathna, M. Schadt, M. Poliks, A. Gowda, J. Erlbaum, H. Xiong, and C. Hitchcock, "Planar SiC Power Module Packaging and Interconnections Using Direct Ink Writing", *IEEE/ECTC Proceedings*, June 2023, pp. 668–675.
35. Kaul, A., M. Hossen, M. Manley, and M. Bakir, "Design Considerations for Power Delivery Network and Metal-Insulator-Metal Capacitor Integration in Bridge-Chips for 2.5-D Heterogeneous Integration", *IEEE/ECTC Proceedings*, June 2023, pp. 985–990.
36. Su, P, J. Pu, D. Ho, and Y. Wang, "FO-EB-T Package Solution for High Performance Computing", *IEEE/ECTC Proceedings*, June 2023, pp. 1038–1042.
37. Pervez, M., and A. Kaur, "Design of a Compact Size Bridge Connected Multiband MIMO Antenna for Automotive 5G and DSRC Communications System", *IEEE/ECTC Proceedings,* June 2023, pp. 1921–1925.
38. Marushima, C., T. Aoki, K. Nakamura, R. Miyazawa, A. Horibe, I. Sousa, and K. Sikka, and T. Hisada, "Dimensional Parameters Controlling Capillary Underfill Flow for Void-Free Encapsulation of a Direct Bonded Heterogeneous Integration (DBHi) Si-bridge Package", *Proceedings of IEEE/ECTC*, May 2022, pp. 586–590.
39. Horibe, A., T. Watanabe, C. Marushima, H. Mori, S. Kohara, R. Yu, M. Bergendahl, T. Magbitang, R. Wojtecki, D. Taneja, M. Godard, C. Cristina, B. Pulido, I. Sousa, K. Sikka, and T. Hisada, "Characterization of Non-Conductive Paste Materials (NCP) for Thermocompression Bonding in a Direct Bonded Heterogeneoussly Integrated (DBHi) Si-Bridge Package", *Proceedings of IEEE/ECTC*, May 2022, pp. 625–630.
40. Horbe, A., C. Marushima, T. Watanabe, A. Jian, E. Turcotte, I. Sousa, K. Sikka, and T. Hisada, "Super Fine Jet Underfill Dispense Technique for Robust Micro Joint in Direct Bonded Heterogeneous Integration (DBHi) Silicon Bridge Packages", *Proceedings of IEEE/ECTC*, May 2022 pp. 631–634.
41. Chowehury, P., K. Sakuma, S. Raghawan, M. Bergendahl, K. Sikka, S. Kohara, T. Hisada, H. Mori, D. Taneja, and I. Sousa, "Thermo-mechanical Analysis of Thermal Compression Bonding Chip-Joint Process", *Proceedings of IEEE/ECTC*, May 2022, pp. 579–585.
42. Qiu, Y., Y. Beilliard, I. Sousa, and D. Drouin, "A Self-Aligned Structure based on V-groove for Accurate Silicon Bridge Placement", *Proceedings of IEEE/ECTC*, May 2022, pp. 668–673.
43. Duan, G., Y. Knaoka, and R. McRee, "Die embedded challenges for EMIB advanced packaging technology," *Proceedings of IEEE/ECTC*, May 2021, pp. 1–7.
44. Sikka, K., R. Bonam, Y. Liu, P. Andry, D. Parekh, A. Jain, M. Bergendahl, et al., "Direct Bonded Heterogeneous Integration (DBHi) Si Bridge", *IEEE/ECTC Proceedings*, June 2021, pp. 136–147.
45. Matsumoto, K., M. Bergendahl, K. Sikka, S. Kohara, H. Mori, and T. Hisada, "Thermal Analysis of DBHi (Direct Bonded Heterogeneous Integration) Si Bridge", *Proceedings of IEEE/ECTC*, May 2021, pp. 1382–1390.
46. Jain, A., K. Sikka, J. Gomez, D. Parekh, M. Bergendahl, J. Borkulo, K. Biesheuvel, R. Doll, and M. Mueller, "Laser vs. Blade Dicing for Direct Bonded Heterogeneous Integration (DBHi) Si Bridge", *Proceedings of IEEE/ECTC*, May 2021, pp. 1125–1130.
47. Cao, L., T. Lee, Y. Chang, S. Huang, J. On, E. Lin, and O. Yan, "Advanced HDFO Packaging Solutions for Chiplets Integration in HPC Application", *Proceedings of IEEE/ECTC*, May 2021, pp. 8–13.
48. Lee, J., G. Yong, M. Jeong, J. Jeon, D. Han, M. Lee, W. Do, E. Sohn, M. Kelly, D. Hiner, and J. Khim, "S-Connect Fan-out Interposer For Next Gen Heterogeneous Integration", *Proceedings of IEEE/ECTC*, May 2021, pp. 96–100.
49. Chong, C., G. Lim, D. Ho, H. Yong, C. Choong, S. Lim, and S. Bhattacharya, "Heterogeneous Integration with Embedded Fine Interconnect", *Proceedings of IEEE/ECTC*, May 2021, pp. 2216–2221.

50. Chiu, C., Z. Qian, and M. Manusharow, "Bridge interconnect with air gap in package assembly," *US Patent* No. 8,872,349, 2014.
51. Mahajan, R., R. Sankman, N. Patel, D. Kim, K. Aygun, Z. Qian, et al., "Embedded multi-die interconnect bridge (EMIB)—a high-density, high-bandwidth packaging interconnect," *Proceedings of IEEE/ECTC*, May 2016, pp. 557-565.
52. Mahajan, R., Q. Zhiguo, R. Viswanath, S. Srinivasan, K. Avgun, W. Jen, S. Sharan, and A. Dhall, "Embedded Multidie Interconnect Bridge - A Localized, High-Density Multichip Packaging Interconnect", *IEEE Transactions on Components, Packaging and Manufacturing Technology*, Vol. 9, Issue 10, October 2019, pp. 1952–1962.
53. Hsiung, C., and A. Sundarrajan, "Methods and Apparatus for Wafer-Level Die Bridge", *US 10,651,126*, date of patent: May 12, 2020, filed on December 8, 2017.
54. Lau, J. H., C. Ko, P. Lin, T. Tseng, R. Tain, and H. Yang, "Package Structure and Manufacturing Method Thereof", *US 11,410,933*, patent date: Aug. 9, 2022, filed on May 7, 2021.
55. Weerasekera, R., S. Bhattacharys, K. Chang, and V. Rao, "Semiconductor Package and Method of Forming the Same", *US 11,018,080*, patent date: May 25, 2021, filed on Mar 17, 2017.
56. Lin, J., C. Key Chung, C. F. Lin, A. Liao, Y. Lu, J. Chen, and D. Ng, "Scalable Chiplet package using Fan-Out Embedded Bridge", *Proceedings of IEEE/ECTC*, May 2020, pp. 14–18.
57. You, J., J. Li, D. Ho, J. Li, M. Zhuang, D. Lai, C. Chung, Y. Wang, "Electrical Performances of Fan-Out Embedded Bridge", *Proceedings of IEEE/ECTC*, May 2021, pp. 2030–2034.
58. Lin, V., D. Lai, and Y. Wang, "The Optimal Solution of Fan-Out Embedded Bridge (FO-EB) Package Evaluation during the Process and Reliability Test", *Proceedings of IEEE/ECTC*, May 2022, pp. 1080–1084.
59. Su, P., D. Ho, J. Pu, and Y. Wang, "Chiplets Integrated Solution with FO-EB Package in HPC and Networking Application", *Proceedings of IEEE/ECTC*, May 2022, pp. 2135–2140.
60. Liu, S., N. Kao, T. Shih, and Y. Wang, "Fan-Out Embedded Bridge Solution in HPC Application", *Proceedings of IEEE/EPTC*, December 2021, pp. 222–225.
61. Podpod, A., J. Slabbekoorn, A. Phommahaxay, F. Duval, A. Salahouedlhadj, M. Gonzalez, et al., "A Novel Fan-Out Concept for Ultra-High Chip-to-Chip Interconnect Density with 20-μm Pitch", *Proceedings of IEEE/ECTC*, May 2018, pp. 370–378.
62. Podpod, A., A. Phommahaxay, P. Bex, J. Slabbekoorn, J. Bertheau, A. Salahouelhadj, E. Sleeckx, A. Miller, G. Beyer, E. Beyne, A. Guerrero, K. Yess, and K. Arnold, "Advances in Temporary Carrier Technology for High-Density Fan-Out Device Build-up", *Proceedings of IEEE/ECTC*, May 2019 pp. 340–345.
63. Sharma, D., G. Pasdast, Z. Qian, and K. Aygun, "Universal chiplet interconnect express (UCIe): an open industry standard for innovations with chiplets at package level," *IEEE Trans. on CPMT*, Vol. 122, No. 9, Sept. 2022, pp. 1423–1431.
64. Lau, J. H., and T. J. Tseng, "CHIP PACKAGE STRUCTURE AND MANUFACTURING METHOD THEREOF", *Patent: TW 1848577*, filed on Oct. 20, 2022.
65. Anand, L., "Constitutive Equations for the Rate-Dependent Deformation of Metals at Elevated Temperatures", *ASME Transactions, Journal of Engineering Materials and Technology*, Vol. 104, January 1982, pp. 12–17.
66. Lau, J. H., and N. C. Lee, *Assembly and Reliability of Lead-Free Solder Joints*, Springer, New York, NY, 2020.
67. Lau, J. H., "State-of-the-Art of Lead-Free Solder Joint Reliability", *ASME Transactions, Journal of Electronic Packaging*, Vol. 143, June 2021, pp. 1–36.
68. Lau, J. H., C. Lin, N. Liu, H. Yang, T. Xia, C. Ko, B. Lin, Y. Chuang, R. Chen, M. Ma, T. Tseng, M. Li, and K. Leung, "Hybrid Substrate with Ultra-Large Organic Interposer for Heterogeneous Integration", *IEEE Transactions on CPMT*, Vol. 13, No. 9, September 2023, pp. 1371–1379.
69. Lau, J. H., G. Chen, C. Yang, V. Teng, A. Peng, J. Huang, N. Liu, Y. Chen, T. Tseng, and M. Li, "Hybrid Substrate for Heterogeneous Integration", *ASME Transactions, Journal of Electronic Packaging*, Vol. 146, June 2024, pp. 1–18.
70. Pang, J., *Lead Free Solder: Mechanics and Reliability*, Springer, New York, NY, 2012.

# Chapter 6
# Fatigue and Fracture Models for Electronic Packaging

**Abstract** This chapter details solid mechanics theories and empirical models relevant to electronic packaging reliability. Coverage includes material behavior (brittle/viscoplasticity/creep) and constitutive models like Chaboche, Anand, and Hyperbolic Sine. Reliability prediction focuses on solder fatigue using the Coffin-Manson, energy density models (Darveaux), and the modified energy density model. Brittle fracture criteria and interfacial delamination analysis using the Modified Virtual Crack Closure Technique (MVCCT) are also presented.

## 6.1 Introduction

Mechanics is a broad field; in this chapter, the discussion will focus specifically on the aspects of solid mechanics that are directly relevant to electronic packaging. The electronic packaging structure plays a vital role in protecting semiconductor devices, ensuring optimal device performance, and mitigating mechanical and thermal stresses during operation. Packaging structures are typically complex assemblies composed of diverse materials, including silicon, organic substrates, metals, and solder alloys. Each of these materials contributes differently to the mechanical behavior of the package. For example, silicon, a core material in integrated circuits, exhibits brittle fracture characteristics. In contrast, organic materials and solder joints, commonly used for electrical interconnects, display viscoelastic or viscoplastic behavior, which is strongly temperature-dependent.

A comprehensive understanding of each constituent material's and mechanical properties is essential to ensure electronic packages' mechanical integrity and long-term reliability under thermal cycling, vibration, and other environmental loads. This chapter will detail the relevant material behaviors, including elasticity, plasticity, nonlinearity, creep, stress relaxation, and temperature sensitivity, as they pertain to electronic packaging applications. Furthermore, critical failure mechanisms, such as solder joint reliability, interfacial delamination, and fracture, will be discussed, along with reliability simulation models. These models, including those for



J. Lau, K.-N. Chiang, *Cu-Interconnects, Glass, and AI-Assisted Simulation for Chiplets and Heterogeneous Integration*,
https://doi.org/10.1007/978-981-95-6891-8_6

solder liquid formation and damage accumulation, are commonly employed to analyze and design robust advanced electronic packaging structures.

## 6.2 Material Characteristic

Figure 6.1 illustrates the different stress-strain responses of materials under loading, showing how various materials behave mechanically when subjected to tensile forces. The x-axis represents strain (ε), which measures the deformation of the material relative to its original length, while the y-axis represents stress (σ), which is the force applied per unit area. The curve labeled "Brittle type" rises steeply, indicating high stiffness and strength but low ductility. This type of plastic can withstand high stress but breaks suddenly with slight deformation. Examples include silicon and glass, which fracture shortly after reaching their maximum stress.

The "Tough & Strong type" curve represents a material, e.g., ceramic, that can withstand high stress and exhibits some ductility before breaking. These materials typically have a higher yield strength and can absorb more energy before failure, making them suitable for applications where both strength and toughness are essential. This curve peaks and then drops slightly, indicating some plastic deformation before fracture. The "Strain Hardening Type" curve, e.g., solder, represents a less strong material than the previous one, but can undergo significant plastic deformation. It does not reach as high a stress level but stretches considerably, showing better toughness and ductility. After an initial yielding, the curve flattens out and

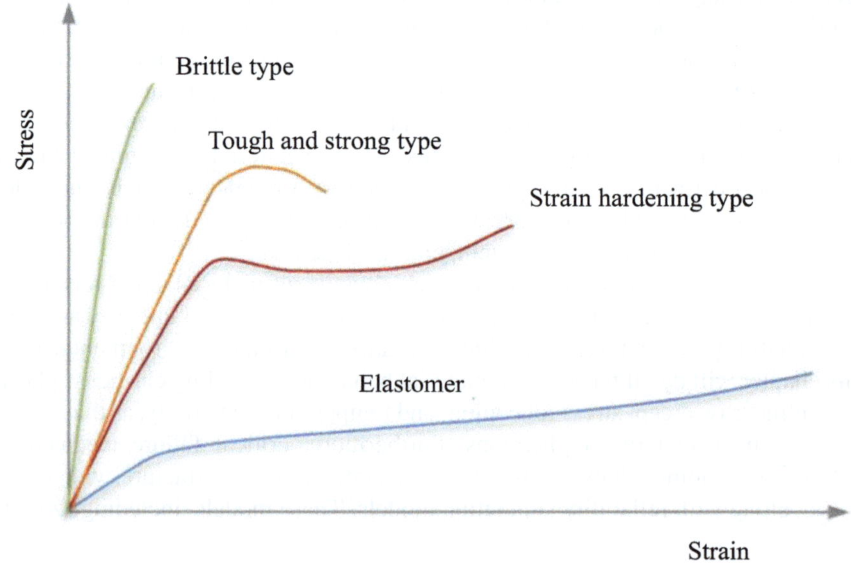


**Fig. 6.1** Material characteristics

rises again, indicating strain hardening, where the material becomes stronger as it is stretched further. At the bottom, the "Elastomer" curve, e.g., solder mask, underfill, is shown. This material, such as rubber, displays much strain under relatively low stress. Its gradual curve does not exhibit a sharp yield or fracture point, which is typical for elastomers. These materials can stretch extensively and return to their original shape, showing excellent elasticity but relatively low strength.

Overall, this graph compares the mechanical behavior of different material types. Brittle material breaks easily without stretching, while tough material can absorb more energy before failure. Elastomers, on the other hand, stretch significantly but don't carry much load. Understanding these characteristics is crucial for selecting the appropriate material for specific electronic packaging applications.

## 6.2.1 *Elasticity [1]*

Material elastic behavior refers to a material's ability to deform under an applied external force and then return to its original size and shape once that force is removed. It's a temporary deformation, meaning no permanent change occurs in the material's structure within its elastic limit. Figure 6.2 presents a typical material under tensile loading.

Figure 6.2 illustrates the stress-strain relationship of a material under tensile loading. The horizontal axis represents strain, which is the deformation or

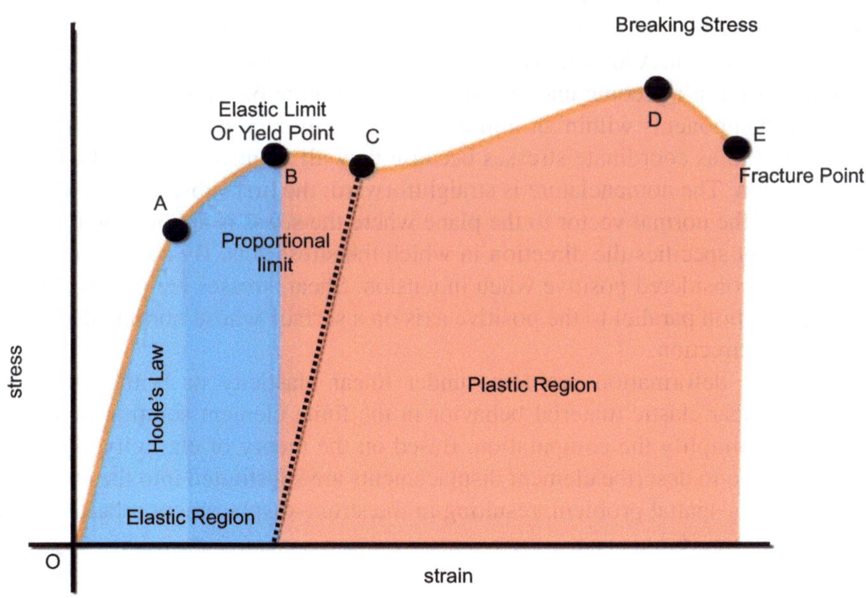


**Fig. 6.2** Stress-Strain curve under tensile loading

elongation of the material relative to its original length. In contrast, the vertical axis represents stress, which is the reaction force per unit area within the material. Initially, the material behaves elastically, as shown in the blue regions labeled the elastic region. The stress and strain are directly proportional within this region, a relationship described by Hooke's Law. This linear behavior continues up to point **A**, known as the proportional limit. Up to this point, if the applied force is removed, the material will return to its original shape without any permanent deformation.

As the stress increases beyond point **A**, the material still exhibits elastic behavior, but the linearity of the stress-strain curve ends. Point **B** marks the elastic limit, also known as the yielding point, which is the maximum stress the material can withstand while returning to its original shape when the load is removed. Once stress exceeds this yielding point, the material enters the plastic region (shown in orange), where permanent deformation occurs. This means that the material will not return to its original shape even if the force is removed.

In the plastic region, the material may continue to endure increasing stress and exhibit a phenomenon known as strain hardening, where it becomes stronger as it is deformed. This continues until it reaches point **D**, the ultimate tensile strength, which is the maximum stress the material can endure. Beyond this point, the material starts to weaken, and the stress required to continue deformation decreases until it finally reaches point **E**, the fracture point, at which the material breaks completely. Overall, this stress-strain curve is crucial for understanding how materials behave under various loads and for determining properties such as elasticity, yield strength, tensile strength, and ductility.

In mechanics, especially for brittle materials, the failure of a structure is fundamentally linked to stress. Specifically, a structure breaks when the applied or the resultant stress exceeds a critical limit. This means stress is currently the most crucial factor in design. Additionally, engineers often consider the strain to ensure that deformations don't become uncomfortably large. Figure 6.3 shown below displays the stress components within an infinitesimally small cube of a continuum. These are referred to as coordinate stresses because they all align with a specific coordinate direction. The nomenclature is straightforward: the first subscript identifies the direction of the normal vector to the plane where the stress is applied, and the second subscript specifies the direction in which the stress acts. By convention, axial stresses are considered positive when in tension. Shear stresses are positive if they act in a direction parallel to the positive axis on a surface whose normal also points in the same direction.

For small deformation problems under linear elasticity or brittle materials, assuming linear elastic material behavior in the finite element solution process is common to simplify the computation. Based on the theory of elasticity, the shape functions used to describe element displacements are substituted into the strain formulation for a spatial problem, resulting in the strain–displacement relationship as expressed in Eq. (6.1).

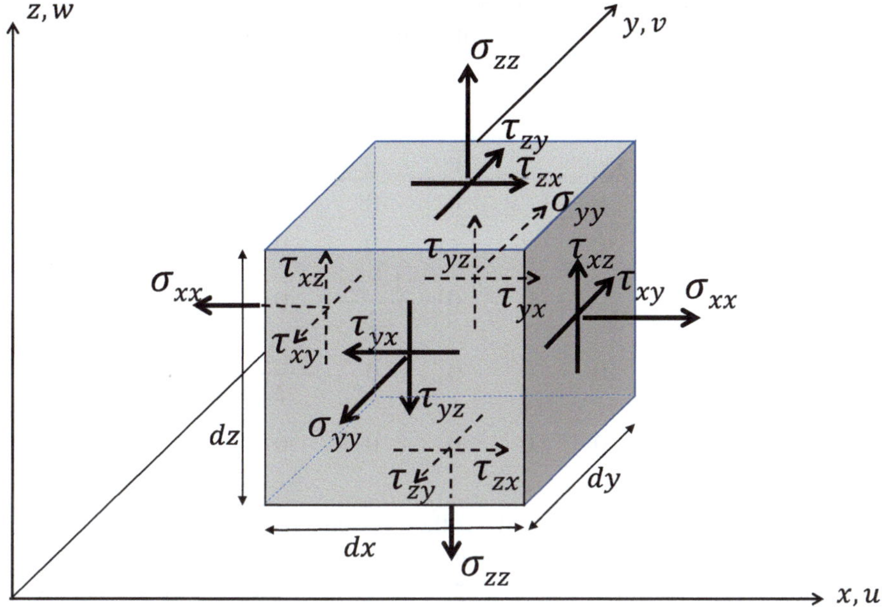


**Fig. 6.3** 3D stress components

$$\begin{Bmatrix} \varepsilon_x \\ \varepsilon_y \\ \varepsilon_z \\ \gamma_{xy} \\ \gamma_{yz} \\ \gamma_{xz} \end{Bmatrix} = \begin{bmatrix} \frac{\partial}{\partial x} & 0 & 0 \\ 0 & \frac{\partial}{\partial y} & 0 \\ 0 & 0 & \frac{\partial}{\partial z} \\ \frac{\partial}{\partial y} & \frac{\partial}{\partial x} & 0 \\ 0 & \frac{\partial}{\partial z} & \frac{\partial}{\partial y} \\ \frac{\partial}{\partial z} & 0 & \frac{\partial}{\partial x} \end{bmatrix} \begin{Bmatrix} u \\ v \\ w \end{Bmatrix} \tag{6.1}$$

where $\varepsilon_x$, $\varepsilon_y$ and $\varepsilon_z$ represent the normal strains and $\gamma_{xy}$, $\gamma_{yz}$ and $\gamma_{xz}$ represent the shear strains. The variables $u$, $v$, and $w$ denote the displacement components in the $x$, $y$, and $z$ directions, respectively, in the Cartesian coordinate system.

If the material is assumed to be isotropic, meaning that $E = E_x = E_y = E_z$ and $\nu_{xy} = \nu_{yz} = \nu_{xz}$, then the stress-strain relationship can be expressed by the following Equation:

$$\begin{Bmatrix} \sigma_x \\ \sigma_y \\ \sigma_z \\ \tau_{xy} \\ \tau_{yz} \\ \tau_{xz} \end{Bmatrix} = \frac{E}{(1+\nu)(1-2\nu)} \begin{bmatrix} 1-\nu & \nu & \nu & 0 & 0 & 0 \\ \nu & 1-\nu & \nu & 0 & 0 & 0 \\ \nu & \nu & 1-\nu & 0 & 0 & 0 \\ 0 & 0 & 0 & \frac{1-2\nu}{2} & 0 & 0 \\ 0 & 0 & 0 & 0 & \frac{1-2\nu}{2} & 0 \\ 0 & 0 & 0 & 0 & 0 & \frac{1-2\nu}{2} \end{bmatrix} \begin{Bmatrix} \varepsilon_x \\ \varepsilon_y \\ \varepsilon_z \\ \gamma_{xy} \\ \gamma_{yz} \\ \gamma_{xz} \end{Bmatrix} \tag{6.1a}$$

where $E$ is the Young's modulus and $\nu$ is the Poisson's ratio.

The principal stresses are the normal stresses acting on planes where the shear stress is zero. Let's use 2D as an example; in two-dimensional stress analysis, the principal stresses are the maximum and minimum normal stresses acting on a particular plane. The principal stresses $\sigma_1$ and $\sigma_2$ can be found using the following equations:

$$\sigma_{1,2} = (\sigma_x + \sigma_y)/2 \pm \sqrt{\left(\frac{\sigma_x - \sigma_y}{2}\right)^2 + \tau_{xy}^{\ 2}} \tag{6.2}$$

where $\sigma_x$ and $\sigma_y$ are the normal stresses in the x and y directions, $\tau_{xy}$ is the shear stress acting on the xy-plane. The principal stresses represent the extreme values of normal stress that a material element experiences, which is crucial for assessing the failure criteria of brittle materials. For example, principal stresses are the criterion for determining whether a silicon chip will be at risk of cracking.

### 6.2.2 Material Nonlinearity

Nonlinearities fall into three categories: material nonlinearity, geometric nonlinearity, and loading nonlinearity (such as follow forces). Material nonlinearity typically breaks down into plastic deformation and viscous behavior. When a material exhibits its plastic behavior, its stress has exceeded its yield strength. The material no longer behaves elastically at this point, and the strain won't return to zero when the external force is removed. We commonly refer to this as permanent strain, as Fig. 6.4 illustrates.

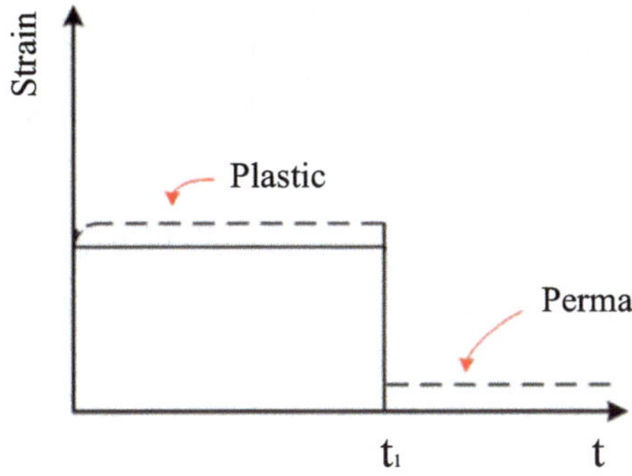


**Fig. 6.4** Plastic deformation vs. time

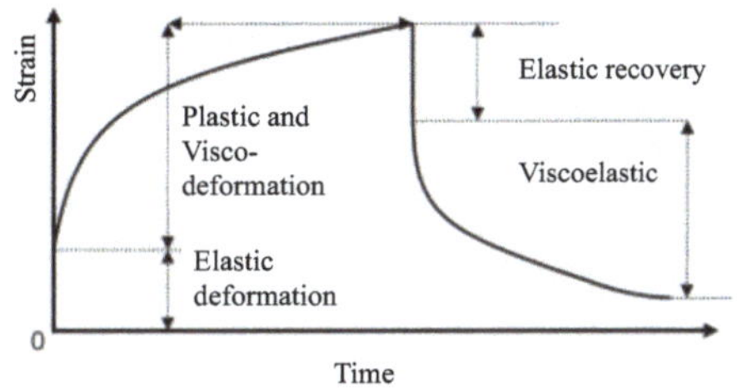


**Fig. 6.5** Viscoplasticity behavior

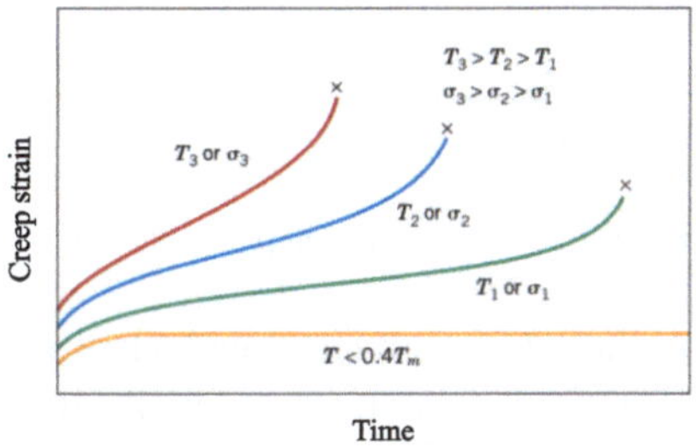


**Fig. 6.6** Creep strain under different stress and temperature vs. time

Viscous behavior can be classified into viscoelasticity and viscoplasticity, which are time-dependent and encompass phenomena such as creep and relaxation. This discussion focuses on viscoplasticity, as illustrated in Figs. 6.5 and 6.6, particularly the creep behavior of materials. Creep refers to the progressive deformation of a material under a constant load over time. As illustrated in Fig. 6.7, the blue solid line represents the strain evolution over time, while the black dashed line shows the strain rate as a function of time during stress application. The figure shows that in the primary creep stage, the strain rate decreases over time due to strain hardening. During the secondary (steady-state) creep stage, the strain increases linearly with time at a relatively constant rate. In the tertiary creep stage, the final stage, the strain rate accelerates rapidly, ultimately leading to fracture. This acceleration is caused by the formation and coalescence of internal micro-voids and cracks, which reduce the material's effective cross-sectional area, eventually leading to material failure.

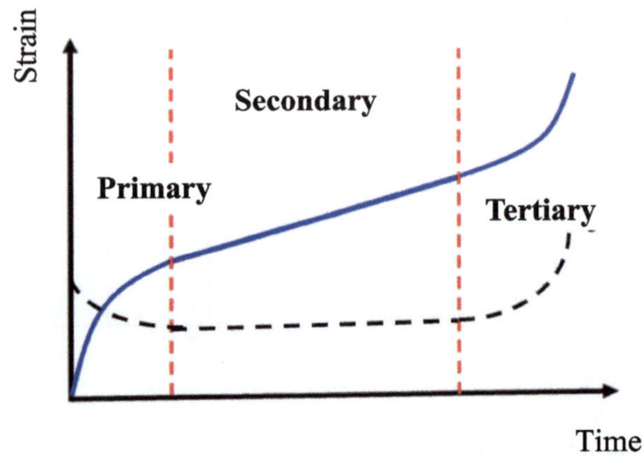


Fig. 6.7 Three stages of material creep behavior

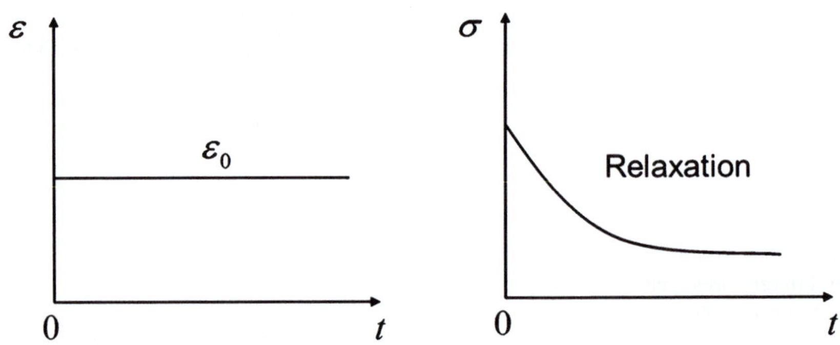


Fig. 6.8 Stress relaxation phenomenon

An initial stress is generated when a constant strain is applied to a material. However, over time, this stress gradually decreases. This phenomenon is known as stress relaxation, as Fig. 6.8 illustrates.

## 6.3 Strain Hardening

Plasticity refers to the ability of a solid material to undergo permanent, non-reversible deformation in response to applied forces; unlike elastic deformation, where a material returns to its original shape once the force is removed, plastic deformation results in a permanent change in shape. The material remains deformed even after the load is removed. The Chaboche model [2, 3] is a widely used constitutive equation for analyzing cyclic plasticity and viscoplasticity in electronic packaging. For example, simulating the mechanical behavior of electronic packaging under accelerated thermal cycling test (ATCT) conditions (Fig. 6.9, JEDEC Standard: JESD22-A140D, Test Condition B, −55 ~ 125 °C). Additionally, hardening rules

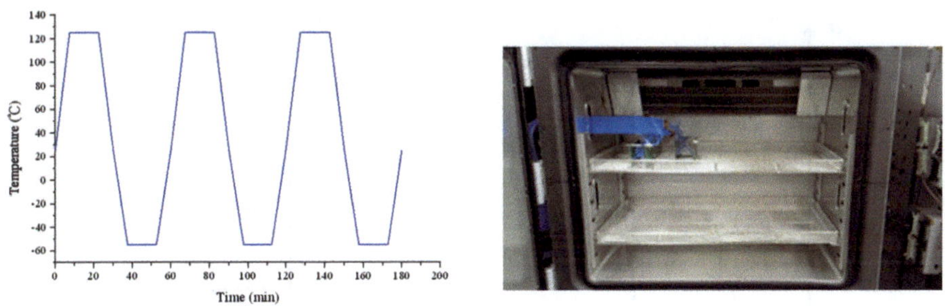


**Fig. 6.9** Accelerated thermal cycling test

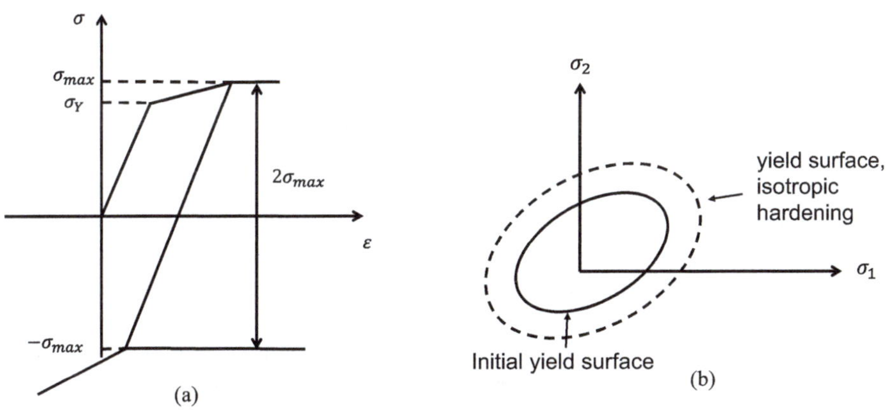


**Fig. 6.10** Isotropic hardening

must also be considered in the analysis. Hardening rules can be categorized into isotropic strain hardening and kinematic strain hardening. The kinematic strain hardening rule is preferred for cyclic loading conditions such as ATCT.

### *6.3.1 Isotropic Strain Hardening*

When a material hardens isotropically (Fig. 6.10), its "yield surface", which defines the boundary between elastic and plastic deformation in multiaxial stress space, expands uniformly in all directions. This uniform expansion means that the material's yield strength increases by the same amount regardless of the direction of subsequent loading. Following plastic deformation in tension, the yield surface undergoes an isotropic expansion such that an identical increase in compressive yield stress accompanies the increase in tensile yield stress. The yield surface does not shift its center or undergo shape distortion. It simply gets bigger. Isotropic hardening is generally a simpler model and is often used for monotonic loading when

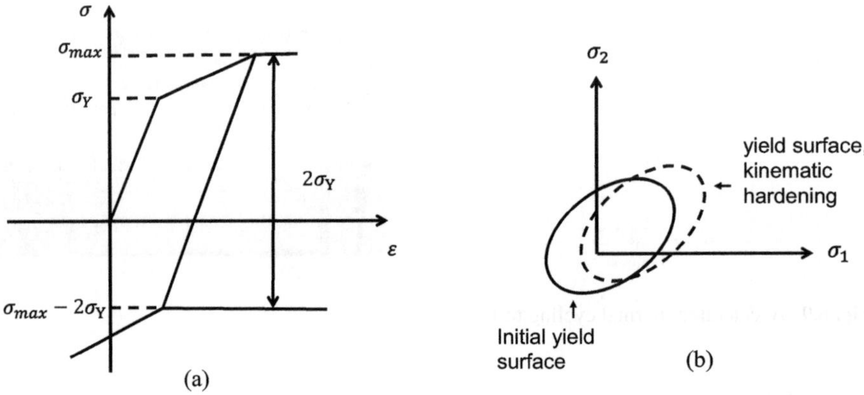


**Fig. 6.11** Kinematic hardening

the load is applied in one general direction and continuously increases (e.g., a simple tensile test to fracture).

While conceptually simple, isotropic hardening has a significant limitation: it does not account for the Bauschinger effect. The Bauschinger effect is a phenomenon observed in many metals where plastic deformation in one direction reduces the yield strength in the opposite direction. Since isotropic hardening predicts an increase in yield strength in all directions, it's often a poor choice for simulating materials subjected to cyclic loading, such as repeated tension-compression cycles, where the Bauschinger effect is prominent. For such cases, kinematic hardening is more appropriate.

### *6.3.2 Kinematic Strain Hardening*

Unlike isotropic hardening, which uniformly expands the yield surface, kinematic hardening (Fig. 6.11) shifts the yield surface without altering its yield size. This shift accounts for the Bauschinger effect, where materials exhibit different yield strengths in tension and compression after prior plastic deformation. Kinematic hardening is commonly employed in cyclic loading conditions, such as thermal cycling, where materials experience cycles of loading and unloading that repeatedly change the direction of stress. It helps predict how materials behave under repeated loading and unloading cycles, making it helpful in assessing the mechanical behavior of electronic packaging and in finite element simulations.

# 6.4 Material Constitutive Models

## 6.4.1 *Chaboche Constitutive Equation*

To accurately capture the nonlinear behavior of solder material such as SAC305 under varying temperature conditions, the Chaboche kinematic hardening model and the Anand/Hyperbolic sine viscoplastic models are widely adopted in finite element analysis. The Chaboche model [2, 3] describes cyclic plasticity without considering time-dependent effects. In contrast, the Anand and hyperbolic sine models focus on creep-dominated viscoplastic behavior, explicitly accounting for the influence of strain rate and temperature on the material response. This section provides a detailed introduction to these constitutive models.

The Chaboche model adopts the von Mises yield criterion, which is expressed by Eq. (6.3) as follows:

$$f(\sigma-\alpha,k)=\sqrt{(s-\alpha)\cdot(s-\alpha)}-k=0 \tag{6.3}$$

where $s$ is the deviatoric stress tensor, and $\alpha$ is the deviatoric back stress tensor, which represents the center of the yield surface. The parameter $k$ defines the initial size of the yield surface, with $k=\sqrt{\frac{2}{3}}\sigma_0$ indicating its initial state, and $\sigma_0$ is the initial yield stress in uniaxial tension. For dynamic hardening, $k$ is considered a constant. The evolution of plastic strain proceeds in the direction of the gradient of the yield surface, as expressed in Eq. (6.4):

$$d\varepsilon^p = d\lambda\frac{\partial f}{\partial \sigma} = d\lambda\frac{\overline{s}}{\sqrt{2\overline{J_2}}} \tag{6.4}$$

where $d\varepsilon^p$ is the incremental plastic strain, and $\lambda$ is the plastic multiplier, which is one of the hardening parameters. The term $\overline{s}=s-\alpha$ denotes relative deviatoric stress and $\sqrt{2\overline{J_2}}=\sqrt{\overline{s}\cdot\overline{s}}$ represents the equivalent deviatoric stress. The function $f$ denotes the yield function.

The back stress $\alpha$ in the Chaboche model can be represented as the sum of $n$ independent components, which can be mathematically expressed as:

$$\{\alpha\}=\sum_{i=1}^{n}\{\alpha_i\} \tag{6.5}$$

$$\{\Delta\alpha_i\}=\frac{2}{3}C_i\{\Delta\varepsilon^{pl}\}-\gamma_i\{\alpha_i\}\Delta\varepsilon_{eq}^{pl}+\frac{1}{C_i}\frac{dC_i}{d\theta}\Delta\theta\{\alpha_i\} \tag{6.6}$$

where $\Delta\varepsilon_{eq}^{pl}$ is the equivalent plastic strain increment, and $\{\Delta\varepsilon^{pl}\}$ denotes the plastic strain increment. $\{\alpha_i\}$ is the back stress tensor associated with the $i_{th}$ component, $\theta$ is the temperature-dependent variable. $C_i$ and $\gamma_i$ are material parameters.

Assuming an isothermal, uniaxial tensile loading condition, the temperature-dependent terms in Eq. (6.6) can be omitted, and the expression can be simplified as follows:

$$\Delta\alpha = \frac{2}{3}C_1\Delta\varepsilon^{pl} - \gamma_1\alpha\Delta\varepsilon_{eq}^{pl} \tag{6.7}$$

Under uniaxial tensile loading, it can be assumed that the plastic strain increment is approximately equal to the equivalent plastic strain increment ($\Delta\varepsilon^{pl} = \Delta\varepsilon_{eq}^{pl}$). Substituting this relationship into Eq. (6.7) and integrating yields the following expression:

$$\sigma = \frac{2}{3}\frac{C_1}{\gamma_1}\left(1 - e^{-\gamma_1\varepsilon_{eq}^{pl}}\right) + \sigma_0 \tag{6.8}$$

The tensile test data of SAC305 solder material were fitted to Eq. (6.8) to obtain the required parameters $C_1$ and $\gamma_1$. This fitting process yields a well-defined nonlinear equation that accurately describes the material's kinematic hardening behavior. Table 6.1 and Fig. 6.12 demonstrate the coefficients and material curve of SAC305.

**Table 6.1** SAC305 Chaboche hardening model parameters at different temperatures

| T(K) | $\sigma_0$ | $C$ | $\gamma$ |
|---|---|---|---|
| 233 | 47.64 | 8994.8 | 639.2 |
| 253 | 38.87 | 8573.3 | 660.0 |
| 313 | 24.06 | 6011.4 | 625.2 |
| 353 | 18.12 | 5804.2 | 697.7 |
| 398 | 14.31 | 4804.6 | 699.9 |

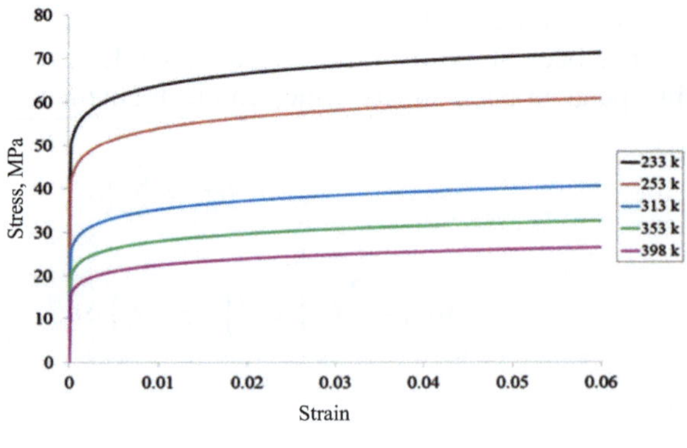


**Fig. 6.12** SAC305 Chaboche material curve at different temperatures

## 6.4.2 Creep Models

Creep is a time-dependent deformation of materials that occurs when they are subjected to a constant load or stress over a long period, especially at high temperatures. When a material is subjected to a constant load or stress, its behavior is influenced by time, strain rate, temperature, and stress, and these factors interact in complex ways. Material creep is one of the most significant factors affecting electronic package structures' long-term reliability and integrity, particularly under elevated temperature conditions. Creep deformation becomes a dominant failure mechanism when materials operate at a high homologous temperature, defined as the ratio of the operating temperature to the material's melting point in absolute terms, $T/T_m$, exceeding 0.3 to 0.4. For many solder alloys used in electronic packaging, such as Sn-Ag-Cu (SAC) formulations, this threshold is easily exceeded during standard operation, making them highly susceptible to creep. The solder balls can exhibit creep behavior when electronic components undergo thermal cycling, such as per JESD22-A104D condition G (which cycles from −40 to 125 °C). Notably, this creep can occur even at the lower end of the spectrum, around −40 °C.

The drive for miniaturization and increased power density in electronics exacerbates creep conditions. The solder joints are the primary failure points within an electronic package due to creep. In operation, electronic devices undergo thermal cycling due to power-on and power-off sequences, as well as fluctuations in ambient temperature. The mismatch in the Coefficient of Thermal Expansion (CTE) between the silicon die, the substrate, and other package components induces significant thermo-mechanical stress on the solder interconnects, e.g., in wafer-level packaging or flip-chip assemblies.

This cyclic stress and elevated operating temperatures cause the solder to deform via creep. The accumulated creep strain with each cycle leads to microstructural degradation, including grain coarsening and the initiation of microcracks. Over time, these cracks propagate through the joint, ultimately causing a complete electrical and mechanical failure. Closely related to creep is stress relaxation, a phenomenon in which the stress within a material diminishes over time under constant strain. The initial thermal stress is gradually converted into inelastic creep strain in a constrained solder joint. This relaxation alters the stress-strain behavior during a thermal cycle, often leading to a larger plastic deformation range and accelerating fatigue damage accumulation.

Creep is not exclusive to metallic alloys. Polymeric materials used in packaging, such as underfills and adhesives, also exhibit significant creep behavior. Underfills are dispensed between the die and the substrate to mechanically couple the two, thereby distributing stress away from the solder joints. However, the viscoelastic nature of underfill materials means they can also creep over time, which can lead to a loss of adhesion (delamination) or a reduction in their stress-distributing capability, re-exposing the solder joints to higher strain and increasing the risk of failure.

To ensure product durability, engineers rely on finite element analysis (FEA) to simulate and predict the long-term reliability of electronic packages. Empirical constitutive models are developed to capture solder's viscoplastic (rate-dependent) behavior accurately.

**Table 6.2** SAC305 Hyperbolic Sine model parameters

| Hyperbolic Sine Model [6] | | | |
|---|---|---|---|
| $\dot{\varepsilon} = A\left[\sinh\left(B\sigma\right)\right]^n e^{\frac{-Q}{RT}}$ | | | |
| $A(1/s)$ | $B(MPa^{-1})$ | $n$ | $Q(kJ/Mol)$ |
| 2631 | 0.0453 | 5.0 | 52.4 |

There are two models commonly used by electronic packaging researchers: the Garofalo-Arrhenius (Hyperbolic Sine, steady-state only) [4] and the Anand [8] (full range creep) creep models. The Hyperbolic Sine model describes the creep behavior only; however, the Anand model provides a framework for describing materials' plasticity and creep behavior at high temperatures without explicitly separating the two phenomena. By implementing the Anand or hyperbolic sine models in FEA simulations, engineers can obtain the accumulation of creep strain and strain energy density in solder joints under various thermal and power cycling scenarios. These results are then used to estimate the fatigue life of the package and guide reliability assessment decisions (Table 6.2).

#### 6.4.2.1 Garofalo-Arrhenius (Hyperbolic Sine) Creep Model

Figure 6.7 shows that when the material is subjected to creep, it undergoes three stages; the second-stage creep (also known as steady-state creep) is the focus of the Garofalo-Arrhenius creep theory. One of the main advantages of the hyperbolic sine model is its ability to accurately represent creep rates across a wide range of applied stress levels. At relatively low stress levels, the hyperbolic sine function approximates a power-law creep (similar to the Norton law), where the creep rate is proportional to the power of the stress. At high stress levels, the hyperbolic sine function exhibits an exponential dependence on stress. This exponential behavior is often observed in materials under high stress. It is commonly implemented in finite element analysis software, such as ANSYS, to simulate the long-term behavior of components subjected to high temperatures and stresses. In summary, the hyperbolic sine creep model is a powerful tool for engineers and material scientists to understand and predict the time-dependent deformation of materials in high-temperature environments, providing a balance between physical realism and simulation tractability.

Creep behavior is influenced by temperature and time and is related to the relationship between the material's melting point and the operating temperature. Homologous temperature $T_h$ is defined as an index, which is the ratio of the operation temperature $T$ and the material melting temperature $T_m$ using Kelvin temperature. When the value of homologous temperature is over a specific value, e.g., 0.3 to 0.4, the creep behavior needs to be considered. In the commercial standard accelerated thermal cycling test, the operating temperature ranges from 233 K to 398 K, and the melting temperature of the solder material Sn95.5Ag4.0Cu0.5 (SAC405) is approximately 490 K. Therefore, creep behavior can be considered in the finite element simulation as needed.

$$T_h = \frac{T}{T_m} \tag{6.9}$$

The creep equation is a function of temperature, strain rate, and stress. The Garofalo-Arrhenius equation is used as Eq. (6.10). In terms of stress-dependent creep, the hyperbolic sine function is used for description. The Garofalo-Arrhenius equation [4] is developed to describe the steady-state creep behavior of a solder alloy.

$$\dot{\varepsilon} = A\left[\sinh\left(B\sigma\right)\right]^n e^{\frac{-Q}{RT}} \tag{6.10}$$

$\dot{\varepsilon}$ : steady-state creep strain rate
*A, B*: material constant
$\sigma$: applied stress
*n*: stress exponent
*Q*: activation energy for creep (J/mol)
*R:* gas constant (8.314 J/mol·K)
*T*: absolute temperature in Kelvin
sinh: hyperbolic sine function

The hyperbolic sine model requires the determination of several material constants (*A, B, n, Q*) through extensive experimental creep testing, which can be time-consuming and costly. The value of "*n*" typically falls within the range of 5 to 7, corresponding to dislocation creep, as shown in the creep deformation map in Fig. 6.13. Figure 6.14 demonstrates the Stress-Strain curve of the Hyperbolic-Sine model for 96.5Sn-3.5Ag solder.

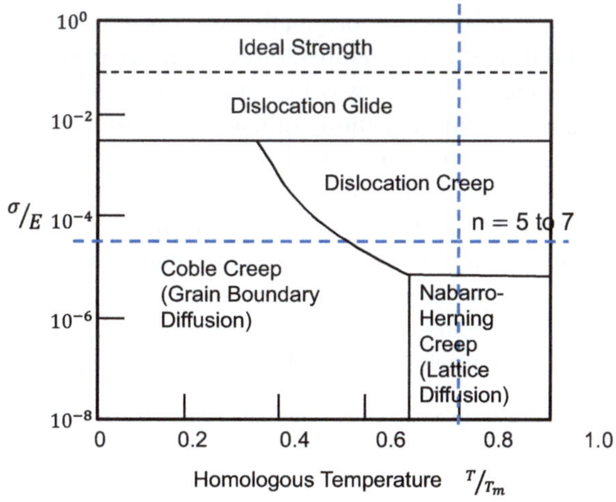


**Fig. 6.13** Creep deformation map [5]

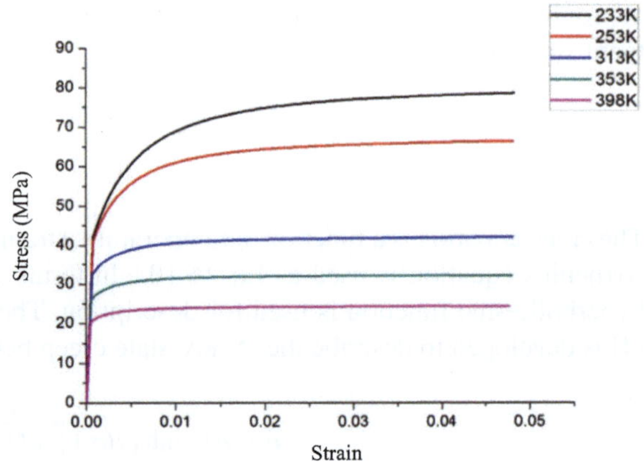


**Fig. 6.14** Stress-Strain curve of Hyperbolic-Sine model for 96.5Sn-3.5Ag solder [6]

### 6.4.2.2 Anand Creep Model

The Chaboche model effectively characterizes a material's cyclic plasticity, delivering stable and accurate results when thermal cycling is consistent. However, the model's inability to incorporate time-dependent deformation mechanisms, such as creep, is a limitation. In certain loading and application conditions, this becomes especially problematic for materials like SAC solder, such as SAC305, which are often exposed to elevated temperatures and sustained loading. Under these conditions, rate-dependent effects become prominent, and neglecting creep can lead to an incomplete or inaccurate representation of the material's behavior. To precisely capture the viscoplastic behavior of SAC solder, particularly during thermomechanical loading, activating the creep option within the simulation framework is essential. The Anand model [8], which encompasses three creep stages, primary, secondary, and tertiary creep, and incorporates both strain rate and temperature dependence, is introduced in this section as a suitable constitutive framework for modeling such behavior.

The Anand model, originally developed to describe the mechanical behavior of materials at elevated temperatures, typically ranging from 0.5 to 0.8 times the melting temperature ($T_m$), was first applied to high-strength aluminum and other structural alloys. Subsequent modifications by Brown [9] expanded its applicability to a broader range of materials. In simulating the thermomechanical behavior of lead-free solder joints in electronic packaging, effective use of the Anand model requires careful calibration of its material-specific constitutive parameters. This parameter identification process constitutes the central focus of many researchers. This constitutive framework adopts a unified formulation to simultaneously characterize solder joints' creep and plastic deformation behaviors, thereby establishing a fully viscoplastic model. Notably, it exhibits two key features [10]:

1. It eliminates the need for an explicit yield condition or distinct loading and unloading criteria.

2. It introduces a single internal state variable, the deformation resistance $s$, which encapsulates the average resistance to macroscopic plastic flow. This resistance reflects the influence of microstructural factors, including dislocation density, solid solution effects, and grain structure.

In the one-dimensional case under uniaxial loading, the equivalent stress is directly proportional to the deformation resistance. This relationship is expressed in Eq. (6.11).

$$\sigma = cs; c < 1; c = c\left(T, \dot{\varepsilon}_p\right) \tag{6.11}$$

where $c$ is a material-dependent parameter that reflects the influence of strain rate and temperature, $s$ is the deformation resistance, $T$ denotes the temperature, and $\dot{\varepsilon}_p$ is the plastic strain rate. Its explicit formulation is given in Eq. (6.12):

$$c = \frac{1}{\xi}\sinh^{-1}\left\{\left[\frac{\dot{\varepsilon}_p}{A}e^{\left(\frac{Q}{RT}\right)}\right]^m\right\} \tag{6.12}$$

where $A$ is the pre-exponential constant, $\xi$ represents the stress multiplier, $Q$ is the activation energy for deformation, $R$ is the universal gas constant, and $m$ characterizes the material's sensitivity to strain rate.

By combining Eqs. (6.11) and (6.12) and performing appropriate rearrangement, the general expression for the plastic strain rate $\dot{\varepsilon}_p$ can be derived. This formulation exhibits the typical characteristics of hyperbolic-type creep behavior and is known as the flow equation of the Anand model. The final form is presented in Eq. (6.13).

$$\dot{\varepsilon}_p = Ae^{-\left(\frac{Q}{RT}\right)}\left[\sinh\left(\xi\frac{\sigma}{s}\right)\right]^{1/m} \tag{6.13}$$

The differential Equation governing the evolution of the internal variable $s$ is shown in Eq. (6.14):

$$\dot{s} = h\left(\sigma, s, T\right)\dot{\varepsilon}_P \tag{6.14}$$

The function $h(\sigma, s, T)$ represents the hardening behavior, incorporating both strain hardening and dynamic recovery effects, and is explicitly given by:

$$h\left(\sigma, s, T\right) = h_0\left|\left(1-\frac{s}{s^*}\right)\right|^a \operatorname{sign}\left(1-\frac{s}{s^*}\right); a \geq 1 \tag{6.15}$$

In Eq. (6.15), $h_0$ denotes the hardening constant, while $a$ reflects the strain rate sensitivity associated with the hardening behavior. The term $s^*$ represents the saturation value of the internal variable $s$. By substituting Eq. (6.14) into Eq. (6.15), the resulting expression for $s^*$ is obtained as follows:

$$\dot{s} = h_0 \left| \left( 1 - \frac{s}{s^*} \right) \right|^a \text{sign}\left( 1 - \frac{s}{s^*} \right) \dot{\varepsilon}_p; \\ s^* = \hat{s} \left[ \frac{\dot{\varepsilon}_p}{A} e^{\left(\frac{Q}{RT}\right)} \right]^n \tag{6.16}$$

where $\hat{s}$ is the coefficient representing the saturation level of deformation resistance, and $n$ denotes the strain rate sensitivity exponent.

By combining Eqs. (6.11), (6.12), and (6.16), Eq. (6.17) can be derived. From Eq. (6.11), it follows that $\sigma^* = cs^*$ and $\sigma_0 = cs_0$, where $\sigma^*$ and $\sigma_0$ denote the saturation and initial values of the equivalent stress, respectively. In the one-dimensional case under uniaxial loading, $\sigma^*$ corresponds to the ultimate tensile strength (UTS), while $\sigma_0$ represents the yield stress.

$$\sigma^* = \frac{\hat{s}}{\xi} \left[ \frac{\dot{\varepsilon}_p}{A} e^{\left(\frac{Q}{RT}\right)} ) \right]^n \sinh^{-1} \left\{ \left[ \frac{\dot{\varepsilon}_p}{A} e^{\left(\frac{Q}{RT}\right)} \right]^m \right\} \tag{6.17}$$

Based on the proportionality $\frac{\sigma}{\sigma^*} = \frac{s}{s^*}$ Equations (6.11) and (6.12) can be used to derive the formulation presented in Eq. (6.18).

$$\frac{d\sigma}{d\varepsilon_p} = ch_0 \left| \left( 1 - \frac{\sigma}{\sigma^*} \right) \right|^a \text{sign}\left( 1 - \frac{\sigma}{\sigma^*} \right); a \geq 1 \tag{6.18}$$

Here, $\sigma_0 = cs_0$. $s_0$ represents the initial value of deformation resistance. Integrating the above Equation gives Eq. (6.19).

$$\sigma = \sigma^* - \left[ \left( \sigma^* - cs_0 \right)^{1-a} + (a-1)\left\{ \left( ch_0 \right) \left( \sigma^* \right)^{-a} \right\} \varepsilon_p \right]^{1/(1-a)} \tag{6.19}$$

At this point, all nine parameters required for the Anand constitutive model, along with the values for SAC305, are presented in Tables 6.3 and 6.4. Figure 6.15 demonstrates the Stress-Strain curve of the Anand model for 96.5Sn-3.5Ag solder.

Both the Anand model and the hyperbolic sine creep model are used to describe the time-dependent deformation of materials at elevated temperatures (creep) and have been widely adopted by electronic packaging researchers in finite element simulations. They have distinct approaches and, therefore, different advantages and

**Table 6.3** Nine parameters of the Anand model

| | Parameter | Description | Unit |
|---|---|---|---|
| 1 | $A$ | Pre-exponential factor | 1/time |
| 2 | $Q$ | Activation energy | Energy |
| 3 | $a$ | Strain rate sensitivity of hardening or softening | Dimensionless |
| 4 | $\xi$ | Multiplier of stress | Dimensionless |
| 5 | $m$ | Strain rate sensitivity of stress | Dimensionless |
| 6 | $h_0$ | Hardening/softening constant | Stress |
| 7 | $\hat{s}$ | Deformation resistance saturation value | Stress |
| 8 | $n$ | Strain rate sensitivity of saturation resistance | Dimensionless |
| 9 | $s_0$ | Initial value of deformation resistance | Stress |

**Table 6.4** SAC 305 Anand model parameters used in simulation [7, 11–13]

| Author | $S_o$(Mpa) | Q/R (k) | $A$ (1/s) | $\xi$ | $m$ | $h_o$(Mpa) | $\hat{S}$(Mpa) | $n$ | $a$ |
|---|---|---|---|---|---|---|---|---|---|
| Chang et al. | 45.9 | 7460 | 5,870,000 | 2 | 0.094 | 9350 | 58.3 | 0.015 | 1.5 |
| Herkommer et al. | 1.06 | 10,413.3 | 1,430,000 | 1.47 | 0.141 | 5023.9 | 20.297 | 0.0324 | 1.120 |
| Mysore et al. | 2.15 | 9970 | 17.994 | 0.35 | 0.153 | 1525.98 | 2.536 | 0.028 | 1.69 |
| Motalab et al. | 21 | 9320 | 3501 | 4 | 0.25 | 180,000 | 30.2 | 0.01 | 1.78 |

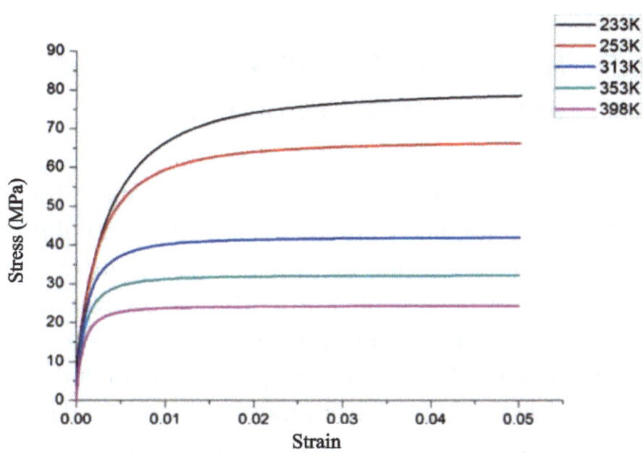


**Fig. 6.15** Stress-Strain curve of Anand model for 96.5Sn-3.5Ag solder [7].

disadvantages. The hyperbolic sine creep model is relatively simple in its mathematical formulation, making it computationally less expensive to implement in finite element simulations. Due to its simplicity and effectiveness for steady-state creep, it has been widely adopted, and its parameters are often available in material databases.

The Anand model is a unified viscoplastic constitutive model. It describes both time-independent plasticity and time-dependent creep as a single, unified phenomenon without separate yield criteria or loading and unloading conditions. A major advantage is its unified approach to plasticity and creep. A single set of equations can describe material behavior under a wide range of strain rates, from quasi-static plasticity to long-term creep. This is particularly useful for materials like solders exhibiting significant viscoplastic behavior. Similar to the hyperbolic sine model, it has proven to be particularly well-suited and widely adopted for modeling the behavior of solder alloys in electronic packaging due to their significant viscoplasticity at operating temperatures. However, the Anand model typically requires nine material parameters compared to a simple hyperbolic sine creep model (4 parameters). Calibrating these parameters accurately from experimental data can be challenging and requires comprehensive tests. The complexity of the equations and the evolution of the internal variable can sometimes lead to numerical challenges and convergence issues in finite element simulations, especially under severe loading conditions.

In summary, if the focus is primarily on predicting steady-state creep and computational efficiency is crucial, the hyperbolic sine creep model is recommended for application. However, for materials exhibiting significant viscoplasticity and complex loading histories, and where capturing both plasticity and all stages of creep is important, the Anand model offers a more comprehensive and robust solution despite its increased complexity and parameter requirements. Both models can accurately predict solder fatigue life for many studies if the finite element mesh size is appropriately selected.

## 6.5 Solder and Traceline Fatigue Life and Fracture Prediction

In electronic packaging, predicting the fatigue life cycles of components, particularly solder joints, is crucial for ensuring reliability and performance. Electronic packages are subjected to various stresses, most commonly thermal cycling due to power on/off cycles, operation/idle or environmental temperature fluctuations, as well as mechanical vibrations and shocks. Some key life cycle prediction models commonly used in electronic packaging include the strain-based Coffin-Manson model, the total strain-life model, the energy density-based Darveaux model, and the modified energy density empirical model. Fatigue is a progressive, localized structural damage that occurs when a material is subjected to repeated or fluctuating stresses or strains. It occurs over time and through many cycles, often initiating microcracks at stress concentration points, and has distinct stages: crack initiation, crack propagation, and final fracture. In a broader sense, failure refers to any situation where a material or component no longer fulfills its intended purpose. This can occur through various mechanisms, including, but not limited to, fatigue. It can occur suddenly or over time, including complete breakage, excessive deformation, or wear.

In materials science and engineering, fracture refers to the separation or fragmentation of a solid material into two or more pieces due to applied stress. It's a fundamental mode of failure in brittle materials such as silicon and glass. Fracture typically occurs when the applied stress on a material, especially at a localized point (like a crack or defect), exceeds the material's ability to resist separation. Brittle fracture typically occurs when a material breaks or fractures with little or no macroscopic plastic deformation. Cracks propagate very rapidly, often at speeds approaching the speed of sound in the material. The fracture surface often appears shiny and flat. In crystalline materials, atomic bonds break along specific crystal planes; in amorphous materials (like glass), it's more random. Brittle fracture absorbs very little energy, and even microscopic flaws can significantly reduce a material's fracture strength. This mechanism makes it a dangerous mode of failure as it can be sudden and catastrophic, with no prior warning. For materials that fail due to brittle fractures, the ***maximum normal stress theory*** is the most suitable theory to apply.

The Coffin-Manson model is by far the most widely used model for predicting the fatigue life of ductile materials, especially for solder joints subjected to thermal cycling. It's a strain-based model focusing on the plastic deformation experienced by the material during each cycle. The original Coffin-Manson model primarily considers low-cycle fatigue (LCF), where plastic strain dominates. The total strain-life model is a well-known modification for trace line fatigue, combining the Coffin-Manson and Basquin equations to provide a comprehensive model for fatigue life across both low-cycle fatigue (LCF) and high-cycle fatigue (HCF) regimes. Allows for more comprehensive fatigue analysis, especially when the transition between plastic and elastic dominance is unclear or when components experience varying load amplitudes. Often employing total strain and accounting for the mean temperature and cycle frequency, it serves as the primary model for the reliability assessment of power modules. Energy density-based models postulate that fatigue damage is related to the dissipated strain energy per cycle, typically the creep or inelastic strain energy density. They can be more robust for complex loading conditions, such as strain-rate-sensitive ones.

Increasingly, data-driven approaches using machine learning algorithms (e.g., Artificial Neural Networks, Random Forests, Support Vector Regression) are being employed. These models learn complex relationships between input parameters (e.g., material properties, geometry, loading conditions) and observed reliability life cycle data. Machine learning and AI models can handle complex, multi-factorial datasets and potentially uncover nonlinear relationships that traditional physics-based empirical models might struggle to handle. Particularly useful for predicting life for new material combinations or complex package architectures where empirical models are difficult to establish.

The choice of the life prediction model for electronic packaging depends heavily on the specific failure mechanism being addressed, the material system (especially the type of solder alloy), the type of loading (thermal, mechanical, or combined), and the desired level of accuracy versus computational effort. A combination of these models and methodologies, coupled with experimental validation and finite element analysis, often provides the most reliable life assessment for electronic packages.

### 6.5.1 Coffin-Manson Model

The Coffin-Manson model [14, 15] is a widely used empirical relationship that describes materials' low-cycle fatigue (LCF) behavior, particularly metals and solder joints. It establishes a relationship between the plastic strain amplitude experienced by a material and the number of cycles to failure under cyclic loading. LCF, this specific type of fatigue occurs when a material is subjected to high stress or strain levels that cause significant plastic (permanent) deformation in each cycle. This typically fails a relatively low number of cycles (often less than 10,000). In contrast, high-cycle fatigue (HCF) involves lower stress or strain levels and primarily elastic deformation, leading to failure after many cycles.

The Coffin-Manson model was initially developed to model thermally induced metal fatigue failures and is widely used for this purpose, particularly in electronic components such as solder joints. When electronic devices are on/off or operating/idle, they undergo temperature fluctuations, resulting in thermal expansion and contraction. If different materials in an assembly have different coefficients of thermal expansion, these temperature changes can induce significant stresses and plastic strains, leading to fatigue failure. The model is particularly valuable in accelerated thermal cycling testing for predicting the lifespan of components under normal operating conditions based on tests conducted at more extreme (accelerated) conditions. Engineers can induce failures more quickly by subjecting products to more rapid and larger temperature cycles, and the acceleration factor model can be used to extrapolate the expected life under typical use conditions. The ATCT can be performed by finite element simulation and coupling with an empirical life cycle prediction model. One of the most common and critical applications of the empirical Coffin-Manson model is assessing solder joints' reliability in electronic packaging. Solder joints are highly susceptible to thermal fatigue due to the coefficient of thermal expansion mismatches between the components they connect. The constants in the Coffin-Manson equation ($\alpha$ and $n$) are material- and geometry-dependent, which can be determined by "life cycles" through fatigue experiment tests and "strain" by simulation. These values provide insights into a material's resistance to plastic deformation and its fatigue behavior.

Incremental equivalent plastic strain is a key parameter in the Coffin-Manson model. It represents the magnitude of the irreversible (plastic) deformation that occurs during each loading cycle. When a material undergoes plastic deformation, its internal structure changes, and this accumulation of damage leads to fatigue. The core of the Coffin-Manson model is an empirical power law relationship, often expressed as:

$$N_f = \alpha\left(\Delta\varepsilon_{eq}^{pl}\right)^{-n} \tag{6.20}$$

Number of Cycles to Failure ($N_f$): This is the total number of load cycles a material can withstand before it fails due to fatigue. Where the empirical constants $\alpha$ is 0.235, and $n$ is 1.75 for SAC305, $\Delta\varepsilon_{eq}^{pl}$ represents incremental plastic strain and can be expressed as follows:

$$\Delta\varepsilon_{eq}^{pl} = \frac{\sqrt{2}}{3}\sqrt{\left(\Delta\varepsilon_{x}^{pl} - \Delta\varepsilon_{y}^{pl}\right)^2 + \left(\Delta\varepsilon_{y}^{pl} - \Delta\varepsilon_{z}^{pl}\right)^2 + \left(\Delta\varepsilon_{z}^{pl} - \Delta\varepsilon_{x}^{pl}\right)^2 + \frac{3}{2}\left(\Delta\gamma^{pl}\right)}$$
$$\Delta\gamma^{pl} = \left(\Delta\gamma_{xy}^{pl}\right)^2 + \left(\Delta\gamma_{yz}^{pl}\right)^2 + \left(\Delta\gamma_{xy}^{pl}\right)^2 \tag{6.21}$$

where $\Delta\varepsilon_{x}^{pl}$, $\Delta\varepsilon_{y}^{pl}$ and $\Delta\varepsilon_{z}^{pl}$ are incremental plastic strains in $x$, $y$, and $z$ directions. $\Delta\gamma_{xy}^{pl}$, $\Delta\gamma_{yz}^{pl}$, and $\Delta\gamma_{xy}^{pl}$ are incremental plastic shear strains in $xy$, $yz$, and $xz$ directions.

## 6.5.2 *Energy Density Model*

When creep analysis becomes necessary, the Darveaux creep model [16–21] and the energy density model [21–23] are highly influential and widely used models. They are especially valuable because they consider the complex creep behavior of solders, which are soft materials that exhibit significant time-dependent deformation even at relatively low temperatures (high homologous temperatures). Creep is the time-dependent plastic deformation of a material under constant stress. Solder alloys are highly prone to creep, particularly at operating temperatures that represent a large fraction of their melting points. This means that the solder will gradually deform over time, even if the stress is below the yield strength. This creep deformation contributes significantly to fatigue damage in solder joints.

### 6.5.2.1 Darveaux Creep Model

The Darveaux model uses a viscoplastic constitutive model to accurately describe the complex stress-strain-temperature-time behavior of solder alloys, including both time-independent plasticity and time-dependent creep. This is crucial because the response of the solder is highly dependent on temperature and strain rate. Instead of solely relying on strain, the Darveaux model typically uses volume-averaged inelastic (or viscoplastic) strain energy density (ΔW) as the driving force for damage accumulation. This parameter represents the energy dissipated within the solder joint per thermal cycle due to plastic deformation and creep. Darveaux proposed that solder joint failure can be divided into two stages: Crack initiation ($N_0$), which is the number of cycles of initial crack. This is related to the inelastic work density. Crack propagation (*da/dN*): The rate at which the initiated crack grows per cycle. This is also related to the inelastic work density. The total number of cycles to failure is then calculated by combining the crack initiation life and the crack propagation life, considering a characteristic crack length ($a$).

The Darveaux model is heavily reliant on finite element analysis. Numerical simulations are used to model the geometry and material properties of the electronic assembly (chip, board, solder joints). Apply the thermal cycling loading profiles to calculate the inelastic strain or creep energy density within the solder joints, which is then used in the damage equations. The mesh size dependency is a critical aspect of using the Darveaux model with FEA. The inelastic strain energy density near stress/strain concentrations (like the corner of a solder joint) can highly depend on the FEA model's element size. Selecting the appropriate mesh size in critical regions is crucial to ensure consistent results. Implementing the Darveaux model requires significant expertise in FEA, materials characterization, mechanics theories, domain knowledge, and understanding of the model's specific parameters. Furthermore, FEA simulations can be computationally intensive, especially with detailed visco-plastic models.

The equations of the Darveaux model can be expressed as follows:

$$N_0 = K_1 \left(\Delta W_{inelastic}\right)^{K_2} \tag{6.22}$$

$$\frac{da}{dN} = K_3 \left(\Delta W_{inelastic}\right)^{K_4} \tag{6.23}$$

where $N_0$ is the initial crack cycle, $a$ is the characteristic crack length, and $\Delta W_{inelastic}$ is the incremental inelastic energy density. $K_1 = 13.77$, $K_2 = -1.64$, $K_3 = 1.708 \times 10^{-3}$, and $K_4 = 1.04$, based on the WLCSP [24]. In sum, the characteristic life (63.2%) $N_f$ is calculated as follows:

$$N_f = N_0 + \frac{a}{da/dN} = N_0 + N_1 \tag{6.24}$$

where $N_1$ is called the crack propagation life.

In summary, the Darveaux creep model offers a robust framework for predicting the thermo-mechanical fatigue life of solder joints by integrating advanced material constitutive models with a two-stage damage accumulation approach driven by inelastic strain energy density. In electronic packaging reliability testing, Daisy Chain structures are commonly used to monitor the electrical resistance of interconnects (e.g., solder joints). When the measured resistance exceeds a predefined threshold, it's typically interpreted as an open circuit or interconnect failure, indicating that at least one joint in the chain has experienced severe damage or cracking. Testers generally cannot determine when an initial crack occurred based solely on resistance measurements, as early-stage (micro) cracks typically do not interrupt the electrical path, especially in ductile materials like SAC305. There's still enough contact area to maintain low resistance. The resistance increase due to an initial crack is often in the micro-ohm range, which is far below the measurement resolution or within the noise level, especially when the total chain resistance may be in the milliohm to ohm range.

Daisy Chain testing cannot determine when an initial crack forms; it only detects failure after the resistance increases significantly. Therefore, it's a reliable electrical indicator of functional failure but not a feasible diagnostic tool for detecting the very early stages of damage. Therefore, for advanced packaging such as wafer-level packaging, 3D, and system-in-package, etc., the initial crack in the Darveaux model is difficult to verify.

#### 6.5.2.2 Modified Energy Density Model

The Modified Energy-Based Method [21–23] as a fatigue life prediction model for electronic packaging. Its mathematical form is similar to that of the Coffin-Manson strain-based approach, with the key difference being the substitution of the incremental inelastic strain energy density obtained from simulation. This relationship is expressed in Eq. (6.25):

$$N_f = C\left(\Delta W_{inelastic}\right)^{-n} \tag{6.25}$$

where $N_f$ represents the fatigue life of the solder joint. $C$ and $n$ are fitting parameters obtained from experimental data. $\Delta W_{inelastic}$ denotes the increment of inelastic strain energy density per thermal cycle, which accounts for both creep and plastic deformation. This value can be obtained from a finite element simulation.

The modified energy-based empirical reliability life prediction model is usually used in conjunction with the Chaboche, Anand, or hyperbolic sine constitutive models. When considering the creep behavior only, it can be expressed in Eqs. (6.26), where $C = 192$ and $n = 1.8$.

$$N_f = C\left(\Delta W_{creep}\right)^{-n} \tag{6.26}$$

The modified energy-based model demonstrates a good prediction of the reliability life of advanced packaging. It is more straightforward than Darveaux's model and is suitable for newer lead-free solder alloys (e.g., SAC305, SAC405, and SAC-Q).

### *6.5.3 Total Strain and Cumulative Damage Model*

In the electronic power module's power cycling test, Figs. 6.16 and 6.17, the current-induced heat will generate thermal stress, causing fatigue fracture after tens of thousands or millions of cyclic load currents. The fatigue life covers an extensive range. Low-cycle fatigue generally corresponds to a fatigue life between $10^4$ and $10^5$ cycles, while high-cycle fatigue is associated with fatigue life exceeding $10^5$ cycles.

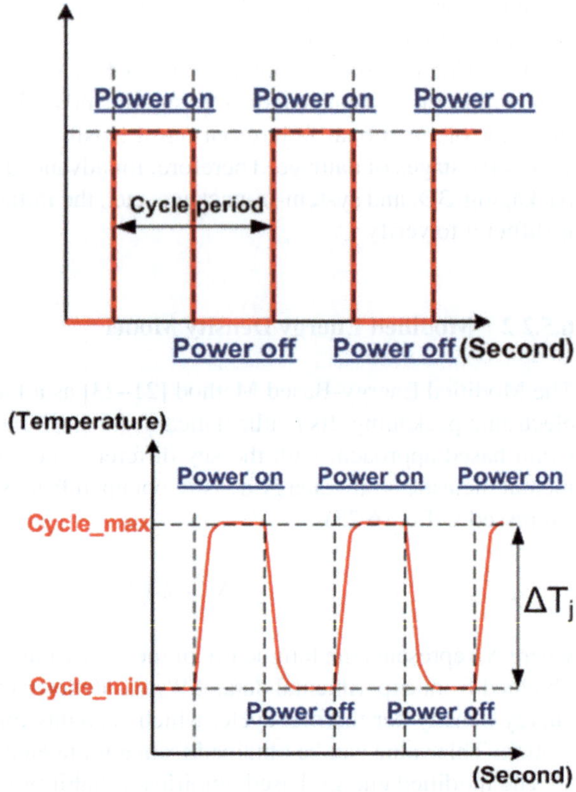


**Fig. 6.16** Schematic of a typical power cycle and junction temperature history [25]

High-cycle fatigue typically occurs within the elastic strain range, where plastic deformation is minimal and does not significantly impact the fatigue process. In contrast, low-cycle fatigue involves substantial plastic strain in each loading cycle, allowing for a more detailed examination of localized yielding, particularly in areas with stress concentrations.

Traditionally, a material's fatigue behavior is represented using an S-N curve, which plots cyclic stress amplitude (S) against the number of cycles to failure (N), as illustrated in Fig. 6.18. For ferrous alloys, if the applied stress falls below a certain threshold, the material can withstand an essentially infinite number of cycles without failure. This threshold is known as the fatigue limit or endurance limit. In contrast, non-ferrous alloys generally do not exhibit a distinct fatigue limit. While various mathematical relationships can describe the S-N curve, they typically reflect statistical trends due to variability in material behavior. The material's ability to resist failure under repeated stress loading, as depicted in the S-N curve, is referred to as its fatigue strength.

The strain-life relationship has been established for a wide range of engineering metals. Morrow and Dowling [26] demonstrated that fatigue strength could be characterized by combining both the elastic and plastic components of strain, referred to as the total strain range (TSR), and the fatigue life is related to the total strain amplitude $\Delta\varepsilon_t/2$ as:

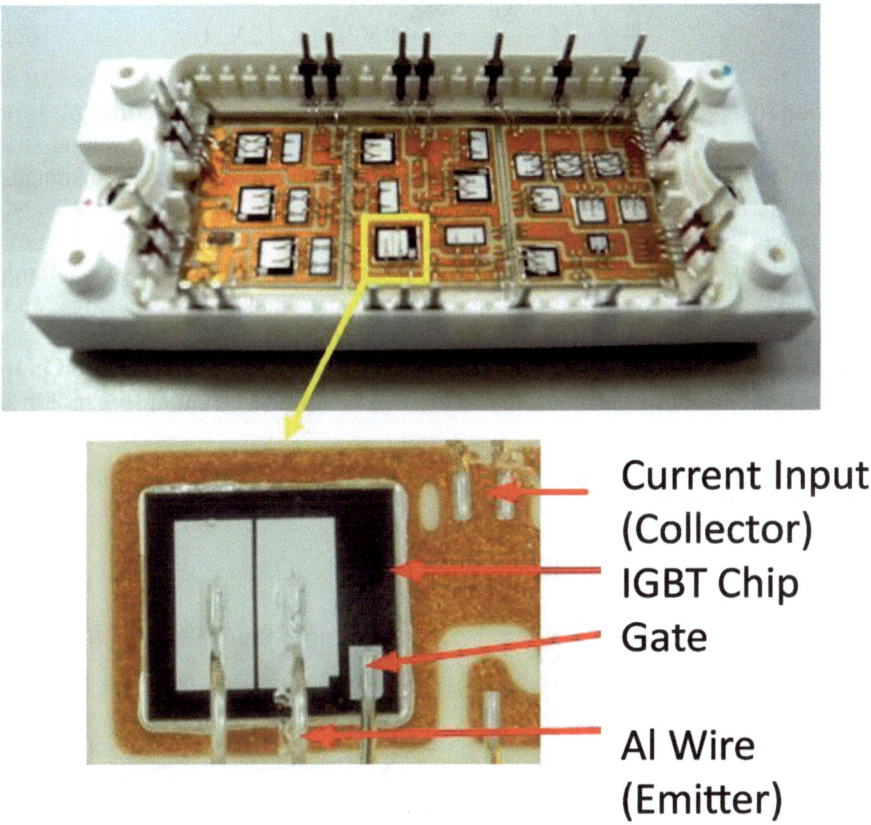


**Fig. 6.17** Power module sample

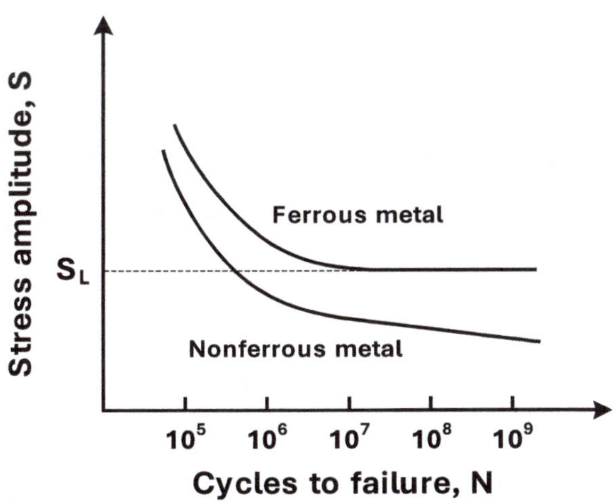


**Fig. 6.18** S-N curve, stress-cycles to failure curve

$$\frac{\Delta\varepsilon_t}{2} = \frac{\Delta\varepsilon_e}{2} + \frac{\Delta\varepsilon_p}{2} = \frac{\sigma_f'}{E}\left(2N_f\right)^b + \varepsilon_f'\left(2N_f\right)^c \tag{6.27}$$

Equation (6.27) indicates that at high total strain amplitudes, fatigue life ($N_f$) is primarily governed by the plastic strain component ($\Delta\varepsilon_p$), while at low total strain amplitudes, it is dominated by the elastic strain component ($\Delta\varepsilon_e$), as illustrated in Fig. 6.19 [27].

In power cycling tests, cyclic stress can lead to crack initiation and propagation due to localized cyclic plastic deformation, ultimately resulting in solder fatigue failure. As cracks grow, the stress amplitude at the crack tip may change, and a range of stress amplitudes significantly affects fatigue behavior. Therefore, accurately predicting fatigue life under varying stress amplitudes, especially in power cycling tests driven by current-induced thermal loading, is essential (Fig. 6.20).

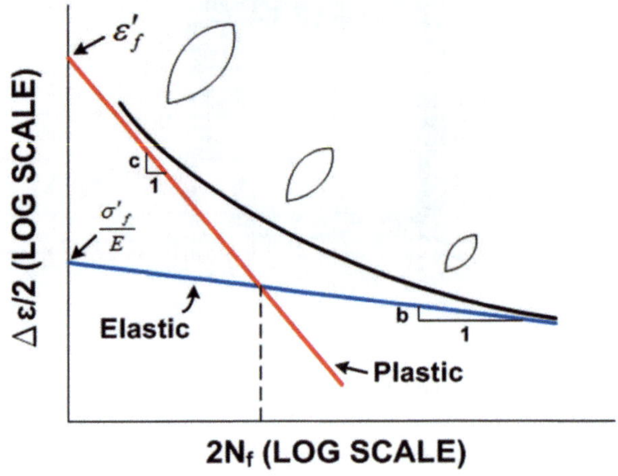


**Fig. 6.19** Superposition of elastic and plastic curves in terms of total strain [26]

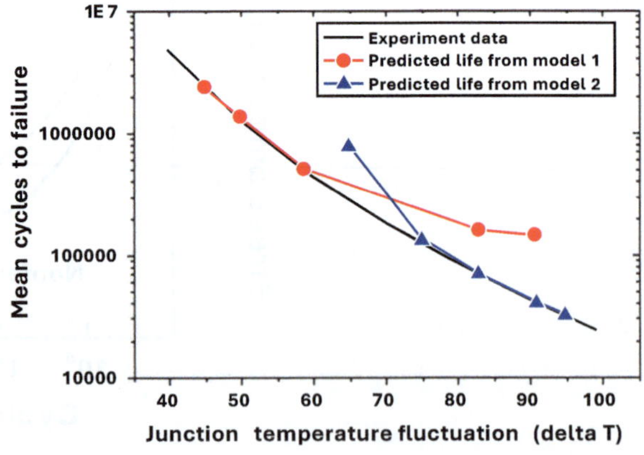


**Fig. 6.20** Bonding wire of power module mean-time-to-failure vs. deltaT [25]

To address this, cumulative damage theories have been developed. These models estimate damage accumulation at different stress levels by assuming that each stress cycle causes permanent damage and that the total damage is the sum of damage increments from individual stress amplitudes. As each fluctuating stress contributes to material degradation, the cumulative damage grows until it reaches a failure threshold. According to these theories, fatigue behavior reflects the progressive exhaustion of a material's inherent life.

Figure 6.21 shows the schematic of the cumulative damage theory for each of the different constant-stress tests. The number of cycles to failure is 300 at stress amplitude $\sigma_1$, whereas it is 600 at stress amplitude $\sigma_2$.

Based on Fig. 6.21, the total life of loading conditions $\sigma_1$ and $\sigma_2$ are 300 and 600 cycles, respectively. According to cumulative damage theory, fatigue life is gradually consumed throughout the stress history (e.g., from point A to B or C to D). For example, if one-third of the fatigue life is depleted during the stress history from A to E under a given stress amplitude, it implies that $\sigma_1$ loading condition has caused one-third of the total damage. If the stress amplitude is then changed from its initial value to a new one, $\sigma_2$, the damage contribution of the new amplitude can be considered equivalent in effect to the previous one, allowing for a direct comparison. Consequently, after the subsequent loading phase from E to F, two-thirds of the fatigue life (e.g., 400 cycles) is estimated to remain, based on the cumulative damage accumulated during the entire loading sequence.

The cumulative damage theory seeks to explain and predict fatigue life under fluctuating stress amplitudes. Palmgren introduced it, later formalized by Miner [28, 29] as the linear damage rule, commonly known as the Palmgren-Miner hypothesis. Over the stress history, each of the stress levels $S_i$ can cause damage fraction $D_i$, and failure is determined as:

$$D_1 + D_2 + \cdots + D_i \geq 1 \tag{6.28}$$

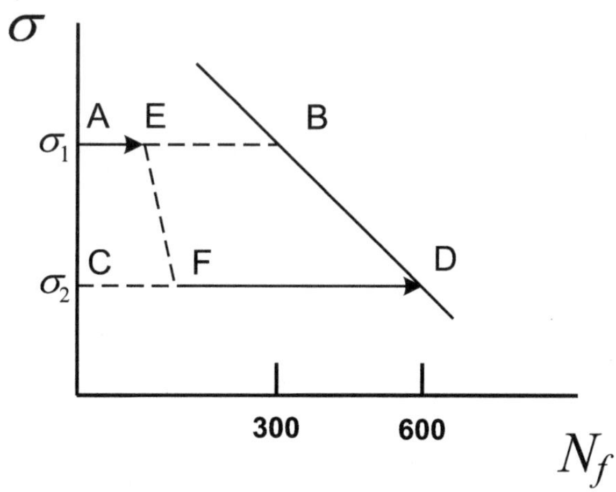


**Fig. 6.21** Damage accumulation in high-to-low load order

According to the linear damage rule, the damage fraction at any stress level $S_i$ is linearly proportional to the ratio of the operating number of cycles $n_i$ to the number of cycles to failure $N_i$:

$$D_i = \frac{n_i}{N_i} \tag{6.29}$$

Combining this with Eq. (6.28) yields a more general equation that applies to spectrum loading:

$$\sum_{i=1}^{m} \frac{n_i}{N_i} = 1 \text{ or } \frac{n_1}{N_1} + \frac{n_2}{N_2} + \cdots + \frac{n_m}{N_m} = 1 \tag{6.30}$$

where $m$ is the number of stress levels in the loading spectrum; $N_i$ is the number of cycles to failure corresponding to stress level $S_i$; and $n_i$ is the actual number of cycles performed at stress level $S_i$.

In linear damage theory, the influence of prior loading history on the damage accumulation rate during subsequent loading is neglected. Each stress level is assumed to contribute to damage independently, regardless of the loading sequence. As a result, the predicted fatigue behavior is the same whether the material is subjected to a low-to-high or high-to-low stress order.

### 6.5.4 Engelmaier Fatigue Life Prediction Model

For copper in electronic packaging structures, such as electrodeposited or rolled copper foil, Engelmaier [30–32] proposed a new method for ductility and flexural fatigue testing of copper foil and flexible printed wiring. The copper foil ductility round-robin study is applied to investigate its fatigue properties. Additionally, experiments have yielded the following correlations, which can be utilized to simplify the determination of parameters.

$$\Delta\varepsilon_e = 0.9\left(\frac{\sigma_u}{E}\right)\left[\frac{\sigma_f \exp\left(D_f\right)}{\sigma_u \cdot 0.36}\right]^{0.1785 \cdot \log\left(\frac{10^5}{N_f}\right)} \tag{6.31}$$

$$\Delta\varepsilon_p = D_f^{0.75} N_f^{-0.6} \tag{6.32}$$

where $\Delta\varepsilon_e$ and $\Delta\varepsilon_p$ are the elastic strain and plastic strain increment, $\sigma_u$ is the ultimate tensile strength, $\sigma_f$ is the fracture strength, $D_f$ is the fatigue ductility, $E$ denotes the modulus of elasticity, and $N_f$ represents the number of cycles to failure. Fatigue life in the low-cycle range depends on the material's ductility, while tensile properties, such as tensile strength, modulus of elasticity, and fracture strength, play a crucial role in the high-cycle range. Thus, the ductility test is usually carried out between $30 \leq N_f \leq 500$ where plastic strain and the exhaustion of ductility dominate. On the other hand, in the higher cycle range, elastic strain dominates the fatigue behavior, and tensile properties can significantly affect ductility determination. For copper, the ultimate tensile and fracture strengths are identical; the total strain increment of the Engelmaier model can be simplified as:

$$\Delta\varepsilon = \Delta\varepsilon_e + \Delta\varepsilon_p = 0.9\left(\frac{\sigma_u}{E}\right)\left[\frac{\exp(D_f)}{0.36}\right]^{0.1785.\log\left(\frac{10^5}{N_f}\right)} + D_f^{0.75} N_f^{-0.6} \tag{6.33}$$

For copper foil, ductility, fatigue life, and cyclic strain range are interdependent and can be derived from each other when combined with the foil's tensile properties. Specifically, ductility can be calculated if fatigue life, cyclic strain range, and tensile properties are known. Similarly, if the applied cyclic strain range, tensile properties, and ductility are available, the fatigue life of the copper foil can be predicted. In electronic package analysis, several studies applied the stable TSR in one thermal cycle from FEM analysis into the Engelmaier model to predict the thermal fatigue life of copper-plated through-holes and micro-vias, and the results show good agreement with experimental data.

## 6.6 Brittle Fracture Failure

Predicting failure and establishing a geometry that will avert failure is a relatively simple matter if the machine part is subjected to a static uniaxial state of stress. It is necessary to have the simple uniaxial stress-strain curve available for the material of interest, which can be readily obtained from one or a few simple tension and compression experiments. For example, if yielding has been established as the governing failure mode for a uniaxially stressed machine part under consideration, failure would be predicted to occur when the maximum normal stress in the part reaches the yield point level determined from the stress-strain curve established in the simple tension test experiment.

Failure prediction becomes considerably more complex when a machine component is subjected to a multiaxial stress state. Yielding can no longer be assumed to occur solely when the maximum normal stress attains the uniaxial tensile yield strength, as additional normal stress components contribute to the onset of yielding. Furthermore, no single experiment can be performed to characterize failure in the multiaxial state of stress. A large number of complex multiaxial tests would be required, in which all stress components would vary over their entire range of values in all possible combinations. Even then, it would be difficult to assess the influence of outside factors such as stress concentration, temperature, and environment. Such a testing program would be prohibitively costly, time-consuming, and perhaps even impossible for certain stress states. When engineers or researchers face a problem of this complexity, their inclination is always to attempt to develop a theory that relates behavior in the complex situation to behavior in a simple, easily evaluated test through some characteristic “modulus”. Specifically, when one desires to predict failure in a machine part subjected to a multiaxial state of stress, it is usual to utilize a theory that relates failure in the multiaxial state of stress to failure by the same mode in a simple tension test through a well-chosen “modulus” such as stress, strain, or energy. To be useful, such moduli must be calculable in the multiaxial state of stress and readily measurable in the simple uniaxial evaluation test.

The basic assumption that constitutes the framework for all combined stress failure theories is that “failure is predicted to occur when the maximum value of the selected mechanical modulus in the multiaxial state of stress becomes equal to or exceeds the value of the same modulus that produces failure in a simple uniaxial stress test using the same material”. Numerous combined-stress failure theories have been proposed to capture these essential aspects. Among them, six are most widely applied: the maximum normal stress theory, the maximum shear stress theory, the maximum normal strain theory, the total strain energy theory, the distortion energy theory, and Mohr’s failure theory. The maximum normal stress criterion is generally recommended for brittle materials such as silicon.

The maximum normal stress theory: Failure is predicted to occur in the multiaxial state of stress when the maximum principal normal stress becomes equal to or exceeds the maximum normal stress at the time of failure in a simple uniaxial stress test using a specimen of the same material. If any of the following conditions become true, then the failure is predicted to occur.

$$\begin{array}{ccc} \sigma_1 \le \sigma_c & & \sigma_1 \ge \sigma_t \\ \sigma_2 \le \sigma_c & \text{or} & \sigma_2 \ge \sigma_t \\ \sigma_3 \le \sigma_c & & \sigma_3 \ge \sigma_t \end{array} \tag{6.34}$$

where $\sigma_1$, $\sigma_2$ and $\sigma_3$ are the principal stresses, $\sigma_t$ is the maximum allowed tensile stress in design, $\sigma_c$ is the maximum allowed compressive stress in design.

## 6.7 Delamination and Interfacial Strength [33–39]

In electronic packaging, especially in microelectronics and semiconductor devices, the interface between different materials, such as metals, polymers, ceramics, and silicon, plays a critical role in ensuring mechanical stability, electrical performance, and long-term reliability. Interfacial strength refers to the mechanical integrity or bonding strength between two materials at their interface. It quantifies how well the materials adhere together under stress, such as thermal, mechanical, or environmental loading. In electronic packaging, Interfaces exist between layers, such as die attach material and substrate, mold compound and leadframe, or underfill and chip.

High interfacial strength is crucial to prevent crack initiation or propagation during thermal cycling or mechanical shocks.

Delamination is the separation or debonding two materials at their interface, where the interfacial strength is insufficient to withstand applied stresses. The primary causes of electronic packaging issues include thermal mismatch resulting from the different coefficients of thermal expansion of materials during temperature cycling, moisture absorption, inadequate surface preparation or contamination, and mechanical stress from bending, vibration, or handling. The delamination may cause electrical failure, degradation in thermal performance, package cracking, or total device failure (Fig. 6.22). Knowing the adhesion strength is crucial, and several methods can be used to measure the delamination between materials, including peel tests, shear tests, pull tests, and four-point bending tests (4PB, Fig. 6.23). Interfacial strength is a key measure of bond integrity between materials. Delamination is a critical failure mode that compromises electrical and mechanical function. In electronic packaging, managing interfacial strength is essential to ensure device reliability under thermal, mechanical, and environmental stress.

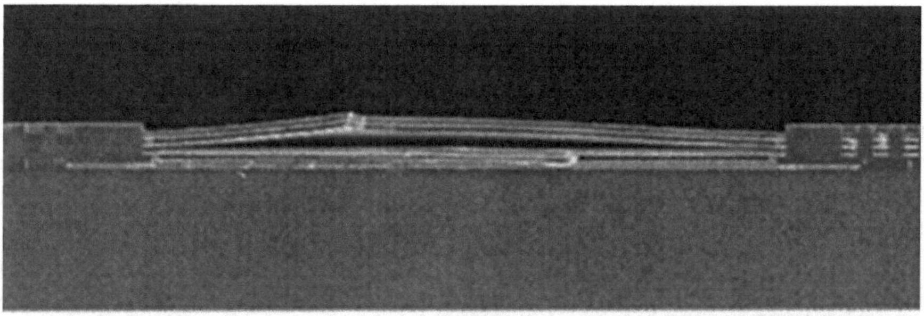

(a)

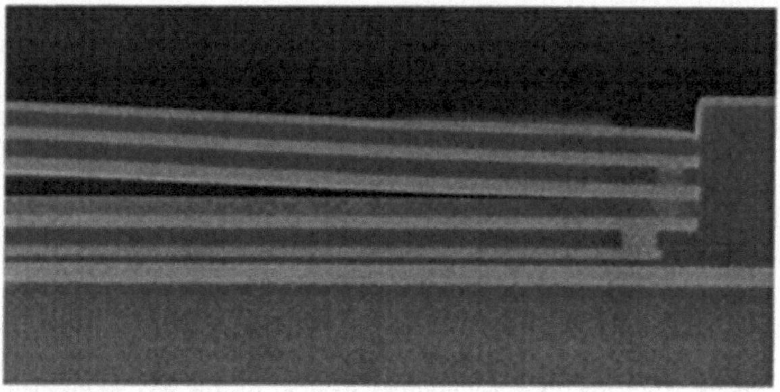

(b)

**Fig. 6.22** Delamination at the bond pad in flip-chip packaging [40]

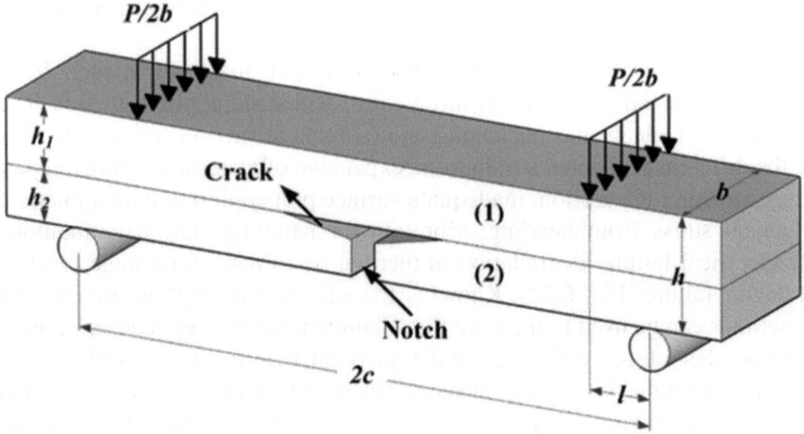


**Fig. 6.23** A notched bimaterial 4 PB specimen

## *6.7.1 Energy Release Rate and Four-Point Bending Test*

The 4PB delamination test, e.g., developed by Charalambides et al. [41, 42] for measuring the fracture resistance, such as fracture toughness and critical energy release rate of bi-material interfaces, is a widely used experimental method for characterizing adhesion in thin-film structures. It can provide reliable delamination parameters for finite element modeling and design validation. The analytical solution for the energy release rate (open mode) can be derived from the 4PB test. For the crack located between the inner loading lines, as shown in Fig. 6.23, the crack is subject to constant moment conditions. Consequently, by analogy with other decohesion problems [43, 44], the strain energy release rate is expected to exhibit steady-state characteristics, at least when the interface crack length is significantly greater than the thickness of the upper layer of the beam [41].

The steady-state energy release rate, $G_{ss}$, can be deduced analytically by recognizing that it is simply the difference in the strain energy in the uncracked and cracked beam. Since the strain energy in the beam above the crack can be neglected, $G_{ss}$ can be deduced from consideration of the energies in the uncracked section and a section of the lower beam beneath the crack. According to Euler-Bernoulli beam theory and plan strain assumptions, the energies can be expressed in terms of the applied moment $M$ as

$$U = \left(1-\upsilon^2\right)\frac{M^2}{2EI} \tag{6.35}$$

where $U$ is the strain energy per unit cross-section, $I$ is the second moment of area per unit width, $E$ is Young's modulus, and $\upsilon$ is Poisson's ratio. Consequently, $G_{ss}$ can be expressed as

$$G_{ss} = \frac{M^2\left(1-\upsilon_2^2\right)}{2E_2}\left(\frac{1}{I_2}-\frac{\lambda}{I_c}\right) \tag{6.36}$$

where

$$\lambda = \frac{E_2\left(1-\upsilon_1^2\right)}{E_1\left(1-\upsilon_2^2\right)}$$

$$I_c = \frac{h_1^3}{12}+\lambda\cdot\frac{h_2^3}{12}+\frac{\lambda h_1 h_2\left(h_1+h_2\right)^2}{4\left(h_1+\lambda h_2\right)} \tag{6.37}$$

$$I_2 = h_2^3/12$$

The subscript $c$ refers to the composite beam, and the subscript 1 or 2 represents each material, as shown in Fig. 6.23. In such a case, the moment per unit width can be presented as $M = Pl/2b$ where $P$ is the total load, $b$ is the specimen width, and $l$ is the spacing between the inner and outer loading. Moreover, the critical strain energy release rate $G_c$ can be further organized into a simplified form at the critical value of applied load $P_c$ in the case of $E = E_1 = E_2$, $\upsilon = \upsilon_1 = \upsilon_2$, and $h_1 = h_2 = h/2$ as

$$G_c = \frac{21P_c^2 l^2\left(1-\upsilon^2\right)}{2Eb^2h^3} \tag{6.38}$$

This formula is the most commonly applied analytical solution for measuring the energy release rate using the 4PB test. Based on the above concepts, Charalambides et al. further derived the solution by considering the frictional effects [42]. Accordingly, the applied moment can be expressed as

$$M = M^P + M^F = \frac{Pl}{2b}+\left(\frac{-\mu P}{2}\right)\frac{h}{b} = \frac{Pl}{2b}\left(1-\mu h/l\right) \tag{6.39}$$

where $\mu$ is the coefficient of static friction. The friction forces exert a bending moment $M^F$ on the specimen, which opposes the applied bending moment $M^P$, and the steady state energy release rate $G_{ss}$ becomes

$$G_{ss}^{\mu} = \frac{3}{2}\cdot\left[\frac{1}{\eta_2^3}-\frac{\lambda}{\eta_1^3+\lambda\eta_2^3+3\lambda\frac{\eta_1\eta_2}{\eta_1+\lambda\eta_2}}\right]\cdot\frac{P^2l^2\left(1-\upsilon_2^2\right)}{E_2b^2h^3}\cdot\left(1-\mu\frac{h}{l}\right)^2 \tag{6.40}$$

where the superscript $\mu$ denotes that frictional effects are considered. The notation $\eta_i$ is given by $\eta_i = h_i/h$. Similarly, geometry and material symmetric specimens, as mentioned above, can be simplified $G_{ss}^{\mu}$ to

$$G_{ss}^{\mu} = \frac{21}{2} \cdot \frac{P^2 l^2 \left(1-\upsilon^2\right)}{Eb^2 h^3} \cdot \left(1-\mu\frac{h}{l}\right)^2 \tag{6.41}$$

The critical strain energy release rate, $G_c^{\mu}$, can be determined by applying the critical value $P_c$ of the applied load at which the crack propagates. Based on this Equation, it is clear that neglecting friction effects will lead to an overestimation of the macroscopic work of fracture, as the additional energy dissipated through friction has not been considered. By plotting the dimensionless normalized parameter $G_c^{\mu=0} / G_c^{\mu}$ versus the loading parameter *l/h* for varying values of the friction coefficient, the effect of the friction force can be observed, as shown in Fig. 6.24. In general, larger applied moment arms, *l,* are preferable since larger values of *l/h* reduce the influence of friction. However, in some cases, the coefficient of friction may vary with the applied normal loading. As a result, quantifying the frictional contribution to the measured fracture energy is generally not practical. Many of these uncertainties will be removed by selecting a larger applied moment arm.

Figure 6.24 shows that a good testing practice is to maintain *l/h* greater than ~4 [45]. Based on the analytical solution of the energy release rate proposed by Charalambides et al. [41, 42] in 1989, the $G_c$ values in thin film structures can be measured simply and quickly by the 4PB test. Volinsky et al. [46] presented a characteristic load-displacement curve for the four-point bending (4PB) test, as shown in Fig. 6.25. The initial, nearly linear portion of this curve, at low loads, represents the stiffness of the notched sample before any delamination extension. A distinct peak or deviation from this initial linearity indicates the kinking of a crack into the interface of interest. As the crack propagates and extends sufficiently far from the pre-notch between the inner loading pins, the applied load stabilizes, reaching a steady-state plateau. This critical load value is then used to calculate the critical energy release rate, $G_c$.

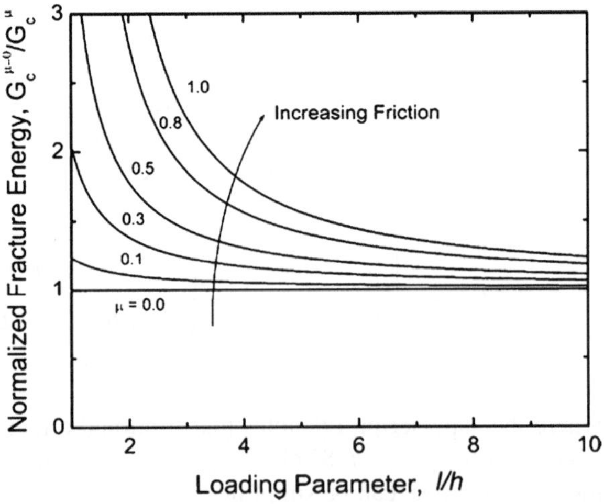


**Fig. 6.24** Normalized fracture energy versus the dimensionless loading parameter l/h for a range of coefficient of friction values [45]

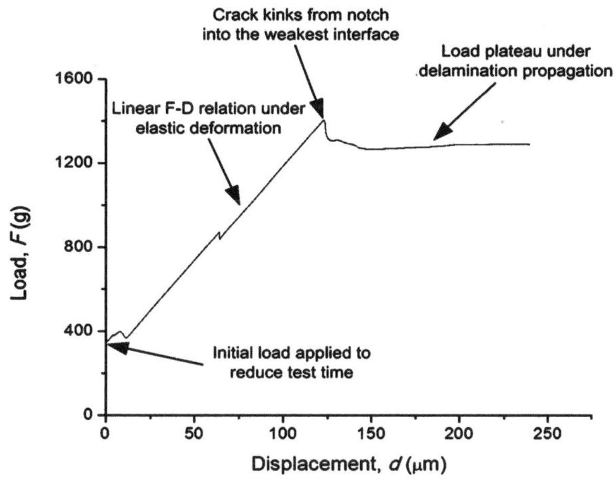


**Fig. 6.25** A typical applied displacement vs. loading during the 4 PB test

## 6.7.2 *Theories of Fracture Mechanics [35–39]*

In engineering design, there are two major approaches to identifying failures in structures or materials. One is the strength of materials approach, and the other is the fracture mechanics approach. In the former case, the anticipated design stress is compared to the flow properties of candidate materials; a material is considered adequate if its strength exceeds the expected applied stress. The fracture mechanics approach is based on three fundamental variables: applied stress, flaw size, and fracture toughness. Compared with the former approach, the additional structural variable is flaw size, and fracture toughness replaces strength as the relevant material property. Fracture mechanics quantifies the critical combinations of these three variables. In fracture analysis, two main methodologies are available: the energy criterion and the stress intensity approach, both of which will be briefly introduced below [47].

### 6.7.2.1 The Energy Criterion

The energy approach to fracture states that a crack will propagate when the energy available for its extension is sufficient to overcome the material's inherent resistance. Griffith initially put forth this energy criterion for fracture [48]. However, the approach we know today was developed mainly by Irwin [49]. A crucial metric in this context is the energy release rate ($G$), which measures explicitly how the potential energy changes as the crack area increases within a linear elastic material. At the moment of fracture, $G = G_c$ is the critical energy release rate, which measures fracture toughness.

For a crack of length $2a$ in an infinite plate subject to tensile stress, as shown in Fig. 6.26, the energy release rate can be described as

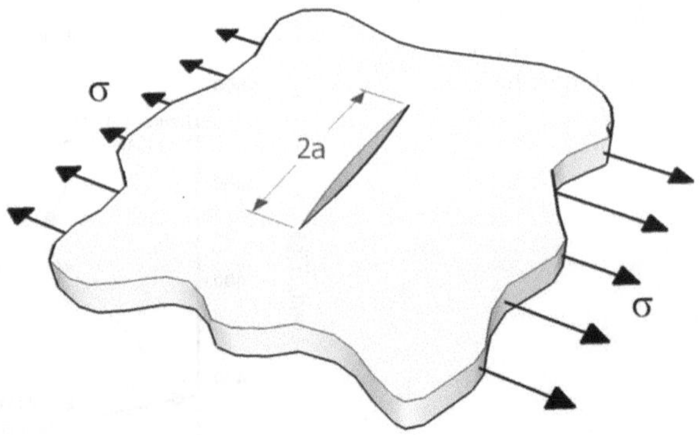


**Fig. 6.26** Through-thickness crack in an infinite plate subject to a remote tensile stress

$$G = \frac{\pi \sigma^2 a}{E} \tag{6.42}$$

where $E$ is Young's modulus, $\sigma$ is the applied stress, and $a$ is the half-crack length. At fracture, $G = G_c$, and Eq. (6.43) describes the critical combinations of stress and crack size for failure as

$$G_c = \frac{\pi \sigma_f^2 a_c}{E} \tag{6.43}$$

In Eq. (6.43), for a constant $G_c$ value, the failure stress, $\sigma_f$, varies with the crack length, $a_c$. The energy release rate, $G$, is the driving force for fracture, while $G_c$ is the material's resistance to fracture. One of the fundamental assumptions of fracture mechanics is that fracture toughness, such as $G_c$, is independent of the size and geometry of the fractured structure. A fracture toughness measure in a laboratory can be applied to a structure. The similitude assumption is valid as long as the material behavior is linear and elastic.

#### 6.7.2.2 The Stress Intensity Approach

The in-plane stresses on an element near the crack tip in an elastic material can be described as (Fig. 6.27)

$$\begin{aligned}
\sigma_{xx} &= \frac{K_I}{\sqrt{2\pi r}} \cos\left(\frac{\theta}{2}\right)\left[1 - \sin\left(\frac{\theta}{2}\right)\sin\left(\frac{3\theta}{2}\right)\right] \\
\sigma_{yy} &= \frac{K_I}{\sqrt{2\pi r}} \cos\left(\frac{\theta}{2}\right)\left[1 + \sin\left(\frac{\theta}{2}\right)\sin\left(\frac{3\theta}{2}\right)\right] \\
\tau_{xy} &= \frac{K_I}{\sqrt{2\pi r}} \cos\left(\frac{\theta}{2}\right)\sin\left(\frac{\theta}{2}\right)\sin\left(\frac{3\theta}{2}\right)
\end{aligned} \tag{6.44}$$

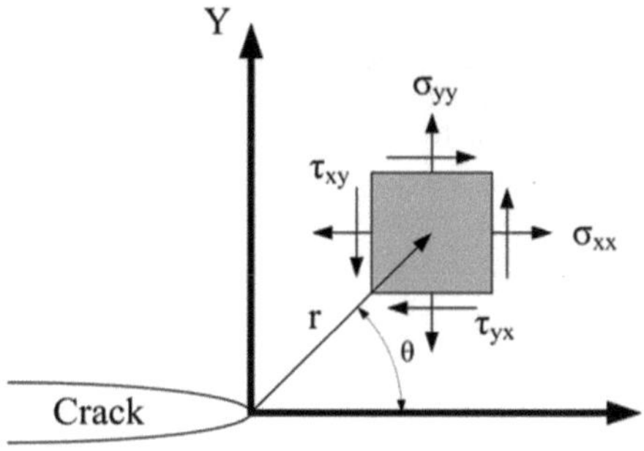


**Fig. 6.27** Stress near the crack tip in an elastic material

In Eq. (6.44), each stress component is proportional to a single constant, $K_I$, which is called the stress intensity factor and completely characterizes the crack tip conditions in a linear elastic material. If $K_I$ is known, the entire stress distribution at the crack tip can be computed with Eq. (6.45). The fracture will occur at a critical stress intensity, $K_{Ic}$, while the material fails locally at some critical combination of stress and strain. For the plate in Fig. 6.26, the stress intensity factor is given by

$$K_I = \sigma\sqrt{\pi a} \tag{6.45}$$

Fracture occurs when the applied stress intensity factor $K_I$ reaches the material's fracture toughness $K_{Ic}$. Here, $K_I$ represents the driving force for crack propagation, while $K_{Ic}$ denotes the material's intrinsic resistance to fracture. Comparing Eqs. (6.42) and (6.45) results in a relationship between $K_I$ and $G$.

$$G = \frac{K_I^2}{E} \tag{6.46}$$

Thus, the energy and stress intensity approach to fracture mechanics are essentially equivalent for linear elastic materials.

Delamination is a common failure mode in classical composite structures, as well as in microelectronic devices and electronic packaging. These systems often experience complex loading conditions that generate intermediate tensile and shear stresses at structural discontinuities, leading to delamination through mode I, mode II, and mode III fracture mechanisms. Due to interlaminar tension, $G_I$, the mode I component is a so-called open mode, where the relative displacement of the fracture surfaces is perpendicular to the direction of crack extension. The mode II component, known as the sliding mode, resulting from interlaminar shear, $G_{II}$, occurs when the foregoing displacement is parallel to the crack propagation direction. The mode III component, $G_{III}$, which dominates as a tear mode, is generated by out-of-plane shear stress, resulting in the relative displacement of the fracture surfaces parallel to the crack front.

A schematic figure is presented to describe the above three modes in Fig. 6.28. The total strain energy release rate, $G_T$, which is obtained from experiments, can be

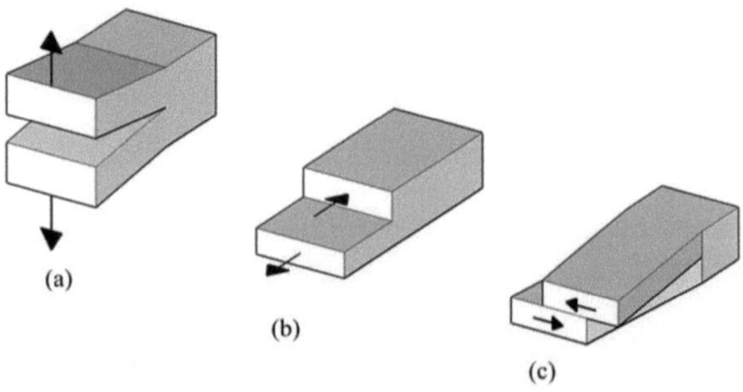


**Fig. 6.28** Three fracture modes: (**a**) Mode I, open mode; (**b**) Mode II, sliding mode; (**c**) Mode III, tear mode

calculated by summing the contributions from the three modes. All fracture problems occurring in practical engineering can be described as using either one or a combination of the above-mentioned fractured modes. Generally speaking, the fracture behavior of structures is quite complicated due to the mixed-mode effect of the fracture driving force. However, the investigation of the opening mode is of great concern, as the condition of crack extension to structural failure is easily compromised by a peeling force.

### *6.7.3 Numerical Methods for Fracture Resistance*

With the rapid advancement of computational technology, numerical mechanics has become a powerful tool for engineers and scientists, offering detailed and multidisciplinary analyses that are often difficult to achieve through experimental methods alone. Several fracture mechanics methods based on the finite element method have been developed to calculate stress intensity factors (SIF, $K$) and strain energy release rates ($G$). These include the J-integral method [50–57], the virtual crack extension (VCE) method [58–63], the modified virtual crack closure technique (MVCCT) [64–72], and the area release energy (ARE) method [73–76]. Using the J-integral method, an integral path must be defined to compute the stress intensity factor (SIF). In two-dimensional FEM models, this path can be easily described; however, the integration process becomes significantly more complex in three-dimensional cases. The virtual crack extension method requires FE analysis to determine the SIF and is capable of evaluating all three fracture modes. Despite its efficiency, it involves additional complex numerical procedures to obtain accurate results. On the other hand, the area release energy method provides a straightforward approach to estimating the energy release rate associated with delamination. It does so by dividing the energy released during crack extension over a specified area, eliminating the need for a predefined crack at a specific location. The ARE method offers an advantage by eliminating the need for a predefined delamination at a specific location. However, due to its area-based formulation, it is not suitable for evaluating crack propagation.

The virtual crack closure technique operates on the principle that the energy released during crack extension is equal to the energy required to close the crack. Consequently, the forces needed to close the crack are equivalent to those acting on the upper and lower surfaces of the crack when it is closed. These forces are obtained from an initial FE analysis with the crack in its closed state, while the corresponding displacements are derived from a second FE analysis after the crack has been extended to its target length.

The modified virtual crack closure technique is an adaptation of the traditional crack closure method, also known as the two-step VCCT [71, 72]. MVCCT is based on the assumption that a small crack extension does not significantly alter the stress and displacement fields near the crack tip. As a result, the displacements behind the extended crack tip are considered approximately equal to those behind the original tip, allowing the use of a single-step FE analysis instead of the two-step process required in VCCT. MVCCT offers a simplified computational approach while maintaining accuracy in calculating the energy release rate ($G$) for all three fracture modes. Its ease of implementation and reliability make it an attractive method in numerical fracture mechanics. The technique has also been successfully applied to model delamination propagation in composite structures, where various crack front patterns have been defined [65]. In this book, MVCCT is employed to analyze fracture behavior numerically.

### *6.7.4 The Modified Virtual Crack Closure Technique (MVCCT) [35–39]*

This section introduces the modified virtual crack closure technique, a method suitable for assessing delamination in electronic packaging. The original virtual crack closure technique, developed by Rybicki and Kanninen [77], was published nearly 25 years ago. The following provides a detailed explanation of the principles underlying VCCT and its application in this context.

The terminology of the virtual crack closure technique (VCCT) in previous technical literature is not clearly defined; it appears appropriate to include a related method: the crack closure method or the two-step virtual crack closure technique [71, 72]. Referring to the VCCT as the crack closure method may be more appropriate because the crack is physically extended or closed during two complete finite element calculations, as shown in Fig. 6.29. Based on Irwin's crack closure integral [49], the released energy of $\Delta E$ is assumed to be identical to the energy required to close the crack between location $l$ and $i$, as shown in Fig. 6.29a, when the crack is extended by $\Delta a$ from $a$ (Fig. 6.29b) to $\Delta a + a$. Index "1" and index "2" indicate the first and second steps, respectively. For a crack model with two-dimensional four-noded elements, as shown in Fig. 6.29, the required energy of $\Delta E$ to close a crack along one element side can be expressed as

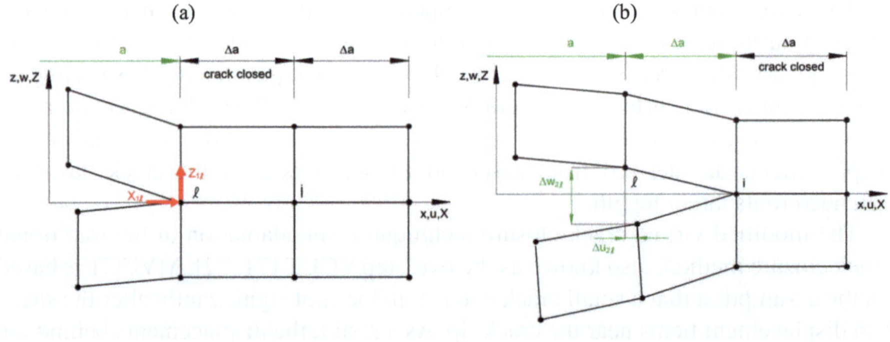


**Fig. 6.29** Crack closure method (Two-step VCCT). (**a**) First step—Crack closed; (**b**) Second step—Crack extended [71]

$$\Delta E = \frac{1}{2}\left[X_{1l} \cdot \Delta u_{2l} + Z_{1l} \cdot \Delta w_{2l}\right] \tag{6.47}$$

where $X_{1l}$ and $Z_{1l}$, which can be obtained at the first step, are the shear and opening forces at node $l$ to be closed, as shown in Fig. 6.29a. Similarly, $\Delta u_{2l}$ and $\Delta w_{2l}$ obtained at the second step indicate the shear and opening displacement at node $l$, as shown in Fig. 6.29b. Because the crack closure method is considered at that initial stage of crack growth, the forces loaded at the upper and lower crack surfaces are identical. The forces $X_{1l}$ and $Z_{1l}$ can be evaluated from a first finite element calculation while the crack is closed, as shown in Fig. 6.29a. The displacements $\Delta u_{2l}$ and $\Delta w_{2l}$ may be obtained from a second finite element analysis while the crack has been extended to its full length $\Delta a + a$ as shown in Fig. 6.29b.

The modified or virtual crack closure method has the same assumptions as the previously described crack closure method. In addition, it also assumes that the driving force to extend a crack is insignificant in changing the state at the crack tip when the crack extends from a + (a (node *i*) to a + 2(a (node *k*)), as shown in Fig. 6.30. Therefore, the displacements behind the crack tip at node *i* are almost the same as the displacements at node *l* behind the original crack tip. In addition, the released energy of $\Delta E$ is equal to the required energy to close the crack between node i and node k, while a crack extends from $a + \Delta a$ to $a + 2\Delta a$. For a crack model with 2D four-noded elements, as shown in Fig. 6.30, the work $\Delta E$ required to close the crack along one element length can be written as follows:

$$\Delta E = \frac{1}{2}\left[X_i \cdot \Delta u_l + Z_{\mathrm{i}} \cdot \Delta w_l\right] \tag{6.48}$$

As shown in Fig. 6.30, $X_i$ and $Z_i$ refer to the shear and opening forces at node *i*. Meanwhile, $\Delta u_l$ and $\Delta w_l$ are the shear and opening displacement at node *l*. By using this method, the energy $\Delta E$ required to close a crack can be obtained through a single finite element analysis. The energy release rate can be calculated by $G = \Delta E/$

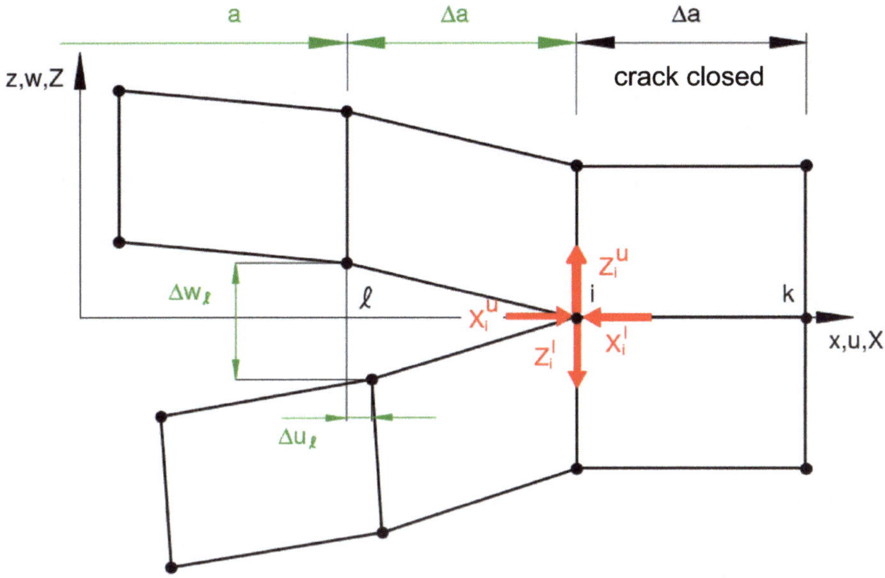


**Fig. 6.30** Modified virtual crack closure technique [72]. (One-step VCCT)

$\Delta A$, where $\Delta A$ is the crack surface area created, and the individual mode components can also be obtained. The following paragraph will introduce the equations for different mode components, element types, and coordinate systems.

## 6.7.5 *Equations for Using the Modified Virtual Crack Closure Technique*

For finite element simulation, based on the assumptions of plane stress or plane strain conditions, the equations for the mixed-mode strain energy release rates using 2D finite element models are introduced. Different solutions for fracture or delamination problems, modeled using plane, shell, or 3D solid models, are discussed in detail. Moreover, corrections for elements with different lengths or widths at the crack tip and procedures for arbitrarily shaped crack front contours will also be introduced.

### 6.7.5.1 Formulae for Two-Dimensional Analysis

As the 2D finite element model with plane stress or plane strain conditions is shown in Fig. 6.31, the crack length "a" is represented by a one-dimensional discontinuity through a line of nodes. Nodes at the top and bottom surfaces of the discontinuity

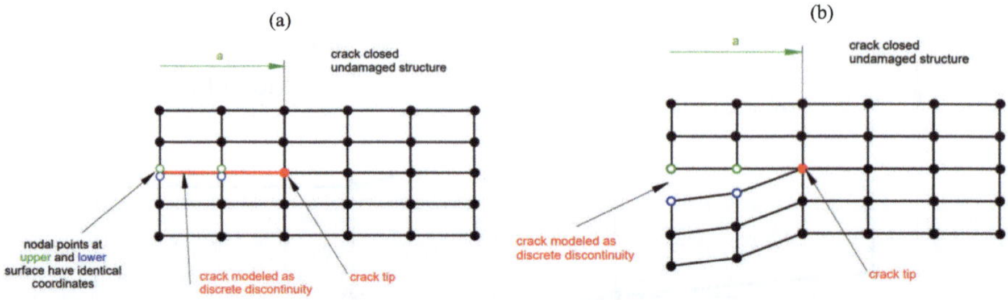


**Fig. 6.31** Two-dimensional finite element crack model [72]

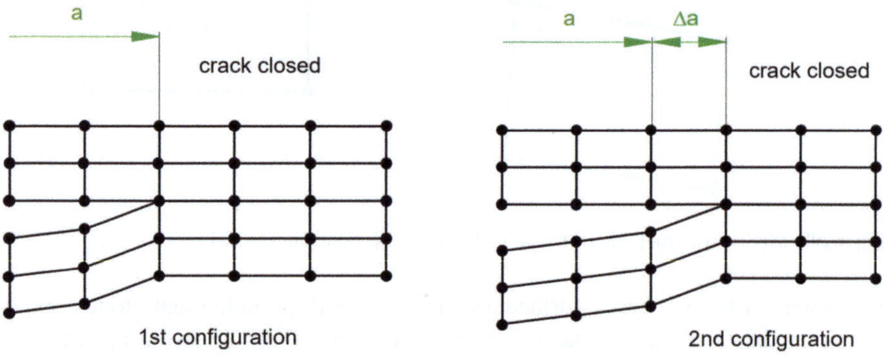


**Fig. 6.32** Crack propagation finite element analysis [72]

have identical coordinate systems but are not connected, as shown in Fig. 6.31a. Therefore, the elements connected to the top surface of the crack can deform independently from those connected to the bottom surface. This allows the crack to open during the finite element calculation, as shown in Fig. 6.31b. The crack tip and the undamaged portion, or the portion where the crack is closed and the elements are still intact, can be modeled using either single nodes or two nodes with the multi-point constraint (MPC) coupling technique during crack propagation analysis. For a crack propagation analysis, as shown in Fig. 6.32, the four-noded element can open or close the crack by sequentially releasing nodes along the crack. Therefore, elements are separated when the crack extension exceeds the entire length of the element.

As shown in Fig. 6.33a, the mode I ($G_I$) and mode II ($G_{II}$) components of the strain energy release rate for four-noded elements can be calculated through the following equations:

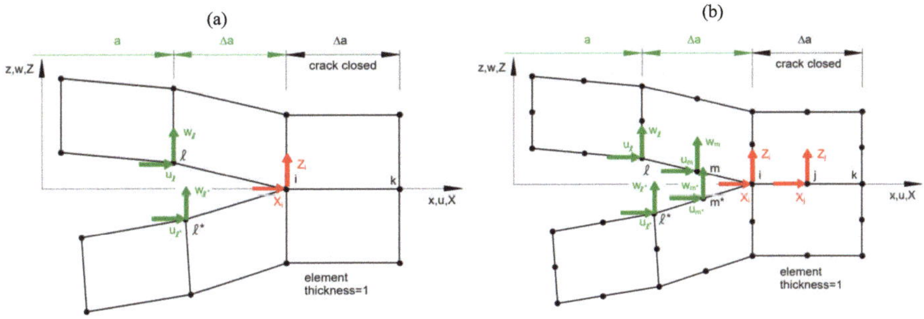


**Fig. 6.33** MVCCT for 2D solid elements. (**a**) Four-noded elements; (**b**) Eight-noded elements [72]

$$G_I = -\frac{1}{2\Delta a} \cdot Z_i \cdot (w_l - w_{l*})$$
$$G_{II} = -\frac{1}{2\Delta a} \cdot X_i \cdot (u_l - u_{l*}) \tag{6.49}$$

where $\Delta a$ is the length of the elements at the crack front, and $X_i$ and $Z_i$ refer to the forces at the crack tip (node $i$). The opening displacement behind the crack tip is calculated by the nodal displacements at the upper surface ($u_l$ and $w_l$) and the lower surface ($u_{l*}$ and $w_{l*}$) of the node $l$. In addition, the fractured surface $\Delta A = \Delta a \times 1$ is computed based on the assumption of a unit thickness defined in a 2D finite element analysis. Moreover, as shown in Fig. 6.33b, the mode I ($G_I$) and mode II ($G_{II}$) of the strain energy release rate for the eight-noded elements are calculated through the following equations:

$$G_I = -\frac{1}{2\Delta a} \left[ Z_i \cdot (w_l - w_{l*}) + Z_j \cdot (w_m - w_{m*}) \right]$$
$$G_{II} = -\frac{1}{2\Delta a} \left[ X_i \cdot (u_l - u_{l*}) + X_j \cdot (u_m - u_{m*}) \right] \tag{6.50}$$

As with the previously described four-noded elements, $\Delta a$ represents the length of the element at the crack front. In addition to the forces *Xi* and *Zi* at the crack tip node *i*, the forces *Xj* and *Zj* at the mid-side node ahead of the crack front (node *j*) are also required. The relative shearing and opening displacements behind the crack tip can be determined using the displacements $u_l$ and $w_l$ at the upper crack face (node *l*), and $u_l*$ and $w_l*$ at the corresponding lower crack face. Moreover, the relative displacements at node *m* and $m^*$ are also calculated by the displacements at the upper fractured surface $u_m$ and $w_m$, as well as the displacements at the lower fractured surface $u_{m*}$ and $w_{m*}$ [78]. The total energy release rate $G_T$ is estimated from the individual mode components as follows:

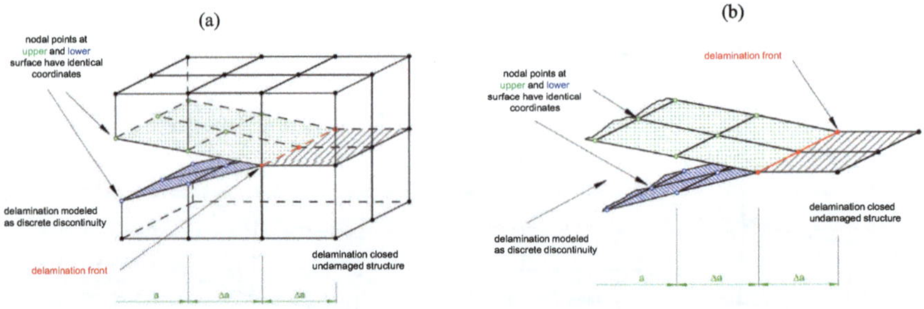


**Fig. 6.34** MVCCT for 3D elements. (**a**) 3D solid elements; (**b**) 3D plate/shell elements [72]

$$G_T = G_I + G_{II} + G_{III} \tag{6.51}$$

where $G_{III} = 0$ for the 2D case.

Since the VCCT proposed by Rybicki and Kanninen [77] does not make any assumptions about the forms of stresses and displacements, the singularity at a crack tip can be ignored. However, the special 2D crack tip elements with quarter-point nodes are applied in some fracture problems. Based on the location of the nodal points, the stress field of the crack tip with a $1/\sqrt{r}$ singularity can be precisely simulated by these quarter-point elements. Several studies reported the application of quarter-point elements [78–80].

#### 6.7.5.2 Formulae for Three-Dimensional Solids

In finite element models, which can be composed of 3D solid or plane/shell-type elements (illustrated in Fig. 6.34), the delaminated length "a" is characterized as a 2D discontinuity between two surfaces. This additional dimension facilitates calculating the energy release rate along the delamination front and determining $G_{III}$. Consistent with the explanation in the preceding section, a delamination propagation analysis can similarly be conducted utilizing a coupling technique, such as multipoint constraints.

Figure 6.35 shows that the region around the crack tip is modeled using eight-noded 3D solid elements. Each component of the energy release rate, the mode I ($G_I$), mode II ($G_{II}$), and mode III ($G_{III}$), is calculated as [81]:

$$\begin{aligned} G_I &= -\frac{1}{2\Delta A} Z_{Li} \cdot \left(w_{Ll} - w_{Ll*}\right) \\ G_{II} &= -\frac{1}{2\Delta A} X_{Li} \cdot \left(u_{Ll} - u_{Ll*}\right) \\ G_{III} &= -\frac{1}{2\Delta A} Y_{Li} \cdot \left(v_{Ll} - v_{Ll*}\right) \end{aligned} \tag{6.52}$$

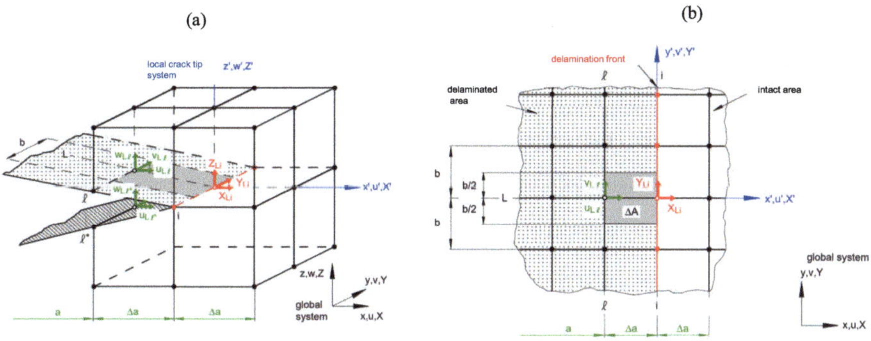


**Fig. 6.35** MVCCT for 3D eight-noded solid elements [72]

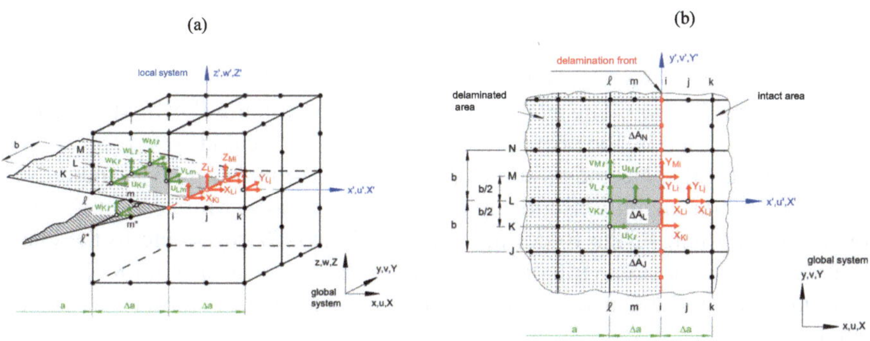


**Fig. 6.36** MVCCT for 3D twenty-noded solid elements [72]

Here, the virtually closed area is $\Delta A = \Delta a \times b$ (Fig. 6.35), where $\Delta a$ refers to the length of the elements at the delamination front, and $b$ is the width of the elements. In addition, $X_{Li}$, $Y_{Li}$, and $Z_{Li}$ individually denote the forces at the delamination front in column $L$, row $i$. The corresponding displacements behind the crack at the upper face node row $l$ are $u_{Ll}$, $v_{Ll}$, and $w_{Ll}$, and the displacements of the lower face node row $l^*$ are $u_{L*l}$, $v_{L*l}$, and $w_{Ll*}$ for the node in column $L$ as shown in Fig. 6.35. The finite element analysis can obtain all forces and displacements in the global coordinate system. A local crack tip coordinate system ($x'$, $y'$, $z'$) has been added to the illustration, defining the normal and tangential coordinate directions at the delamination front in the deformed configuration. This local coordinate system can be applied in the geometrically nonlinear analyses, which will be introduced later. Moreover, for high-order 3D solid elements such as twenty-noded solid elements, the equations for the strain energy release rate components at element corner nodes (location $Li$) can be expressed as [82] (See Fig. 6.36):

$$G_I = -\frac{1}{2\Delta A_L}\cdot \left[\frac{1}{2}Z_{Ki}\cdot(w_{Kl} - w_{Kl*}) + Z_{Li}\cdot(w_{Ll} - w_{Ll*}) + Z_{Lj}\cdot(w_{Lm} - w_{Lm*}) + \frac{1}{2}Z_{Mi}\cdot(w_{Ml} - w_{Ml*})\right]$$
$$G_{II} = -\frac{1}{2\Delta A_L}\cdot \left[\frac{1}{2}X_{Ki}\cdot(u_{Kl} - u_{Kl*}) + X_{Li}\cdot(u_{Ll} - u_{Ll*}) + X_{Lj}\cdot(u_{Lm} - u_{Lm*}) + \frac{1}{2}X_{Mi}\cdot(u_{Ml} - u_{Ml*})\right]$$
$$G_{III} = -\frac{1}{2\Delta A_L}\cdot \left[\frac{1}{2}Y_{Ki}\cdot(v_{Kl} - v_{Kl*}) + Y_{Li}\cdot(v_{Ll} - v_{Ll*}) + Y_{Lj}\cdot(v_{Lm} - v_{Lm*}) + \frac{1}{2}Y_{Mi}\cdot(v_{Ml} - v_{Ml*})\right] \quad (6.53)$$

where $\Delta A_L = \Delta a \times b$, $X_{Ki}$, $Y_{Ki}$, and $Z_{Ki}$ denote the applied forces at the delamination front in column $K$, row $i$. The relative displacements at column $K$ behind the delamination are calculated by the displacements at the lower face node row $l*$ as $u_{Kl*}$, $v_{Kl*}$, and $w_{Kl*}$, and the displacements at the upper face node row $l$ as $u_{Kl}$, $v_{Kl}$, and $w_{Kl}$ (See Fig. 6.36). Similar definitions are applicable in other nodes. Moreover, it should be noted that only half of the forces at locations $Ki$ and $Mi$ contribute to the energy required to close the area virtually $\Delta A_L$. Half of the forces at location $Ki$ contribute to the adjacent area $\Delta A_J$ and half of the forces at location $Mi$ contribute to the adjacent area $\Delta A_N$ (Fig. 6.37).

#### 6.7.5.3 Formulae for Plate/Shell Elements

The application of four-node plate elements for evaluating the mixed-mode strain energy release rate has been documented by Whitcomb and Shivakumar [83]. Wang and Raju [84] also developed equations for utilizing the VCCT with both four- and nine-noded plate elements. The subsequent paragraph will briefly present the equations for calculating the strain energy rate components for four-noded rectangular plate elements (Fig. 6.38).

$$\begin{aligned} G_I &= -\frac{1}{2\Delta A}\cdot Z_{Li}\cdot(w_{Ll} - w_{Ll*}) \\ G_{II} &= -\frac{1}{2\Delta A}\cdot X_{Li}\cdot(u_{Ll} - u_{Ll*}) \\ G_{III} &= -\frac{1}{2\Delta A}\cdot Y_{Li}\cdot(v_{Ll} - v_{Ll*}) \end{aligned} \quad (6.54)$$

The Equation is similar to that of the eight-noded 3D solid elements. The difference is that $\Delta A = \Delta a \times (b_1 + b_2)/2$ is the crack surface closed as shown in Fig. 6.38.

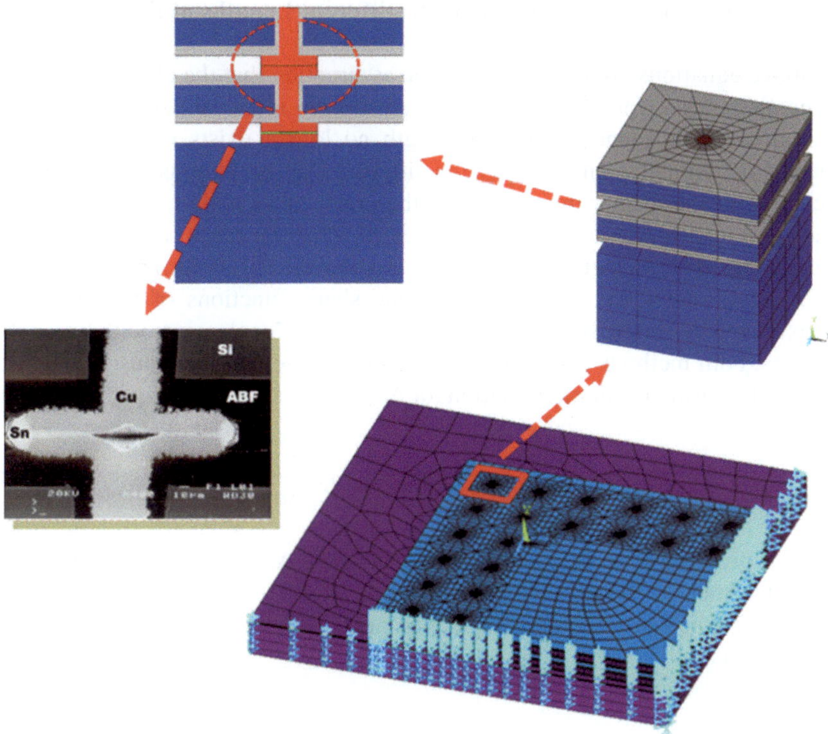


Fig. 6.37 Delamination 3D finite element modeling with MPC

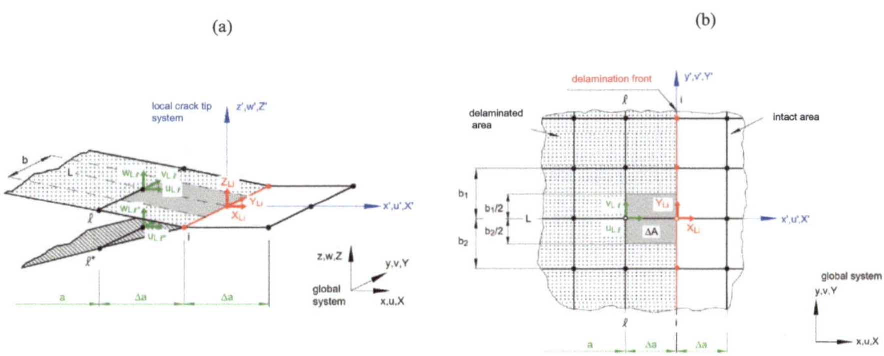


Fig. 6.38 MVCCT for 3D plate/shell elements [72]

### 6.7.5.4 Corrections for Elements with Different Lengths at the Crack Tip

The above equations are derived under the assumption that the element lengths $\Delta a$ for the elements before or behind the crack tip are identical. Corrections will be required once the identical element length no longer exists in models. In two-dimensional finite element analysis, the different element lengths are accounted for by correcting the displacements behind the crack tip at node $l$, which were computed for a length $\Delta a_1$, to match the forces $Xi$, $Zi$ at the crack tip at node $i$, which were computed for an element length $\Delta a_2$, as shown in Fig. 6.39. The displacements can be adjusted by taking into account the shape functions of the elements or approximated by simple linear interpolation [71, 72]. For elements with linear shape functions, both methods are identical. The displacement at locations $\tilde{l}$ and $\tilde{l}^*$ are calculated by using linear interpolation for $\Delta a_1 > \Delta a_2$ as shown in Fig. 6.39, yielding.

$$\tilde{u}_l = u_l \cdot \frac{\Delta a_2}{\Delta a_1} \text{ and } \tilde{u}_{l^*} = u_{l^*} \cdot \frac{\Delta a_2}{\Delta a_1} \tag{6.55}$$

The same procedure can be applied to the case while $\Delta a_1 > \Delta a_2$, and the crack opening displacement $w$. Based on this relationship of the displacements, the corrected equations for the energy release rate components for two-dimensional analysis can be expressed as

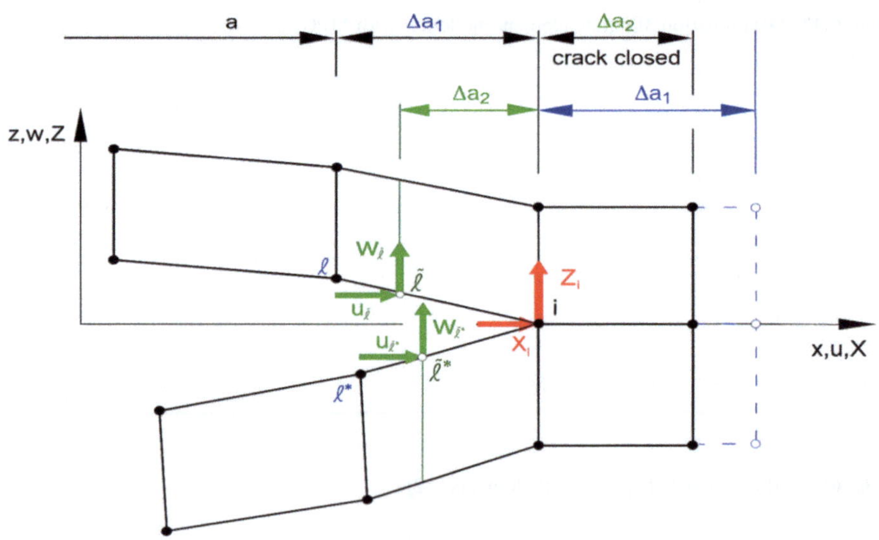


**Fig. 6.39** MVCCT for different lengths in front and behind the crack tip [72]

$$G_I = -\frac{1}{2\Delta a_2} \cdot Z_{Li} \cdot (w_l - w_{l*}) \cdot \frac{\Delta a_2}{\Delta a_1}$$
$$G_{II} = -\frac{1}{2\Delta a_2} \cdot X_{Li} \cdot (u_l - u_{l*}) \cdot \frac{\Delta a_2}{\Delta a_1} \tag{6.56}$$

Besides the above approach, Rybicki and Kanninen [77] reported the application of the $1/\sqrt{r}$ singularity of the stress field at the crack tip to derive the corrected equations. However, the former method is less restrictive, as the results depend only on the finite element discretization at the crack tip.

#### 6.7.5.5 Corrections for Elements with Different Widths Along the Delamination Front

In the previous derivation of the equations, the crack front is assumed to be straight, and the element width $b$ is also assumed to be constant along the front. In the case of free meshing, the element length $\Delta a$ and width $b$ may vary along the front. Therefore, corrections for such a condition need to be considered. The correction is straightforward for models with four-node plate/shell elements or eight-node solid brick elements, as shown in Fig. 6.40. In the condition of a variation with only the element length $\Delta a$, the correction can be derived from the previous corrections for the different length case as

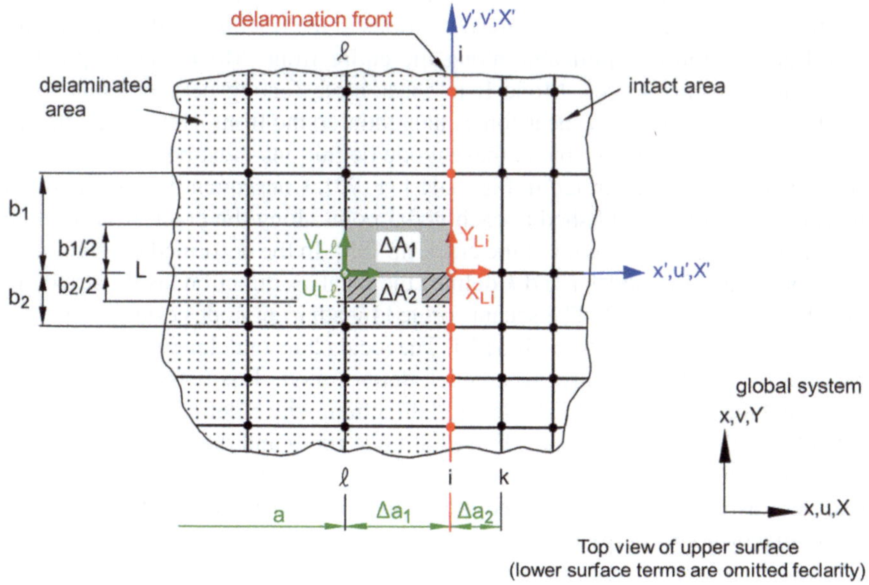


**Fig. 6.40** MVCCT for different widths along the crack front [72]

$$
\begin{aligned}
G_I &= -\frac{1}{2\Delta a_2 \cdot b} \cdot Z_{Li} \cdot (w_{Ll} - w_{Ll*}) \cdot \frac{\Delta a_2}{\Delta a_1} \\
G_{II} &= -\frac{1}{2\Delta a_2 \cdot b} \cdot X_{Li} \cdot (u_{Ll} - u_{Ll*}) \cdot \frac{\Delta a_2}{\Delta a_1} \\
G_{III} &= -\frac{1}{2\Delta a_2 \cdot b} \cdot Y_{Li} \cdot (v_{Ll} - v_{Ll*}) \cdot \frac{\Delta a_2}{\Delta a_1}
\end{aligned}
\tag{6.57}
$$

The additional variation of the element width $b$ requires the separate calculation of the contributing element surfaces $\Delta A_1 = \Delta a_2 \times b_1/2$ and $\Delta A_2 = \Delta a_2 \times b_2/2$. In such a condition, the corrected equations for the energy release rate components are

$$
\begin{aligned}
G_I &= -\frac{1}{2(\Delta A_1 + \Delta A_2)} \cdot Z_{Li} \cdot (w_{Ll} - w_{Ll*}) \cdot \frac{\Delta a_2}{\Delta a_1} \\
G_{II} &= -\frac{1}{2(\Delta A_1 + \Delta A_2)} \cdot X_{Li} \cdot (u_{Ll} - u_{Ll*}) \cdot \frac{\Delta a_2}{\Delta a_1} \\
G_{III} &= -\frac{1}{2(\Delta A_1 + \Delta A_2)} \cdot Y_{Li} \cdot (v_{Ll} - v_{Ll*}) \cdot \frac{\Delta a_2}{\Delta a_1}
\end{aligned}
\tag{6.58}
$$

#### 6.7.5.6 Procedures for Arbitrarily Shaped Crack Front Contours

The equations presented earlier were derived under the assumption of a straight crack front. In this simplified case, the definitions of the fracture modes are straightforward and uniformly applicable along the entire front: Mode I corresponds to out-of-plane opening forces, Mode II to shear forces acting perpendicular to the crack front, and Mode III to shear forces tangential to the front. However, when the crack front has an arbitrary shape, these mode definitions must be adapted locally along the contour, as illustrated in Fig. 6.41 [71, 72]. Consequently, a local coordinate system must be established at each nodal point along the crack front to accurately characterize the behavior at the crack tip [85]. In the deformed configuration, the vector connecting nodes $i$ and $k$ defines the local $x*$ direction, as illustrated for a sample node in Fig. 6.41. The secant formed by adjacent nodes establishes the $y'$ direction, which corresponds to Mode III. The local plane of delamination is defined by $x^*$ and $y'$. The local $z'$ direction, which defines mode I, is perpendicular to the plane of delamination. Finally, the local $x'$ direction, which defines mode II, is perpendicular to the $y'$ and $z'$ planes. Forces at node row $i$ and displacements at node row $l$ need to be transformed to the local $x'$-$y'$-$z'$ system at the tip. The transformed forces and displacements are utilized in the previously derived equations to calculate the mixed-mode energy release rates.

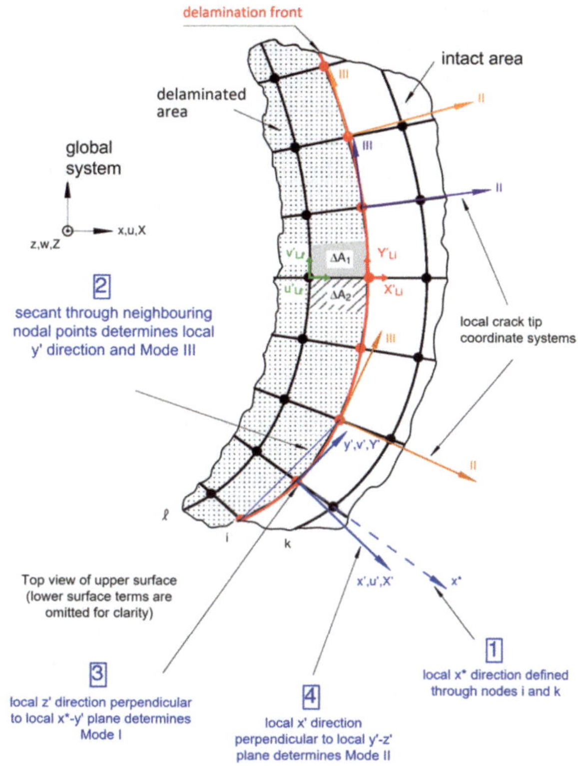


**Fig. 6.41** MVCCT for arbitrarily shaped crack front contours [72]

## 6.8 Summary and Recommendations

This chapter provides a comprehensive overview of the mechanics relevant to electronic packaging, covering material characteristics, failure mechanisms, constitutive models, life prediction methodologies, thermal analysis, and interfacial strength considerations. The goal is to ensure electronic packages' mechanical integrity and long-term reliability under various environmental loads, like thermal cycling and vibration. Mechanics, specifically solid mechanics, are crucial for electronic packaging because they help protect semiconductor devices, ensure optimal performance, and lessen mechanical and thermal stresses during operation.

Electronic packaging structures are often complex assemblies composed of diverse materials, including silicon, organic substrates, metals, and solder alloys. Each material behaves differently; for instance, silicon exhibits brittle fracture, while organic materials and solder joints show viscoelastic or viscoplastic behavior, which depends heavily on temperature. Understanding these material and mechanical properties is essential for electronic packages' integrity and long-term reliability under various loads, such as thermal cycling and vibration. Simulation models, such as those for solder liquid formation and damage accumulation, are used to analyze and design robust packaging structures. A comprehensive understanding of each constituent material's material and mechanical properties is crucial for ensuring the mechanical integrity and long-term reliability of electronic packages under various loads, including thermal cycling and vibration.

The maximum normal stress criterion for brittle materials such as silicon is generally recommended. For creep analysis, both the Anand and hyperbolic sine models can predict reliability life within an acceptable accuracy range. However, the hyperbolic sine model is recommended for electronic packaging applications if computational efficiency is critical.

## References


1. S. P. Timoshenko and J. N. Goodier, Theory of Elasticity, McGraw-Hill Book Company, 1951.
2. J. L. Chaboche, "Constitutive equations for cyclic plasticity and cyclic viscoplasticity," International Journal of Plasticity, vol. 5, 3, pp. 247–302, 1989.
3. J. L. Chaboche, "On some modifications of kinematic hardening to improve the description of ratchetting effects," International Journal of Plasticity, vol. 7, no. 7, pp. 661–678, 1991.
4. H. Ma, "Constitutive models of creep for lead-free solders," Journal of Materials Science, vol. 44, pp. 3841–3851, 2009.
5. H. Ma, and J. C. Suhling, "A review of mechanical properties of lead-free solders for electronic packaging," Journal of Materials Science, vol. 44, pp. 1141–1158, 2009.
6. X. Chen, G. Chen, and M. Sakane, "Prediction of Stress-Strain Relationship with an Improved Anand Constitutive Model for Lead-Free Solder Sn-3.5Ag," IEEE Transactions on Components and Packaging Technologies, Vol. 28, pp. 111–116, 2005.
7. J. Chang, L. Wang, J. Dirk, and X. Xie, "Finite Element Modeling Predicts the Effects of Voids on Thermal Shock Reliability and Thermal Resistance of Power Device," Welding Journal, Vol. 85, pp. 63–70, 2006.
8. L. Anand, "Constitutive equations for hot-working of metals," International Journal of Plasticity, vol. 1, no. 3, pp. 213–231, 1985.
9. S. B. Brown, K. H. Kim, and L. Anand, "An internal variable constitutive model for hot working of metals," International Journal of Plasticity, vol. 5, no. 2, pp. 95–130, 1989.
10. Q. Wang, "Experimental determination of Anand constitutive model parameters for microelectronics lead-free material SnAgCu and solder joint life prediction," Zhejiang University of Technology, 2007.
11. D. Herkommer, J. Punch and M. Reid, "Constitutive Modelling of Joint-Scale SAC305 Solder Shear Samples", IEEE Transactions on Components Packaging Manufacturing Technology, Vol.3, pp.275–281, 2013.
12. M. Pei and J. Qu, "Constitutive Modelling of Lead-Free Solders", 10th International Symposium on Advanced Packaging Materials: Processes, Properties and Interfaces, Irvine, CA, USA, March 16–18, 2005.
13. M. Motalab, Z. Cai, J. C. Suhling, and P. Lall, "Determination of Anand Constants for SAC Solders Using Stress-Strain or Creep Data", 13th Intersociety Conference on Thermal and Thermomechanical Phenomena in Electronic Systems (ITherm), San Diego, CA, USA, May 30–June 1, 2012.
14. L. F. Coffin, "A Study of the Effects of Cyclic Thermal Stress on a Ductile Metal," *Transactions ASME,* Vol. 76, pp. 931–950, 1954.
15. S. S. Manson, Thermal stress and low cycle fatigue, New York: McGraw-Hill, 1966.
16. R. Darveaux and K. Banerji, "Fatigue analysis of flip chip assemblies using thermal stress simulations and a Coffin-Manson relation," *41st Electronic Components and Technology Conference (ECTC)*, Atlanta, GA, USA, May 11–16, 1991.

17. R. Darveaux, K. Banerji, A. Mawer, and G. Dody, Reliability of plastic ball grid array assembly, New York, McGraw-Hill, 1995.
18. R. Darveaux, "Effect of simulation methodology on solder joint crack growth correlation," *50th Electronic Components and Technology Conference (ECTC)*, Las Vegas, NV, USA, May 21–24, 2000.
19. R. Darveaux, "Effect of simulation methodology on solder joint crack growth correlation and fatigue life prediction," *Journal of Electronic Packaging*, vol. 124, pp. 147–154, 2002.
20. K. C. Wu, S. Y. Lin, T. Y. Hung, and K. N. Chiang, "Reliability Assessment of Packaging Solder Joints Under Different Thermal Cycle Loading Rates", *IEEE Transactions on Device and Materials Reliability*, Volume 15, Issue 3, Pages 437–442, September 2015.
21. V. Ramachandran and K. N. Chiang, "Feasibility Evaluation of Creep Model for Failure Assessment of Solder Joint Reliability of Wafer Level Packaging", IEEE Transactions on Device and Materials Reliability, Volume PP, Issue 99, Pages 1–1, September 2017.
22. C. Y. Tsou, T. N. Chang, K. C. Wu, P. L. Wu and K.N. Chiang, "Reliability Assessment using Modified energy-based model for WLCSP Solder Joints", ICEP2017, Yamagata, Japan, April 19–22, 2017
23. P. H. Wang, Y. C. Lee, C. K. Lee, H. H. Chang, and K. N. Chiang, "Solder Joint Reliability Assessment and Pad Size Studies of FO-WLP with Glass Substrate", *IEEE Transactions on Device and Materials Reliability*, Vol. 21, No. 1, pp. 96–101
24. S. C. Chaparala *et al.*, "Effect of geometry and temperature cycle on the reliability of WLCSP solder joints," *IEEE Transactions on Components and Packaging Technologies,* vol. 28, no. 3, pp. 441–448, 2005.
25. T. Y. Hung, C. C. Wang, and K. N. Chiang, "Bonding Wire Life Prediction Model of the Power Module under Power Cycling Test", 2013 EuroSimE, Wroclaw, Poland, April 14–17, 2013
26. J. Morrow, J. F. Martin, and N. E. Dowling, "Local Stress-Stain Approach to Cumulative Fatigue Damage Analysis," Final Report, T & A. M. Report No. 379, Department of Theoretical and Applied Mechanics, University of Illinois, Urbana, 1974.
27. J. M. Hu, M. Pecht, and A. Dasgupta, "A probabilistic approach for predicting thermal fatigue life of wire bonding in microelectronics," Transactions of the ASME. Journal of Electronic Packaging, vol. 113, Issue 3, pp. 275–285, 1991.
28. A. Palmgren, "Durability of ball bearings," Zeitschrift des Vereins Deutscher Ingenieure (ZVDI), vol. 68, pp. 339–341, 1924.
29. M. A. Miner, "Cumulative damage in fatigue," Transactions ASME, vol. 67, pp. 159–164, 1945.
30. W. Engelmaier, "A New Ductility and Flexural Fatigue Test Method for Copper Foil and Flexible Printed Wiring," IPC-TP-204, Institute for Interconnecting and Packaging Electronic Circuits, Evanston, USA, 1978.
31. R. V. Pucha, "Modeling spatial strain gradient effects in thermo-mechanical fatigue of copper microstructures," International Journal of Fatigue, vol. 26, pp. 947–957, 2004.
32. W. Engelmaier and A. Wagner, "Fatigue Behaviour and Ductility Determination for Roller Annealed Copper Foil and Flex Circuits on Kapton," Circuit World, vol. 14, pp. 30–38, 1988.
33. S. Balakumar, X. T. Chen, Y. W. Chen, T. Selvaraj, B. F. Lin, R. Kumar, T. Hara, M. Fujimoyo, and Y. Shimura, "Peeling and Delamination in Cu/SiLK Process During Cu-CMP," Thin Solid Films, Vol. 462–463, pp. 161–167, Sep. 2004.
34. W. D. van Driel, M. A. J. van Gils, R. B. R. van Silfhout, and G. Q. Zhang, "Prediction of Delamination Related IC & Packaging Reliability Problems," Microelectronic Reliability, Vol. 45, No. 9–11, pp. 1633–1638, Sep.-Nov. 2005.
35. C. J. Wu, "Application of Interfacial Strength Simulation in Three-Dimensional Chip Stacking Electronic Packaging and Four-Point Bending Delamination Test", Ph.D. Thesis Dissertation, National Tsing Hua University, Taiwan, 2010.
36. C. C. Lee, "Investigation of Micro Electronic Devices Reliability Using Interfacial Crack Growth Prediction Methodology", Ph.D. Thesis Dissertation, National Tsing Hua University, Taiwan, 2006.

37. C. J. Wu, M. C. Hsieh, C. C. Chiu, M. C. Yew, and K. N. Chiang, "Interfacial delamination investigation between copper bumps in 3D chip stacking package by using the modified virtual crack closure technique", Microelectronic Engineering, Volume: 88, pp. 739–744, 2011.
38. C. J. Wu, M.C. Hsieh, and K.N. Chiang, "Strength evaluation of silicon die for 3D chip stacking packages using ABF as dielectric and barrier layer in through-silicon via," Microelectronic Engineering, vol. 87, pp. 505–509, 2010.
39. C. C. Lee, Y. F. Su, C. S. Wu, and K. N. Chiang, "Investigation of interconnect design on interfacial cracking energy of Al/TiN barriers under a flexural load," Thin Solid Films, Volume 544, Pages 530–536, October 2013.
40. L. L. Mercador, V. Sarihan, Y. Guo, and A. Mawer, "Impact of Solder Pad Size on Solder Joint Reliability in Flip Chip PBGA Packages," *IEEE Transactions on Advanced Packaging*, Vol. 23, pp. 415–420, Aug. 2000.
41. P. G. Charalambides, J. Lund, A. G. Evans, and R. M. McMeeking, "A test specimen for determining the fracture resistance of bimaterial interfaces," *Journal of Applied Mechanics-Transactions of the Asme,* vol. 56, pp. 77–82, 1989.
42. P. G. Charalambides, H. C. Cao, J. Lund, and A. G. Evans, "Development of a Test Method for Measuring the Mixed-Mode Fracture-Resistance of Bimaterial Interfaces," *Mechanics of Materials,* vol. 8, pp. 269–283, 1990.
43. M. D. Drory, M. D. Thouless, and A. G. Evans, "On the Decohesion of Residually Stressed Thin-Films," *Acta Metallurgica,* vol. 36, pp. 2019–2028, 1988.
44. M. D. Thouless, A. G. Evans, M. F. Ashby, and J. W. Hutchinson, "The Edge Cracking and Spalling of Brittle Plates," *Acta Metallurgica,* vol. 35, pp. 1333–1341, 1987.
45. D. M. Gage, K. Kim, C. S. Litteken, and R. H. Dauskard, "Role of friction and loading parameters in four-point bend adhesion measurements," *Journal of Materials Research,* vol. 23, pp. 87–96, 2008.
46. A. A. Volinsky, N. R. Moody, and W. W. Gerberich, "Interfacial toughness measurements for thin films on substrates," *Acta Materialia,* vol. 50, pp. 441–466, 2002.
47. T. L. Anderson, Fracture mechanics: fundamentals and applications/T. L., 2nd ed., CRC Press, 1994.
48. A. A. Griffith, "The Phenomena of Rupture and Flow in Solids," *Philosophical Transactions of the Royal Society of London. Series A,* vol. 221, pp. 163–198, 1921.
49. G. R. Irwin, "Fracture Dynamics," *Fracturing of Metals, American Society for Metals, Cleveland,* 1948.
50. C. C. Chiu, H. H. Chang, C. C. Lee, C. C. Hsia, and K. N. Chiang, "Reliability of interfacial adhesion in a multi-level copper/low-k interconnect structure," Microelectronics Reliability, vol. 47, pp. 1506–1511, 2007.
51. C. C. Chiu, C. C. Lee, T. L. Chou, C. C. Hsia, and K. N. Chiang, "Analysis of Cu/Low-k structure under back end of line process," Microelectronic Engineering, vol. 85, pp. 2150–2154, 2008.
52. C. C. Lee, C. C. Chiu, C. C. Hsia, and K. N. Chiang, "Interfacial Fracture Analysis of CMOS Cu/Low-k BEOL Interconnect in Advanced Packaging Structures," IEEE Transactions on Advanced Packaging, vol. 32, pp. 53–61, 2009.
53. C. C. Lee, T. L. Chou, C. C. Chiu, C. C. Hsia, and K. N. Chiang, "Cracking energy estimation of ultra low-k package using novel prediction approach combined with global-local modeling technique," Microelectronic Engineering, vol. 85, pp. 2079–2084, 2008.
54. C. C. Lee, T. C. Huang, C. C. Hsia, and K. N. Chiang, "Interfacial fracture investigation of low-k packaging using J-integral methodology," IEEE Transactions on Advanced Packaging, vol. 31, pp. 91–99, 2008.
55. M. Chiarelli and A. Frediani, "A Computation of the 3-Dimensional J-Integral for Elastic-Materials with a View to Applications in Fracture-Mechanics," Engineering Fracture Mechanics, vol. 44, pp. 763–788, 1993.
56. W. H. Chen and C. W. Wu, "On the J-Integral for a Pressurized Crack in Bonded Materials," International Journal of Fracture, vol. 16, pp. R47-R51, 1980.

57. W. H. Chen and Y. H. Huang, "J-Integral for Cracked Structure with Inclusions," International Journal of Fracture, vol. 15, pp. R73-R76, 1979.
58. P. W. Claydon, "Maximum Energy-Release Rate Distribution from a Generalized 3D-Virtual Crack Extension Method," *Engineering Fracture Mechanics,* vol. 42, pp. 961–969, 1992.
59. T. K. Hellen, "A Substructuring Application of the Virtual Crack Extension Method," *International Journal for Numerical Methods in Engineering,* vol. 19, pp. 1713–1731, 1983.
60. S. C. Lin and J. F. Abel, "Variational Approach for a New Direct-Integration Form of the Virtual Crack Extension Method," *International Journal of Fracture,* vol. 38, pp. 217–235, 1988.
61. D. M. Parks, "Virtual Crack Extension Method for Non-Linear Material Behavior," *Computer Methods in Applied Mechanics and Engineering,* vol. 12, pp. 353–364, 1977.
62. T. K. Hellen, "On the Method of Virtual Crack Extension," *International Journal for Numerical Methods in Engineering,* vol. 9, pp. 187–207, 1975.
63. D. M. Parks, "Stiffness Derivative Finite-Element Technique for Determination of Crack Tip Stress Intensity Factors," *International Journal of Fracture,* vol. 10, pp. 487–502, 1974.
64. A. Leski, "Implementation of the virtual crack closure technique in engineering FE calculations," *Finite Elements in Analysis and Design,* vol. 43, pp. 261–268, 2007.
65. A. C. Orifici, R. S. Thomson, R. Degenhardt, C. Bisagni, and J. Bayandor, "Development of a finite-element analysis methodology for the propagation of delaminations in composite structures," *Mechanics of Composite Materials,* vol. 43, pp. 9–28, 2007.
66. C. Schuecker and B. D. Davidson, "Evaluation of the accuracy of the four-point bend end-notched flexure test for mode II delamination toughness determination," *Composites Science and Technology,* vol. 60, pp. 2137–2146, 2000.
67. G. C. Tsai, "Parametric study of mode I energy release rate of soldered joints," *Finite Elements in Analysis and Design,* vol. 38, pp. 671–689, 2002.
68. K. S. Venkatesha, B. Dattaguru, and T. S. Ramamurthy, "Finite element analysis of an interface crack with large crack-tip contact zones," *Engineering Fracture Mechanics,* vol. 54, pp. 847–860, 1996.
69. K. S. Venkatesha, T. S. Ramamurthy, and B. Dattaguru, "Generalized modified crack closure integral (GMCCI) and its application to interface crack problems," *Computers & Structures,* vol. 60, pp. 665–676, 1996.
70. G. S. Palani, B. Dattaguru, and N. R. Iyer, "Numerically integrated modified virtual crack closure integral technique for 2-D crack problems," *Structural Engineering and Mechanics,* vol. 18, pp. 731–744, 2004.
71. R. Krueger, "Virtual crack closure technique: History, approach, and applications," *Applied Mechanics Reviews,* vol. 57, pp. 109–143, 2004.
72. R. Krueger, "The Virtual crack closure technique: History, approach, and applications." NASA/CR-2002-211628, ICASE Report No. 2002-10, 2002.
73. M. A. J. van Gils, O. van der Sluis, G. Q. Zhang, J. H. J. Janssen, and R. M. J. Voncken, "Analysis of Cu/low-k bond pad delamination by using a novel failure index," *Microelectronics Reliability,* vol. 47, pp. 179–186, 2007.
74. O. van der Sluis, R. A. B. Engelen, R. B. R. van Silfhout, W. D. van Driel, and M. A. J. van Gils, "Efficient damage sensitivity analysis of advanced Cu/low-k bond pad structures by means of the area release energy criterion," *Microelectronics Reliability,* vol. 47, pp. 1975–1982, 2007.
75. C. A. Yuan, O. van der Sluis, W. D. van Driel, and G. Q. Zhang, "The need for multi-scale approaches in Cu/low-k reliability issues," *Microelectronics Reliability,* vol. 48, pp. 833–842, 2008.
76. Z. W. Zhong, "Wire bonding of low-k devices," *Microelectronics International,* vol. 25, pp. 19–25, 2008.
77. E. F. Rybicki and M. F. Kanninen, "Finite-Element Calculation of Stress Intensity Factors by a Modified Crack Closure Integral," *Engineering Fracture Mechanics,* vol. 9, pp. 931–938, 1977.
78. I. S. Raju, "Calculation of Strain-Energy Release Rates with Higher-Order and Singular Finite-Elements," *Engineering Fracture Mechanics,* vol. 28, pp. 251–274, 1987.

79. K. B. Narayana, B. Dattaguru, T. S. Ramamurthy, and K. Vijayakumar, "Modified Crack Closure Integral Using 6-Noded Isoparametric Quadrilateral Singular Elements," *Engineering Fracture Mechanics,* vol. 36, pp. 945–955, 1990.
80. R. Sethuraman and S. K. Maiti, "Finite-Element Based Computation of Strain-Energy Release Rate by Modified Crack Closure Integral," *Engineering Fracture Mechanics,* vol. 30, pp. 227–231, 1988.
81. K. N. Shivakumar, P. W. Tan, and J. C. Newman, "A Virtual Crack-Closure Technique for Calculating Stress Intensity Factors for Cracked 3-Dimensional Bodies," *International Journal of Fracture,* vol. 36, pp. R43-R50, 1988.
82. I. S. Raju, R. Sistla, and T. Krishnamurthy, "Fracture mechanics analyses for skin-stiffener debonding," *Engineering Fracture Mechanics,* vol. 54, pp. 371–385, 1996.
83. J. D. Whitcomb and K. N. Shivakumar, "Strain-Energy Release Rate Analysis of Plates with Postbuckled Delaminations," *Journal of Composite Materials,* vol. 23, pp. 714–734, 1989.
84. J. T. Wang and I. S. Raju, "Strain energy release rate formulae for skin-stiffener debond modeled with plate elements," *Engineering Fracture Mechanics,* vol. 54, pp. 211–228, 1996.
85. R. Dauskardt, M. Lane, Q. Ma, and N. Krishna, "Adhesion and debonding of multi-layer thin film structures," *Engineering Fracture Mechanics,* vol. 61, pp. 141–162, 1998.

# Chapter 7
# AI and Machine Learning Algorithms

**Abstract** In electronic packaging (EP) applications, machine learning provides a data-driven approach for rapid reliability prediction and parameter inference. This technology offers a significant advantage over conventional methods. While accelerated thermal cycling tests (ATCT) take months and finite element (FE) simulations require days or weeks, ML models can deliver reliability predictions in seconds to minutes, substantially accelerating the design cycle. Furthermore, AI-assisted Design-on-Simulation (DoS) technology ensures consistent prediction outcomes, mitigating inconsistencies often attributed to human factors in traditional simulations. The primary application in EP is supervised regression-type machine learning, which focuses on predicting continuous numerical values. These tasks include forecasting the fatigue life (reliability life cycles) for structures like wafer-level packaging and 3D packaging and extracting material parameters for complex constitutive models (e.g., Garofalo, Chaboche, Anand models). The general ML workflow involves several critical steps: Data Preprocessing is necessary to standardize input and output data ranges to improve modeling performance. Cross-validation (often k-fold) is employed to select robust parameters and prevent overfitting. Grid Search systematically evaluates defined ranges to find the optimal configuration of hyperparameters.

## 7.1 Introduction

Artificial intelligence (AI) and machine learning (ML) represent rapidly advancing fields that are becoming increasingly crucial for complex decision-making, with extensive applications across various scientific and industrial sectors, including the electronics industry. Machine learning is a fundamental discipline within AI and statistical inference. It leverages theoretical principles of AI in conjunction with large datasets to guide computers through training and learning, ultimately developing a simplified AI model that maps input to output relationships to aid in design decisions. The primary objective of machine learning is to enable computers to



J. Lau, K.-N. Chiang, *Cu-Interconnects, Glass, and AI-Assisted Simulation for Chiplets and Heterogeneous Integration*,
https://doi.org/10.1007/978-981-95-6891-8_7

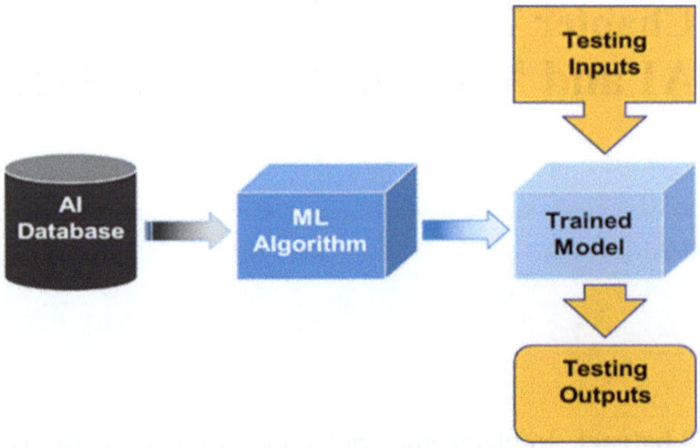


**Fig. 7.1** AI training flow chart

automatically learn from data without requiring human involvement. This learning process relies on observing data, such as examples, past experiences, or instructions, to help the system create standardized patterns for improving future decisions, as shown in Fig. 7.1. Machine learning aims to enhance a system's ability to generalize and solve previously unknown problems by learning from prior data, referred to as "experience."

## 7.2 Overview of Machine Learning

Solomonoff introduced the concept of machine learning in 1957 [1]. However, its practical widespread adoption was initially hampered by the limited computing power available at the time. Today, driven by the exponential growth in computational hardware and the vast availability of data, machine learning has been successfully applied across numerous fields, yielding highly accurate predictive models. The general process flow of machine learning involves constructing various training datasets, employing testing data to evaluate performance, and fine-tuning hyperparameters during the training phase [2–6]. A typical machine learning workflow begins with building a training dataset, which is then fed into various algorithms [3]. This process leads to the development of a hypothesis or model, which subsequently receives input data to test and predict the output based on the given test information. Before model development, it is crucial to preprocess the data to standardize the input and output datasets, as their values may not be in the same range, which improves modeling performance. Cross-validation, often employing the k-fold method, is a crucial technique for selecting parameters to prevent overfitting by dividing the dataset into training and validation sets [7]. Grid search is employed to find the best hyperparameters for the training model, systematically searching a defined range of values. Selecting an appropriate algorithm and configuring it optimally, including data preprocessing and hyperparameter tuning, is crucial for achieving high accuracy and efficiency. Machine learning encompasses a variety of

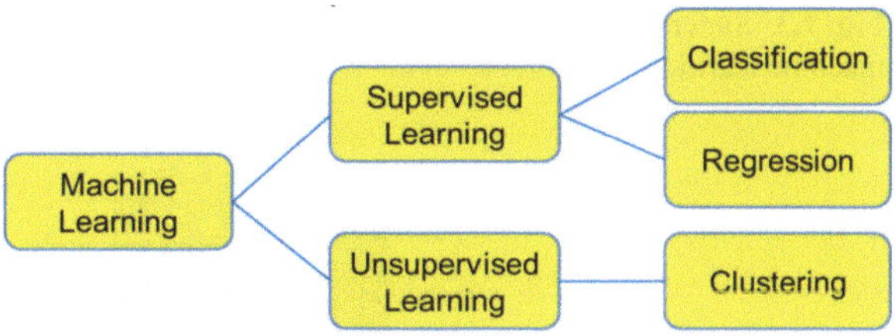


**Fig. 7.2** Machine learning paradigms

algorithms, each with distinct methodologies and applications, particularly in predicting the reliability of advanced packaging.

## 7.3 Classification of Machine Learning

Machine learning algorithms are generally categorized based on their learning paradigms or their specific learning objectives. Based on learning paradigms, the algorithms (Fig. 7.2) include supervised learning (Fig. 7.3), unsupervised learning (Fig. 7.4), and semi-supervised learning [4]. Reinforcement learning is not included here, as it constitutes an interaction-based paradigm rather than a data-driven one. Based on learning objectives, the algorithms include regression [8–10], classification [11–13], ranking, clustering, and generation (GANs). In electronic packaging applications, machine learning provides a data-driven approach for rapid prediction and parameter inference related to reliability. This technology enables the estimation of the reliability life cycles of new packaging structures without requiring extensive simulations or experiments, thereby reducing the time and cost typically associated with traditional methods. Furthermore, AI-assisted design-on-simulation (DoS) technology offers a viable solution to the challenge of inconsistent simulation results among researchers, a problem often attributed to human factors, ensuring more consistent prediction outcomes.

### *7.3.1 Regression [2–7]*

As illustrated in Fig. 7.5, regression in machine learning is a fundamental aspect of artificial intelligence that focuses on predicting continuous target values by modeling the relationship between various input features and a numerical output. This approach is widely utilized for forecasting and time-series prediction problems where the output variable is continuous. Regression algorithms aim to fit a mathematical function to data points, thereby establishing a relationship that can predict outcomes for new, unseen inputs. The electronic packaging (EP) community has increasingly embraced AI-assisted design-on-simulation technology to address critical challenges in ensuring reliable and consistent simulation results. Traditional

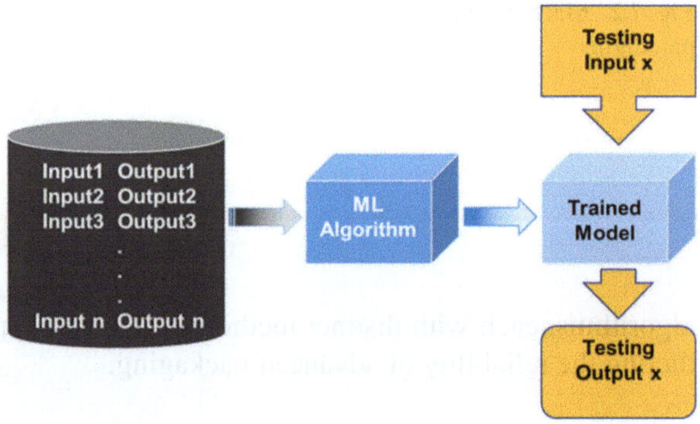


**Fig. 7.3** Supervised machine learning

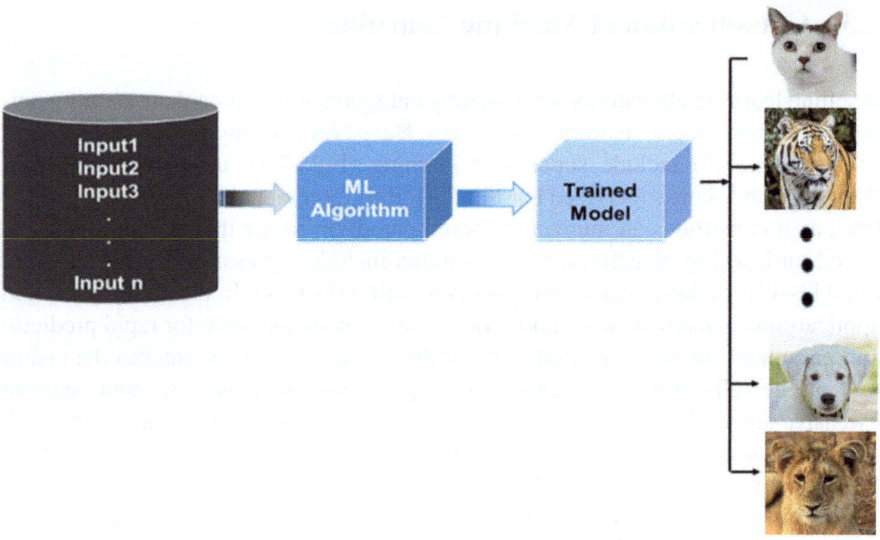


**Fig. 7.4** Unsupervised machine learning

experimental methods, such as accelerated thermal cycling tests (ATCT), are notably costly and time-consuming, often requiring months to generate essential reliability data. While finite element simulations can reduce this to days or weeks, machine learning can provide predictions in mere seconds to minutes, significantly accelerating the design cycle.

Furthermore, conventional FE simulation results can be inconsistent due to variations in domain knowledge, theoretical backgrounds, and modeling techniques among researchers. AI-assisted DoS technology offers a viable solution to this, providing consistent and reliable prediction results. Machine learning also proves beneficial for optimizing design parameters in packaging, where establishing explicit empirical equations for complex and nonlinear interactions is challenging. ML models, especially neural networks, excel at capturing these intricate nonlinear relationships commonly found in engineering systems.

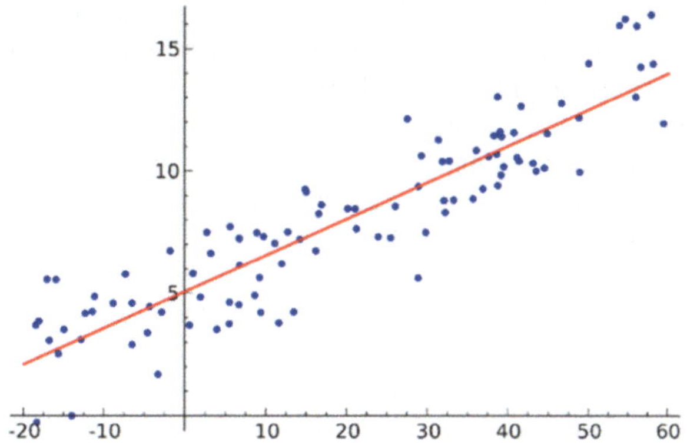


**Fig. 7.5** Regression illustration

Regression-based ML models are primarily applied to several key areas in electronic packaging. For the reliability assessment, ML models are trained to forecast the fatigue life or reliability life cycles, such as the number of thermal cycles to failure, for diverse packaging structures, including wafer-level packaging, wafer-level chip scale packaging (WLCSP), heterogeneous packaging, fanout packaging, and 3D packaging. Another crucial use is material parameters extraction, where AI-assisted methods aid in deriving parameters for complex material constitutive models, such as the Garofalo, Chaboche, and Anand models, which are vital for accurate FE simulations. ML models can also predict warpage packaging structures, such as panel-level fanout packaging. Additionally, AI-driven workflows can be used to estimate equivalent material properties of complex structures, such as printed circuit boards (PCBs), including equivalent coefficient of thermal expansion (CTE) and Young's modulus, thereby simplifying subsequent packaging simulations.

A range of regression algorithms is well-suited for AI-assisted DoS technology in reliability and warpage prediction. These include, but are not limited to, Artificial Neural Network (ANN), known for its ability to model complex nonlinear relationships, Polynomial Regression (PR), utilized when the response variable exhibits nonlinear behavior, and Gaussian Process Regression (GPR), offering a probabilistic approach that is advantageous in scenarios with limited data. Furthermore, the Recurrent Neural Network (RNN) is employed for time-series data and fatigue life prediction. Ensemble learning techniques such as bagging, boosting, and stacking are also applied to combine multiple base models, including ANN, GPR, and RNN, thereby enhancing prediction accuracy and robustness, particularly when data is limited. Hybrid models, such as K-means with KRR (K-K) and KNN, are also explored to boost performance for nonlinear datasets and improve the training efficiency of prediction models.

The typical workflow for reliability prediction using ML in electronic packaging begins with data generation. An extensive database of reliability results is typically created using validated finite element simulations, which often replace costly and time-consuming physical experiments. The input features for these models usually consist of the package's geometric design parameters, such as chip thickness, upper pad diameter, lower pad diameter, and stress buffer layer (SBL) thickness. The output corresponds to the reliability life cycle, often expressed as the number of thermal cycles to failure.

Before training, the data undergoes preprocessing. Scaling techniques such as Standard Scaler, Robust Scaler or MinMax Scaler are applied to ensure all input and output datasets fall within a uniform range. Preprocessing prevents certain features from dominating the model, thereby enhancing its performance. Cross-validation, particularly the K-fold method, is widely used for parameter selection and to mitigate overfitting issues by dividing the dataset into multiple parts for training and validation. Grid search is another technique used to find optimal hyperparameter configurations, such as the number of neurons, hidden layers, kernel types, and regularization factors, which helps improve accuracy and prevent overfitting. Parallel computing, leveraging both CPUs and GPUs, can significantly reduce the time required for hyperparameter tuning. Once the model is trained, its performance is evaluated using metrics such as mean absolute error (MAE) and maximum absolute error, comparing predicted values against FEA or experimental results. Finally, with a robust regression model in place, designers can simply input the geometric data of a new package and obtain the reliability life cycle almost instantaneously.

### *7.3.2 Classification*

Classification in machine learning, as shown in Fig. 7.6, is a type of supervised learning where the goal is to train a model that can assign input data to predefined categories or classes. For example, given a set of features, such as temperature, size, shape, or image data, the model learns patterns from labeled examples and then uses that knowledge to classify new, unseen inputs into one of the categories it has learned. This approach is widely used in tasks like image recognition, medical diagnosis, and quality inspection.

In the context of electronic fabrication, classification plays an important role in quality control and process automation. By applying classification techniques to data collected from sensors, cameras, or testing equipment, manufacturers can

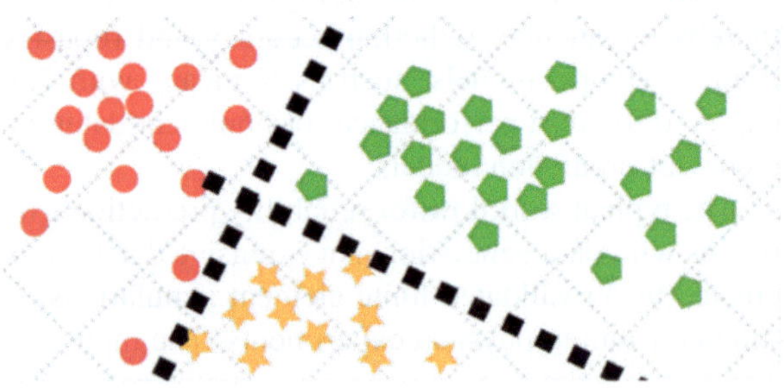

**Fig. 7.6** Classification illustration

automatically detect and categorize defects in electronic components. For instance, based on optical images, machine learning models can be trained to recognize various types of surface defects, such as cracks, voids, or delamination. Similarly, process-related data, such as pressure, temperature, and time during soldering or bonding, can determine whether a manufacturing process will likely yield a pass or fail result.

The use of classification models in electronic packaging helps reduce human error, increase throughput, and maintain consistent product quality. It also supports predictive maintenance and process optimization by identifying patterns that lead to failures, allowing engineers to intervene before producing defective products. In this way, machine learning-based classification becomes a valuable tool in advancing smart manufacturing within the electronics industry. Some algorithms frequently employed for classification problems include Support Vector Machines (SVM), which work by finding the hyperplane that best separates data points into classes, defined by the support vectors, and maximizes the margin between them. With kernel functions, it can handle nonlinear and complex decision boundaries. K-Nearest Neighbors (KNN) is another versatile algorithm capable of classification, where the predicted category for a new data point is determined by a majority vote among its K-nearest neighbors. Similarly, Random Forest (RF), an ensemble method, combines multiple decision trees for robust classification, summarizing predictions across all trees. Convolutional Neural Network (CNN) is also recognized as a supervised learning algorithm that can be applied to classification tasks, particularly for high-dimensional, structured data like images.

Although various machine learning algorithms, including those capable of classification, such as convolutional neural networks (CNN), KNN, and RF, are mentioned, their applications within these references [2–7] for electronic packaging are primarily framed as regression tasks for estimating continuous properties. For instance, a proposed AI-driven workflow aims to estimate equivalent material properties of printed circuit boards (PCBs), such as equivalent coefficient of thermal expansion and Young's modulus, by training a CNN model on PCB patterns. This task yields continuous numerical outputs, aligning with a regression objective, rather than classifying PCB patterns into discrete categories.

### 7.3.3 *Specific Electronic Packaging Applications and Algorithms*

In the domain of electronic packaging, machine learning is primarily applied to supervised regression problems, focusing on predicting continuous numerical values rather than discrete categories. This emphasis on continuous outcomes is crucial for tasks such as predicting reliability life or material properties. The key applications of machine learning within electronic packaging, as detailed in the electronic packaging-related applications, include:

- Reliability Prediction [2–7]: Machine learning models are trained using geometric design parameters of the package as inputs and the number of thermal cycles to failure as the output, enabling rapid reliability prediction for wafer-level packaging structures. This significantly reduces the time cost compared to traditional experiments and simulations. The Random Forest method, for instance, has been applied to predict WLP reliability life, showing good agreement with simulation results. Other algorithms, such as K-nearest neighbors, Kernel Ridge Regression (KRR), Gaussian Process Regression (GPR), and Polynomial Regression (PR), have also been used for WLP reliability prediction.
- Material Parameter Extraction [14]: An Artificial Neural Network approach has been employed to fit the stress-strain behavior of complex materials, such as SAC305 solder balls. This approach enables the development of predictive models for material parameters governed by constitutive models, such as the Garofalo, Chaboche, and Anand models. This method effectively mitigates sensitivity to initial values, a common issue in traditional fitting methods, such as the least squares method.
- Equivalent Material Property Prediction: The study proposes a workflow using Generative Adversarial Networks (GAN) to synthesize substrate or printed circuit board (PCB)-like patterns from a limited number of real PCB images. Subsequently, a Convolutional Neural Network (CNN) is trained to predict the equivalent material properties (e.g., coefficient of thermal expansion (CTE), Young's modulus, Poisson's ratio) from these generated patterns.
- Data Reduction and Sampling [2]: Small-data machine learning methodologies are explored to overcome the limitation of rapidly increasing training data requirements with more input features. An efficient data sampling strategy, such as K-medoids clustering, is introduced for data reduction and prototype selection to enhance sampling efficiency in training ML models, especially under small-data conditions.
- Ensemble Learning [15, 16]: This technique involves combining multiple base models, such as ANN, GPR, and Recurrent Neural Network (RNN), to improve prediction accuracy and robustness, particularly under limited data conditions or with small datasets. Ensemble learning is considered a "training strategy" rather than a specific algorithm. Bagging, boosting, and stacking are three ensemble learning approaches mentioned.

Common machine learning algorithms suitable for or applied in AI-assisted DoS technology for electronic packaging, as mentioned across the references [2–7]. Some studies have also explored hybrid models, such as Gaussian mixture model-KMeans-Passive Aggressive Regression (GMM-KMeans-PAR), and K-medoids clustering for specific tasks. The general machine learning process involves building various training datasets using validated simulation methodologies [2–7], applying test data to evaluate performance, and adjusting hyperparameters during training, without resorting to a grid search function in some cases. A brief description of these commonly used ML algorithms for electronic packaging is as follows: the detailed description will be illustrated in the next few sections.

**Artificial neural network (ANN)** is a computational model inspired by the structure of biological neural systems and is typically implemented as a multilayer feedforward network to approximate complex target functions. They are well-suited for modeling complex nonlinear relationships. They are primarily used for regression problems to predict continuous values, such as the reliability life cycles of WLP. ANN adjusts weights, deviations, and hidden layers through algorithms like backpropagation, offering strong learning capabilities. The accuracy of ANN models generally improves with larger datasets. However, increasing the number of neurons and hidden layers, while enhancing nonlinearity, can lead to high computational complexity, overfitting, and longer training times. Additionally, ANN can serve as a base model in ensemble learning approaches, such as bagging, to enhance prediction accuracy and robustness for small datasets.

**Recurrent Neural Network (RNN)** is a type of neural network that models "time-like" series data and adopts a nonlinear structure, where connections between nodes form a directed network in a sequence, allowing the output from a previous layer to feed back into the input for time-dependent modeling. RNN, particularly Long Short-Term Memory (LSTM) network, excels at handling sequential data and capturing long-range dependencies, which helps solve the vanishing gradient problem. RNN is used for predicting WLP reliability. While RNN can predict reliability more precisely than ANN in some cases, they generally require significantly more training time when using a CPU environment. However, utilizing a GPU environment can reduce RNN's training time to an acceptable range while maintaining performance trends.

**Support Vector Regression (SVR)** is a machine learning algorithm that evolved from Support Vector Machine (SVM) and is well-suited for high-dimensional features. It transforms data into a high-dimensional feature space and applies an $\varepsilon$-insensitive loss function to perform linear regression, aiming to minimize the norm of weights to avoid overfitting. SVR provides global optimal solutions and is widely adopted in various fields. It can utilize different kernel functions, such as linear, polynomial, Radial Basis Function (RBF), and Sigmoid, with the RBF kernel often exhibiting better accuracy and CPU time performance. Overall, SVR generally offers good accuracy comparable to KNN, KRR, and RF.

**K-Nearest Neighbors Regression (KNN)** is a simple, non-parametric, instance-based (lazy) learning algorithm that stores all available cases and predicts new data based on similarity measures, often distance. It approximates the function locally by considering surrounding neighborhoods. KNN can be used for both regression and classification problems. Although sometimes considered more suitable for classification, it demonstrated good regression accuracy for the reliability of the life cycle prediction of WLP [7].

**Kernel Ridge Regression (KRR)** combines ridge regression with the "kernel trick" to learn linear functions in a feature space induced by the kernel, allowing it to handle nonlinear functions in the original data space. KRR is often simpler and faster to train than SVR due to its closed-form solution, and can achieve better performance. It utilizes various kernel functions, including RBF, Sigmoid, and polynomial. The RBF kernel generally provides good accuracy for KRR. KRR has been extensively used to predict WLP reliability. KRR, when combined with K-means clustering (K-K hybrid model), can significantly reduce training time while maintaining similar prediction accuracy, making it an extremely efficient approach. K-means partitions the entire dataset into K subsets. A separate KRR model is then trained on each subset. When a new input arrives, it is first assigned to a cluster by the K-means model, and then the corresponding KRR predictor is applied. The K–K framework essentially reduces the size of the training data used in each individual model, thereby decreasing CPU time, especially in the case of large training datasets.

**Random Forest (RF)** is an ensemble learning method that builds a collection of decision trees, typically fully grown and unpruned Classification and Regression Trees (CARTs). It employs the bagging method to create bootstrapped training subsets for each tree, randomly selecting features for splitting, which helps reduce correlation between trees and improve overall accuracy. RF then summarizes the prediction results from all decision trees. RF is known for its multivariable predicting accuracy without excessive computational effort and is less prone to overfitting, performing relatively stably even with missing values or non-equilibrium data. It has been applied to WLP reliability prediction. The calculation time for RF is generally very fast, making it suitable for analyzing large datasets. Under certain conditions, Extra Trees (ET) can outperform RF, which may be attributed to its increased level of randomness in node splitting. Some researchers note that RF's accuracy in regression tasks may not be as high as that of ANN, RNN, SVR, or KRR, as it is considered more suitable for classification purposes.

**Polynomial Regression (PR)** is a model used for nonlinear relationships between input and response variables, where the scatter plot shows a curvilinear structure. It can be considered an extended version of the linear regression model. PR is noted for being a simplified model that is very fast and reduces CPU time. It can transform lower-dimensional feature spaces to higher dimensions, which impacts performance and reduces computational time. PR has been utilized to predict the reliability of electronic packaging. While slightly more complex and requiring more CPU time than simple linear regression due to its multivariable attributes and higher-order polynomials, PR's accuracy for WLP reliability is comparable to KRR and GPR, and better than KNN. Its CPU time is generally very small, especially for smaller datasets. When combined with an unsupervised learning algorithm like K-means, PR can produce more accurate predictions and significantly reduce training time, particularly for large datasets, allowing simpler models to perform well.

**Gaussian Process Regression (GPR)** is a generic supervised learning method designed for regression and probabilistic classification problems. It implements a Gaussian process where the prior mean is typically zero or the mean of the training data, and the covariance is defined by a kernel object, with hyperparameters optimized by maximizing the log-marginal likelihood. GPR has been applied to predict WLP reliability [7]. It is noted for exhibiting the best accuracy among the compared algorithms in some studies, with good accuracy stemming from its use of kernel functions for nonlinear datasets and its simplicity as a model. However, a significant drawback of GPR is that it can take a considerable amount of time to train the model compared to other algorithms, often having the highest CPU time among the ones compared.

**Convolutional Neural Network (CNN)** is a variant of ANN and falls under supervised learning methods. They are particularly adept at image recognition and image classification tasks, excelling at extracting spatial features from image data. CNN is ideal for predicting material properties from images, such as estimating the equivalent material properties of PCB patterns (including coefficient of thermal expansion, Young's modulus, and Poisson's ratio).

**K-medoids and K-means Clustering** are unsupervised learning algorithms used for clustering, where they partition a dataset into distinct groups based on similarity measures, typically distance. A key distinction from K-means is that K-medoids chooses actual data points as cluster centers, while K-means computes centroids that may not correspond to any real sample. This algorithm is primarily used for data reduction and sampling and for improving the spatial distribution of training datasets, which is especially beneficial for small-data training. These two clustering algorithms serve entirely different purposes. K-Medoids reduces the data volume by using the medoid of each cluster as the representative data point, where the number of clusters K essentially corresponds to the number of sampled points. In contrast, K-Means partitions the entire dataset into K subsets. E.g., for KRR, a separate KRR model is then trained on each K subset. When a new input arrives, it is first assigned to a cluster by the K-means model, and then the corresponding KRR predictor is applied. The K–K framework essentially reduces the size of the training data used in each individual model, thereby decreasing CPU time, especially in the case of large training datasets.

**Generative Adversarial Network (GAN)** are unsupervised learning techniques and prominent models in the field of generative AI. Their primary purpose is to synthesize new, realistic samples from an underlying data distribution, such as PCB-like patterns based on a small set of real PCB images. This capability is useful for augmenting training datasets in challenging data acquisition scenarios.

## 7.4 Generalization, Overfitting, and Underfitting

In the context of machine learning for electronic packaging, concepts such as generalization, overfitting, and underfitting are critical for developing robust and reliable predictive models. The references [2–7] emphasize these aspects, particularly in the context of predicting reliability life cycles of wafer-level packaging and extracting material parameters, which primarily involve supervised regression problems.

### 7.4.1 Generalization

Generalization refers to a machine learning model's ability to accurately predict outcomes for new, unseen data that was not part of its training set. In electronic packaging, achieving good generalization is paramount because models are often trained on large datasets generated through finite element simulations. However, their ultimate purpose is to predict the reliability or material properties of real-world physical packages or designs. References [2–7] highlight efforts to assess generalization by:

- Applying testing data: After training, models are evaluated on separate testing datasets that fall within the feature boundary but were not used during training, to assess their performance on unseen data.
- Analyzing average testing error/difference: The goal is for this metric to decrease and converge as the training dataset size increases gradually. This indicates improved generalization.
- Cross-validation: This technique helps to evaluate model performance and identify optimal parameters by splitting the dataset into multiple subsets for training and validation, ensuring that the model generalizes well to new data and avoids overfitting.

### 7.4.2 Overfitting

Overfitting occurs when a machine learning model learns the training data, including noise and outliers, too precisely, leading to excellent performance on the training set but poor performance on new, unseen data. References [2–7] extensively discuss identifying and preventing overfitting. Symptoms of overfitting:

- A large difference between training and testing error.
- Models showing "perfect" training difference (e.g., zero error) often necessitate checking for overfitting.
- High computational complexity and excessive hidden layers or neurons in models like ANN or RNN can lead to overfitting. While good for high-dimensional features, SVR may not be suitable for small datasets and could lead to overfitting without proper regularization. Polynomial Regression with an excessively high degree can decrease the model's generalization ability and lead to overfitting.

Mitigation Strategies:

- Cross-validation: K-fold cross-validation is commonly used to select parameters and prevent overfitting by ensuring the model is tested on data it hasn't seen during a specific training fold.
- Regularization: Techniques like ridge regression (in KRR) inherently incorporate regularization (e.g., parameter $\lambda$ or $\alpha$) to penalize large weights and prevent

overfitting, even when converting a lower-dimensional feature space to a higher-dimensional space.

- Data Quantity and Quality: Increasing the training dataset size generally helps reduce overfitting and improve performance, as shown for various algorithms.
- Adaptive Sampling: Strategies such as K-medoids clustering are used to improve the distribution of limited training data by focusing on "high-interest regions" (e.g., near feature boundaries where prediction errors are higher), thereby enhancing model performance without exponentially increasing the data volume.
- Ensemble Learning: Techniques such as bagging, boosting, and stacking combine multiple base models to enhance overall accuracy and robustness, effectively reducing variance and the risk of overfitting in individual models, especially with small datasets.
- Choosing appropriate algorithms: Certain algorithms, like Random Forest (RF), are noted for being less prone to overfitting due to their internal randomness and ensemble nature.

### *7.4.3 Underfitting*

Underfitting occurs when a machine learning model is too simplistic to capture the underlying patterns and relationships in the training data, resulting in poor performance on both the training set and new data. This means the model hasn't learned enough from the available data. While the sources don't explicitly use the term "underfitting" as often as "overfitting," they address the concept indirectly through:

- Ensuring sufficient model complexity: The goal is to achieve a low "average training difference" or "training error," indicating that the model has successfully learned the trend of the training data.
- For an Artificial Neural Network (ANN), this involves selecting an appropriate number of hidden layers and neurons. Insufficient complexity can lead to poor training performance.
- For Polynomial Regression (PR), selecting an appropriate polynomial degree ensures the model can capture nonlinear relationships. A very low degree might result in underfitting.
- Data preprocessing methods, such as MinMax Scaler or Robust Scaler, scale feature values, which can optimize model performance by preventing certain features from dominating the learning process due to large value ranges. This helps the model better capture underlying trends and avoid underfitting.

The ultimate aim in electronic packaging reliability prediction is to find a balance that yields a model with strong generalization ability, successfully avoiding both overfitting (where the model is too complex and sensitive to training data noise) and underfitting (where the model is too simple to learn the data's true patterns). This balance is achieved through careful data generation, sampling, algorithm selection, hyperparameter tuning, and validation strategies.

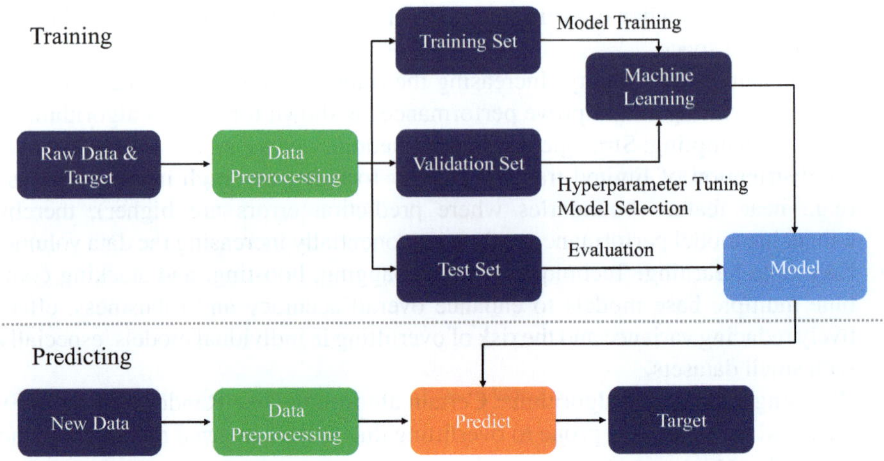


**Fig. 7.7** Methodology flow chart

## 7.5 Training Methodology

AI training methodology systematically prepares data, trains models, and evaluates their performance to achieve accurate and reliable predictions. This process is crucial for applications such as predicting the reliability of wafer-level packaging [7] or predicting warpage in fanout panel-level packaging. The methodology flowchart is illustrated in Fig. 7.7.

### *7.5.1 Data Preprocessing*

To identify the most effective machine learning model, one first needs to train the model with the input data using, e.g., Python, which offers robust support for machine learning. Before developing the model, verifying whether data preprocessing has been applied is important, as both input and output datasets need to be scaled. This is crucial because the values in the dataset often vary in scale. Proper data preprocessing ensures that the data is scaled, which can significantly improve model performance. This study employs various preprocessing techniques, such as Min-Max Scaler, Robust Scaler, Max Absolute Scaler, and Standard Scaler, to prepare the data for regression-based machine learning models. Figure 7.8 illustrates the various methods for rescaling and preprocessing a dataset [17]. The figure also displays a synthetic two-class classification dataset with two characteristics in the first plot. The first characteristic (the x-axis value) ranges from 10 to 15. The y-axis value (the second feature) is somewhere between 1 and 9. After that, the different preprocessing methods [18] will be discussed as follows:

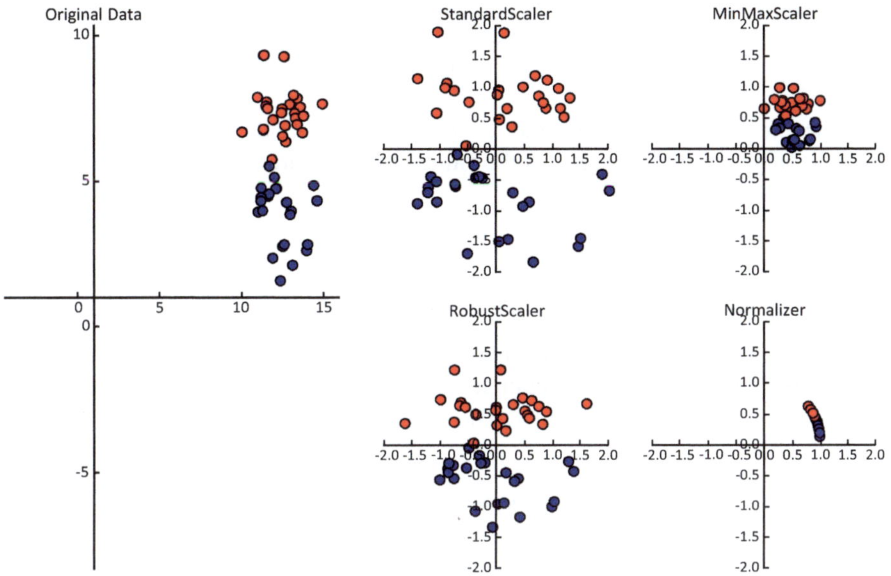


**Fig. 7.8** Various methods for rescaling and preprocessing a dataset [17]

#### 7.5.1.1 Standard Scaler Methods

The Standard Scaler ensures that the mean of each feature is zero and the variance is one, thereby standardizing the scale of all features to the same level. This scaling, however, does not guarantee that the features have specific minimum and maximum values. The mathematical formula for the standard scalar is as follows:

$$x' = \frac{x - \mu}{\sigma} \tag{7.1}$$

Where $x$ is the value of the feature, $\mu$ and $\sigma$ are the mean and standard deviation values of the feature; $x^{'}$ is the value of the feature after feature scaling.

#### 7.5.1.2 Robust Scaler Method

Like the Standard Scaler, the Robust Scaler scales features to bring them onto a comparable scale. However, unlike the Standard Scaler, which uses the mean and standard deviation, the Robust Scaler relies on the median and interquartile range (IQR). This makes it less sensitive to outliers, which are extreme values that can distort the scaling process in other methods. By focusing on the central distribution of the data, the Robust Scaler effectively minimizes the impact of anomalies such as measurement errors. The formula for the Robust Scaler is:

$$x' = \frac{x - Q_2}{Q_3 - Q_1} \tag{7.2}$$

Where $x$ is the value of the feature; $Q_1$, $Q_2$, and $Q_3$ are the first, second, and third quartile values of the feature; $x'$ is the value of the feature after feature scaling.

#### 7.5.1.3 Minmax Scaler Method

The MinMax Scaler transforms data such that all features are scaled to a specific range, typically between 0 and 1. This means that for a two-dimensional dataset, all data points will fall within a rectangle defined by the x-axis, ranging from 0 to 1, and the y-axis, ranging from 0 to 1. The equation for the Min-Max Scaler is:

$$x' = \frac{x - x_{\min}}{x_{\max} - x_{\min}} \tag{7.3}$$

where $x$ is the value of the feature; $x_{\min}$ and $x_{\max}$ are the minimum and maximum values of the feature; $x'$ is the value of the feature after feature scaling.

#### 7.5.1.4 MaxAbs Scaler Method

The MaxAbs Scaler is similar to the MinMax Scaler, but it scales data based on the maximum absolute value of each feature. The output range depends on the presence of negative or positive values: if the data contains only positive values, the range is (0, 1); if it contains only negative values, the range is (−1, 0); and if both positive and negative values are present, the range becomes (−1, 1). Both MaxAbs Scaler and MinMax Scaler behave similarly when applied to strictly positive data. However, like MinMax Scaler, the MaxAbs Scaler is still sensitive to large outliers, which can distort the scaling. The formula for MaxAbs Scaler is:

$$x' = \frac{x}{\max\left(\mathrm{Abs}\left(x\right)\right)} \tag{7.4}$$

Where $x$ is the value of the feature; Max(Abs($x$)) is the maximum of the absolute value of the feature; $x'$ is the value of the feature after feature scaling.

#### 7.5.1.5 Normalizer Method

The Normalizer performs a distinct type of scaling by rescaling each data point (i.e., row) to have a unit norm. Specifically, it transforms each feature vector so that its Euclidean norm equals 1. In geometric terms, this projects each data point onto the surface of a unit circle (or a unit sphere in higher dimensions). This process involves multiplying each data point by the inverse of its length. Normalization is particularly useful when the data's direction (or angle) is more important than its magnitude. The formula for the Normalizer is:

$$x' = \frac{x}{\|x\|_p} \tag{7.5}$$

where $x$ is the value of the feature; $\|x\|_p$ is the p-norm value; $x'$ is the value of the feature after feature scaling, and p-norm is calculated by the following equation:

$$\|x\|_p = \left( \sum_{i=1}^{n} |x_i|^p \right)^{1/p}, p = 1, 2, \ldots \tag{7.6}$$

where $n$ is the number of features, the 1-norm and 2-norm are most commonly used.

Hence, among all preprocessing methods, selecting one that can help obtain the best model and a more accurate predicted output from the input dataset is essential.

## 7.5.2 *Cross-Validation*

Cross-validation is an indispensable technique for parameter selection across various machine learning algorithms. Its fundamental principle involves reserving a portion of the dataset, not for training, but for evaluating model performance under different parameter configurations. This allows researchers to assess how well the model generalizes to unseen data. In essence, the training data is used to fit the model with various parameter configurations. Each configuration is then evaluated on a separate validation set to identify the most accurate and robust model parameters. Several types of cross-validation methods exist, which will be discussed in further detail.

### 7.5.2.1 K-Fold Cross-Validation

Cross-validation is a statistical technique for evaluating a model's generalization performance more robustly and comprehensively than a simple train–test split. Cross-validation repeatedly partitions the data instead of using a single split and trains multiple models to assess performance. The most commonly used method is K-fold cross-validation, where the dataset is divided into K equal (or nearly equal) parts or "folds." Typical values for K are 5 or 10. In five-fold cross-validation, for example, the data is split into five folds. Each iteration uses one fold as the validation set, while the remaining four folds are used for training. This process repeats five times, with each fold serving as the test set once. As a result, five models are

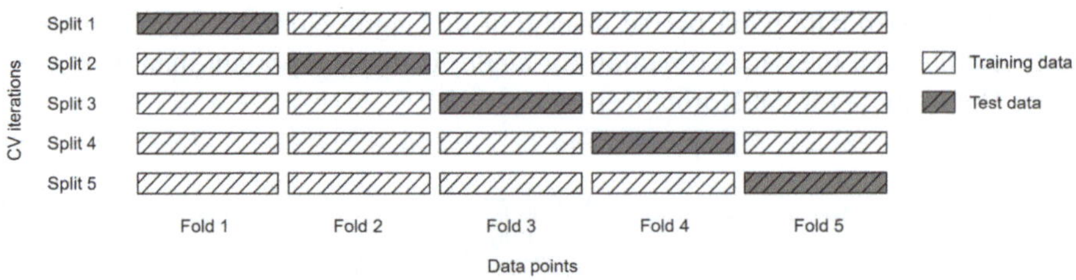


**Fig. 7.9** Five-fold cross-validation data splitting

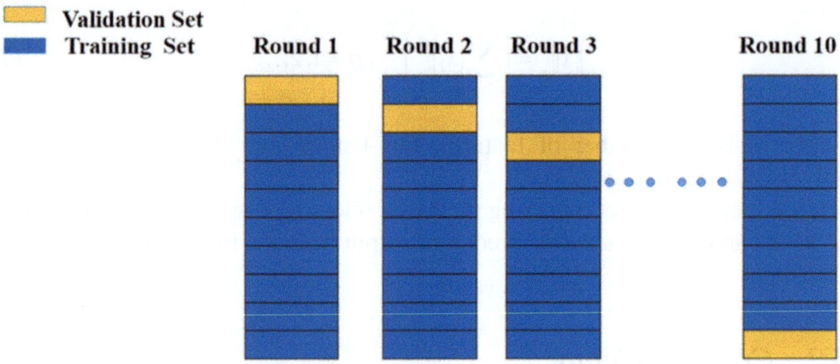


**Fig. 7.10** Ten-fold cross-validation model diagram

trained and evaluated, producing five accuracy scores. The average of these scores gives a more reliable estimate of the model's performance.

This process is illustrated in Fig. 7.9, which shows how the data is split during five-fold cross-validation. Figure 7.10 demonstrates ten-fold cross-validation, dividing the dataset into ten parts. Each part serves as a validation or training set at different stages. These validation sets can also be used to estimate how well the model is likely to perform on unseen data [17].

#### 7.5.2.2 Leave-One-out Cross-Validation

Leave-one-out is another popular cross-validation strategy. Leave-one-out cross-validation is similar to K-fold cross-validation in that each fold represents a single sample. One can select a single data point as the validation set for each split. This process can take considerable time, especially for large datasets, but it can yield better results on smaller datasets [17].

#### 7.5.2.3 Shuffle-Split Cross-Validation

Shuffle-split cross-validation is another flexible approach to cross-validation. Each split randomly divides the dataset into a training and a disjoint validation set,

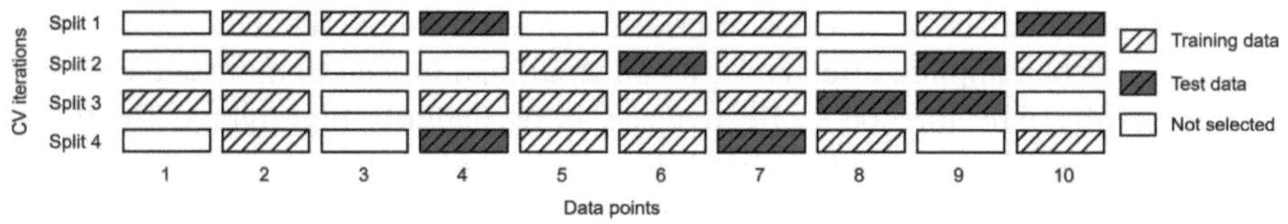


**Fig. 7.11** Shuffle-split with 10 points cross-validation

repeating the process n times. Figure 7.11 shows four iterations of partitioning a 10-point dataset into a training set of 5 points and test sets of 2 points each (researchers can use integers for train and test sizes to use absolute sizes for these sets, or floating-point values to use fractions of the entire dataset) [17].

Shuffle-split cross-validation enables you to specify the number of iterations independently of the training and validation sizes, which may be beneficial in certain cases. It also permits only a portion of the data to be used in each iteration by supplying train and validation sizes that do not sum up to one. Experimenting with huge datasets can be made easier by subsampling the data in this fashion [17]. Due to the significant computational burden of the shuffle-split and leave-one-out methods, many researchers often choose the K-fold method. After the data preprocessing method was confirmed, the cross-validation technique will usually be employed to avoid the overfitting problem in the machine learning model.

### *7.5.3 Grid Search*

Grid search is a fundamental and widely used method in machine learning for hyperparameter optimization, aiming to find the optimal combination of hyperparameters for a given model. This technique systematically evaluates different hyperparameter configurations to enhance a model's performance, particularly in improving the accuracy of model predictions [2–7, 14–16]. The mechanism of grid search involves a systematic and exhaustive exploration of a predefined hyperparameter space. The model builder first specifies a search range for each hyperparameter. Grid search then divides these ranges into several discrete parts and runs the model with every possible combination of hyperparameter values within these segments. For instance, when tuning Support Vector Regression, grid search can iteratively refine the optimal values for parameters such as gamma ($\gamma$) and epsilon ($\varepsilon$) by uniformly subdividing their defined ranges [18]. Similarly, for Artificial Neural Network, it helps determine the optimal number of neurons per layer, alongside other hyperparameters like the activation function, optimizer, learning rate, and training epochs. It can also be employed to find optimal initial coefficients for complex constitutive models, such as those presented by Anand and Chaboche [14]. The GridSearchCV method, available in scikit-learn for Python, is noted as a suitable tool for this purpose.

One of the primary advantages of grid search is its simplicity and ease of implementation. When applied effectively, it can significantly lead to more accurate prediction results. The method's effectiveness is particularly highlighted for Radial Basis Function (RBF) kernels in SVR due to their reliance on only one parameter ($\gamma$), which simplifies the search space. In some cases, its application can also speed up the overall process of obtaining accurate results. Despite its simplicity, grid search has notable limitations. It can be quite time-consuming, especially when dealing with a large number of hyperparameters or when the defined search ranges are wide. The time required for grid search also increases with the size of the dataset being processed. Some studies mentioned in the sources chose not to apply grid search and instead fixed hyperparameters during the training process, likely to mitigate the extensive computation time. However, it is generally preferred over manual hyperparameter tuning. To address the time consumption issue, particularly in complex models, grid search can be combined with parallel computing architectures, leveraging both Central Processing Units (CPUs) and Graphics Processing Units (GPUs) to reduce the search time and enhance efficiency. Parallel processing provides faster grid search times, and the difference becomes more significant as the dataset size increases.

Grid search finds extensive application in predicting Wafer-Level Packaging reliability, where it is used across various machine learning algorithms, including ANN, RNN, SVR, Kernel Ridge Regression, Gaussian Process Regression, Polynomial Regression, and GMM-KMean-Passive Aggressive Regression. For instance, it has been utilized to select the optimal number of neurons for ANN models and to define optimal initial values for parameters in Anand's constitutive model.

## *7.5.4 Error Estimation Methods [7, 18]*

Similar to classification, regression performance can be assessed by analyzing overpredictions and underpredictions relative to the target. In most cases, however, using the default $R^2$ scoring method in most regressors is sufficient. While the Mean Squared Error (MSE) is a prevalent choice for regression loss functions, AI researchers employ a range of additional metrics to evaluate regression models. These often include the Mean Absolute Error, the Maximum Absolute Error, the Mean Absolute Percentage Error, and the Mean Squared Logarithmic Error. The following are the error metrics that AI researchers employ:

**1. Mean Absolute Error**

The mean absolute error function calculates mean absolute error, a risk metric equal to the predicted value of the absolute error loss, also known as $l_1$ norm loss.

$$\text{MAE}(y,\hat{y}) = \frac{1}{n}\sum_{i=1}^{n}|y_i - \hat{y}_i| \tag{7.7}$$

where, $\hat{y}_i$ is the predicted value of the $i^{th}$ sample, and $y_i$ is the corresponding true value, $n$ is the number of samples, and MAE is a metric used to estimate the average magnitude of errors in predictions.

### 2. Mean Squared Error

The mean squared error function calculates the expected value of a squared (quadratic) mistake or loss, which is a risk metric and is defined as:

$$\text{MSE}(y,\hat{y}) = \frac{1}{n}\sum_{i=1}^{n}(y_i - \hat{y}_i)^2 \tag{7.8}$$

where, $\hat{y}_i$ is the predicted value of the $i^{th}$ sample, and $y_i$ is the corresponding true value, $n$ is the number of samples, and MSE, or Mean Squared Error, is a metric used to estimate the average of the squared differences between predicted and actual values.

### 3. Max Error

The maximum residual error, a metric that represents the worst-case error between the predicted and actual values, is computed by the max error function.

$$\text{Max Error}(y,\hat{y}) = \text{Max}(|y_i - \hat{y}_i|) \tag{7.9}$$

where, $\hat{y}_i$ is the predicted value of the $i^{th}$ sample, and $y_i$ is the corresponding actual value, $n$ is the number of samples, and Max Error is estimated.

### 4. Mean Absolute Percentage Error

The mean absolute percentage error (MAPE), also known as the mean absolute percentage deviation (MAPD), is a key evaluation metric for regression problems. This statistic is designed to be sensitive to relative errors.

$$\text{MAPE}(y,\hat{y}) = \frac{1}{n}\sum_{i=1}^{n}\frac{|y_i - \hat{y}_i|}{\max(\epsilon, |y_i|)} \tag{7.10}$$

where, $\hat{y}_i$ is the predicted value of the $i^{th}$ sample, and $y_i$ is the corresponding true value, $n$ is the number of samples, and MAPE is the Mean absolute percentage error. To avoid undefined results when $y$ is zero, $\in$ is an arbitrary small but strictly positive number.

**5. $R^2$ Score:**

It denotes the fraction of $y's$ variance that can be explained by the model's independent variables. The proportion of explained variance indicates the model's goodness of fit, which is a measure of how well unseen samples are likely to be predicted by the model.

$$R^2\left(y,\hat{y}\right)=1-\frac{\sum_{i=1}^{n}\left(y_i-\hat{y}_i\right)^2}{\sum_{i=1}^{n}\left(y_i-\overline{y}\right)^2} \tag{7.11}$$

where, $\hat{y}_i$ is the predicted value of the $i^{th}$ sample, and $y_i$ is the corresponding actual value, $n$ is the number of samples $\overline{y}$ is the mean value of $y_i$.

## 7.6 Machine Learning Algorithms

Many machine learning algorithms can be applied to electronic packaging areas, including ANN, RNN, Random Forest, Extra Trees, and SVR. This section provides a detailed introduction to machine learning procedures and theories.

### *7.6.1 Artificial Neural Networks*

Artificial Neural Networks (ANNs) emerged from early attempts to model the human nervous system, with McCulloch and Pitts [19] proposing the foundational concept in 1943 by introducing a logical calculus method for simulating biological neural networks. This was followed by Rosenblatt's [20] introduction of the Perceptron in 1957 and 1958, which significantly advanced the overall architecture of ANN. Krizhevsky et al. [21] applied backpropagation with GPU acceleration to develop AlexNet. The backpropagation enables ANNs to solve more complex problems and has become one of today's most widely used algorithms.

The theoretical basis of ANN lies in their design to mimic the information processing capabilities of the human nervous system. A typical ANN, often referred to as a multi-layer perceptron (MLP), is structured with three primary types of layers: an input layer, one (Fig. 7.12) or multiple hidden layers (Fig. 7.13), and an output

Fig. 7.12 Structure of single hidden layer ANN

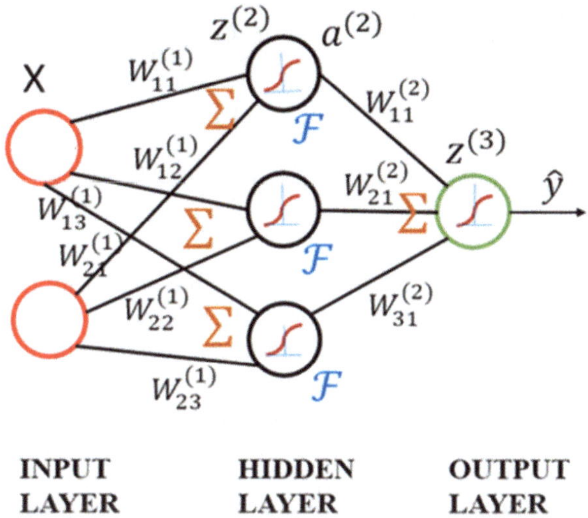


Fig. 7.13 Structure of a multiple hidden layer ANN

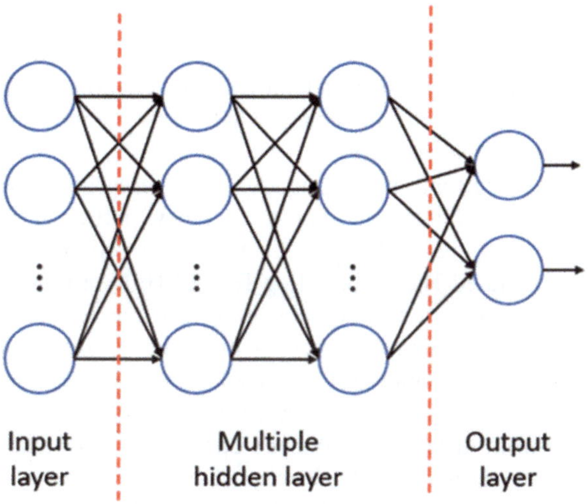


layer [22, 23]. The input layer receives the raw feature values of the training data. The hidden layer(s), situated between the input and output layers, perform computations by multiplying the received data by weights and linking them to individual neuron nodes. As shown in Fig. 7.14, the basic unit of a neural network receives numerical inputs, each associated with an individual weight. These inputs are combined via a weighted summation within the neuron and subsequently passed through an activation function to generate the final output.

The crucial optimization of these weights is achieved through the backpropagation algorithm. This algorithm efficiently calculates the gradient of the loss function

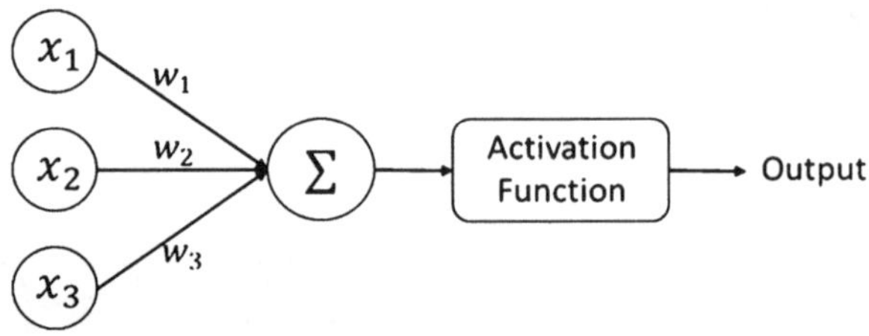


**Fig. 7.14** Schematic of a single neuron

with respect to all weights, using the chain rule to propagate gradients backward from the output layer to the input layer to facilitate updates that minimize the loss. Increasing the number of hidden layers and neurons within them enhances the network's nonlinearity, making it more adept at fitting complex data, although this also increases the risk of overfitting if not properly regularized. The output layer then takes the computed results from the last hidden layer, transforms them using an activation function, and produces the final result. Key components also include activation functions such as Sigmoid, Hyperbolic Tangent (Tanh), and Rectified Linear Unit (ReLU), which introduce nonlinearity, enabling the network to process more complex models. A loss function, such as Mean Squared Error (MSE) for regression, quantifies the difference between the model's predicted output and the actual data, with the training goal being to minimize it. Optimizers, such as Gradient Descent, Stochastic Gradient Descent (SGD), AdaGrad, RMSProp, and Adam, are algorithms used to update the network's parameters (weights and biases) to minimize the loss function.

In electronic packaging, ANN is increasingly employed to address traditional design challenges and boost efficiency, particularly for reliability assessment and prediction [2–7, 22, 23]. Conventional reliability assessment methods, such as Accelerated Thermal Cycling Tests (ATCT) or Finite Element Method (FEM) simulations, are often time-consuming and costly. FEM results can also vary among researchers due to differences in model construction and solution procedures. AI/Machine Learning provides a solution by integrating these varying results and statistically deriving optimal regression models, which can then unify the errors resulting from different personnel's operations. For instance, ANN is applied to predict the reliability of wafer-level package structures by incorporating data from FEM simulations. This enables the quick estimation of packaging structure reliability within a reasonable size range by utilizing an AI model derived from the regression model. This capability is especially valuable for predicting continuous values (regression problems), such as the life cycles of packaging structures with various geometric parameters, particularly when the relationship between these parameters is complex and nonlinear, where linear regression might fail.

Beyond reliability prediction, ANN is also vital for material model fitting. The accuracy of material models and their coefficients significantly influences simulation outcomes [14, 25]. ANN can be used to fit precise material model curves, such as the Chaboche model, based on reliable experimental data (e.g., Anand model data), thereby enhancing the reliability of FEM analyses by providing accurate material parameters. By establishing a database of reliability data from validated FE simulations of diverse packaging structures, ANN can be trained to rapidly predict reliability, significantly shortening development time and mitigating discrepancies arising from different research approaches [2–7, 22, 23]. For example, studies have focused on using ANN to predict reliability based on structural parameters like upper pad size, lower pad size, chip thickness, and SBL thickness, optimizing hidden layer neuron counts and layers to find the best regression model.

Furthermore, ANN can achieve accurate predictions with relatively small amounts of data compared to other machine learning methods requiring extensive datasets. Since grid search may require significantly more computing time than training, some research also explores reducing the time required for hyperparameter searching in ANN. This is achieved by leveraging techniques such as Grid Search, combined with parallel computing (CPU and GPU), to improve reliability prediction efficiency further.

#### 7.6.1.1 Activation Functions

In Fig. 7.12, the ANN model, $a_i^l$ is the $i_{th}$ activation element of the $l_{th}$ layer in the hidden layer. $b_i^l$ is the bias, $z_i^l$ is equal to the input value times the weight $W_{ji}^l$ and add the bias in Eq. (7.12) [24]:

$$z_j^{l+1} = \sum_{i=1}^{n} w_{ij}^l a_i^l + b_j^l \tag{7.12}$$

From a calculation point of view, the input layer data is first combined with the bias and weights to obtain a value. That calculated input value, i.e., $a_i^l$ is substituted into different types of activation functions, such as Sigmoid, rectified linear unit (ReLU), Hyperbolic tangent (Tanh), and softmax, to convert them into nonlinear types before entering the next layer. After that, we can find $a_i^{l+1}$ and go to the next layer. These functions are highlighted as follows [24]:

$$a_i^{l+1} = \phi\left(z_i^l\right) \tag{7.13}$$

where, the activation function $\phi$ is shown in Fig. 7.15.

Sigmoid Activation Function

The sigmoid function plays a significant role in the Artificial Neural Network primarily as an activation function. It is described as a smooth S-shaped curve that outputs values between 0 and 1. In the context of ANN calculations, an input value is transformed by the Sigmoid activation function into a nonlinear output before being passed to the next layer. The relationship gives the Sigmoid function:

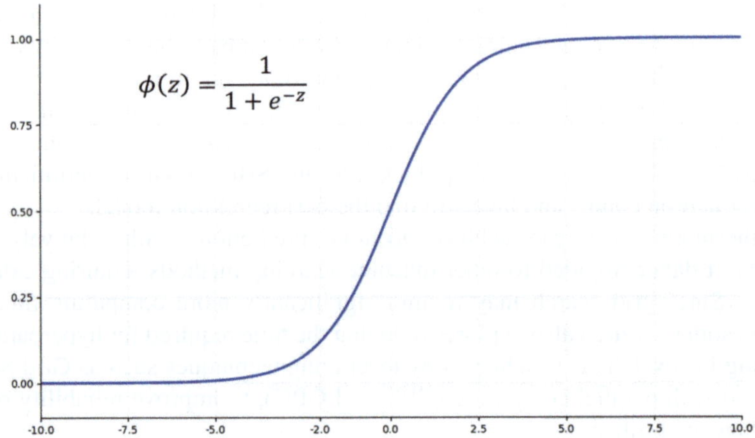


**Fig. 7.15** Sigmoid activation function

$$\varnothing(z) = \frac{1}{1+e^{-z}} \tag{7.14}$$

where, $z$ is the input value and $\varnothing(z)$ is sigmoid activation function.

Rectified Linear Unit (ReLU) Function

The rectified linear unit (ReLU) function is a commonly used activation function in the Artificial Neural Network. It produces values greater than or equal to 0. Key characteristics and comparisons of the ReLU function include that it outputs zero if the input is negative and the input value itself if the input is positive. This means its output value has a lower bound of 0 but no upper limit. Like other activation functions, ReLU introduces nonlinearity into the network's computations, enabling the model to handle complex problems.

A significant advantage of ReLU is that it avoids the "gradient vanishing" problem, which can occur with functions like Sigmoid and Tanh when inputs fall outside a certain interval and their derivatives approach zero. Due to avoiding gradient vanishing, ReLU typically offers faster convergence during the training process. A potential issue with ReLU is that if the inputs to certain neurons are negative, the output and its derivative will be zero, which can prevent parameters from being updated (known as "dead neurons"). In a real application case [23], when comparing activation functions for fitting Chaboche curves, ReLU was noted to have smaller errors in certain temperature ranges and offered faster convergence. Unlike

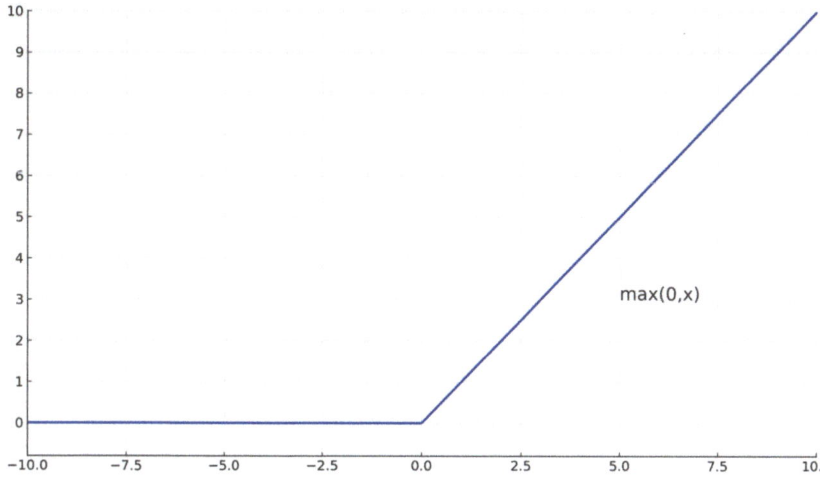


**Fig. 7.16** ReLU activation function

the Sigmoid function, which outputs values between 0 and 1, and the Hyperbolic Tangent (Tanh) function, which outputs values between −1 and 1, ReLU has no upper bound. While Sigmoid and Tanh can suffer from gradient vanishing and slower convergence due to exponential calculations or non-zero-centered outputs (Sigmoid), ReLU generally avoids these issues, leading to faster training. ReLU and Tanh functions performed similarly in a comparison of training cycles, with ReLU showing slightly better performance across most temperatures.

The ReLU activation function applies a threshold operation to each input element, setting any values less than zero to zero. Thus, the ReLU is given by.

$$f(x) = \max(0,x) = \begin{cases} x_i, & \text{if } x_i \geq 0 \\ 0, & \text{if } x_i < 0 \end{cases} \tag{7.15}$$

where, $x$ is the input value, and $f(x)$ is the ReLU activation function (Fig. 7.16).

## Tanh Activation Function

The Hyperbolic Tangent (Tanh) function is also a commonly used activation function in Artificial Neural Network (ANN). It is characterized as a smooth S-shaped curve with output values between −1 and 1. A visual representation of the Tanh function and its derivative is provided in Fig. 7.17 of the sources. Similar to other activation functions, Tanh introduces nonlinearity into the network's calculations, enabling the model to address more complex problems. A significant advantage of the Tanh function is that its output is zero-centered. This property contributes to faster parameter updates during the training process. A notable limitation of the

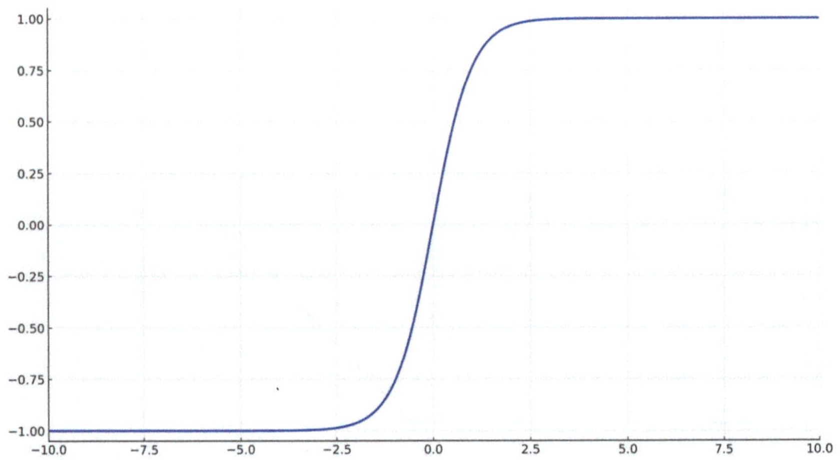


**Fig. 7.17** Tanh activation function

Tanh function is its susceptibility to the "gradient vanishing" problem. When input values fall outside a certain interval, its derivative tends to approach zero, which can impede effective parameter updates, similar to the Sigmoid function.

Tanh and ReLU functions generally offer faster convergence during training than the Sigmoid function. This is attributed to Tanh's zero-centered output and ReLU's ability to avoid the gradient vanishing problem in its active range. The tanh function, also known as the hyperbolic tangent function, is a smoother zero-centered function with a range of −1 to 1, and its output of the Tanh function is given by.

$$f(x) = \frac{e^{x} - e^{-x}}{e^{x} + e^{-x}} \tag{7.16}$$

where, $x$ is the input value, and $f(x)$ is the Tanh activation function.

Compared to the sigmoid function, the tanh function became favored because it provides higher training performance for multi-layer neural networks. The tanh function, on the other hand, alleviates the vanishing gradient problem compared to the sigmoid function. The function's key benefit is that it creates zero-centered output, which aids the backpropagation process. Figure 7.17 shows the picture of the Tanh activation function.

#### 7.6.1.2 The Gradient Descent Algorithm

The Gradient Descent algorithm, also referred to as the steepest descent algorithm, is a first-order optimization algorithm widely used in ANNs. Its primary purpose is minimizing the loss function, quantifying the difference between a model's

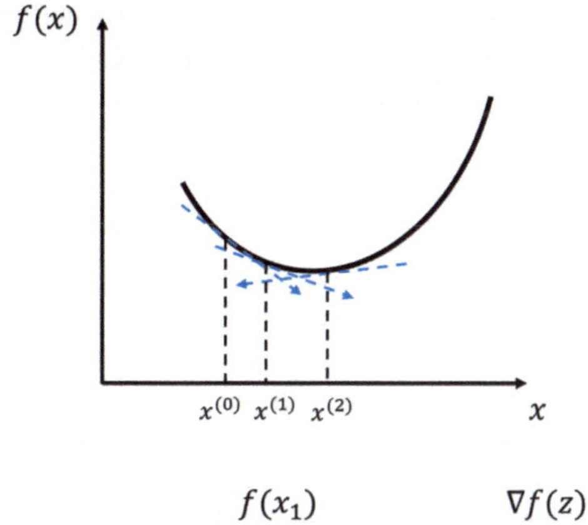


**Fig. 7.18** An example of a gradient search for a stationary point

predicted outputs and target values. Doing so facilitates updating network parameters (weights) to improve the model's predictions. Gradient Descent works by iteratively adjusting the network's weights in the direction opposite to the gradient of the loss function. The gradient indicates the direction of the steepest ascent of the loss function. Therefore, moving in the gradient's negative (opposite) direction helps to descend towards a minimum. One option is starting at an arbitrary point, moving along the gradient at that point to the next point, and repeating this process until (ideally) converging to a stationary point. This is depicted in Fig. 7.18. This process is often part of the backpropagation algorithm, where the error is propagated backward through the network to compute these gradients for each connection weight. Gradient descent is one of the most widely used optimization algorithms and is by far the most frequent method for optimizing neural networks. At the same time, almost every current Deep Learning package will implement several gradient descent optimization methods (e.g., Scikit-learn, Caffe, and Keras library). However, the limited availability of practical analyses detailing their respective advantages and limitations often leads to these algorithms being employed as "black-box" optimizers [26].

In general, a search for a stationary point can be divided into two parts: the direction and the step size. The step size specifies how far we move in a particular direction, and the direction determines the next direction we search. Such methods can be generally described as starting at some arbitrary point $x^{(0)}$ and then at every step $k \geq 0$, iteratively moving in a direction $\Delta x^{(k)}$ by step size $t_k$ to the next point $x^{(k+1)} = x^{(k)} + t_k. \Delta x^{(k)}$. In gradient descent, the direction we search is the negative gradient at the point, i.e., $\Delta x = -\nabla f(x)$. Thus, the iterative search of gradient descent can be described through the following recursive rule:

$$x^{(k+1)} = x^{(k)} - t_k \nabla f\left(x^{(k)}\right) \tag{7.17}$$

Given that the search for a stationary point is currently at a certain point $x^{(k)}$, and step size $t_k$. Since our objective is to minimize the function, one reasonable approach is to choose the step size in a manner that will minimize the value of the new point, i.e., find the step size that minimizes $f(x^{(k+1)})$. Since $x^{(k+1)} = x^{(k)} - t \nabla f(x^{(k)})$ the step size $t_k^*$ of this approach:

$$t_k^* = \text{argmin}_{t \geq 0} f\left(x^{(k)} - t\nabla f\left(x^{(k)}\right)\right) \tag{7.18}$$

A gradient descent method is suggested for unconstrained minimization, assuming that each selected point lies within the feasible region S and is defined with a precision $\epsilon > 0$. The stopping condition $\|\nabla f(x^{(k)})\| \leq \epsilon$ does not a priori guarantee that the point $x^{(k)}$ is within $\epsilon$ of the optimal solution, i.e., that $f(x^{(k)}) - \min_{x \in S} f(x) \leq \epsilon$. Finally, computing the step size as shown here is referred to as exact line search.

The learning rate is a crucial hyperparameter that determines the step size taken during each parameter update. An appropriate learning rate allows the model to converge efficiently to the minimum of the loss function. If the learning rate is too small, the updates will be slow, requiring many iterations to reach the minimum. If the learning rate is too large, the updates might overshoot the minimum, causing the loss function value to oscillate or even increase. The steepest descent method provides a straightforward approach for gradient-based optimization, offering stable convergence by utilizing full-batch gradients, although it may not be the most computationally efficient compared to modern optimizers. While Gradient Descent is foundational, many advanced optimizers or "solvers" have been developed to improve convergence speed and stability, particularly in complex neural networks. These algorithms determine how the gradients (computed by backpropagation) are used to update model parameters. Common examples include:

- Stochastic Gradient Descent (SGD): This variant performs updates using randomly selected subsets of the training data, rather than the entire dataset. This approach can be faster for large datasets, but may lead to oscillations in the training process.
- Adaptive Gradient (AdaGrad): It adapts the learning rate for each parameter individually based on the historical sum of squared gradients. However, a drawback is that the learning rate can become very small over time, potentially causing updates to stop prematurely.
- Momentum Optimizer: Inspired by the concept of momentum in physics, this optimizer incorporates information from previous gradients to accelerate learning in consistent directions and dampen oscillations.
- Root Mean Squared Propagation (RMSProp): This method improves upon AdaGrad by using an exponentially decaying average of squared gradients, which helps to prevent the learning rate from becoming too small too quickly.
- Adam (Adaptive Moment Estimation): Considered a highly effective and often default choice among solvers, Adam combines the benefits of both Momentum and RMSProp. It offers adaptive learning rates for each parameter, incorporates momentum, and includes bias correction, leading to smoother and faster convergence during training.

- L-BFGS (Limited-memory Broyden–Fletcher–Goldfarb–Shanno): This quasi-Newton method approximates second-order derivatives (Hessian matrix) without explicitly storing it. L-BFGS uses a limited history of gradient and parameter differences to guide parameter updates more effectively. It is known for offering faster convergence for smaller datasets or problems where high accuracy is desired.

Both Adam and L-BFGS have been adopted explicitly as solvers for ANN models in relevant studies [2–7].

#### 7.6.1.3 The Backpropagation Algorithm

Backpropagation [27, 28], as shown in Fig. 7.19, is a fundamental algorithm used in neural networks, particularly in training multi-layer feedforward Artificial Neural Network (ANN). Its primary purpose is to efficiently update the network's parameters by propagating errors in reverse. This algorithm's development significantly enhanced ANN's learning capabilities for various real-world applications. In the backpropagation algorithm, supervised learning is applied by supplying the network with input–output pairs and calculating the error as the difference between the observed and predicted outcomes. The goal of the backpropagation algorithm is to minimize this error until the ANN has learned all the training data. The training begins with random weights, which are then fine-tuned to minimize error.

There is a two-phase training process for the ANN algorithm. First, forward propagation involves processing data sequentially, layer by layer, from the input layer through the hidden layers to the output layer, thereby transmitting information. After the model generates predictions, these are compared with the actual target values. This triggers the backward propagation phase, during which the error is propagated backward through the network. It recursively evaluates how each

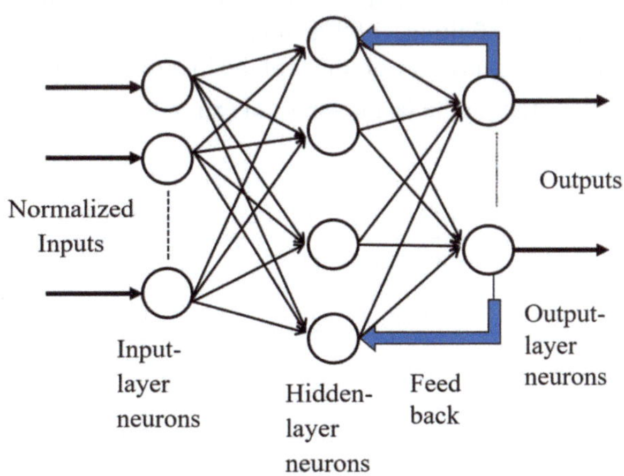


**Fig. 7.19** Training in a backpropagation neural network

artificial neuron in the final hidden layer contributes to the overall error observed at the output layer. This process continues layer by layer, tracing backward to determine the error contribution of neurons in preceding layers, all the way to the input layer. During this backward phase, the gradient of the error for each connection weight is precisely computed. These gradients are then used to update the network parameters. The algorithm calculates the gradient of the loss function (e.g., Mean Squared Error (MSE) for regression tasks) with respect to the weights using the chain rule. In Equation-Informed Neural Networks (EINNs) [14], specific operations within the network may be excluded from backpropagation updates. Weight or coefficient updates are computed for each training sample using gradient-descent-based backpropagation, while sample-specific ratios can be incorporated to emphasize particular data instances and enable selective learning.

In Fig. 7.20, there is an input layer with $n$ dimensions and an output layer with $m$ dimensions, comprising a total of $L$ layers. Each layer $l$ contains $n^l$ neurons. $o_i$ denotes the $i$-th component of the output vector, and $t_i$ denotes the $i$-th component of the target vector. The loss function is defined as follows:

$$C = \frac{1}{2}\sum_{k=1}^{m}\left(t_k - o_k\right)^2 = \frac{1}{2}\sum_{k=1}^{m} C_k \tag{7.19}$$

The relevant parameters for the $j$-th neuron in the $l$-th layer of the neural network are defined as follows:

In Fig. 7.21, the weight between the $i$-th neuron in the previous layer and the $j$-th neuron in layer $l$ is denoted as $w_{ij}^{l}$, $Z_j^l$ represents the input to this neuron, and $f\left(z_j^l\right) = o_j^l$ is the result after applying the activation function, which is also the output of the neuron. The definition of $Z_j^l$ is as follows:

$$z_j^l = \sum_{i=1}^{n^{l-1}} w_{ij}^l f\left(z_j^{l-1}\right) \tag{7.20}$$

The backpropagation algorithm can be introduced based on the above structure and the definitions of the related symbols. To update the parameters, gradient descent requires the gradient of the loss function with respect to each parameter, i.e., $\frac{\partial C}{\partial w_{ij}^l}$. Taking the last layer as an example:

$$\begin{aligned} \frac{\partial C}{\partial w_{ij}^L} = \frac{\partial C}{\partial o_j^L}\frac{\partial o_j^L}{\partial z_j^L}\frac{\partial z_j^L}{\partial w_{ij}^L} = \frac{\partial\left(\frac{1}{2}\sum_{k=1}^{m}\left(t_k - o_k\right)^2\right)}{\partial o_j^L} \\ \frac{\partial\left(f\left(z_j^L\right)\right)}{\partial z_j^L}\frac{\partial \sum_{k=1}^{n^{L-1}} w_{ij}^L f\left(z_j^{L-1}\right)}{\partial w_{ij}^L} = \left(o_j - t_j\right)\cdot f'\left(z_j^L\right)\cdot f\left(z_i^{L-1}\right) \end{aligned} \tag{7.21}$$

Taking the $l$-th layer (not the last layer) as an example, and applying the chain rule, we obtain the following equation:

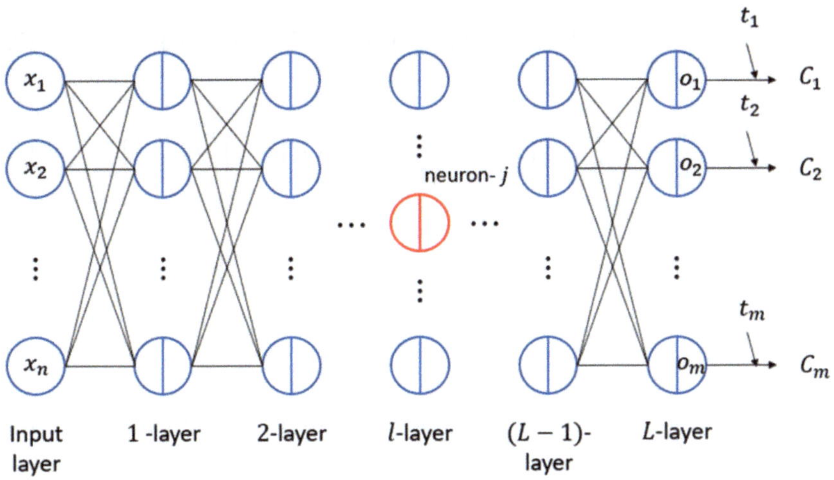


**Fig. 7.20** Schematic diagram of an Artificial Neural Network structure

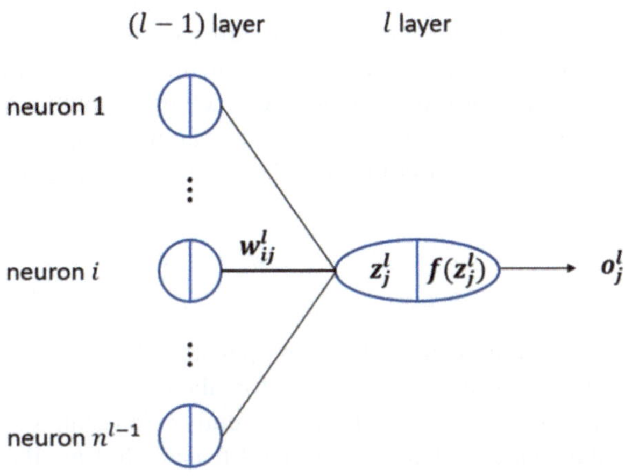


**Fig. 7.21** Schematic diagram of the configuration for the $j$-th neuron in the $l$-th

$$\frac{\partial C}{\partial w_{ij}^l} = \frac{\partial C}{\partial z_j^l}\frac{\partial z_j^l}{\partial w_{ij}^l} = \delta_j^l \cdot f\left(z_i^{l-1}\right) \tag{7.22}$$

$$\delta_j^l = \frac{\partial C}{\partial z_j^l} = \sum_{k=1}^{n^{l+1}} \frac{\partial C}{\partial z_j^{l+1}}\frac{\partial z_j^{l+1}}{\partial o_j^l}\frac{\partial o_j^l}{\partial z_j^l} = \sum_{k=1}^{n^{l+1}} \delta_k^{l+1} \cdot w_{jk}^{l+1} \cdot f'\left(z_j^l\right) \tag{7.23}$$

By rearranging the above equation, we can derive the formula for the gradients of each layer as follows:

$$\frac{\partial C}{\partial w_{ij}^{L}} = \delta_j^L \cdot f^{'(z_j^L)} \cdot f\left(z_i^{L-1}\right) \text{ where } \delta_j^L = \left(t_j - o_j\right) \tag{7.24}$$

$$\frac{\partial C}{\partial w_{ij}^{L-1}} = \delta_j^{L-1} \cdot f\left(z_i^{L-2}\right) = \sum_{k=1}^{n^L} \delta_k^L \cdot w_{jk}^L \cdot f'\left(z_j^{L-1}\right) \cdot f\left(z_i^{L-2}\right) \tag{7.25}$$

$$\frac{\partial C}{\partial w_{ij}^{L-2}} = \delta_j^{L-2} \cdot \left(z_i^{L-3}\right) = \sum_{k=1}^{n^{L-1}} \delta_k^{L-1} \cdot w_{jk}^{L-1} \cdot f'\left(z_j^{L-2}\right) \cdot f\left(z_i^{L-3}\right) \tag{7.26}$$

$$\vdots$$

$$\frac{\partial C}{\partial w_{ij}^{l}} = \delta_j^{l} \cdot f\left(z_i^{l-1}\right) = \sum_{k=1}^{n^{l+1}} \delta_k^{l+1} \cdot w_{jk}^{l+1} \cdot f'\left(z_j^{l}\right) \cdot f\left(z_i^{l-1}\right) \tag{7.27}$$

From Eqs. (7.24) to (7.27), it can be seen that calculating the gradient for each layer requires the $\delta$ of the subsequent layer as well as the corresponding weights. Therefore, the backpropagation algorithm is divided into two parts to obtain the necessary parameters. The forward pass is used to compute $z_i^l$ and $w_{ij}^l$ for each neuron, while the backward pass calculates the $\delta^L$ for the final layer and then propagates backward to compute $\delta$ for each preceding layer. After obtaining all required values and computing all gradients, parameter updates are performed using gradient descent or other optimizers. The parameter update formula is as follows:

$$w_{ij,new}^{l} = w_{ij}^{l} - \eta \cdot \frac{\partial C}{\partial w_{ij}^{l}} \tag{7.28}$$

The significance of backpropagation lies in its efficiency. Training them was computationally impractical before the widespread adoption of multi-layered neural networks in the 1980s. By intelligently calculating gradients layer by layer, it avoids redundant calculations, making it feasible to train the complex deep learning models we see today.

#### 7.6.1.4 Training Methodology

The training process for Artificial Neural Network (ANN) is a systematic approach to developing a model capable of making accurate predictions, typically involving several key stages from data handling to model evaluation.

1. Data Preparation and Dataset Establishment: The initial and crucial step in ANN training is to prepare and establish the dataset, as reliable data is fundamental for accurate predictions.

2. Data Division: Data is generally categorized into three subsets: a training set used to build the learning model's algorithm, a validation set for examining the model and adjusting parameters to enhance its generalization ability, and a testing set to evaluate the algorithm's predictive performance after training and validation.
3. Feature Selection: Researchers usually select the most sensitive parameters as input variables for training and learning. For example, features like upper pad diameter, lower pad diameter, chip thickness, and stress buffer layer (SBL) thickness are often utilized in wafer-level packaging reliability prediction.
4. Preprocessing: Data preprocessing is vital for improving convergence and model performance. As mentioned above, common methods include the Standard Scaler, Minimax Scaler, Robust Scaler, and Normalizer. Preprocessing can also involve filling missing values and standardizing data.
5. Model Design and Structure Establishment: ANN is designed to simulate how biological neurons transmit information by connecting inputs and outputs.
6. Layers: An ANN typically comprises three main types of layers: (a) Input Layer: Receives the feature values from the training data, (b) Hidden Layers: Perform calculations on the input data. Increasing the number of hidden layers and neurons can enhance the network's nonlinearity, allowing it to fit more complex data. However, too many can lead to high computational complexity, overfitting, and poor performance. (c) Output Layer: Receives the final calculated results from the last hidden layer and outputs them after transformation by an activation function.
7. Neurons: Basic units within the ANN, represented as circles in diagrams. When neurons reach a certain number, the cross-validation score tends to saturate, suggesting that an optimal number for computational efficiency is gradually achieved.
8. Hyperparameter Configuration: Hyperparameters are crucial settings that govern the training process and the model's structure.
9. Activation Functions: Introduce nonlinearity into the network's computations, enabling the model to handle complex problems. (a) Sigmoid Function: Outputs values between 0 and 1, but can suffer from "gradient vanishing" and slower convergence due to exponential calculations. (b) Hyperbolic Tangent (Tanh) Function: Outputs values between $-1$ and 1. Also susceptible to gradient vanishing, but can converge faster than Sigmoid due to its zero-centered output. (c) Rectified Linear Unit (ReLU) Function: Outputs values greater than or equal to 0. It avoids the gradient vanishing problem and offers faster convergence, but can result in "dead neurons" if inputs are negative.
10. Loss Functions: Measure the discrepancy between the model's predictions and actual data, with smaller values indicating higher accuracy. For regression tasks, common types include: (a) Mean Squared Error (MSE): Calculates the average of the squared errors, (b) Mean Absolute Error (MAE): Calculates the average of the absolute errors.
11. Optimizers: Algorithms that update parameters (weights and biases) to minimize the loss function and bring predictions closer to actual data. Examples

include: (a) Gradient Descent: Iteratively updates weights based on the gradient of the loss function, (b) Stochastic Gradient Descent (SGD): Uses random training data for faster computation, though it can lead to oscillations, (c) Adaptive Gradient (AdaGrad): Adapts learning rates for different parameters, but the learning rate can become too small, prematurely stopping updates, (d) Momentum: Accelerates learning by considering the direction of past gradients, providing an "inertia" to the updates, (e) Root Mean Squared Propagation (RMSProp): Adjusts learning rates using an exponentially decaying average of squared gradients, improving upon AdaGrad, (f) Adam (Adaptive Moment Estimation): Combines the benefits of Momentum and RMSProp, providing adaptive learning rates and smoother, faster updates, (g) Limited-memory Broyden–Fletcher–Goldfarb–Shanno (L-BFGS): A quasi-Newton method approximating second-order derivatives, offering faster convergence for some scenarios, especially with small datasets.

12. Learning Rate: Determines the step size for parameter updates; an appropriate learning rate is crucial for efficient convergence.
13. Epochs (Training Cycles): The number of times the entire training dataset is passed forward and backward through the neural network.
14. Grid Search: A common method to find the best hyperparameter combination by exhaustively testing all combinations within a defined range. This can be time-consuming for large datasets. Parallel computing can be combined with grid search to reduce the time needed to find optimal hyperparameters.

Training Process (Backpropagation) Backpropagation (BP) is the fundamental algorithm for training neural networks.

1. Forward Pass: Input values propagate through the network, multiplied by weights and added with biases, then transformed by activation functions.
2. Backward Pass: The error is calculated at the output layer and then propagated backward through the network, using the chain rule to calculate the gradient of the loss function with respect to each weight in every layer. An optimizer then uses these gradients to update the weights and reduce the value of the loss function.

Model Evaluation: The performance of AI models is assessed using various criteria.

1. Performance Metrics include maximum and average training differences (indicating underfitting) and maximum and average testing differences (indicating overfitting). Mean Absolute Error (MAE) and Maximum Absolute Error (MaxAE) are commonly used for evaluation.
2. Cross-validation: A common technique to avoid overfitting, where the dataset is divided into multiple parts, each serving as a validation or training set in different steps.

Generalization Capability: The ability of a trained model to accurately predict outcomes on new, unseen data.

1. CPU Time: The computational time required for training is a key factor in efficiency. Using GPU environments can significantly reduce training time compared to CPU environments. For instance, ANN training on a GPU showed significantly shorter wall times compared to CPU.

Advanced Techniques

1. Ensemble Learning: Combines multiple trained models to improve prediction accuracy and generalization. Bagging is a method where multiple ANNs are trained, and their predictions are averaged, which can reduce the number of inaccurately predicted data points.
2. Adaptive Sampling Methods: Used to build machine learning models with smaller, more optimally distributed datasets, reducing computational costs while maintaining prediction performance. K-medoids is an example of a clustering algorithm used for adaptive data selection. This approach helps avoid the issue of exponentially increasing data needed for higher design parameters with traditional space-filling methods.

### *7.6.2 CNN Algorithm*

The Convolutional Neural Network (CNN), Fig. 7.22, [29–32] is well-suited for handling high-dimensional, structured data, such as images, making it widely used in image and natural language processing. The CNN machine learning algorithm has recently been applied to warpage prediction of the fan-out panel-level packaging (FO-PLP). The development of CNN has significantly advanced the field of deep learning, breaking through previous limitations of neural networks.

A typical CNN architecture is composed of several key components:

1. Convolutional Layer: This is the primary distinguishing feature of CNN from traditional ANN. It is designed to perform "feature extraction". In this layer, a set of learnable filters, also known as kernels, is applied across the input data (e.g., an image, electronic packaging warpage fringe pattern). This operation involves performing an inner product between the filter and local regions of the input, generating an "Activation Map" or "Feature Map". Each filter is designed to

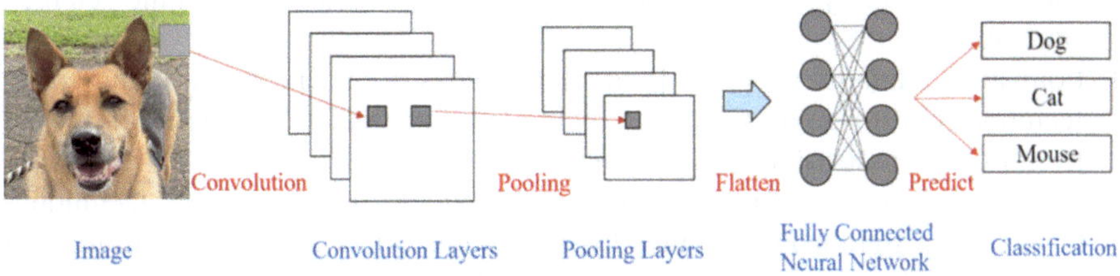


**Fig. 7.22** Schematic diagram of a Convolutional Neural Network

detect a specific local feature, such as edges or textures, within the input. For instance, specific filters can be used to detect vertical or horizontal lines in a grayscale image. The process often includes padding, such as zero padding, to maintain the spatial resolution of the feature maps and prevent the loss of border information, especially when multiple convolution operations are stacked. The stride parameter determines the step size of the filter's movement, which in turn influences the downsampling of the output size.
2. Pooling layer: Typically following a convolutional layer, the pooling layer's main objective is to downsample the spatial dimensions of feature maps. This process reduces computational cost and achieves a degree of translation invariance, meaning the network becomes less sensitive to slight shifts in the input. Standard pooling methods include maximum pooling (which retains the maximum value in a region and is useful for detecting boundaries or textures) and average pooling (which calculates the average value and is suitable for observing brightness changes and contrast).
3. Fully Connected Layer: After the feature extraction by convolutional and pooling layers, the processed feature maps are typically flattened into a one-dimensional vector. This vector then serves as the input to one or more fully connected neural network layers, which perform the final classification or regression task based on the extracted features.

CNN learn by iteratively adjusting its internal parameters (weights and biases). This learning process is fundamentally driven by the Backpropagation method, which calculates the gradients of the loss function with respect to each parameter. These gradients are then used by optimization algorithms, such as Gradient Descent and its advanced variants (e.g., Adam, L-BFGS), to update the parameters in a direction that minimizes the discrepancy between the model's predictions and actual target values. Applications of CNNs are widespread, such as:

1. Image Recognition and Classification: They have shown excellent performance in tasks like handwritten digit recognition and image classification competitions.
2. Warpage Prediction: In the context of fan-out panel-level packaging (FO-PLP), CNNs or modified CNN architectures are utilized to estimate warpage based on structural dimensions, allowing for correction to explore the influence of these dimensions on warpage. For instance, they have been applied to predict the warpage of FO-PLP structures by learning the relationship between panel-level package geometry and warpage from data generated by finite element simulations.
3. Equivalent Material Property Prediction: CNNs are adept at extracting spatial features from image data, making them ideal for predicting material properties from PCB or redistribution layer images.
4. Computational Efficiency: By incorporating feature extraction techniques, CNNs offer significantly higher learning capabilities and computational efficiency for image processing compared to other machine learning algorithms.

Historically, the concept of neural networks dates back to McCulloch and Pitts [19] in 1943, with significant advancements in CNN seen from LeCun's [33, 34] work in 1989 and Krizhevsky's [21] AlexNet in 2012. This research further explores the use of CNN-based models, sometimes with modified architectures tailored to specific problems, such as warpage prediction, demonstrating their utility in engineering and materials science applications. Application to Warpage Prediction: In the context of electronic packaging, CNNs have been specifically applied to estimate the warpage of fan-out panel-level packaging structures. The primary goal is to learn the complex relationship between panel-level package geometry and warpage from a generated database (often derived from finite element method (FEM) simulations). Once trained, the CNN model can quickly predict the corresponding warpage by simply inputting the geometric dimensions, thereby significantly reducing the time and potential user error associated with traditional simulation methods.

#### 7.6.2.1 CNN Layers

Figure 7.23 illustrates the difference between convolutional and traditional fully connected layers. As shown on the left, each pair of connected neurons has its own distinct weight in a fully connected layer. In contrast, on the right side, the convolutional layer divides neurons into different local regions, where a single set of shared weights is used for all neurons within that region. The output value of the convolutional layer can be expressed as shown in Eq. (7.29):

$$a_i^{(l)} = \mathrm{act}^{(l)}\left(\left[a_{i-K^{(l)}/2}^{(l-1)},\ldots,a_i^{(l-1)},\ldots,a_{i+K^{(l)}/2}^{(l-1)}\right]\begin{bmatrix} w_1 \\ \cdots \\ w_{K^{(l)}} \end{bmatrix} + b^{(l)}\right) \tag{7.29}$$

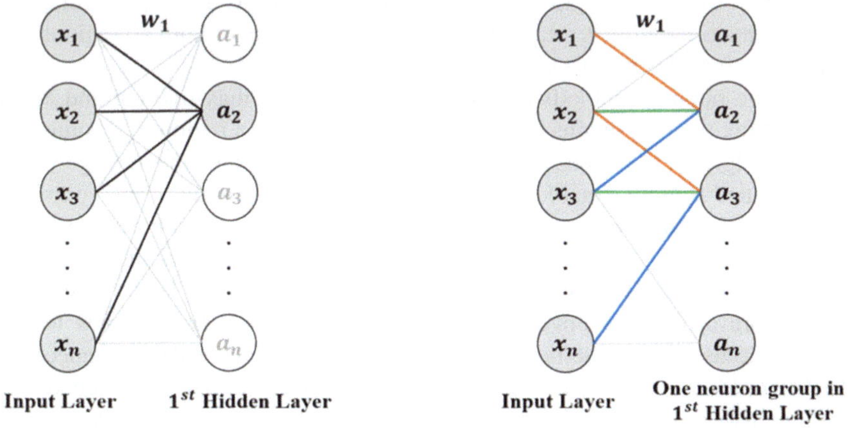


**Fig. 7.23** Schematic diagram of a fully connected layer (left) and a convolutional layer (right)

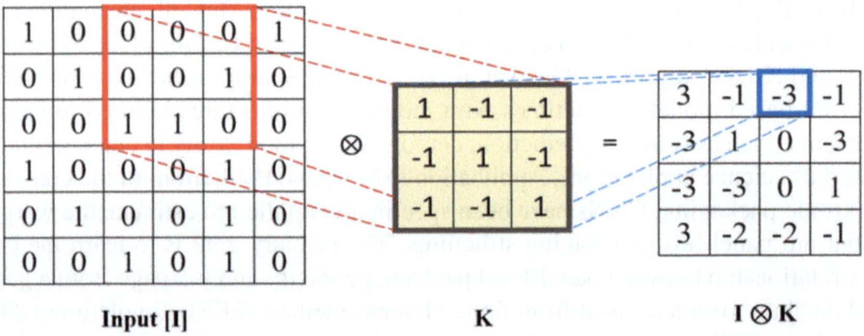


**Fig. 7.24** Convolutional layer diagram

Figure 7.24 is a schematic diagram illustrating the computation process of a convolutional layer. The input values are multiplied by the weights through an inner product operation, which serves to detect local features within a region of size K. Each fixed set of weights is designed to detect a specific type of feature, and such a set of weights is referred to as a *filter* or *kernel*. While detecting a single output value $a_n$ may not carry much meaning on its own, performing detection across a range allows us to determine whether a certain feature is present. The values obtained from these inner products are referred to as the *feature map* or *activation map*.

The pooling layer is typically placed after the convolutional layer. Its primary purpose is to perform downsampling on the feature map, aiming to reduce the number of inputs and computational cost by compressing the data while preserving the overall features and discarding less important information. A schematic diagram of a pooling layer is shown in Fig. 7.25. After obtaining the feature map from the convolutional layer (as shown on the left), the matrix is compressed to a size of 2 × 2 by retaining only the local maximum values. This means that subsequent computations only need to process a 2 × 2 matrix. The detailed process involves using a 2 × 2 filter to scan the feature map on the left. If max pooling is applied, the maximum value within each region is retained. This operation continues until all regions of the feature map have been processed.

Common pooling methods in pooling layers include max pooling and average pooling. Max pooling retains the maximum value within a region. It is suitable for detecting edges and textures, while average pooling computes the average value of the region and is more suitable for detecting changes in brightness and contrast.

The overall architecture of a Convolutional Neural Network is shown in Fig. 7.26. When an image is used as input, it is first passed through a series of convolutional and pooling layers. The number of convolutional and pooling layers is not fixed; having more of these layers allows the network to learn more complex features. However, it also means the network filters out many details it deems unnecessary. After completing *feature extraction*, the resulting feature map is flattened into a one-dimensional vector. This vector then serves as the input to the fully connected layers. Through neural network training, an output is produced. In classic CNN, the

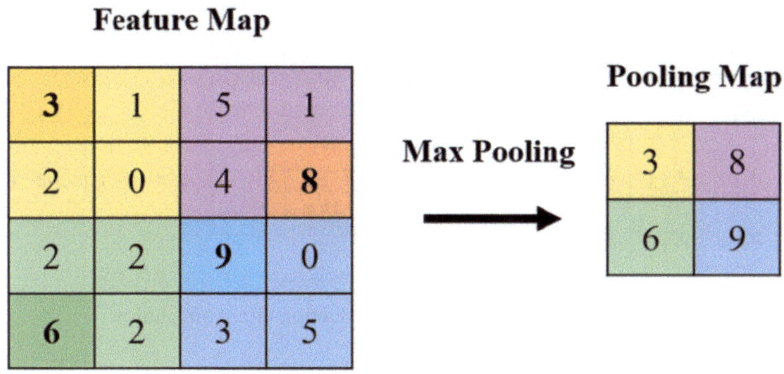


**Fig. 7.25** Pooling layer diagram

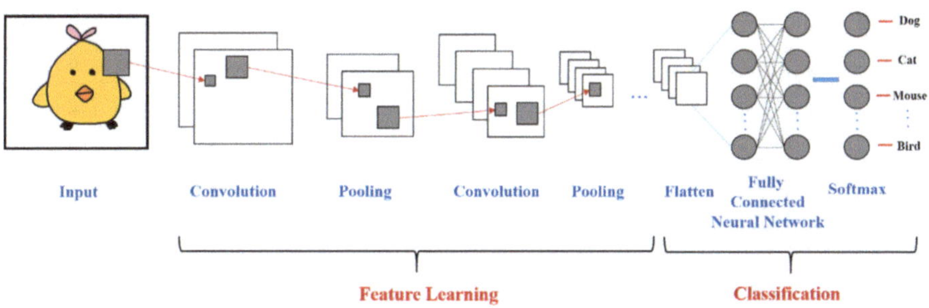


**Fig. 7.26** Schematic diagram of Convolutional Neural Network architecture

weights used in the convolutional layers as well as in the fully connected layers are all learned automatically by the model.

#### 7.6.2.2 Zero Padding

During the convolution operation in a Convolutional Neural Network, matrix operations cause the resulting feature map to become smaller than the original image, because the convolution kernel cannot fully cover the image edges, as shown in Fig. 7.27. The zero padding method addresses this by adding a border of zeros around the original image, allowing the size of the feature map to remain the same as the original image after the convolution, as illustrated in Fig. 7.28.

The advantage of using zero padding is that it allows the feature map to maintain the same size as the original input. This prevents loss of features during convolution operations and ensures that sufficient information is preserved even after multiple convolutional layers. Additionally, since the values added by zero padding are all zeros, there is no risk of noise contamination.

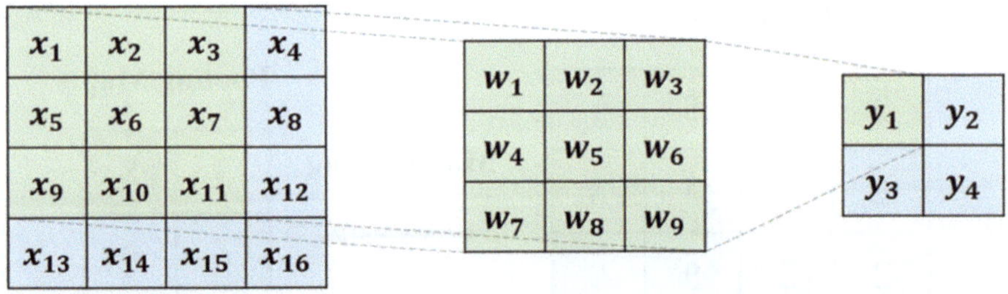


**Fig. 7.27** Schematic diagram showing output size reduction after convolution

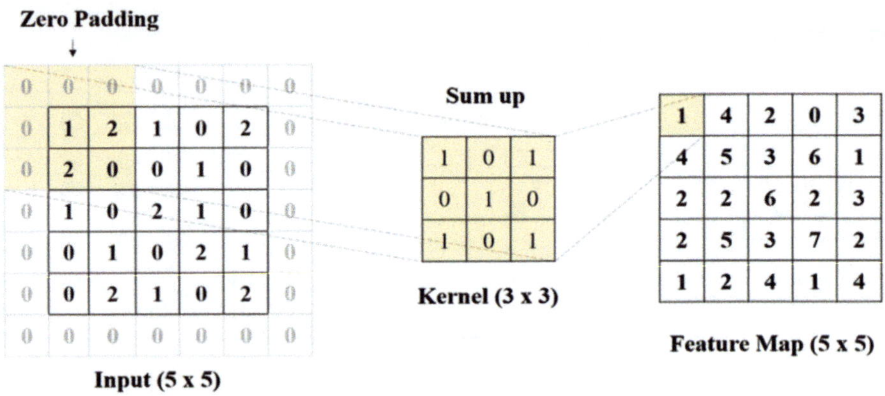


**Fig. 7.28** Illustration of zero padding, maintaining the output matrix size

In conclusion, the Convolutional Neural Network is a powerful type of Artificial Neural Network that excels at processing high-dimensional, structured data, such as images. It can be applied to various electronic packaging applications, including the prediction of temperature distribution and warpage.

#### 7.6.2.3 Training Methodology

The training involves several distinct strategies and components:

1. Core Architecture and Operation: CNN is composed primarily of convolutional layers, pooling layers, and fully connected layers. (1) Convolutional Layer: This layer performs "feature extraction" by applying learnable filters (also called kernels) that slide across the input. Each filter is designed to detect specific local features, such as horizontal or vertical lines. The operation involves an inner product between the input and the filter, producing a "Feature Map" or "Activation Map". Weight sharing within a range of neurons significantly reduces the number of parameters compared to fully connected layers. During convolutional operations, output matrices can become smaller than the original

input. Zero padding involves adding a border of zero values around the input image to maintain the original size of the feature map, preventing loss of border information and ensuring consistency in output dimensions, especially in deep architectures. (2) Pooling Layer: Typically placed after convolutional layers, pooling layers perform "down-sampling" on the feature maps. Their primary purpose is to reduce the amount of information and computational time while preserving important features. Common pooling strategies include max pooling (retaining the maximum value in a region, useful for detecting specific boundaries or textures) and average pooling (taking the average, suitable for observing brightness and contrast).

2. Data Preparation and Preprocessing: (1) Input Data, input data is often high-dimensional structured image data for CNN, especially in applications like predicting mechanical properties from PCB/substrate images. (2) Feature Scaling/Standardization: Similar to other machine learning models, CNN benefits from data preprocessing to ensure that input features fall within a similar range. StandardScaler (standardizing data to have zero mean and unit standard deviation) and MinMax Scaler (scaling features to a specific range) are common methods. The Robust Scaler is also used, as it is preferred for its stability and reduced sensitivity to outliers compared to other normalization techniques.
3. Data Reduction for Training: To reduce computational time, strategies such as edge detection can be applied to feature maps to identify and retain only boundary points, thereby removing redundant data with similar information. This can significantly reduce the number of training data points and accelerate training speed by three to four times. The threshold value used in edge detection affects the number of feature maps and keypoints selected, influencing training performance.
4. Optimizer: The Adam (Adaptive Moment Estimation) optimizer is frequently used. It is a first-order optimization algorithm that combines the advantages of momentum and adaptive learning rate techniques, making it efficient and robust for training deep networks. L-BFGS is also mentioned as an optimizer for smaller datasets, offering faster convergence by approximating second-order derivatives.
5. Activation Function: ReLU (Rectified Linear Unit) is a common choice. It introduces nonlinearity, is computationally efficient, and helps mitigate vanishing gradient problems often seen with Sigmoid or Tanh functions.
6. Loss Function: The Mean Squared Error (MSE) is commonly used for regression tasks.
7. Epochs (Maximum Iterations): A fixed number of iterations, such as 5000, can be set. It's noted that the model's progress may flatten beyond a certain number of iterations, indicating diminishing returns in performance improvement for increased training time.
8. Hyperparameters: Hyperparameters like the number of neurons per layer can be determined using grid search methods to find optimal configurations.

9. Backpropagation: Like other neural networks, CNN uses backpropagation (an adaptation of the backpropagation algorithm for sequential data) for updating parameters, despite its distinct structure.
10. Cross-Validation: N-fold cross-validation (e.g., ten-fold) is used to avoid overfitting and help in parameter selection by validating model parameters on unseen data subsets.
11. Overfitting and Regularization: Overfitting occurs when a model performs well on training data but poorly on unseen test data, often due to excessive model complexity. Regularization methods (e.g., adding a penalty term to the loss function) are used to combat overfitting, although their effectiveness can vary depending on the model and dataset.
12. Incremental Learning: This strategy is used to handle continually growing datasets by using parameters from a previously trained model as starting parameters for new training data, rather than retraining the entire model from scratch. This can lead to faster convergence and reduced computation time, potentially saving 17–30% of training time.
13. Transfer Learning: While less directly tied to CNN training for reliability prediction in the sources, transfer learning is mentioned as a solution for scarce training data, where a model pre-trained on a large "source data" set is fine-tuned with a smaller "target data" set.
14. Ensemble Learning: CNN can be part of an ensemble learning framework to enhance prediction accuracy and robustness, especially with small datasets. For instance, CNN is used to predict the equivalent mechanical properties of composite materials within an AI-driven workflow that includes Generative Adversarial Networks (GANs) for image generation.

Furthermore, like ANN, training CNN models can be significantly faster in a GPU environment than in a CPU environment.

### *7.6.3 RNN Algorithm*

A Recurrent Neural Network (RNN) [35–40] is a distinct type of Artificial Neural Network (ANN) that is specifically engineered to process data with sequential or temporal dependencies. Unlike a traditional feedforward ANN, which treats each input independently, an RNN is characterized by its internal feedback loops and hidden states, allowing it to retain contextual information from prior steps within a sequence. This internal state functions as a form of memory, enabling the network to leverage knowledge acquired from past inputs to inform the processing of the current input. A key operational aspect of RNN is that each hidden layer shares the same parameters (weights) across different time steps during the training phase. The core idea is to enhance a basic ANN regression model by incorporating a structure that depends on the output from the previous time step, introducing temporal dependencies. This results in a looped structure where nodes are connected directionally

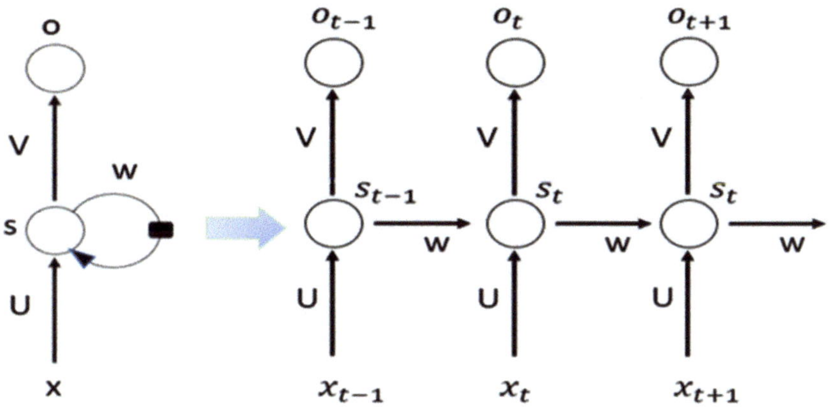


**Fig. 7.29** RNN unfolded along the time dimension

to form a cyclic neural network, as illustrated to the left of the arrow in Fig. 7.29. This structure includes an activation function and weight parameters. The hidden state of the RNN functions similarly to memory, allowing it to behave as a dynamic model that evolves over time, unlike feedforward neural networks (FNNs), which are inherently static.

By unfolding this cyclic neural network along the time axis, as shown on the right side of the arrow in Fig. 7.29, the RNN model is able to retain information from the previous time step and use it in computing the current output. At any given time step $t$, the hidden state $s_t$ and the output $o_t$ can be expressed by Eqs. (7.30) and (7.31), respectively. From these equations, it can be seen that the input to a hidden layer at time $t$ includes the output from the input layer (or the previous hidden layer) and the output from the same hidden layer at the previous time step. Each hidden layer shares the same parameters during training across all time steps. This is why only three weight matrices, W, U, and V, appear in Fig. 7.29.

$$s_t = W * s_{t-1} + U * x_t + b \tag{7.30}$$

$$o_t = g\left(V * s_t\right) \tag{7.31}$$

RNN is highly versatile and can be configured in various architectural patterns depending on the nature of the sequential data and the desired output. These structures include one-to-one (similar to a standard ANN), one-to-many (a single input yielding multiple outputs, such as generating an image caption), many-to-one (multiple sequential inputs leading to a single output, for instance, in sentiment analysis of a sentence), and many-to-many (multiple inputs producing multiple outputs, which can be synchronized, like in language translation, or non-synchronized), see Fig. 7.30. The many-to-one structure has been utilized for specific applications,

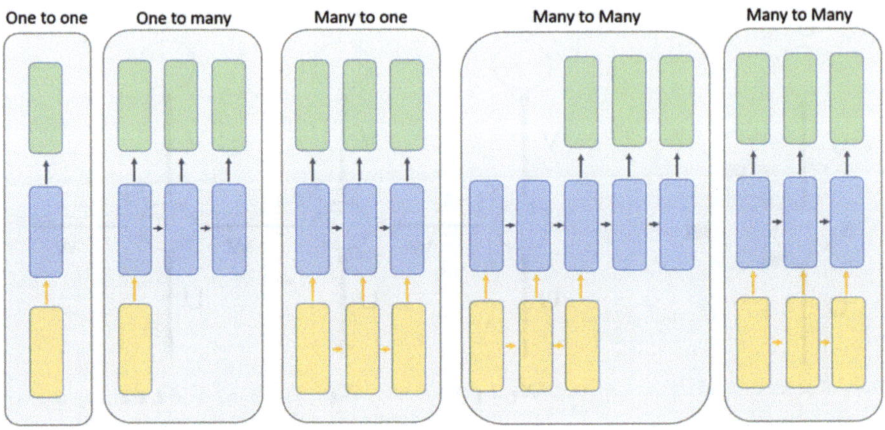


**Fig. 7.30** Illustration of different types of RNNs

such as predicting wafer-level package reliability, where multiple input parameters lead to a single reliability prediction.

The training of RNNs relies on a method known as Backpropagation Through Time (BPTT), which is an extension of the standard backpropagation algorithm adapted to account for the temporal dimension of sequential data. However, basic RNN architectures often face significant challenges, notably the vanishing or exploding gradients problem, particularly when learning long-term dependencies within extended sequences. These gradient issues can severely hinder the network's ability to effectively learn from data points far apart in a sequence. To overcome these limitations, more advanced variants of RNNs have been developed. For example, the Long Short-Term Memory (LSTM) network addresses the vanishing gradient problem through a more sophisticated memory structure that incorporates input, output, and forget gates. These gates enable the LSTM to store or discard information selectively, making it highly effective at capturing long-range dependencies, albeit at the cost of higher computational requirements. A simpler yet still powerful alternative is the Gated Recurrent Unit (GRU), which streamlines some of the LSTM's gates to reduce the number of parameters, potentially leading to faster convergence while maintaining much of its ability to manage long-term information. This variant specifically features an update gate and a reset gate.

#### 7.6.3.1 Long Short-Term Memory, LSTM

Long Short-Term Memory (LSTM) was proposed by Hochreiter and Schmidhuber [41] in 1997. As shown in Fig. 7.31, the original architecture mainly includes memory-capable Cells, Input Gates, and Output Gates. In 1999, Gers et al. [42] introduced the element of Forget Gates, which allows LSTM to reset its state. Later, they added peephole connections [37], enabling the Gates to receive input from the cell state. Cho et al. [43] also simplified the Input Gates and Forget Gates in the LSTM architecture into a single Update Gate, a structure known as GRU (Gated

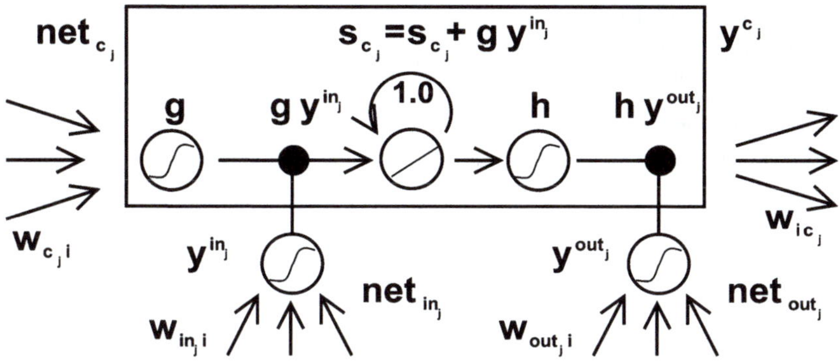


**Fig. 7.31** Schematic diagram of the original LSTM architecture

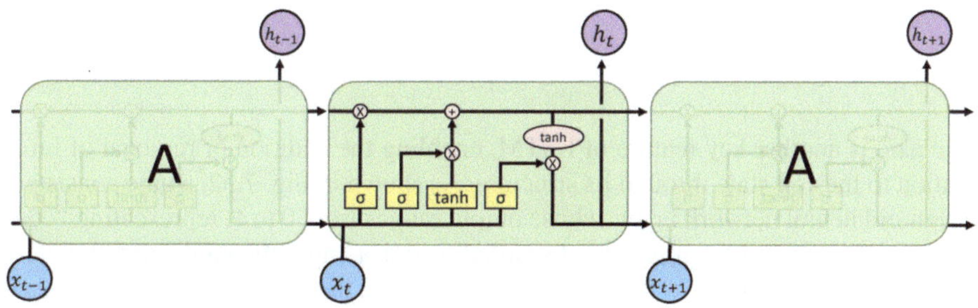


**Fig. 7.32** Actual architecture diagram of LSTM

Recurrent Unit). All these are considered variants and historical developments of LSTM.

All RNN architectures possess a recurrent neural network module arranged in a chain-like form. In a standard RNN structure, the repeated module can be as simple as a single layer of neurons with an activation function and a bias term. In contrast, an LSTM also consists of such repeating modules. Still, each has a more complex structure, containing four interacting neural network layers that communicate with each other in a special way, as shown in Fig. 7.32.

In Fig. 7.33, each black line denotes the transmission of an entire vector, which may function as either the input or output of a node. When multiple lines converge, they represent the connection of vectors; when they diverge, they indicate duplication of contents rather than splitting. The pink circles represent element-wise operations, such as vector addition or multiplication. The yellow rectangles typically denote activation functions, which can be regarded as layers in the neural network.

One of the key features of LSTM is the cell state, which runs through the top of the LSTM module as a horizontal line. Its function is similar to a conveyor belt operating directly along the chain, allowing information to flow with only minor linear interactions. This enables the data to remain more unchanged compared to a standard RNN structure. $C_{t-1}$、 $C_t$ represent the cell states at time steps *t-1* and *t*, respectively.

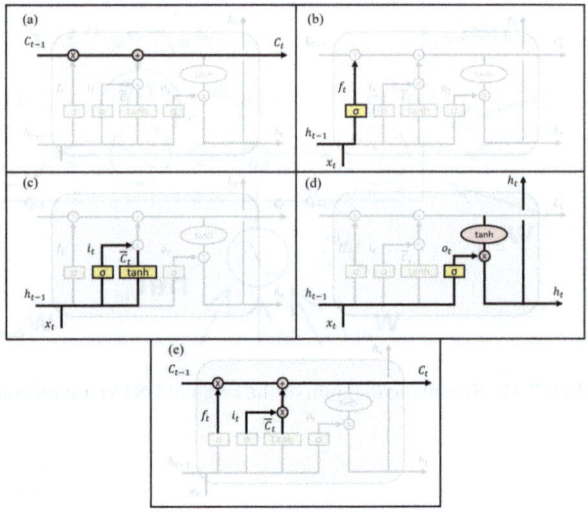


**Fig. 7.33** Flowchart of the operation of the LSTM structure. (**a**) Cell State, (**b**) Forget Gate, (**c**) Input Gate, (**d**) Output Gate, (**e**) Update Cell State

The gate is another key feature of LSTM, enabling the addition or removal of information to the cell state through its structure. As shown in Fig. 7.32, a gate consists of a sigmoid neural network layer (whose output ranges from 0 to 1, representing "completely retained" to "completely discarded") and a multiplication operation. The LSTM architecture used in this study has three gates: the input gate, output gate, and forget gate, which control the cell state. The weights in the LSTM and the timing and extent to which each gate opens are automatically learned by the machine.

Figure 7.33 is a flowchart illustrating the operation of the LSTM structure. The process begins by determining how much information to discard, which the forget gate accomplishes. The outputs from $h_{t-1}$ (the output of the previous time step) and $X_t$ (the input at the current time step) are passed through a sigmoid function to produce a value $f_t$ between 0 and 1. This value, together with the multiplication symbol on the cell state chain, represents the degree to which information is discarded. Next, the process determines what new information to add, which the input gate controls. This part also consists of two steps. First, the input is passed through a sigmoid function to generate a value between 0 and 1. The multiplication symbol indicates how much new information is acquired. Second, the input is passed through another activation function (in this case, a tanh layer) to create a new candidate value vector $\tilde{C}_t$, which serves as a potential value to be added to the cell state.

Next is the stage of updating the previous cell state, where $C_{t-1}$ is updated to $C_t$. As shown in Eq. (7.32), the previous state $C_{t-1}$ is multiplied by $f_t$ to discard information, and then $i_t$ multiplied by the candidate value $\tilde{C}_t$ is added. This results in the new cell state $C_t$.

$$C_t = f_t * C_{t-1} + i_t * \tilde{C}_t \tag{7.32}$$

Finally, the LSTM needs to determine its output value based on the updated cell state. A sigmoid function is used to determine the extent to which the cell state is

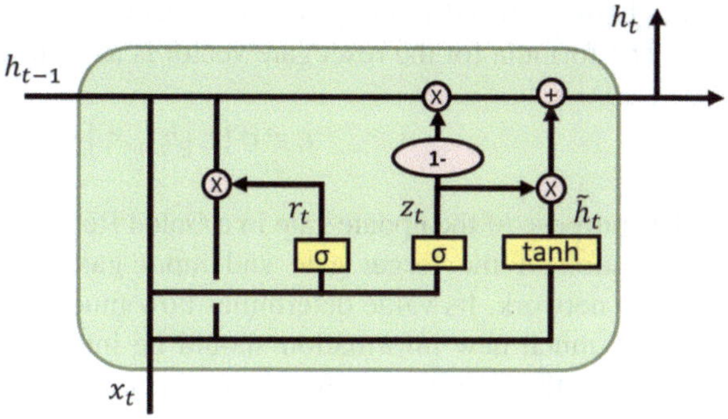


**Fig. 7.34** Actual architecture diagram of GRU

output $o_t$; this part is known as the output gate. In addition, the updated cell state is passed through another activation function (in this case, a tanh layer), and then multiplied by the output of the output gate $o_t$. This determines the output $h_t$ of the LSTM module at the current time step, which also serves as one of the inputs for the next time step. According to the above, it can be seen that recurrent neural networks are more complex than conventional artificial neural networks. In addition to introducing parameters for time series and considering the effects of different input sequences, the neural network units can operate with various types of cells, such as long short-term memory (LSTM) units.

#### 7.6.3.2 Gate Recurrent Unit, GRU

Although Long Short-Term Memory (LSTM) networks are well-suited for handling and predicting important events with very long intervals and delays in time series, their drawback lies in the complexity of their structure. The model's parameters are significantly higher than those of a simple recurrent neural network (Vanilla RNN), resulting in longer training times. To address this, Cho et al. [43] proposed a compromise in 2014: the Gated Recurrent Unit (GRU). The GRU architecture is an improvement based on LSTM, where the forget gate and input gate in LSTM are replaced with a single update gate, and the cell state is merged with the hidden state ($h_t$). The method for calculating new information differs from that of LSTM. For this reason, GRU is also known as a simplified version of LSTM. Its internal structure is shown in Fig. 7.34.

Like in LSTM, in a GRU unit, each black line represents the direction of transmission for a complete vector, which can serve as the output or input of any node. When the black lines are merged, it represents the combination of vectors; when the black lines are separated, it refers to copying for repeated operations, rather than splitting. The pink circles represent operations performed by each node on the vectors, such as vector addition or multiplication. The yellow rectangles usually represent activation functions. The computational steps for the GRU are as follows: First, the reset gate and update gate vectors are calculated. The reset gate determines how much of the previous information is retained. Since the sigmoid function outputs values between 0 and 1, it is used here to decide the retention percentage of information. The formula for the reset gate vector is as follows:

$$r_t = \sigma\left(r_z\left[h_{t-1}, x_t\right] + b_r\right) \tag{7.33}$$

The purpose of the update gate in a Gated Recurrent Unit (GRU) is similar to the combination of the forget gate and input gate in a Long Short-Term Memory (LSTM) network. Its value determines how much information should be discarded and how much new information should be introduced. Since it also represents a percentage, the sigmoid function is used here to achieve this purpose. The formula is as follows:

$$Z_t = \sigma\left(w_z\left[h_{t-1}, x_t\right] + b_z\right) \tag{7.34}$$

The next step is to calculate the candidate value for the hidden state $\tilde{h}_t$, also known as the *h* candidate. The previous hidden state is passed through the reset gate to determine how much information is retained, and then it is combined with the input from the previous layer. The formula is as follows:

$$\tilde{h}_t = \tanh\left(w \cdot \left[r_t \cdot h_{t-1}, x_t\right]\right) \tag{7.35}$$

When determining the output of a GRU unit, the main factor is the value of the update gate, which decides the proportion of the candidate value $\tilde{h}_t$ to be retained. The remaining part is supplemented by the hidden state from the previous time step $h_{t-1}$. The formula is as follows:

$$h_t = \left(1 - Z_t\right) \cdot h_{t-1} + Z_t \cdot \tilde{h}_t \tag{7.36}$$

Chung et al. [38] used an experimental approach in their paper to compare Long Short-Term Memory (LSTM) units and Gated Recurrent Units (GRU) by applying them to tasks such as chord music modeling and speech signal modeling. Their experiments demonstrated that these advanced recurrent units, such as LSTM and GRU, outperform traditional ones. Moreover, they pointed out that although GRU has a simpler structure, it still achieves performance comparable to that of LSTM.

In conclusion, when comparing performance, while RNNs, including their LSTM and GRU variants, can achieve good prediction accuracy for WLP reliability (with average testing errors observed within approximately 10 cycles for datasets of around 432 samples), they generally demand significantly more training time than ANN when operating on CPUs. In some cases, RNN training can take nearly ten times longer than ANN training in a CPU environment. However, leveraging a GPU environment can substantially reduce training time for both RNN and ANN, bringing it within an acceptable range while maintaining similar performance trends. The suitability of RNNs can also depend on the nature of the data's sequential aspect; if the time sequence in the data is not inherently long or highly relevant to the problem, simpler RNN architectures (like Vanilla RNN) or even an ANN might yield comparable or superior performance. Increasing the size of the training dataset generally leads to improved model performance, although this comes at the expense of increased training time. There might be a point where the performance gain from additional data saturates, while training time continues to climb. Importantly, in various comparison studies, RNN has demonstrated the capability to predict reliability with greater precision than ANN. Optimal performance for RNNs in these specific applications is often achieved with RobustScaler for data preprocessing and Mean Absolute Percentage Error (MAPE) as the chosen loss function.

#### 7.6.3.3 Training Methodology

A Recurrent Neural Network (RNN) is specifically designed to handle sequential or temporal data, and its training involves distinct strategies and considerations compared to traditional Artificial Neural Network (ANN). RNN is trained using Backpropagation Through Time (BPTT), an adaptation of the backpropagation algorithm for sequential data that accounts for the time dimension. A significant challenge with simple RNN architectures is their susceptibility to vanishing or exploding gradients, particularly when dealing with long sequences. This issue can prevent the network from learning long-term dependencies effectively. To mitigate these gradient problems, more advanced RNN variants have been developed:

1. Long Short-Term Memory (LSTM) networks introduce a more complex memory structure with input, output, and forget gates. These gates enable the model to selectively retain or discard information, making them effective at capturing long-range dependencies.
2. Gated Recurrent Unit (GRU) is a simplified version of LSTM that merges some gates to reduce parameters, potentially offering faster convergence while still managing long-term information. However, one study noted that GRU did not consistently improve prediction performance and often failed when using more hidden layers, especially for the WLP reliability problem, where the time sequence might not be long or highly relevant.
3. Model Configuration and Hyperparameters: In some studies, hyperparameters for machine learning models, including RNN, were fixed during the training process rather than using a grid search function. However, other contexts demonstrate that a grid search method was employed to find optimal configurations for parameters such as the number of neurons. Specific hyperparameter settings and choices for RNN include: (1) Activation Function: The Rectified Linear Unit (ReLU) is commonly selected due to its computational efficiency and ability to mitigate vanishing gradients compared to Sigmoid functions, especially for deeper network structures. (2) Optimizer: The Adam (Adaptive Moment Estimation) optimizer is frequently used. Adam is an advanced variant of RMSprop that incorporates momentum and adaptive learning rates, providing more stable training. (3) Learning Rate: An initial learning rate of 0.01 is often

set. If performance is unsatisfactory, it can be adjusted to 0.001. (4) Hidden Layers: The number of hidden layers can vary, with three layers being a common choice. Studies have also explored 1 to 8 layers or specific ranges like 2, 5, 10, and 20 layers. (5) Neuron Number: Per layer, neuron counts such as 100 or 200 are used. Other studies explored 10, 20, 50, 100, 200, and 500 neurons. Increasing the number of neurons often leads to improved performance, but also increases the risk of overfitting. (6) Epochs (Maximum Iterations): A maximum of 2000 epochs is noted for some RNN models, while others specify up to 10,000 iterations. (7) Loss Function: Mean Absolute Percentage Error (MAPE) is generally found to be the optimal choice for both ANN and RNN models, as it provides a lower percentage error and avoids being dominated by large numerical values, especially compared to Mean Squared Error (MSE) or Mean Squared Logarithmic Error (MSLE).

4. RNN Unit Types: For ensemble learning, both Simple RNN (also known as Vanilla RNN) and LSTM units are employed to ensure diversity among the base models. For RNNs, input data is typically structured in three dimensions: batch size, time step, and number of features. This differs from the two-dimensional input used by an ANN.
5. Common input-output configurations for RNNs include: (1) Many-to-one models: Multiple sequential inputs (e.g., four WLP geometric parameters like upper pad diameter, lower pad diameter, chip thickness, and stress buffer layer thickness) lead to a single output (reliability life cycle). The specific order of these inputs was studied, and in one case, input order did not significantly impact training results. (2) Synchronized Many-to-many models: Multiple inputs lead to multiple outputs, where outputs directly correspond to inputs at each time step. For WLP, this involved changing one feature value over time while keeping others constant.
6. Data Preprocessing: RobustScaler is commonly adopted as the preprocessing method for RNNs. It is preferred because it performs more stably and is less sensitive to outliers compared to other methods, such as Standardization, MinMax Scaler, and MaxAbs Scaler.
7. The data is typically shuffled before training: (1) Cross-validation, specifically K-fold cross-validation (e.g., ten-fold), is widely used to ensure smooth performance, avoid overfitting, and help in parameter selection by validating model parameters on unseen data subsets. (2) Early stopping is employed to prevent overfitting, halting training if the validation loss does not improve after a certain number of iterations (e.g., 500 steps).
8. Data Quantity, Distribution, and Sampling Strategies: The training dataset's size significantly impacts the RNN's performance. Increasing the training data generally improves prediction accuracy. However, there is often a point where the performance gain from additional data saturates, while training time continues to increase linearly. The distribution of feature values within the training data is crucial. Training with uniformly distributed feature values yields better results than using randomly selected ones, as it ensures broader coverage of the feature space. Often combined with algorithms like K-medoids, adaptive sampling tech-

niques are utilized to efficiently reduce the required training dataset size while maintaining prediction accuracy, particularly important for small-data learning tasks. The boosting concept can be incorporated by feeding mispredicted testing samples back into the training dataset for retraining, which helps to enhance model accuracy.

9. Ensemble Learning with RNN: RNN can serve as base models within ensemble learning frameworks such as Bagging, Boosting, and Stacking. Ensemble learning is employed to enhance prediction accuracy and robustness, particularly when training on small datasets where individual models may be unstable or prone to overfitting. Studies have shown that an ensemble of ANN and RNN models achieves the best overall performance. The effectiveness of ensemble learning heavily relies on its base models' diversity and accuracy.

The computational environment has a substantial impact on training time. Training RNN models on a CPU environment takes significantly longer (e.g., almost ten times longer than an ANN). Utilizing a GPU environment can drastically reduce training time for both RNNs and ANN, bringing it into an "acceptable range," while maintaining similar performance trends.

### *7.6.4 Random Forest and Extra Trees Algorithms*

Random Forest (RF) and Extremely Randomized Trees (ERT), also known as Extra Trees (ET) [44–49], are both powerful ensemble learning algorithms that use multiple Decision Trees to make predictions. They are effective for regression and classification problems and recognized for their computational speed and suitability for analyzing large-scale data. Both algorithms aim to enhance model prediction performance by reducing both bias and variance. For regression tasks, the final prediction from both algorithms is obtained by averaging the predictions made by all the individual decision trees within the ensemble. Despite these similarities, their primary distinctions lie in how they sample data for individual trees and select split points within them. A decision tree is a machine learning algorithm that uses a tree-like structure to progressively split data for prediction and analysis. It operates by segmenting data from a large dataset into multiple branches using a binary split. Random Forest and Extra Trees can be applied to both regression and classification problems. The program automatically splits the input database into training and testing sets. From the training set, it selects a subset of feature vector values and a portion of the data to generate a decision tree, which is usually a CART (Classification and Regression Tree). This process is repeated multiple times to create decision trees with diverse structures. The predictions from all trees are then combined, resulting in the final output through an ensemble method.

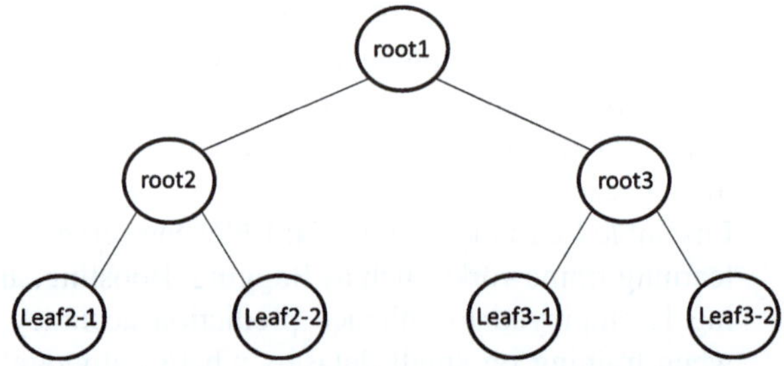


**Fig. 7.35** CART decision tree model

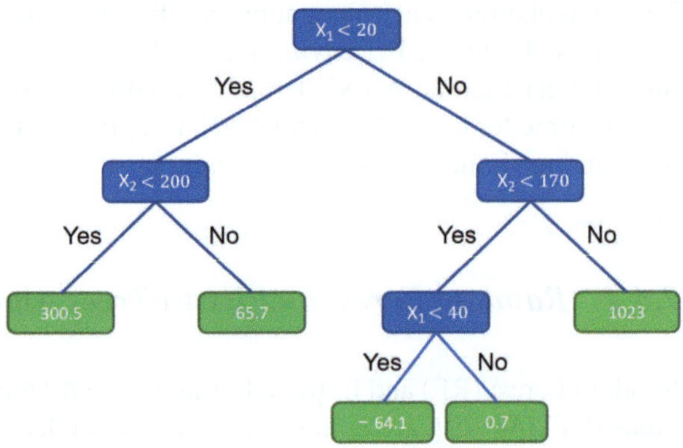


**Fig. 7.36** Example of binary split

#### 7.6.4.1 Classification and Regression Tree

The CART decision tree concept encompasses both classification and regression trees, utilizing binary splitting (as illustrated in Figs. 7.35 and 7.36). In each decision process step, the observed variables are split into two groups, which differs from other decision trees that may split into multiple branches. The generation process of a CART tree is divided into "CART generation" and "CART pruning," and the methods for generating CART classification trees and CART regression trees are different.

The generation of a CART regression tree uses the least squares method for optimal splitting. The computational steps are as follows: (1) Traverse all variables $j$ (feature vectors) to find the optimal split point s (feature vector value), and then perform the split using the optimal pair $(j, s)$, (2) Use the selected $(j, s)$ to partition the region and determine the corresponding output values: The expression is shown as follows:

$$\min_{j,s}\left[\min_{c_1}\sum_{x_i\in R_1}(y_i-c_1)^2+\min_{c_2}\sum_{x_i\in R_2}(y_i-c_2)^2\right] \tag{7.37}$$

$$R_1(j,s)=\left\{x|x^{(j)}\le s\right\},R_2(j,s)=\left\{x|x^{(j)}>s\right\}, \tag{7.38}$$

$$c_m=\frac{1}{N_m}\sum_{x_i\in R_{m(j,s)}}y_i,x\in R_m,\ m=1,2 \tag{7.39}$$

Equation (7.37) illustrates that the entire dataset is divided into $R_1$ and $R_2$ regions by the value of $x^{(j)}$ at $s$. All data points in $R_1$ satisfy $x^{(j)} \le s$, while data points in $R_2$ fall within the range $x^{(j)} > s$.

Continue to apply steps 1 and 2 to $R_1$ and $R_2$, respectively, until the stopping criteria are met. The training set space is then divided into $m$ regions, $R_1$, $R_2$, ..., $R_m$, thereby concluding the "CART generation" process.

$$f(x)=\sum_{m=1}^{M}\widehat{c_m}I(x\in R_m) \tag{7.40}$$

In Eq. (7.40), $I$ is the identity matrix, and $\widehat{c_m}$ is the mean output matrix for each region $R_m$. This equation indicates that each region has its own specific output value $c_m$, which serves as the $f(x)$ of the CART decision tree.

To prevent the decision tree from overfitting the samples, each feature vector has been fully considered after CART generation, resulting in very pure leaf nodes in each CART tree. This leads to strong performance on the training samples, but poor generalization to unseen samples. Therefore, CART pruning can be regarded as a way to simplify the decision tree to avoid overfitting. CART pruning consists of two sub-steps: Pre-pruning and post-pruning. Pre-pruning is performed during the CART generation process by setting stopping criteria to prevent the decision tree from growing indefinitely and causing overfitting. Common stopping criteria include: (1) All sample data in the subtree belong to the same class, (2) Limit the maximum tree depth, (3) The number of samples in the subtree node is less than a certain threshold or below a specified proportion, (4) The decision value after optimal splitting is less than a specific error value, such as specifying the least squares error term.

Post-Pruning: After the CART is generated, some subtrees are removed to achieve simplification. The post-pruning of CART adopts cost-complexity pruning (CCP), with its loss function defined as follows:

$$C_\alpha(T)=C(T)+\alpha|T| \tag{7.41}$$

Let $T$ be any subtree, $T_0$ a fully grown tree, $C(T)$ the prediction error (squared error for regression trees) used to measure how well the training data is fitted, and $|T|$ the number of leaf nodes in subtree $T$, representing the complexity of the tree. $\alpha$

is a parameter that balances the degree of fit and the complexity of the tree. The goal is to find a suitable $\alpha$ that achieves the best balance between fit and complexity. During the process, for each $\alpha$, the optimal subtree $T(\alpha)$ that minimizes $C_\alpha(T)$ can be found. When $\alpha = 0$, the original fully grown tree $T_0$ is the optimal subtree; when $\alpha = \infty$, a single node $J$ is the optimal subtree.

CCP (Cost-Complexity Pruning) first generates a sequence of subtrees by pruning the child nodes under an internal node back to that internal node, making it a leaf node. For any internal node $t$, before pruning, there are $|T_t|$ leaf nodes, and the prediction error is $C(T_t)$. After pruning, only one leaf node (the internal node itself) remains, and the prediction error becomes $C(t)$. Substituting the above into eq. (7.42) gives the loss function for the subtree rooted at node t before pruning.

$$C_\alpha(T_t) = C(T_t) + \alpha|T_t| \tag{7.42}$$

In this way, only the local loss change of the internal node remains, so when calculating the overall error, it is sufficient to calculate the loss function of the internal node before and after pruning. Breiman [45] proved that as $\alpha$ increases from zero, $0 = \alpha_0 < \alpha_1 < \ldots < \alpha_n < +\infty$, in each interval $\alpha_i,\ \alpha_{i+1}$, the subtree $T_i$ is optimal within that interval. There always exists an $\alpha$ such that $C_\alpha(T_t) = C_\alpha(t)$; this value is:

$$\alpha = \frac{C(t) - C(T_t)}{|T_t| - 1} \tag{7.43}$$

Therefore, for the current node $t$, as long as $\alpha$ is greater than this value, $C_\alpha(t) < C_\alpha(T_t)$ will always hold, meaning that pruning this node yields a better result than not pruning it. Thus, each optimal subtree corresponds to an interval in which it remains optimal. Next, for each internal node t in $T_t$, the reduction in the overall loss function is calculated and denoted as g($t$):

$$g(t) = \frac{C(t) - C(T_t)}{|T_t| - 1} \tag{7.44}$$

Here, $t$ is a variable. The $\alpha$ mentioned above is an expression with respect to the constant $t$. Finally, by using independent validation data or cross-validation, the squared error for each subtree is calculated, and the one with the smallest error is selected as the optimal subtree in the sequence. This completes the CART decision tree model.

#### 7.6.4.2 RF and ET

Random Forest (RF, Fig. 7.37) is categorized as a Bagging (Bootstrap aggregating) ensemble method (Fig. 7.38). It creates multiple training subsets by randomly sampling from the original training dataset with replacement. This process means that each tree is trained on a slightly different subset of data, and some data points might

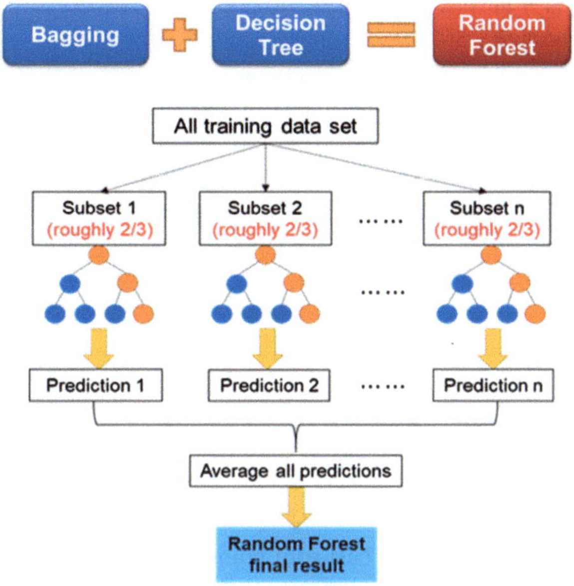


**Fig. 7.37** Random Forest structure

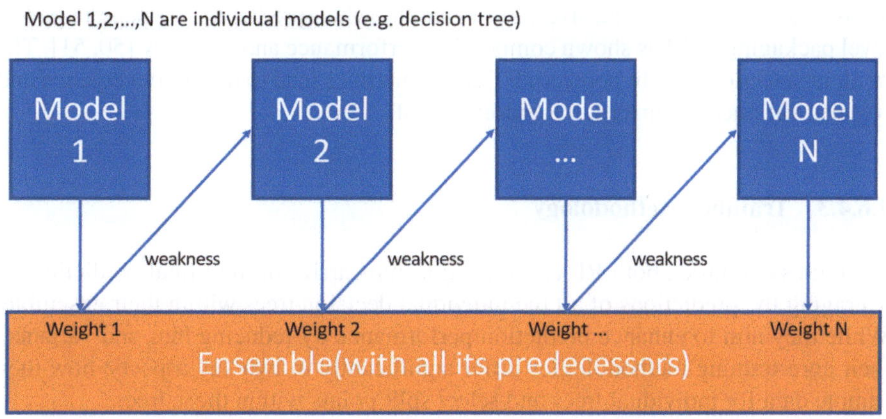


**Fig. 7.38** Ensemble learning process procedure

appear multiple times or not at all in a specific subset. This approach helps to avoid local optima during training. The sampled data typically amounts to about two-thirds (~63.2% unique samples) of the total data. The algorithm calculates and selects the "best split point" for data partitioning at each internal node. Regression tasks typically involve minimizing the mean squared error (MSE).

For Extra Trees (Fig. 7.39), a key distinction of ET is that it uses all the input data directly as the training set for each decision tree, rather than employing Bootstrap sampling. This can help avoid local optima during training. Extra Trees randomly selects split points for data partitioning at each node, instead of exhaustively

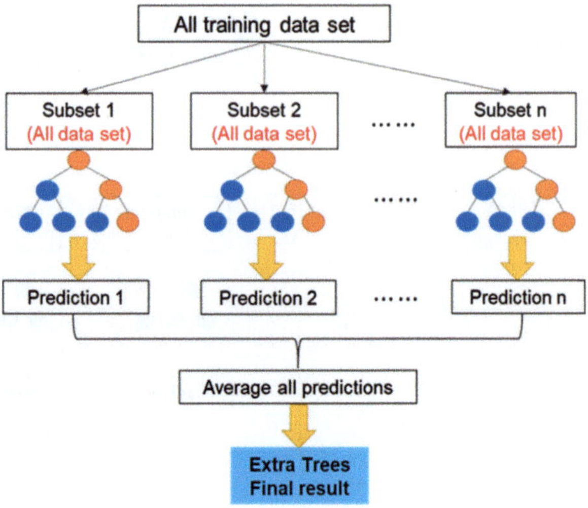


**Fig. 7.39** Extra Trees structure

searching for the "best split point". Extra Trees has been observed to demonstrate superior prediction performance over Random Forest for specific datasets. In some comparative studies, ET's average testing error was consistently lower than RF across various training data sizes. For example, when predicting warpage in panel-level packaging, ET has shown competitive performance and accuracy [50, 51]. The random selection of split points in ET can sometimes lead to faster convergence and reduced variance compared to Random Forest.

#### 7.6.4.3 Training Methodology

For regression tasks, both RF and ET algorithms arrive at their final prediction by averaging the predictions of all the individual decision trees within their ensemble. While they aim to enhance prediction performance by reducing bias and variance, their core training methodologies differ significantly in two key aspects: how they sample data for individual trees and select split points within those trees.

Random Forest operates as a Bagging (Bootstrap aggregating) ensemble method. The training process for RF involves (1) Data Sampling (Bagging), for each decision tree in the "forest," RF creates a randomly sampled subset of the original training data, with replacement. This means that some data points from the original dataset may appear multiple times in a subset, while others may not appear at all. This sampling process, known as Bagging, helps increase the model's generalization ability and avoid local optima during training. Typically, the sampled data for each tree amounts to about two-thirds of the total data. (2) Split Point Selection, at each internal node of a decision tree, the algorithm calculates and selects the "best split point" for partitioning the data. For regression problems, this "best split" is determined by minimizing the Mean Squared Error of the resulting sub-regions. This involves searching through various features and their potential split points to find the one that yields the minimum MSE. (3) Tree Growth, RF builds a "forest" by combining multiple Classification and Regression Trees (CART). The process of finding the best split point is repeated for each new node until the MSE of the branched nodes is zero or a stopping condition is met. The individual trees in Random Forest are typically grown to their full depth without pruning, relying on the ensemble to reduce overfitting.

Extremely Randomized Trees, a variant of Random Forest, also uses an ensemble of decision trees but introduces greater randomness in its training process. The training process for ET involves (1) Data Sampling (No Bagging), unlike Random Forest, Extra Trees uses all of the original input data directly as the training set for each decision tree in the ensemble, rather than employing Bootstrap sampling (Bagging). This approach can also help to avoid local optima during training. (2) Split Point Selection (Randomization), a critical distinction for ET is that at each internal node, it randomly selects split points for data partitioning, rather than exhaustively searching for the "best split point" (i.e., the one that minimizes the mean squared error, or MSE). This deliberate randomization makes the algorithm faster and can sometimes lead to faster convergence and reduced variance compared to Random Forest. (3) Tree Growth, similar to Random Forest, ET also uses multiple decision trees. The prediction results from each decision tree are then averaged to obtain the final prediction.

In summary, the key differences in training, the fundamental differences in their training methodologies, can be summarized as follows:

1. Data Subsampling: Random Forest uses bootstrap aggregating (Bagging), meaning each tree is trained on a random sample of the original data with replacement. On the other hand, Extra Trees uses all of the original training data for each tree, without any resampling.
2. Split Point Calculation: Random Forest searches for the optimal split point at each node by calculating and minimizing a criterion (e.g., MSE for regression). Extra Trees randomly selects split points at each node, rather than exhaustively searching for the best one.

These methodological differences lead to varying performance characteristics. Due to their random split selection, Extra Trees generally exhibit faster training times than Random Forest for the same data size. In some cases, Extra Trees has also shown superior prediction performance and reduced variance compared to Random Forest. Interestingly, while Random Forest can benefit from data preprocessing, such as standardization, studies suggest that typical preprocessing methods, including normalization or scaling, do not significantly improve Extra Trees' prediction accuracy.

### 7.6.5 SVR Algorithms

Support Vector Regression (SVR) [52–58] is a robust supervised machine learning algorithm specifically designed to address regression problems, where the goal is to predict continuous output values. It is derived from the theoretical foundations of Support Vector Machine (SVM), which is primarily used for classification tasks. The core concept behind SVR involves identifying a hyperplane in a high-dimensional feature space that best represents the relationship between input variables and their corresponding continuous output values. Instead of trying to fit a curve to the data points, SVR aims to find a function that deviates from the training data by a margin, referred to as the tolerance value ($\varepsilon$), while keeping the function as flat as possible. The $\varepsilon$-insensitive loss function means that errors smaller than $\varepsilon$ are tolerated and do not contribute to the loss function. The size of $\varepsilon$ directly impacts the number of support vectors; a larger $\varepsilon$ leads to fewer support vectors and a less complex model, while a smaller $\varepsilon$ results in more support vectors and a more complicated model.

SVR is formulated as a constrained quadratic programming (QP) problem. This characteristic means that SVR is a convex problem, ensuring that the solution found is a global optimal solution, rather than merely a local optimum. The primary objective of the SVR optimization problem is to minimize the structural risk (SRM), which strikes a balance between model complexity and accuracy. This contrasts with empirical risk minimization (ERM), which is used by algorithms such as neural networks. For complex, non-linear datasets, SVR utilizes the "Kernel Trick". This technique enables SVR to implicitly map the input data from a lower-dimensional input space to a higher-dimensional (or even infinite) feature space, where a linear relationship can be identified. The Radial Basis Function (RBF) kernel is the most commonly used kernel. The RBF kernel only has one gamma ($\gamma$) parameter, making tuning easier. SVR performs well even when dealing with numerous features, particularly with the use of kernel functions. Due to its ability to automatically control model complexity and its $\varepsilon$-insensitive loss function, SVR can handle noise in the data. As it solves a convex optimization problem, SVR guarantees finding a global optimal solution. SVR has been widely adopted in various fields, including biological and behavioral research, image analysis, and medical research. Recently, SVR has been used explicitly to predict the reliability life of a wafer-level package. This involves learning the relationship between design parameters (such as chip thickness and pad diameters) and the predicted reliability life obtained from validated finite element simulations.

#### 7.6.5.1 Theory of SVR

Unlike the support vector machines, which seek to find the optimal hyperplane for linearly separating input data, traditional support vector regression aims to find the optimal hyperplane to represent the input data distribution. Support vector

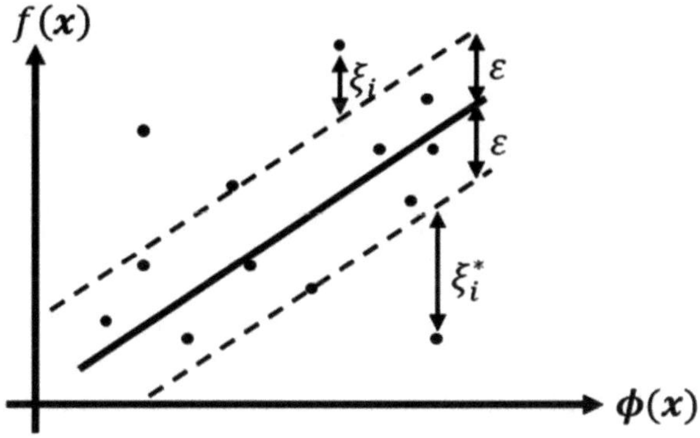


**Fig. 7.40** Support Vector Regression model

regression determines the size of the margin by setting an error tolerance value $\varepsilon$, which is known as $\varepsilon$-SVR, as shown in Fig. 7.40.

When using Support Vector Regression (SVR), a linear model, as shown in Eq. (7.45), is suitable if the relationship between data points is linear. However, real-world data is often non-linear. To handle non-linear relationships, we typically introduce a mapping function, which is a vector function. This transforms the original data into a higher-dimensional space where a linear relationship can then be found. The non-linear SVR model is represented by Eq. (7.46).

Equation (7.47) illustrates the connection between the mapping and kernel functions. The kernel function is a scalar function that calculates the dot product of the mapped vectors without explicitly performing the mapping itself. This is a crucial optimization: in SVR, we define the kernel function instead of the mapping function because the kernel function handles the necessary calculations more efficiently. Defining the mapping function directly would significantly increase computational complexity.

$$y = \langle w, x \rangle + b \tag{7.45}$$

$$y = \langle w, \phi(x) \rangle + b \tag{7.46}$$

$$K(x_i, x_j) = \langle \phi(x_i), \phi(x_j) \rangle \tag{7.47}$$

In Eqs. (7.45) to (7.47), $w$ is the weight, $b$ is the bias, $x$ is the feature, $y$ is the output, $K(x_i, x_j)$ is a kernel function, $\phi(x_i)$ is a mapping function, $x_i$ and $x_j$ stand for two data, $\langle \bullet, \bullet \rangle$ stands for dot product.

The commonly used kernel functions are the polynomial kernel, sigmoid kernel, and radial basis function (RBF) kernel, as shown in Eqs. (7.48) to (7.50). We can observe that the polynomial kernel has three parameters, the sigmoid kernel has two parameters, and the RBF kernel only has one parameter. When we perform parameter setting, we often use a grid search. The RBF kernel is strong enough to perform a grid search due to its single parameter. The RBF kernel can also transform features into an infinite number of dimensions. In this study, we will focus on the RBF kernel.

$$K(x_i, x_j) = \left(\gamma \langle x_i, x_j \rangle + b\right)^d \tag{7.48}$$

$$K(x_i, x_j) = \tanh\left(\gamma \langle x_i, x_j \rangle + b\right) \tag{7.49}$$

$$K(x_i, x_j) = \exp\left(-\gamma \left\| x_i - x_j \right\|^2\right), \gamma = 1/2\sigma^2 \tag{7.50}$$

In Eqs. (7.48) to (7.50), $x_i$ and $x_j$ stand for two data, $\langle \bullet, \bullet \rangle$ stands for dot product, $\|x_i - x_j\|$ is the Euclidean distance between $x_i$ and $x_j$.

The primal problem in SVR is presented in Eq. (7.51), which is an optimization problem with inequality constraints. The $\langle \bullet, \bullet \rangle$ stands for dot product. The superscript * represents the other side of the model. The $\xi_i$ and $\xi_i^*$ are called slack variables.

$$\begin{aligned} &\text{minimize } \frac{1}{2}\|w\|^2 + C\sum_{i=1}^{l}\left(\xi_i + \xi_i^*\right) \\ &\text{subject to} \begin{cases} y_i - \langle w, \phi(x_i) \rangle - b \le \varepsilon + \xi_i \\ \langle w, \phi(x_i) \rangle + b - y_i \le \varepsilon + \xi_i^* \\ \xi_i, \xi_i^* \ge 0 \end{cases} \end{aligned} \tag{7.51}$$

From the primal problem, we learn that we need to minimize the square of the weight and $\xi_i$. And the $\xi_i$ is the value of the loss function on $x_i$, shown in Eq. (7.53). The loss function used in SVR is called $\varepsilon$-insensitive loss function, shown as Eq. (7.52).

$$L(x) = \begin{cases} 0 & \mathit{if} \left| y - f(x) \right| \le \varepsilon \\ \left| y - f(x) \right| - \varepsilon & \text{otherwise} \end{cases} \tag{7.52}$$

$$\xi_i = L(x_i) \tag{7.53}$$

Since solving the primal problem directly is challenging, one can transform it into its dual form using the Lagrange multiplier method. This transformation requires that the Karush-Kuhn-Tucker (KKT) conditions be satisfied. The resulting dual problem in Support Vector Regression (SVR) is presented in Eq. (7.54). The $\alpha_i$ and $\alpha_i^*$ are called Lagrange multipliers. It's clear that the variables in the dual problem are $\alpha_i$, $\alpha_i^*$ corresponding to each data point $x_i$. Before starting training the model, one needs to set the kernel function, $\varepsilon$, and $C$. Where $y_i$ is the result. So the left variables are $\alpha_i$, $\alpha_i^*$ corresponding to each data point $x_i$. The variables $\alpha_i$, $\alpha_i^*$ can be solved using Eq. (7.54). Many ways to solve $\alpha_i$, $\alpha_i^*$, and the most famous one is sequential minimal optimization (SMO).

$$\text{Maximize} -\frac{1}{2}\sum_{i,j=1}^{l}\left(\alpha_i - \alpha_i^*\right)\left(\alpha_j - \alpha_j^*\right)K\left(x_i, x_j\right)$$

$$-\varepsilon\sum_{i=1}^{l}\left(\alpha_i + \alpha_i^*\right) + \sum_{i=1}^{l} y_i\left(\alpha_i - \alpha_i^*\right) \tag{7.54}$$

$$\text{subject to} \sum_{i=1}^{l}\left(\alpha_i - \alpha_i^*\right) = 0 \text{ and } \alpha_i, \alpha_i^* \in [0, C]$$

$$w = \sum_{i=1}^{l}\left(\alpha_i - \alpha_i^*\right)\phi\left(x_i\right) \tag{7.55}$$

$$y = \sum_{i=1}^{l}\left(\alpha_i - \alpha_i^*\right)K\left(x_i, x\right) + b \tag{7.56}$$

From one of the KKT conditions, we derive Eq. (7.55). This Equation reveals that each data point $x_i$ corresponds to a pair $\left(\alpha_i - \alpha_i^*\right)$. However, not all data points contribute to the weight vector computation. Specifically, if $\left(\alpha_i - \alpha_i^*\right)$ equal to zero, the corresponding data point has no contribution to the weight. In contrast, data points for which $\left(\alpha_i - \alpha_i^*\right)$ not equal to zero, these do contribute to the weight calculation and are referred to as support vectors. Substitute Eq. (7.55) into Eq. (7.46) to obtain Eq. (7.56), which is the final model of SVR.

#### 7.6.5.2 Parameter Selection of SVR

When training the model, it is necessary to set specific parameters. As indicated by the dual formulation, the parameters that need to be defined are $\gamma$, $\varepsilon$, and $C$, where $\gamma$ is the parameter in the kernel function. For example, for parameter $C$, one can apply the approach proposed by Cherkassky and Ma [59] shown in Eq. (7.57). $\overline{y}$ and $\sigma_y$ are the average and standard deviation of the results of the training data. *max* stands for selecting the maximum value.

$$C = \max\left(\left|\overline{y} + 3\sigma_y\right|, \left|\overline{y} - 3\sigma_y\right|\right) \tag{7.57}$$

Grid search is applied to determine the parameter $\varepsilon$, which requires defining a search range. The method proposed by Cherkassky and Ma [59] is also employed, as shown in Eq. (7.58), where $\sigma$ denotes the standard deviation of the noise and $n$ is the number of training samples. Equation (7.58) provides a reference value for $\varepsilon$, and the search range is defined from 0 to twice this value. For instance, if the computed value from Eq. (7.58) is 5, the corresponding search range for $\varepsilon$ is 0 to 10.

$$\varepsilon = 3\sigma\sqrt{\frac{\ln n}{n}} \tag{7.58}$$

To determine the parameter $\gamma$, grid search is applied, which requires specifying a search range. The method proposed by Cherkassky and Ma [59], shown in Eqs. (7.59) and (7.60), is used to define this range. In these equations, $x$ represents the input features. For example, if *range*($x$) = 1 and the corresponding *range*($\gamma$) is from 2 to 50. Thus, Eqs. (7.59) and (7.60) provide a systematic way to define an appropriate search range for $\gamma$.

$$p \sim (0.1 - 0.5) * \text{range}(x) \tag{7.59}$$

$$\gamma = \frac{1}{2p^2} \tag{7.60}$$

The initial range is uniformly divided into ten intervals in the grid search for selecting $\varepsilon$ and $\gamma$. The two values yielding the lowest average testing error are chosen to form a narrower range. This new range is again uniformly divided into ten intervals, and the process is repeated to identify the two values that perform best based on the average testing error. This iterative refinement continues until the optimal parameter settings are determined.

#### 7.6.5.3 Training Methodology

Unlike Support Vector Machines (SVMs), which aim to find an optimal hyperplane for classification, SVR seeks an optimal hyperplane that best represents the input data distribution. Key aspects of SVR include: $\varepsilon$-insensitive loss function, which defines a "margin" ($\varepsilon$) around the hyperplane. Errors within this margin are not penalized, promoting a smoother model. The goal is to find a hyperplane that is as flat as possible, with data points remaining within the $\varepsilon$-margin. However, for data points that fall outside this $\varepsilon$-margin, slack variables ($\xi$ and $\xi^*$) are introduced. These variables quantify the deviation of these samples from the boundaries defined by the $\varepsilon$-loss function. Essentially, they allow the SVR model to tolerate some errors or noise in the data without penalizing them excessively, making the model more robust.

Hyperparameters include $C$, $\varepsilon$, and $\nu$. (1) $C$ (Penalty Factor), this constant determines the trade-off between model smoothness (minimizing the weight vector '$w$') and the amount of acceptable error (penalizing slack variables). A larger $C$ value means less error tolerance, potentially leading to overfitting. (2) $\varepsilon$ (Epsilon) defines the width of the insensitive zone (margin) around the regression line. Its value also impacts the number of support vectors. (3) $\nu$ (Nu-SVR): A newer SVR technique, also known as Nu-SVR, introduces the parameter $\nu$ to replace $\varepsilon$. This enables the algorithm to adaptively adjust the value of $\varepsilon$ during the training process, resulting in potentially more accurate predictions and improved generalization capabilities by controlling the number of support vectors. Nu-SVR often achieves the most accurate predictions due to its adaptive $\varepsilon$-tuning capability and control over the support vector count, leading to minimized structural risk.

Kernel Functions: These functions implicitly transform input features from the input space to a higher-dimensional feature space without explicitly knowing the transformation equation. This "Kernel Trick" reduces computational complexity. Common kernel functions include: (1) Radial Basis Function (RBF) Kernel: Widely used due to its ability to map data to an infinite-dimensional feature space and its fewer parameters to tune, making it easier to optimize. It is often preferred for optimal prediction performance after data preprocessing. (2) Polynomial Kernel, (3) Sigmoid Kernel, (4) Matérn Kernel, (5) Rational Quadratic Kernel, and (5) Exponential Sine Squared Kernel.

Support Vectors: These are the data points that either lie on the boundary or outside the $\varepsilon$-margin and contribute to the calculation of the model's weight. The number of support vectors influences the prediction performance; generally, more support vectors can lead to better performance and increase training time.

In summary, the application of SVR techniques, especially the newer Nu-SVR, is similar to other machine learning algorithms, such as ANN and RNN, providing an efficient and effective approach for front-end designers to quickly assess the feasibility of their WLP designs by predicting reliability. This reduces reliance on time-consuming and costly experimental and purely FEA-based approaches.

### *7.6.6 GPR Algorithm*

Gaussian Process Regression (GPR) [60–62] is a sophisticated, kernel-based, non-parametric Bayesian approach designed to solve regression and probabilistic classification problems. As a non-neural network algorithm used for training sub-models, the inclusion of GPR enhances model diversity. GPR captures relationships between data points through a defined covariance function known as a kernel. GPR has been applied explicitly for the reliability life prediction of wafer-level packaging structures, using large datasets generated by finite element method simulations. Gaussian Process Regression is a supervised machine learning technique. GPR can apply the prior Gaussian Process to make a prediction. The prior distribution can be specified by the mean and covariance function. Therefore, the covariance function plays a crucial role in GPR models. The kernel trick can be used to determine the covariance function. During the fitting process, the hyperparameter of the kernel is optimized by maximizing the margin likelihood using a gradient descent optimizer.

Unlike parametric models, GPR directly models the underlying function probabilistically, as shown in Fig. 7.41. One of its key advantages is the ability to quantify predictive uncertainty via posterior distribution. The core assumption of GPR is that

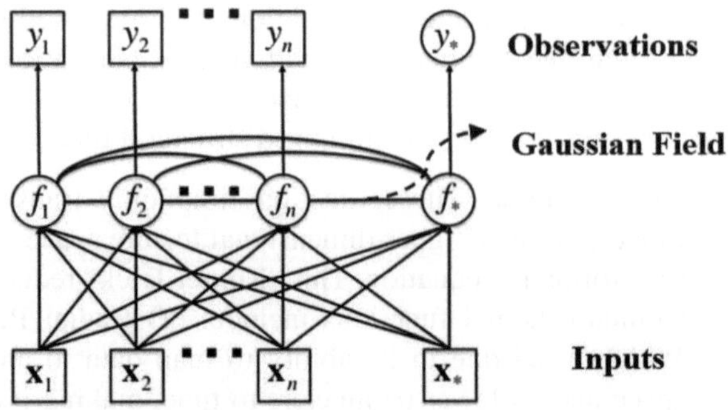


**Fig. 7.41** Schematic of GPR

the target function $f(x)$ is governed by a Gaussian Process prior, while the noise in observations is normally distributed, as represented in Eq. (7.61).

$$\begin{aligned} y &= f(x) + \epsilon \\ f(x) &\sim \mathrm{GP}\left(m(x), k(x, x')\right) \\ \epsilon &\sim N\left(0, \sigma_n^2\right) \end{aligned} \tag{7.61}$$

where $m(x)$ denotes the mean function, which is typically assumed to be zero, and $k(x, x')$ represents the kernel function that defines the covariance between the input points $x$ and $x'$.

Let the training inputs be denoted as $X$ with corresponding observed outputs $y$, and the test inputs as $X_*$ with their predicted outputs $y_*$ obtained through the latent function $f_*$. The joint distribution of the training and test outputs is given in Eq. (7.62).

$$\begin{bmatrix} y \\ f_* \end{bmatrix} \sim N\left[0, \begin{matrix} K(X,X) + \sigma_n^2 I & K(X,X_*) \\ K(X_*,X) & K(X_*,X_*) \end{matrix}\right] \tag{7.62}$$

Using the properties of conditional Gaussian distributions, the predictive mean and covariance of the test outputs can be derived, as shown in Eq. (7.63).

$$\begin{aligned} \overline{f_*} &= K(X_*,X)\left[K(X,X) + \sigma_n^2 I\right]^{-1} y, \\ \operatorname{cov}(f_*) &= K(X_*,X_*) - K(X_*,X)\left[K(X,X) + \sigma_n^2 I\right]^{-1} K(X,X_*) \end{aligned} \tag{7.63}$$

This framework not only adapts to the underlying structure of the data but also optimizes its hyperparameters through maximum likelihood estimation, allowing for flexible and accurate modeling of complex functions. On the other hand, the performance of GPR is highly dependent on the choice of the kernel function. The kernel functions used in this study are the Matérn kernel and the Radial Basis Function (RBF) kernel, as shown in Eqs. (7.64) and (7.65), respectively.

$$k\left(x_i, x_j\right) = \frac{1}{\Gamma(v)2^{v-1}}\left(\frac{\sqrt{2v}}{l}d\left(x_i, x_j\right)\right)^v K_V\left(\frac{\sqrt{2v}}{l}d\left(x_i, x_j\right)\right) \tag{7.64}$$

$$k\left(x_i, x_j\right) = \exp\left(\frac{-d\left(x_i, x_j\right)^2}{2l^2}\right) \tag{7.65}$$

Where $\Gamma(\bullet)$ is the gamma function, $v$ is the smoothness parameter that controls the differentiability of the function, $K_V(\bullet)$ is the modified Bessel function of the second kind, $l$ is the length-scale parameter, and $d(\bullet, \bullet)$ is the Euclidean distance between inputs.

#### 7.6.6.1 Cluster Analysis

Cluster analysis is a data analysis technique introduced by anthropologists Driver and Kroeber [63] in 1932 to study cultural groups. Used in fields such as statistics and machine learning, it's not a specific algorithm but a general term for a family of methods. The core purpose is to divide data into meaningful clusters by calculating the correlation between data points, often using their features, distribution, or density as a guide. Cluster analysis is a kind of unsupervised learning in machine learning. The main purpose of using cluster analysis with the GPR is to explore the relationship between the time complexity of the Gaussian process regression model and the amount of data. Analyze the K-means algorithm [64, 65] in Cluster Analysis to divide the data. The amount of data in the Gaussian Process Regression model will significantly impact the training time of a single stroke, and its time complexity is O ($n^3$), where $n$ is the number of data points to be substituted. It can be understood that the training time of the Gaussian Regression model will increase exponentially with the amount of data. Using the K-means algorithm, large volumes of data and test samples are grouped into several clusters based on the Euclidean distance between their feature vectors. Each cluster is then trained in batches to generate individual training and prediction results. These results are combined using a weighted average to produce the entire dataset's overall training and prediction outcomes. The K-means clustering process is illustrated below [65].

$$\sum_{i=0}^{n}\sum_{j=0}^{n_c}\min\left(x_i - \mu_j^2\right) \tag{7.66}$$

$x_i$ is each data point 、 $\mu_j$ is the cluster centroid 、 $n_c$ is the cluster number、 $||x_i - \mu_j||$ is Euclidean distance. K-means procedure illustrated as follows:

1. Select the number of clusters to be divided, which is called the K value.
2. Randomly add K new centroids to the original data.

3. Calculate the Euclidean distance between each data point and each K centroid feature size. The data points with the closest distance to each centroid are grouped into the same cluster, with their centroids serving as the cluster centers. This method updates the overall clustering.
4. Average the data points of each cluster to obtain the new centroid of each cluster.
5. Iteratively recalculate the new centroids and regroup the data closest to the new centroids. Update the new centroids of each cluster, and repeat steps 3 and 4 until the centroids of each cluster remain unchanged.

#### 7.6.6.2 Training Methodology

The training methodology for Gaussian Process Regression (GPR) models, as described in the reference [62], is a structured process designed to predict the reliability life of wafer-level packaging structures. It integrates data generation, preprocessing, hyperparameter tuning, and specialized strategies to enhance performance and manage computational costs. The research follows a structured machine learning process that includes building different training datasets using a validated simulation methodology, evaluating algorithm performance with testing data, and comparing performance and training time across various data sizes. The machine learning GPR regression models are estimated using a flow that involves data preprocessing and hyperparameter tuning.

Data preprocessing is a crucial step to standardize input and output datasets. This process helps ensure that all dataset values are within the same range, which improves training effectiveness and model accuracy. It also addresses data inconsistencies, such as outliers. Various preprocessing methods are evaluated, including Min-Max Scaler, Robust Scaler, Max Absolute Scaler, and Standard Scaler. The Standard Scaler consistently yields better training and testing accuracy for GPR and is therefore chosen. This method standardizes data by subtracting the mean and dividing by the standard deviation, resulting in a mean of 0 and a standard deviation of 1. The Robust Scaler is also mentioned for providing good performance and being less sensitive to outliers.

In practice, grid search is utilized to find the best hyperparameters for GPR before the main training runs. Both exhaustive and bisection methods of grid search can be used for GPR. The choice of kernel function has a significant impact on GPR performance. Various kernels are explored, including the Radial Basis Function (RBF), Matérn, Rational Quadratic (RQ), and Exponential Sine Squared kernels. The Matérn kernel is often chosen for its superior accuracy, even if it might take slightly longer than RBF. A significant drawback of GPR is its high computational cost, with training time increasing substantially as the dataset size grows larger. For example, training time for GPR can range from 1.4 s for 500 data points to 864 s for 9000 data points, making it considerably slower than other models, such as KNN or PR. This is mainly due to the complexity of calculating the inverse of the kernel matrix. K-means clustering can be applied to address the high computational cost of GPR, especially with large datasets. This strategy involves partitioning large datasets into smaller clusters based on Euclidean distance. The GPR model is then trained in batches on these individual clusters. The overall prediction is obtained through weighted averaging of the cluster results. This strategy effectively reduces training time without a significant loss of accuracy. Future work considerations for GPR Future improvements for GPR model training include further exploration of multi-kernel approaches to enhance prediction accuracy.

### 7.6.7 KRR Algorithm

Arthur and Robert [66] proposed the idea of Ridge Regression (RR) to address the issue of multicollinearity in high-dimensional data. With the evolutionary progress of the algorithm, the terminology "Kernel Ridge Regression" (KRR) was proposed in 2000 by Cristianini and Shawe-Taylor [67]. The essence of KRR lies in the combination of RR and the kernel trick. K-means clustering can perform data partitioning to reduce the workload of KRR. This would significantly reduce the training time required for the prediction model.

#### 7.6.7.1 Least Squares Method

The Least Squares Method (LSM) is commonly used in statistics to solve regression problems. It is a mathematical optimization technique that determines the best-fitting function for a dataset by minimizing the sum of squared errors. The squared loss function for LSM is presented in Eq. (7.67).

$$L(\beta)=(y-X\beta)^T(y-X\beta) \tag{7.67}$$

In Eq. (7.67), $X$ is the matrix expression of the data input. The row of the matrix is the number of data samples. The column of the matrix is the data dimension. $y$ is the output. $\beta$ is the equation coefficient.

In LSM, our target value $\hat{\beta}$ is the value that minimizes $L(\beta)$. To do this, we need to solve the partial derivative of $L(\beta)$ to $\beta$. The resulting expression of $\hat{\beta}$ is shown as Eq. (7.68).

$$\hat{\beta}=\left(X^TX\right)^{-1}X^Ty \tag{7.68}$$

Training models using LSM enables them to fit known sample points quickly and accurately; however, it may also cause overfitting. In Eq. (7.68), when there is multicollinearity in the data dimension, $X^TX$ is no longer a full-rank matrix, and it would be tough to solve its inverse matrix directly. RR is an algorithm that is proposed to solve this problem. RR is essentially an improved LSM. By relinquishing the unbiasedness of the LSM, a more realistic mathematical model is obtained, albeit at the expense of losing some information and reducing accuracy. Equations (7.69) and (7.70) show the new loss function and target value expression.

$$J(\beta) = (y - X\beta)^T (y - X\beta) + k\beta^2 \tag{7.69}$$

$$\beta = \left(X^T X + kI\right)^{-1} X^T y \tag{7.70}$$

In Eqs. (7.69) and (7.70), $k$ is the ridge parameter, $I$ is the identity matrix. As $k$ increases, the undetermined coefficient $\beta$ would stabilize, and we were looking for the smallest $k$ value under the condition that the coefficient is stable.

KRR is the combination of RR and kernel tricks. When a linear model cannot learn the data in the input space, the data can be converted to other dimensions of space through a transformed function and then solved using the inner product. It is found that the same effect can be achieved directly by defining a function $K$, which is called the kernel function. There are three commonly used kernel functions: the polynomial kernel, the sigmoid kernel, and the radial basis function (RBF) kernel, as shown in Eqs. (7.71) to (7.73). As we can see, the polynomial kernel has three parameters. The sigmoid kernel has two parameters, whereas the RBF kernel has only one parameter. This is a key strength of the RBF kernel. Using the RBF kernel reduces the hyperparameter search space; the dominant computational cost of KRR still comes from kernel-matrix operations. In this section, we focus on the RBF kernel.

$$K\left(x_i, x_j\right) = \left(\gamma \left\langle x_i, x_j \right\rangle + b\right)^d \tag{7.71}$$

$$K\left(x_i, x_j\right) = \tanh\left(\gamma \left\langle x_i, x_j \right\rangle + b\right) \tag{7.72}$$

$$K\left(x_i, x_j\right) = \exp\left(-\gamma \left\| x_i - x_j \right\|^2\right) \tag{7.73}$$

In Eqs. (7.71) to (7.73), $x_i$ and $x_j$ represent two data points, $\langle \bullet, \bullet \rangle$ represents the dot product, $\|x_i - x_j\|$ is the Euclidean distance between $x_i$ and $x_j$.

We need to express the RR solution as an inner product to introduce the kernel function. Converting the original formula to a particular form requires the matrix inversion lemma [68].

Consider a general partitioning matrix $M = \begin{pmatrix} E & F \\ G & H \end{pmatrix}$. Assuming that both $E$ and $H$ are invertible, we have:

$$\left(E - FH^{-1}G\right)^{-1} = E^{-1} + E^{-1}F\left(H - GE^{-1}F\right)^{-1} GE^{-1} \tag{7.74}$$

$$\left(E - FH^{-1}G\right)^{-1} FH^{-1} = E^{-1}F\left(H - GE^{-1}F\right)^{-1} \tag{7.75}$$

$$\left|E - FH^{-1}G\right| = \left|H - GE^{-1}F\right|\left|H^{-1}\right|\left|E\right| \tag{7.76}$$

Use Eq. (7.75) to simplify the optimal solution of $\beta$. Let $H^{-1} \triangleq k^{-1}I$, $F \triangleq X^T$, $G \triangleq -X$, $E \triangleq I$, then we obtain a new expression shown as Eq. (7.77).

$$\beta = \left(X^T X + kI\right)^{-1} X^T y = X^T \left(kI + XX^T\right)^{-1} y \tag{7.77}$$

Now, it is very close to kernelization. The remaining task is to predict $y^*$ when a new sample point $x^*$ comes in. We write $\beta$ in the form of a vector summation and let $\alpha \triangleq (kI + X^T X)^{-1} y$. Then we rewrite Eq. (7.77) to Eq. (7.78).

$$\beta = X^T \alpha = \sum_{i=1}^{N} \alpha_i x_i \tag{7.78}$$

$\beta$ is a weighted average of all samples. Thus, the predicted value for a new sample is:

$$y^* = \beta^T x^* = \sum_{i=1}^{N} \alpha_i x_i^T x^* = \sum_{i=1}^{N} \alpha_i K\left(x^*, x_i\right) \tag{7.79}$$

The predicted value is the weighted average of the inner product of the new and old samples. In this way, we have completed the conversion of the original formula to the inner product form and chosen different kernel functions K to simplify the operation.

#### 7.6.7.2 Training Methodology

The training methodology for Kernel Ridge Regression models, as described in references [4, 5], for predicting wafer-level packaging reliability is a comprehensive process involving data generation, preprocessing, hyperparameter tuning, and specialized strategies to optimize performance and manage computational demands. Data preprocessing is a crucial step to standardize input and output datasets. This standardization ensures that all values fall within the same range, thereby improving training effectiveness and model accuracy and addressing inconsistencies such as outliers. Various preprocessing methods are evaluated for KRR, including Min-Max Scaler, Robust Scaler, Max Absolute Scaler, and Standard Scaler. For KRR, the Min-Max Scaler consistently yields better testing accuracy and is therefore preferred in some studies; however, the Standard Scaler and Robust Scalers also show good training performance, depending on the dataset. In practice, grid search is utilized to find the best hyperparameters for KRR before the main training runs. Bayesian optimization has also been explored as an alternative tuning method. The choice of kernel function significantly impacts KRR performance, particularly with nonlinear datasets.

K-means clustering is applied as a preprocessing step to address the high computational cost of KRR, especially with large datasets. This strategy, known as the K-K (K-means—KRR) hybrid model, partitions large datasets into smaller clusters based on Euclidean distance. The KRR model is then trained in batches on these individual clusters, effectively replacing a single global model with multiple regional sub-models. The overall prediction for new data is obtained by first classifying the data into a cluster using K-means, then using the corresponding sub-model. This strategy significantly reduces training time without compromising accuracy. For instance, the case shown, using K-means with 32 clusters, can reduce KRR training time from 1340 to 7 s for a 9000-point dataset.

## *7.6.8 Polynomial Regression Algorithm*

Polynomial Regression (PR) is a machine learning model used to describe the relationship between independent and dependent variables by fitting a polynomial equation. It is a type of supervised learning algorithm, meaning it operates on datasets where the input variables are labeled with known output values, making it suitable for predicting continuous outputs, such as reliability life cycles. To pursue a more efficient algorithm, polynomial regression [69, 70] is introduced in this section for model training. The implemented model includes several components: the dependent variable, independent variables, interaction terms involving all features, and a necessary bias term. This approach delivers acceptable accuracy in most scenarios.

Polynomial Regression (PR) is a sort of regression that is used to find the dependence between independent and dependent variables.

$$\begin{aligned} y &= \beta_0 + \beta_1 x + \beta_2 x^2 + \beta_3 x^3 \\ &+ \ldots + \beta_n x^n + \varepsilon \end{aligned} \tag{7.80}$$

Where $y$ is the function of $x$, $y$ is the dependent variable, $x$ is the independent variable, $\beta$ means the coefficient, $n$ represents the highest degree in the formula (Degree), which is different from the order (Order), and $\varepsilon$ represents the residual value (Residual). Although this value is a constant like $\beta_0$, but it cannot be combined. The specific reason is that $\beta_0$ represents the *0th* power coefficient of $x$, not a simple constant.

In Fig. 7.42, the blue dots represent the original data, while the orange line illustrates the result of linear regression. When a quadratic curve (yellow line) is applied, it becomes apparent that the added curvature allows the model to better capture the overall trend compared to a straight line. As the degree of the polynomial increases, the model can pass through more data points. This suggests that higher-order terms act as correction terms, progressively refining the fit. With each increase in

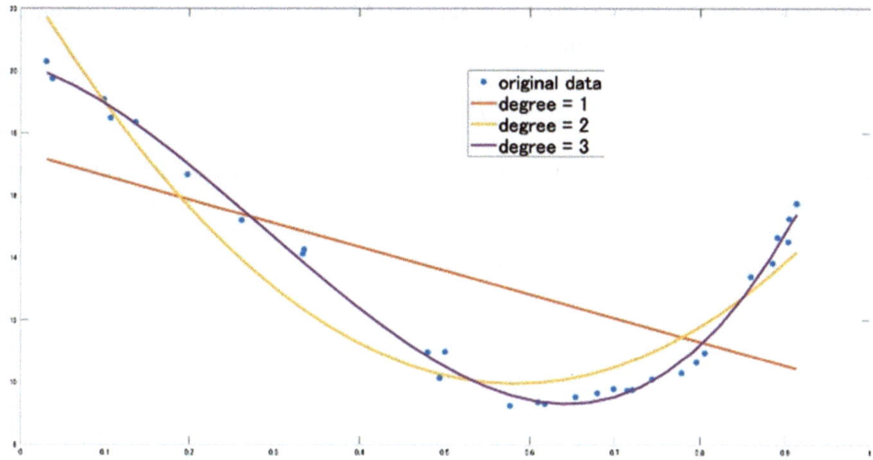


**Fig. 7.42** Polynomial regression

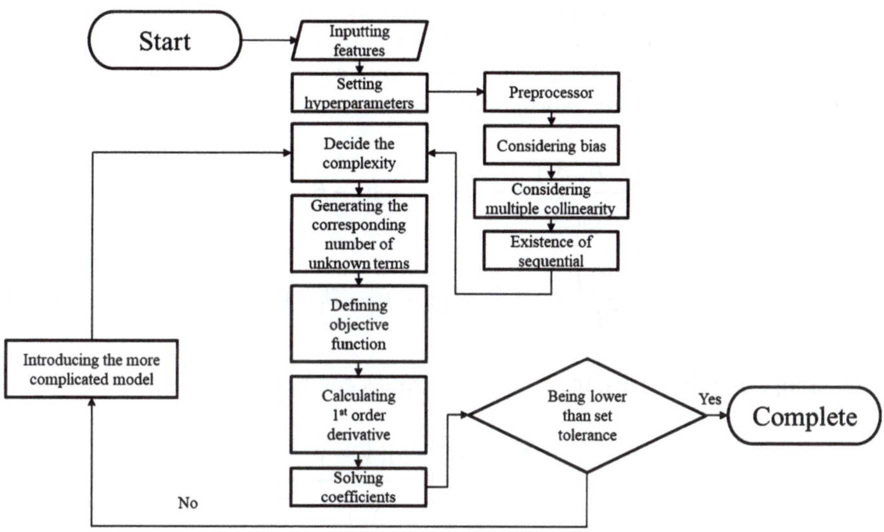


**Fig. 7.43** Polynomial regression solution flowchart

polynomial degree, the model more closely approximates the underlying pattern of the data. The flowchart of Polynomial regression is shown in Fig. 7.43.

It is important to address a key issue: balancing underfitting and overfitting. The model should avoid underfitting, where it fails to capture the underlying trend, but it must also prevent overfitting, where it becomes too closely tailored to the training data. Overfitting occurs when the model is overly complex, fitting not only the general pattern but also the noise and outliers in the data [69]. A regression model typically comprises independent variables, dependent variables, residuals, and coefficients. The aim is to estimate the values of all the coefficients. The most common approach for this is the Least Squares Method. Below is a basic example. Consider a linear regression model:

$$y = \beta_0 + \beta_1 x + \varepsilon \tag{7.81}$$

Before performing the calculation, it is required that the independent variable is not a constant. The next step is to define the objective function. The third step involves squaring the difference between the actual value and the estimated value, then dividing by the number of data points:

$$Q(\beta_0,\beta_1) = \frac{1}{n}\sum_{i=1}^{n}(y_i - \beta_0 - \beta_1 x_i)^2 = \frac{1}{n}\sum_{i=1}^{n}(U_i)^2 \tag{7.82}$$

Where $Q$ represents the objective function, $n$ represents the amount of data, $y$ is the dependent variable, $x$ is the independent variable, $\beta_0$ and $\beta_1$ are coefficients of different powers, and $U$ represents the difference. The reason for squaring the differences before summing them is to prevent the positive and negative values from canceling each other out, which would result in an inaccurate measure of the difference. By squaring the differences, all values become positive, accurately representing the total error. Taking the first derivative with respect to the coefficients yields two equations, known as the Normal Equations:

$$\frac{\partial Q(\beta_0,\beta_1)}{\partial \beta_0} = \frac{-2}{n}\sum_{i=1}^{n}(y_i - \beta_0 - \beta_1 x_i) = 0 \tag{7.83}$$

$$\frac{\partial Q(\beta_0,\beta_1)}{\partial \beta_1} = \frac{-2}{n}\sum_{i=1}^{n}(y_i - \beta_0 - \beta_1 x_i)x_i = 0 \tag{7.84}$$

After comparing the terms, the coefficient values can be solved as follows:

$$\hat{\beta}_1 = \frac{\sum_{i=1}^{n}\left(x_i - \overline{x_n}\right)\left(y_i - \overline{y_n}\right)}{\sum_{i=1}^{n}\left(x_i - \overline{x_n}\right)^2} \tag{7.85}$$

$$\hat{\beta}_0 = \overline{y_n} - \hat{\beta}_1 \overline{x_n} \tag{7.86}$$

The profile of the model is obtained, and recalling the preset condition, that is, the independent variables cannot be the same constant, or $\hat{\beta}_1$ cannot be solved because of the zero in its denominator.

Polynomial Regression primarily relies on the least squares method to compute the coefficients of each term. Consequently, the most significant factor influencing CPU time is the number of terms, which depends on the number of features and the degree of the polynomial. Since polynomial regression includes interaction terms between different features, the number of terms can grow rapidly, increasing computational time exponentially. To address this, cluster analysis can be used to reduce the number of data points, simplifying the overall model. This makes it more feasible to apply polynomial regression to represent the original dataset effectively.

$$\text{Number of Items} = \frac{(n+d)!}{n!d!} \tag{7.87}$$

Where $n$ represents the number of data features to be estimated, and $d$ is the set degree.

#### 7.6.8.1 Training Methodology

Polynomial Regression (PR) is a machine learning model designed to fit a polynomial equation to data, primarily used for describing nonlinear or curvilinear relationships between independent and dependent variables. Preprocessing is vital in standardizing the input data, ensuring that all values are within a similar range. This improves training effectiveness and model accuracy, especially when features have vastly different scales. Common preprocessing methods evaluated for PR include Standard Scaler, Min-Max Scaler, Robust Scaler, and Max Absolute Scaler. The Min-Max Scaler is often chosen due to its consistent performance and ability to compress values from 0 to 1.

Critical hyperparameters need to be configured for the PR model: Increasing the degree allows the model to fit complex trends better, as higher-order terms act as correction terms. However, a higher degree also increases model complexity and the risk of overfitting. As the amount of data increases, a model with a higher degree is generally required to describe the trend accurately. Including a bias term is necessary for better model performance. It serves as a correction term, addressing issues that can arise with specific preprocessing methods, particularly when scaled minimum values become zero. If this setting is enabled, it restricts the model from including interaction terms between different features, excluding terms where a variable is multiplied by itself. This can lead to suboptimal results, as it removes important higher-order correction terms, causing the model's performance to plateau with increasing degrees of freedom. Therefore, it is generally advised not to use this setting for optimal PR performance.

K-means clustering can be applied as a preprocessing step to further optimize PR, particularly for large datasets. This hybrid approach involves partitioning the dataset into smaller clusters and then training a separate PR model on each cluster. This strategy reduces training time significantly and allows for using simpler PR models (lower degrees) without sacrificing accuracy, sometimes even achieving better results. K-means helps to address the challenge of increased computational time and model complexity that comes with larger datasets in pure PR by reducing the amount of data needed for each PR model. The choice of preprocessing for K-means is crucial to ensure all clusters contain data.

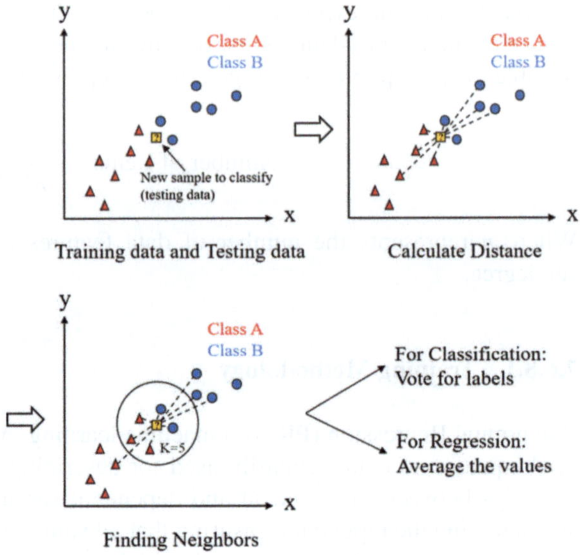


**Fig. 7.44** The step of KNN

## 7.6.9 *KNN Algorithm*

The K-Nearest Neighbors (KNN) [71–73] algorithm is a simple, non-parametric instance-based learning algorithm classified as supervised learning. Unlike many other algorithms, KNN is considered a "lazy algorithm" because it does not explicitly train a model or find any functions beforehand; instead, it stores all available training data and defers computation until a new data point needs to be classified or regressed.

The K-Nearest Neighbors algorithm follows a straightforward process. First, it calculates the distance between the new and existing data points. Once the value of K is specified, the algorithm identifies the K nearest neighbors based on the defined distance metric. A majority voting approach is used for classification tasks, where the class most frequently represented among the K neighbors is assigned to the new point. For regression problems, the algorithm predicts the outcome by averaging the values of the K nearest data points. The steps and flowchart of the KNN algorithm are illustrated in Figs. 7.44 and 7.45.

### 7.6.9.1 Distance Definition

K-Nearest Neighbors is a simple yet effective algorithm that performs classification or regression based on the similarity between new data points and existing training data. This similarity is typically defined in terms of the distance between data points. Three commonly used distance metrics are introduced below: Euclidean distance, Manhattan distance, and Minkowski distance.

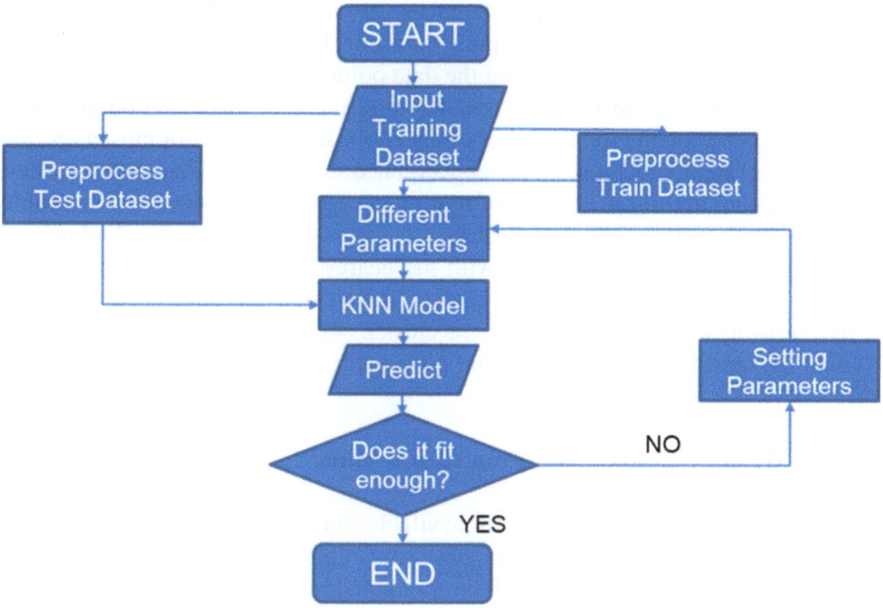


**Fig. 7.45** KNN flowchart

Euclidean distance, the most widely used metric, represents the straight-line distance between two points and is defined in Eq. (7.88). Manhattan distance, illustrated in Eq. (7.89), can be interpreted as the distance traveled along a grid, calculated by summing the absolute differences across all features. Minkowski distance generalizes these metrics, as shown in Eq. (7.90). When the parameter P = 1, the Minkowski distance is equivalent to the Manhattan distance; when P = 2, it becomes the Euclidean distance.

$$D = \sqrt{\sum_{i}^{n} (x_i - y_i)^2} \tag{7.88}$$

$$D = \sum_{i}^{n} |x_i - y_i| \tag{7.89}$$

$$D = \left( \sum_{i}^{n} |x_i - y_i|^p \right)^{\frac{1}{p}} \tag{7.90}$$

Cluster analysis is a data simplification technique that groups samples with high similarity based on their attributes. Typically, similarity is measured by the distance between data points; the smaller the distance, the greater the similarity. Conversely, a larger distance indicates lower similarity. After clustering, intra-group differences are minimized, while inter-group differences are maximized. K-means is a widely used clustering algorithm that partitions data into K clusters based on similarity, measured by distance. The average of the data points within each cluster is referred to as the centroid. As an unsupervised learning method, the time complexity of K-means depends on the total number of data points, the number of clusters K, and the number of iterations required for convergence.

The K-means algorithm consists of four main steps: (1) Initialization: Randomly select $K$ data points as the initial cluster centers. (2) Assignment: Assign each remaining data point to the cluster with the nearest center, based on the shortest distance (i.e., highest similarity). (3) Update: Recalculate the centroid of each cluster by averaging all data points within the cluster. (4) Iteration: Repeat steps (2) and (3) until the cluster centers stabilize and no longer change positions.

#### 7.6.9.2 Weight Definition and Empirical K Value

In KNN, two weighting strategies are available: uniform weighting and distance-based weighting. For regression tasks, uniform weighting assigns equal importance to all $K$ nearest neighbors, meaning the predicted value is the simple average of these neighbors. In contrast, distance-based weighting assigns weights that are inversely proportional to the distance between the new data point and each neighbor; thus, closer neighbors have a greater influence on the prediction. K-Nearest Neighbors lacks an explicit training model, which makes its performance highly sensitive to the choice of the K value. A slight change in K can significantly impact the overall results. When introducing a new data point, it can be challenging to make an accurate prediction without prior tuning. In such cases, based on past experience or validation, an empirically determined K value can be used to improve prediction reliability. The formula is shown in Eq. (7.91).

$$\mathrm{K} = \frac{\sqrt{n}}{2} \tag{7.91}$$

Where "K" is the empirical K value and "$n$" is the number of training data points. Although not optimal, this empirical K typically yields acceptable performance and can be obtained within milliseconds. It is because the empirical K value is directly given as a K value. The Flowchart of KNN is shown below.

### *7.6.10 K-Means and K-Medoids Clustering Algorithm*

Clustering analysis is a method used to simplify data by grouping samples with high similarity together to form a cluster. It is classified as unsupervised learning, meaning that the training data does not contain predefined labels or categories. Instead, the algorithm explores the data to uncover underlying patterns or structures within the dataset. The primary goal of clustering is to make the differences within each group (cluster) smaller and the differences between different groups larger. This is typically achieved by quantifying similarity based on the distance between data points; points closer together are considered more similar, while points farther apart are less similar. Section 7.6.6.1 Cluster Analysis mentions that the clustering algorithm can be coupled with various machine learning algorithms, such as GPR, KRR, PR, and KNN, to improve prediction accuracy and training efficiency. Two specific clustering algorithms are K-means (Section 7.6.6.1) and K-medoids.

K-means is a commonly used clustering algorithm that divides n data samples into K sets (clusters). The core concept is to minimize the sum of squares of the distance between each data point and its assigned cluster center, making data points within the same group more similar. A crucial parameter is the K value, which represents the predefined number of cluster centers or groups. The choice of K has a significant impact on the algorithm's results. Scholars note that K-means is sensitive to the initial value of K. The proportion of K to the total number of data points (n) should be within a reasonable range, such as 3% to 20%, to avoid under- or over-clustering. The K-means procedure has been detailed and illustrated in Section 7.6.6.1.

Similar to K-means, K-medoids [74, 75] operates in a manner very similar to K-means in its overall procedure. It also partitions a dataset into distinct groups based on a defined similarity measure, typically distance. The main difference lies in how cluster centers are selected and updated. Unlike K-means, which computes centroids that may not correspond to any real data point, K-medoids chooses actual data points from the dataset as its cluster centers (called "medoids"). In K-medoids, the medoid for each cluster is identified as the data point that minimizes the total distance to all other points within that same cluster. This means the medoid is the most representative feature within its local region. K-medoids is utilized for data reduction and prototype selection, enhancing sampling efficiency in training machine learning models. It is particularly suitable for identifying representative samples from an existing dataset to serve as a training set.

Since clustering is an unsupervised method, it lacks explicit labels for direct accuracy assessment. Several techniques and metrics are used to evaluate clustering results. (1) Within-Cluster Sum of Squares (WCSS) / Sum of Squared Errors (SSE), this metric calculates the sum of the squared distances between each data point and its assigned cluster center within its cluster. Lower SSE values indicate better clustering. However, SSE tends to decrease as the number of clusters (K) increases, which can lead to a bias towards over-partitioning. (2) The Elbow Method uses the SSE (or WCSS) to help determine the optimal K value. It involves plotting the SSE against different K values and looking for an elbow point in the curve, where the rate of decrease in SSE significantly slows down. (3) Adjusted Sum of Squared Errors (Adjusted SSE), it is proposed to address the limitations of standard SSE, especially with increasing K. It incorporates a penalty term (n-K in the denominator, where n is the number of data points and K is the number of clusters) to provide a more accurate reflection of the average deviation between each data point and its cluster center, thereby helping to prevent underestimation of error as K increases. (4) Silhouette Plot Approach, which is used for the analysis and validation of data clusters. (5) Dendrogram Technique is also used to identify optimal clusters.

Numerous electronic packaging research studies have demonstrated that clustering analysis, particularly K-means and K-medoids, plays a crucial role in enhancing the efficiency and consistency of machine learning models for predicting the reliability life of wafer-level packages and warpage in fan-out panel-level packaging. Clustering helps reduce the training data by selecting representative points, thereby simplifying model complexity and saving computational time. This is crucial for large datasets generated by finite element simulations. For algorithms like K-Nearest Neighbors, which lack a training phase, clustering can help reduce the error between the optimal K value and an empirical K value, offering a "model-like" concept for KNN. It ensures that KNN algorithms can operate normally by preventing scenarios where a cluster lacks sufficient data points. Clustering can be integrated into hybrid models, such as KRR combined with K-means (K-K model), and GMM-KMean-PAR, or Random Forest combined with cluster analysis and Artificial Neural Networks. This integration leverages the strengths of different algorithms, enabling efficient prediction while maintaining accuracy. In scenarios with small datasets or numerous design variables, clustering (e.g., K-medoids in adaptive sampling) helps ensure a more even distribution of data, particularly near feature boundaries where prediction errors are often higher. This strategy enables models to acquire more relevant information, thereby reducing local prediction errors and enhancing generalization.

## 7.7 Parallel Computing and Computing Speed of CPU and GPU

AI training, particularly for complex machine learning models, significantly benefits from the interplay of parallel computing [56, 76–79], Graphics Processing Units (GPUs), and Central Processing Units (CPUs) [80, 81]. This integration is crucial for efficiently building predictive models, such as those used for reliability life prediction in electronic packaging. Machine learning involves using AI theory and large datasets to train computers to learn and develop prediction models. This training phase, especially for deep learning models like Artificial Neural Networks and Recurrent Neural Networks, inherently involves many matrix and tensor computations. The goal is to establish a regression model by analyzing input data (e.g., packaging geometry) and labeled output data (e.g., reliability life cycle). Once trained, these AI models can immediately provide reliability predictions by simply inputting new design parameters.

A significant challenge in AI training is the hyperparameter search, a computationally exhaustive process that involves finding the optimal set of parameters for a model. This process, often conducted via grid search, can significantly increase the overall time required for model development, particularly with larger datasets and more complex models. Traditionally, CPUs were used for model training. A CPU is designed for latency-oriented computing, focusing on maximizing the processing speed of individual tasks. It has a few powerful cores, each capable of independently handling different instructions (Multiple Instruction, Multiple Data Stream—MIMD architecture). In a multi-core CPU setup, parallel computing can be achieved by simultaneously assigning different models or tasks to different cores. While effective for some tasks, CPUs typically have lower parallel computing efficiency for large-scale data and computations than GPUs due to hardware limitations and fewer cores. When training an ANN, for instance, CPU training time increases proportionally with the number of hidden layers and exponentially with the number of neurons, as these increase the dimensionality and complexity of matrix operations.

GPUs are designed for throughput-oriented computing, excelling at performing a large number of simple, repetitive operations simultaneously. They feature a massive number of simpler cores (often thousands), enabling Single Instruction Multiple Data (SIMD) architecture. This architecture is highly beneficial for AI training because many core operations, like matrix multiplications in neural networks, are inherently parallelizable. GPUs can process multiple data streams using the same instruction, simultaneously updating many parameters, which significantly reduces training time. NVIDIA's CUDA (Compute Unified Device Architecture) is a prominent platform that allows developers to write parallel computing tasks specifically for GPUs, leveraging their hierarchical architecture of grids, blocks, and threads.

### *7.7.1 CPU Vs. GPU Performance*

GPUs provide significant speed-ups in AI training, particularly for large datasets and complex neural network models. For example, training an RNN in a CPU environment can take almost ten times longer than in a GPU environment. Electronic packaging research has shown that for large training datasets (e.g., 9000 data points) in reliability prediction of WLP, GPU training time for ML models like ANN can be as low as 82 s, compared to CPU time of 345 s for ANN. For K-Nearest Neighbors (KNN), which has very fast computation due to local input considerations, GPU times ranged from 0.03 to 0.21 s for training datasets of 500 to 9000. GPUs show clear advantages over larger datasets. However, for smaller datasets, a CPU's single-core efficiency can sometimes lead to faster training times than a GPU's single-model training.

In many situations, Hybrid (CPU-GPU) Parallel Computing demonstrates the best of both worlds. A heterogeneous parallel computing approach combines the strengths of both CPUs and GPUs to further optimize AI training, especially for time-consuming processes such as hyperparameter search. In this setup, CPUs manage task assignment (task parallelism) by distributing different model configurations within the grid search to GPUs, which handle each model's data-level parallel computations (data parallelism). When a GPU finishes training one model, the CPU assigns the next, maximizing throughput. This hybrid method has been demonstrated to be the most efficient for hyperparameter search, significantly reducing total search time. For instance, in one study [78], this approach was over 77% faster than a CPU-based grid search and approximately 50% faster than a GPU-based grid search for larger datasets.

In conclusion, the fundamental relationship is that AI training requires massive computational power, particularly for operations such as matrix multiplication. While CPUs offer powerful sequential processing, their limited number of cores makes them less efficient for the highly parallel nature of AI workloads. With their architecture optimized for parallel processing, GPUs are exceptionally well-suited for accelerating these computations, especially with large datasets. Furthermore, combining CPU-driven task management with GPU-driven data processing in a hybrid parallel computing environment provides the most efficient strategy for not only training individual AI models but also for the critical, time-consuming process of hyperparameter optimization, thereby greatly expediting the development of AI-assisted design and simulation technologies.

## 7.8 Summary and Recommendations

This Chapter comprehensively outlines the fundamentals of Artificial Intelligence (AI) and Machine Learning (ML), with a particular focus on their applications in electronic packaging (EP), common algorithms, training methodologies, and the crucial role of parallel computing (CPU and GPU). The core Concepts of AI and ML are as follows: AI involves simulating human learning using computers for complex decision-making. Machine learning is a core discipline within AI that enables computers to learn from large datasets and develop predictive models without human intervention. The widespread adoption of ML is driven by exponential growth in computational hardware and data availability. The typical ML workflow includes data preprocessing, training with algorithms, evaluating performance, and fine-tuning hyperparameters. Data preprocessing is crucial for standardizing input/output and improving modeling performance. Cross-validation (e.g., K-fold) is a key technique for parameter selection and preventing overfitting. Grid search is a fundamental method for systematically finding optimal hyperparameter combinations.

Utilize clustering techniques (e.g., K-means, K-medoids) as a preprocessing step, especially for algorithms with high computational costs like GPR, KRR, or PR, to reduce training time and maintain accuracy. Implement hybrid models to leverage the strengths of combined algorithms and enhance efficiency. Recognize that hyperparameter configuration is crucial for model performance. While manual tuning is an option, grid search is a fundamental and often preferred method for finding optimal hyperparameter combinations. Combining it with parallel computing is crucial for efficiency.

Applications in electronic packaging utilize ML to provide a data-driven approach for rapid prediction and parameter inference, particularly in the reliability life cycle assessment. It can estimate lifetime significantly faster (seconds to minutes) compared to traditional experimental methods (months) or finite element simulations (days to weeks). AI-assisted Design-on-Simulation (DoS) technology, which will be introduced in the following two chapters, ensures consistent prediction outcomes and addresses inconsistencies arising from human factors in traditional simulations. Regression ML is primarily used to predict continuous values such as reliability life (e.g., thermal cycling to failure) and warpage for various packaging structures. It also supports extracting material parameters and predicting equivalent material properties for complex structures, such as printed circuit boards, substrates, and redistribution layers. Classification ML is important for quality control and process automation in EP, enabling automatic detection and categorization of defects (e.g., cracks, voids, delamination). It is highly recommended that AI-assisted design-on-simulation technology be actively leveraged for reliability life prediction, warpage prediction, material parameter extraction, and other potential applications in electronic packaging. Compared to traditional methods, this approach significantly lowers both time and cost and ensures greater consistency in prediction results.

## References


1. R. J. Solomonoff, "An inductive inference machine," in IRE Convention Record, Section on Information Theory, 1957, vol. 2: Institute of Radio Engineers New York, pp. 56–62.
2. Q. Su, C. Yuan, and K.-N. Chiang, "A Small Database with an Adaptive Data Selection Method for Solder Joint Fatigue Life Prediction in Advanced Packaging," Materials, vol. 17, no. 16, p. 4091, 2024.
3. S. K. Panigrahy, Y.-C. Tseng, B.-R. Lai, and K.-N. Chiang, "An overview of AI-Assisted design-on-Simulation technology for reliability life prediction of advanced packaging," Materials, vol. 14, no. 18, p. 5342, 2021.
4. Q. Su and K.-N. Chiang, "Predicting Wafer-Level Package Reliability Life Using Mixed Supervised and Unsupervised Machine Learning Algorithms," Materials, vol. 15, no. 11, p. 3897, 2022.
5. S. K. Panigrahy and K.-N. Chiang, "Study on an Artificial Intelligence Based Kernel Ridge Regression Algorithm for Wafer Level Package Reliability Prediction," in 2021 IEEE 71st Electronic Components and Technology Conference (ECTC), 2021: IEEE, pp. 1435–1441.
6. H. Hsiao and K. Chiang, "AI-assisted reliability life prediction model for wafer-level packaging using the random forest method," Journal of Mechanics, vol. 37, pp. 28–36, 2020.
7. S. K. Panigrahy, "Study on AI-Assisted Design-On-Simulation Technology for Reliability Life Prediction of Wafer Level Package", Ph.D. Dessertation, National Tsing Hua University, Taiwan, 2022.
8. P. Sinha, "Multivariate polynomial regression in data mining: methodology, problems and solutions," International Journal of Scientific and Engineering Research, vol. 4, no. 12, pp. 962–965, 2013.
9. M. Welling, "Kernel ridge regression," Max Welling's classnotes in machine learning, pp. 1–3, 2013.

10. Y. Hamed, A. I. Alzahrani, Z. Mustaffa, M. C. Ismail, and K. K. Eng, "Two steps hybrid calibration algorithm of support vector regression and K- 151 nearest neighbors," Alexandria Engineering Journal, vol. 59, no. 3, pp. 1181–1190, 2020
11. R. M. Mohana, C. K. K. Reddy, P. Anisha, and B. R. Murthy, "Random forest algorithms for the classification of tree-based ensemble," Materials Today: Proceedings, 2021.
12. A. C. Braun, U. Weidner, and S. Hinz, "Classification in high-dimensional feature spaces—Assessment using SVM, IVM and RVM with focus on simulated EnMAP data," IEEE Journal of Selected Topics in Applied Earth Observations and Remote Sensing, vol. 5, no. 2, pp. 436–443, 2012.
13. G. Ortac and G. Ozcan, "Comparative study of hyperspectral image classification by multidimensional Convolutional Neural Network approaches to improve accuracy," Expert Systems with Applications, p. 115280, 2021.
14. C. A. Yuan, Q. H. Su, K. N. Chiang, "Coefficient Extraction of SAC305 Solder Constitutive Equations Using Equation-Informed Neural Networks", Materials 2023, 16 ,4922, July 10 2023.
15. Q. Su and K. N. Chiang, "Applying Ensemble Learning with Multiple Machine Learning Algorithms to Predict the Reliability of Wafer-Level Packaging," EuroSimE 2025, Utrecht, Netherlands, April. 6–9, 2025
16. Q. H. Su and K. -N. Chiang, "Multi-Algorithm Ensemble Learning Framework for Predicting the Solder Joint Reliability of Wafer-Level Packaging", Materials, vol. 18, no. 17, p. 4074, 2025.
17. A. C. Müller and S. Guido, *Introduction to machine learning with Python: a guide for data scientists*. "O'Reilly Media, Inc.", North Sebastopol, CA, USA., 2016.
18. F. Pedregosa *et al.*, "Scikit-learn: Machine learning in Python," *the Journal of machine Learning research,* vol. 12, pp. 2825–2830, 2011.
19. W. S. McCulloch and W. Pitts, "A logical calculus of the ideas immanent in nervous activity," *The bulletin of mathematical biophysics,* vol. 5, no. 4, pp. 115–133, 1943.
20. F. Rosenblatt, "The Perceptron, a Perceiving and Recognizing Automaton Project Para", Cornell Aeronautical Laboratory, 1957
21. A. Krizhevsky, I. Sutskever, and G. E. Hinton, "Imagenet classification with deep convolutional neural networks," *Advances in neural information processing systems,* vol. 25, 2012.
22. P. H. Chou, "Reliability Assessment of Wafer Level Package using Artificial Neural Network Regression Model", MS Thesis, National Tsing Hua University, 2019.
23. C. E. Lee, "Using Wafer-Level Packaging as a Case Study of Reducing Artificial Neural Network Hyperparameter Searching Time", MS Thesis, National Tsing Hua University, 2024.
24. P. Chou, K. Chiang, and S. Y. Liang, "Reliability assessment of wafer level package using artificial neural network regression model," Journal of Mechanics, vol. 35, no. 6, pp. 829–837, 2019.
25. C. Nwankpa, W. Ijomah, A. Gachagan, and S. Marshall, "Activation functions: Comparison of trends in practice and research for deep learning," *arXiv preprint arXiv:1811.03378,* 2018.
26. S. Ruder, "An overview of gradient descent optimization algorithms," arXiv preprint arXiv:1609.04747, 2016.
27. C. Gershenson, "Artificial neural networks for beginners," *arXiv preprint cs/0308031,* 2003.
28. T. Tieleman and G. Hinton, "Lecture 6.5-rmsprop, coursera: Neural networks for machine learning," *University of Toronto, Technical Report,* vol. 6, 2012.
29. S, W, Liu, "Warpage Prediction of Fan-out Panel Level Packaging Using Convolution Neural Network Model", MS Thesis, National Tsing Hua University, 2019.
30. M. Y. Chen, "Using Modified Convolutional Neural Network Method to Estimate the Warpage of the Fan-Out Panel Level Packaging", MS Thesis, National Tsing Hua University, 2020.
31. T. Wang, Y. Chen, M. Qiao, and H. Snoussi, "A fast and robust convolutional neural network-based defect detection model in product quality control," The International Journal of Advanced Manufacturing Technology, vol. 94, no. 9-12, pp. 3465–3471, 2018.
32. A. Krizhevsky, I. Sutskever, and G. E. Hinton, "Imagenet classification with deep convolutional neural networks," in Advances in neural information processing systems, 2012, pp. 1097–1105.

33. Y. LeCun et al., "Backpropagation applied to handwritten zip code recognition," Neural computation, vol. 1, no. 4, pp. 541–551, 1989.
34. Y. LeCun, L. Bottou, Y. Bengio, and P. Haffner, "Gradient-based learning applied to document recognition," Proceedings of the IEEE, vol. 86, no. 11, pp. 2278–2324, 1998.
35. Y. C. Tseng, "Research on the influence of Different Data applied to Recurrent Neural Network Regression Model for Wafer-Level Package Reliability Assessment", MS Thesis, National Tsing Hua University, 2020.
36. Y. C. Lee, "Reliability Assessment of Wafer Level Package using Recurrent Neural Network Regression Model", National Tsing Hua University, 2019.
37. F. A. Gers and E. Schmidhuber, "LSTM recurrent networks learn simple context-free and context-sensitive languages," IEEE Transactions on Neural Networks, Vol. 12, no. 6, pp. 1333–1340, Nov. 2001.
38. J. Chung, C. Gulcehre, K. Cho, and Y. Bengio, "Empirical evaluation of gated recurrent neural networks on sequence modeling," *arXiv preprint arXiv:1412.3555,* 2014.
39. M. Schuster and K. K. Paliwal, "Bidirectional recurrent neural networks," *IEEE transactions on Signal Processing,* vol. 45, no. 11, pp. 2673–2681, 1997.
40. G. Petneházi, "Recurrent neural networks for time series forecasting," *arXiv preprint arXiv:1901.00069,* 2019.
41. S. Hochreiter and J. Schmidhuber, "Long short-term memory," Neural computation, vol. 9, no. 8, pp. 1735–1780, 1997.
42. F. A. Gers, J. Schmidhuber, and F. Cummins, "Learning to forget: Continual prediction with LSTM," 1999.
43. K. Cho et al., "Learning phrase representations using RNN encoder-decoder for statistical machine translation," arXiv preprint arXiv:1406.1078, 2014.
44. X. H. Li, "Using Random Forest for Classification and Regression", Chinese Journal of Applied Entomology, vol. 50, no. 4, pp. 1190–1197, 2013.
45. L. Breiman, Classification and Regression Trees, 1st ed, Routledge, 2017.
46. T. K. Ho, "Random Decision Forests", IEEE transactions on pattern analysis and machine intelligence, vol. 20, no. 8, pp. 832–844, 1998.
47. D. R. Cutler, T. C. Edwards, K. H. Beard, A. Cutler, K. T. Hess, J. Gibson, and J. J. Lawler. "Random forests for classification in ecology ", Ecology, vol. 88, 11, pp. 2783–2792, 2007.
48. Grömping, Ulrike. "Variable importance assessment in regression: linear regression versus random forest", The American Statistician, vol. 63, no. 4, pp. 308–319, 2009.
49. H. Y. Hsiao, "Reliability Assessment of Wafer-Level Packaging using Random Forest Regression Model", MS Thesis, National Tsing Hua University, 2019.
50. B. S. Wang, "Using Extra Trees to Predict Warpage of Panel Level Packaging and Discuss Asymmetric Warpage", MS Thesis, National Tsing Hua University, 2022.
51. T. H. Tasi, "The Research of WLCSP Reliability Prediction with Data Distribution in Random Forest Regression Model", MS Thesis, National Tsing Hua University, 2020.
52. C. Cortes and V. Vapnik, "Support-vector networks," Machine learning, vol. 20, no. 3, pp. 273–297, 1995.
53. V. Vapnik, S. Golowich, and A. Smola, "Support vector method for function approximation, regression estimation and signal processing," Advances in neural information processing systems, vol. 9, 1996.
54. A. J. Smola and B. Schölkopf, "A tutorial on support vector regression," Statistics and Computing, vol. 14, no. 3, pp. 199–222, 2004.
55. H. C. Kuo, B. R. Lai, and K. N. Chiang, "Study on Reliability Assessment of Wafer Level Package Using Design-on-Simulation with Support Vector Regression Techniques," *15th International Microsystems, Packaging, Assembly and Circuits Technology Conference (Impact 2020),* pp. 192–195, 2020.
56. C. Y. Chang, "Using Grid Search Methods and Parallel Computing to Reduce AI Training Time for Reliability Lifetime Prediction of Wafer-Level Packaging", MS Thesis, National Tsing Hua University, 2023.

57. B. R. Lai, "Research on Reliability Assessment of Wafer Level Package by using Multiple Kernel SVR and New Support Vector Regression", MS Thesis, National Tsing Hua University, 2020
58. Y. T. Shen, "Reliability Assessment of Wafer Level Package using Support Vector Regression Model", MS Thesis, National Tsing Hua University, 2019.
59. V. Cherkassky and Y. Ma, "Practical selection of SVM parameters and noise estimation for SVM regression," Neural Networks, vol. 17, no. 1, pp. 113–126, 2004.
60. J. Wang, "An intuitive tutorial to Gaussian process regression," *Computing in Science & Engineering,* vol. 25, no. 4, pp. 4–11, 2023.
61. C. K. Williams and C. E. Rasmussen, Gaussian processes for machine learning (no. 3). MIT Press, Cambridge, MA, 2006.
62. H. M. Chang, "Predict Reliability Life of Wafer Level Packaging Using GPR with Cluster Analysis", MS Thesis, National Tsing Hua University, 2022.
63. H. E. Driver and A. L. Kroeber, "Quantitative expression of cultural relationships" (no. 4). Berkeley: University of California Press, 1932.
64. K. Krishna and M. N. Murty, "Genetic K-means algorithm," IEEE Transactions on Systems, Man, and Cybernetics, Part B (Cybernetics), vol. 29, no. 3, pp. 433–439, 1999.
65. G. Hamerly and C. Elkan, "Learning the k in k-means," Advances in neural information processing systems, vol. 16, pp. 281–288, 2003.
66. Arthur E. Hoerl and Robert W. Kennard, "Ridge Regression: Biased Estimation for Nonorthogonal Problems," Technometrics, vol. 12, 1, pp. 55–67, 1970.
67. N. Cristianini, and J. Shawe-Tylor, An introduction to support vector machines and other kernel-based learning methods. Cambrige university press, 2000.
68. D. J. Tylavsky and G. R. Sohie, "Generalization of the matrix inversion lemma," Proceedings of the IEEE, vol. 74, no. 7, pp. 1050–1052, 1986.
69. E. Ostertagová, "Modelling using polynomial regression," *Procedia Engineering,* vol. 48, pp. 500–506, 2012.
70. H. H. Liao, "Study on Prediction Reliability Life of Wafer Level Packaging Using Polynomial Regression with Cluster Analysis", MS Thesis, National Tsing Hua University, 2022.
71. N. S. Altman, "An introduction to kernel and nearest-neighbor nonparametric regression," The American Statistician, vol. 46, no. 3, pp. 175–185, 1992.
72. M. Sarkar and T.-Y. Leong, "Application of K-nearest neighbors algorithm on breast cancer diagnosis problem," in Proceedings of the AMIA Symposium, 2000: American Medical Informatics Association, p. 759.
73. B. S. Chen, "Predict the Reliability Life of Wafer Level Packaging using K-Nearest Neighbors algorithm with Cluster Analysis", MS Thesis, National Tsing Hua University, 2022.
74. S. A. Abbas et al., "K-means and k-medoids: Cluster analysis on birth data collected in city Muzaffarabad, Kashmir," IEEE Access, vol. 8, pp. 151847–151855, 2020.
75. W. Utomo, "The comparison of k-means and k-medoids algorithms for clustering the spread of the covid-19 outbreak in Indonesia," ILKOM Jurnal Ilmiah, vol. 13, no. 1, pp. 31–35, 2021.
76. G. Bathla, "Stock Price prediction using LSTM and SVR," in 2020 Sixth International Conference on Parallel, Distributed and Grid Computing (PDGC), Waknaghat, India, 2020: IEEE, pp. 211–214.
77. C. Chang et al., "Using grid search methods and parallel computing to reduce AI training time for reliability lifetime prediction of wafer-level packaging," in 2023 24th International Conference on Thermal, Mechanical and Multi-Physics Simulation and Experiments in Microelectronics and Microsystems (EuroSimE), Graz, Austria, 16–19 Apr 2023: IEEE, pp. 1–5.
78. Y. T. Su, "Research on Reliability Prediction of Wafer-Level Packaging by Combining Uncertainty Analysis and Parallel Accelerated Neural Networks", MS Thesis, National Tsing Hua University, 2025.

79. C. E. Lee, "Study on Reducing Artificial Neural Network Parameter Searching Time and Reliability Prediction Time for Wafer-Level Packaging", MS Thesis, National Tsing Hua University, 2025.
80. D. B. Kirk and W. H. Wen-Mei, Programming massively parallel processors: a hands-on approach. Morgan Kaufmann, 2016.
81. S. Che et al., "A performance study of general-purpose applications on graphics processors using CUDA," Journal of Parallel and Distributed Computing, vol. 68, no. 10, pp. 1370–1380, 2008.

# Chapter 8
# Modeling and Design-on-Simulation Technology

**Abstract** The electronic packaging (EP) community has transitioned from the costly, time-consuming, and physically limited Design-on-Experiment (DoE) methodology to the more strategic Design-on-Simulation (DoS) technology, thereby securing trustworthy finite element simulation results. This transition reduces physical experiments, shortens design cycles, and lowers costs. The core simulation tool is the Finite Element Method (FEM), which is used to analyze complex physical phenomena, including thermal-induced strains and stresses arising from the mismatch in the coefficient of thermal expansion among components, which can lead to fatigue failure in solder joints. A major inherent challenge of traditional FEM is the inconsistency and subjectivity of results, which are heavily influenced by the individual researcher's expertise and the unavoidable dependency on mesh size.

DoS technology is a modeling verification procedure or standardized validation step executed before full-scale simulation is performed. This procedure ensures consistent and trustworthy predictions by iteratively tuning and verifying theories, material properties, empirical equations, and solution procedures until simulation results match physical test outcomes or literature data. To optimize computational efficiency, especially for complex 3D models, which are computationally intensive, both 2D and 3D FEM are utilized. Crucially, the Multipoint Constraints (MPC) method is employed in 3D simulations to manage complex models by establishing coupling relationships between nodes of varying mesh densities within a single model, drastically reducing element number and computation time while being suitable for nonlinear analysis.

Accurate reliability assessment also requires a precise prediction of the post-reflow shape of solder joints, as their geometry strongly influences fatigue life. The primary prediction methods include the geometric Truncated Sphere Method and the Force-Balanced Analytical Solution, both suited for lightweight packages and circular pads. The Energy-Based Method (e.g., Surface Evolver) is noted for its superior accuracy, ability to account for gravitational effects in larger solder balls, and applicability to pads with arbitrary shapes.

Further addressing the mesh-dependency issue in FEM, particularly near stress concentration areas, the Finite Volume-Weighted Method characterizes strain



J. Lau, K.-N. Chiang, *Cu-Interconnects, Glass, and AI-Assisted Simulation for Chiplets and Heterogeneous Integration*,
https://doi.org/10.1007/978-981-95-6891-8_8

response by averaging state variables over a targeted zone. The resulting stabilized equivalent plastic strain can be used consistently as input for reliability life empirical formulations.

For highly complex, multi-scale electronic package architectures, equivalent methods simplify detailed structures into homogenized representations to reduce modeling and computational burden. While the Rule of Mixture (ROM) and Inverse Rule of Mixture (IROM) offer fast estimations of equivalent material properties (like Young's Modulus and CTE) based on volume fraction, they fail to account for differing structural layouts with the same material proportions, often requiring correction factors.

Finally, AI-assisted methodologies enhance DoS, notably through the Equation-Informed Neural Networks (EINNs) combined with Bayesian inference (BI). EINN efficiently extracts and refines coefficients for complex, nonlinear constitutive equations (such as Garofalo, Anand, and Chaboche models) based on measurement data. This ensures accurate material curves for simulations and allows prioritized learning of specific data regions. The combination of validated simulation (including 2D/3D FEM) with AI-assisted data generation can also create the massive datasets needed for training predictive AI models, enabling reliability life cycles to be predicted in milliseconds.

## 8.1 Introduction

The electronic packaging (EP) community has historically relied on various methodologies for designing and developing new packaging structures. Initially, Design-on-Experiment (DoE) methods were widely employed for reliability estimation. However, DoE is notoriously time-consuming, costly, and energy-intensive, often requiring months or even a year to complete for new structures, which hinders the ability to meet market demands and lacks sufficient physical information for structural optimization.

To overcome these significant challenges, the electronic packaging industry introduced Design-on-Simulation (DoS) technology [1–7] more than a decade ago. DoS represents a strategic approach or broader methodology aimed at securing the finite element simulation results, and therefore can be used to reduce the number of costly and time-consuming physical experiments, thereby shortening design cycles and reducing overall costs in product development. The core simulation technology for electronic packaging is the Finite Element Method (FEM). Finite element modeling is a specific computational technique or tool used to analyze complex physical phenomena in electronic packaging structures numerically. It involves dividing a complex geometry into smaller, simpler elements to approximate the behavior of materials under various conditions, and calculating parameters such as stress, strain, and temperature distribution. The quality and trustworthiness of FEM results in the traditional design procedure are highly dependent on the individual researcher's expertise, including their domain knowledge, understanding of mechanics theory, and FEM background. This often leads to inconsistent analysis results between different researchers performing the same simulation. Therefore, a standardized validation of the simulation theories, material properties, and solution procedures is critical. This validation procedure, known as DoS technology, ensures that the results are located within an accurate range.

## 8.2 Simulation Technology

Finite element simulation [1–10] is a widely adopted and essential technology in the electronic packaging (EP) community for designing new packaging, as shown in Figs. 8.1 and 8.2, particularly in assessing the reliability of new packaging structures. This approach is crucial for shortening design cycles, reducing the need for numerous physical experiments, such as accelerated thermal cycling testing (ATCT), and thus meeting time-to-market demands while also reducing power consumption. FEM is used to evaluate the reliability performance of packaging solder joints. It helps analyze thermal-induced strains and stresses that arise from the mismatch in the coefficient of thermal expansion (CTE) among components and the printed circuit board (PCB). These mismatches can lead to fatigue failure, often seen as cracking at the interface between solder joints and components. FEM assists in optimizing package structures and identifying critical failure points, such as the corner solder joint with the largest Distance from Neutral Point (DNP). This enables virtual prototyping and analysis prior to physical fabrication. By simulating thermal cycling tests, FEM can significantly reduce the number of costly and time-consuming physical experiments, such as Accelerated Thermal Cycling Tests (ATCT). A single ATCT run can take more than a month to complete. FEM enables a detailed study of complex material behaviors, such as creep and viscoplasticity, particularly in solder joints at high homologous temperatures. Creep is a major failure mechanism

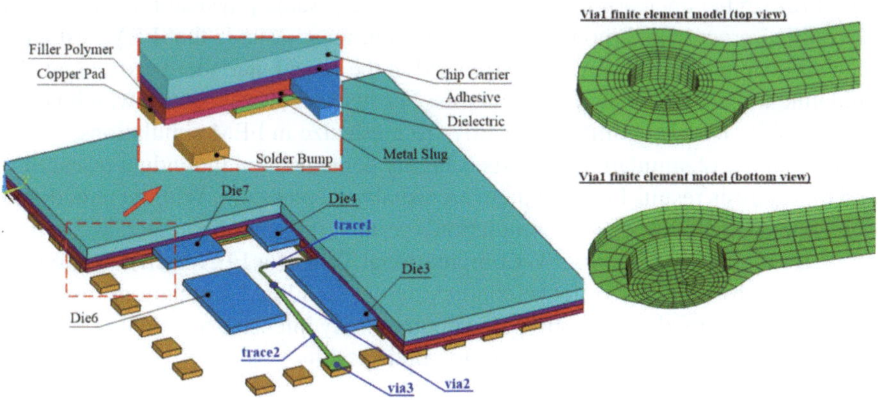


**Fig. 8.1** 3D SiP finite element modeling

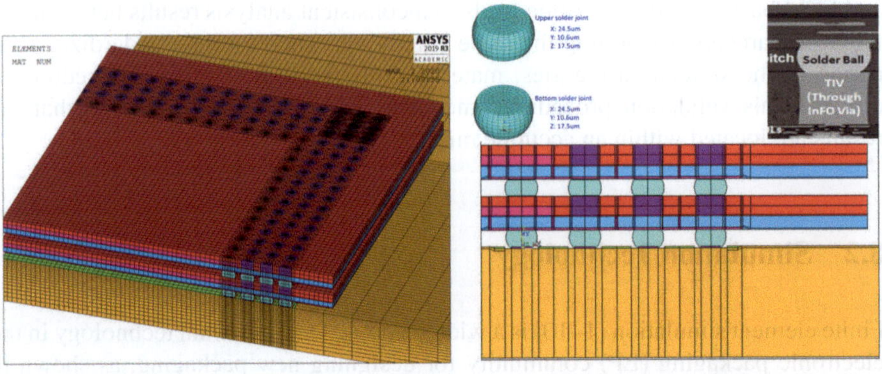


**Fig. 8.2** A stacked 3D quarter symmetric model of on-board wafer-level packaging

in lead-free solder materials due to their high homologous temperatures. However, a key challenge of simulation is achieving consistent and trustworthy simulation results, as analysis quality can vary significantly between researchers due to a lack of guidelines, especially regarding mesh size.

## *8.2.1 Conventional Finite Element Simulation Procedure*

Despite the advantages of finite element simulation over DoE in terms of speed and cost reduction, traditional simulation technology, which is heavily reliant on FEM, still faces its own set of critical challenges stemming from the nature of FEM itself such as (1) inconsistency and subjectivity, the quality and trustworthiness of FEM results in traditional simulation are highly dependent on the individual researcher's expertise, including their domain knowledge, understanding of mechanics theory, and FEM background. This often leads to inconsistent analysis results between different researchers performing the simulation of the same physical model. (2) Lack of certified process and modeling work. A significant issue is that FEM results are inevitably dependent on the mesh size used in the model, and there is often no clear guideline to help researchers address this. For instance, creep strain, which is critical for solder joint reliability, is sensitive to mesh size in FEM simulations.

The quality of simulation analysis is highly dependent on individual researchers, and the analysis results between them are usually inconsistent. Without a methodology to consistently predict the simulation's results, it cannot be treated as a reliable replacement for experiments. (3) Computational Time (for FEA itself): While much faster than physical experiments, conducting detailed FEM simulations, especially for complex 3D models, can still be computationally intensive and time-consuming, often requiring several days or weeks to obtain results.

## 8.2.2 Design-on-Simulation Technology

It is a well-known fact that the results of finite element simulations are inevitably dependent on the mesh size. A suitable and fixed element size is crucial for consistently predicting reliability life. A mesh that is too fine can lead to stress/strain concentration, as well as convergence issues, while a mesh that is too coarse may underestimate values. These inherent limitations of traditional FEM led to the development of Design-on-Simulation (DoS, Fig. 8.3), a modeling verification procedure that should be executed, which aims to overcome simulation instability, deliver consistent and trustworthy predictions, and further reduce design cycles. DoS technology is a broader methodology or design paradigm that actively incorporates computational simulations, predominantly FEM, as an integral part of the design process. Its goal is to replace or significantly reduce the need for physical prototyping and testing by predicting performance and reliability virtually.

In Fig. 8.3, it is explicitly shown that the DoS procedure includes a "Model Validation" step. This step occurs after the initial "Design" phase and before the "Build Small Quantity of Prototypes" phase. In real design work, the modeling and simulation procedure can be verified using literature results instead of conducting experimental work. Designers can iteratively tune and verify the theories, material properties, empirical equation models, mesh sizes, and solution procedures until the simulation results match those of real physical tests. Once the items mentioned above are verified, the designer can proceed to the main "Simulation" phase, where they will design the target electronic packaging structure and build a small quantity sample for real-world testing.

For example, Figs. 8.4 and 8.5 demonstrate a landing gear subjected to a static loading. These figures indicate that the maximum von Mises stress of the simulation varies with mesh size; it is difficult to determine which mesh size provides a correct result without verification from an experiment. Therefore, the designer should use a simple model to verify which mesh size can yield a similar result to the experiment

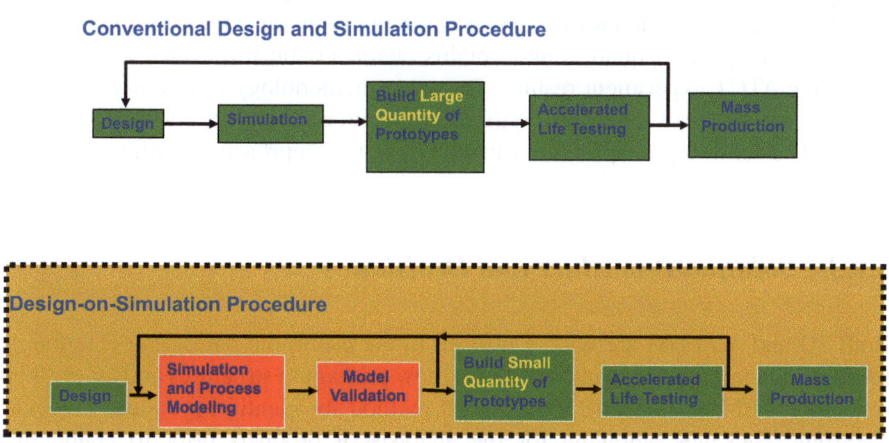


**Fig. 8.3** Conventional design and simulation procedure vs. DoS

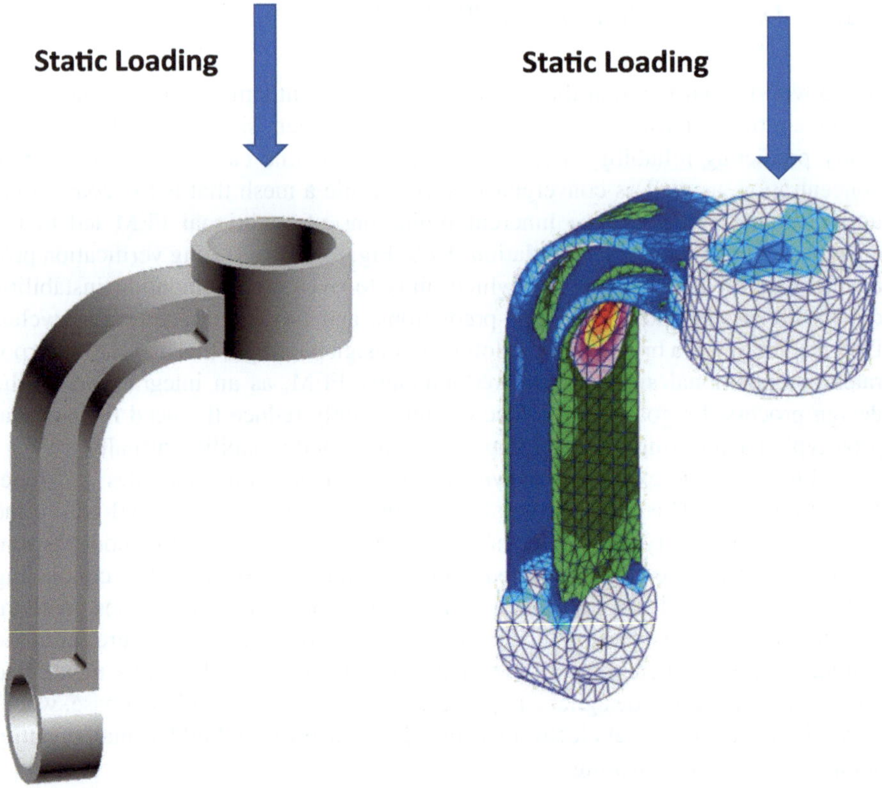


**Fig. 8.4** Landing gear subjected to a static loading

before proceeding with the detailed full-scale design. Once the theories, empirical equations, and mesh size, etc., are verified, the same modeling procedure can be applied to other similar landing gear designs. Similarly, for the solder joint of wafer-level packaging, e.g., with a solder diameter between 80 $\mu m$ and 600 $\mu m$, one should be able to find a suitable mesh size that consistently predicts the strain for the reliability life empirical equation and obtains characteristic life cycles (63.2%) that match the ATCT experiment results. The same methodology can be used to determine the mesh size of the critical region for traceline failure and interlayer delamination. The following chapter will illustrate how this approach is applied.

### 8.2.3 Solution Procedure and Decision

Both 2D and 3D FEM are widely utilized in the electronic packaging community for DoS technology, which aims to design new packaging structures and assess their reliability. FEM is a reliable and efficient method for identifying failure modes in electronic packages, allowing for the analysis of products prior to conducting

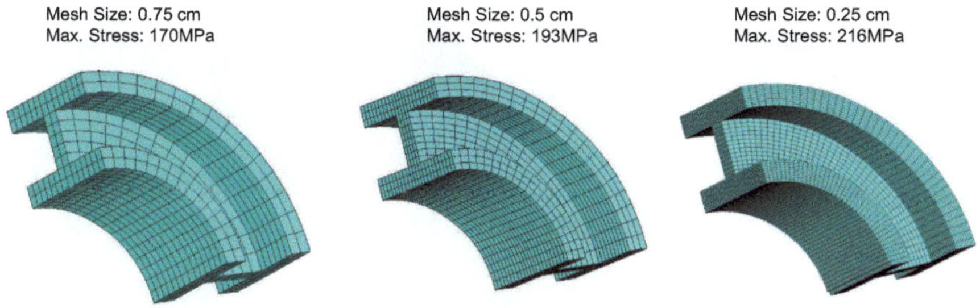


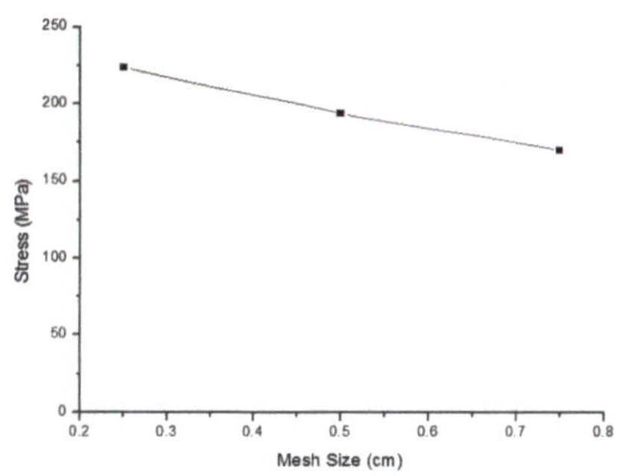


| Mesh Size (cm) | von Mises (MPa) |
|---|---|
| 0.25 | 216 |
| 0.5 | 193 |
| 0.75 | 170 |

**Fig. 8.5** Mesh size effect on the stress results of landing gear

reliability tests. The use of FEM helps reduce the number of experiments and shorten product design cycles, addressing the critical need for faster time-to-market and reduced power consumption in the development of new electronic packaging structures. 2D FEM models require a significantly smaller number of elements, thereby reducing simulation time compared to 3D models. For symmetric-type electronic packaging structures, their lower computational cost makes 2D models a necessity for extensive parametric analysis to optimize packaging structures, especially for complex or 3D stacked configurations, such as Package-on-Package (PoP) structures. Developers and researchers often adopt a plane strain assumption, which means they cannot truly represent the actual cylindrical geometries of solder balls. However, the numerous results indicate that a 2D finite element model can still yield accurate reliability results with a suitable mesh size selection. One thing to keep in mind is that 2D modeling is limited to symmetric package sizes and solder joint arrangements.

3D finite element models provide a more accurate representation of the actual packaging structure compared to 2D models, as they can accurately represent asymmetric packaging structures and cylindrical solder features. In contrast, the challenge of 3D finite element modeling is generating a large number of elements, resulting in high CPU time consumption, which can render comprehensive parametric studies infeasible. Reducing computational time is a critical issue for 3D finite element simulations. The effort includes reducing the total number of elements, applying multipoint constraints (MPC) for local-global analysis (Fig. 8.6), and utilizing equivalent material properties [11–13], among other measures. The MPC

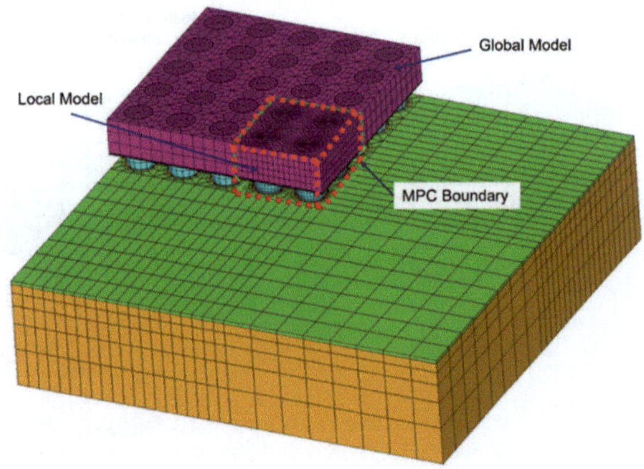


**Fig. 8.6** Multipoint constraints 3D model

technology can be used to reduce the total number of elements in a 3D model, thereby decreasing computational time. In the CoWoS and wafer-level packaging cases, the MPC models have been shown to effectively reduce elements and computing time while maintaining accuracy comparable to full fine mesh models.

In summary, both 2D and 3D FEM play crucial roles in electronic packaging design and reliability assessment. The goal is to ensure that both 2D and 3D FE models can obtain reliable life results consistent with experimental results within an accurate range. Their simulation results should be validated by comparing them with physical tests, such as ATCT. The quality of simulation analysis is highly dependent on individual researchers' domain knowledge, mechanics theory, and FEM background. Furthermore, regardless of whether the simulation is 2D or 3D, the results inevitably depend on the mesh size, and choosing an appropriate and fixed element mesh size is crucial for consistent and accurate predictions. Of course, the combination of mesh size, material property determination, plasticity or creep strain selection, and coupling with strain-based or energy-density-based reliability prediction empirical equations is critical for the success of simulation work, and this requires an expert to handle. The simulation procedure for decision trees is shown in Fig. 8.7.

AI-assisted DoS combines machine learning (ML) with trustworthy finite element simulation. A key step involves generating a large database, which is much easier to achieve through validated simulation (including 2D/3D FEM) than through experiments alone. This validated simulation procedure can replace experiments for creating the massive data needed for AI training. The aim is to overcome traditional simulation instability issues and deliver consistent and trustworthy prediction results. Once a final AI model is obtained, it can predict reliability life cycles from packaging geometries within milliseconds, a significant improvement over traditional methods that can take days, weeks, or months.

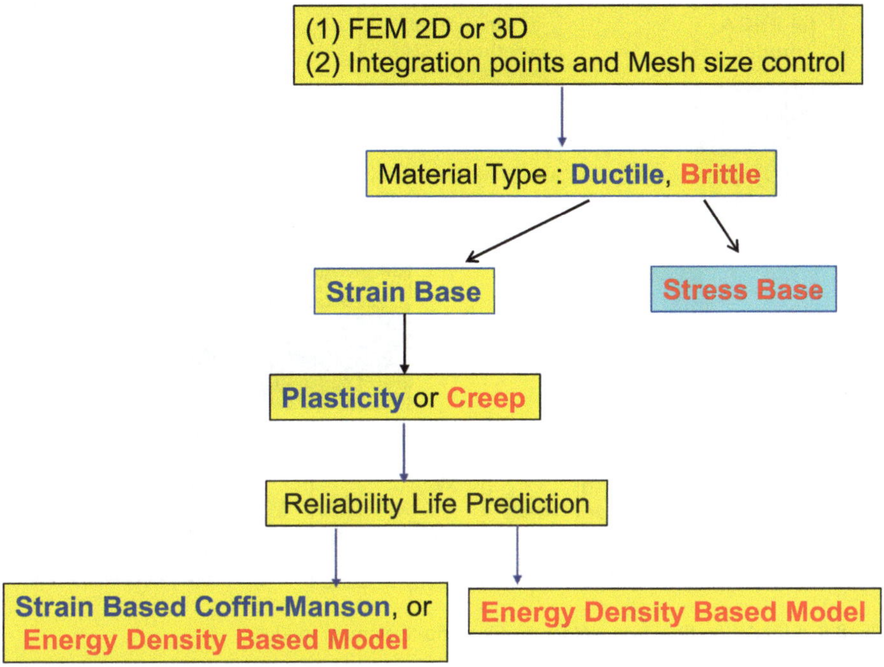


**Fig. 8.7** Simulation procedure decision tree

## 8.3 Solder Liquid Formation

The continuous push for high performance and smaller electronic packages has led to the adoption of area array packaging technologies, including flip-chip (FC), wafer-level packaging, and system-in-package (SiP). As these advanced packaging methods become more prevalent, increasing emphasis has been placed on solder joint reliability and surface mount process yield. Studies have shown that solder joint reliability is strongly influenced by both the geometry of the joints and the properties of the solder material. Consequently, accurately predicting the shape of solder reflow has become crucial for assessing solder joint reliability, and the primary purpose of solder reflow prediction is to address critical issues in electronic packaging design, particularly those concerning solder joint reliability and assembly yield. As electronic packages become smaller and more sophisticated with higher density inputs/outputs (I/Os), the geometry of solder joints becomes a crucial factor.

In electronic packaging, the substrate and the chip are electrically and thermally connected, as well as mechanically integrated, through solder balls. Due to the mismatch in the CTE between different materials, the upper and lower surfaces of solder joints are often subjected to repeated stresses and bending moments during thermal cycling, leading to fatigue failure. Therefore, the fatigue life of solder joints is commonly used as a key metric for evaluating the reliability of electronic

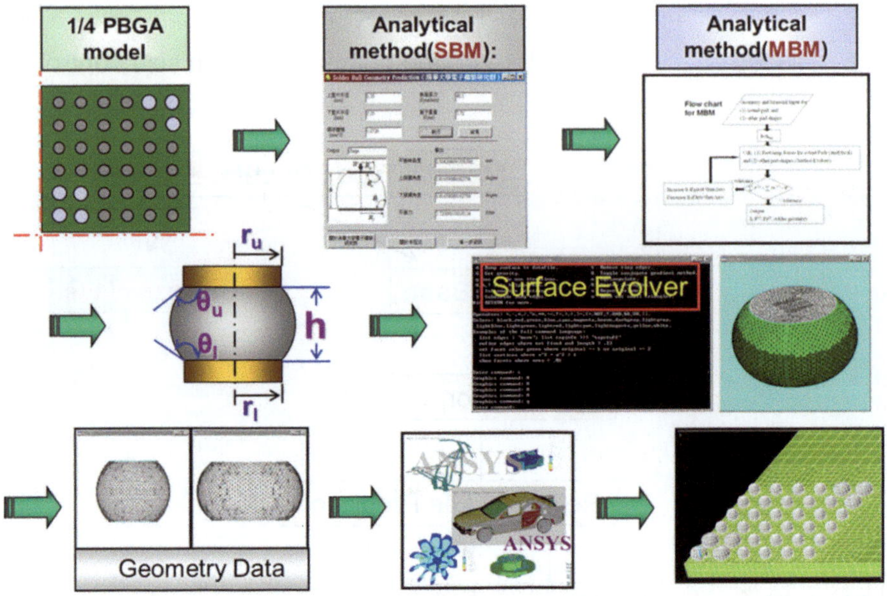


**Fig. 8.8** Simulation flow from solder profile to finite element model

packages. The fatigue life of solder joints is closely related to both their material properties and geometric profile. As such, accurately predicting the post-reflow shape of solder bumps can significantly enhance the accuracy of reliability assessments for the entire packaging structure. The detailed geometrical information of the solder joint obtained from prediction algorithms can be transferred to finite element programs (Fig. 8.8), such as MSC/NASTRAN, ANSYS, or ABAQUS, to analyze the stress, strain, and reliability of the package.

The primary methods for predicting solder bump geometry are the truncated sphere method, the force-balanced analytical solution, and the energy-based algorithm [14–19]. Each method has specific characteristics and limitations:

## *8.3.1 Truncated Sphere Method*

This method assumes that the molten solder bump forms a spherical segment truncated by the chip and substrate, as shown in Fig. 8.9. This is a purely geometrically based algorithm that assumes the solder bump after reflow takes the shape of a truncated sphere. The dimensions of the solder bump, pad, and solder height are interrelated, and the standoff height can be estimated based on the volume of the solder ball, as well as the upper and lower pads. For an area array package on a printed circuit board (PCB), the solder joint forms a "double truncated sphere" between the upper and lower pads. Based on the geometric relationship between the

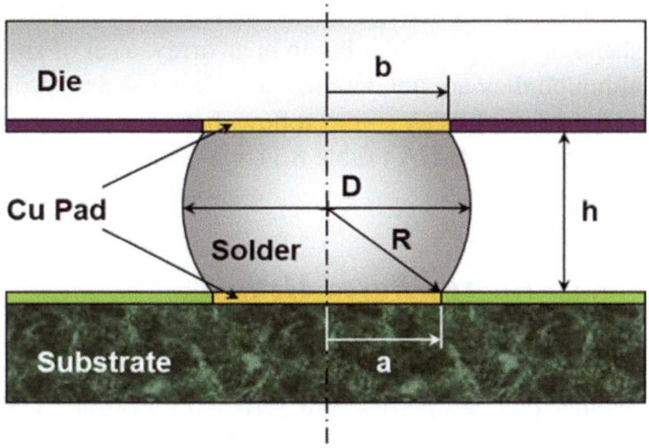


**Fig. 8.9** Truncated sphere geometry

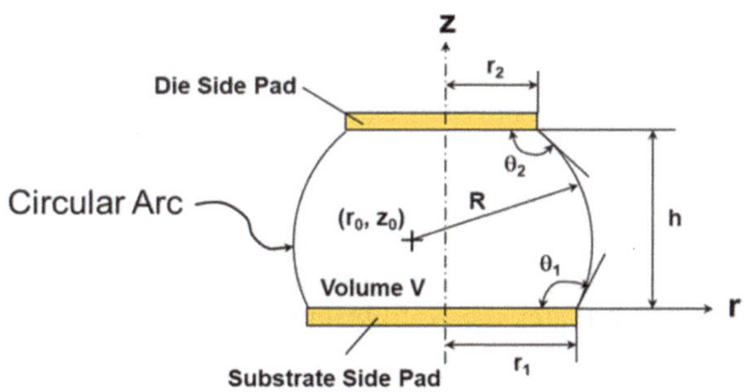


**Fig. 8.10** Force-balanced method solder geometry

bump and the substrate before reflowing, the equilibrium height and maximum radius of the solder ball (approximated as a perfect sphere) can be derived. The truncated sphere method can accurately predict the standoff height and maximum solder width of the solder ball, particularly for lightweight packages. It has also demonstrated accurate predictions for micro-bump shapes, flip-chip bump shapes, wafer-level packages, and BGA solder ball shapes.

### *8.3.2 Force-Balanced Analytical Solution*

Introduced by Heinrich et al. in 1996 [15], this method is based on the principle of mechanical equilibrium of the solder bump in solidification stage, as shown in Fig. 8.10. This method includes the surface tension effect and calculates solder joint contours using a Laplace equation. It assumes that the solder joint reaches static equilibrium upon solidification, the solder pad is circular and perfectly aligned, the free surface is axisymmetric, and a circular arc can approximate the meridian defining the free surface. It also assumes the solder pad is completely covered and perfectly wettable. The solder ball shape is modeled as a rotational surface formed by revolving a circular arc about a central axis. To analyze the restoring force and shape of a single solder bump in an array, a set of simplifying assumptions is made, although they are not detailed here. It has shown promising results in predicting the shape of the PBGA solder reflow. It also yields good results for lightweight packages, such as PBGA (Table 8.1), wafer-level packages, and flip chips.

### *8.3.3 Energy-Based Method*

The truncated sphere method and force-balanced analytical solution are limited to circular pad geometries. For pads with arbitrary shapes, such as elliptical, square, and rectangular, the energy-based method remains the only applicable approach. This method takes into account surface tension, gravitational effects, and internal and external pressures. Solder joint equilibrium is achieved when the forces due to these factors are balanced, and the total energy of the liquid body, including surface tension energy, gravitational energy, and energy from external forces, is minimized. The Surface Evolver can examine fully 3-D problems by discretizing an initial surface and iterating it towards a minimal energy configuration [17–19]. Among the available models, this approach has been widely validated for its superior accuracy. Brakke et al. [17–19] developed the software Surface Evolver based on this principle. It simulates the equilibrium shape of a liquid surface by minimizing the total energy, assuming that the lowest-energy configuration corresponds to the minimal surface area under force balance. Through variational analysis, the total energy $E_{total}$ of the solder ball can be expressed in the following variational form:

$$\begin{aligned}
\delta E_{total} &= \partial E_{surface\ tension} + \partial E_{gravity} + \partial E_{external\ force} \\
\delta E_{surface\ tension} &= T_S \iint_s \left(\mathrm{div}\vec{h} - \vec{n}\cdot D\vec{h}\cdot\vec{n}\right) dA \\
\delta E_{gravity} &= \rho g \iint_s \left(\left(\mathrm{div}\frac{z^2}{2}\vec{k}\right)\vec{h} - \mathrm{curl}\left(\vec{h}\times\frac{z^2}{2}\vec{k}\right)\right)\cdot d\vec{A} \\
\delta E_{external\ force} &= -P\iint_s \vec{h}\cdot d\vec{A}
\end{aligned} \tag{8.1}$$

**Table 8.1** PBGA Solder profile analytical solution vs. experiment

| Joint dimension (mm) | Experiment (16 samples) Average | Experiment (16 samples) Standard deviation | Heinrich model (Circular-arc revolution) |
|---|---|---|---|
| Stand-off Height, h | 0.52 | 0.02 | 0.51 |
| Maximum Width, w | 0.86 | 0.02 | 0.87 |

Since the total energy $E_{total}$ is a function of the solder ball height $H$, taking the first partial derivative of Eq. (8.1) with respect to $H$ yields the expression for the restoring force of the solder ball, as shown below:

$$F_r = \frac{\partial E_{total}}{\partial H} = \frac{\partial E_{surface\ tension} + \partial E_{gravity} + \partial E_{external\ force}}{\partial H} \tag{8.2}$$

Where $T_S$ denotes the surface tension of the solder ball, $P$ is the externally applied pressure, $A$ represents the area of an elemental surface on the solder ball, $z$ is the height of the element in the direction of gravity, $\vec{k}$ is the unit vector in the direction of gravity, $\vec{n}$ is the unit normal vector to the elemental surface, and $F_r$ is the restoring force.

Each component of the restoring force $F_r$ in Eq. (8.2) can be reformulated as shown in Eqs. (8.3) to (8.5):

$$\frac{\partial E_{surface\ tension}}{\partial H} = T_S \iint_s \left(\nabla \cdot \vec{h} - \vec{n} \cdot D\vec{h} \cdot \vec{n}\right) dA \tag{8.3}$$

$$\frac{\partial E_{gravity}}{\partial H} = \rho g \iint_s \left( \left( \nabla \cdot \frac{z^2}{2} \vec{k} \right) \vec{h} - \nabla \times \left( \vec{h} \times \frac{z^2}{2} \vec{k} \right) \right) \cdot d\vec{A} \tag{8.4}$$

$$\frac{\partial E_{external\ force}}{\partial H} = -P \frac{\partial V}{\partial H} = -P \iint_s \vec{h} \cdot d\vec{A} \tag{8.5}$$

where $\rho$ is the density of the solder, $g$ is the gravitational acceleration, and $V$ is the volume of the solder ball. In addition, $h$ represents the perturbation function, which can be expressed as:

$$\vec{h} = \left[ \frac{z_{top} - z}{z_{top} - z_{base} - H} \right] \vec{k} \tag{8.6}$$

where $z_{top}$ *and* $z_{base}$ represent the upper and lower boundary conditions of the solder ball, respectively.

One can determine the restoring force of the solder ball in the direction of gravity by applying a small upward or downward displacement at the pad. When the restoring force equals the gravitational force acting on the solder ball, the system reaches a state of static equilibrium in the molten condition. Solving the resulting equilibrium equation yields the final standoff height and geometric shape of the solder ball.

By fixing the volume of the solder ball after reflowing and using the Truncated Sphere Method to estimate the initial height $h$ and radius $R$, the estimated height $h$ can be set as the initial range for the standoff height input into Surface Evolver. By incorporating the surface tension of the solder material, the pressure acting on each solder ball is computed, and the corresponding geometric shape can then be

**Fig. 8.11** Comparison of three solder profile prediction methods

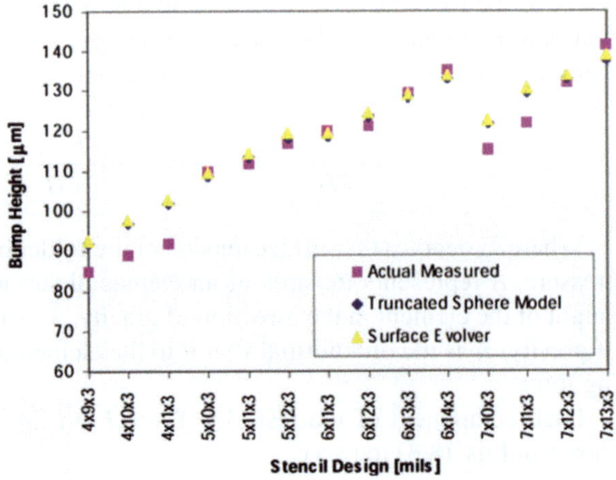


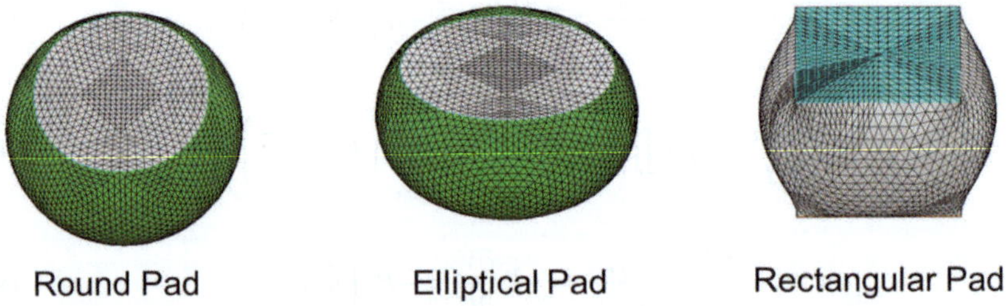


**Fig. 8.12** Surface evolver solder profile prediction under different pad geometry

determined. It has been successfully applied for predicting the final shape of BGA joints after reflow, considering various pad sizes and shapes, solder volumes, specific solder heights, and surface tensions. The force-balanced analytical solution, energy-based method, and truncated sphere theory can accurately predict micro-bump shapes. The energy-based Surface Evolver is considered the best among the simulation methods for predicting the geometry of both bumps and solder on a board.

In summary, all three methods can predict solder reflow shapes with reasonable accuracy (Fig. 8.11). However, the energy-based theory generally yields the most reliable results, particularly when accounting for gravitational effects in larger solder balls. The truncated sphere method, though purely geometrical, performs well for lightweight packages and micro-bumps but loses accuracy with heavier components. Similarly, the force-balanced analytical solution is suitable for lightweight packages but is highly sensitive to both package weight and pads geometric assumptions. Notably, only the energy-based method is applicable to non-circular solder pads, as illustrated in Fig. 8.12.

## 8.4 Finite Volume-Weighted Method

The finite volume-weighted method, also known as the volume-weighted averaging technique, is a methodology developed to characterize strain response in specific areas, particularly in electronic packages. This method was proposed to resolve the strain of mesh size-dependent problems in finite element method (FEM) analysis. It is critical for characterizing stress/strain information near interfaces or singular geometry regions, such as those between the solder joint and the under-bump metallurgy (UBM) or solder pads. These interfaces often exhibit abrupt geometry changes that form sharp corners, resulting in singular (concentrated) stress/strain responses. The volume-weighted averaging technique aims to effectively characterize the strain response in the vicinity of these singular regions.

An enhanced volume-weighted averaging technique [20, 21] is proposed to more accurately characterize the strain response at the corner of the solder joint. Rather than averaging state variables across the entire material domain, the method introduces a targeted zone within which the averaging is performed, as detailed below:

$$\tilde{\varepsilon} = \sum_{e=1}^{n} \int_{\Omega_e} \varepsilon_e d\Omega / \sum_{e=1}^{n} \int_{\Omega_e} d\Omega \tag{8.7}$$

Where $n$ is the total number of elements in the zone, $\varepsilon_e$ is the strain of the $e$-th element, $\Omega e$ is the volume (area) of the $e$-th element, and $\tilde{\varepsilon}$ is the volume-weighted averaging strain of the specific zone. The dimension of the finite zone is determined empirically to be small enough to capture the maximum strain field, yet large enough to obtain a converging solution as the mesh density increases. To explore the dimension of the particular finite zone, several different finite element models, each with varying mesh densities in the area of observation, should be examined in the FEM study. The simulation results should then be validated against the experiment.

In the research described [21], the volume-weighted criterion is used to examine the equivalent plastic strain in the critical area of solder bumps. The resulting stabilized equivalent plastic strain is then applied as an input for the Coffin-Manson empirical formulation to predict the reliability life of solder bumps. For validation, models with different mesh densities (122, 212, 364, and 435 elements, Fig. 8.13) were subjected to a uniformly distributed shear force. The studies concluded that for a solder bump with a 150 μm pad diameter, the equivalent plastic strain averaged over an area of 5 μm radius became irrelevant to the mesh density (Fig. 8.14). In this case, the designer can use the average volume-weighted zone larger than 5 μm to determine an appropriate radius for the Coffin-Manson or energy-density model to predict the reliability life cycles of the solder joint, and compare the result with the experiment. However, in 3D finite element simulations, calculating the average strain over a volume-weighted zone becomes less practical. Therefore, it is recommended to use the average strain from a single critical element, typically the corner element of the upper pad. As previously mentioned, the size of this element can be carefully controlled to both minimize singularity effects and accurately capture the

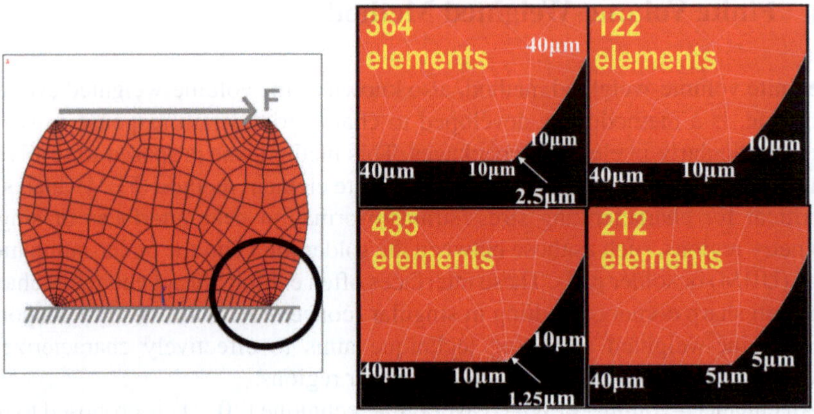


**Fig. 8.13** Illustration of volume-weighted zone

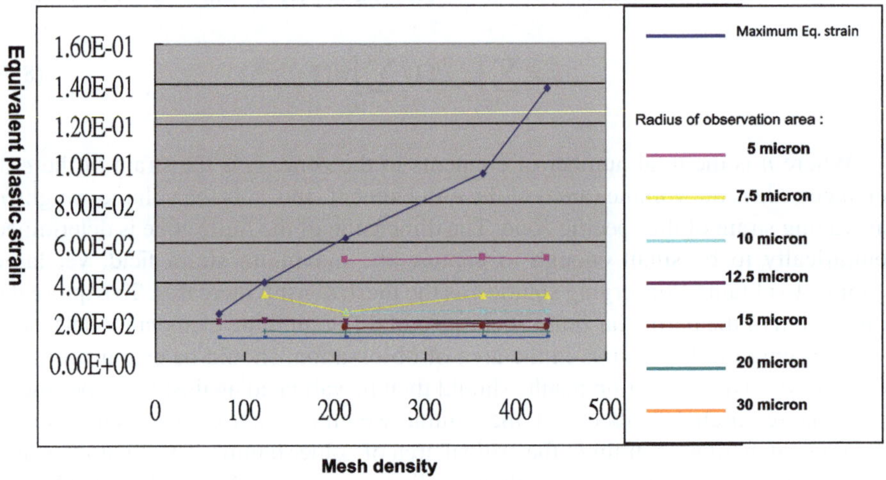


**Fig. 8.14** Strain vs. mesh density under different volume-weighted zone

strain value required for empirical models used in life cycle prediction. It is an art; the size of this particular element will need to be verified through several ATCT results to ensure it is robust enough to handle different sizes of solder balls.

## 8.5 AI-Assisted Coefficient Extraction

Solder material curves are nonlinear, strain-rate, stress, and temperature-dependent; accurate material curves play a crucial role in finite element simulations. For such complex, multiple material curves, AI can assist in fitting experimental data into the material's constitutive equation. AI-assisted coefficient extraction is an efficient methodology, primarily utilizing Equation-Informed Neural Networks (EINNs) [22, 23] or ANNs, to extract coefficients from constitutive equations based on measurement data. A similar approach is the Physics-Informed Neural Networks (PINNs), a type of artificial neural network that integrates physical laws, described by differential equations, directly into its learning process. Unlike traditional ANNs that learn solely from data, a PINN is trained to satisfy both the given data and the underlying physical equations. This section will focus on EINN; this approach is particularly valuable for modeling nonlinear material properties, such as those found in electronic packaging.

### 8.5.1 *Coefficient Extraction Using EINN*

The equation-informed neural network (EINN) [22] is developed as an efficient method for extracting the coefficients of constitutive equations. Subsequently, numerical Bayesian inference (BI) iterations were applied to estimate the distribution of these coefficients, thereby refining them further. One could generate coefficients optimally aligned with the targeted application scenario by carefully adjusting pre-processing mapping parameters and identifying dataset preferences. Leveraging graphical representation techniques, the EINN formulation is implemented in temperature- and strain-rate-dependent hyperbolic Garofalo, Anand, and Chaboche constitutive models to extract the corresponding coefficients for lead-free SAC305 solder material. The performance of the EINN-based extracted coefficients, obtained from the experimental results of SAC305 solder material, is comparable to that of existing studies. The methodology offers the dual advantage of providing the coefficients' value and distribution against the training dataset.

Solder, a key component in electronic packaging, is often associated with potential fatigue failures. The research [22] has developed the Equation-Informed Neural Network (EINN) method, synergistically incorporating the Bayesian inference (BI) iteration technique to extract the coefficients of constitutive equations from measurement data. As visualized in Fig. 8.15, the foundational concept of EINN begins with constructing an artificial neural network to embody the constitutive equation $f$, where the coefficient $\alpha_k$ designates as their respective weights. Subsequently, EINN deploys a pre-processing mapping technique on the input/output data pairs, a strategy rooted in neural network learning theory, which enables in-depth exploration of coefficient fitting across various domains.

This neural network can be incrementally trained using input and output data pairs, facilitating the simultaneous approximation of coefficient $\alpha_k$. Theoretically, the steepest descent algorithm of the neural network backpropagation enhances computational efficiency and facilitates the selective learning of data pairs. The post-processing conversion obtains the final coefficients. Utilizing the coefficients obtained by EINN as initial values, Bayesian Inference (BI) is applied to obtain the

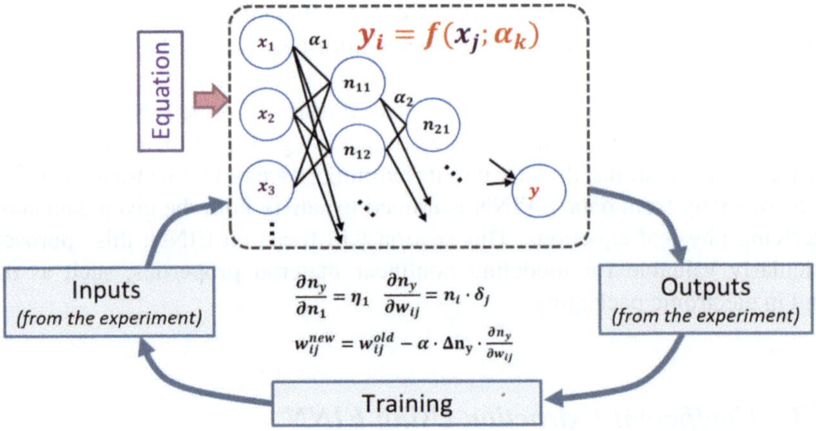


**Fig. 8.15** The concept of equation-informed neural networks (EINS)

distribution of the coefficients against the training datasets and further enhance the accuracy of coefficient extraction.

#### 8.5.1.1 The Framework of the Equation-Informed Neural Network (EINN

Assume the function gives a constitutive equation:

$$y_i = f\left(x_j;\alpha_k\right) \tag{8.8}$$

where $y_i$, $x_j$ and $\alpha_k$ are vectors in real space with dimensions $i$, $j$, and $k$, respectively. The $x_j$ and $y_i$ represent the input and output of the functions, while $\alpha_k$ refers to the coefficients. Design pre-processing mapping functions:

$$\mathcal{M}_x\left(x_j\right) = X_j \text{ and } \mathcal{M}_y\left(y_i\right) = Y_i, \tag{8.9}$$

which serve to effectively modify the domains of $x_j$ and $y_i$ to optimize the precision of coefficient extraction. Consequently, a new function can be formulated as $Y_i = F(X_j, A_k)$. Meanwhile, the corresponding neural network representation of $Y_i$ is formulated, and the coefficients $A_k$ is assigned as the weightings.

The learning process of the neural network involves continuous adjustment of these weights, or coefficients. These adjustments can be computed for each known data pair using steepest-descent-based backpropagation as $A_k^{new} = A_k^{new} - \eta\Delta_y \frac{\partial y_i}{\partial A_k}$. Since these updates are independent of each data pair, the computationally expensive matrix multiplication and inversion inherent in the least squares-based approaches can be avoided. Furthermore, incorporating ratios into the adjustments enables users to emphasize specific data pairs. This can be implemented as $A_k^{new} = A_k^{new} - \eta \sum_l r_l \cdot \left( \Delta_y \frac{\partial y_i}{\partial A_k} \right)^{(l)}$, where $l$ is the coefficient adjustment from each data pair and $\sum_l r_l = 1$.

Following several learning iterations with satisfactory accuracy, the coefficient $A_k$ of the constitutive equation can be obtained. However, due to the pre-processing mapping Eq. (8.9) being applied, counteractions are required to reverse its effect. Therefore, we define the post-processing conversion functions as follows.

$$a_k = g_k\left(x_j, y_i, A_k\right). \tag{8.10}$$

Through the combined application of pre-processing mapping functions and post-processing coefficient conversion, the EINN framework gains an additional degree of freedom, thereby bolstering the accuracy of coefficient extraction. Additionally, the steepest descent method offers a unique opportunity to prioritize specific data pairs while maintaining high computational efficiency.

The Mean Squared Error (MSE) function of Eq. (8.8) for the coefficients $\alpha_k$, is defined as:

$$\epsilon\left(\alpha_k\right) = \sum_l \left\| y_{t,i}^{(l)} - y_i\left(x_j^{(l)}; \alpha_k\right) \right\|^2, \tag{8.11}$$

where $x_j^l$ and $y_{t,i}^l$ denote the input and ground truth of the $l$-th datapair, respectively.

Assume that the distribution of the data pairs $y_{t,i}^{(l)}$ and $x_j^{(l)}$ are normal, and so is the error function $\epsilon(\alpha_k)$, denoted as $\epsilon(\alpha_k) \sim N(\mu, \tau)$. Because the parameter $\tau$ cannot be negative, we assume it follows the gamma distribution, so that $\tau \sim G(a_0, b_0)$, where $a_0$ and $b_0$ are the gamma distribution parameters of $\tau$. Moreover, assume that all the coefficients follow the normal distribution, say $\alpha_k \sim N(\mu_k, \tau_k)$, and $\mu_k$ and $\tau_k$ are the average and precision, respectively. The posterior distribution after the BI remains a normal distribution. In practice, we set $\mu_k$ equals $\alpha_k$.

Consequently, the probabilities of the coefficient $\tau$ and $\alpha_k$ can be derived as $\mathcal{P}(\tau) = \frac{b_0^{a_0} \tau^{a_0 - 1} e^{-b_0 \tau}}{\Gamma(a_0)}$ and $\mathcal{P}(\alpha_k) = (2\pi)^{-\frac{1}{2}} \tau_k^{\frac{1}{2}} e^{-\frac{1}{2}\tau_k(\alpha_k - \mu_k)^2}$, respectively. The likelihood with respect to the coefficient $\tau$ and $\alpha_k$ is $\mathcal{L}(\alpha_k, \tau) \equiv P(data|\alpha_k, \tau) = \prod_{l=1}^{L} \frac{1}{\sqrt{2\pi}} \tau^{\frac{n}{2}} e^{-\frac{\left(y_{t,i}^{(l)} - y_i^{(l)}\right)^2}{2\sigma^2}} = (2\pi)^{-\frac{n}{2}} \tau^{\frac{n}{2}} e^{-\frac{\tau}{2}\epsilon(\alpha_k)}$ [24].

The posterior of the $\tau$ distribution can be updated by the gamma-gamma conjugate:

$$a_0^{new} = a_0^{old} + \frac{n}{2} \text{ and } b_0^{new} = b_0^{old} + \frac{1}{2} \in \quad (8.12)$$

As Eq. (8.8) is not always a linear function, the posterior of the coefficient $\alpha_k$ cannot always be computed by conjugate. Therefore, a numerical integration approach is applied:

$$\int_0^{\infty} \mathcal{L}(\alpha_k) \cdot P(\alpha_k) d\alpha_k \sim \sum_{n=1}^{n=N} \mathcal{L}\left(\alpha_k^{(0)} + n \cdot \Delta\alpha_k\right) \cdot P\left(\alpha_k^{(0)} + n \cdot \Delta\alpha_k\right) \cdot \Delta\alpha_k \quad (8.13)$$

where $\alpha_k^{(0)}$ is the minimal value of $\alpha_k$ and $n$ is the number of equal splits between the assigned maximum and minimum $\alpha_k$ with a total of $N$ splits. The posterior can then be obtained using an approximation based on the normal distribution.

Employ the Markov Chain Monte Carlo (MCMC) method to compute large hierarchical models requiring integrations over many parameters. By applying the Gibbs sampling, the $\tau$ distribution parameters $a_0$ and $b_0$ are first updated through the conjugate (Eq. 8.12), and a new $\tau$ value will be sampled from the gamma distribution. Each $\alpha_k$ will be updated sequentially, and the new value will be accepted. Following thousands of iterations, every $\alpha_k$ exhibits a normal distribution. The mean value of this distribution is computed and assigned as the updated value for $\alpha_k$.

The development of Equation-Informed Neural Network (EINN) formulations for the hyperbolic Garofalo, nine-parameter Anand, and Chaboche models, including pre-processing mapping and post-processing coefficient functions, is outlined as follows.

#### 8.5.1.2 The Hyperbolic Garofalo Model

The conventional hyperbolic Garofalo constitutive equation can be written as:

$$\dot{\varepsilon}_p = C_1 \cdot \left[\sinh\left(C_2\sigma\right)\right]^{C_3} \cdot e^{-\frac{Q}{RT}}, \quad (8.14)$$

where $\dot{\varepsilon}_p$, $\sigma$, $Q$, $R$, and $T$ represent the plastic strain rate, stress, activation energy, gas constant, and temperature, respectively. The $C_1$, $C_2$ and $C_3$ are the coefficients that need to be extracted from the experimental data.

We introduce $e = \dot{\varepsilon}_p \cdot e^{\frac{Q}{RT}}$ and accumulate the datapairs of $\{e\}$ and $\{\sigma\}$ from the experimental results. The pre-processing matching functions are defined as follows:

$$\mathcal{M}_x(\sigma) = \frac{\sigma - \sigma_m}{\Delta\sigma} b + a = x \text{ and}$$
$$\mathcal{M}_y(e) = \frac{e - e_m}{\Delta e} b + a = y, \quad (8.15)$$

where $\sigma_m$ and $\Delta\sigma$ represent the minimal and maximum different values of set $\{\sigma\}$, and $e$ and $\Delta e$ correspond to set $\{e\}$. Parameters $a$ and $b$ are parts of the pre-processing mapping, and $a = 0.001$ and $b = 1$ are assigned for this case. The values after the pre-processing are defined as $x$ and $y$, respectively. Subsequently, a new function can be derived as:

$$y = C\left[\sinh Ax\right]^n, \tag{8.16}$$

The corresponding neural network can be defined in Fig. 8.16. The definition of the neurons is given in Table 8.2.

Accordingly, the post-processing conversion of the coefficients can be approximated as $C_1 = C\Delta e \cdot r_2^n$, $C_2 = A \bullet b/\Delta\sigma$ and $C_3 = n$, where $r_2 = \left(\frac{Ab}{\Delta\sigma} - Aa\right)$.

#### 8.5.1.3 The Anand Model

Anand et al. [25] proposed a set of viscoplastic constitutive equations for the rate-dependent deformation of metals. Recently, the Anand model has been extensively applied to microelectronic solders exhibiting large viscoplastic deformations. In addition to the activation energy, the Anand model includes eight coefficients. A two-step approach is commonly employed to extract these eight coefficients.

The governing equation for the first step of the Anand model, including the ultimate tensile stress ($\sigma^*$), plastic strain rate ($\dot{\varepsilon}_p$), activation energy ($Q$), and temperature ($T$), is expressed in Eq. (8.17). The $\hat{s}$, $\xi$, $A$, $n$, and $m$ are the coefficients that need to be extracted.

$$\sigma^* = \frac{\hat{s}}{\xi}\left(\frac{\dot{\varepsilon}_p}{A} \cdot e^{\frac{Q}{RT}}\right)^n \sinh^{-1}\left[\left(\frac{\dot{\varepsilon}_p}{A} e^{\frac{Q}{RT}}\right)^m\right], \tag{8.17}$$

Utilizing the same method as in the previous section, we assume $e_0 = \dot{\varepsilon}_p \cdot e^{\frac{Q}{RT}}$. Since the value of the strain rate is relatively small compared to other input parameters, a

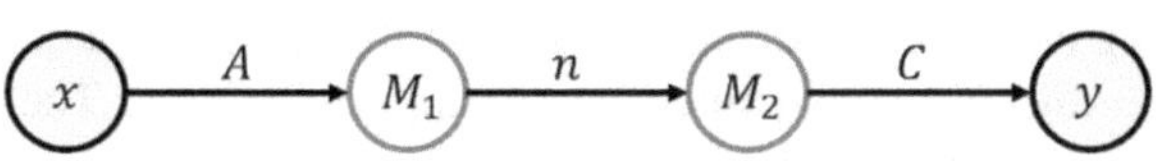


**Fig. 8.16** The EINS for the hyperbolic Garofalo model

**Table 8.2** The neuron definition of the EINS representation of the hyperbolic Garofalo equation

| Neuron | Net value | Activation |
|---|---|---|
| $M_1$ | $M_{1,net} = A \bullet x$ | $M_1 = \ln(\sinh M_{1,net})$ |
| $M_2$ | $M_{2,net} = n \bullet M_1$ | $M_2 = e^{M_{2,net}}$ |
| $y$ | $y_{net} = C \bullet M_2$ | $y = y_{net}$ |

scaling factor $R$ is applied, such that $e = \frac{e_0}{R}$. For consistency within this paper, the same activation function as in the previous section is assumed. The data pair of $\{e\}$ and $\{\sigma^*\}$ are collected from the Motalab et al. [26, 27]. Additional $y$ and $x$ are introduced to represent the output and input parameters, and the pre-processing mapping functions are defined as:

$$x = \frac{e - e_m}{\Delta e} b_e + a_e \text{ and } y = \frac{\sigma^* - \sigma_m^*}{\Delta \sigma^*} b_\sigma + a_\sigma, \tag{8.18}$$

$$y = \beta^* \left( \frac{x}{A^*} \right)^{n^*} \sinh^{-1} \left[ \left( \frac{x}{A^*} \right)^{m^*} \right], \tag{8.19}$$

where $e_m$ and $\Delta e$ are the minimal and maximum difference among set $\{e\}$, and so does the $\sigma_m^*$ and $\Delta\sigma^*$ in $\{\sigma^*\}$. The $a_e$, $b_e$, $a_\sigma$ and $b_\sigma$ are the mapping coefficients. By defining $\beta = \frac{\hat{s}}{\xi}$, the new function can be written as:

Based on Eq. (8.19), the EINN representation can be formulated as shown in Fig. 8.17; the definitions of this network are listed in Table 8.3.

By defining $r = \sigma_m^* - a_\sigma \frac{\Delta \sigma^*}{b_\sigma}$, the post-processing of the coefficients can be written as follows:

$$\begin{aligned} &\frac{1}{A} = \frac{1}{A^*} \left( \frac{b b_e}{\Delta e} + \frac{a_e}{avg(y)} \right) \cdot \frac{1}{R}, n = n^*, m = m^* \\ &\text{and } \beta = \left[ \frac{\Delta \sigma}{b_\sigma} \cdot \beta^* + \frac{r}{\beta^* \cdot avg(x)} \right] = \frac{\hat{s}}{\xi}, \end{aligned} \tag{8.20}$$

where $avg(x)$ and $avg(y)$ are the averges of $\{e\}$ and $\{\sigma^*\}$.

The governing equation of the second step of the Anand model is shown in Eq. (8.21), and $s_0$, $a$ and $h_0$ are the three remaining coefficients. The parameter $c$ is defined in Eq. (8.22), and the $\xi$ is defined as the smallest positive real number to keep $c < 1$.

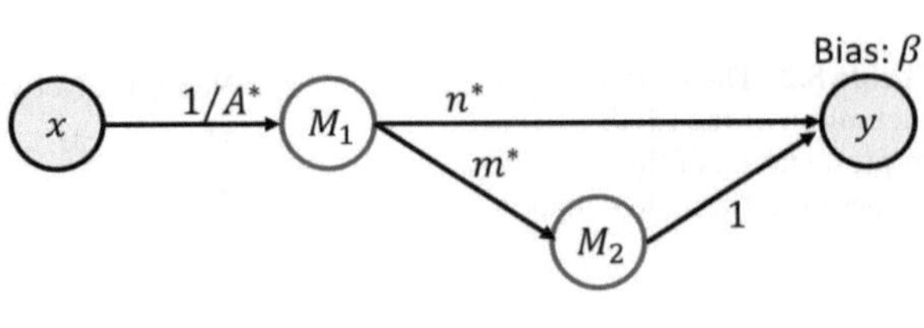


**Fig. 8.17** The EINS for Step 1 Anand Equation

**Table 8.3** The neuron definition of the EINS representation of the step 1 Anand equation

| Neuron | Net value | Activation |
|---|---|---|
| $M_1$ | $M_{1,net} = 1/A^* \bullet x$ | $M_1 = \ln(M_{1,net})$ |
| $M_2$ | $M_{2,net} = m^* \bullet M_1$ | $M_2 = \ln\left(\sinh^{-1} e^{M_{2,net}}\right)$ |
| $y$ | $y_{net} = n^* \bullet M_1 + M_2 + \beta$ | $y = e^{y_{net}}$ |

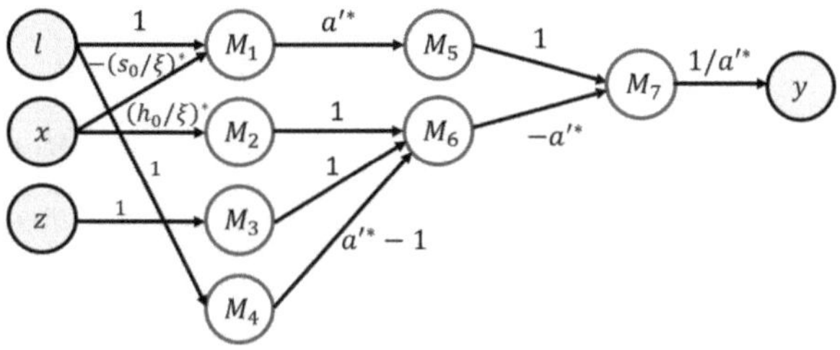


**Fig. 8.18** The EINS for Step 2 Anand Equation

$$\sigma = \sigma^* - \left[\left(\sigma^* - cs_0\right)^{1-a} + (a-1)\left\{\left(ch_0\right)\left(\sigma^*\right)^{-a}\right\}\varepsilon_p\right]^{1/(1-a)}, \tag{8.21}$$

$$c = \frac{1}{\xi}\sinh^{-1}\left(\frac{\dot{\varepsilon}_p}{A} e^{\frac{Q}{RT}}\right)^m, \tag{8.22}$$

We assume that $x = \sinh^{-1}\left[\left(\frac{\dot{\varepsilon}}{A} e^{\frac{Q}{RT}}\right)^m\right]$, $y = (\sigma^* - \sigma)$, $l = \sigma^*$ and $z = \varepsilon_p$, and the pre-processing mapping functions are defined as:

$$y = \frac{y - y_m}{\Delta y} b_y + a_y, l = \frac{l - l_m}{\Delta l} b_l + a_l, x = \frac{x - x_m}{\Delta x} b_x + a_x \text{ and } z = \frac{z - z_m}{\Delta z} b_z + a_z, \tag{8.23}$$

By assuming $1 - a = a'$, the new function can be written as:

$$y = \left[\left(\overline{l} + \left(-\frac{s_0}{\xi}\right)^* \cdot x\right)^{a'^*} - a'^* \left\{\left(\left(\frac{h_0}{\xi}\right)^* \cdot\right)(l)^{a'^*-1}\right\} \cdot \overline{z}\right]^{1/a'^*}, \tag{8.24}$$

Based on Eq. (8.24), the EINN representation can be formulated as shown in Fig. 8.18; the definitions of this network are listed in Table 8.4. Moreover, the post-processing of coefficients can be derived as:

**Table 8.4** The neuron definition of the EINS representation of the step 2 Anand equation

| Neuron | Net value | Activation |
|---|---|---|
| $M_1$ | $M_{1,net} = l + \left(-\frac{s_0}{\xi}\right)^* x$ | $M_1 = \ln(M_{1,net})$ |
| $M_2$ | $M_{2,net} = \left(\frac{h_0}{\xi}\right)^* \cdot M_1$ | $M_2 = \ln(M_{2,net})$ |
| $M_3$ | $M_{3,net} = z$ | $M_3 = \ln(M_{3,net})$ |
| $M_4$ | $M_{4,net} = l$ | $M_4 = \ln(M_{4,net})$ |
| $M_5$ | $M_{5,net} = a'^* \bullet M_1$ | $M_5 = e^{M_{5,net}}$ |
| $M_6$ | $M_{6,net} = M_2 + M_3 + (a'^* - 1) \bullet M_4$ | $M_6 = e^{M_{6,net}}$ |
| $M_7$ | $M_{7,net} = M_5 - a'^* \bullet M_6$ | $M_7 = \ln(M_{7,net})$ |
| $y$ | $y_{net} = \frac{1}{a'^*} \cdot M_2$ | $y = e^{y_{net}}$ |

$$a'^* = a', s_0' = \frac{r_y}{r_l} \cdot (s_0), h_0' = \frac{r_y}{r_l} \cdot r_z (h_0), \tag{8.25}$$

where $r_l = \frac{b_l}{\Delta l}$, $r_x = \frac{b_x}{\Delta x}, r_y = \frac{b_y}{\Delta y}$ and $r_z = \frac{b_z}{\Delta z}$.

#### 8.5.1.4 The Chaboche Model

The Chaboche model [4, 28] is often used to present metallic materials exhibiting the Bauschinger effect under cyclic loading. The original function can be written as:

$$\alpha = \frac{C}{\gamma}\left(1 - e^{-\gamma \cdot \varepsilon_p}\right) + \sigma_0 \tag{8.26}$$

where $\alpha$ and $\varepsilon_p$ are the back tensile stress and the plastic strain; $\sigma_0$ is the initial yielding stress, and $C$ and $\gamma$ are the fitting coefficients. To simplify the equation, we substitute and $C/\gamma$ as $\beta$. Let $x = \varepsilon_p$, $y = \alpha$, as the parameters, with the pre-processing mapping functions:

$$x = \frac{\varepsilon_p - \varepsilon_{p,m}}{\Delta \varepsilon_p} \text{ and } y = \frac{\alpha - \alpha_m}{\Delta \alpha}, \tag{8.27}$$

where $\varepsilon_{p,m}$ and $\Delta\varepsilon_p$ are the minimal and maximum differences among the set $\{\varepsilon_p\}$, and so does the $\alpha_m$ and $\Delta\alpha$ in $\{\alpha\}$, and $s = \sigma_0$. Hence, the new function can be rewritten as:

$$y = \beta^* \left(1 - e^{-\gamma^* \cdot x}\right) + s, \tag{8.28}$$

With the EINN formulation shown in Fig. 8.19 and the neuron definition listed in Table 8.5.

Furthermore, the post-processing of coefficients can be derived as:

$$\sigma_0 = x_m + s^* \cdot \Delta\alpha, = \frac{\gamma^*}{\Delta\varepsilon_p}, \text{and } C = \frac{\Delta\alpha}{\Delta\varepsilon_p} \cdot \left(\beta^* \cdot \gamma^*\right) \tag{8.29}$$

#### 8.5.1.5 Applications

Building on the EINN formulation and Bayesian Inference (BI) iteration described in the preceding section, this chapter discusses the extraction of coefficients from the Hyperbolic Garofalo, nine-parameter Anand, and Chaboche models for the SAC305 solder material.

Hyperbolic Garofalo Model

The experimental dataset is drawn from Xiao and Armstrong [29]. To determine the coefficient $C$ in Eq. (8.16), we employ a grid search combined with a bisection optimization technique, whereas the EINN structure for coefficients $A$ and $n$ is addressed using standard backpropagation. To emphasize coefficient extraction for low temperatures (both 318 and 353 K) and low strain rates, ratios are assigned to the datapairs, as shown in Table 8.6. Utilizing the post-processing conversion formula, the EINN coefficients, $C_1$, $C_2$ and $C_3$, are obtained and presented in the middle column of Table 8.7. The hyperbolic model, when compared to the experimental data, is depicted in Fig. 8.20. The data at 388 K exhibits a more significant difference than the others, primarily due to the ratio setting outlined in Table 8.6.

A total of 1000 Bayesian inference interactions were performed to obtain the distribution of the coefficients. The distributions are displayed in Figure 8.21a–c, represented as the ratio of each coefficient value to the average, and are expanded

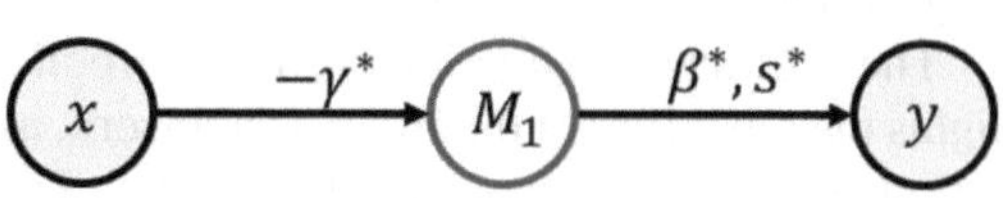


**Fig. 8.19** The EINS formulation for the Chaboche model

**Table 8.5** The neuron definition of the EINS representation of the Chaboche equation

| Neuron | Net value | Activation |
|---|---|---|
| $M_1$ | $M_{1,\,net} = (-\gamma^*) \bullet x$ | $M_1 = 1 - e^{M_{1,net}}$ |
| $y$ | $y_{net} = \beta \bullet M_1 + s^*$ | $y = y_{net}$ |

**Table 8.6** The ratios applied to the datapair to emphasize the preference

| Data ID | Temperature (K) | Plastic strain rate ( $\dot{\varepsilon}_p$,$10^{-9}$ $1/s$ ) | Stress (MPa) | Ratio |
|---|---|---|---|---|
| 1 | 318 | 1.39 | 10.54 | 18 |
| 2 | 318 | 4.01 | 12.3 | 6 |
| 3 | 318 | 13.6 | 14.25 | 5 |
| 4 | 318 | 43.9 | 15.92 | 0.5 |
| 5 | 353 | 6.6 | 10.43 | 18 |
| 6 | 353 | 13.1 | 11.59 | 6 |
| 7 | 355 | 21.2 | 12.4 | 5 |
| 8 | 353 | 57.8 | 13.8 | 3 |
| 9 | 353 | 95 | 14.46 | 1 |
| 10 | 388 | 13.4 | 8.73 | 15 |
| 11 | 388 | 19.8 | 9.35 | 12 |
| 12 | 388 | 34.2 | 10.07 | 6 |
| 13 | 388 | 50.5 | 10.67 | 6 |
| 14 | 388 | 166 | 12.41 | 1 |
| 15 | 423 | 90.1 | 7.51 | 1 |
| 16 | 423 | 165 | 8.63 | 1 |
| 17 | 423 | 315 | 9.82 | 1 |
| 18 | 423 | 454 | 10.54 | 0.5 |
| 19 | 423 | 810 | 11.52 | 0.5 |

**Table 8.7** The comparison of the extracted coefficients

| | Xiao and Armstrong | EINS | EINS+BI |
|---|---|---|---|
| Q(kJ/mol) | 62,000 | 65,000 | 65,000 |
| $C_1$ | 0.184 | 0.539 | 0.443 |
| $C_2$ | 0.221 | 0.473 | 0.482 |
| $C_3$ | 2.89 | 1.055 | 1.073 |
| MSE | 37,188.5 | 11,794.9 | 10,967.4 |

by the precision $\tau$ of the error function. As denoted by the dashed lines in Figure 8.21a–c, which signify a 5% difference, stable distributions of coefficients $C_2$ and $C_3$ are observed, while the large variation in $C_1$ is attributed to the ratio setting, which induces a higher discrepancy among the 388 K data.

The coefficient extraction of the hyperbolic Garofalo constitutive equation highlights the flexibility of the EINN framework, as it allows for assigning ratios to datapairs to prioritize specific data. The fitting accuracy of the EINN results demonstrates a significant improvement compared to the original reports [29], as indicated by the Mean Squared Error (MSE) in Table 8.7, which is further supported by Eq. (8.11). Although the distribution of the $C_1$ coefficients demonstrate a small faction of the outliners from BI integration, both $C_2$ and $C_3$ show statistical robustness within ±5% difference. Over 1000 iterations, only 58 instances of $C_1$ shows more than ±5%

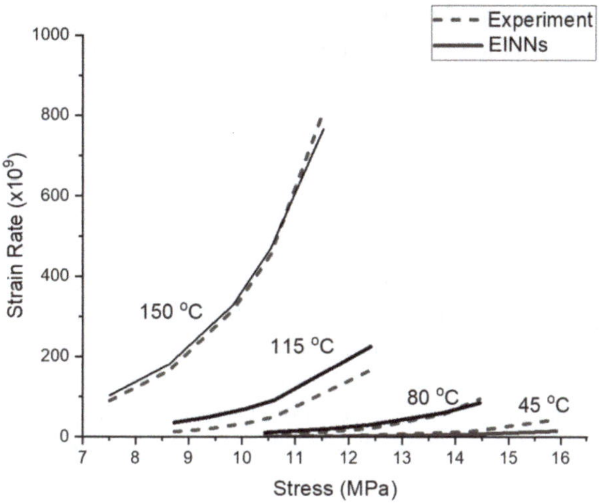


**Fig. 8.20** The obtained hyperbolic Garofalo curves for different temperatures

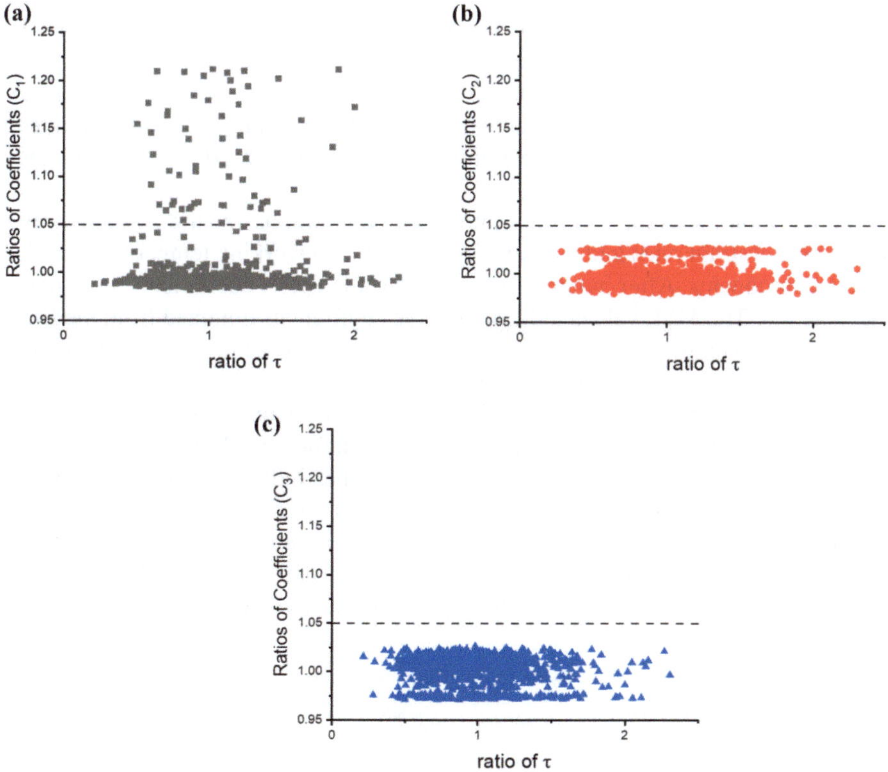


**Fig. 8.21** The distribution of the coefficients from the BI iteration. (**a**), (**b**) and (**c**) are coefficients $C_1$, $C_2$, and $C_3$, respectively

difference of the average value. Consequently, a robust set of coefficients for the hyperbolic Garofaolo constitutive model is achieved.

### Anand Model

In this section, the Anand constitutive model coefficients are extracted for the lead-free SAC305 solder. The same activation energy as in the previous section is applied for the sake of research consistency. To extract the remaining eight coefficients of the Anand constitutive model, the first step involves using temperature- and strain-rate-dependent ultimate tensile stresses to determine the initial four coefficients. Subsequently, the second step defines the remaining parameters based on temperature- and strain-rate-dependent stress and plastic strain.

The experimental data are sourced from Motalab et al. [26, 27]. The EINN formulation, following Eq. (8.19), is applied with the pre-processing mapping coefficients $a_e$, $b_e$, $a_\sigma$ and $b_\sigma$ (Eq. 8.20) are 0.8, 0.15, 0.9, and 0.1. It is vital to note that the selection of these mapping coefficients depends on the numerical characteristics of the dataset, and it is essential for preventing numerical errors during the backpropagation-based machine learning of the EINN formulation.

During the learning phase of the EINN formulation, the coefficients $n^*$, $m^*$, $A^*$, and $\beta$ of Eq. (8.19) and Fig. 8.17 are constrained to be positive. A grid search technique is employed to identify the optimal initial values based on the experimental data.

Furthermore, learning ratios are implemented to emphasize the learning preference for low strain rates and temperatures close to the working temperature of electronic components. After hundreds of iterations, the EINN coefficients are reported in Table 8.8. The MSE values indicate that the coefficients obtained from the EINN formulation exhibit similar accuracy to those obtained using conventional methods. The obtained step 1 Anand model is plotted in Fig. 8.22.

The EINN formulation coefficients serve as initial inputs for Bayesian Inference (BI) to analyze the statistical distribution of the coefficients. Figure 8.23 illustrates the distribution of the coefficients, with dashed lines indicating differences within ±5%. Due to its low value, the coefficient $n$ was not examined. Both coefficients $A$ and $\beta$ exhibit a distribution within ±5% difference. Out of 1500 values, only 61 cases of coefficient m exceed ±5% difference, which can be attributed to the preference settings during the EINN learning process. The average coefficients obtained

**Table 8.8** The comparison of the extracted coefficients of step 1 Anand

| | Motalab et al. | EINN | EINN+BI |
|---|---|---|---|
| A | 3501 | 1650 | 1649 |
| $n$ | 0.01 | 0.000154 | 0.000164 |
| $m$ | 0.25 | 0.54 | 0.53 |
| $\beta = \hat{s} / \xi$ | 7.55 | 4.11 | 4.16 |
| MSE | 17.03 | 15.96 | 15.78 |

**Fig. 8.22** Step 1 Anand model at different t temperatures

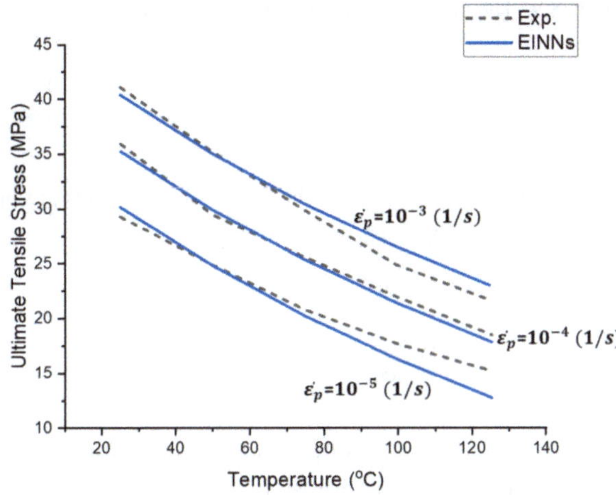


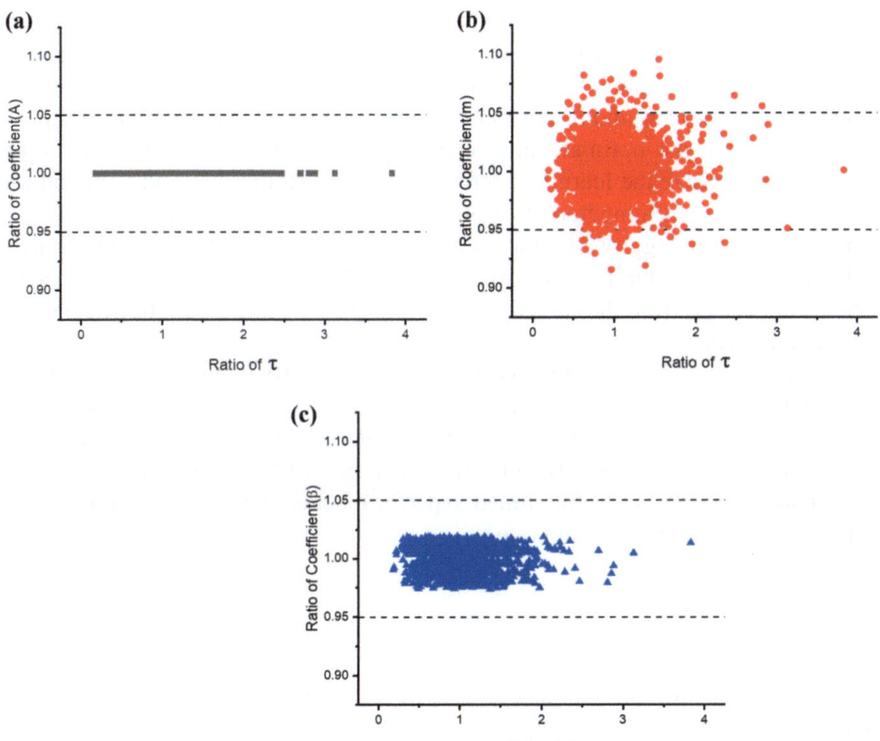


**Fig. 8.23** The distribution of the coefficients from the BI iteration. (**a**), (**b**) and (**c**) are coefficients A, m, and β, respectively

**Table 8.9** The pre-processing mapping parameters

| | $y$ | $l$ | $x$ | $z$ |
|---|---|---|---|---|
| a | 1 | 0.8 | 0.1 | 0.8 |
| $b$ | 0 | 0.4 | 0.05 | 0.1 |

**Table 8.10** The comparison of the extracted coefficients of step 2 Anand model

| | Motalab et al. | EINN | EINN+BI |
|---|---|---|---|
| $\xi$ | 4 | 17.66 | 17.66 |
| $s_0$ | 21 | 55.96 | 57.45 |
| $a$ | 1.78 | 2.30 | 2.26 |
| $h_0$ | 18,000 | 828,822 | 828,822 |

from BI are presented in the last column of Table 8.8 and are utilized for the subsequent coefficient extraction step in the Anand model.

The temperature- and strain-rate-dependent stress-strain curves are obtained from Motalab et al. [26, 27]. The EINN formulation of the step 2 Anand model, as indicated in Eq. (8.24) and Fig. 8.18, is applied with the pre-processing mapping parameters shown in Table 8.9, based on Eq. (8.23). While the EINN learning procedure, the values of $s_0$ and $h_0$ are forced to be positive. A grid search technique is applied to define the optimal initial coefficients. The learning ratios are implemented to emphasize the learning preference for low strain rates and temperatures close to the working temperature of electronic components, following the coefficient extraction strategy of Motalab et al. [26, 27]. With Eq. (8.25), the optimized coefficients can be obtained, as listed in Table 8.10. Figure 8.24 illustrates the stress-strain curves from the Anand model at various strain rates, plotted against the experimental data.

The dataset with high preference is applied to the BI iteration to mitigate the large coefficient shifting. Figure 8.25 plots the MSEs of EINN and EINN with BI against the Anand coefficient obtained by Motalab et al. [26, 27], under different temperatures and strain rates. By adjusting the ratio of EINN network learning, the coefficient extraction can be fine-tuned to perform better at the working temperature in the room, even at a low strain rate.

### Chaboche Model

To study the lifetime of ball-grid-array-type advanced electronic packaging, the Chaboche material model is often applied [23, 30]. The Chaboche model and its coefficients can be extracted from the temperature-dependent stress-strain curves at a given strain rate. Unlike the previous sectors, this section investigates the extraction of Chaboche coefficients from the Anand model.

The Anand coefficients from Tables 8.7 and 8.8, adjusted via Bayesian Inference, are utilized to generate inputs for the Chaboche model. A strain rate of $10^{-5}$ (1/s) is maintained, given that the Anand coefficients have been optimized for lower strain

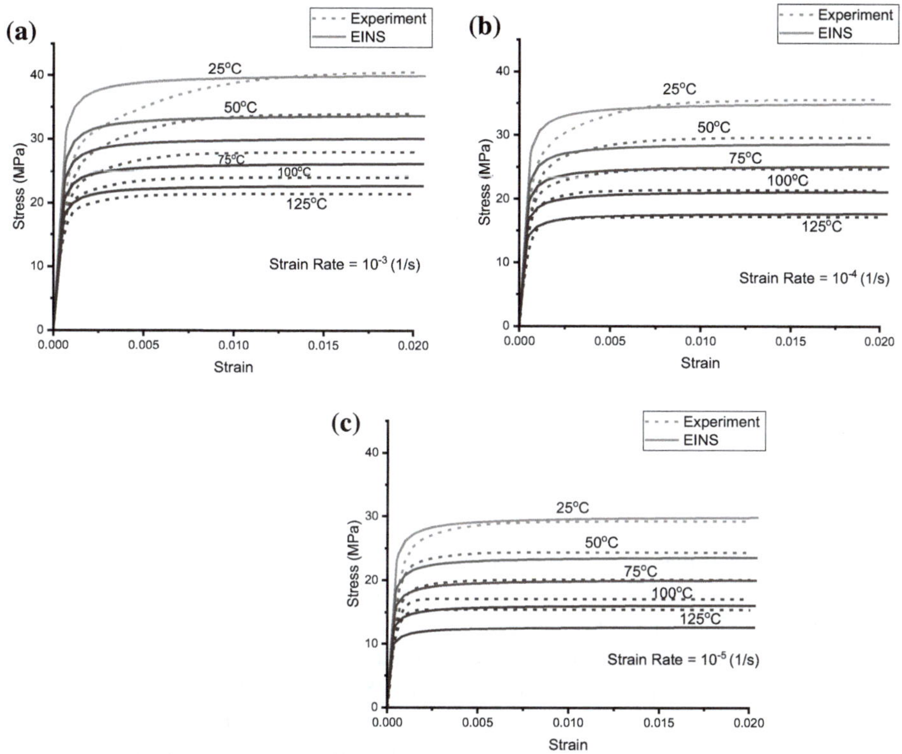


**Fig. 8.24** The obtained step 2 Anand model curves for different temperatures. (**a**), (**b**) and (**c**) are the obtained Anand model with strain rates of $10^{-3}$, $10^{-4}$ and $10^{-5}$ (1/sec)

**Fig. 8.25** The MSE Plot of step 2 Anand model coefficient extraction results

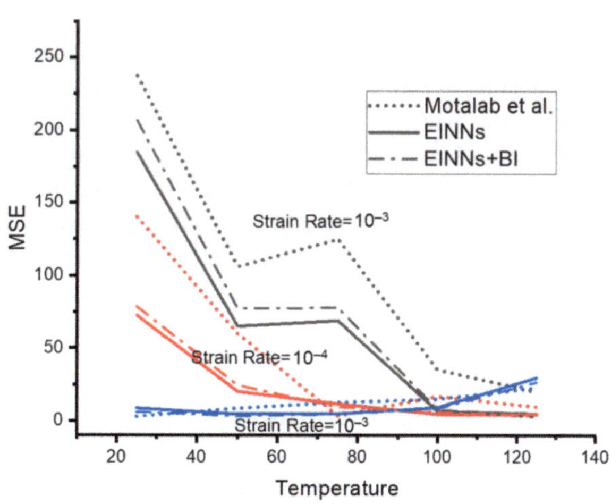

**Table 8.11** Temperature-dependent Chaboche coefficients of EINN

| Temperature | $\sigma_0$ | $C$ | $\gamma$ | MSE |
|---|---|---|---|---|
| 233 K | 39.32 | 9174.1 | 1004.7 | 1.11 |
| 253 K | 33.80 | 7535.7 | 964.0 | 0.84 |
| 313 K | 21.35 | 4216.0 | 840.0 | 0.50 |
| 353 K | 15.43 | 2988.0 | 824.3 | 0.28 |
| 395 K | 10.50 | 1886.5 | 759.1 | 0.18 |

**Table 8.12** Temperature-dependent Chaboche coefficients of EINN and BI

| Temperature | $\sigma_0$ | $C$ | $\gamma$ | MSE |
|---|---|---|---|---|
| 233 K | 39.30 | 9174.1 | 1004.7 | 1.11 |
| 253 K | 33.78 | 7535.7 | 964.0 | 0.84 |
| 313 K | 21.39 | 4216.0 | 840.0 | 0.49 |
| 353 K | 15.45 | 2988.0 | 824.3 | 0.27 |
| 395 K | 10.53 | 1886.5 | 759.1 | 0.17 |

rates, as demonstrated in the previous section. Stress-strain curves can be generated using the Anand model (as Eqs. (8.21) and (8.22)) for each temperature point, including 233, 253, 313, 353, and 395 K.

The temperature-dependent stress-strain data serve as the training datasets. With the pre-processing mapping established by Eq. (8.27), we apply the EINN formulation to the Chaboche model, as shown in Eq. (8.28). Following this, the steepest-descent coefficient optimization is applied to the EINN formulation with the neural definitions outlined in Table 8.5. The post-processing of the coefficients in Eq. (8.29) enables the acquisition of Chaboche coefficients at various temperatures. The resultant data are documented in Table 8.11, with the Mean Squared Errors (MSE) compared to the input dataset.

The coefficients derived from the EINN formulation are subsequently incorporated into BI iterations for the temperature-dependent Chaboche model. Table 8.12 contains the averaged coefficient post-BI.

While variations in all coefficients lie within a ±5% difference, a larger variety, coupled with a lower MSE, as listed in Tables 8.11 and 8.12, is evident at higher temperatures. This suggests a reduced coefficient sensitivity at these elevated temperatures. By introducing Young's modulus obtained by linear extrapolation from the experiment, the temperature-dependent stress-strain curves are plotted in Fig. 8.26.

## 8.6 Modeling Techniques

The increasing demand for lighter, thinner, high-efficiency, and multifunctional electronic products, especially those incorporating advanced packaging technologies such as heterogeneous integration and chiplets, has led to incredibly complex package architectures. The high integration density and complexity of modern packages, especially those with multiple dies and numerous micro-bumps, lead to

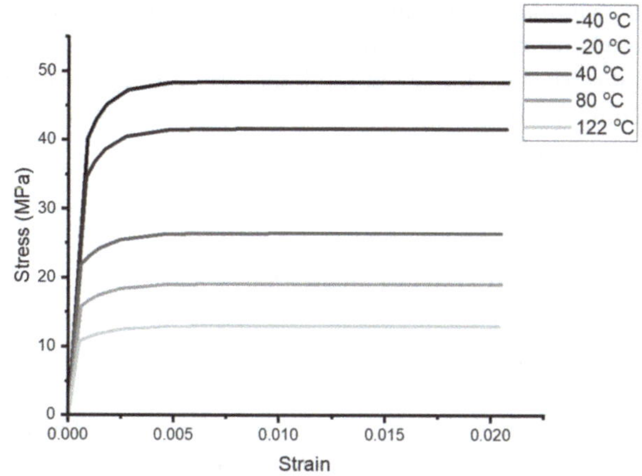


**Fig. 8.26** The temperature-dependent stress-strain curves

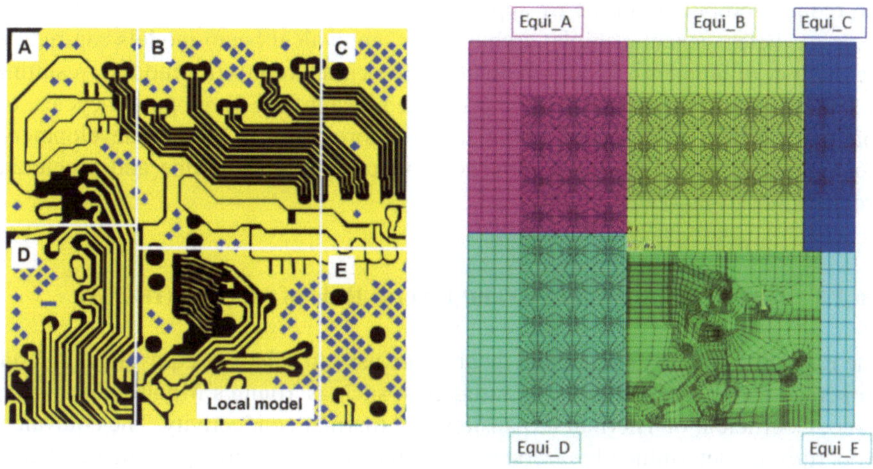


**Fig. 8.27** The equivalent model partition of the top layer traceline

multi-scale issues, highly complex modeling, and exceptionally long computation times for simulations. Asymmetric architectures, common in heterogeneous integration, particularly increase computation time if a full 3D model is used. This complexity poses significant challenges in simulation; therefore, it is critical to apply the equivalent method to reduce the modeling size. Equivalent methods in electronic package design refer to advanced simulation strategies employed within finite element analysis to streamline the modeling of complex structures. These techniques involve substituting detailed geometries or multi-layered components with simplified, homogenized representations that retain equivalent mechanical, thermal, or other physical properties. For example, in the local-global simulation process, the global model (Fig. 8.27) allows the copper layer of the PCB to use equivalent material coefficients. This will reduce the overall computational load of the simulation. Because the material composition of the entire copper layer is not uniform, it can be segmented into several parts based on the material distribution ratio for this equivalent modeling. By doing so, engineers can significantly reduce computational effort while maintaining the accuracy required for reliable assessments and performance predictions.

## 8.6.1 *Equivalent Methods*

Equivalent methods are crucial because they directly address the challenges mentioned above, offering a range of significant benefits: (1) Significant reduction in computational resources and time. They drastically reduce the number of model elements and overall computation time, often to a small fraction of the original. In wafer-level packaging simulations, equivalent beam models combined with MPC can reduce computation time by approximately 80% compared to full models. For complex, heterogeneous chiplet architectures, equivalent beam models can reduce the number of elements by approximately 50% compared to full models, resulting in substantial efficiency gains. (2) Accelerated design and development cycles, by enabling efficient analysis and optimization before physical product development, these methods significantly shorten the product research timeline and accelerate product time-to-market.

### 8.6.1.1 Rule of Mixture (ROM) and Inverse Rule of Mixture (IROM)

The ROM and IROM fundamental theories are used to calculate equivalent material properties for composite layers or structures, such as Young's Modulus, Poisson's Ratio, Coefficient of Thermal Expansion, and thermal conductivity. The equivalent properties are determined by accounting for the volume or area percentage of each constituent material. These approaches are particularly effective for simplifying complex patterned layers, such as Redistribution Layers (RDL) or joint layers containing micro-bumps and underfill, into homogeneous equivalent layers. The rule of mixtures and the inverse rule of mixtures are the simplest models used to estimate the effective material properties of composite materials, and they are widely adopted in various research papers. The concept of the rule of mixtures is based on the volume or area fraction of each constituent material, assuming that the effective properties are the linear summation of those of the individual materials. In a bi-material composite containing fibers, the effective material properties along the fiber length direction are typically dominated by the properties of the fibers. Assuming perfect bonding between the matrix and the fibers, this composite model can be simplified into a basic model consisting of two stacked material blocks, with proportions matching those of the original model, as illustrated in Fig. 8.28.

If this simplified model is subjected to an external force applied parallel to the fiber length direction, assuming that the axial strain is uniform throughout the entire

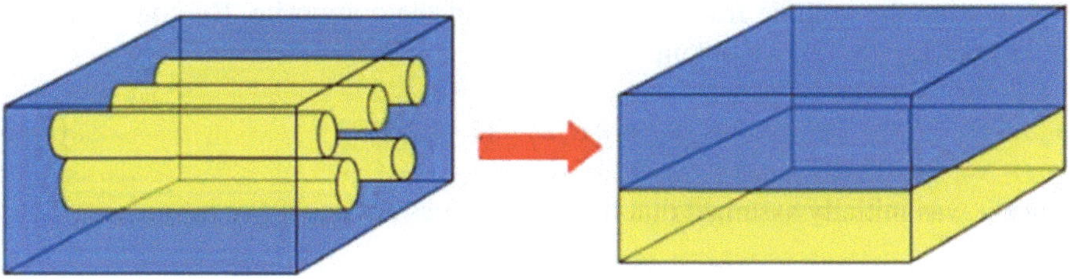

Fig. 8.28 A simple Bi-materials equivalent model

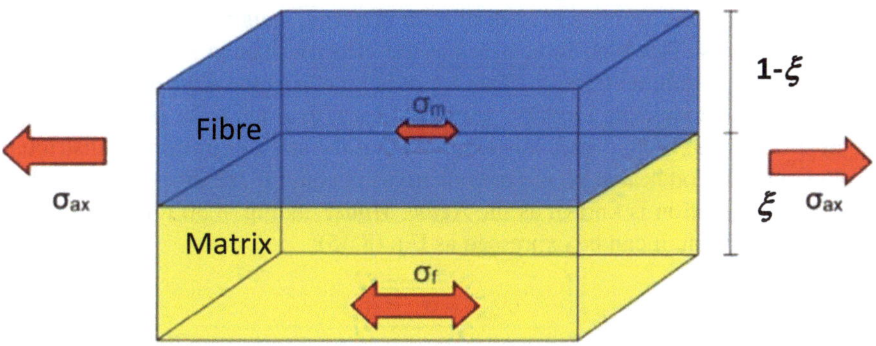


Fig. 8.29 The Voigt model

model. The strain in the fibers will be equal to the strain in the matrix, which is also equal to the overall strain of the composite material. This is known as the Voigt model [31], as illustrated in Fig. 8.29. Based on this assumption, it can be expressed as Eq. (8.30):

$$\varepsilon_{ax} = \varepsilon_f = \varepsilon_m \tag{8.30}$$

$\varepsilon_{ax}$ represents the overall strain of the composite material, while $\varepsilon_f$ and $\varepsilon_m$ denote the strain values of the fiber and the matrix, respectively. Since strain is equal to stress divided by Young's modulus, Eq. (8.30) can be rewritten as:

$$\varepsilon_{ax} = \varepsilon_f = \frac{\sigma_f}{E_f} = \varepsilon_m = \frac{\sigma_m}{E_m} \tag{8.31}$$

$\sigma_f$, $\sigma_m$ and $E_f$, $E_m$ represent the stress values and Young's modulus of the fiber and matrix, respectively. Since stress is the force applied per unit area, the total stress can be expressed as the weighted sum of the stress values of each material:

$$\sigma_{ax} = \xi\sigma_f + (1-\xi)\sigma_m \tag{8.32}$$

$\xi$ represents the volume fraction of the fiber in the composite. By combining Eq. (8.31) and Eq. (8.32), we obtain:

$$E_{ax}\varepsilon_{ax} = \xi E_f \varepsilon_f + (1-\xi) E_m \varepsilon_m \quad (8.33)$$

Since it was initially assumed that the strain values are equal, we have:

$$E_{ax} = \xi E_f + (1-\xi) E_m \quad (8.34)$$

Therefore, the effective Young's modulus in the fiber direction can be obtained as the linear sum of the Young's moduli of each material, weighted by their respective volume fractions. This also indicates that the effective Young's modulus depends only on the fiber volume fraction. This result is known as the ***Rule of Mixtures***.

Suppose the composite model is subjected to an external force perpendicular to the fiber direction. In that case, it is assumed that the stress is uniformly distributed throughout the model; that is, the overall stress is equal to the stress in each material. This assumption is known as the ***Reuss Model***, as Fig. 8.30 illustrates. Based on this assumption, it can be expressed as Eq. (8.35):

$$\sigma_{trans} = \sigma_f = \sigma_m \quad (8.35)$$

Since stress equals strain multiplied by Young's modulus, and the total strain can be expressed as the weighted sum of the strain values of each material, we have:

$$\sigma_{trans} = \sigma_f = E_f \varepsilon_f = \sigma_m = E_m \varepsilon_m \quad (8.36)$$

$$\varepsilon_{trans} = \xi \varepsilon_f + (1-\xi) \varepsilon_m \quad (8.37)$$

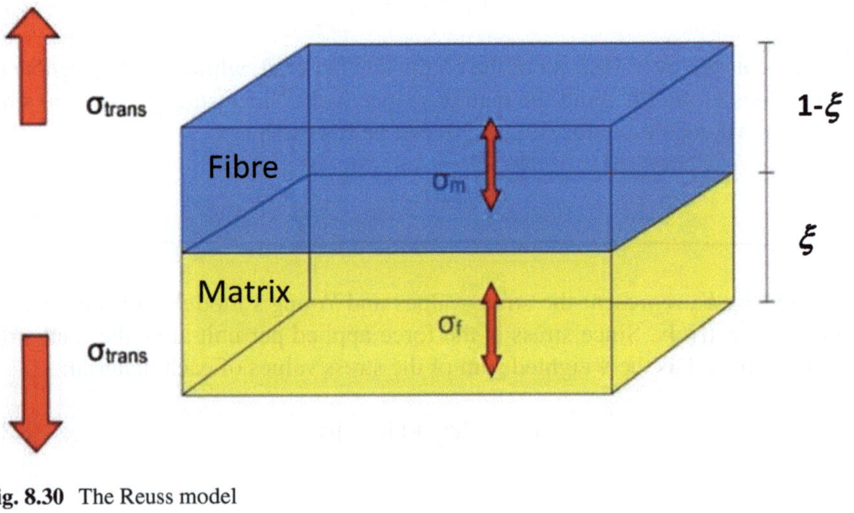


**Fig. 8.30** The Reuss model

Based on the above two Equations, we can have:

$$\frac{1}{E_{trans}}=\frac{\varepsilon_{trans}}{\sigma_{trans}}=\frac{\xi\varepsilon_f+(1-\xi)\varepsilon_m}{\sigma_f}=\frac{\xi}{E_f}+\frac{(1-\xi)}{E_m} \tag{8.38}$$

When considering the Poisson's ratio effect, where $\nu_f$ and $\nu_m$ are the Poisson's ratio of the fiber and matrix, Eq. (8.38) can be rewritten as Eq. (8.39)

$$E_x=E_y=\left[\frac{\xi\left(1-\nu_f^2\right)}{E_f}+\frac{(1-\xi)\left(1-\nu_m^2\right)}{E_m}\right]^{-1} \tag{8.39}$$

Therefore, the effective Young's modulus in the direction perpendicular to the fiber length can be obtained as the reciprocal of the sum of the volume fractions divided by the respective Young's moduli of each material. This also indicates that the effective Young's modulus in the transverse direction depends only on the fiber volume fraction. This result is known as the ***Inverse Rule of Mixtures***.

The effective Young's moduli derived from the **Voigt Model** and **Reuss Model** can be extended to multi-phase composite materials. For a composite consisting of $n$ different materials, the effective Young's modulus in the directions parallel and perpendicular to the fiber length can be expressed as:

$$E_{axial}=\xi_1E_1+\xi_2E_2+\ldots+\xi_nE_n=\sum_i^n\xi_iE_i \tag{8.40}$$

$$\frac{1}{E_{transverse}}=\frac{\xi_1}{E_1}+\frac{\xi_2}{E_2}+\ldots+\frac{\xi_n}{E_n}=\sum_i^n\frac{\xi_i}{E_i} \tag{8.41}$$

The same derivation method can be extended to calculate the effective shear modulus and Poisson's ratio. Therefore, the formulas for the effective shear modulus and Poisson's ratio are:

$$\frac{1}{G_{ROM}}=\frac{\xi_1}{G_1}+\frac{\xi_2}{G_2}+\ldots+\frac{\xi_n}{G_n}=\sum_i^n\frac{\xi_i}{G_i} \tag{8.42}$$

$$\nu_{ROM}=\xi_1\nu_1+\xi_2\nu_2+\ldots+\xi_n\nu_n=\sum_i^n\xi_i\nu_i \tag{8.43}$$

$G_{ROM}$ and $\nu_{ROM}$ represent the effective shear modulus and Poisson's ratio calculated using the rule of mixtures. $G_i$ and $\nu_i$ denote each material's shear modulus and Poisson's ratio, where $i = 1, 2, ..., n$.

In the derivation of the effective coefficient of thermal expansion for a two-phase composite, several assumptions are made [32]:

1. When deriving the out-of-plane effective thermal expansion, the in-plane stress is assumed to be zero.
2. All materials are perfectly bonded to each other under all conditions.
3. The materials are considered linearly elastic, meaning the strain is assumed to be small.

Based on these assumptions, when the bi-material structure is subjected to a uniform temperature load, the matrix is constrained by the fiber, resulting in a compressive force $F_m$ in the Z-direction. Meanwhile, the fiber experiences a tensile force $F_f$, as illustrated in Fig. 8.31. At this point, the thermal deformation of both the matrix and the fiber can be expressed by Eq. (8.44).

$$\delta_i = \alpha_i (\Delta T) L_i \, (i = x,y,z) \tag{8.44}$$

$\delta_i$ represents the thermal deformation in a given direction, $\alpha_i$ is the coefficient of thermal expansion in that direction, $\Delta T$ is the temperature change, and $L_i$ is the original length. The total elongation of the structure in the Z-direction can be expressed as:

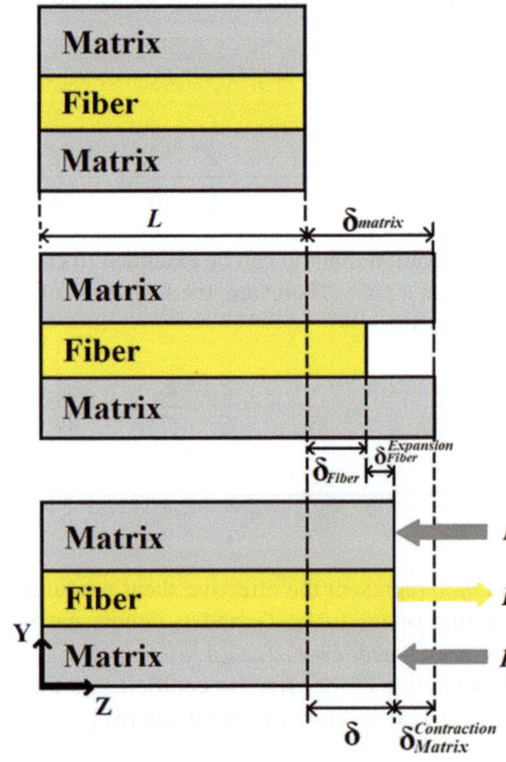


**Fig. 8.31** Axial thermal deformation of a Bi-material model under uniform temperature loading

$$\delta^F = \delta_m - \delta_m^{Contraction} = \delta_f + \delta_f^{Expansion} \tag{8.45}$$

According to the principles of mechanics of materials, the contraction of the matrix $\delta_m^{Contraction}$ and the expansion of the fiber $\delta_f^{Expansion}$ can be alternatively expressed regarding the compressive and tensile forces they experience. Therefore, Eq. (8.45) can be rewritten as:

$$\frac{F_m L}{E_m A_m} + \frac{F_f L}{E_f A_f} = \alpha_m (\Delta T) L - \alpha_f (\Delta T) L \tag{8.46}$$

$E_m/E_f$ and $A_m/A_f$ represent the Young's modulus and cross-sectional area of the matrix and fiber, respectively. $\alpha_m$, $\alpha_f$ are the thermal expansion coefficients of the matrix and fiber. The total cross-sectional area $A^T$ in the fiber direction is the sum of the individual cross-sectional regions of the two materials, that is:

$$A^T = A_m + A_f \tag{8.47}$$

If no external loading, the net stress in the structure should be zero, i.e., $F_m$=$F_f$. By rearranging Eq. (8.46) under this condition, we obtain:

$$F_m = \frac{\Delta T\left(\alpha_m - F_f\right) E_m A_m E_f A_f}{E_m A_m + E_f A_f} = F_f \tag{8.48}$$

Substituting (8.48) into (8.45), the final elongation can be expressed as:

$$\delta^F = \frac{\Delta T\left(\alpha_m E_m A_m + \alpha_f E_f A_f\right) L}{E_m A_m + E_f A_f} \tag{8.49}$$

The out-of-plane effective coefficient of thermal expansion for this bi-material composite can be expressed as:

$$\alpha_Z = \frac{\alpha_m E_m A_m + \alpha_f E_f A_f}{E_m A_m + E_f A_f} \tag{8.50}$$

For $A_m/A^T = \xi$, and $A_f/A^T = 1 - \xi$, Eq. (8.50) can be expressed as:

$$\alpha_Z = \frac{\alpha_m E_m \xi + \alpha_f E_f (1-\xi)}{E_m \xi + E_f (1-\xi)} \tag{8.51}$$

When considering Poisson's ratio effect [34, 36], where $v_f$ and $v_m$ are the Poisson's ratio of the fiber and matrix, the equivalent CTE along the x and y direction can be written as:

$$\alpha_x = \alpha_y = \xi\left(1+v_m\right)\alpha_m + \left(1-\xi\right)\left(1+v_f\right)\alpha_f - \alpha_z v_{xz} \tag{8.52}$$

$$\alpha_x = \alpha_y = \xi\left(1+v_m\right)\alpha_m + \left(1-\xi\right)\left(1+v_f\right)\alpha_f - \hat{\alpha}_a \bar{v} \tag{8.53}$$

$$\alpha_z = \hat{\alpha}_a = \frac{\alpha_m E_m \xi + \left(1-\xi\right)\alpha_f E_f}{\xi E_m + \left(1-\xi\right)E_f} \tag{8.54}$$

The subscripts *m* and *f* represent matrix and fiber, respectively. Therefore, the only difference between the two equations lies in the final terms, $\nu_{xz}$ and $\bar{v}$, where $\nu_{xz}$ can be expressed by Eq. (8.55), and $\bar{v}$ represents the average Poisson's ratio of the materials.

$$v_{xz} = v_{yz} = \xi v_m + \left(1-\xi\right)v_f \tag{8.55}$$

The effective coefficient of thermal expansion, excluding Poisson's ratio, for a multi-material composite can be derived similarly. The equivalent coefficient of thermal expansion for a multi-material composite can be derived using the same concept. If there are a total of n materials, the out-of-plane equivalent coefficient of thermal expansion can be expressed as:

$$\alpha_Z = \sum_i^n \alpha_i E_i \xi_i \,/\, \sum_i^n E_i \xi_i \tag{8.56}$$

#### 8.6.1.2 Comparison of ROM, IROM, and Finite Element Approach

In packaging simulations, the increasing complexity of packaging structure designs in recent years, along with the possibility of layouts containing intricate structural patterns, has made it both difficult and time-consuming to create finite element models. Even minor changes in the model's structure or geometry may require the entire finite element model to be rebuilt, further increasing the time needed for model construction. In previous studies [33–35], equivalent material formulations have been used to quickly calculate the equivalent material parameters of structures with complex layouts. Introducing these equivalent materials into simulation analyses can significantly reduce the computational and modeling time required for simulations.

The advantage of ROM and IROM is that they are convenient and fast to compute. However, their most significant limitation is that when the material proportions are the same but the structure layout differs, ROM and IROM cannot distinguish between them, so the result remains unchanged. This drawback does not affect the finite element model approach, which can account for such differences. The difference between ROM, IROM, and the finite element model approach will be discussed in this section.

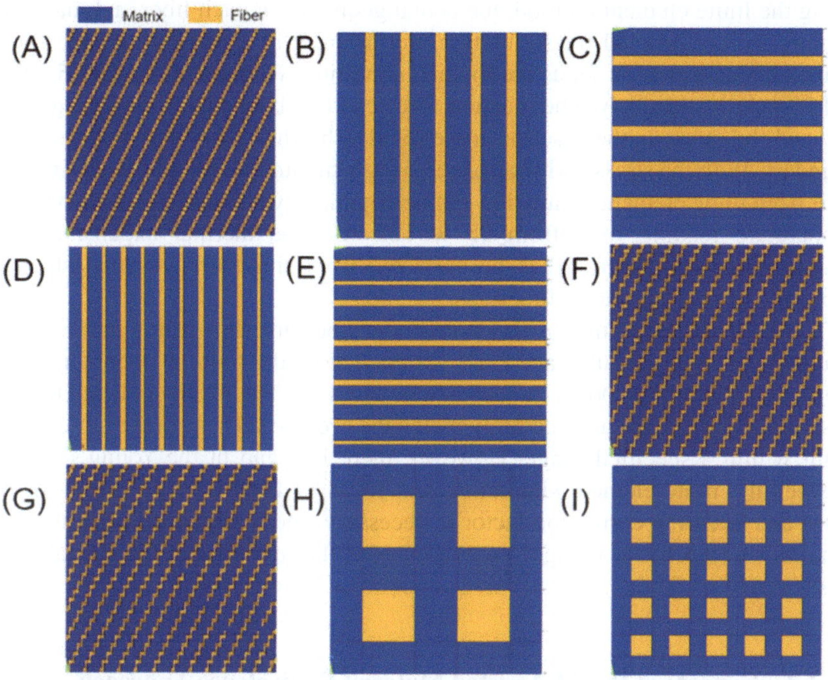


**Fig. 8.32** Nine Matrix-Fiber patterns, (**a**) Continuous Inclined Pattern; (**b**) Uniform Vertical Pattern; (**c**) Uniform Horizontal Pattern; (**d**) Non-Uniform Vertical Pattern; (**e**) Non-Uniform Horizontal Pattern; (**f**) Zigzag Inclined Pattern; (**g**) Discontinuous Inclined Pattern; (**h**) 2×2 Square Array; (**i**) 5×5 Square Array

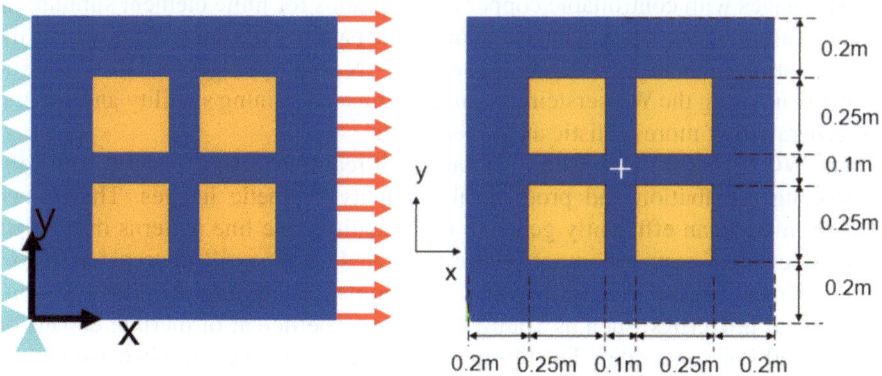


**Fig. 8.33** The Matrix-Fiber loading condition

This section presents nine different layout test models, as shown in Fig. 8.32. Although the fiber shapes and distributions vary among the models, the matrix-to-fiber ratio in all models is fixed close to 4:1. This eliminates the influence of material proportion, allowing the analysis to focus solely on the effects caused by fiber patterns. The boundary conditions and loads for all test models are also identical. The prescribed displacement with the boundary conditions set as shown in Fig. 8.33.

Using the finite element method, the actual geometry of each fiber and matrix pattern will be constructed for simulation. For Young's modulus estimation, the prescribed displacement is set as the loading condition, with magnitudes of 5, 10, 15, and 20 mm, respectively. The reaction force can be obtained at the loading boundary, and the stress is the reaction force divided by the cross-sectional area of the boundary. With the stress and strain, the Young's modulus is therefore obtained. The CTE from finite element simulation can be obtained by a similar method, where the temperature loading rises from 25 to 125 °C. The thermal expansion in the x-direction and y-direction can be extracted from simulation results, which can define the CTE along those directions.

The material properties of the matrix and fiber are shown in Table 8.13. The results of Young's modulus and CTE obtained from the finite element simulation, ROM, and IROM are shown in Tables 8.14 and 8.15. The results indicate that both Eqs. (8.52) and (8.53) can accurately match the average CTE from the finite element result within a specified range. However, there is a gap in the Young's modulus between the finite element simulation and ROM or IROM. In this case, for both ROM and IROM, a correction factor is necessary, and it must also account for the material's geometry. It is highly recommended that users build their database for these correction factors.

#### 8.6.1.3 Concept of the AI-Assisted Material Equivalence Approach

Figure 8.34 presents the workflow of the AI-assisted material equivalence approach [37]. This approach employs a data augmentation strategy based on Wasserstein Generative Adversarial Networks (WGANs) to synthetically generate diverse PCB layout images with controllable copper content ratios for finite element simulation, which is used to generate a database of trace line patterns. Figure 8.35 illustrates the architectural details of the WGAN model. WGAN employs a generator and a critic network, utilizing the Wasserstein distance to enhance training stability and support the generation of more realistic and diverse data.

The WGAN model is trained on a dataset of real trace line patterns to learn the underlying distribution and produce high-fidelity synthetic images. The trained WGAN model can efficiently generate high-quality trace line patterns that reflect varying levels of copper content. These generated images will be transferred to a finite element model for subsequent FE simulations to extract corresponding mechanical properties, such as Young's modulus, coefficient of thermal expansion (CTE), and Poisson's ratio. The ultimate goal is to create a comprehensive dataset of image-material property pairs, facilitating the rapid prediction of equivalent

**Table 8.13** Material properties of matrix and fiber

| Material | Young's modulus (GPa) | Poisson ratio | CTE (ppm/°C) | Ratio |
|---|---|---|---|---|
| Matrix | 3.033 | 0.382 | 60 | 0.75 |
| Fiber | 67.568 | 0.2 | 4.9 | 0.25 |

**Table 8.14** Comparison of CTE estimation results for finite elements, ROM, and IROM

| | A | B | C | D | E | F | G | H | I |
|---|---|---|---|---|---|---|---|---|---|
| x-direction CTE | 76.29 | 88.51 | 25.79 | 92.35 | 20.01 | 74.29 | 74.01 | 60.42 | 60.41 |
| y-direction CTE | 34.03 | 25.79 | 88.51 | 20.01 | 92.32 | 37.69 | 38.43 | 60.42 | 60.41 |
| Average CTE (FEM) | 55.16 | 57.15 | 57.15 | 56.18 | 56.17 | 55.99 | 56.22 | 60.42 | 60.41 |
| (8.52) | 59.81 | | | | | | | | |
| (8.53) | 60.33 | | | | | | | | |

**Table 8.15** Comparison of Young's modulus estimation results for finite elements, ROM, and IROM

| | A | B | C | D | E | F | G | H | I |
|---|---|---|---|---|---|---|---|---|---|
| x-direction Young's Modulus | 4.41 | 4.43 | 19.17 | 4.49 | 19.17 | 4.63 | 4.61 | 4.84 | 4.95 |
| y-direction Young's Modulus | 7.36 | 19.17 | 4.43 | 19.17 | 4.49 | 7.35 | 7.24 | 4.84 | 4.95 |
| Average Young's Modulus (FEM) | 5.89 | 11.80 | 11.80 | 11.83 | 11.83 | 5.99 | 5.93 | 4.84 | 4.95 |
| ROM | 19.17 | | | | | | | | |
| IROM (Eq. 8.39) | 3.98 | | | | | | | | |

material properties from trace line layouts using machine learning techniques, e.g., ANN.

## 8.6.2 *Local-Global Simulation Technology*

Specific Boundary Constraints (SBC) and Multipoint Constraints (MPC) [13, 38–40] are both techniques used in finite element analysis (FEA) to manage complex models, particularly in the context of "global-local" methods, which aim to strike a balance between computational efficiency and accuracy. While they aim to simplify models, their approaches and mechanisms differ significantly.

### 8.6.2.1 Specific Boundary Constraints (SBC)

SBC, also known as the sub-modeling method, focuses on obtaining highly accurate results in specific regions of interest within a larger structure while simultaneously saving computation time. This method requires the creation of two distinct models. The global model, specifically with a coarser mesh, is utilized for the entire structure to obtain an overall, less detailed solution, which helps reduce initial

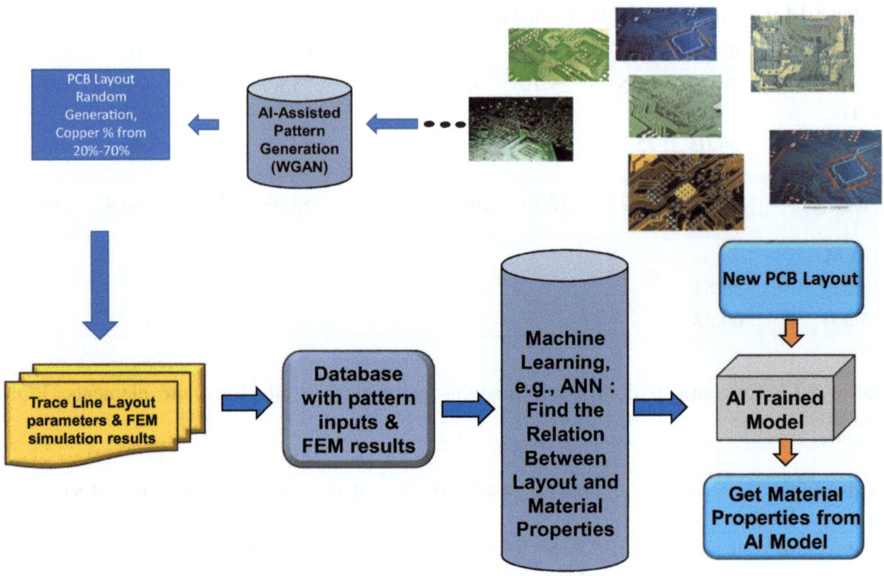


**Fig. 8.34** Workflow of AI-assisted material equivalence

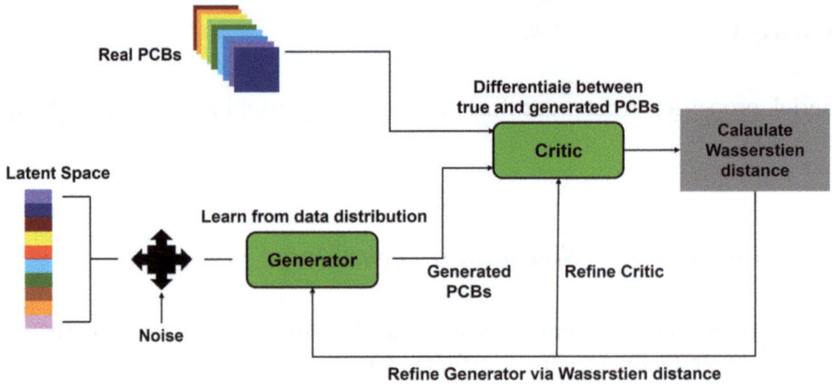


**Fig. 8.35** Architecture of WGANs

computation time. A finer, more detailed mesh is explicitly created for the region of particular interest (e.g., areas of high stress concentration). The crucial link between these two models is the "cut boundary." The results (such as displacements or forces) obtained from the coarser global model are applied as boundary conditions to the finer submodel. This transfer is typically achieved by interpolating the global model's solution onto the submodel's cut boundary. SBC can significantly reduce the computational time and storage space required compared to a single, uniformly fine-meshed model. The accuracy of SBC is highly dependent on the location of the cut boundary. If this boundary is too close to areas where stress changes dramatically (stress concentrations), interpolation errors can arise, potentially affecting the precision of the submodel's results. However, SBC is unsuitable for nonlinear analysis; therefore, this method is inadequate for electronic packaging simulation.

#### 8.6.2.2 Multipoint Constraints (MPC)

In contrast, MPC directly establishes coupling relationships between nodal degrees of freedom (DOF) within a single, integrated finite element model. It allows for varying mesh densities within different regions of that single model. Unlike SBC, MPC utilizes a single finite element model that incorporates regions of varying mesh densities. For instance, critical areas can have a fine mesh, while less critical areas can have a coarse mesh. The connections between nodes (even those with different DOFs or belonging to different mesh densities) are defined by linear constraint equations. These equations mathematically link the degrees of freedom of various nodes, creating a "coupling" between them. So, MPC allows the linking of elements with different numbers of DOFs (e.g., solid elements with 3 DOFs to shell elements with 6 DOFs), which would otherwise cause computational issues. MPC is also capable of connecting different mesh densities. This primary use enables fine and coarse meshes to coexist in different parts of a single model, thereby improving efficiency. MPC significantly improves computational efficiency by drastically reducing the total number of elements and nodes. Studies show that it can reduce elements by approximately 70% and computation time to as little as 13–35% of the original, making it especially beneficial for complex packaging designs, such as Chiplets. When properly implemented, it can maintain high accuracy. However, one important point to note is that the area of applying MPC should be minimized as much as possible, because MPC itself also requires significant computation time. Applying too many MPCs can diminish the intended effect of reducing overall computation time.

Similar to SBC, the accuracy of MPC is sensitive to the placement of the transition boundary between coarse and fine meshes. Placing too close to areas with sharp stress changes can lead to discontinuous stress distributions and computational errors. MPC is suitable for nonlinear analysis; therefore, it is the primary simulation method used for complex electronic packaging structures, such as 3D, SiP, and Chiplet.

The Multipoint Constraints method's core concept lies in establishing coupling relationships between element nodes to ensure continuous transfer of forces or displacements when dealing with elements that have incompatible degrees of freedom or mismatched boundaries. This method aims to overcome local discontinuities present in the model, thereby improving the accuracy of simulation calculations. Specifically, the MPC method can define linear relationships between the degrees of freedom of one or multiple nodes, ensuring that these nodes remain consistent in motion and deformation. In general mathematical form, it can be expressed as Eq. (8.57):

$$\sum_{i=1}^{N}\left[Coefficient\left(i\right)\cdot D.O.F\left(i\right)\right]=Constant \tag{8.57}$$

Here, $N$ is the number of terms in the equation, *DOF* ($i$) denotes the degree of freedom of the $i$-th node, and *Constant* is a constant whose value is usually zero. Based on the above formula, a multipoint constraint equation can be derived, which contains an arbitrary set of nodal degrees of freedom and is used to describe the linear relationships between these degrees of freedom. The MPC method can be applied to many problems related to element connectivity, with three common applications: connecting meshes of different densities, connecting different element types, and creating rigid regions. These will be explained in detail below.

#### 8.6.2.3 Different Mesh Connection

When performing finite element model analysis, elements are often meshed with varying densities to effectively shorten the computation time required for simulation. For example, fine meshes are used for critical regions where the stress gradient changes significantly, while coarser meshes are applied to the remaining areas. However, the simulation may encounter modeling problems if the mesh sizes are mismatched. To address this issue, the Multipoint Constraints method can be introduced to establish constraint relationships between different nodes, ensuring the continuity and stability of the model.

As shown in the example of Fig. 8.36, Elements I, II, and III are all four-node plane elements. The schematic reveals that node 2 is not directly associated with Element III. To establish a relationship between them, the concept of the finite element shape function can be applied, using the shape function of Element III to derive the constraint equation between the nodes.

Introduce the following constraint equation at the interface, and we can have:

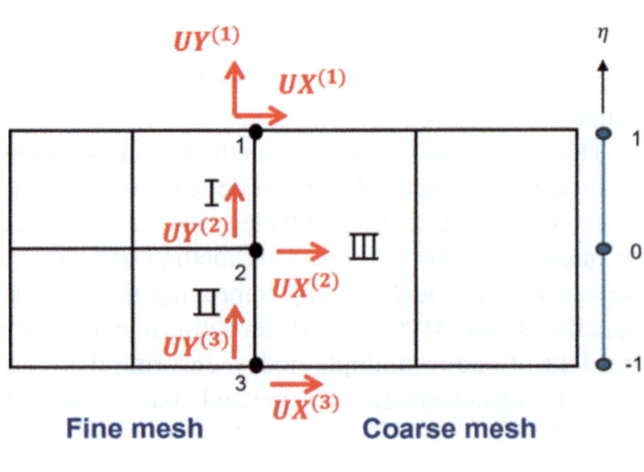


**Fig. 8.36** Schematic diagram of connecting meshes with different densities

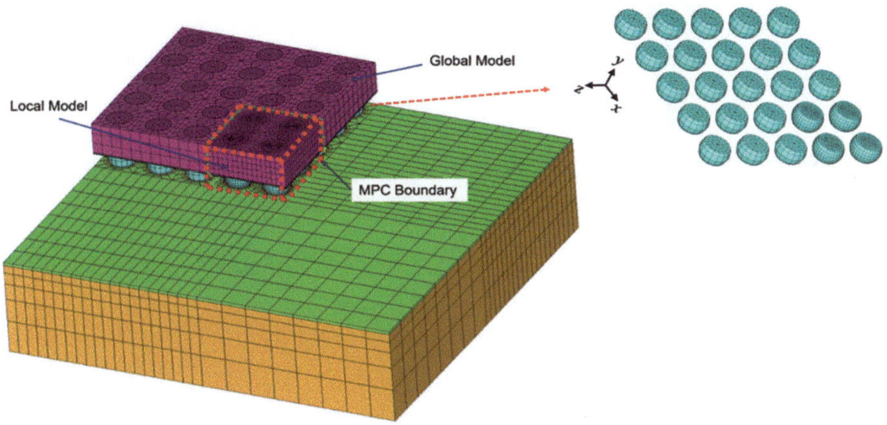


**Fig. 8.37** 2 × 2 solder ball MPC example of a 3D WLP model

$$UX = \frac{1}{2}(1+\eta)\cdot UX^{(1)} + \frac{1}{2}(1-\eta)\cdot UX^{(3)} \tag{8.58}$$

$$UY = \frac{1}{2}(1+\eta)\cdot UY^{(1)} + \frac{1}{2}(1-\eta)\cdot UY^{(3)} \tag{8.59}$$

Let $\eta = 0$ at node 2,

$$UX^{(2)} = \frac{1}{2}UX^{(1)} + \frac{1}{2}UX^{(3)} \tag{8.60}$$

$$UY^{(2)} = \frac{1}{2}UY^{(1)} + \frac{1}{2}UY^{(3)} \tag{8.61}$$

Finally, the linear relationship between the degrees of freedom of the nodes can be obtained and expressed in terms of the field function $\phi$ as follows:

$$\phi^{(2)} = \frac{1}{2}\phi^{(1)} + \frac{1}{2}\phi^{(3)} \tag{8.62}$$

The field function $\phi$ describes the displacement components $u$ and $v$ in the context of elasticity.

There are generally two methods for introducing constraint equations in finite element analysis. The first is the penalty function method, in which, after solving the system equations, the constraint is applied to obtain an approximate solution. The second method is to directly introduce the constraint equation after forming the stiffness matrices and load vectors of Elements I and II, Fig. 8.36, eliminating the degree of freedom $\phi^{(2)}$ in Eq. (8.62) so that it is not included in the system matrix for

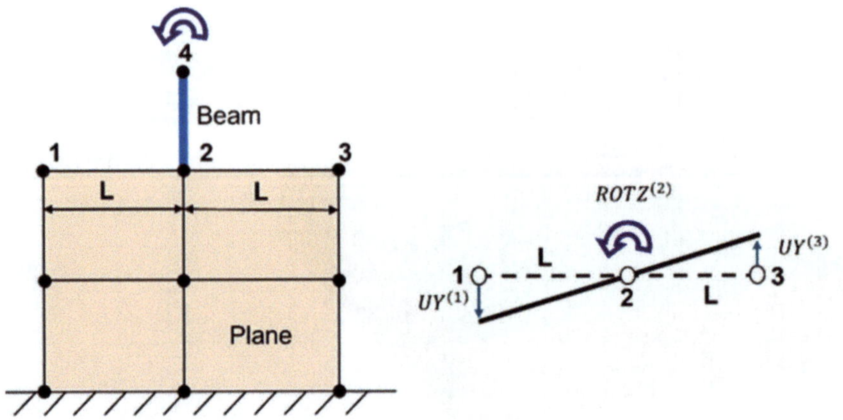


**Fig. 8.38** Schematic diagram of connecting different element types

solution. This approach provides a more precise computation. Furthermore, even if the elements used in the simulation are not the four-node elements mentioned above, but elements with more nodes, such as eight-node quadrilateral elements (Fig. 8.37), the same principle can still be applied to obtain the constraint equations between nodes. This treatment not only preserves computational accuracy but also improves the overall reliability of the model.

#### 8.6.2.4 Different Element Type Connection

In a finite element model, when different types of elements are connected in the same region, the mismatch in nodal degrees of freedom between these elements may hinder the smooth transfer of loads. In such cases, the Multipoint Constraints method can be used to establish linear relationships between the degrees of freedom of different elements, enabling effective load transfer between them. Figure 8.38 illustrates an example of applying the MPC method to connect different types of elements, where the beam element has rotational degrees of freedom, whereas the plane element does not. A constraint equation must be introduced to allow the moment to be transferred at node 2.

The moment transfer at node 2 can be expressed using Eqs. (8.63) and (8.64):

$$-UY^{(1)} = \sin\left(ROTZ^{(2)}\right)L \tag{8.63}$$

$$+UY^{(3)} = \sin\left(ROTZ^{(2)}\right)L \tag{8.64}$$

Summation of Eqs. (8.63) and (8.64), we can obtain Eq. (8.65):

$$0 = UY^{(3)} - UY^{(1)} - 2L \cdot \sin\left(ROTZ^{(2)}\right) \tag{8.65}$$

Considering small rotations, Eq. (8.65) can be simplified to Eq. (8.66), which represents the linear constraint equation between the beam element and the plate element, as follows:

$$0 = UY^{(3)} - UY^{(1)} - 2L \cdot ROTZ^{(2)} \tag{8.66}$$

#### 8.6.2.5 The WLP Application

To demonstrate the feasibility of the global–local finite element model for three-dimensional wafer-level packaging structures established in this section, using the Multipoint Constraints method, a model of a test vehicle with the specified geometry is constructed, as detailed in Table 8.16. The linear, nonlinear, and SAC305 Chaboche material properties are shown in Tables 8.17 and 8.18. The simulation results of this test vehicle are then compared with the experimental value. When the differences fall within an acceptable range, the feasibility of this method can be verified.

A quarter-symmetric WLP fine mesh finite element model, shown on the left side of Fig. 8.39, and a 1 × 1 MPC model, on the right side of Fig. 8.39, are presented. An MPC is applied to the outermost solder joint, which is located at the largest distance from the neutral point. The results show that the number of modeling elements can be reduced from 145,275 to 35,404, significantly reducing CPU time from 103 to 25 min. With suitable critical mesh control, both the full model and the MPC model (with different mesh size connections) can achieve a good agreement with experimental results in predicting solder joint reliability. The principle of MPC modeling is to position the MPC boundary at a certain distance from the region of

**Table 8.16** Geometry sizes of WLP

| Unit (mm) | TV2 |
|---|---|
| Chip | 4.0 × 0.33 |
| Lower Pad | 0.22 × 0.025 |
| RDL | 0.25 × 0.004 |
| SBL1 | 0.0075 |
| SBL2 | 0.004 |
| Low-k | 0.005 |
| Upper Pad | 0.24 × 0.0086 |
| PCB | 8 × 1 |
| Ball diameter | 0.25 |
| Pitch | 0.4 |
| Ball number | 100 |
| Ball height | 0.16 |
| MTTF | **1013** |

**Table 8.17** Linear and nonlinear material properties of WLP

| **Linear material** | **Young's modulus (Gpa)** | **Poisson's ratio** | **CTE (ppm/°C)** |
|---|---|---|---|
| Chip | 150 | 0.28 | 2.62 |
| Low-k | 10 | 0.16 | 5 |
| SBL | 2 | 0.33 | 55 |
| Cu | 68.9 | 0.34 | 16.7 |
| Solder mask | 6.87 | 0.35 | 19 |
| PCB | 18.2 | 0.19 | 16 |
| **Nonlinear material** | **Young's modulus (GPa)** | **Poisson's ratio** | **CTE (ppm/°C)** |
| SAC 305 | 38.7–0.176 T | 0.35 | 25 |

**Table 8.18** The Chaboche kinematic hardening model parameters

| T(K) | $\sigma_0$ | $C$ | $\gamma$ |
|---|---|---|---|
| 233 | 47.64 | 8894.8 | 639.2 |
| 253 | 38.87 | 8573.3 | 660.0 |
| 313 | 24.06 | 6011.4 | 625.2 |
| 353 | 18.12 | 5804.2 | 697.7 |
| 398 | 14.31 | 4804.6 | 699.9 |

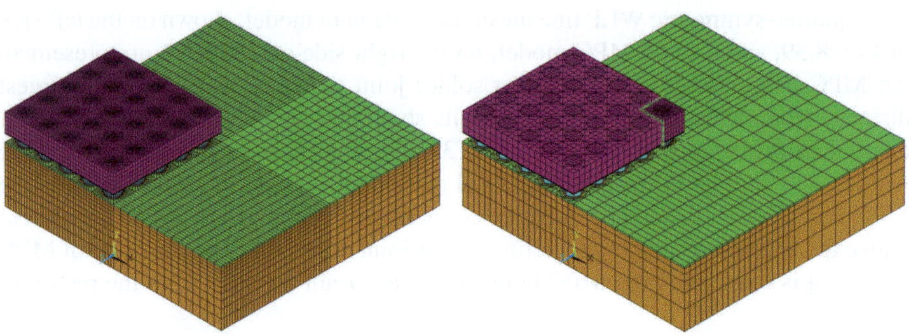

**Fig. 8.39** The 1/ 4 fine mesh model and The 1 x 1 Solder Ball MPC model

interest. For example, Fig. 8.37 (2 × 2 solder ball MPC) illustrates a better MPC setup than Figure 39 (1 × 1 solder ball MPC). Table 8.19 describes the total element number, CPU time, and reliability life cycles for fine mesh, 1 × 1, and 2 × 2 solder ball MPC models, along with the experimental reliability results. In this WLP case, both MPC configurations achieved good agreement with experimental results when an appropriate mesh size control was applied. It demonstrates the importance of mesh size control in the critical region.

**Table 8.19** Comparison of fine mesh model and MPC models

| | Total element number | Simulation CPU time for 8 cycles | Reliability life cycles (simulation) | MTTF (experiment) |
|---|---|---|---|---|
| 1x1 MPC model | 35,404 | 25 min | 1092 | 1013 |
| 2x2 MPC Model | 48,524 | 35 min | 1065 | |
| Full Fine Mesh Model | 145,275 | 103 min | 1010 | |

## 8.7 Summary and Recommendations

The electronic packaging industry has transitioned from Design-on-Experiment, which was time-consuming, costly, and lacked sufficient physical information for optimization, to Design-on-Simulation technology. DoS is a strategic approach that can address the uncertainty issue in the traditional approach of finite element simulations. It primarily uses the validated finite element method to secure simulation results, reduce physical experiments, shorten design cycles, and lower costs in product development. The challenges in the traditional FEM approach are that FEM results were highly dependent on individual researcher expertise, leading to inconsistencies, and there was a lack of clear guidelines, especially regarding mesh size, which significantly impacts accuracy and computational time. DoS was developed to address these challenges by providing a modeling verification procedure that ensures consistent and trustworthy predictions, further reducing design cycles. This includes iteratively tuning and verifying theories, material properties, empirical equations, mesh sizes, and solution procedures against literature or experimental results.

Both 2D and 3D FEM can be used for electronic packaging simulation. 2D models are computationally less expensive and suitable for symmetric structures and extensive parametric analysis; however, they simplify and package structures, as well as the cylindrical geometries of the solder joint. 3D models offer more accurate representations but are computationally intensive, requiring techniques like Multipoint Constraints to reduce element numbers and computation time. The finite volume-weighted method aims to resolve mesh size-dependent strain problems in FEM analysis, particularly in singular regions such as interfaces between solder joints and under-bump metallurgy. It involves averaging strain over a targeted finite zone to capture maximum strain and obtain converging solutions. For 2D and 3D simulations, using the average strain from a single critical corner element is recommended, with its size carefully controlled and verified through experiments. To address modern electronic packages' complexity and long computation times (e.g., heterogeneous integration, SiP, 3D, and chiplets), equivalent methods and local-global simulation technologies are crucial. The Rule of Mixture (ROM) and Inverse Rule of Mixture (IROM) are fundamental theories for calculating equivalent material properties based on volume/area fractions. However, they cannot distinguish between different structural layouts with the same material proportions and may require correction factors. More importantly, simulation processes must be validated against physical tests before applying the full-scale simulation to the packaging structure.

## References


1. K. C. Wu, S. Y. Lin, T. Y. Hung and K. N. Chiang, "Reliability Assessment of Packaging Solder Joints Under Different Thermal Cycle Loading Rates", IEEE Transactions on Device and Materials Reliability, Volume 15, Issue 3, Pages 437–442, September 2015.
2. C. H. Lee, K. C. Wu and K. N. Chiang, "A Novel Acceleration-Factor Equation for Packaging-Solder Joint Reliability Assessment at Different Thermal Cyclic Loading Rates", Journal of Mechanics, Volume 31, Issue 1, Pages 35–40, February 2017.
3. V. Ramachandran and K. N. Chiang, "Feasibility Evaluation of Creep Model for Failure Assessment of Solder Joint Reliability of Wafer Level Packaging", IEEE Transactions on Device and Materials Reliability, Volume PP, Issue 99, Pages 1–1 , September 2017.
4. P. H. Wang, Y. C. Lee, C. K. Lee, H. H. Chang, and K. N. Chiang, "Solder Joint Reliability Assessment and Pad Size Studies of FO-WLP with Glass Substrate", IEEE Transactions on Device and Materials Reliability, Vol. 21, No. 1, pp. 96–101, March 2021.
5. C. H. Tsai, S. W. Liu and K. N. Chiang, "Warpage analysis of fan-out panel-level packaging using equivalent CTE", DOI 10.1109/TDMR.2019.2956049, IEEE Transactions on Device and Materials Reliability, Volume 20, Issue 1, pp. 51–57, March 2020.
6. P. L. Wu, P. H. Wang and K. N. Chiang, "Empirical Solutions and Reliability Assessment of Thermal Induced Creep Failure for Wafer Level Packaging", IEEE Transactions on Device and Material Reliability, Volume 19, Issue 1, Pages 126–132, March 2019.
7. L. L. Liao, T. Y. Huang, C. K. Liu, W. Li, M. J. Dai, and K. N. Chiang, "Electro-thermal finite element analysis and verification of power module with aluminum wire", Microelectronic Engineering, Volume 120, Pages 114–120, May 2014.
8. J. Chang, L. Wang, J. Dirk, and X. Xie, "Finite Element Modeling Predicts the Effects of Voids on Thermal Shock Reliability and Thermal Resistance of Power Device", Welding Journal, Vol. 85, pp. 63–70, 2006.
9. R. D. Cook, D. S. Malkus, M. E. Plesha, and R. J. Witt, Concepts and Applications of Finite Element Analysis: Wiley; 4th edition 2001.
10. F. Feustel, S. Wiese, and E. Meusel, "Time-Dependent Material Modeling for Finite Element Analyses of Flip Chips", 50th Electronic Components and Technology Conference, Las Vegas, NV, USA, May 21–24, 2000.
11. J. H. Zhu et al., "A multipoint constraints based integrated layout and topology optimization design of multi-component systems," Structural and Multidisciplinary Optimization, vol. 51, pp. 397–407, 2015.
12. M. Y. Chan, "An MPC-Based Method with Mesh Compensation and Equivalent Beam Approach for WLP Reliability Prediction", MS Thesis, National Tsing Hua University, Taiwan, 2025.
13. Y. W. Huang, "Study on Design and Reliability Assessment of Heterogeneous Chiplets Packaging Using Design-on-Simulation Technology", Ph.D. Dissertation, National Tsing Hua University, Taiwan, 2025.
14. M. J. Pfeifer, "Solder Bump Size and Shape Modeling and Experimental Validation," IEEE Transactions on Components, Packaging and Manufacturing Technology—Part B, Vol. 20, No.4, pp. 452–457,1997.

15. S. M. Heinrich, M. Schaefer, S. A. Schroeder and P. S. Lee, "Prediction of Solder Joint Geometries in Array-Type Interconnects," ASME Journal of Electronic Packaging, Vol. 118, pp. 114–121,1996
16. K. N. Chiang and W. L. Chen, "Electronic Packaging Reflow Shape Prediction for the Solder Mask Defined Ball Grid Array," Transactions on Journal of Electronic Packaging, Vol. 120, No. 2, pp,175–178, 1998.
17. K. A. Brakke, The Surface Evolver, Experimental Mathematics, Vol.1, No.2, pp. 141–165, 1992.
18. K. A. Brakke, Surface Evolver Manual, version 2.01 Minneapolis, MN: The Geometry Center, 1996.
19. K. A. Brakke, "The Surface Evolver and the Stability of Liquid Surfaces" Philosophical Transactions: Mathematical Physical and Engineering Sciences, Vol. 354, pp. 2143–2157, 1996.
20. H. C. Cheng, K. N. Chiang, and C. K. Chen, "Parametric Analysis of Thermally Enhanced BGA Reliability Using a Finite-Volume-Weighted Averaging Technique", Conference International Mechanical Engineering Congress and Exposition, The Winter Annual Meeting of ASME, 1998.
21. Y. T. Lin, "Design and Reliability Analysis of Wafer Level / Flip Chip Packages", MS Thesis, National Tsing Hua University, Taiwan, 2000.
22. C. A. Yuan, Q. H. Su, K. N. Chiang, "Coefficient Extraction of SAC305 Solder Constitutive Equations Using Equation-Informed Neural Networks", Materials, 16, 4922, July 2023.
23. Y. T. Su, "Research on Reliability Prediction of Wafer-Level Packaging by Combining Uncertainty Analysis and Parallel Accelerated Neural Networks", MS Thesis, National Tsing Hua University, Taiwan, 2025.
24. S. Watanabe, Mathematical Theory of Bayesian Statistics; Chapman & Hall: 2020.
25. L. Anand, Constitutive equations for hot-working of metals. International Journal of Plasticity 1985, 1, 213–231.
26. M. Motalab, Z. Cai, J. C. Suhling, P. Lall, Determination of Anand constants for SAC solders using stress-strain or creep data. In Proceedings of the 13th InterSociety Conference on Thermal and Thermomechanical Phenomena in Electronic Systems (ITHERM), San Diego, CA, USA, 30 May 2012–01 June 2012, 2012.
27. M. Motalab, M. Mustafa, J. C. Suhling, J. Zhang, J. Evans, M. J. Bozack, P. Lall, Thermal Cycling Reliability Predictions for PBGA Assemblies that Include Aging Effects. In Proceedings of the ASME 2013 International Technical Conference and Exhibition on Packaging and Integration of Electronic and Photonic Microsystems (InterPACK2013), Burlingame, CA, USA, July 16–18, 2013, 2013.
28. J. L. Chaboche, On some modifications of kinematic hardening to improve the description of ratchetting effects. International Journal of Plasticity 1991, 7, 661–678.
29. Q. Xiao, W. D. Armstrong, Tensile creep and microstructural characterization of bulk Sn3. 9Ag0. 6Cu lead-free solder. Journal of Electronic Materials 2005, 34, 196–211.
30. Y. W. Huang, K. N. Chiang, Study of shear locking effect on 3D solder joint reliability analysis. Journal of Mechanics 2022, 38, 176–184.
31. W. Voigt, "Ueber die Beziehung zwischen den beiden Elasticitätsconstanten isotroper Körper," Annalen der physik, vol. 274, no. 12, pp. 573–587, 1889.
32. H.-C. Cheng, R.-S. Li, S.-C. Lin, W.-H. Chen, and K.-N. Chiang, "Macroscopic mechanical constitutive characterization of through-silicon-via-based 3-D integration," IEEE Transactions on Components, Packaging and Manufacturing Technology, vol. 6, no. 3, pp. 432–446, 2016.
33. H.-C. Cheng, L.-C. Tai, and Y.-C. Liu, "Theoretical and experimental investigation of warpage evolution of flip chip package on packaging during fabrication," Materials, vol. 14, no. 17, p. 4816, 2021.
34. G. Ye, X. Fan, and G. Zhang, "Practical aspects of thermomechanical modeling in electronics packaging: A case study with a SiC power package," Microelectronics Reliability, vol. 132, 114514, 2022.

35. H. C. Cheng, K. N. Chiang, and M. H. Lee, "An effective approach for three-dimensional finite element analysis of ball grid array typed packages,", Journal of Electronic Packaging, vol. 120, no. 2, pp. 129–134, Jun 1998.
36. R. A. Schapery, "Thermal expansion coefficients of composite materials based on energy principles," *Journal of Composite Materials,* vol. 2, no. 3, pp. 380–404, 1968.
37. Q. H. Su, Y. T. Su, and K. N. Chiang, "Using WGAN-Based Data Augmentation Machine Learning Algorithm to Estimate the Equivalent Material Properties of Electronic Packaging", EPTC 2025, Singapore, Dec. 2–5, 2025.
38. J. Y. Wang, "Applicability Investigation of Equivalent Theory and Global-Local Finite Element Method in Stress Analysis of Full Model Package", MS Thesis, National Tsing Hua University, Taiwan, 2023.
39. M. -Y. Chan and K. -N. Chiang, "Study on the Applicability of Multi-Point Constraints for 3D WLP Finite Element Model," EuroSimE 2025, Utrecht, Netherlands, April. 6–9, 2025.
40. Wang, P.-H.; Huang, Y.-W.; Chiang, K.-N. Reliability Evaluation of Fanout Type 3D Packaging-On-Packaging. Micromachines 2021, 12, 295.

# Chapter 9
# Integrating AI with Design-on-Simulation for Advanced Packaging

**Abstract** The AI-DoS methodology begins by validating finite element simulations against mechanics theories and experimental results, ensuring their accuracy. This validated simulation procedure is then used to efficiently generate the large labeled datasets required for machine learning (ML) training, linking packaging geometries (inputs) to reliability life cycles (outputs). Strategies such as adaptive data sampling and ensemble learning are employed to enhance predictive robustness and accuracy, particularly when data volumes are constrained.

This Chapter evaluates several supervised ML algorithms, including Artificial Neural Networks (ANN), Random Forest (RF), Kernel Ridge Regression (KRR), and Gaussian Process Regression (GPR), for Wafer-Level Packaging (WLP) reliability prediction. While fast algorithms like K-Nearest Neighbor (KNN) and Polynomial Regression (PR) are efficient, models such as ANN, GPR, and KRR demonstrate superior accuracy, consistently achieving very low average testing errors (often below 1%). Ensemble learning further improves stability and suppresses extreme deviations. Once the AI model is established and trained, it can predict reliability life within milliseconds, enabling rapid structural optimization and transforming the design and development cycles.

## 9.1 Introduction

The electronic packaging (EP) community has long relied on finite element simulation technology to develop new packaging structures. However, this traditional approach has faced significant challenges, particularly in ensuring the trustworthiness and consistency of simulation results, which can be highly dependent on the individual researcher's domain knowledge, understanding of mechanics theory, and finite element method (FEM) background. This lengthy design cycle hinders the ability to meet critical time-to-market demands for increasingly complex packaging, such as system-in-package (SiP), Chiplet, and heterogeneous packaging.



J. Lau, K.-N. Chiang, *Cu-Interconnects, Glass, and AI-Assisted Simulation for Chiplets and Heterogeneous Integration*,
https://doi.org/10.1007/978-981-95-6891-8_9

To overcome these limitations and achieve more consistent, trustworthy predictions, as well as further reduce design cycles, Artificial Intelligence (AI) and Machine Learning (ML) algorithms have been introduced, leading to AI-assisted Design-on-Simulation (AI-DoS) technology. AI is a broader field encompassing machine learning, which focuses on enabling systems to generalize and solve previously unknown problems by learning from prior data. In the context of electronic packaging, this involves using an AI approach combined with large datasets, often generated by validated FEM simulations, to train models that establish the relationship between packaging geometry (input) and reliability life cycle (output). Once trained, these AI models can predict reliability life within seconds by simply inputting geometric data, making structural optimization feasible.

The primary advantages of AI-Assisted Design-on-Simulation include:

1. Ensuring trustworthy and consistent prediction results for packaging reliability and mechanical behavior by ensuring that estimation results are within an accurate range when compared with experimental data.
2. Significantly shortening design cycles and reducing time-to-market, this enables reliability prediction within milliseconds to seconds once the AI model is established, providing a stark contrast to the lengthy experimental methods. This addresses a critical issue in the development of new electronic packaging.
3. Reducing costs by minimizing the need for extensive physical experiments.
4. Packaging structures, such as wafer-level packaging, System-in-Package, Chiplet, heterogeneous packaging, and panel-level fanout packaging, are utilized to achieve optimal component combinations, thereby improving reliability, reducing interface delamination, and minimizing warpage.
5. Lowering the training threshold for simulation expertise, making advanced design analysis more accessible, and unifying results across different developers.

The methodology typically involves four key steps: (1) Validating finite element models and approaches against mechanics theories and experimental results to ensure simulation accuracy. (2) Require generating large labeled datasets if the adaptive data sampling technique is not applied. Obtaining data from validated simulations is far more efficient than experimental data collection, but still requires extensive CPU time. (3) Training and testing various machine learning algorithms (such as Artificial Neural Network (ANN), Support Vector Regression (SVR), Kernel Ridge Regression (KRR), K-Nearest Neighbor (KNN), Recurrent Neural Network (RNN), Random Forest (RF), Gaussian Process Regression (GPR), Polynomial Regression (PR), and Convolutional Neural Network (CNN)) on this data to establish a robust regression model, and (4) Deploying the trained AI model for design purposes, allowing designers to quickly obtain reliability predictions by simply inputting new packaging geometries. This transformative technology offers a feasible and robust solution to current electronic packaging design challenges. Figure 9.1 illustrates the key steps and flow of AI-Assisted DoS.

A significant challenge for AI-assisted DoS is the lack of sufficient labeled data, as machine learning generally requires large datasets for robust training. This has driven research into small-data machine learning methodologies. Key strategies

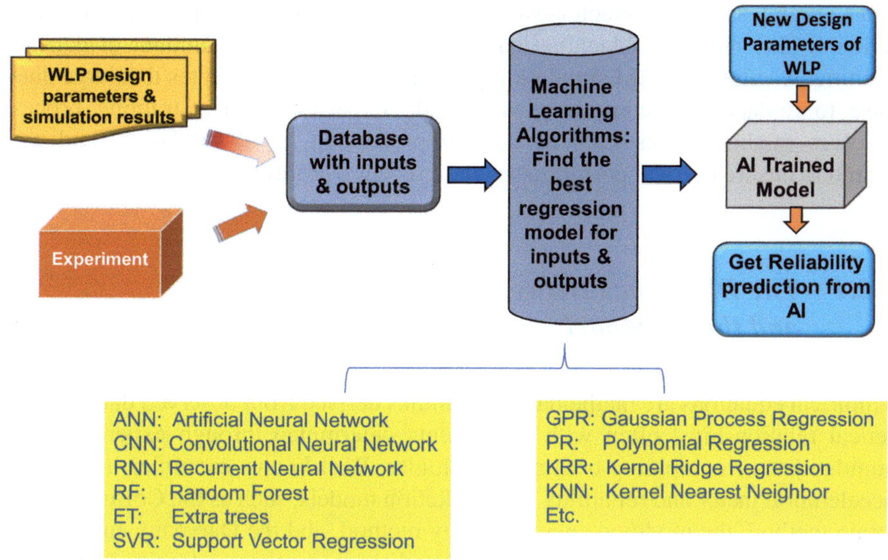


**Fig. 9.1** Illustration of the AI-assisted DoS

include: (1) Efficient data sampling strategies, Such as adaptive sampling, which strategically positions data points in "high-interest areas" (e.g., feature boundaries where prediction errors are concentrated, such as using K-medoids K-means to focus on "high-interest areas," like feature boundaries where prediction errors are concentrated) rather than relying solely on uniform space-filling. This improves data quality and model performance with fewer samples. (2) Ensemble learning: A "training strategy" that combines multiple trained models (weak learners) to improve overall prediction accuracy and robustness, especially beneficial for small datasets by reducing variance and risk of overfitting. Common approaches include bagging, boosting, and stacking, all of which have been incorporated in this research framework. Ensemble learning has shown significant potential to improve prediction accuracy when training with small datasets. Several machine learning algorithms, e.g., ANN, RF, and GPR, are selected for demonstration purposes in this Chapter; other ML algorithms introduced in Chap. 7 can be applied similarly.

## 9.2 Empirical Model Vs. AI Model

The empirical and AI models can be treated as a transformation box between inputs and outputs. Empirical equations are typically applied after finite element simulations to predict reliability life. While FEM itself reduces the need for extensive physical experiments, generating simulation data for complex structures is still a detailed and potentially time-consuming process. The quality of these predictions can be highly sensitive to factors such as mesh size; a validated finite element simulation is also required. Machine learning requires a massive amount of data for training. A validated FEM simulation procedure typically generates this large database to replace time-consuming and costly experiments. The training process involves methodologies such as data preprocessing, cross-validation, and grid search to ensure optimal performance and prevent overfitting.

### 9.2.1 The Difference Between the Empirical Model and the AI Model

Empirical equations are mathematical formulas derived from observed data or theoretical relationships, often with coefficients determined through experiments or simulations, or combining both methodologies. Well-known examples include the acceleration factor and reliability life prediction models, such as the Coffin-Manson strain method, the modified energy-density method, and the Darveaux energy density method. These models utilize calculated strain or energy density from finite element method (FEM) simulations to predict fatigue life. These equations are considered "transformation boxes" that have a direct physical meaning, such as transforming strain or energy density into reliability life cycle. On the other hand, the AI approach involves using machine learning algorithms combined with large datasets to enable computers to learn and train automatically, eventually developing a simpler AI model that defines input-output relationships. While an AI model can also be viewed as a "transformation box" between input and output, it doesn't necessarily reveal their explicit physical relationship. The acceleration factor is also a good example to illustrate the empirical model; the solution procedure is presented in the following section.

### 9.2.2 Empirical Model Case Study: Acceleration Factor

The acceleration factor (AF) [1–3] is a crucial concept in microelectronics reliability assessment, serving as a rapid method to predict product lifetime under field-use conditions based on accelerated thermal cycling test (ATCT) results and loading conditions. Essentially, it's a ratio that compares the fatigue life cycle in a field environment to that in a test environment. The Norris-Landzberg (N-L) AF empirical equation [1–3], Eq. (9.1), is a well-known and widely used formula developed based on the ATC test. It was initially established using SnPb eutectic solder material. The initial N-L equation incorporates three ATC loading factors: the temperature variation between operating temperatures, the maximum temperature, and frequency, which combine ramp rate and dwell time through the concept of cycle time. According to this Equation, a shorter cycle time would result in a longer

**Table 9.1** Test results vs. N-L prediction results under various ramping rate conditions

| Packaging Type | Material | T (℃) | Dwell Time (s) | Ramp Rate (℃ / min.) | Exp. Results | AF | N-L |
|---|---|---|---|---|---|---|---|
| WLP | SnAg | -40 - 125 | 600 | 11 | 287 | 0.843 | 172 |
| | | | 600 | 16.5 | 246 | 0.909 | 185 |
| | | | 600 | 33 | 204 | 1(Base) | 204 |

predicted lifetime; even if the shorter time is associated with a higher ramp rate, the N-L AF equation predicts a longer lifetime when the ramp rate is higher, which contradicts experimental test data. Experimental data consistently show that solder joints have a shorter test lifetime at higher ramp rates (Table 9.1). This discrepancy arises because the N-L AF equation combines both ramp rate and dwell time factors into a single frequency term. Consequently, it does not adequately reflect the effect of the solder strain rate.

$$AF = \frac{N_{field}}{N_{test}} = \left(\frac{\Delta T_{test}}{\Delta T_{field}}\right)^{1.9} \left(\frac{f_{field}}{f_{test}}\right)^{1/3} \exp\left(\frac{1414}{T_{field}^{MAX}} - \frac{1414}{T_{test}^{MAX}}\right) \tag{9.1}$$

where $AF$ is the acceleration factor, $N$ is the fatigue life cycle, $\Delta T$ is the difference between the maximum and minimum temperature of the thermal loading, $f$ is the frequency of thermal loading, and $T^{MAX}$ is the maximum operating temperature (K). The subscripts "*test*" and "*field*" represent the acceleration test environment and the product field-use environment, respectively. A modified AF equation [3–5] has been proposed to separate the effects of ramp rate and dwell time in AF prediction, thereby addressing the discrepancy between predicted and experimental results, as expressed in Eq. (9.2).

$$AF = \frac{N_{field}}{N_{test}} = \left(\frac{\Delta T_{test}}{\Delta T_{field}}\right)^{1.9} \left(\frac{R_{test}}{R_{field}}\right)^{a} \left(\frac{t_{dwell,test}}{t_{dwell,field}}\right)^{b} \exp\left(\frac{1414}{T_{field}^{MAX}} - \frac{1414}{T_{test}^{MAX}}\right) \tag{9.2}$$

where $t_{dwell}$ is the dwell time and $R$ is the ramp rate per cycle (C/min). Parameter $a$ is the ramp rate-related coefficient, and $b$ is the dwell time-related coefficient. The values of $a$ and $b$ can be determined from fatigue life cycles, derived either from experimental data or simulation results. Once these parameters are obtained, the modified AF model can be employed as a general empirical equation.

#### 9.2.2.1 Modified Acceleration Factor

The FE models of the selected packages are built based on the dimensions and material information described in the literature [3]. The material properties of the Sn/Ag solder joints follow the hyperbolic sine creep equation and stress-strain curve to describe the deformation behavior of the solder joints. The related information is listed in Tables 9.2, 9.3 and 9.4, and Fig. 9.2. To derive the values of *a* and *b* of Eq. (9.2), three thermal cycling profiles are selected in Table 9.5. The value of *a* is 0.14, and that for *b* is 0.08 based on the fatigue life cycle from simulation, and the value of *a* is 0.39, and that for *b* is 0.25 based on the fatigue life cycle from test data. The corresponding new equations are presented below.

$$AF = \frac{N_{field}}{N_{test}} = \left(\frac{\Delta T_{test}}{\Delta T_{field}}\right)^{1.9}\left(\frac{R_{test}}{R_{field}}\right)^{0.14}\left(\frac{t_{dwell,test}}{t_{dwell,field}}\right)^{0.08}\exp\left(\frac{1414}{T_{field}^{MAX}}-\frac{1414}{T_{test}^{MAX}}\right) \tag{9.3}$$

$$AF = \frac{N_{field}}{N_{test}} = \left(\frac{\Delta T_{test}}{\Delta T_{field}}\right)^{1.9}\left(\frac{R_{test}}{R_{field}}\right)^{0.39}\left(\frac{t_{dwell,test}}{t_{dwell,field}}\right)^{0.25}\exp\left(\frac{1414}{T_{field}^{MAX}}-\frac{1414}{T_{test}^{MAX}}\right) \tag{9.4}$$

The variations in the values of *a* and *b* between Eqs. (9.3) and (9.4) arise from the different datasets used in their calculation. Specifically, Eq. (9.3) is derived from simulation results, while Eq. (9.4) is based on experimental data. These two empirical equations are compared across different types of electronic packaging structures subjected to various thermal loading conditions [6–10]. The corresponding life cycle predictions are summarized in Tables 9.6 and 9.7.

**Table 9.2** Dimension info of WLP package

| Material | Dimension (mm) | Thickness/height (μm) |
|---|---|---|
| Silicon Chip | 8x8 | 200 |
| Polyimide | 0.18 | 5 |
| Al Pad | 0.25 | 2 |
| Cu UBM | 0.24 | 1.8 |
| Solder | 0.26 (after reflow) | 280 |
| PCB Cu Pad | 0.24 | 40 |

**Table 9.3** Material Properties Used in WLP Model

| Material | Young's modulus (GPa) | Poisson's ratio | CTE (ppm) |
|---|---|---|---|
| Silicon | 150 | 0.278 | 2.8 |
| Polyimide | 3.5 | 0.35 | 35 |
| UBM | 150 | 0.33 | 14.5 |
| Copper | 68.9 | 0.34 | 16.7 |
| PCB | 26 | 0.39 | 13 |

**Table 9.4** Sn/Ag Hyperbolic Sine Creep Model

$$\dot{\varepsilon} = A\left[\sinh\left(B\sigma\right)\right]^{n} e^{\frac{-Q}{RT}}$$

| Parameter | Unit | Value |
|---|---|---|
| A | 1/s | 23.17 |
| B | $MPa^{-1}$ | 0.0509 |
| n | – | 5.04 |
| Q | kJ/mol | 41.6 |

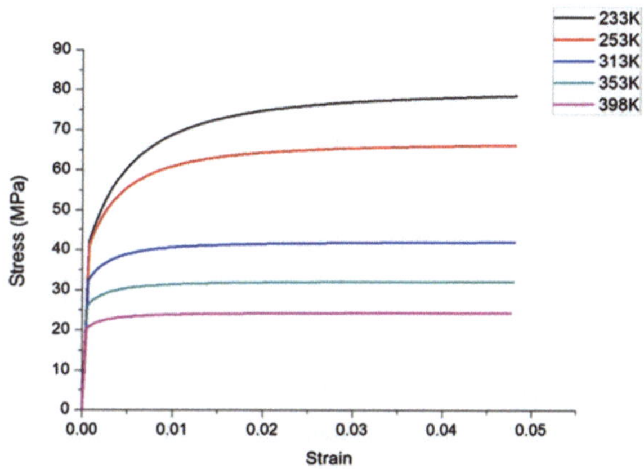


**Fig. 9.2** Sn/Ag stress-strain curve

**Table 9.5** Three sets of simulation/test data used to determine *a* and *b*

| Ramp time (s) | Dwell time (s) | Fatigue cycle (simulation) | Fatigue cycle (experiment) |
|---|---|---|---|
| 600 | 600 | 245 | 246 |
| 900 | 600 | 258 | 288 |
| 600 | 300 | 259 | 293 |

In addition to the WLP test data used for curve fitting parameters *a* and *b*, various package types reported in the literature [6–10], including BGA and chip-scale packages (CSP), are also used to compare and validate the proposed AF equation. The results show that Eq. (9.3), with *a* and *b* derived from simulation data, demonstrates a stronger capability for life cycle prediction than Eq. (9.4), which is based on experimental data. For WLP test legs, however, Eq. (9.4) exhibits smaller differences, since the values of *a* and *b* are directly obtained from these test data. The reduced accuracy of the AF lifetime prediction using Eq. (9.4) stems from uncertainties inherent in the testing process, which vary from case to case. Such factors include tolerances in temperature, ramp time, and dwell time during thermal cycling tests. By contrast, these uncertainties are excluded in simulations. Therefore, it is recommended that the new AF terms in Eq. (9.2) be paired with *a* and *b* coefficients derived from simulation data, allowing test-related uncertainties to be estimated on a case-by-case basis.

**Table 9.6** AF results based on Eq. (9.3)

| Package type | Temp Range (°C) | Ramp (min) | Dwell (min) | Life cycle (test) | AF | Life cycle (AF) | Difference |
|---|---|---|---|---|---|---|---|
| WLP | −40–125 | 5 | 10 | 204 | 0.91 | 223 | 9% |
| | | 10 | 10 | 246 | 1 | – | – |
| | | 15 | 10 | 288 | 1.06 | 260 | 10% |
| | | 10 | 5 | 293 | 1.06 | 260 | 11% |
| | | 18 | 12 | 297 | 1.07 | 263 | 11% |
| CBGA [6] | 0–100 | 7.5 | 7.5 | 1286 | 1.17 | 1165 | 9% |
| | | 7.5 | 52.5 | 997 | 1 | – | – |
| CBGA [7] | 0–100 | 10 | 10 | 1040 | 1.33 | 865 | 17% |
| | | 10 | 350 | 651 | 1 | – | – |
| BGA [8, 9] | 0–100 | 15 | 10 | 1040 | 1.41 | 916 | 12% |
| | | 10 | 350 | 651 | 1 | – | – |
| CSP [7] | 0–100 | 10 | 10 | 1025 | 1.15 | 908 | 11% |
| | | 10 | 60 | 787 | 1 | – | – |
| CSP [10] | −40–125 | 20.6 | 30 | 8797 | 1.16 | 8128 | 8% |
| | | 6.8 | 30 | 6970 | 1 | – | – |

**Table 9.7** AF results based on Eq. (9.4)

| Package type | Temp range (°C) | Ramp (min) | Dwell (min) | Life cycle (test) | AF | Life cycle (AF) | Difference |
|---|---|---|---|---|---|---|---|
| WLP | −40–125 | 5 | 10 | 204 | 0.76 | 188 | 8% |
| | | 10 | 10 | 246 | 1 | – | – |
| | | 15 | 10 | 288 | 1.17 | 288 | 0% |
| | | 10 | 5 | 293 | 1.19 | 293 | 0% |
| | | 18 | 12 | 297 | 1.2 | 296 | 0.4% |
| CBGA [6] | 0–100 | 7.5 | 7.5 | 1286 | 1.63 | 1622 | 26% |
| | | 7.5 | 52.5 | 997 | 1 | – | – |
| CBGA [7] | 0–100 | 10 | 10 | 1040 | 2.43 | 1583 | 52% |
| | | 10 | 350 | 651 | 1 | – | – |
| BGA [8, 9] | 0–100 | 15 | 10 | 1040 | 2.85 | 1855 | 78% |
| | | 10 | 350 | 651 | 1 | – | – |
| CSP [7] | 0–100 | 10 | 10 | 1025 | 1.57 | 1232 | 20% |
| | | 10 | 60 | 787 | 1 | – | – |
| CSP [10] | -40–125 | 20.6 | 30 | 8797 | 1.53 | 10,698 | 22% |
| | | 6.8 | 30 | 6970 | 1 | – | – |

Based on the proposed AF equation (Eq. 9.3), an excellent agreement is observed between AF lifetime predictions and FE simulation results, as shown in Table 9.8. The solder fatigue life cycles predicted by the AF equation closely match those obtained from the energy-based lifetime equation, $N_f = C(\Delta W_{inelastic})^{-n}$, where $N_f$ represents the fatigue life of the solder joint. $\Delta W_{inelastic}$ denotes the increment of inelastic strain energy density per thermal cycle, which accounts for both creep and

**Table 9.8** WLP solder ball fatigue life: simulation results and AF results

| Ramp (min) | Dwell (min) | Life cycle (simulation) | AF | Life cycle (AF) | Difference |
|---|---|---|---|---|---|
| 2 | 5 | 208 | 0.84 | 207 | 0.6% |
| 5 | | 237 | 0.96 | 235 | 0.6% |
| 10 | | 259 | 1.06 | 259 | 0% |
| 15 | | 273 | 1.12 | 274 | 0.4% |
| 2 | 10 | 196 | 0.79 | 196 | 0.1% |
| 5 | | 223 | 0.91 | 223 | 0.3% |
| 10 | | 245 | 1 (Base) | 245 | 0% |
| 15 | | 259 | 1.06 | 260 | 0.2% |
| 2 | 15 | 189 | 0.77 | 190 | 0.1% |
| 5 | | 216 | 0.88 | 216 | 0.1% |
| 10 | | 237 | 0.97 | 237 | 0.2% |
| 15 | | 251 | 1.02 | 251 | 0.2% |
| 2 | 30 | 179 | 0.73 | 179 | 0.4% |
| 5 | | 203 | 0.83 | 204 | 0.3% |
| 10 | | 224 | 0.91 | 225 | 0.3% |
| 15 | | 237 | 0.96 | 238 | 0.4% |

plastic deformation. $C$ and $n$ are fitting parameters obtained from simulation and experimental data, $C$ is 78, and $n$ is defined as 2.8 in ball grid array type packaging. The results suggest that the modified AF equation efficiently reduces simulation time under varying thermal cycling conditions.

## 9.3 AI-Assisted Design-on-Simulation for Reliability Prediction of WLP

With the increasing complexity of packaging structures, manufacturing reliability test vehicles, and conducting ATCT experiments have become time-consuming and expensive processes. Consequently, the Design-on-Experiment (DoE) methodology for packaging design is becoming infeasible. As a result of the widespread adoption of finite element analysis [11–17], accelerated thermal cycling tests have been significantly reduced in the semiconductor industry, leading to a decrease in package development time and cost. In a 3D WLP model, Liu [18] applies the Coffin-Manson life prediction empirical model to predict the reliability life of a solder joint within an accurate range. However, the results of finite element simulations are highly dependent on the mesh size, and there are no guidelines to help researchers address this issue. Therefore, Chiang et al. [19] proposed the concept of "volume-weighted averaging" to determine the local strain, especially in critical areas. Tsou [20] successfully predicted packaging reliability through finite element simulation with a fixed mesh size in the critical area of the WLP structure. However, simulation analysis results are highly dependent on the individual researcher, and the results are usually inconsistent between simulations. To overcome this problem, the present work comparatively reviews an artificial intelligence approach in which electronic packaging design is performed using a machine learning algorithm [21, 22]. Using an AI model for analyzing electronic packaging reliability is the most effective way to obtain a reliable prediction result and meet the time-to-market demand.

AI-assisted Design-on-Simulation technology (Fig. 9.1, [24]) offers numerous advantages for predicting the reliability of electronic packaging, especially advanced electronic packaging. These benefits primarily stem from its ability to enhance efficiency, accuracy, consistency, and cost-effectiveness compared to traditional methods. Traditional experimental methods, such as Accelerated Thermal Cycling Tests, are time-consuming and resource-intensive, often taking months to a year to yield results. Finite element analysis simulations reduce this to days or weeks. However, AI-assisted DoS can predict the reliability life cycle of a new WLP structure in a few seconds after the model is established. This significantly shortens the design and development cycle. Traditional FEA results can vary significantly depending on the researcher's experience, domain knowledge, and modeling assumptions. AI-assisted DoS provides consistent and trustworthy prediction results regardless of who performs the analysis. The DoS solution procedure is illustrated in Fig. 9.1, and validation plays an essential role in the DoS procedure. In summary, AI-assisted DoS provides a fast, accurate, consistent, and cost-effective framework for reliability prediction in advanced electronic packaging, fundamentally transforming the design and development cycle. The following sections will demonstrate the finite element validation and AI training procedures.

### *9.3.1 WLP Solder Joint Reliability Prediction and Verification Procedure*

If the simulation consistently predicts the experiment's result [23, 24], then the simulation can be considered an experiment. In this case, the experiment work can be replaced by a validated simulation procedure, allowing for creating a large database for AI training and then developing a small and accurate AI model for reliability life cycle prediction. Once the AI model is obtained, developers can simply input the WLP geometries for a new WLP structure, and the life cycle can be obtained within a few milliseconds. Figure 9.1 illustrates this procedure. A validated finite element simulation procedure is critical for the success of the AI-Assisted Design-on-Simulation approach. Next, we will use WLP as an example to illustrate the validation process.

### 9.3.2 *Mesh Size Control and Empirical Life Prediction Model*

Because a large amount of data is required to build the AI training model, a two-dimensional finite element method (FEM) model is used for simulation in this case. Since a reliable database is critical for AI training, this work validated the simulation results using five WLP test vehicles (TVs), as shown in Tables 9.9 and 9.10. All of the sizes and specifications for different materials and the mean-times-to-failure (MTTF) of the test vehicles are presented in the tables (Table 9.9 and 9.10) [25, 26]. The simulation method primarily employed a fixed mesh size at a critical location, determined through appropriate mechanical behaviors, and an empirical life prediction equation was used to validate the reliability of the life cycles of all test vehicles. This work utilized a fixed mesh size for the solder joint at the maximum distance from the neutral point of the WLP to establish the modeling pattern and simulation procedure, as proposed by Tsou et al. [20]. As shown in Fig. 9.3, the fixed mesh size

**Table 9.9** Dimension of WLP TV1 and TV2

| Material | TV1 (mm) | TV2 (mm) |
|---|---|---|
| Si Chip | 5.3 × 5.3 × 0.33 | 4 × 4 × 0.33 |
| Cu RDL | 0.26 × 0.008 | 0.26 × 0.008 |
| UBM | – | 0.24 × 0.0086 |
| Cu Pad | 0.22 × 0.025 | 0.22 × 0.025 |
| SBL1 | 5.3 × 5.3 × 0.0075 | 4 × 4 × 0.0075 |
| SBL2 | 0.01 | 0.004 |
| PCB | 10.6 × 10.6x1 | 8 × 8 × 1 |
| Low-k | 5.3 × 0.005 | 4 × 0.005 |
| Ball diameter (mm) | 0.25 | 0.25 |
| Ball pitch (mm) | 0.4 | 0.4 |
| Ball counts | 121 | 100 |
| **MTTF (cycles)** | **318** | **1013** |

**Table 9.10** Dimension of WLP TV3, TV4, and TV5

| Material | TV3 (mm) | TV4 (mm) | TV5 (mm) |
|---|---|---|---|
| Si Chip | 4 × 4 × 0.33 | 4 × 4 × 0.33 | 6 × 6 × 0.33 |
| Cu RDL | 0.18 × 0.004 | 0.2 × 0.004 | 0.25 × 0.0065 |
| UBM | 0.17 × 0.0086 | 0.19 × 0.0086 | 0.24 × 0.0075 |
| Cu Pad | 0.22 × 0.025 | 0.22 × 0.025 | 0.22 × 0.04 |
| SBL1 | 4 × 4 × 0.0075 | 4 × 4 × 0.0075 | 6 × 6 × 0.008 |
| SBL2 | 0.004 | 0.004 | 0.0065 |
| PCB | 8 × 8 × 1 | 8 × 8 × 1 | 12 × 12 × 1 |
| Low-k | 4 × 0.005 | 4 × 0.005 | – |
| Ball Diameter (mm) | 0.18 | 0.2 | 0.25 |
| Ball pitch (mm) | 0.3 | 0.3 | 0.4 |
| Ball counts | 144 | 144 | 196 |
| **MTTF (cycles)** | **587** | **876** | **904** |

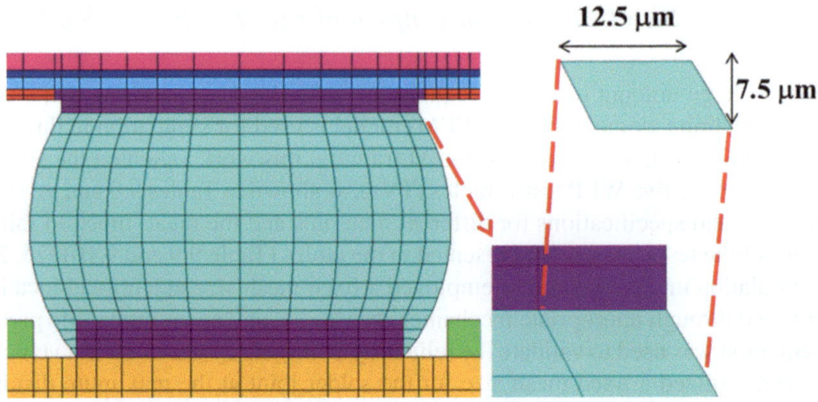


**Fig. 9.3** Critical region mesh size

**Table 9.11** Material properties of WLP

| Material | Young's modulus (Gpa) | Poisson's ratio | CTE (ppm/°C) |
|---|---|---|---|
| Solder joint | Temperature dependent | 0.35 | 25 |
| Silicon chip | 150 | 0.28 | 2.62 |
| Copper | 68.9 | 0.34 | 16.7 |
| Low-k | 10 | 0.16 | 5 |
| Solder mask | 6.87 | 0.35 | 19 |
| PCB | 18.2 | 0.19 | 16 |

in the critical region had a width and height of 12.5 and 7.5 μm, respectively. The solder joint geometry was generated using the Surface Evolver [27] and then converted into a 2D profile.

In the simulation process, the solder material was a nonlinear plastic material. Therefore, PLANE182, which exhibits good convergence characteristics and can handle large deformations, was utilized as the solder ball element. This work used PLANE42 for other components, which had linear material properties. Table 9.11 presents the list of individual material properties. The Young's moduli of the solder joint at different temperatures are listed in Table 9.12. Figure 9.4 shows the stress-strain curve for a Sn-Ag-Cu, SAC305, solder joint. The stress-strain curve [28] obtained from tensile testing, along with the Chaboche kinematic hardening model, was used to describe the tensile curves at different temperatures. Once the model is built, boundary conditions and external thermal loading are applied for the WLP simulation.

Electronics packaging geometry is usually symmetrical; therefore, half of the 2D structure was modeled along the diagonal in this case, as shown in Fig. 9.5. The X-direction displacement at each node was fixed to zero along the Y-direction of the central line. To prevent rigid body motion, the node at the lowest point of the neutral axis, located on the printed circuit board (PCB), has all degrees of freedom fixed.

**Table 9.12** The Young's modulus under different temperatures

| Temperature | Young's modulus (GPa) |
|---|---|
| 233 K | 45.74 |
| 253 K | 42.22 |
| 313 K | 31.66 |
| 353 K | 24.62 |
| 398 K | 16.70 |

**Fig. 9.4** Stress-strain curve for SAC305 solder

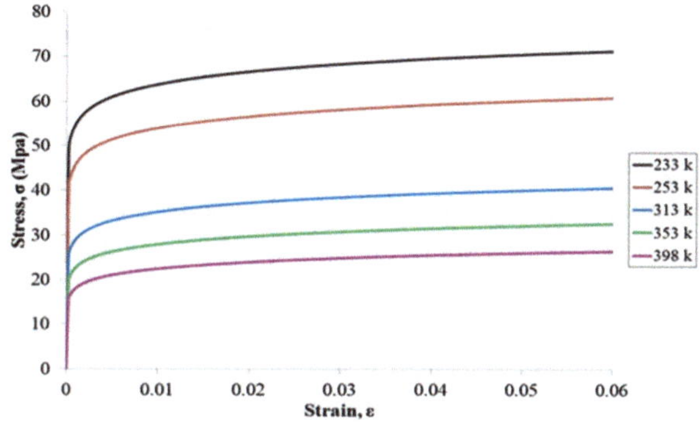


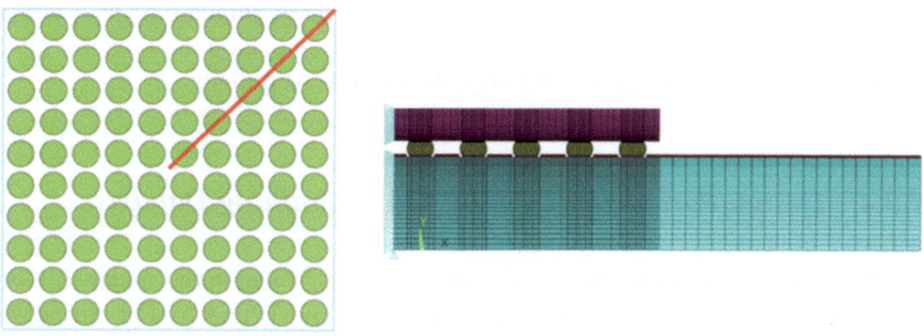

**Fig. 9.5** Schematic of symmetrical WLP solder ball geometry

The thermal loading condition used in this research was the JEDEC JESD22-A104D condition G [29], with a temperature range of −40 to 125 °C. The ramp rate was set at 16.5 °C/min, and the dwell time was 10 min. A qualified design should have an MTTF higher than 1000 thermal cycles. After the simulation process is completed, typically involving eight thermal cycles, the incremental equivalent plastic strain in the critical zone is substituted into the strain-based Coffin-Manson model to predict the reliability life cycle. This strain-based method is as accurate as the energy-based empirical model for a fixed temperature ramp rate, but with significantly less CPU time.

The empirical formula for the Coffin-Manson equivalent plastic strain model is shown as $N_f = C\left(\Delta\varepsilon_{eq}^{pl}\right)^{-n}$, where $N_f$ is the mean cycle to failure, $C$ and $n$ are empirical constants, and $\Delta\varepsilon_{eq}^{pl}$ is the incremental equivalent plastic strain. For SAC305 solder, $C$ and $n$ are 0.235 and 1.75 [30, 31].

### *9.3.3 Comparison of Simulation Results and Experiment Results*

Table 9.13 presents the predicted reliability life cycles of the WLP structures. The results show that the difference between the FEM-predicted life cycle and the experiment result is within a small range. Multiple test vehicles (TVs) with varying structural dimensions are used for validation, and the simulation results consistently align with experimental data. The difference between FEA-predicted life cycles and experimental MTTF is less than 10%, which is considered acceptable. In many cases, the error percentage is even lower. Therefore, ATCT experiments can be replaced by this validated FEM simulation to minimize cost and time. Compared to the experimental approach, one ATCT test can take several months. This validated FEM simulation procedure, however, can provide large amounts of data within much less time. This validation is crucial for treating simulations as a reliable alternative to experiments for generating large datasets for AI training.

### *9.3.4 Design Parameters of WLP for AI Training Database*

The WLP structure consists of several components: the solder mask, solder ball, I/O pad, UBM, stress buffer layer, PCB size, and silicon chip (Fig. 9.6). For illustration purposes, the four most influential parameters were selected to build the AI model and predict the reliability life cycles of new WLP structures, such as silicon chip thickness, stress buffer layer thickness, upper pad diameter, and lower pad diameter. These four design parameters were used to generate both training and testing datasets for AI machine learning algorithms. Tables 9.14, 9.15, 9.16, 9.17, 9.18 and 9.19

**Table 9.13** WLP finite element results for five test vehicles

| Test vehicle | Experimental reliability (cycles) | Simulation reliability (cycles) | Difference |
|---|---|---|---|
| TV1 | 318 | 313 | −5 |
| TV2 | 1013 | 982 | −31 |
| TV3 | 587 | 587 | 0 |
| TV4 | 876 | 804 | 72 |
| TV5 | 904 | 885 | 19 |

| Design Parameters | |
|---|---|
| Chip Thickness | Chip Size |
| Upper Pad Thickness | Upper Pad Diameter |
| Lower Pad Thickness | Lower Pad Diameter |
| SBL Thickness | PCB Thickness |
| Solder Diameter | Pitch |

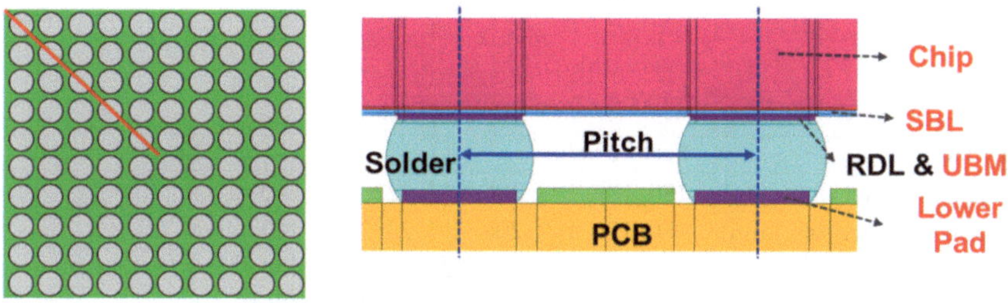


**Fig. 9.6** Design parameters of WLP

**Table 9.14** Training data 256 of FEM for four input features

| Feature name | Feature value | Level |
|---|---|---|
| Upper pad diameter (unit: μm). | 180, 200, 220, 240 | 4 |
| Lower pad diameter (unit: μm). | 180, 200, 220, 240 | 4 |
| Chip thickness (unit: μm). | 150, 250, 350, 450 | 4 |
| SBL layer thickness (unit: μm). | 5, 14.17, 23.33, 32.5 | 4 |
| Number of dataset | 256 | |

**Table 9.15** Training data 625 of FEM for four input features

| Feature name | Feature value | Level |
|---|---|---|
| Upper pad diameter (unit: μm) | 180, 195, 210, 225, 240 | 5 |
| Lower pad diameter (unit: μm) | 180, 195, 210, 225, 240 | 5 |
| Chip thickness (unit: μm). | 150, 225, 300, 375, 450 | 5 |
| SBL layer thickness (unit: μm). | 5, 11.88, 18.75, 25.63, 32.5 | 5 |
| Number of dataset | 625 | |

**Table 9.16** Training data 1296 of FEM for four input features

| Feature Name | Level (μm) |
|---|---|
| Upper pad diameter | 180, 190, 200, 210, 220, 230 |
| Lower pad diameter | 180, 190, 200, 210, 220, 230 |
| Chip thickness | 200, 250, 300, 350, 400, 450 |
| Stress buffer layer thickness | 7.5, 12.5, 17.5, 22.5, 27.5, 32.5 |
| Total number | 1296 |

**Table 9.17** Training data 2401 of FEM for four input features

| Feature Name | Feature Value | Level |
|---|---|---|
| Upper pad diameter (unit: μm) | 180, 190, 200, 210, 220, 230, 240 | 7 |
| Lower pad diameter (unit: μm) | 180, 190, 200, 210, 220, 230, 240 | 7 |
| Chip thickness (unit: μm) | 150, 200, 250, 300, 350, 400, 450 | 7 |
| SBL layer thickness (unit: Mm) | 5, 9.58, 14.17, 18.75, 23.33, 27.92, 32.5 | 7 |
| Number of dataset | 2401 | |

**Table 9.18** Training data 4096 of FEM for four input features

| Feature name | Feature value | Level |
|---|---|---|
| Upper pad diameter (unit: μm) | 180, 189, 197, 206, 214, 223, 231, 240 | 8 |
| Lower pad diameter (unit: μm) | 180, 189, 197, 206, 214, 223, 231, 240 | 8 |
| Chip thickness (unit: μm) | 150, 193, 236, 279, 321, 364, 407, 450 | 8 |
| SBL layer thickness (unit: μm) | 5, 8.93, 12.86, 16.79, 20.71, 24.64, 28.57, 32.5 | 8 |
| Number of dataset | 4096 | |

**Table 9.19** Training data 9000 of FEM for four input features

| Feature name | Size (unit: mm) | Level |
|---|---|---|
| Upper pad diameter | Includes 25 distinct values (prediction values are derived based on combinations of the four parameters) | 25 |
| Lower pad diameter | | |
| Chip thickness | | |
| SBL layer thickness | | |
| Total data | 9000 | |

show the generated training dataset of different sizes of design parameters obtained through FEM simulation. First, the number of training features generated in this research was 256 (Tables 9.14) to 9000 (Table 9.19). For testing, a few hundred features were randomly selected from the interpolation of the above training dataset, which helps build the AI training model. Increasing the training dataset will improve AI performance; however, the CPU/GPU time will also increase.

### 9.3.5 Machine Learning Algorithms for Training

In recent years, AI has been widely used in various research domains. Machine learning is a subset of AI that uses provided data for training and learning. Eventually, a simple AI model with well-defined input and output relationships will be developed to assist researchers in making informed design decisions [32–35]. Machine learning [36–39] can be applied to regression or classification models using supervised or unsupervised learning. The input datasets are labeled in this section, so the learning algorithm used to predict reliability life falls under supervised machine learning. Several machine learning algorithms exist, including artificial neural networks (ANN), support vector regression (SVR), K-nearest neighbor (KNN), kernel ridge regression (KRR), recurrent neural networks (RNN), random forests (RF), and convolutional neural networks (CNN).

The ANN is one of the most common machine learning methods. McCulloch and Pitts [40] proposed an ANN that mimics biological neural structures, using different numbers of hidden layers and neurons to construct different neural network structures. Denoeux [41] also explored data classification using a neural network algorithm. In this algorithm, the data are entered at the input layer and then calculated in the hidden layer. Once the calculation is complete, the result will be displayed in the output layer. RNN is a class of ANN algorithms in which connections between nodes form a directed network along a temporal sequence, making them more suitable for deep learning [42, 43] when many nonlinear datasets are available [44, 45].

SVR, proposed by Cortes and Vapnik [46], is suitable for high-dimensional features but is not well-suited for small datasets. The concept of SVR is similar to the support vector machine (SVM), which is used to solve classification problems. SVM finds the best classifier by searching for the hyperplane with the largest margin [47]. Along with SVR, KRR is among the most popular kernel-based methods. Kernel-based methods are useful for nonlinear structural datasets [48]. Due to its closed-form solution, KRR is simpler and faster to train and can outperform SVR. Non-parametric calibration models eliminate the normality assumption and can represent almost any type of data, whether scattered or following a certain trend. However, the models can exhibit only one type of prediction behavior, i.e., they cannot combine local and general prediction. Local calibration models, such as KNN interpolation, consider the surrounding neighborhood as input to obtain information about the output [49, 50]. KNN is suitable for both classification and regression problems. The KNN classification output is determined by the nearest neighborhood, which depends on the number of nearest neighbors that belong to that class. In contrast, the regression output is determined by the average of the nearest neighbor values [51, 52].

KNN and RF are more suitable for classification tasks; these two algorithms also show prominent performance for regression tasks. The RF algorithm was proposed in 2001 by Breiman [53]. An RF is formed by combining multiple classifications and regression trees (CART); it analyzes [54, 55] the data features and data distribution to generate multiple decision trees with different structures, and finally summarizes the prediction results of all decision trees. The CNN is a machine learning algorithm for image recognition and image classification. In 2012, AlexNet, proposed by Krizhevsky et al. [56], achieved the highest accuracy in the ImageNet competition; consequently, the CNN has become a focus in academia and has developed rapidly [57, 58].

This Chapter will demonstrate the use of ANN, RF, and KRR to predict WLP's life cycle reliability. The aim is to develop and establish a regression model for the relationship between packaging geometries (input) and life cycle results (output).

Before implementing the above AI algorithms, an AI training database was generated using a finite element simulation combined with the Coffin–Manson empirical equation for predicting WLP reliability life cycles. This standard simulation procedure was experimentally verified before generating the massive simulation database.

## 9.4 Selected AI Training Models

Several regression algorithms are effective for AI-assisted DoS in reliability and mechanical behavior prediction. Artificial Neural Networks (ANN) capture complex nonlinear relationships, Polynomial Regression (PR) addresses nonlinear responses, and Gaussian Process Regression (GPR) provides a probabilistic framework suitable for limited data. Recurrent Neural Networks (RNN) are applied to time-like series and fatigue life prediction, while ensemble methods such as bagging, boosting, and stacking enhance accuracy and robustness by combining models like ANN, GPR, and RNN. Hybrid approaches, such as K-means with Kernel Ridge Regression (K-K), further enhance efficiency and performance on nonlinear datasets. Chapter 7 comprehensively introduces various machine learning algorithms that can be applied to electronic packaging. Most machine learning algorithms are a tradeoff among accuracy, data size, and computational efficiency. This Chapter selects a few representative machine learning algorithms for demonstration purposes; other ML algorithms can be applied similarly.

### *9.4.1 Artificial Neural Network (ANN)*

Artificial Neural Networks (ANN), as described in Chap. 7, are utilized in the case of wafer-level packages to predict their reliability life cycle. The ANN model for WLP is a multi-layer neural network. It comprises three fundamental layers. (1) Input Layer: This layer receives the design parameters of the WLP structure. The primary influential parameters used as inputs are silicon chip thickness, stress buffer layer thickness, upper pad diameters, lower pad diameters, and their related reliability life cycles. (2) Hidden Layer: After data enters the input layer, it is processed and calculated within one or more hidden layers. Increasing the number of hidden layers and neurons in these layers enhances the network's capacity to handle nonlinearity in the WLP data. (3) Output Layer: This layer displays the predicted reliability life cycle of the WLP. The flowchart of the ANN training procedure is shown in Fig. 9.7.

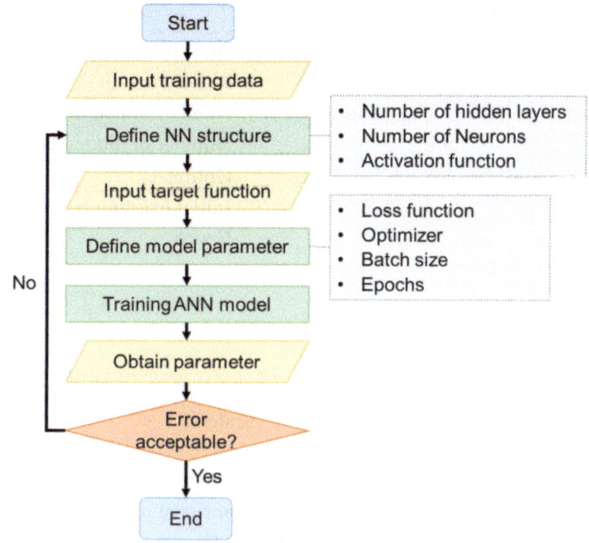


**Fig. 9.7** The ANN training flowchart

#### 9.4.1.1 ANN Settings and Training

The training procedure for an ANN used in the reliability prediction of WLP is a systematic process designed to accurately establish a regression model between WLP design parameters and their reliability life cycle. The primary goal is to learn and establish a regression model that represents the relationship between WLP packaging geometries (input) and their corresponding reliability life cycle (output). This model, once trained, allows designers to immediately obtain the predicted reliability life cycle by simply inputting the WLP geometric data.

Data from the input layer (WLP design parameters), which are features and results, are combined with bias and weights to produce a calculated value. This calculated input value is then transformed into a nonlinear form using an activation function, such as the ReLU, Sigmoid, Tanh, or other functions, before being passed to the next layer. Table 9.20 illustrates one of the hyperparameter settings of the ANN training. This ANN training process allows the network to learn complex, nonlinear relationships in the WLP design data. The ANN learns and establishes a regression model that maps the relationship between the WLP packaging geometries and their corresponding reliability life cycles. If adaptive data selection is not applied, this learning process relies on a large training and testing dataset, which is generated using finite element simulation. This validated simulation procedure effectively reduces the time and cost constraints associated with physical experiments in generating extensive datasets. For example, training datasets ranging from 256 to 9000+ entries and corresponding testing datasets are used.

**Table 9.20** The ANN hyperparameters setting for training

| | Tested hyperparameter values |
|---|---|
| Activation function | ReLU |
| Scaler | Robust |
| Number of neurons | 100 |
| Number of hidden layers | 6 |
| Optimizer | Adam |
| Initial learning rate | 0.001 |
| Epoch | 1000 |

**Table 9.21** ANN hyperparameters grid search candidates

| | Grid Search of Hyperparameters |
|---|---|
| Activation function | ReLU, Sigmoid, Tanh |
| Scaler | MinMax, MaxAbs, Standrad, Normalizer, Robust |
| Number of neurons | 20, 50, 100, 200, 400 |
| Number of hidden layers | 2, 3, 4, 5, 6 |
| Optimizer | Adam, SGD |
| Initial learning rate | 0.001 |
| Epoch | 200, 500, 1000, 2000, 4000 |
| Total combinations | 3750 |

#### 9.4.1.2 Hyperparameter Setting and Grid Search

Data preprocessing methods, such as Minimax scalar, are applied to standardize the input and output datasets, which improves the model's performance. Cross-validation (e.g., ten-fold method) is employed to select parameters and prevent overfitting. The grid search technique is used to find the optimal hyperparameters, such as the number of neurons and hidden layers, to achieve the best model performance. The grid search method is a widely used approach to find the hyperparameters of a machine learning model. Its concept involves dividing the hyperparameter range into several parts and exhaustively running all possible combinations within the range to find the best hyperparameter combination. The grid search is a highly time-consuming process, requiring significantly more time than training. For example, the grid search candidates on the hyperparameters listed in Table 9.21 resulted in 3750 combinations. The best hyperparameter combination will be used for the ANN final training work.

#### 9.4.1.3 ANN Performance on Different Sizes of Datasets

In general, the ANN algorithm demonstrated excellent accuracy. ANN demonstrates a clear relationship between the size of the training dataset, its predictive accuracy, and the computational resources required, particularly in the case of WLP reliability

**Table 9.22** ANN hyperparameters, dataset sizes, training, and testing results

| | 256 | 625 | 1296 | 2401 | 4096 |
|---|---|---|---|---|---|
| Scaler | MinMax | Standard | Standard | Robust | Robust |
| Number of neurons | 400 | 400 | 400 | 100 | 400 |
| Number of hidden layers | 4 | 4 | 4 | 4 | 4 |
| Max. Training diff. | 36.1 (3.9%) | 22.3 (2.0%) | 31.2 (3.1%) | 43.1 (2.5%) | 23.5 (2.4%) |
| Ave. training diff. | 4.3 (0.4%) | 3.4 (0.3%) | 5.7 (0.6%) | 3.1 (0.3%) | 1.9 (0.2%) |
| Max. testing diff. | 46.5 (4.1%) | 32.0 (2.8%) | 38.9 (4.7%) | 27.2 (2.7%) | 31.5 (3.2%) |
| Ave. testing diff. | 9.8 (1.0%) | 6.3 (0.6%) | 7.0 (0.7%) | 5.1 (0.5%) | 3.3 (0.3%) |
| Grid search GPU time | 77.3 s | 207.2 s | 377.5 s | 604.0 s | 783.3 s |

prediction. The accuracy of the ANN model for WLP reliability prediction improves significantly as the training dataset size increases. For the training dataset consisting of 256 entries, with the selection of MinMax (256 dataset) or Robust scalar (4096 dataset) preprocessing, ReLU activation function, and Adam optimizer, the ANN produced maximum and average training differences of 36.1 cycles (3.9%) and 4.3 cycles (0.4%), respectively. In contrast, the maximum testing difference was 46.5 cycles (4.1%), with an average testing difference of 9.8 cycles (1.0%). An effective AI model should maintain high accuracy across both training and testing datasets, with the target objective being to achieve training and testing average errors below 1% and avoid overfitting conditions, as demonstrated by the maximum difference.

As the dataset size increases to 4096 entries, the ANN achieves maximum and average training differences of 23.5 cycles (2.4%) and 1.9 cycles (0.2%), respectively. The corresponding maximum and average testing differences decrease to 31.5 cycles (3.2%) and 3.3 cycles (0.3%). These results were obtained using an optimal configuration of four hidden layers with 400 neurons. The detailed hyperparameter settings and training and testing outcomes are summarized in Table 9.22, while the accuracy trend is illustrated in Fig. 9.8. The performance accuracy of the ANN can vary with different hyperparameter settings as well as the choice of GPU or CPU. Table 9.23 presents the grid search and training times corresponding to the optimal hyperparameter set. Generally, with proper hyperparameter tuning, accuracy improves as the dataset size increases; however, this improvement comes at the expense of longer CPU/GPU computation time, highlighting the tradeoff between accuracy and computational efficiency.

Using the WLP with design parameters of upper pad diameter (240 μm), lower pad diameter (220 μm), chip thickness (330 μm), and stress buffer layer thickness (7.5 μm) as an example, the differences between the simulation and AI model are compared with the experiment result in Table 9.24. Once the AI model is trained and finalized, it enables the rapid evaluation of design parameter variations while ensuring consistency and accuracy in prediction results.

Despite higher computational time, the ANN model consistently showed the best accuracy (lowest MAE) when compared against KNN, KRR, Polynomial Regression (PR), GPR, and RF models for WLP reliability prediction. Once the ANN model is

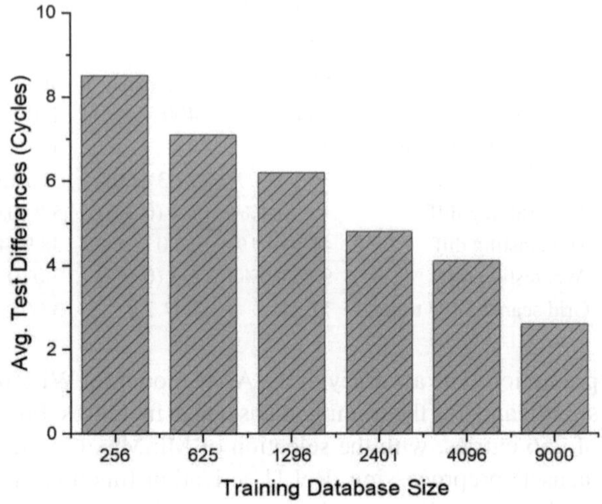


**Fig. 9.8** ANN training accuracy across different dataset sizes

**Table 9.23** CPU/GPU ANN training performance

| | | 256 | 625 | 1296 | 2401 | 4096 | 9000 |
|---|---|---|---|---|---|---|---|
| Best parameter from CPU/GPU | Scaler | MinMax | MinMax | Standard | Robust | Standard | Robust |
| | Neuron | 400 | 400 | 400 | 100 | 100 | 400 |
| | Hidden layer | 4 | 4 | 4 | 4 | 4 | 4 |
| | Solver | adam | adam | adam | adam | adam | adam |
| | Initial learning rate | 0.001 | 0.001 | 0.001 | 0.001 | 0.001 | 0.001 |
| | Epoch | 2000 | 2000 | 2000 | 2000 | 2000 | 2000 |
| | Activation function | ReLU | ReLU | ReLU | ReLU | ReLU | ReLU |
| Best result from CPU/GPU | Maximum training difference | 34.7 (3.8%) | 30.5 (2.6%) | 32.7 (3.1%) | 40.0 (2.4%) | 35.7 (2.1%) | 30.7 (3.4%) |
| | Average training difference | 4.0 (0.4%) | 6.0 (0.6%) | 3.9 (0.4%) | 3.0 (0.3%) | 2.2 (0.2%) | 2.3 (0.2%) |
| | Maximum testing difference | 43.1 (3.8%) | 29.9 (2.6%) | 36.7 (3.2%) | 25.9 (2.6%) | 26.5 (2.7%) | 34.0 (2.7%) |
| | Average testing difference | 8.5 (0.9%) | 7.1 (0.7%) | 6.2 (0.6%) | 4.8 (0.5%) | 4.1 (0.4%) | 2.6 (0.3%) |
| CPU time for this parameter(s) | | 10.5 | 27.2 | 55.1 | 16.8 | 25.0 | 345.0 |
| GPU time for this parameter(s) | | 3.3 | 8.9 | 17.0 | 24.0 | 31.3 | 81.7 |
| CPU time for grid search(s) | | 115.1 | 292.6 | 598.1 | 1006.4 | 1653.8 | 3587.8 |
| GPU time for grid search(s) | | 82.3 | 214.5 | 382.9 | 611.6 | 794.7 | 2086.9 |

Table 9.24 Performance comparison of the simulation and AI models

| Method | Prediction (cycles) | Difference from experiment (cycles) |
|---|---|---|
| Experiment | 1013 | – |
| Simulation | 982 | 31 (3.1%) |
| ANN machine learning | 1029 | 16 (1.6%) |

established and trained, designers can simply input the WLP geometric data, and the predicted reliability life cycle can be obtained almost immediately. This significantly shortens the design cycle, reduces costs, and optimizes packaging structures effectively.

#### 9.4.1.4 GPU/CPU Training Time

The ANN grid search and training time will increase dramatically when the dataset, hidden layers, and number of neurons increase. CPUs are designed for latency-oriented processing, which means they prioritize maximizing the efficiency and processing capability of individual cores. They are traditionally optimized for sequential computation, processing one instruction at a time. While modern multi-core CPUs can handle some parallel tasks (MIMD architecture), their overall efficiency for large-scale data and massive calculations is generally lower compared to GPUs. For example, the training time for an ANN on a CPU can increase exponentially with the number of neurons. However, GPUs offer significant advantages over CPUs for training the ANN, primarily due to their parallel computing architecture and processing paradigm. GPUs are designed for throughput-oriented processing, which emphasizes performing large numbers of identical operations simultaneously. They feature a massive number of simpler processing units organized into Streaming Multiprocessors (SMs). This architecture supports Single Instruction Multiple Data Stream (SIMD) or Single Instruction Multiple Threads (SIMT) processing, enabling the concurrent execution of thousands of threads.

ANN training involves extensive matrix and vector operations, including linear combinations, activation function calculations, forward propagation, backward propagation, and gradient calculations. These mathematical operations are inherently highly parallelizable. GPUs excel at data parallelism, which allows them to process multiple data points or elements of a matrix simultaneously. This capability is particularly beneficial in batch processing, where calculations for different samples and matrix elements are independent. The ability of GPUs to perform massive parallel computations significantly reduces the training time required for an ANN. This acceleration is vital for developing and optimizing complex deep neural networks. The speedup offered by GPUs becomes more significant as the dataset size increases and the neural network models become more complex (e.g., with more neurons or hidden layers). Unlike CPUs, where training time may increase

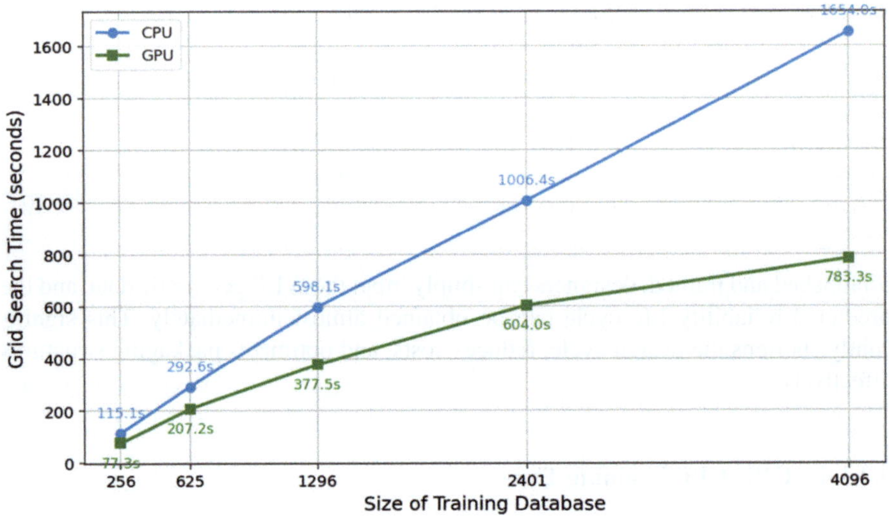


**Fig. 9.9** The CPU/GPU grid search time vs. dataset size

**Table 9.25** The differences between RF and ET

| Comparison | Random forest | Extra trees |
|---|---|---|
| Number of trees | Many | Many |
| Subset of training datasets | Bagging (roughly 2/3) | All datasets |
| Splits nodes | Best split | Random split |

exponentially with the number of neurons, GPUs can maintain a more stable or linear increase in training time as the number of neurons increases within certain ranges, due to their effective parallel handling of larger matrix dimensions. As mentioned above, these advantages make GPUs the dominant and preferred hardware for accelerating neural network training. Figure 9.9 illustrates the CPU and GPU computing performance under different dataset sizes. It is clear that the GPU's advantage lies in handling larger datasets.

### *9.4.2 Random Forest (RF) and Extra Tree (ET)*

Both Random Forest and Extremely Randomized Trees, also known as Extra Tree, as described in Chap. 7, are powerful ensemble learning algorithms used for machine learning tasks, particularly regression and classification. A key difference is that Extra Tree uses all of the original input data for each decision tree, rather than a bootstrapped subset (it does not use Bagging), Table 9.25 and Fig. 9.10. The most significant differentiator is how split points are chosen. Instead of searching for the optimal split point that minimizes a criterion (like Mean Squared Error, MSE) for

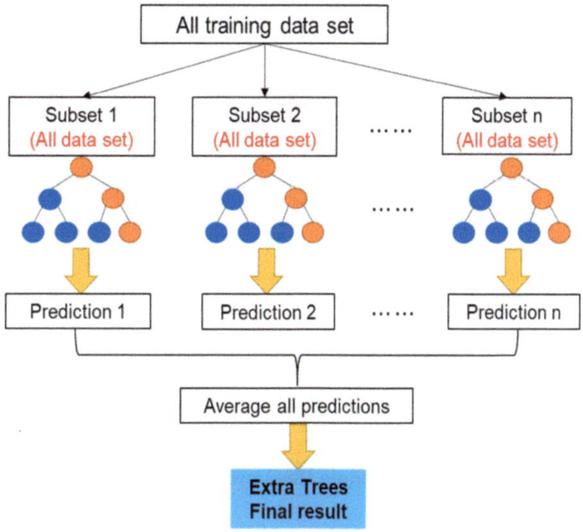


**Fig. 9.10** Schematic of the Extra Tree processes

each feature, Extra Tree randomly selects a split point from within the range of the chosen feature. RF and ET are built upon the foundation of decision trees and leverage collective intelligence to improve predictive performance. RF and ET are highly resistant to overfitting due to the random sampling of data and features and the aggregation of multiple diverse trees. Its generalization error tends to converge to an upper bound as the number of decision trees increases, meaning it does not easily overfit. RF is generally faster to train than complex deep learning models, such as Artificial Neural Networks. For example, for a 1296 dataset, RF took approximately 4 s, whereas ANN required around 24 min for training and prediction. Both Random Forest and Extra Trees have been applied to wafer-level packaging to address critical issues, such as reliability prediction and warpage analysis, providing efficient and accurate predictive models. Like ANN, the primary goal of RF/ET is to quickly and accurately predict the packaging lifetime of WLP, which traditionally requires time-consuming thermal cycling tests or finite element simulations. This helps save time and costs associated with model building and verification, and avoids inconsistent simulation results by different researchers.

#### 9.4.2.1 RF and ET Settings and Training

Random Forest training processes are illustrated in Fig. 9.11. RF parameter settings are crucial for optimizing model performance in tasks such as wafer-level packaging reliability prediction. These parameters control various aspects of the ensemble, including the number of trees and the construction of each tree. Number of decision trees (n_estimators): This parameter determines the number of individual decision trees built in the forest. A higher number of trees generally improves the model's

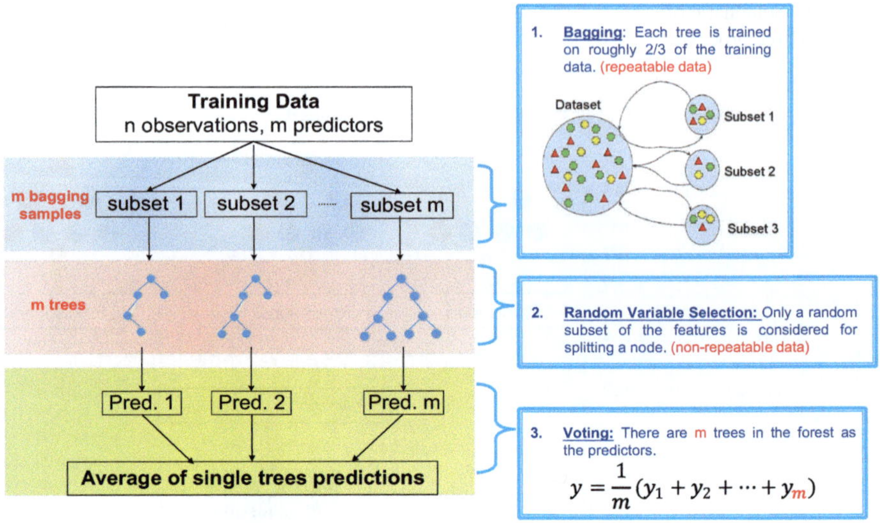


**Fig. 9.11** Schematic of The Random Forest training processes

overall prediction accuracy and stability, as it averages out the individual tree biases and variances. However, beyond a certain point, increasing the number of estimators yields diminishing returns and increases computational cost. For WLP reliability prediction, studies have shown that prediction accuracy (*R2* score, Eq. 9.5) tends to saturate after approximately 50 trees, Fig. 9.12.

$$R2 = 1 - \frac{\sum_{i=1}^{m}\left(y_{true} - y_{pred}\right)^2}{\sum_{i=1}^{m}\left(y_{true} - \overline{y_{true}}\right)^2} \tag{9.5}$$

The above Equation indicates that there are *m* training targets, where $y_{true}$ represents the actual target values, $y_{pred}$ represents the predicted target values, and $\overline{y_{true}}$ denotes the mean of all target values. Therefore, a value of *R2* closer to 1 indicates a better prediction performance that corresponds to the actual situation. This study uses this evaluation metric to describe the training results of the random forest model.

Maximum Tree Depth (max_depth): This parameter limits how deep each decision tree in the forest can grow. A deeper tree can capture more complex patterns. An overly deep tree can lead to overfitting, while a very shallow tree might be too simplistic to capture the underlying relationships in the data. For WLP reliability, the *R2* score was observed to saturate at a depth of 12, indicating that the decision trees could fully grow and capture the data's complexity within this depth, given the dataset size. The optimal parameter setting often uses "None" for max_depth, allowing trees to grow fully unless other constraints are met. Minimum

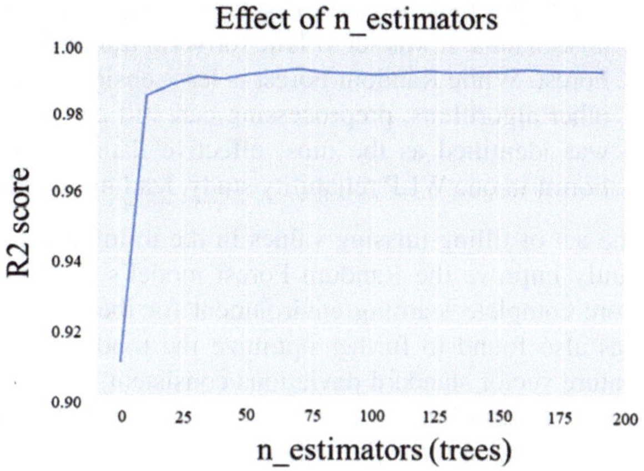


**Fig. 9.12** The number of estimators vs. R2 score

Samples to Split (min_samples_split): This sets the minimum number of samples required to be present at an internal node before it can be split further. Increasing this value makes the model simpler and less prone to overfitting; however, a too high value might prevent the tree from learning meaningful splits. For WLP reliability, the default value of 2 was chosen, as increasing it led to a decrease in the *R2* score, implying that making the model simpler worsened performance for the dataset used. Minimum Samples per Leaf (min_samples_leaf): This defines the minimum number of samples that must be present in a leaf node. Similar to min_samples_split, it limits tree growth and helps in regularization. The default value of 1 was selected, as increasing it (making the model simpler) decreased the *R2* score. Number of Features to Consider for Best Split (max_features): At each split point during tree construction, only a random subset of max_features is considered to find the best split. This randomness helps to decorrelate the individual trees in the forest, which is key to Random Forest's robustness against overfitting. Considering more features can lead to faster convergence. With four input features (Upper Pad Diameter, Lower Pad Diameter, Chip Thickness, Stress Buffer Layer Thickness), max_features was set to 4 (the maximum possible). This setting demonstrated faster convergence in the *R2* score, especially when the number of trees was low.

Splitting Criterion: This is the function used to measure the quality of a split at each node. For regression problems, such as WLP reliability prediction, Mean Squared Error (MSE) is commonly used and adopted in the case studies. Random State (random_state): This parameter seeds the random number generator, ensuring reproducibility of results across multiple runs. Bootstrap Aggregating (bootstrap): This specifies whether bootstrap samples (random subsets of data with replacement) are used for training each decision tree. Using bootstrapping (Bagging) is a core component of Random Forest, as it creates diversity among trees and helps prevent overfitting. It was set to True for WLP training. Out-of-Bag Score (oob_score): This option enables the use of out-of-bag samples (data points not included in a tree's bootstrap sample) to estimate the *R2* score on unseen data. It was set to True for WLP training. Data Preprocessing for Random Forest: While Random Forest is less sensitive to data scaling compared to some other algorithms, preprocessing can still enhance performance. MinMaxScaler was identified as the most effective data preprocessing method for Random Forest in one WLP reliability study, leading to reduced prediction errors.

The act of filling missing values in the training dataset was also found to significantly improve the Random Forest model's prediction accuracy, as it provides a more complete learning environment for the decision trees. Data standardization was also found to further optimize the model's predictive capability by making feature vector standard deviations consistent, even though its impact was less pronounced than filling missing values. The optimal combination of these parameters, often identified through methods such as Grid Search, enables the Random Forest and Extra Tree models to achieve high accuracy and efficiency in predicting WLP reliability.

#### 9.4.2.2 RF and ET Performance on Different Sizes of Datasets

This section aims to apply RF machine learning to WLP reliability analysis. Figure 9.13 illustrates a flowchart of the training procedure and data preprocessing, which includes filling in missing values and data standardization. In the process of building the RF model, parameter estimation was achieved using a grid search (Table 9.26 with 648 hyperparameter combinations) with cross-validation. The best prediction results after hyperparameter grid search.

Random Forest determines outcomes by constructing an ensemble of decision trees during the training process. This WLP case study explored 648 hyperparameter combinations through a grid search, covering scalers, estimators, maximum depth, minimum samples per split, minimum samples per leaf, and maximum features. With the dataset expanded to 9000 entries, the average testing difference was reduced to 1.0%. While RF is generally robust to noise and demonstrates strong generalization across many applications, its accuracy can be limited when the sample size is small. This limitation arises from its data-driven partitioning approach, which restricts smooth extrapolation and reduces accuracy in sparse regions. Moreover, the sampling rules prevent RF from fully leveraging the complete dataset, and the effectiveness of splits is inherently constrained under low-data conditions. Expanding the dataset size alleviates these restrictions and improves performance.

Table 9.27 presents the results of a Random Forest grid search study across different dataset sizes (256, 625, 1296, 2401, 4096, and 9000 datasets). It reports hyperparameter configurations (scaler, number of estimators, maximum depth, minimum samples per split, minimum samples per leaf, maximum features) and corresponding performance outcomes (training/testing differences, both maximum and average, in cycles and percentages), as well as grid search computation time in seconds. The MinMax or Robust normalization method was applied for the

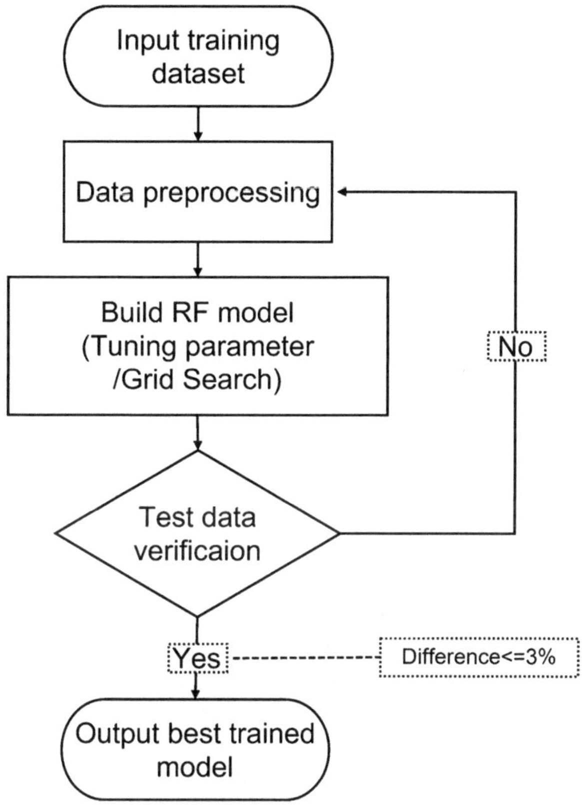


**Fig. 9.13** Flowchart of Random Forest

**Table 9.26** The hyperparameters grid search combination of RF

| Parameters | Values |
|---|---|
| Scaler | Robust, MinMax, Standard |
| N estimator | 100, 200, 300 |
| Max depth | 5, 10, 15 |
| Min sample split | 2, 3, 4 |
| Min sample leaf | 1, 2 |
| Max features | 1, 2, 3, 4 |

combination of hyperparameters, depending on the dataset size. The number of decision trees ranges from 100 to 300 across different dataset sizes. Max depth varies between 10 and 15, indicating the maximum depth of each tree. The minimum number of samples required to split an internal node ranges between 2 and 3. The minimum number of samples needed for a leaf node ranges from 1 to 2, and either 1 or 4, defining the maximum number of features considered at each split.

The training and testing performance of RF shows a maximum training difference ranging from 154.3 cycles (9.1%) with the smallest dataset (256) to 30.8 cycles (2.6%) with the largest dataset (9000). This shows a general trend of decreasing

**Table 9.27** Best prediction results of RF on different datasets after grid search

| | | 256 | 625 | 1296 | 2401 | 4096 | 9000 |
|---|---|---|---|---|---|---|---|
| Parameter | Scaler | MinMax | MinMax | Robust | MinMax | MinMax | Robust |
| | N estimators | 300 | 300 | 200 | 100 | 100 | 300 |
| | Max depth | 10 | 15 | 10 | 15 | 15 | 15 |
| | Min simples split | 2 | 2 | 2 | 3 | 2 | 2 |
| | Min simples leaf | 2 | 2 | 2 | 1 | 1 | 1 |
| | Max features | 1 | 1 | 1 | 1 | 1 | 4 |
| Result (cycles) | Maximum training difference | 154.3 (9.1%) | 141.7 (8.3%) | 106.2 (6.2%) | 84.0 (4.9%) | 40.5 (2.3%) | 30.8 (2.6%) |
| | Average training difference | 41.3 (4.5%) | 29.0 (3.2%) | 23.7 (2.5%) | 13.1 (1.4%) | 6.7 (0.7%) | 3.9 (0.4%) |
| | Maximum testing difference | 146.2 (11.6%) | 116.2 (13.8%) | 101.6 (1.01%) | 140.8 (18.9%) | 114.1 (15.1%) | 64.2 (6.1%) |
| | Average testing difference | 45.2 (4.3%) | 35.4 (3.5%) | 34.6 (3.6%) | 37.3 (3.8%) | 31.9 (3.2%) | 10.1 (1.0%) |
| Grid search time(s) | | 138 | 156 | 199 | 299 | 409 | 873 |

maximum training error as the dataset size increases. The average training difference decreases from 41.3 cycles (4.5%) at 256 samples to just 3.9 cycles (0.4%) at 9000 samples, indicating improved model fitting with larger datasets. For the testing case, the maximum testing difference fluctuates more compared to the training differences: the highest is 146.2 cycles (11.6%) at 256 samples and 140.8 cycles (18.9%) at 2401 samples, and the lowest is 64.2 cycles (6.1%) at 9000 samples. It suggests that testing performance is more sensitive to dataset size and possibly hyperparameter tuning. The average testing difference gradually decreases from 45.2 cycles (4.3%) at 256 samples to 10.1 cycles (1.0%) at 9000 samples. Confirms that larger datasets generally yield more reliable predictions. Figures 9.14 and 9.15 illustrate the average testing differences and the grid search time versus the training dataset size of RF.

The same set of datasets and hyperparameter grid search is also utilized in ET for both training and testing. Table 9.28 summarizes the grid search optimization results for the Extra Trees (ET) model across different dataset sizes: 256, 625, 1296, 2401, 4096, and 9000 entries. It contains: Hyperparameter settings (scaler type, number of estimators, tree depth, minimum samples per split/leaf, maximum features). Performance results (training/testing differences in cycles and percentages). Grid search computation times (seconds). This illustrates how dataset size and hyperparameter choices affect ET model accuracy and efficiency. Maximum Training Difference: Lowest: 3.1 cycles (0.2%) at 1296 samples. Highest: 30.5 cycles (2.9%) at 9000 samples. Most dataset sizes show very low training differences (≤1%), suggesting a strong fit with minimal overfitting. The average training difference is extremely low across all sizes: 0.0 cycles (0.0%) at 625 and 1296 (perfect fitting). Peaking at 2.7 cycles (0.3%) for 9000 samples. The above results confirm that ET achieves excellent training accuracy. Maximum testing difference ranges from 40.9 cycles (3.6%) at 4096 entries (lowest) to 64.0 cycles (6.8%) at 256 entries

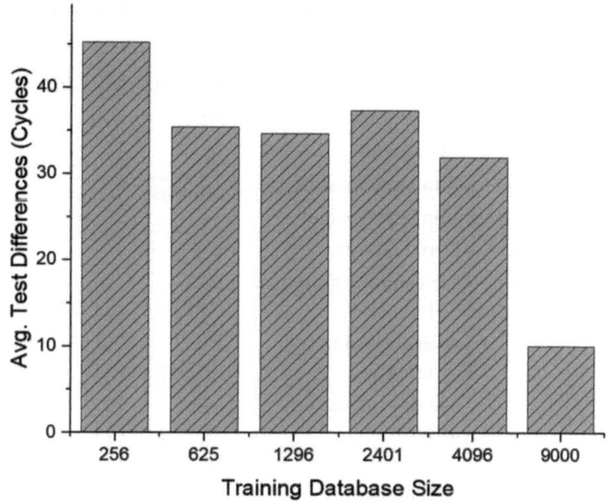


**Fig. 9.14** RF average testing difference vs. training data size

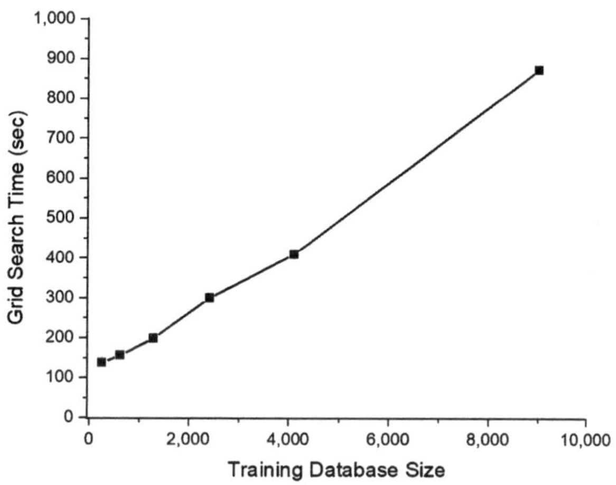


**Fig. 9.15** RF grid searching time vs. training data size

(highest). Errors reduce significantly as the dataset size increases, indicating better generalization. The average testing difference decreases consistently with larger datasets, from 16.5 cycles (1.6%) at 256 samples to 5.5 cycles (0.5%) at 4096 samples. Slight rise again to 8.8 cycles (0.8%) at 9000, possibly due to noise sensitivity. Suggests that ET generalizes well, with the lowest average testing errors ranging from 0.5% to 0.8%. Larger datasets provide stronger accuracy and stability, but grid search times scale significantly, peaking at 610 s.

The Extra Trees model exhibits very strong training accuracy (with an average error of $\leq$0.3%) and excellent testing accuracy ($\leq$1.6% average error, typically closer to 0.5–0.8%), demonstrating better performance than RF. Increasing dataset size improves generalization, although very large datasets (9000 samples) introduce a slight performance tradeoff. As with other ensemble methods, computation time

**Table 9.28** Best prediction results of ET on different datasets after grid search

| | | 256 | 625 | 1296 | 2401 | 4096 | 9000 |
|---|---|---|---|---|---|---|---|
| Parameter | Scaler | MinMax | MinMax | Robust | MinMax | MinMax | Robust |
| | N estimators | 300 | 300 | 200 | 100 | 100 | 300 |
| | Max depth | 10 | 15 | 10 | 15 | 15 | 15 |
| | Min simples split | 2 | 2 | 2 | 3 | 2 | 2 |
| | Min simples leaf | 2 | 2 | 2 | 1 | 1 | 1 |
| | Max features | 1 | 1 | 1 | 1 | 1 | 4 |
| Result (cycles) | Maximum training difference | 12.8 (1.0%) | 4.5 (0.5%) | 3.1 (0.2%) | 7.4 (0.5%) | 11.6 (0.9%) | 30.5 (2.9%) |
| | Average training difference | 1.6 (0.2%) | 0.0 (0.0%) | 0.1 (0.0%) | 0.4 (0.0%) | 1.1 (0.1%) | 2.7 (0.3%) |
| | Maximum testing difference | 64.0 (6.8%) | 55.3 (4.8%) | 52.9 (4.6%) | 45.7 (4.0%) | 40.9 (3.6%) | 56.4 (4.6%) |
| | Average testing difference | 16.5 (1.6%) | 8.5 (0.8%) | 7.1 (0.7%) | 6.1 (0.5%) | 5.5 (0.5%) | 8.8 (0.8%) |
| Grid search time(s) | | 106 | 138 | 185 | 248 | 345 | 610 |

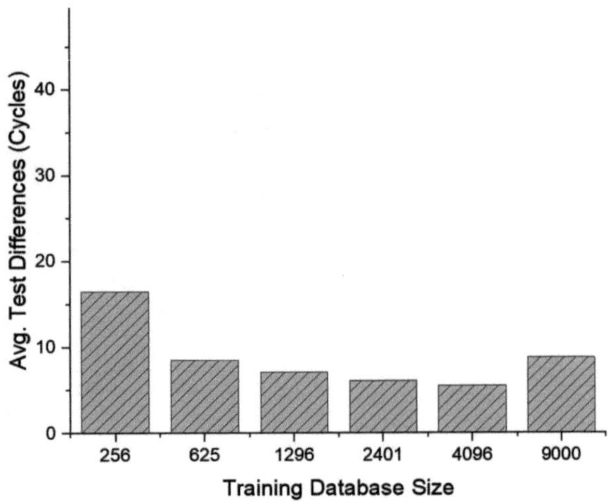


**Fig. 9.16** ET average testing difference vs. training data size

grows rapidly with dataset size, highlighting the balance between accuracy and efficiency. Figures 9.16 and 9.17 illustrate the average testing differences and the grid search time versus the training dataset size of ET.

Using the WLP with design parameters of upper pad diameter (240 μm), lower pad diameter (220 μm), chip thickness (330 μm), and stress buffer layer thickness (7.5 μm) as an example, the differences between the simulation and AI model are compared with the experiment result in Fig. 9.18. RF demonstrates good prediction accuracy compared with the simulation result.

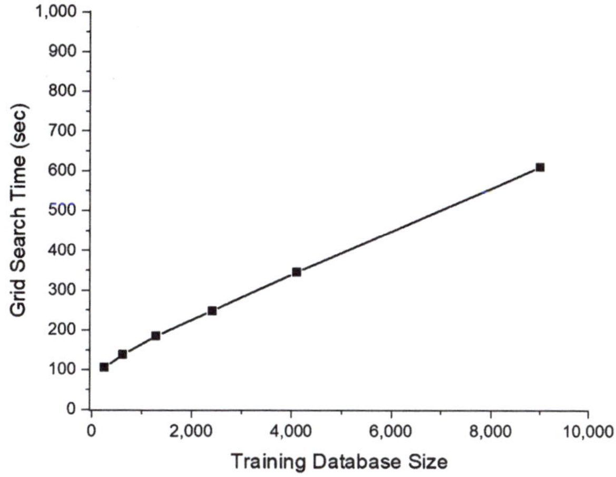


**Fig. 9.17** ET grid searching time vs. training data size

| | TCT experiment | Simulation | Random Forest |
|---|---|---|---|
| Lifetime (cycles) | 1,013 | 982 | 976 |

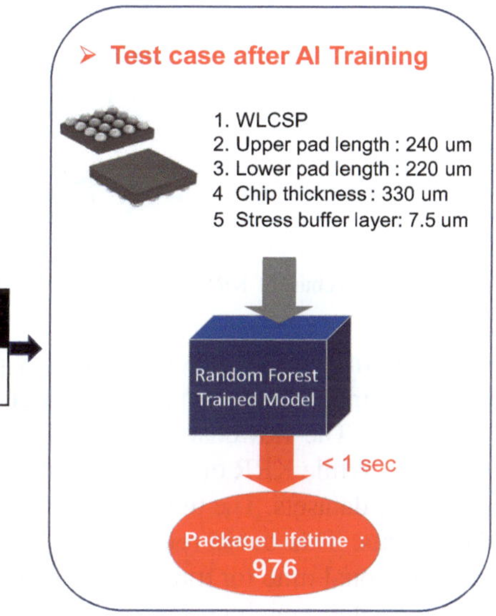


**Fig. 9.18** Reliability prediction of WLP Using RF

### *9.4.3 Kernel Ridge Regression (KRR)*

This section describes how Kernel Ridge Regression is used to predict the reliability of WLPs. KRR is an advanced machine learning algorithm that combines Ridge Regression (RR) with the kernel trick, as described in Chap. 7, to effectively handle complex and nonlinear regression problems. The kernel trick is a technique that allows KRR to process complex, nonlinear data without explicitly transforming the data into a high-dimensional feature space, which would be computationally

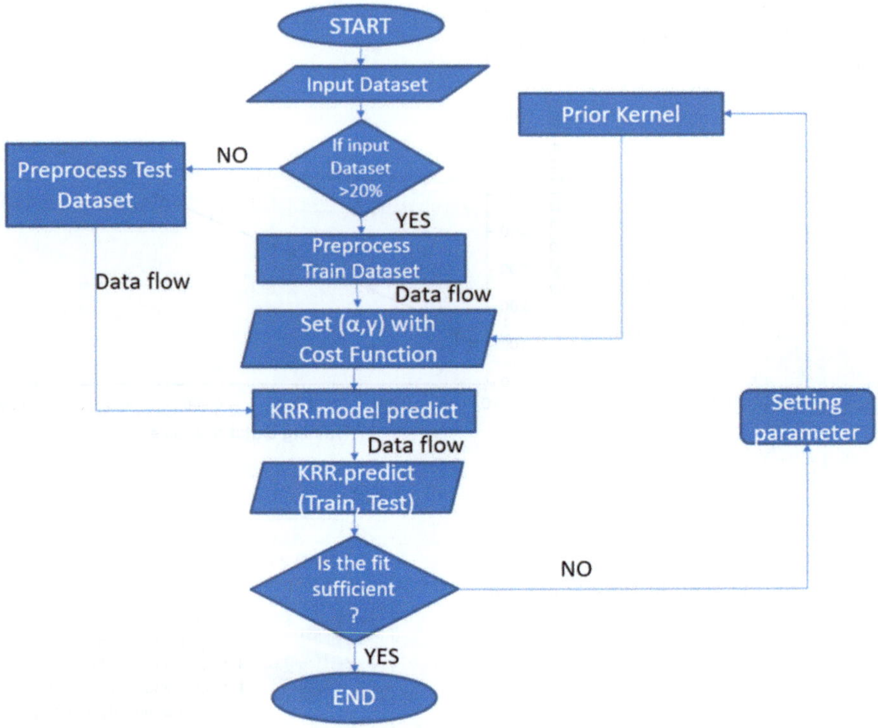


**Fig. 9.19** Flowchart of KRR

intensive. In the case of WLP reliability prediction, KRR is employed to develop a predictive model that estimates the fatigue life of solder balls under thermal cycling conditions. The flowchart of the KRR machine learning procedure is shown in Fig. 9.19. While KRR offers significant advantages, it faces challenges, particularly with large datasets. The time complexity of KRR is $O(n^3)$, where $n$ is the number of data points. This cubic relationship means that as the dataset size increases, the computational time for training the KRR model grows sharply. To mitigate this, the hybrid model, which combines K-means clustering with KRR (K-K model), is a good approach for reducing training time.

#### 9.4.3.1 Hyperparameter Setting and Grid Search

The hyperparameter settings for KRR are crucial to its performance and are typically determined through optimization methods such as Grid Search or Bayesian Optimization. The process also involves data preprocessing to scale features effectively. KRR combines Ridge Regression with the kernel trick to handle complex, nonlinear data. The core hyperparameters of KRR include: Alpha, which is the ridge parameter in Ridge Regression. KRR utilizes kernel functions to implicitly map

**Table 9.29** The hyperparameters grid search combination of KRR

| Parameters | Values |
|---|---|
| Scaler | MinMax, Standard, Robust |
| Kernal | Poly, RBF |
| Alpha | 0, 0.1, 0.01, 0.001, 0.0001 |
| Gamma | 0.1, 1, 10 |
| Degree | 5–9 |

data into higher-dimensional feature spaces, reducing computational complexity. Standard kernel functions and their associated hyperparameters include the Radial Basis Function (RBF) Kernel, the Polynomial Kernel, and the Sigmoid Kernel. Gamma ($\gamma$): This parameter controls the influence of individual training samples. For RBF, it's the only parameter that simplifies computation. In the WLP case, gamma values explored ranged from 0.01 to 100. The degree of the polynomial, values explored included from 5 to 9.

Before training the KRR model, data preprocessing is essential to reduce prediction errors. Standard methods include MinMaxScaler, which is used to transform features by scaling them to a specified range, typically between 0 and 1. RobustScaler scales features using robust statistics (median and interquartile range), making it less sensitive to outliers. MaxAbsScaler scales features by dividing by the maximum absolute value, fitting data into the $[-1, 1]$ range. StandardScaler standardizes features by removing the mean and scaling to unit variance (mean = 0, standard deviation = 1), assuming a normal distribution. In the WLP case, MinMaxScaler, StandardScaler, and RobustScaler are frequently employed as preprocessing methods for KRR models. RobustScaler was consistently found to be the most optimal preprocessing method when using the Radial Basis Function Kernel in KRR. This combination was noted for its excellent prediction performance and efficient training time. In the Grid Search, when applied to datasets of various sizes (256, 1296, 2401, and 4096 data points), RobustScaler was chosen as the preprocessing method for the KRR model. Table 9.29 illustrates the total number of combinations, 450, from the hyperparameter grid search for each dataset.

#### 9.4.3.2 KRR Performance on Different Sizes of Datasets

Kernel Ridge Regression is utilized in the wafer-level package reliability prediction to model the relationship between package design parameters and their fatigue life. Its performance, in terms of both prediction accuracy and computational time, varies significantly with the size of the dataset used for training. Studies have explored KRR performance across various dataset sizes ranging from 256 to 9000 data points. The performance of KRR models was evaluated using two different kernel functions, Radial Basis Function and Polynomial, along with various preprocessing methods and optimized hyperparameters via Grid Search across six dataset sizes: 256, 625, 1296, 2401, 4096, and 9000 data points. Table 9.30 presents the training and testing results of the KRR model under different dataset sizes (256 to 9000

**Table 9.30** Best prediction results of KRR on different datasets after grid search

| | | 256 | 625 | 1296 | 2401 | 4096 | 9000 |
|---|---|---|---|---|---|---|---|
| Parameter | Scaler | Robust | Standard | Robust | Robust | Robust | MinMax |
| | Kernel | Poly | Poly | Rbf | Rbf | Rbf | Rbf |
| | Alpha | 0 | 0.001 | 0.0001 | 0.0001 | 0.0001 | 0.0001 |
| | Gamma | 0.1 | 0.1 | 1 | 1 | 1 | 10 |
| | Degree | 8 | 8 | NaN | NaN | NaN | NaN |
| Result (cycles) | Maximum training difference | 17.3 (1.5%) | 42.7 (3.6%) | 33.2 (3.0%) | 44.9 (2.7%) | 38.2 (2.9%) | 37.4 (2.2%) |
| | Average training difference | 3.4 (0.3%) | 5.8 (0.5%) | 3.7 (0.3%) | 4.6 (0.4%) | 3.4 (0.3%) | 1.8 (0.2%) |
| | Maximum testing difference | 46.1 (4.0%) | 36.4 (2.5%) | 26.9 (1.8%) | 31.9 (2.8%) | 27.8 (2.4%) | 28.9 (2.3%) |
| | Average testing difference | 8.7 (0.8%) | 6.4 (0.6%) | 5.5 (0.5%) | 5.0 (0.4%) | 4.7 (0.4%) | 3.0 (0.3%) |
| Grid search time(s) | | 1.7 | 7.9 | 37.7 | 253 | 1467.3 | 13,783 |

samples) using various scalers, kernels, and hyperparameters. Two normalization methods were tested: Robust, Standard, and MinMax. Both polynomial (poly) and radial basis function kernels were applied. The regularization parameter Alpha decreases with larger datasets (from 0.001 to 0.0001), thereby controlling model complexity. Gamma adjusts the influence of a single training sample, ranging from 0.1 to 10, depending on the dataset size. Degree used only for polynomial kernels (degree = 8 at small dataset sizes, not applicable for RBF).

Training results (Cycles) are as follows: The maximum training difference ranges from 17.3 cycles (1.5%) at 256 entries to 44.9 cycles (2.7%) at 2401 entries. The largest dataset (9000) shows 37.4 cycles (2.2%), indicating relatively stable training errors across scales. The average training difference is very low across all dataset sizes, ranging from 0.2% to 0.5%. Best performance at 9000 entries: 1.8 cycles (0.2%). Suggests KRR consistently achieves excellent training accuracy regardless of dataset size. Regarding the testing results (Cycles), the maximum testing difference starts at 46.1 cycles (4.0%) for 256 entries. Decreases with larger datasets, reaching as low as 26.9 cycles (1.8%) at 1296 entries. Remains stable around 28–32 cycles (2.3%–2.8%) for datasets 2401–9000, indicating that generalization improves with larger datasets but stabilizes beyond ~1296 samples. The average testing difference shows steady improvement as the dataset size increases. 8.7 cycles (0.8%) at 256 → reduced to 3.0 cycles (0.3%) at 9000. It confirms that KRR achieves strong predictive consistency with more data. Figures 9.20 and 9.21 illustrate the average testing differences and the grid search time versus the training dataset size of KRR.

Using the WLP with design parameters of upper pad diameter (240 μm), lower pad diameter (220 μm), chip thickness (330 μm), and stress buffer layer thickness (7.5 μm) as an example, the differences between the simulation and AI model are compared with the experiment result in Fig. 9.22. KRR demonstrates good prediction accuracy compared with the simulation result.

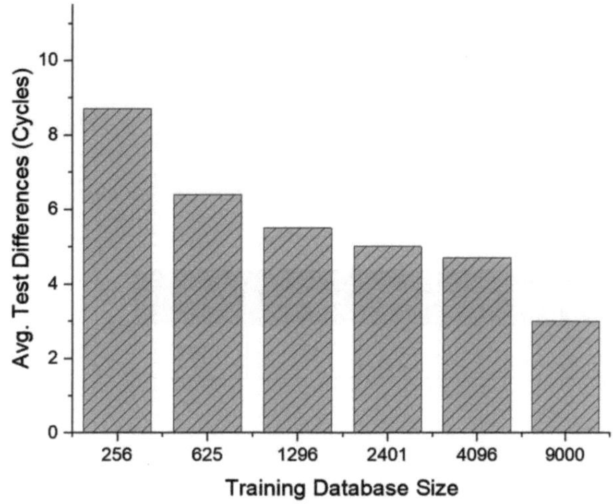


**Fig. 9.20** KRR average testing difference vs. training data size

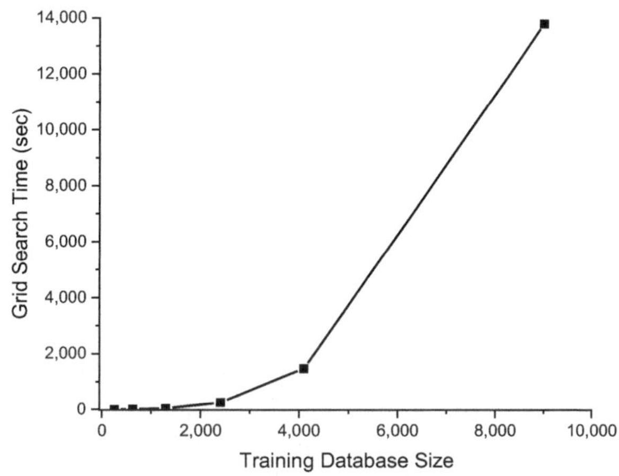


**Fig. 9.21** KRR grid searching time vs. training data size

## *9.4.4 Gaussian Process Regression (GPR)*

Gaussian Process Regression is a supervised machine learning algorithm used in this section to predict WLP reliability. It is a kernel-based, non-parametric Bayesian approach that probabilistically models the underlying function directly. GPR applies a prior Gaussian Process to make predictions and provides uncertainty quantification alongside its predictions. The target function is assumed to follow a Gaussian Process prior, and noise in observations is normally distributed. The model involves a mean function (often assumed to be zero) and a kernel function (or covariance function). The kernel function defines the covariance or relationship between input data points. The choice of kernel function is crucial for the performance of GPR. Examples include the Radial-Basis function (RBF) kernel, the Matern kernel,

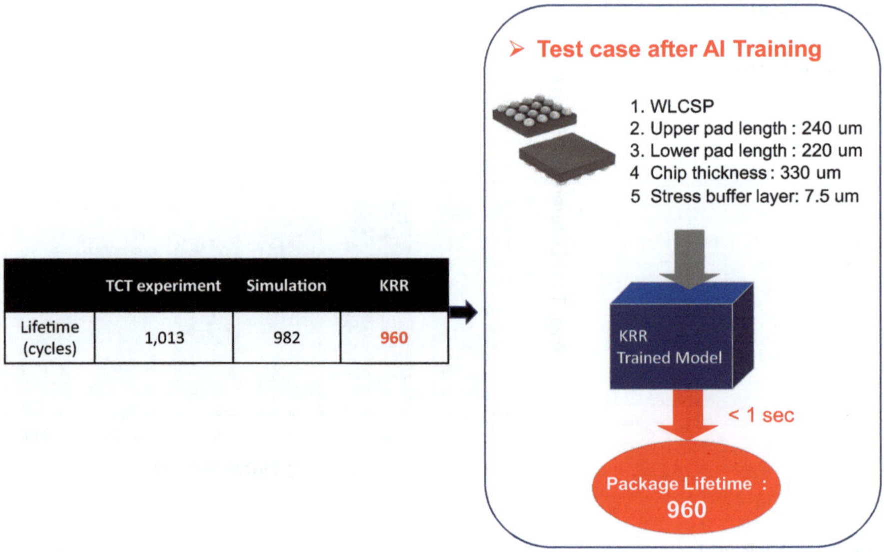


**Fig. 9.22** Reliability prediction of WLP using KRR

**Table 9.31** The hyperparameters grid search combination of GPR

| Parameters | Values |
|---|---|
| Scaler | MinMax, Standard, Robust |
| Kernel | RBF, Matern, Rational Quadratic, |
| Alpha | 1E-10, 1E-8, 1E-6, 1E-4, 1E-2, 1 |

a generalization of the RBF that is noted for its good performance, Rational Quadratic (RQ) kernel, Exp-Sine-Squared (EQ) kernel, and Double kernel functions (e.g., constant kernel multiplied by another) can also be used, sometimes offering better results for small datasets, but increasing computation time.

#### 9.4.4.1 Hyperparameter Setting and Grid Search

The hyperparameter settings for Gaussian Process Regression are crucial for achieving accurate prediction results. The optimization of these hyperparameters, particularly those related to the kernel function, is typically done by maximizing the margin likelihood using a gradient descent optimizer. Table 9.31 illustrates the total number of 54 hyperparameter combinations for grid search, using 256, 625, 1296, 2401, 4096, and 9000 data points. In the preprocessing, similar to KRR, MinMaxScaler, StandardScaler, and RobustScaler are frequently employed as preprocessing methods for GPR models. The performance of GPR models was evaluated using three different kernel functions: Radial Basis Function, Matern kernel, and Rational Quadratic kernel, along with various preprocessing methods and an alpha parameter

**Table 9.32** Best prediction results of GPR on different datasets after grid search

| | | 256 | 625 | 1296 | 2401 | 4096 | 9000 |
|---|---|---|---|---|---|---|---|
| Parameter | Scaler | Robust | MinMax | MinMax | Robust | Standard | Robust |
| | Kernel | Matern | Matern | Matern | Matern | Matern | Matern |
| | Alpha | 0.01 | 0.01 | 0.01 | 0.01 | 0.01 | 0.0001 |
| Result (cycles) | Maximum training difference | 12.1 (1.0%) | 20.6 (1.2%) | 17.7 (1.7%) | 19.6 (1.5%) | 25.1 (1.4%) | 1.1 (0.1%) |
| | Average training difference | 2.6 (0.2%) | 2.3 (0.2%) | 2.0 (0.2%) | 1.7 (0.2%) | 1.5 (0.1%) | 0.0 (0.0%) |
| | Maximum testing difference | 43.3 (3.1%) | 35.9 (2.8%) | 24.0 (1.6%) | 22.8 (1.6%) | 28.6 (2.9%) | 27.9 (2.3%) |
| | Average testing difference | 11.9 (1.1%) | 7.6 (0.7%) | 4.9 (0.5%) | 4.2 (0.4%) | 3.9 (0.4%) | 2.6 (0.2%) |
| Grid search time(s) | | 3 | 21 | 107 | 521 | 1852 | 11,052 |

ranging from 1 to $10^{-10}$ via Grid Search across six datasets. Table 9.32 illustrates the optimal combination of hyperparameters for various datasets.

#### 9.4.4.2 Training and Testing Performance

Gaussian Process Regression is utilized in this study for wafer-level package reliability prediction to model the relationship between package design parameters and their corresponding reliability life cycles. Its performance, in terms of training, testing accuracy, and computational time, varies significantly with the size of the dataset used for training. Table 9.32 has explored GPR performance across various dataset sizes ranging from 256 to 9000 points. The Table summarizes the results of a grid search conducted to optimize the parameters for a machine learning model, given the use of a Matern kernel and an alpha parameter. The grid search tested different combinations of parameters on datasets of varying sizes: 256, 625, 1296, 2401, 4096, and 9000.

The Table summarizes the model's performance across both training and testing datasets. For the model trained on 9000 data points, the average training error is 0.0, indicating a perfect fit to the training data. The average error on unseen test data serves as a critical metric for evaluating the model's generalization capability. In this case, the model trained on 9000 data points achieves an average testing error of 2.6 cycles—the lowest among all configurations—demonstrating superior performance on unseen data. This result slightly outperforms the 3.9-cycle error observed with the 4096-point training set. Furthermore, the small maximum differences in both training and testing results indicate that overfitting is not a significant issue for the trained GPR model.

The grid search time (sec) has a strong scaling effect, increasing from 357.84 s at 256 samples to 3696 s at 9000 samples. Highlights a tradeoff between accuracy and computational efficiency, with GPR becoming computationally expensive for large datasets. Figures 9.23 and 9.24 illustrate the average testing differences and the grid search time versus the training dataset size of GPR.

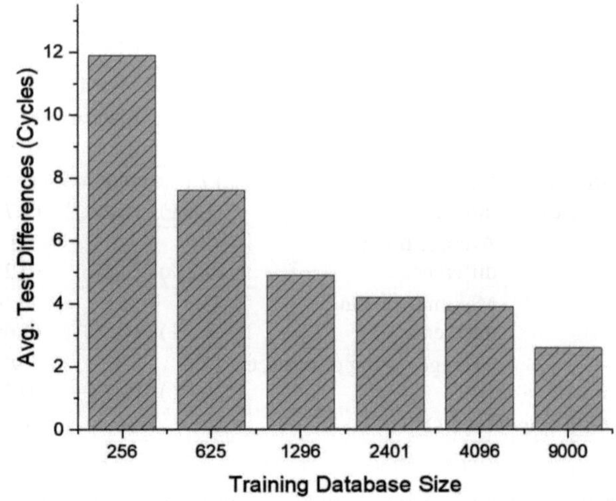


**Fig. 9.23** GPR average testing difference vs. training data size

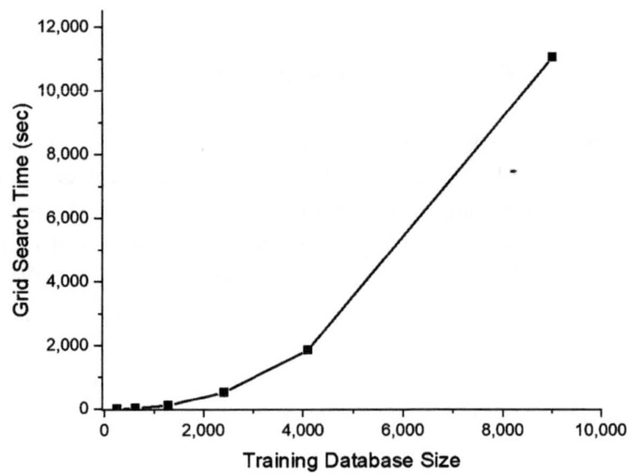


**Fig. 9.24** GPR grid searching time vs. training data size

## 9.5 Ensemble Learning

Ensemble learning is a machine learning strategy that integrates multiple predictive algorithms and frameworks to enhance predictive robustness and improve prediction accuracy and stability. It is considered a "training strategy" rather than a specific method or algorithm. In situations where training on small datasets is necessary due to the high time cost of data generation, individual Artificial Intelligence (AI) models can exhibit unstable predictive performance and a higher risk of overfitting. Ensemble learning addresses these issues by leveraging the complementary

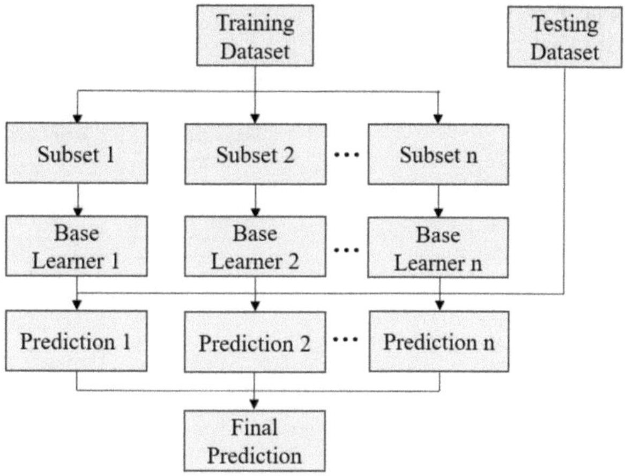


**Fig. 9.25** Schematic of Bagging

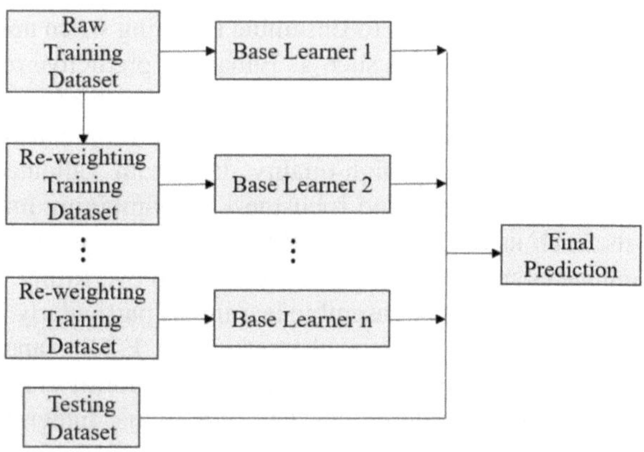


**Fig. 9.26** Schematic of Boosting

strengths of different algorithms and frameworks, enabling accurate predictions even with constrained training data. There are three main ensemble strategies: Bagging (Bootstrap Aggregating, Fig. 9.25) improves prediction by averaging the outputs of multiple base learners, trained on different data subsets. For regression tasks, optional weighting can be applied. This method enhances stability. Boosting (Fig. 9.26) works by assigning higher importance to data points that were previously mispredicted (samples with large prediction errors) during successive training rounds. It then aggregates predictions through a weighted average, with subsequent models focusing on harder cases. Stacking (Stacked Generalization, Fig. 9.27) combines the outputs of several base learners as inputs to a meta-model (or secondary model/meta-learner), which is then trained to generate the final prediction.

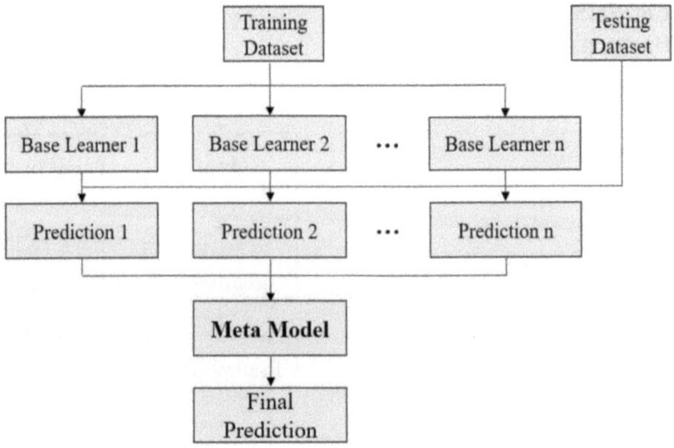


**Fig. 9.27** Schematic of Stacking

## 9.5.1 *WLP Case Study*

There are advantages to Ensemble Learning when applied to wafer-level packaging reliability prediction, such as Enhanced predictive reliability in small-data scenarios. Ensemble learning is an essential approach to enhance predictive reliability when working with small datasets, a common constraint due to the computational cost of generating high-quality data from validated finite element simulations. Improved accuracy and robustness by combining multiple machine learning models, such as ANN, RNN, and GPR. The ensemble framework improves accuracy, enhances stability, and reduces the risk of overfitting. Algorithmic complementarity, the effectiveness of ensemble learning is particularly evident when combining models with complementary characteristics. For instance, the ensemble of ANN and RNN models has shown good results, demonstrating the value of algorithmic complementarity in addressing data limitations. Integrating multiple algorithms helps offset individual model weaknesses, especially in outlier cases where high-error samples differ across algorithms. Ensemble learning is a promising approach for reducing prediction errors and enhancing generalization. The optimal ensemble model in the study achieved an average test error of only seven cycles, along with a notable reduction in the maximum test difference. Ensembles reduce the average difference in predictions and effectively suppress extreme deviations, leading to more reliable and consistent model performance.

The effectiveness also depends on how models are organized and combined, with different ensemble frameworks showing comparable predictive performance. A second-level ensemble, combining two ensemble frameworks, further improved prediction performance. However, it is also noted that simply increasing the number of base models or adding more algorithms does not consistently lead to improved predictive performance. The effectiveness of ensemble learning relies not only on the diversity of models but also on the predictive quality (accuracy) of each base model. For example, in the study, GPR models, despite being theoretically well-suited for small-data scenarios, reduced ensemble accuracy due to their relatively poor individual performance.

**Table 9.33** Dataset configuration and model selection criteria

| Item | Step 1 | Step 2 | Step 3 |
|---|---|---|---|
| Training data | 144 | 144 + 14 | 432 |
| Testing data | 144 | 144 + 144 | – |
| Optimal criteria | Test difference | Test difference | Cross-validation score |

Before data generation, two critical aspects must be addressed to ensure the quality of the training dataset: feature selection and sample distribution. Feature selection relies on domain knowledge or numerical analysis to identify highly relevant variables to the prediction target, thereby enhancing both model accuracy and interpretability [59]. Meanwhile, in small-data scenarios, the distribution of samples plays a pivotal role. A well-structured dataset that adequately covers the input space is essential for improving generalization and ensuring reliable predictive performance.

This study focuses on constructing machine learning datasets based on the four influential parameters: upper pad diameter (U), lower pad diameter (L), SBL thickness (S), and chip thickness (C). For data distribution, the training datasets are constructed using a combination of space-filling and adaptive sampling methods [60]. The base models are trained following the standard ensemble neural network procedure outlined by Su et al. [61].

Each base model is trained using different data volumes and hyperparameter configurations to promote model diversity. Inspired by the concept of boosting, a three-step training process is designed, as summarized in Table 9.33. Three datasets, each containing 144 samples, are prepared. In Step 1, Dataset 1 is used for training, and Dataset 2 serves as the test set. In Step 2, mispredicted samples from the previous test set are added to the training set, while Dataset 2 and Dataset 3 are used for testing. Step 3 combines all available data for training without a separate test set. The sampling details of the three datasets were established based on the adaptive sampling procedure proposed by Su et al. [61].

The grid search method [62] is employed at each step, resulting in multiple candidate models. Only the best-performing model is selected from each step. The selection criteria are based on the minimum average test difference in Steps 1 and 2. Due to the absence of testing data, the cross-validation (CV) score is used as the optimal criterion in Step 3. Finally, the generalization capability of the trained models is evaluated using a dataset consisting of 9601 samples [61]. The three algorithms, namely ANN, RNN, and GPR, are preliminarily combined based on the ensemble structure presented in Fig. 9.28.

The term "ULCS" denotes a set of four geometric structural parameters used as input features in the machine learning model. The optimal models from Steps 1, 2, and 3 are assigned as base models 1, 2, and 3–4, respectively. To enhance model diversity, base model 5 is constructed as a stacking model, which uses the outputs from base models 1 to 3 as its inputs. Tables 9.34, 9.35 and 9.36 present the key hyperparameter settings used for training the predictive models with the three algorithms. As previously mentioned, to enhance diversity among base models, two

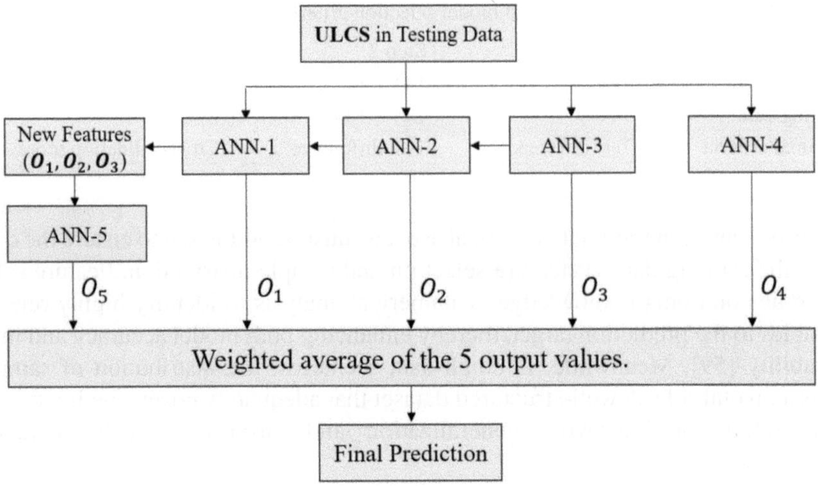


**Fig. 9.28** Ensemble structure for a single algorithm

**Table 9.34** ANN hyperparameter settings

| Hyperparameter | Setting |
|---|---|
| Activation function | ReLU |
| Solver | L-BFGS/Adam |
| Learning rate | Adaptive |
| Initial learning rate | 0.001 |
| Hidden layers | 3 |
| Neuron number | Grid search |
| Max_iter | 5000 |
| Loss function | MSE |

**Table 9.35** RNN hyperparameter settings

| Hyperparameter | Setting |
|---|---|
| Activation function | ReLU |
| Directions | Single |
| Unit | LSTM/Simple RNN |
| Learning rate | Adaptive |
| Initial learning rate | 0.001 |
| Hidden layers | 3 |
| Neuron number | 100/200 |
| Epochs | 2000 |
| Loss function | MAPE |

**Table 9.36** GPR hyperparameter settings

| Hyperparameter | Setting |
|---|---|
| Kernel function | Matérn/RBF |
| Matérn/RBF | Grid search (1e-10 ~ 1) |
| Optimizer | fmin_l_bfgs_b |
| N_restarts | Grid search (0 ~ 20) |

**Table 9.37** Training results of five ANN-based models

| Item | ANN-1 | ANN-2 | ANN-3 | ANN-4 | ANN-5 |
|---|---|---|---|---|---|
| Training data | 144 | 158 | 432 | 432 | 432 |
| Feature | ULCS | ULCS | ULCS | ULCS | Reliability (3) |
| Solver | L-BFGS | L-BFGS | Adam | L-BFGS | L-BFGS |
| Neuron number | 88-112-96 | 104-72-76 | 112-112-52 | 88-64-60 | 56-96-44 |
| Maximum training difference | 0<br>(0) | 0<br>(0) | 61<br>660/721<br>(9.2%) | 6<br>1327/1333<br>(0.5%) | 31<br>903/872<br>(3.4%) |
| Average training difference | 0<br>(0) | 0<br>(0) | 8.3<br>(0.8%) | 0.7<br>(0.1%) | 4.8<br>(0.5%) |
| Maximum testing difference | 71<br>1396/1467<br>(5.1%) | 66<br>1311/1377<br>(5.0%) | – | – | – |
| Average testing difference | 9.7<br>(1.0%) | 9.0<br>(0.9%) | – | – | – |

solvers (Adam and L-BFGS) are employed for the ANN, two types of recurrent units (LSTM and Simple RNN) are selected for the RNN, and two kernels (Matern and RBF) are adopted for the GPR. For example, Tables 9.37 present the key hyperparameter settings and corresponding model performance for the five base models trained using the ANN algorithms. The item "Neuron number" indicates the number of neurons in the three hidden layers. In "Maximum difference", the listed values represent, in order, the absolute difference, target/prediction, and percentage error. The RNN and GPR are applied similarly.

The base models from the three algorithms are combined following the structure shown in Fig. 9.28, where the assigned weights are proportional to their respective training set sizes. In this framework, model weights are determined in proportion to the size of their respective training datasets, such that learners trained on larger datasets contribute more strongly to the final prediction. The ensemble models from each algorithm are then equally weighted and further aggregated into a second-level ensemble. The outcomes are presented in Table 9.38. Thus far, the performance of all ensemble models has been fully demonstrated. The optimal ensemble model achieves an average test error of only seven cycles, resulting in a notable reduction in the maximum test difference. These results demonstrate the advantage of ensemble learning in improving the accuracy and stability. The ensemble reduces the

**Table 9.38** The performance of ensemble models

| Ensemble | Maximum testing difference | Average testing difference |
|---|---|---|
| ANN | 66 (4.1%) | 8.1 (0.8%) |
| RNN | 82 (8.9%) | 8.0 (0.8%) |
| GPR | 69 (6.3%) | 11.3 (1.0%) |
| ANN + RNN | 63 (6.8%) | 7.0 (0.7%) |
| ANN + GPR | 70 (6.4%) | 8.7 (0.9%) |
| RNN + GPR | 77 (5.7%) | 8.8 (0.9%) |
| ALL | 67 (5.9%) | 7.9 (0.8%) |

**Table 9.39** Performance comparison of the simulation and ensemble AI Model

| Method | Prediction (cycles) | Difference from experiment (cycles) |
|---|---|---|
| Experiment | 1013 | – |
| Simulation | 982 | 31 (3.1%) |
| Ensemble machine learning | 1029 | 16 (1.6%) |

average difference and effectively suppresses extreme deviations, resulting in more reliable and consistent model performance.

Finally, the ensemble model demonstrating the best overall performance was selected. Using the WLP with design parameters of upper pad diameter (240 μm), lower pad diameter (220 μm), chip thickness (330 μm), and stress buffer layer thickness (7.5 μm) as an example, the differences between the simulation and AI model are compared in Table 9.39. Once the AI model is trained and finalized, it enables the rapid evaluation of design parameter variations while ensuring consistency and accuracy in prediction results.

## 9.6 Comparison of Machine Learning Algorithms

In this section, various machine learning algorithms were compared for their ability to predict the reliability life of WLP structures. This comparison focused on prediction accuracy and CPU/GPU time consumption, using consistent training data and calculation environments. The reliability prediction is part of an AI-assisted design-on-simulation technology that combines artificial intelligence and simulation to reduce the time and cost associated with traditional reliability testing.

The machine learning process involved generating training datasets ranging from 500 to 9000 data points using a validated finite element simulation methodology. After training, testing data within the feature boundary were applied to evaluate each algorithm's performance. The hyperparameters for each ML model were fixed during training, as a grid search function was not applied for the initial comparisons in some studies, but was used in others to find optimal parameters. Algorithms compared include ANN, RF, ET, KRR, SVR series models, RNN, KNN, PR, and

GPR. The performance of ANN, RF, ET, and KRR has been discussed in previous sections. The performance of other AI algorithms is shown below.

The SGD Regressor and single kernel SVR showed an almost identical average testing difference across each training dataset. For 9000 data points, the SGD Regressor had an average testing difference of 2.64 cycles, while single kernel SVR had a difference of 2.51 cycles. Increasing training data generally decreased the average testing difference. SVR is noted to be suitable for high-dimensional features but not for small datasets. For small training data (500 and 1000), SVR was faster than SGD Regressor (e.g., at 500 data points, SVR: 0.4 s vs. SGD: 1.5 s). However, for large training datasets (2000, 4000, 6000, and 9000), the SGD Regressor was faster than SVR, e.g., at 9000 data points, SVR took 1799 s vs. SGD took 306 s.

The RNN can predict reliability more precisely than an ANN. For example, at 9000 data points using CPU, RNN had an average testing difference of 1.46 cycles, while ANN had 4.18 cycles. The ANN algorithm exhibited the best accuracy in comparing 576 and 1296 training datasets, with MAEs of 4 and 2 cycles, respectively. RNN, SVR, and KRR also showed good accuracy, with their testing errors being very close to each other when comparing all algorithms. RNNs take significantly longer to train than ANNs, approximately ten times longer than the training time of an ANN in a CPU environment. For 9000 data points, ANN took 469 s, while RNN took 4601 s. Using a GPU environment significantly reduced both RNN and ANN's training times to an acceptable range, while maintaining the same performance trend. The average difference remained similar when comparing CPU and GPU environments for both ANN and RNN. For example, with 9000 data points, the ANN CPU training time is 469 s, and the GPU training time is 277 s. RNN CPU training time: 4601 s; GPU training time: 829 s. For RNN, 2000 training data points were sufficient, as increasing the data to 4000, 6000, and 9000 showed almost the same performance. ANN and RNN models required more CPU time than KRR, SVR, KNN, and RF models due to the usage of more hidden layers and a greater number of neurons.

For KNN, PR, and GPR. KNN had relatively higher average testing differences, though it improved with more data (9.3 cycles at 9000 data points with four features). With five features (one extra design parameter, substrate thickness) and 5000 datasets, KNN had an MAE of 9.3 cycles. KNN is considered more suitable for classification purposes, but it also showed good accuracy for regression tasks. For KNN, the Manhattan distance metric performed better than Euclidean distance for both four and five input features. KNN was extremely fast, with training times around 0.04 to 0.21 s across all data sizes for four features. With five features, KNN training time ranged from 0.04 to 0.16 s.

PR also showed a decreasing average difference with increasing data, with 4.2 cycles at 9000 data points and four features. With five features and 5000 datasets, PR had an MAE of 4.8 cycles. PR was also relatively fast but increased with data size (PR: 0.05 to 28 s for four features). With five features, PR training time ranged from 0.22 to 2.9 s. GPR took significantly longer to train than KNN, KRR,

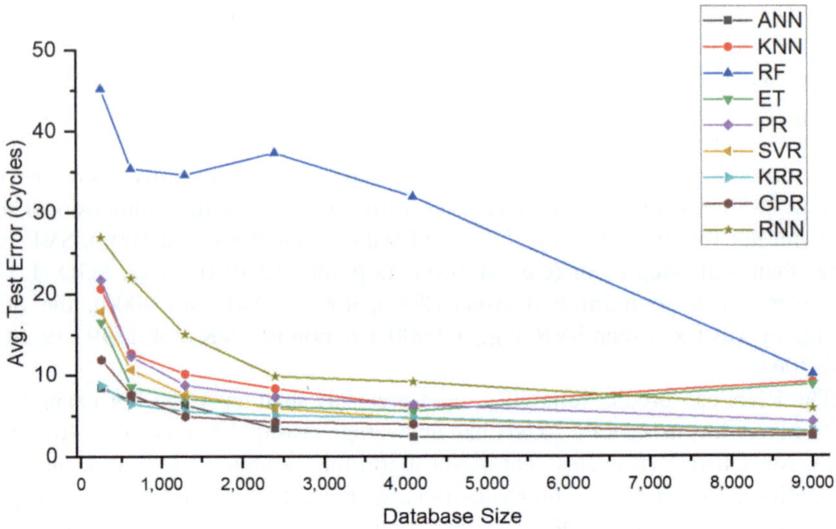


**Fig. 9.29** Average testing error vs. dataset size on different AI algorithms

and PR, especially with larger datasets (e.g., 864 s for 9000 data points with four features).

GPR generally showed low average testing differences, reaching 2.6 cycles at 9000 data points with four features. With five features and 5000 datasets, GPR achieved an MAE of 2.8 cycles. GPR exhibited the best accuracy among all algorithms compared in one of our studies. With five features, GPR training time ranged from 18 to 64 s. GPR's higher CPU time is attributed to the use of complex models, more frequent restarts for optimization, and larger hyperparameter ranges.

Overall Comparison and Conclusions,

1. Accuracy (Fig. 9.29): The GPR algorithm generally exhibited the best accuracy, followed closely by KRR and PR. ANN also showed excellent accuracy in specific comparisons. KNN and RF were noted to be less accurate for regression tasks compared to others, as they are often better suited for classification.
2. Training Time (Fig. 9.30): KNN, KRR, and PR regression models generally required less than 35 s of CPU time, with KNN being the fastest. ANN, RNN, and SVR typically needed 150 to 1800 s of CPU time, with RNN being the most time-consuming. GPR models also consumed significant training time, ranging from 64 to 864 s, depending on the features and dataset size.
3. Data Size Impact: Increasing the training dataset size for all algorithms generally led to improved testing performance. For RNN, 2000 data points were sufficient, as further increases did not significantly improve performance.
4. CPU/GPU Environment: Utilizing a GPU environment significantly reduced the training time for both ANN and RNN, while maintaining similar performance.

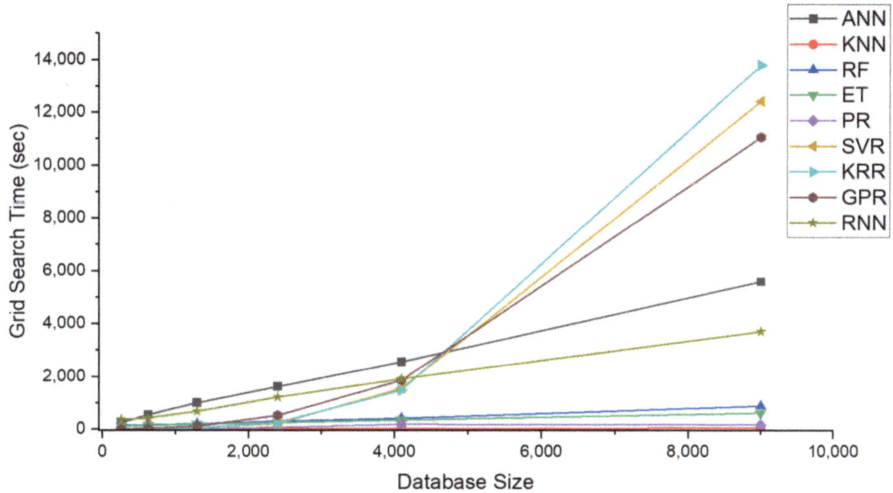


**Fig. 9.30** Grid search time vs. dataset size on different AI algorithms

## 9.7 Summary and Recommendations

The electronic packaging community traditionally relies on design-on-simulation to develop new packaging structures. Still, this approach faces challenges in ensuring the trustworthiness and consistency of simulation results, which are often highly dependent on individual researchers' expertise. It also involves lengthy design cycles that hinder time-to-market demands for complex packaging such as System-in-Package, 3D packaging, Chiplet, and heterogeneous packaging. AI-assisted Design-on-Simulation technology has been introduced to address these limitations, integrating an Artificial Intelligence approach with Machine Learning algorithms and DoS. AI-DoS aims to provide consistent and trustworthy predictions, significantly reducing design cycles.

Key advantages of AI-DoS include ensuring trustworthy and consistent prediction results for packaging reliability and warpage, with estimations within an accurate range compared to experimental data. This approach significantly shortens design cycles and reduces time-to-market, enabling reliability prediction in milliseconds to seconds once the AI model is established, in contrast to the lengthy experimental methods. AI-DoS reduces costs by minimizing the need for extensive physical experiments, lowering the training threshold for simulation expertise, making advanced design analysis more accessible, and unifying results across different developers.

The typical methodology for AI-DoS involves four steps:

1. Validating finite element simulation against mechanics theories and experimental results to ensure simulation accuracy.

2. Generating large labeled datasets, primarily from validated FEM simulations, which is more efficient than experimental data collection, but still CPU-intensive.
3. Training and testing various machine learning algorithms (e.g., ANN, SVR, KRR, KNN, RNN, RF, GPR, PR) on this data to establish robust regression models.
4. Deploying the trained AI model for design purposes, allowing quick reliability predictions by inputting new packaging geometries.

However, a significant challenge for AI-DoS is the lack of sufficient labeled data, as ML generally requires large datasets. Even this database can be generated from a validated finite element simulation; however, this training database will increase significantly as the design parameters increase. This has led to research in small-data ML methodologies, including efficient data sampling strategies like adaptive sampling, which focuses on "high-interest areas" to improve data quality with fewer samples. Ensemble learning, a "training strategy" that combines multiple trained models (weak learners) to improve overall prediction accuracy and robustness, is particularly beneficial for small datasets by reducing variance and overfitting.

AI and empirical models can be considered "transformation boxes" between inputs and outputs. However, empirical equations are mathematical formulas derived from observed data and theoretical relationships (e.g., Coffin-Manson, modified Norris-Landzberg Acceleration Factor). They typically have a direct physical meaning, but not all physical behaviors can be described by a simple empirical equation. In contrast, AI models learn patterns from large datasets without necessarily revealing explicit physical relationships. This Chapter notes that a modified Acceleration Factor equation derived from simulation data demonstrates a stronger capability for life cycle prediction than one based on experimental data, primarily due to the exclusion of inherent testing uncertainties in simulations.

This Chapter details the performance of several ML algorithms for wafer-level package reliability prediction:

- ANN: Demonstrate excellent accuracy, which improves significantly with larger datasets. However, they require substantial computational time, especially for grid searches and training, but GPUs can dramatically reduce this.
- RF and ET: Powerful ensemble algorithms resistant to overfitting and generally faster to train than complex deep learning models like ANN. They show good prediction accuracy, with ET often demonstrating superior performance to RF. Their accuracy can be limited with smaller datasets due to data-driven partitioning.
- KRR: Effective for complex, nonlinear regression problems by combining Ridge Regression with the Kernel Trick. KRR achieves excellent training and testing accuracy, but its computational time scales cubically with dataset size, making it computationally expensive for large datasets.
- Ensemble Learning: A strategy that integrates multiple algorithms to enhance predictive robustness, accuracy, and stability, particularly valuable for small datasets. It leverages complementary strengths of different models (Bagging,

Boosting, Stacking) and can significantly reduce prediction errors and suppress extreme deviations.

Based on the insights from this Chapter, the following recommendations are provided for utilizing AI-assisted Design-on-Simulation. AI-DoS can be applied for complex, advanced electronic packaging structures (e.g., SiP, Chiplet, heterogeneous packaging) where traditional design cycles are too long and experimental costs are prohibitive. Recognize that AI-DoS offers a transformative framework for faster, more accurate, consistent, and cost-effective reliability prediction. The validated finite element simulation procedure serves as the cornerstone for generating large, reliable datasets required for AI training, effectively replacing time-consuming and costly physical experiments.

## References


1. K. C. Norris, and A. H. Landzberg, "Reliability of controlled collapse interconnections," IBM Journal of Research and Development, vol. 13, pp. 266–271, 1969.
2. A. Syed, "Limitations of Norris-Landzberg equation and application of damage accumulation based methodology for estimating acceleration factors for Pb free solders," 11th International Conference on Thermal, Mechanical & Multi-Physics Simulation, and Experiments in Microelectronics and Microsystems (EuroSimE), Bordeaux, France, Apr. 26–28, 2010.
3. K. C. Wu, "Reliability Assessment of Advanced Packaging Solder Joints under Different Thermal Cycling Ramp Rates", Ph.D. Dissertation, National Tsing Hua University, Taiwan, 2016.
4. C. H. Lee, K. C. Wu and K. N. Chiang, "A Novel Acceleration-Factor Equation for Packaging-Solder Joint Reliability Assessment at Different Thermal Cyclic Loading Rates", Journal of Mechanics, Volume 31, Issue 1, Pages 35–40, February 2017.
5. K. C. Wu, and K. N. Chiang, "Characterization on Acceleration-Factor Equation for Packaging-Solder Joint Reliability", Microelectronics Reliability, Volume 65, Pages 167–172, October 2016.
6. S. K. Kang, P. Lauro, D. Y. Shih, D. W. Henderson, T. Gosselin, J. Bartelo, S. R. Cain, C. Goldsmith, K. J. Puttlitz, and T. K. Hwang, "Evaluation of thermal fatigue life and failure mechanisms of Sn-Ag-Cu solder joints with reduced Ag contents," 54th Electronic Components and Technology Conference (ECTC), Las Vegas, NV, USA, June 1–4, 2004.
7. N. Pan, G. A. Henshall, F. Billaut, S. Dai, M. J. Strum, E. Benedetto, and J. Rayner, "An acceleration model for Sn-Ag-Cu solder joint reliability under various thermal cycle conditions," SMTA International Conference, Chicago, IL, USA, Sept. 25–29, 2005.
8. X. Dai, N. Pan, A. Castro, J. Culler, M. Hussain, R. Lewis, and T. Michalka, "High I/O glass ceramic package Pb-free BGA interconnect reliability," 55th Electronic Components and Technology Conference (ECTC), Lake Buena Vista, FL, USA, May 31–Jun. 3, 2005.
9. R. Shih, S. Dai, N. Ramos, F. Billaut, L. Gopalakrishnan, M. Hu, K. Hubbard, S. Teng, N. Lee-Kim, and S. Bezuk, "Reliability of HITCE ceramic ball grid array package," 38th International Symposium on Microelectronics, Philadelphia, PA, USA, Sept. 25–29, 2005.
10. J. Hokka, T. T. Mattila, H. Xu, and M. Paulasto-Kröckel, "Thermal cycling reliability of Sn-Ag-Cu solder interconnections part 1: effects of test parameters," Journal of Electronic Materials, vol. 42, pp. 1171–1183, 2013.
11. Chen, C.; Yu, D.; Wang, T.; Xiao, Z.; Wan, L. Warpage prediction and optimization for embedded silicon fanout wafer-level packaging based on an extended theoretical model. IEEE Trans. Compon. Packaging Manuf. Technol. 2019, 9, 845–853.

12. Lau, J.H.; Ko, C.-T.; Tseng, T.-J.; Yang, K.-M.; Peng, T.C.-Y.; Xia, T.; Lin, P.B.; Lin, E.; Chang, L.; Liu, H.N. Panel-level chip-scale package with multiple diced wafers. IEEE Trans. Compon. Packaging Manuf. Technol. 2020, 10, 1110–1124.
13. Lau, J.H.; Li, M.; Yang, L.; Li, M.; Xu, I.; Chen, T.; Chen, S.; Yong, Q.X.; Madhukumar, J.P.; Kai, W. Warpage measurements and characterizations of fanout wafer-level packaging with large chips and multiple redistributed layers. IEEE Trans. Compon. Packaging Manuf. Technol. 2018, 8, 1729–1737.
14. Qin, C.; Li, Y.; Mao, H. Effect of Different PBO-Based RDL Structures on Chip-Package Interaction Reliability of Wafer Level Package. IEEE Trans. Device Mater. Reliab. 2020, 20, 524–529.
15. Qin, F.; Zhao, S.; Dai, Y.; Yang, M.; Xiang, M.; Yu, D. Study of warpage evolution and control for six-side molded WLCSP in different packaging processes. IEEE Trans. Compon. Packaging Manuf. Technol. 2020, 10, 730–738.
16. Wang, P.-H.; Huang, Y.-W.; Chiang, K.-N. Reliability Evaluation of Fanout Type 3D Packaging-On-Packaging. Micromachines 2021, 12, 295.
17. Yang, C.; Su, Y.; Liang, S.Y.; Chiang, K. Simulation of wire bonding process using explicit FEM with ALE remeshing technology. J. Mech. 2020, 36, 47–54.
18. Liu, C.-M.; Lee, C.-C.; Chiang, K.-N. Enhancing the reliability of wafer level packaging by using solder joints layout design. IEEE Trans. Compon. Packag. Technol. 2006, 29, 877–885.
19. Chiang, K.-N.; Chen, W.-H.; Cheng, H.-C. Large-scale three-dimensional area array electronic packaging analysis. J. Comput. Model. Simul. Eng. 1999, 4, 4–11.
20. Tsou, C.; Chang, T.; Wu, K.; Wu, P.; Chiang, K. Reliability assessment using modified energy-based model for WLCSP solder joints. International Conference on Electronics Packaging (ICEP), Yamagata, Japan, April 19–22, 2017, 7–15.
21. Liu, S.; Panigrahy, S.; Chiang, K. Prediction of fanout panel level warpage using neural network model with edge detection enhancement. IEEE 70th Electronic Components and Technology Conference (ECTC), 2020, 1626–1631.
22. Yuan, C.C.; Lee, C.-C. Solder joint reliability modeling by sequential artificial neural network for glass wafer level chip scale package. IEEE Access 2020, 8, 143494–143501.
23. S. K. Panigrahy, Y. C. Tseng, B. R. Lai, and K. N. Chiang, "An Overview of AI-Assisted Design-on-Simulation Technology for Reliability Life Prediction of Advanced Packaging", Materials, 14, 5342, Sept. 2021.
24. S. K. Panigrahy, "Study on AI-Assisted Design-On-Simulation Technology for Reliability Life Prediction of Wafer Level Package", Ph.D. Dissertation, National Tsing Hua University, Taiwan, 2022.
25. Hsieh, M.-C. Modeling correlation for solder joint fatigue life estimation in wafer-level chip scale packages. International Microsystems, Packaging, Assembly and Circuits Technology Conference (IMPACT), Taipei, Taiwan, October 21–23, 2015, 65–68.
26. Rogers, B.; Scanlan, C. Improving WLCSP reliability through solder joint geometry optimization. International Symposium on Microelectronics (IMAPS), Orlando, Florida, USA, 30 September–3 October, 2013, 546–550.
27. Chiang, K.-N.; Yuan, C.-A. An overview of solder bump shape prediction algorithms with validations. IEEE Trans. Adv. Packag. 2001, 24, 158–162.
28. Chang, J.; Wang, L.; Dirk, J.; Xie, X. Finite element modeling predicts the effects of voids on thermal shock reliability and thermal resistance of power device. Weld. J. 2006, 85, 63s-70s.
29. Association, J.S.S.T. JEDEC Standard JESD22-A104D, Temperature Cycling. Jedec. Org 2005, 11, 2009.
30. Yanjun, X.; Liquan, W.; Fengshun, W.; Weisheng, X.; Hui, L. Effect of interface structure on fatigue life under thermal cycle with SAC305 solder joints. International Conference on Electronic Packaging Technology (ICEPT), Dalian, China, August 11–14, 2013, 959–964.
31. Wang, P.; Lee, Y.; Lee, C.; Chang, H.; Chiang, K. Solder Joint Reliability Assessment and Pad Size Studies of FO-WLP with Glass Substrate. IEEE Trans. Device Mater. Reliab. 2021, 21, 96–101.

32. Gupta, S.; Al-Obaidi, S.; Ferrara, L. Meta-Analysis and Machine Learning Models to Optimize the Efficiency of Self-Healing Capacity of Cementitious Material. Materials 2021, 14, 4437.
33. Jianliang, S.; Mengqian, S.; Hesong, G.; Yan, P.; Jiang, J.; Lipu, X. Research on Edge Surface Warping Defect Diagnosis Based on Fusion Dimension Reduction Layer DBN and Contribution Plot Method. J. Mech. 2020, 36, 889–899.
34. Kuschmitz, S.; Ring, T.P.; Watschke, H.; Langer, S.C.; Vietor, T. Design and Additive Manufacturing of Porous Sound Absorbers—A Machine-Learning Approach. Materials 2021, 14, 1747.
35. Song, H.; Ahmad, A.; Ostrowski, K.A.; Dudek, M. Analyzing the Compressive Strength of Ceramic Waste-Based Concrete Using Experiment and Artificial Neural Network (ANN) Approach. Materials 2021, 14, 4518.
36. Farooq, F.; Czarnecki, S.; Niewiadomski, P.; Aslam, F.; Alabduljabbar, H.; Ostrowski, K.A.; Śliwa-Wieczorek, K.; Nowobilski, T.; Malazdrewicz, S. A Comparative Study for the Prediction of the Compressive Strength of Self-Compacting Concrete Modified with Fly Ash. Materials 2021, 14, 4934.
37. Huang, X.; Wasouf, M.; Sresakoolchai, J.; Kaewunruen, S. Prediction of Healing Performance of Autogenous Healing Concrete Using Machine Learning. Materials 2021, 14, 4068.
38. Salazar, A.; Xiao, F. Design of Hybrid Reconstruction Scheme for Compressible Flow Using Data-Driven Methods. J. Mech. 2020, 36, 675–689.
39. Song, S.-H. A Comparison Study of Constitutive Equation, Neural Networks, and Support Vector Regression for Modeling Hot Deformation of 316L Stainless Steel. Materials 2020, 13, 3766.
40. McCulloch, W.S.; Pitts, W. A logical calculus of the ideas immanent in nervous activity. Bull. Math. Biophys. 1943, 5, 115–133.
41. Denoeux, T. A neural network classifier based on Dempster-Shafer theory. IEEE Trans. Syst. Man Cybern. Syst. 2000, 30, 131–150.
42. Tandel, G.S.; Biswas, M.; Kakde, O.G.; Tiwari, A.; Suri, H.S.; Turk, M.; Laird, J.R.; Asare, C.K.; Ankrah, A.A.; Khanna, N. A review on a deep learning perspective in brain cancer classification. Cancers 2019, 11, 111.
43. Schmidhuber, J. Deep learning in neural networks: An overview. Neural Netw. 2015, 61, 85–117.
44. Kulawik, A.; Wróbel, J.; Ikonnikov, A.M. Model of the Austenite Decomposition during Cooling of the Medium Carbon Steel Using LSTM Recurrent Neural Network. Materials 2021, 14, 4492.
45. Yuan, C.C.; Fan, J.; Fan, X. Deep machine learning of the spectral power distribution of the LED system with multiple degradation mechanisms. J. Mech. 2020, 37, 172–183.
46. Cortes, C.; Vapnik, V. Support-vector networks. Mach. Learn. 1995, 20, 273–297.
47. Braun, A.C.; Weidner, U.; Hinz, S. Classification in high-dimensional feature spaces—Assessment using SVM, IVM and RVM with focus on simulated EnMAP data. IEEE J. Sel. Top. Appl. Earth Obs. Remote sens. 2012, 5, 436–443.
48. Drineas, P.; Mahoney, M.W.; Cristianini, N. On the Nyström Method for Approximating a Gram Matrix for Improved Kernel-Based Learning. J. Mach. Learn. Res. 2005, 6.
49. Hamed, Y.; Alzahrani, A.I.; Mustaffa, Z.; Ismail, M.C.; Eng, K.K. Two steps hybrid calibration algorithm of support vector regression and K-nearest neighbors. Alex. Eng. J. 2020, 59, 1181–1190.
50. Tan, S. An effective refinement strategy for KNN text classifier. Expert syst. Appl. 2006, 30, 290–298.
51. Ho, W.; Yu, F. Chiller system optimization using k nearest neighbour regression. J. Clean. Prod. 2021, 303, 127050.
52. Xia, J.; Zhang, J.; Wang, Y.; Han, L.; Yan, H. WC-KNNG-PC: Watershed clustering based on k-nearest-neighbor graph and Pauta Criterion. Pattern Recognit. 2021, 121, 108177.
53. Breiman, L. Random forests. Mach. Learn.2001, 45, 5–32.

54. Hsiao, H.; Chiang, K. AI-assisted reliability life prediction model for wafer-level packaging using the random forest method. J. Mech. 2020, 37, 28–36.
55. Mohana, R.M.; Reddy, C.K.K.; Anisha, P.; Murthy, B.R. Random forest algorithms for the classification of tree-based ensemble. Materials Today: Proceedings 2021.
56. Krizhevsky, A.; Sutskever, I.; Hinton, G.E. Imagenet classification with deep convolutional neural networks. Adv. Neural Inf. Process. Syst. 2012, 25, 1097–1105.
57. Ortac, G.; Ozcan, G. Comparative study of hyperspectral image classification by multidimensional Convolutional Neural Network approaches to improve accuracy. Expert syst. Appl. 2021, 115280.
58. Petneházi, G. Quantile convolutional neural networks for Value at Risk forecasting. Machine Learning with Applications 2021, 6, 100096.
59. P. Dhal and C. Azad, "A comprehensive survey on feature selection in the various fields of machine learning," Applied intelligence, vol. 52, no. 4, pp. 4543–4581, 2022.
60. J. N. Fuhg, A. Fau, and U. Nackenhorst, "State-of-the-art and comparative review of adaptive sampling methods for kriging," Archives of Computational Methods in Engineering, vol. 28, pp. 2689–2747, 2021.
61. Q. Su, C. Yuan, and K. Chiang, "Utilizing Ensemble Learning on Small Database for Predicting the Reliability Life of Wafer-Level Packaging," in 2024 IEEE 26th Electronics Packaging Technology Conference (EPTC), Singapore, Dec. 03–06 2024: IEEE, pp. 1248–1252.
62. M. Ogunsanya, J. Isichei, and S. Desai, "Grid search hyperparameter tuning in additive manufacturing processes," Manufacturing Letters, vol. 35, pp. 1031–1042, 2023.

# Index





J. Lau, K.-N. Chiang, *Cu-Interconnects, Glass, and AI-Assisted Simulation for Chiplets and Heterogeneous Integration*,
https://doi.org/10.1007/978-981-95-6891-8

## G



## H



## I



## J



## K

MIX
Papier aus verantwortungsvollen Quellen
Paper from responsible sources
FSC® C105338

If you have any concerns about our products,
you can contact us on
**ProductSafety@springernature.com**

In case Publisher is established outside the EU,
the EU authorized representative is:
**Springer Nature Customer Service Center GmbH**
**Europaplatz 3, 69115 Heidelberg, Germany**

Printed by Libri Plureos GmbH
in Hamburg, Germany